Springer Series in Materials Science

Volume 320

The Springer Series in Materials Science covers the complete spectrum of materials research and technology, including fundamental principles, physical properties, materials theory and design. Recognizing the increasing importance of materials science in future device technologies, the book titles in this series reflect the state-of-the-art in understanding and controlling the structure and properties of all important classes of materials.

More information about this series at https://link.springer.com/bookseries/856

Benny E. Raahauge · Fred S. Williams
Editors

Smelter Grade Alumina from Bauxite

History, Best Practices, and Future Challenges

Editors
Benny E. Raahauge
FLSmidth A/S
Retired, Deceased 6/21
Valby, Staden København, Denmark

Fred S. Williams
Alcoa - Retired
Longmont, CO, USA

ISSN 0933-033X ISSN 2196-2812 (electronic)
Springer Series in Materials Science
ISBN 978-3-030-88585-4 ISBN 978-3-030-88586-1 (eBook)
https://doi.org/10.1007/978-3-030-88586-1

This Springer imprint is published by the registered company Springer Nature Switzerland AG
The registered company address is: Gewerbestrasse 11, 6330 Cham, Switzerland

MEMORIAL

This book is published in the memory of its co-editor, Benny Raahauge who passed away on June 2, 2021.

Benny was undefeatable in his efforts to organize the book contents, line up the authors and push for the completion of their manuscripts.

Benny will be fondly remembered by the many associations he made in the alumina industry at technical conferences. Benny had many technical contributions and publications. He was always very gracious friendly, enthusiastic and open in technical discussions with colleagues in the industry.

His presence at future technical meetings will be truly missed.

Fred S. Williams, Co-Editor

Preface

This book details the recovery of alumina from bauxite and gives an extensive description of the principal process used in that recovery. That process, patented in 1888 by Karl Joseph Bayer, is known as the Bayer process. The four steps of the Bayer process carry names used by the ancient alchemists, digestion, clarification, precipitation and calcination. The tie back to alchemy is appropriate. The use of this Bayer process was immediately coupled with the 1886 patents of the Hall–Heroult smelting process. This fortunate combination turned aluminum from a precious metal to an abundant, versatile, inexpensive commodity.

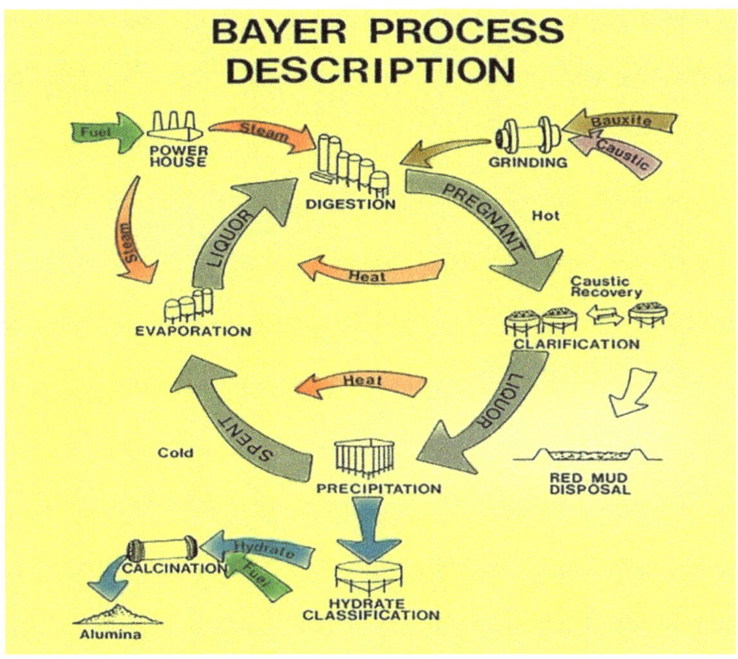

The raw material for the Bayer process, bauxite, is a general term used for mineral deposits that have the hydrated form of alumina present as the major mineral constituent. It turned out that there were a large number of extensive and easily recoverable deposits of bauxite primarily in the tropical regions of the world. Because of the availability of bauxite, the Bayer process has been the primary process used for over 130 years for the recovery of high-purity alumina. This book outlines the development of the Bayer process—past, present and future.

During the 2016 TMS meeting in Nashville, Tennessee, Springer contacted Benny Raahauge and asked if he would edit a new Springer book about alumina. Benny accepted the challenge with Fred Williams as Co-Editor. The formal title of the book is: *Smelter Grade Alumina from Bauxite—History, Best Practices and Future Challenges.*

The first step was to develop a Table of Contents for the new book. Once the TOC was developed, potential chapter authors were contacted. Because there are Bayer plants operating in many countries, experts from around the world were contacted to find those willing to voluntarily contribute manuscripts for the book chapters. This up-to-date published Springer book on Alumina from Bauxite is only made possible by all the many very competent authors who contributed with their exceptional efforts.

Thanks to Springer for their initial offer and patience over the years in finalizing our manuscript for submission.

Hopefully, you the reader will enjoy the content and find inspiration to your own future work to meet today's challenges facing the global alumina industry moving forward.

Longmont, USA

Benny E. Raahauge
Fred S. Williams

Contents

Editors and Contributors

About the Editors

Benny E. Raahauge received his M.Sc. Chemical Engineering from the Danish Technical University in November 1972. He started working with programming minicomputers controlling the raw material mixing for production of Cement Clinker and in December 1972 in Process Technical Department, FLSmidth, Cement Division, Copenhagen, Denmark.

From 1975 to 1976, he worked as a process and plant engineer in a Danish sugar factory before re-joining FLSmidth as R&D engineer and later as R&D Manager in the Mining Division in Copenhagen. He was assigned the task to develop the stationary gas suspension technology for Smelter Grade Alumina (SGA) to replace the rotary kilns.

After successful commissioning of the first 1000 tpd gas suspension calciner (GSC) unit for alumina at Hindalco Industries, India, in 1986, he worked in the role as General Manager-Pyro & Alumina Technology based in Copenhagen. This assignment included the responsibility for design, marketing, sales and commissioning of GSC units for alumina, culminating with commissioning of 3 × 4500 tpd GSC units at Queensland Alumina 2004–2005, the world's largest stationary calciners.

During +44 years employment with FLSmidth & Co., he contributed with many papers on calcination to international ICSOBA, Alumina Quality Workshop and TMS meetings until retiring in 2018.

Together with Don Donaldson, he was Lead-Editor to "Essential Readings in Light Metals—Volume 1—Alumina & Bauxite" with Fred Williams as Co-Editor, Published by Wiley & Sons. Inc. Copyright © 2013 by The Minerals, Metals & Materials Society (TMS).

In 2018, he was the editor and contributor to Chap. 12.1 Alumina, in the *SME Mineral Processing & Extractive Metallurgy Handbook*, Published by Society for Mining, Metallurgy & Exploration (SME), Copyright © 2019.

Fred S. Williams graduated from the University of Kansas with a BS Ch.E. in 1958 and MS Ch.E. in 1960. Upon graduation, he joined Alcoa Research in E. St. Louis, IL, and continued to work for Alcoa in alumina and chemical research and production assignments in the USA and Suriname for 42 years. Among other assignments, he was Manager of Alumina and Chemical Research at the Alcoa Technical Center, New Kensington, PA, Technical Manger Refining and Chemicals, Point Comfort, TX, Production Manager Refining and Chemicals, Point Comfort, TX and Technical Manager Refining and Smelting, Suralco, Suriname.

After retiring in 2002, he continued as a consultant to the alumina and bauxite industry for an additional 10 years.

He is the holder of several patents and has published a number of technical papers. He has been Alumina Subject Chairman three times and many times Session Chairman at TMS annual meetings.

He currently lives with his wife, Anne, at 4102 Heatherhill Circle, Longmont Colorado 80503, USA

Contributors

Dennis R. Audet, PE Director Audet Process Audit, Brisbane, Australia

Quentin D. Avery FLSmidth, INC. Regional Product Line Manager, Filter Press Technologies, Westwood, NJ, USA

Manfred Bach ENAr, Sales Manager Alumina and Pyromet Technology, Taunusstein, Germany

György (George) Bánvölgyi Bán-Völgy Ltd, Budapest, Hungary

Stephan Beaulieu Aughinish Alumina Ltd, Limerick, Ireland

David Cooling Alcoa World Alumina - Retired, Consultant Civil Engineering, Dawesville, Western Australia, Australia

Raphael Costa Hydro Bauxite and Alumina, Rio de Janeiro, Brazil

Carsten Duwe REEL Mollet Gmbh, Pinneberg, Germany

Ken Evans Rio Tinto/Alcan/British Alumina Group-Retired, Consultant - Alumina and Bauxite Residue, Chalfont St. Peter, UK

Anthony T. Filidore FLSmidth, Copenhagen, Denmark

Brady Haneman Hatch, Brisbane, Australia

Steven J. Healy Alumina Technology Consultant, Brisbane, QLD, Australia

Dr. George Komlóssy Rio Tinto Exploration Pty - Retired, Bauxite Consulting Technologist, Budapest, Hungary

Robert LaMacchia Alcoa World Alumina, Perth, Australia

Pascal Lavoie Alcoa Aluminum Center of Excellence, Director Smelting Manufacturing Excellence, Quebec, Canada

Stephen J. Lindsay Hatch p Smelting Specialist Consultant Centre of Excellence, Knoxville, USA

Benny E. Raahauge (Deceased)

Fred Schoenbrunn FL Smidth, Director for Thickeners, Cottonwood Heights, UT, USA

Peter-Hans ter Weer TWS Services & Advice – Bauxite & Alumina Consultancy, Owner-Director., Huizen, The Netherlands

Caio van Deursen Nexa Resources - Innovation Manager, São Paulo, Brazil

Michael Wanyo FLSmidth, Copenhagen, Denmark

Fred S. Williams Alcoa - Retired, Longmont, CO, USA

Chapter 1
Introduction: Primary Aluminum–Alumina–Bauxite

Benny E. Raahauge and Fred S. Williams

Abstract This book centers on the production of purified alumina (Al_2O_3) that is used as a feedstock to smelting cells that convert alumina to purified aluminum (Al). The smelting process used universally around the world for the production of aluminum is known as the Hall-Heroult process. This process was independently discovered and patented in1886 in the United States by Charles Martin Hall and in France by Paul Heroult. For the process to be commercial, however an inexpensive supply of high purity alumina was necessary. Fortuitously, in 1888, the Austrian chemist, Karl Joseph Bayer patented a simple caustic based chemical process for recovery of purified alumina using bauxite as a raw material.

Bauxite turned out to be a very abundant raw material, although, primarily in the tropical regions of the world. As a result, the combination of these two processes remain to this date the most economical method of producing purified aluminum metal.

This Chapter gives an overview of the Bayer Process as a lead into the following chapters where all aspects of the history and improvements to the original process are discussed. Also in this chapter is a brief discussion of alternative aluminum containing raw materials and processes that have been researched and, in a few cases, commercially used to recover alumina.

In 2020 the production of Smelter Grade Alumina (SGA) reached 126.7 Mton with China accounting for 53.3%. The corresponding global average energy intensity was about 10.7 GJ/ton SGA. The global demand growth for primary aluminum in 2021 is estimated to increase about 7% per annum, driving the demand for SGA subject to changes in stored inventory and any oncoming new capacity commissioned.

By the end of 2020, the price of SGA was 305 US$/ton, up from 275 US$/ton the year earlier. The decoupling of SGA prices as a percentage of the 3-month LME primary aluminum price continues.

Benny E. Raahauge is deceased.

B. E. Raahauge
FLSmidth - Retired, Deceased, Haslev, Denmark

F. S. Williams (✉)
Alcoa - Retired, Longmont, CO, USA

B. E. Raahauge and F. S. Williams (eds.), *Smelter Grade Alumina from Bauxite*, Springer Series in Materials Science 320,
https://doi.org/10.1007/978-3-030-88586-1_1

The focus on the Environmental Footprints is changing from best practice bauxite residue management to reduction of "green-house" gases, notably CO_2, which appear to be a tremendous challenge for the global alumina industry for many years to come.

1.1 Smelter Grade Alumina—The Property Driven Industrial Raw Material

This book centers on the manufacture of alumina (Al_2O_3) as a feedstock for producing aluminum by the electrolytic smelting process that is used universally around the world. Alumina has the property of readily dissolving in the molten cell bath (smelting) freeing the aluminum ion up to be electrolytically converted to aluminum metal.

Alumina also has a second very useful property utilized by the aluminum smelting plants. The exhaust gases from the smelting cell contains low concentrations of HF gas molecules. The exhaust gases are passed through a gas treatment center utilizing fluidized beds of alumina or alumina in gas suspension reactors for contacting HF in the exhaust gas with the virgin/primary alumina. The reacted/secondary alumina with HF is introduced into the smelting cell via a smelting cell/pot feeding system (see Chap. 12), that efficiently returns the HF molecules into the molten cell bath.

Not covered in this book are the many other unique properties created by special manufacturing process conditions to create a wide variety of crystallized forms of aluminas as well as the hydrated forms of alumina that enjoy a wide variety of industrial applications. References for discussions on these many products are by Misra [1] and Hart [2].

1.2 The Abundance of Aluminum in the Earth's Crust

Aluminum is never found naturally in the earth's crust as a free metal. The element Al is, however, the third most prevalent of any the periodic table ions and totals 8.13% [1] by weight of that crust. Only oxygen and silica are more numerous. Therefore, aluminum is the most abundant metal ion in the crust and there are a large number of minerals that contain the aluminum ion that can be considered for processing and conversion to aluminum metal via alumina.

The most common raw material that has been used for the production of aluminum is bauxite and that raw material is the subject of Chap. 2 in this book.

Bauxite is not a mineral but a raw material ore that comes in many forms. A raw material is classified as Bauxite if it has the specific quality that the majority of the Al element in the ore is in the mineral form of Gibbsite ($Al_2O_3 \cdot 3H_2O$) or in combination with the mono hydrated mineral forms, Boehmite or Diaspore ($Al_2O_3 \cdot H_2O$).

Commercial grades of bauxite are normally processed by a chemical process known as the Bayer Process (discussed briefly in this chapter and in detailed discussions of the steps in that process in later chapters in this book). The Bayer Process produces the purified alumina that is the raw material for the aluminum smelting process.

Bauxites are graded in quality by the percentage of Al present in the above favored mineral forms. Lower grade bauxites have the aluminum content diluted by minerals containing silica, iron, calcium, magnesium, titanium and many more of the elements in the periodic table. Al can be present or absent in those diluting minerals.

The quality of the bauxites used in the Bayer Process is often measured by the amount of alumina recoverable from the ore. Commercial bauxites vary from 30 to 60% by weight rccoverable Al_2O_3.

1.3 The Bayer Process

The Bayer Process was patented in 1888 in Germany by the Austrian chemist, Karl Joseph Bayer [3]. The simple chemistry of the process is that the hydrated forms of aluminum in bauxite, readily dissolve in heated caustic (NaOH) solutions (**the DIGESTION step**, see Chap. 4). The advantage is that nearly all of the minerals in bauxite containing other metallic ions do not dissolve and so can be discarded with the residue (**the separation or CLARIFICARION step**, see Chap. 5). The cooled solution of sodium aluminate ($NaAlO_2$) can then be seeded with Gibbsite ($Al_2O_3{\cdot}3H_2O$) crystals and additional Gibbsite recovered from the cooled solution that now has a lower equilibrium sodium aluminate concentration (**the PRECIPITATION step**, see Chap. 7). The sodium aluminate solution is returned to digestion and the separated gibbsite crystals heated in a furnace to dehydrate the Gibbsite to alumina (Al_2O_3) (**the CALCINATION step**, see Chap. 10).

A simplified picture of the steps of the Bayer Process is shown in Fig. 1.1. Note: Fig. 1.1 is an enhanced version of a graphic used by Don Donaldson in his TMS short course [4].

Also shown on Fig. 1.1 is that Bayer plants have some type of steam generation plant to supply the needed heat for the process. Often this is a cogeneration steam powerhouse, where high pressure steam is used to run turbines generating electricity for the processing plant and lower pressure steam out of the turbines is used for heating in the process. Most of that heating is done with tubular heat exchangers so that condensate can be returned to the powerhouse for steam generation.

A second feature of a Bayer Process shown in Fig. 1.1 is that the bauxite is ground with caustic solutions. In most plants, the ground bauxite slurry is held in stirred vessels for the time needed for the majority of the silica, contained in the bauxite being ground, to form the insoluble desilication product (DSP) ($Na_2O{\cdot}Al_2O_3{\cdot}2SiO_2{\cdot}xH_2O$). Allowing the DSP to form before digestion, is necessary to reduce DSP scale formation in the process equipment and to keep the silica

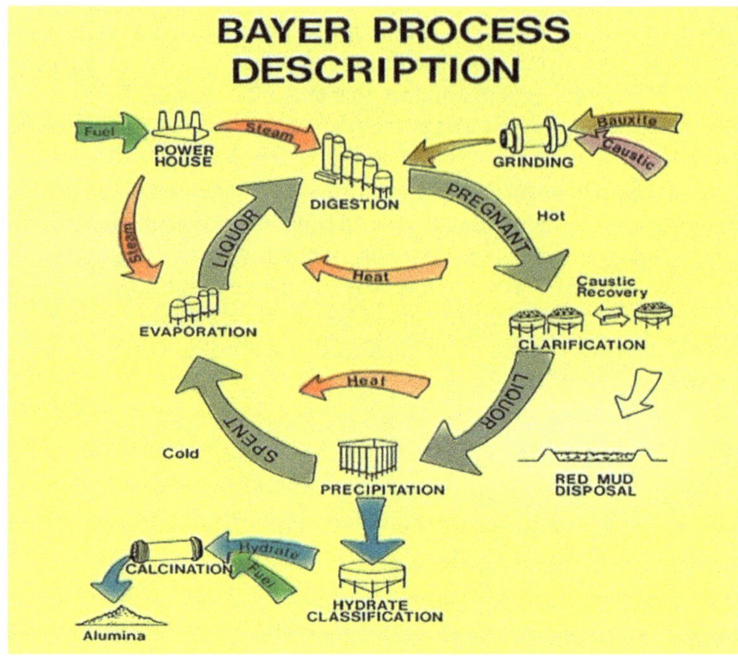

Fig. 1.1 Bayer process description [4]

from being a more significant contaminate in the final alumina product. Grinding and desilication are further discussed in Sects. 4.1 and 4.2.

A third feature shown in Fig. 1.1 is that there needs to be some way of disposing of the large volume of alkaline residue created by the Bayer Process. Residue disposal is further covered in Sect. 6.3.

In the over one hundred years since the patenting of the Bayer Process, the basic chemistry hasn't changed and the process is still the dominant process for recovery of alumna.

The later chapters in this book will discuss the improvements that have been made to the process in modern plants. These improvements have resulted in reduced energy consumption, improved alumina recovery, reduced caustic losses and improved alumina quality, see Chap. 11).

1.4 The Technical Enigma of Expressing Concentrations in Bayer Process Solutions

Because the Bayer Process has been utilized worldwide, some differences do occur in how the chemical composition of Bayer Process solutions are expressed.

This discussion on these the differences used to quantify Bayer liquor composition is based on the TMS short course given by Donaldson [4] (retired from Kaiser Aluminum Company) at the 2009 TMS Annual Meeting as well as to a number of worldwide interested Bayer processing groups in the 2010 to 2014 period.

The first area where reading technical articles on the Bayer Process can be confusing is in the differences employed to describe the free NaOH and total sodium centration that is paired with other anions in plant liquors. More specifically, the free concentration of NaOH and the total concentration of sodium OH and other anions are expressed and symbolized in several different ways. The caustic, or the sodium hydroxide (NaOH), present in solution is what dissolves the available alumina in bauxite.

Bayer plants in Europe, China and Japan, and their allied international plants, use different conventions for expressing concentrations than those plants in North America and their allied international plants. Sometimes the symbols are the same but the meaning is different. To reduce confusion in this discussion a subscript $_{(E)}$ will be attached to the similar symbol used in European and $_{(NA)}$ for that symbol used in North America.

In Europe, the free caustic concentration ($C_{(E)}$) is converted to Na_2O and expressed as g/l Na_2O. In North American the free caustic concentration is symbolized as $C_{(NA)}$ or TC and is converted to Na_2CO_3 and expressed as g/l Na_2CO_3. The total sodium concentration in Europe is symbolized by $S_{(E)}$ and again us Na_2O g/l. In North America the total sodium is symbolized by $S_{(NA)}$ or TS and is Na_2CO_3 g/l.

The Al in solution is universally expressed as alumina (Al_2O_3) g/l. The amount of Al that is in solution relative to the NaOH, however, may again be expressed differently. In France, it is expressed as $RP = Al_2O_3/Na_2O$ g/l. In other locations, it may be expressed as a reversed molar ratio, $MR = Na_2O/Al_2O_3$ mol/l. In North America, the ratio is expressed as $A/C = Al_2O_3/Na_2CO_3$ g/l.

Another concentration term that is used is causticity, which refers to the amount of caustic (NaOH) in solution relative to the amount of sodium combined with (OH) and other ions. In Bayer liquors, the principal ion that combine with sodium, in addition to (OH), is carbonate (CO_3), but other anions such as sulfate (SO_4), Chloride (Cl) fluoride (F) and others may be present. These combined forms of sodium are harmful because they make the sodium unavailable for dissolution of Al in bauxite. The operational efficiency of a Bayer plant is dependent on the solution causticity.

In North America, causticity is referred to as $C_{(NA)}/S_{(NA)}$ or TC/TA. All symbols are still in $Na_2(CO)_3$ g/l. In European plants causticity is expressed as $C_{(E)}/S_{(E)}$ with the symbols again in Na_2O g/l. Sometimes in European plants causticity is called percent carbonate and is the reverse expression $S_{(E)}/C_{(E)}$, again with each in Na_2O g/l.

Table 1.1 below summarizes and compares all of these liquor concentration terms using the base concentrations of Al_2O_3 90 g/l, NaOH of 100 g/l and total soda of Na_2CO_3 of 165 g/l.

Table 1.1 Conventions used to express bayer liquor concentrations

Free caustic (NaOH)			Total sodium		
NaOH g/l	$C_{(NA)}$ or TC Na_2CO_3 g/l	$C_{(E)}$ Na_2O g/l	$S_{(NA)}$ or TA Na_2CO_3 g/l	$S_{(E)}$ Na_2O g/l	
100	132.5	77.5	165	96.51	
Using the above numbers and $Al_2O_3 = 90$ g/l					
Causticity		% Carbonate	Ratio of Al_2O_3 to caustic		Molar ratio
$C_{(NA)}/S_{(NA)}$ or TC/TA	$C_{(E)}/S_{(E)}$	$S_{(E)}/C_{(E)}$	$A/C_{(NA)}$ or A/TC	RP = $A/C_{(E)}$	$C_{(E)}/Al_2O_3$ mol/l
0.803	0.803	1.245	0.679	1.161	1.417

1.5 Alternative Raw Materials and Processes to Recover Alumina

Though this book will entirely focus on bauxite ore as the raw material for making alumina, other ores have also been considered over time and some used to make alumina. These other ores are potentially useable for making alumina and deserves a short review for the sake of completeness.

> "To know what you do not know is kind of knowing it all" by Piet Hein, Danish engineer and poet.

1.5.1 The Lime-Soda-Sinter or Combination Process

Bauxites containing more than a minimum amount of silica in the form of the clay mineral, Kaolinite, ($Al_2Si_2O_5(OH)_4$) are not economical to process with the Bayer Process alone. The Kaolinite dissolves in the heated caustic solution and quickly re-precipitates out as a desilication product (DSP) ($Na_2O{\cdot}Al_2O_3{\cdot}2SiO_2{\cdot}xH_2O$). The formation of DSP is harmful not only because it removes Al_2O_3 from solution but more importantly it removes caustic from solution. The precipitation of DSP on the walls of process equipment can also reduce heat transfer efficiencies and lead to blockage of slurry flow through the process equipment.

High silica bauxites have been used commercially with the Lime-Soda-Sinter Process, also known also as, the Combination Process. In this process, the high silica bauxite is ground with spent sodium aluminate liquor, stored for up to 24 h in heated atmospheric pressure tanks to allow the majority of the desilication process to be completed in the thick slurry and then digested in Bayer Process digesters. The residue solids are then separated from the sodium aluminate liquor. Controlled amounts of limestone and sodium carbonate are added and ground with the residue to prepare it for sintering in rotary kilns. At high temperature (typically 1050 °C), the silica reacts with calcium to form insoluble calcium silicate (Ca_2SiO_4) and Sodium Carbonate

(Na_2CO_3) reacts with Al to form soluble sodium aluminate ($NaAlO_2$). The recovered sodium aluminate solution is then returned to the Bayer Process precipitation step for alumina recovery.

The extra cost of sintering obviously makes the resulting sodium aluminate solution more expensive than from the standard Bayer Process, but the advantage is that the hydrated and calcined aluminas that can be created from this processing feedstock have higher purity and premium qualities in any industrial applications.

1.5.2 Acid Processes for Recovery of Alumina from Clay

The extensive worldwide deposits of kaolinitic clays would seem to be an attractive source for alumina recovery because these potential raw materials can contain up to 39.5% Al_2O_3. A large amount of research has gone into using various acids to dissolve the Al. The advantage of using acids to dissolve alumina in clays is that under acidic conditions the silica is not soluble. However, most of the metallic ions in the other mineral impurities that may be associated with a clay deposit are soluble in acids. The additional construction cost of acid process equipment, additional processing cost to recover purified alumina, and the cost of treating and disposal of acidic wastes have resulted in none of these acidic processes being commercially viable at this time.

1.5.3 Anorthosite

Extensive igneous origin deposits of anorthosite are found around the world. When the deposit formed, they were a molten solution of two minerals: anorthite ($CaAl_2SiO_3$) and albite ($NaAlSi_3O_3$). The Al_2O_3 content of anorthosite is usually in the 26–30% range. Comingled iron and titanium minerals are generally very low in these deposits.

A commercial process developed by the US Bureau of Mines during World War II operated briefly for two years in Wyoming after the war was over. In this process, the anorthosite was ground with limestone and sodium carbonate and the mixture kiln calcined at 1250–1300 °C. The cooled kiln product was then leached with water, and recycled solutions, to recover a sodium aluminate solution. Kiln gases containing CO_2 were used to neutralize the sodium aluminate solution to force the precipitation of alumina trihydrate and the hydrate calcined to form alumina. The energy intensive nature of this process makes it non-competitive with the Bayer Process.

1.5.4 Aluminum Phosphate

Aluminum Phosphate deposits in Florida, United States were researched in the 1950s for recovery of alumina and uranium oxide. The deposits contained 10–27% alumina. Various attempts at using the Bayer Process and the lime–soda process was eventually dropped as being uneconomical.

1.5.5 Nepheline and Nepheline Syenite

Nepheline ($Na_3KAl_4Si_4O_{16}$) and Nepheline Syenite (Nepheline with additional alkaline mineral content) igneous deposits are found in a number of regions worldwide.

The concentrates from these deposits can contain around 26–29% alumina and are utilized as raw material feedstocks for alumina plants in Russia [5]. The ores are calcined with limestone and leached with recycled alkaline liquor. The recovered aluminate solution is desilicated, precipitated by carbonation to recover alumina trihydrate and calcined to recover alumina. The recovered residue is further sintered to produce cement.

1.5.6 Ferruginous Laterite

Mineral deposits known as ferruginous laterites are naturally formed by weathering conditions similar to those that form bauxites. Large deposits do occur worldwide. These deposits, like bauxite are highly variable and may contain 20–40% alumina, 20–40% iron and 5–30% reactive silica.

A process known as the Pederson Process has been extensively researched to treat ferruginous laterite ores. In this process, the laterite ore is mixed with coke and limestone (or lime) and charged into a vertical shaft furnace. Molten iron is tapped out of the bottom of the furnace and processed for iron recovery. Molten calcium aluminate and calcium silicate is tapped out of the furnace at a higher level and allowed to cool and solidify as a block. The formed block on cooling fractures into many small pieces and these then become the feedstock to a leaching step with a sodium carbonate solution. The separated week sodium aluminate solution formed from the leaching step is carbonated to precipitate alumina tri-hydrate and this calcined to form alumina. The leached residue can then be processed in kilns to form cement. Nearly ten tons of cement can be recovered for each ton of alumina. This process has been demonstrated in commercial sized equipment but never continued as a commercial process.

1.5.7 *Alunite*

Extensive Alunite ($KAl_3(SO_4)_2(OH)_6$) deposits can be found in a number of places in the world. The advantage of this raw material is that it can be processed to recover sulfuric acid and potassium sulfate as well as alumina. Commercial processes using this raw material for recovery of alumina have been run in Azerbaijan [1].

1.5.8 *Dawsonite*

Dawsonite ($NaAlCO_3(OH)_2$) has been investigated as a source of alumina in connection with the recovery of oil from oil shales. Dawsonite mineral beds can be associated with the oil shale beds for example in western Colorado. Heating oil shale to drive off the kerogen hydrocarbons, results in water soluble sodium aluminate which could be used for alumina recovery. No commercial process has been placed in operation for alumina recovery.

1.5.9 *Coal Wastes and Fly Ash*

Coal wastes and fly ash from coal fired electric generation plants can contain up to 35% alumina content. Research into recovering alumina from these wastes have included the lime-soda sinter process and processes utilizing hydrochloric acid or sulfuric acid. None of the researched process have come to economic fruition.

Two references by Hudson [6] and O'Connor [7], give a much more extensive discussion on the combination process as well as the alternative aluminum bearing raw materials and processes for recovery of alumina from those raw materials.

1.6 Alumina/Aluminum Supply–Demand Balance and Forecast

In general, the global growth in demand in kg per Capita for metals, and many other materials, is primarily driven by the increase in world GDP per Capita. In addition to this growth in material needs, the growth in aluminum demand is further driven by its very favorable physio-chemical properties, which comprises:

- Light weight.
- High strength per unit weight.
- High electrical conductivity per unit weight.
- Non-corrosive.
- Re-cyclable.

Table 1.2 Global growth of primary aluminum and SGA production in Kmetric ton/annum

Commodity	Primary aluminum [8]				Smelter grade alumina (SGA [9]			
Year	2000		2020		2000		2020	
Total	24.657	100.0%	65.295	100.0%	54.060	100.0%	126.682	100.0%
China	2.794	11.3%	37.337	57.2%	5.549	10.3%	67.463	53.3%
RoW	21.863	88.7%	27.958	42.8%	48.511	89.7%	59.219	46.7%

The global demand for Smelter Grade Alumina (SGA) is directly driven by the global demand for Primary aluminum. This global demand for SGA is illustrated in Table 1.2, where the development in China and the Rest of the World (RoW) is compared.

From 2000 until 2020, the global primary aluminum production has, on average, grown (CAGR) about 4.8%/annum, while production in China has, on average, grown about 13.3%/annum. For SGA, a similar global average growth in production of about 4.2%/annum has taken place, while production in China on average has grown about 10.3%/annum.

The increase in global demand for primary aluminum in 2021 is estimated to be about 7%/annum [10] or about 4.571 Kmton. The estimated increase in demand does not translate directly into an increase in primary aluminum production due to the current oversupply of primary aluminum in inventory which is estimated to be about 0.5 Mmton in China, and 1.2–1.4 Mmton [11] in the Rest of World (RoW). The oversupply of primary aluminum is expected to remain in 2021 [11]. The market balance for SGA is relatively stable at about 550–850 Kton [11]. The SGA supply growth for India and Indonesia is expected to match the increased smelter production in (RoW). The bauxite market remains oversupplied [11].

1.7 Alumina Cost Drivers and Pricing

1.7.1 Cost Drivers

There are two general cost drivers, capital investment expenditures (CAPEX) and operational expenditures (OPEX) for bauxite mines, alumina refineries and primary aluminum smelters. In this chapter only the major cost drivers will be discussed to provide the reader with an overview right from the beginning. A much more in-depth presentation and discussion of the subject is given in Chap. 14 Economics of Smelter Grade Alumina Projects.

In commodity markets like bauxite, alumina, and primary aluminum, the focus is on achieving a low cost (OPEX) position on the global supply/cash cost curve. This is because low-cost producers are more profitable than high-cost producers at all points in the business cycle. Furthermore, low-cost producers are less likely to have to curtail production in periods of low prices during recession or oversupply situations. Unfortunately, low OPEX often involves relatively higher CAPEX and a longer time for new mine or plant capacity to reach the market. The delay in coming up to production capacity can come during an oversupply situation, and thus result in a lower market share than anticipated at the time the capital investment was decided.

1.7.2 CAPEX for Alumina Refineries

Based on real 2013 US$, the installed cost per annual ton of alumina capacity, ranges from $600–$1900 US$/ton covering an installed capacity range from 0.250 to 1.750 M ton alumina capacity per year. This is a wide range including a wide scope. In Chap. 14, Sect. 14.1.4.2 the data do show a correlation between CAPEX and annual capacity [12].

When considering the different process facilities comprising a modern alumina refinery, using high-capacity tube digestion technology, to process boehmitic or diasporic bauxite at 250 °C digestion temperature, Hatch [13] has presented the below estimate of CAPEX contributions (Table 1.3):

The last column provides a reference to other chapters in this book for further discussion of the Facility and their role in the alumina production. See also Table 14.13 in Chap. 14 for a similar table.

Hatch [13] claims a reduction in capital expenditure of about 7% obtained through a combination of reduction in evaporation and boiler plant capacity in addition to a simplified flowsheet. The latter may incur increased risk of operational stability, capacity utilization and/or product quality issues [14].

Table 1.3 Relative capital contribution by facility (2012) [13]

Facility	Capital contribution, %	Chapter reference
Digestion	12	4
Steam generation	10	14
Precipitation	5	7
Hydrate filtration	5	7
Calcination	4.5	10
Mud settling washing	4	6
Bauxite grinding and …	2	3
Security filtration	1.5	5
Evaporation	1.5	9
Classification	1.5	7

1.7.3 OPEX for Bauxite, Smelter Grade Alumina and Primary Aluminum

For bauxite, smelter grade alumina and primary aluminum the major operational cost drivers are raw material, energy/power. All other more minor cost drivers are lumped together, see Table 1.4.

The large spread in Raw Material cost is for bauxite mining, reflecting differences in state tax or other state levies paid for by the mining company. The spread is small for Smelter Grade Alumina producers despite the advantage of owning an integrated bauxite mine with the alumina refinery in some cases. The same may apply for Primary Aluminum producers with respect to owing an alumina refinery but located far away.

Fuel/Power costs differences vary within each commodity as a result of having the advantage of cheap natural gas for some Smelter Grade Alumina producers. World Aluminum Statistics reports that the world average specific energy consumption for producing SGA is 10.736 MJ/ton on 22 June 2020, and the specific power consumption for producing Primary aluminum to be 13.354 kWh/ton on 5th August 2020.

The different levels in Other Cost reflects the differences in allocation of cost elements (i.e., caustic versus anode-carbon) and type of items being reported (i.e., conversion versus fixed cost).

Despite short to medium term fluctuations in the above OPEX Elements, their relative impact on OPEX will change little over time unless a new break-through technology emerges. An example would be if the carbon free electrolytic cell for making Primary Aluminum now being developed by Rio Tinto Alcan and Alcoa is proven successful.

The world average specific energy consumption for producing SGA reported above covers a large spread among refineries as shown in Table 1.5.

The specific energy consumption for producing SGA is dependent on the quality of bauxite being used by a plant and the type of Production Method as shown above in Table 1.5 and Chap. 2. The average specific energy consumption for producing SGA in China using Gas Suspension Calciner (GSC) units has improved to 11.6 GJ/ton SGA [17]. However, calcination fuel also has some minor impact. GSC units using natural gas have a specific energy/heat consumption of 2.7 GJ/ton, while GSC units using coal gas as in China have a specific energy/heat consumption of 3.1 GJ/ton [17].

Table 1.4 Major OPEX elements 2009, 2017 and 2020 [various sources]

Commodity	Raw material, %	Fuel/Power, %	Other cost, %
Bauxite mining	0–10	22–23	68–77
Smelter grade alumina	30–32	22–28	41–46
Primary aluminum	34–38	25–32	~33

Table 1.5 Energy intensity of selected alumina refineries

Alumina refinery	Bauxite type	Production method	Calciner fuel	GJ/ton SGA
Zheng Zhou, PRC [15]	Diaspore	Combined	Heavy fuel oil	34.15
Pinguo, PRC [15]	Gibbsite and diaspore	Bayer	Coal gas	15.10
Gardanne, France [15]	Gibbsite and soft diaspore	Bayer	Heavy fuel oil	13.52
Pinjarra, Australia [15]	Gibbsite	Bayer	Natural gas	11.21
Shennigola, Greece [15]	Diaspore	Bayer	Heavy fuel oil	14.86
Stade, Germany [15]	Gibbsite	Bayer	Natural gas	9.60
Alunorte, Brazil [16]	Gibbsite	Bayer	Heavy fuel oil	8.00

In general, stationary calciners accounts for 2.7–3.1 GJ/ton of the overall refinery energy intensity with the balance taken up by the Bayer process (see Table 1.7).

1.7.4 OPEX Contributions in Modern Alumina Refinery

The relative contributions to OPEX in a modern alumina refinery [18] are shown in Table 1.6.

The last column provides a reference to other chapters in this book for further discussion of the OPEX Elements and their role in the alumina production. The Energy cost can be further subdivided as shown in Table 1.7.

Table 1.6 Relative contributions to OPEX in a modern alumina refinery [18]

OPEX element	Operational cost contribution, %	Chapter reference
Energy	20–40	4, 9, 10, 14
Bauxite	15–30	2, 14
Caustic	10–20	4
Labor/Payroll	5–10	14
Spare and maintenance	5–10	14
Sundry charges	5	14
Lime	<5	4
Water	<5	9
Flocculants and reagents	<5	6

Table 1.7 Relative energy cost distribution in modern alumina refinery [13]

Facility or type	Relative cost, %
Digestion	52
Calcination	36
Electricity	6
Others	6

1.7.5 Pricing of SGA

The alumina price, FOB WA (Free On Board Western Australia) was 305 US$/ton by the end of 2020, up from 275 US$/ton a year earlier [10]. For comparison, the price of Chinese produced alumina (Shanxi) was 360 US$/ton, up from 352 US$/ton a year earlier [10], reflecting the higher cost of alumina produced in China.

Traditionally the SGA price has been linked as a percentage to the 3-month LME pricing of primary aluminum.

While spot prices are driven by market fundamentals, the long-term contracts were negotiated as a percentage of the LME aluminum price. Spot prices have tended to fluctuate between 10 and 25%, and long-term prices historically have been settled at around 11–15% of the LME price [19].

Alumina suppliers have always preferred to move away from contracts linked to primary aluminum LME prices. The move to market alumina with an index pricing system is motivated and supported by the following factors: [19]:

- The disconnect between alumina cost (rising over time) and aluminum prices.
- The change of the structure of the market to being less integrated.
- A shift of power with more new independent alumina refineries being built.

On this last factor, outside China, the non-integrated share of the alumina market has risen from 19% in 2005 to 23% in 2014 and will continue to rise.

There has been no new LME based contracts signed since the start of Alumina Price Index. Alumina Spot price publications by CRU, Metal Bulletin and Platts are aligned [20]. The RoW index prices (Platts alumina, FOB WA (US$/ton)) reflects alumina fundamentals [21] and the shift for alumina sales to index-based pricing continues at full speed. That pricing method is estimated to approaching 85% of alumina sales in 2020 [22].

1.8 Environmental Footprints and Challenges

The bauxite residue area was until a few years ago, the major environmental footprint of an alumina refinery. Especially, after the incident of the red mud pond dam failure at Ajka (Hungary) in October 2010 [23], has safe storage of bauxite residue or red mud been a focus area, see Chap. 6.

However, with the emergence and increasing understanding of the global climate change issue, emission of CO_2 and its reduction, has been the new focus area in addition to other gas emissions from alumina refineries as discussed in Chap. 13.

In the short- to medium-term, a significant reduction in refinery emission of "green-house" gasses, notably CO_2, can be achieved outside China by replacing coal in coal fired boilers and heavy fuel oil fired calciners with natural gas.

China accounts for more than 50% of the world alumina production, see Table 1.2, using coal gas ($CO + H_2$) as fuel. Coal gas is not easily replaced with natural gas, nor is it easy to replace coal used in the boilers supplying steam and electricity to their alumina refineries.

Long-term all current fuels used in boilers and calciners needs to be replaced with Green-H_2 produced from electrolysis plants powered with power from wind turbines and or solar panels. The technology is available today but not yet developed enough for the cost to be competitive.

References

1. C. Misra, *Industrial Alumina Chemicals, ACS Monograph 184* (American Chemical Society, Washington, 1986)
2. D. Hart, in *Alumina Chemicals Science and Technology Handbook* (The American Ceramic Society Inc. Westerville, Ohio, 1990)
3. F. Habashi, A hundred years of bayer process for alumina production, in *Light Metals*, ed. by L.G. Boxall (Phoenix, AZ, 1988), pp. 3–12
4. D. Donaldson, Alumina refinery fundamentals and practice, in *138th TMS Annual Meeting and Exhibition 15 February, 2009, San Francisco, CA*
5. V. Smirnov, Alumina production in Russia Part 1: historical background. JOM **48**(8), 24–26 (1996)
6. L.K. Hudson, in *Alumina Production* (Aluminum Company of America, 1982)
7. D.J. O'Connor, *Alumina Extraction from Non-Bauxitic Materials* (Aluminium-Verlag GMbH, Dusseldorf, 1988)
8. World Aluminium—Primary Aluminium Production Data Sheet
9. World Aluminium—Metallurgical Alumina production Data sheet
10. ALCOA 4Q2020 Presentation
11. Hydro Capital Market Day 10 December 2020
12. B. Haneman, S. Coackley, Energy and innovation in alumina production—key to success, in *20th Bauxite & Alumina Seminar, Metal Bulletin Events, Miami, 24–26 February, 2014*
13. B. Haneman et al., High-capacity tube digestion technology reduces capital and operating cost for refineries, in *Proceedings of XIX international Conference ICSOBA 2013, Krasnoyarsk, Russia, 4–6 September, 2013*
14. D.R. Audet, R. Little, Stabilization of product quality at Yarwun alumina refinery (AQW, Perth, 2012)
15. L. Liu et al., Analysis of the overall energy intensity of alumina refinery process using unit process energy intensity and product ratio method. Energy **31**, 1167–1176 (2006)
16. R. Wischnewski et al., Alunorte global energy efficiency, in *Light Metals* (2011), pp.179–184
17. S. Gu, J. Wu, Review of the energy saving technologies applied in bayer process in China (AQW, Perth, 2012)
18. L. Hendrickson, The need for energy efficiency in bayer refining, in *Light Metals* (2010)
19. M. Haller, in *Metal Bulletin Research Presentation on 20th Bauxite & Alumina Conference, 24–26 February 2014, Miami*

20. T. Reyes, in *Alcoa Presentation on 20th Bauxite & Alumina Conference, 24–26 February 2014, Miami*
21. A. Wod, Keeping the bauxite and alumina industry profitable—challenges and opportunities, in *20th Bauxite & Alumina Conference, 24–26 February 2014, Miami*
22. Hydro 1Q2019 Presentation
23. G. Bánvölgyi, The red mud pond dam failure at Ajka (Hungary) and subsequent developments. Presentation at ICSOBA Red Mud Seminar, Goa, India, 2011

Benny E. Raahauge Deceased—Owner—Director Raahauge-SGA ApS, Denmark

Benny received his M.Sc. Chemical Engineering from the Danish Technical University in November 1972. Start working with programming minicomputers controlling the raw material mixing for production of Cement Clinker, December 1972 in Process Technical Department, FLSmidth, Cement Division, Copenhagen, Denmark.

From 1975–76, Benny worked as process and plant engineer in a Danish Sugar Factory before re-joining FLSmidth as R&D engineer and later R&D Manager in the Mining Division in Copenhagen. Benny was assigned the task to develop the stationary Gas Suspension Technology for Smelter Grade Alumina (SGA) to replace the rotary kilns.

After successful commissioning of the first 1000 tpd Gas Suspension Calciner (GSC) unit for alumina at Hindalco Industries, India, in 1986, Benny worked in the role as General Manager-Pyro & Alumina Technology based in Copenhagen. This assignment included the responsibility for design, marketing, sales and commissioning of GSC units for Alumina, culminating with commissioning of 3 x 4500 tpd GSC units at Queensland Alumina 2004–05, the world's largest stationary calciners.

During +44 years employment with FLSmidth & Co. contributed with many papers on calcination to international ICSOBA, Alumina Quality Workshop and TMS meetings until retiring in 2018.

Together with Don Donaldson, Benny was Lead-Editor to "Essential Readings in Light Metals—Volume 1—Alumina & Bauxite" with Fred Williams as Co-Editor, Published by Wieley & Sons. Inc. Copyright © 2013 by The Minerals, Metals & Materials Society (TMS).

In 2018 Benny was editor and contributor to Chapter 12.1 Alumina, in the "SME Mineral Processing & Extractive Metallurgy Handbook", Published by Society for Mining, Metallurgy & Exploration (SME), Copyright © 2019.

Fred S. Williams Retired Alcoa World Alumina, Longmont, CO, USA. Email: fsw4102@gmail.com

Fred S. Williams graduated from the University of Kansas with a BS Ch.E. in 1958 and MS Ch.E. in 1960. Upon graduation he joined Alcoa Research in E. St. Louis, IL and continued to work for Alcoa in alumina and chemical research and production assignments in the US and Suriname for 42 years. Among other assignments, Fred was Manager of Alumina and Chemical Research at the Alcoa Technical Center, New Kensington, PA, Technical Manger Refining and Chemicals, Point Comfort, TX, Production Manager Refining and Chemicals, Point Comfort, TX and Technical Manager Refining and Smelting, Suralco, Suriname.

After retiring in 2002, Fred continued as a consultant to the alumina and bauxite industry for an additional 10 years.

Fred is the holder of several patents and has published a number of technical papers. He has been Alumina Subject Chairman three times and many times session chairman at TMS annual meetings.

Fred currently lives with his wife, Anne, at 4102 Heatherhill Circle, Longmont Colorado 80503, USA.

Chapter 2
Bauxite: Geology, Mineralogy, Resources, Reserves and Beneficiation

George Komlóssy, Caio van Deursen, and Benny E. Raahauge

Abstract In this chapter the Authors summarize the typical features of bauxite which are believed to play roles in mining, beneficiation and mainly in alumina processing, that is, the industrial value of the raw material. Geological curiosities are neglected. General physio-chemical conditions are outlined in order to understand the ore formation being determinant of the quality (chemistry and mineralogy) and quantity of the bauxite ore. Special emphasise is given to its chemical and mineralogical composition make up and their alterations are shown in selected examples, indicating the possible inhomogeneity of the mine product processed in the refineries. In deposit geology laterite, karst and sedimentary/paralic deposits are distinguished requiring significant differences in mining methods and technology applied in processing. In laterite deposits three main types are introduced namely: (1) plateau type deposits developed on morphological terraces and low land interfluves, (2) whale back plateaus and (3) dome shaped hills confined by hilltops and bauxites on hill tops with associated slope bauxites. For estimating the possible supply furnishing of operating refineries or for establishment of green-field plants an overview is given on global bauxite reserves and resources with their grade and mineralogy, amended with their further potential established with the aid of remote sensing technics introduced by the Author in 2006. Principles of physical bauxite beneficiation of the Run of Mine (ROM) bauxite is presented together with the most common unit operations applied. A flowchart example is shown integrating many of the unit operations. To minimize transportation cost and be able to supply more than one refinery, the beneficiation takes place at the mine site before the beneficiated bauxite is exported to the refinery. Principles of thermo-chemical bauxite beneficiation/activation is presented for three distinctive purposes: (1) Reduce organic carbon in the ROM bauxite before feeding

---

Benny E. Raahauge is deceased.

G. Komlóssy (✉)
Rio Tinto - Retired, Budapest, Hungary

C. van Deursen
Nexa Resources - Innovation Manager, São Paulo, Brazil

B. E. Raahauge
FLSmidth A/S - Retired, Deceased, Valby, Staden København, Denmark

B. E. Raahauge and F. S. Williams (eds.), *Smelter Grade Alumina from Bauxite*, Springer Series in Materials Science 320,
https://doi.org/10.1007/978-3-030-88586-1_2

the refinery and thus reducing the organic carbon input to the Bayer process liquor before digestion; (2) Activate the content of boehmite in the ROM bauxite to enable low temperature digestion with the side benefit of significantly reducing the organic carbon in the ROM bauxite, and (3) Pyro-genic attack or sintering of the ROM bauxite with limestone and/or sodium carbonate, when the bauxite is composed of mainly mono hydrate bauxites (boehmite or diaspore) and low to high silica content in order to maximize yield of alumina, recovery of caustic and reduced energy consumption. Only the soda-lime sintering process has gained commercial status over time, and especially in China. All the thermo-chemical bauxite beneficiation/activation processes are planned to be built at the refinery site, and in the case of the soda-lime sintering process integrated with the Bayer process in the overall process flowsheet.

2.1 Definitions

Bauxite: as petrographic term is a residual sedimentary rock product of tropical chemical weathering in which the free aluminium bearing minerals (alumina hydroxide and oxy-hydroxides) such: gibbsite, boehmite and diaspore attain min. 50% as shown in Fig. 2.1. There are bauxitic deposits, such as, Darling Range—Western Australi—which does satisfy the petrographic term, even if it is so, they are valuable ore in the alumina industries.

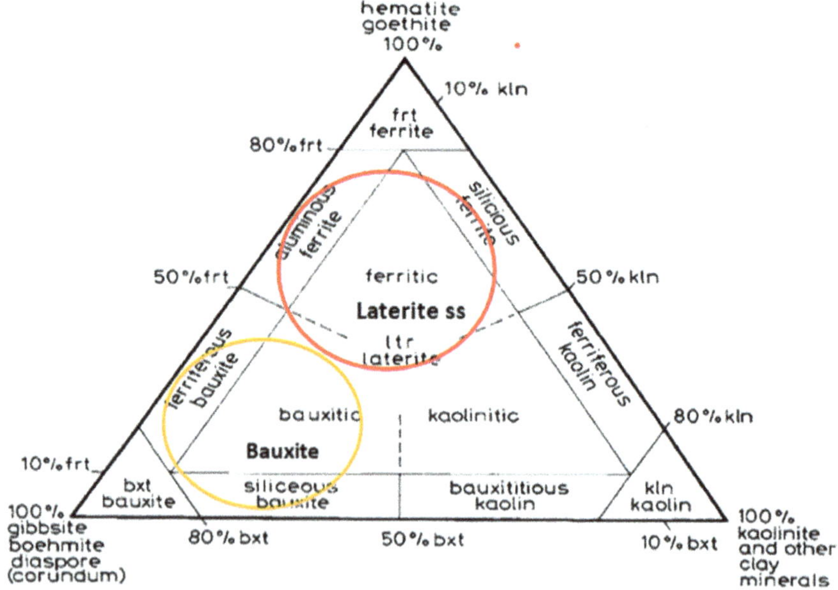

Fig. 2.1 Triangle diagram of the rock family of the laterites (Bárdossy and Aleva [1]), completed by: Komlóssy [2, 3]

Bauxite deposit: geometrically defined space, geological unit composed of the bauxite body and its associated formations. In practice they are the subjects of the mining operation.

Duricrust—cuirassé: Iron rich (Fe_2O_3% attains as much as 60–80%) layer developed on the topmost level of the laterite section in 0.2–0.3 m thickness It is mostly laminar or scoriaceous in texture.

Ferruginous laterite: part of the laterite in which Fe_2O_3 content more than 30–35%.

Industrial (commercial, cut off) grade ore: that part of the deposits from which alumina can be extracted economically.

Laterite sensu lato (sl) the whole weathered profile from the surface to the fresh rock involving all of the formations are shown in Fig. 2.1.

Laterite sensu stricto (ss) TAl_2O_3 changes between 20 and 40% $TSiO_2$: 30 and 45% Fe_2O_2 varies roughly between 30 and 35% (Fig. 2.1).

Protolith: parent rock of laterite cappings.

Regolith: weathering product of the solid rocks, that is, the whole laterite (sl). profile.

Saprolite (lithomarge): lowermost layer of the bauxite bearing laterite section between the bauxite and the altered fresh parent rocks (protoliths). It is clay, composed of kaolin, in its upper level bauxite nodules and lower level zone weathered fragments of the basement are found.

2.2 Chemical and Mineralogical Composition

Only that part of the bauxitic sequence is taken into account which is regarded to be industrial grade raw material. There are elemental and mineralogical concentrations or depletions in small non-minable masses. They are left out of consideration, along with special curiosities which might be characteristic for several deposits, but do not play significant role in the bauxite-alumina industries.

2.2.1 Major, Accessory and Trace Elements

2.2.1.1 Major Elements and Their Distribution in the Bauxites

Main elements are the aluminium, silicon, iron and titanium. The chemically combined (crystal) water is a principal constituent, too. These components are regularly analysed in each sampling interval by different methods not to be covered in this book. The elements are given in oxides, as a convention. The grade of bauxites is basically determined by the concentration of the alumina and silica contents. As a closer approach the available alumina and reactive silica are to be determined. In the reserve/resource estimate the term of the industrial grade ore is established by cut off values set up for minimum alumina and maximum silica contents. The cut-off values are different for different deposits depending on economic considerations. When it is relevant cut-off values are also set up for other contaminant accessories e.g., $\sum SO_3$ or $\sum CO_3$ or for both. Based on the chemical makeup bauxites are used for different purposes:

- Metallurgical grade ore (estimated to be more than 95% of the mined bauxites)
- Chemical grade ore: used in the production of various aluminium salts: TAl_2O_3 min. 55%, SiO_2 max.12%. Fe_2O_3 max. 2%.
- Abrasive grade ore: production of fused alumina, as suitable for natural corundum, in the manufacturing of grinding and polishing equipment TAl_2O_3 min. 55%, SiO_2 max. 7%, Fe_2O_3 max. 6%.
- Refractory grade ore: used to make bricks to line smelting and calcining furnaces Al_2O_3 min.60%, SiO_2 max. 10%, Fe_2O_3 max. 2% [4].

A. Alumina

Routine chemical analyses determine the total alumina (TAl_2O_3) in each sampled interval. The alumina is bound in the hydrated alumina minerals (gibbsite, boehmite and diaspore), in clay minerals, dominantly in kaolinite, and also in iron minerals (goethite and hematite) as isomorph substitution. The available alumina ($AvAl_2O_3$) is analysed in laboratories typically by bomb digestion tests carried out at different temperatures, retention times, caustic molar (or A/C) ratios. Some 92–95% of available alumina can be extracted ($ExtAl_2O_3$) in the refineries to calculate the *specific bauxite consumption* (bauxite tons to alumina ton)

The TAl_2O_3 content of the industrial grade ore in deposit average varies typically between 40% and 60%. In extreme cases it may be even more, such as at Shandong deposits, China: 68% [5, 6] or at Dobri pod deposit in Montenegro: 67.3% [7].

The TAl_2O_3 content most frequently is between 46 and 52% in laterite bauxites and that of in karst bauxites 52–55%. In the premium grade bauxites types e.g., at Sangaredi, Guinea, Onverdacht, Suriname it attains as much as 58% as deposit average. Similar high concentration is gained in beneficiated concentrate at Paragominas, Brazil [1].

In the bauxite ore the gibbsite occurs either in the form of more or less finely disseminated mineral and in structural elements (fissure and pore space fillings, pisoids and ooids). At some places it concentrates in nodules (e.g., Suriname, Brazil, Mexico) in cm or in boulders even in m magnitude (e.g., Gujarat, India, Manica Land, Tanzania). In these cases, the nodules and boulders consist of almost pure gibbsite with 62–65% of TAl_2O_3.

A part of alumina being bound in chamosite (Sect. 2.3.1) is extractable only by high temperature digestion (240–260 °C).

Alumina substitution in iron minerals in extreme cases, in individual samples, or part of ore body, it may attain 20 mol% e.g. at Wigton or even more e.g. at Partville 24 mol% in Jamaican karst bauxites [8]. At Timan bauxite (Russia) it varies between

8-20 mol% [9]. In the bauxites of the Mediterranean Province the alumina substitution attains 10 mol % or even more e.g. Montenegro [7]. In laterite bauxites, based on thousands of analytical data, the substitution may attain 10% in some cases 20% (Sajó, I. personal . communication 2017). At Moengo in Suriname it is 20-25% [10].

In most cases the alumina substitution in goethite is much higher than in the hematite, the ratio varies typically between 10:1 and 3:1

As a general rule vertically the highest alumina concentration is in the middle of the deposits, both in laterite and karst type of ores, although in case of laterites high grade ores can also be found along the scarps of the plateaux or on the surface (see Fig. 2.30 in Sect. 2.5.1.1.).

The highest alumina concentration horizontally indicates the drainage conditions of the basement, the more efficient leaching the higher alumina concentration. See Fig. 2.3.

B. Silica

The silica can be found in clay minerals (dominantly in kaolinite) and occurs in the form of quartz, as well. The chamosite, a typical iron mineral in sedimentary/paralic bauxites, also bears silica. The quartz content in karst bauxites is negligible. The quartz in gibbsitic bauxites (processed with low temperature digestion at 105–140 °C) is non-reactive. For that reason, beside the total silica concentration the determination of the reactive silica ($rSiO_2$) belongs to the routine analyses. In laterite bauxites the total silica content ($TSiO_2$) of industrial grade crude bauxite ore, and in the case of its beneficiation concentrates, in most cases, varies between 2 and 3%. such as In some Guinean bauxites e.g. at Sangaredi and at Fria, in Ghana at Awaso and Kibi, in Indian bauxites e.g. in Eastern Ghats and it is more than 10%. e.g. at Cataguases in Brazil (18%), at Los Pijiguaos, Venezuela (11%), at Darling Range in Australia (22%), at Manantenina, in Madagascar (32%). In these cases, the $rSiO_2$ concentration can be

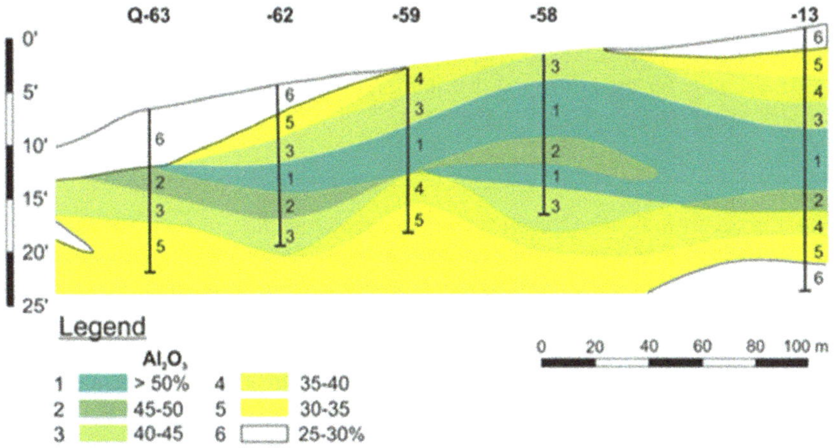

Fig. 2.2 Distribution of alumina in section—Western Ghats, Goa India [11]

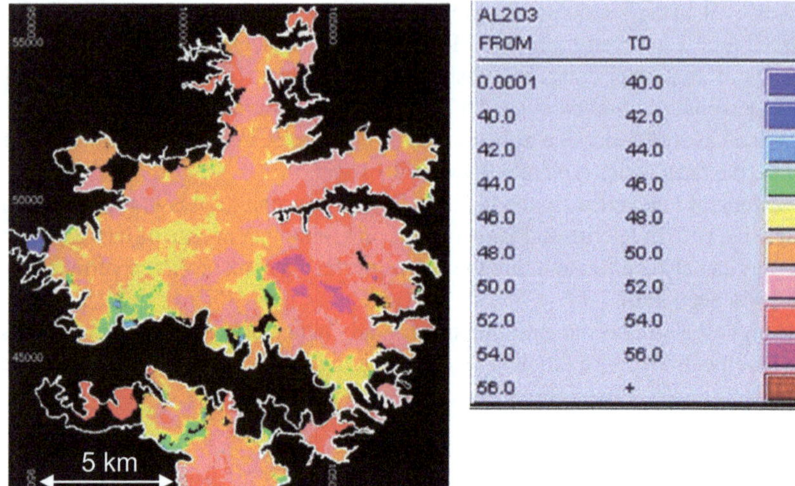

Fig. 2.3 Horizontal distribution of alumina in % at the Gove deposit – Australia r Rio Tinto Expl. Ltd

kept usually below 4%. In laterite bauxites the average total silica is estimated to be around 4%. In karst bauxites, the lowermost total silica values have been revealed in the Caribbean Province, e.g., in Jamaica it is 2–3% [8]. That of in the Mediterranean Province it is 5–7% in general. The lowest total silica content was found in the Distomon—Ghiona ore bodies, between 2 and 3%. In East European Province in some ore bodies of Timan its content exceeds 10% [9]. In the East Asian Province high total silica concentration occurs in the Chinese bauxites (bound both in kaolinite and chamosite). $TSiO_2$ in the run of mine ore of Henan deposits in 2011 was 11–12%(Wang Feihong et al. 2012) [5]. The silica distribution in a typical laterite section is shown in Fig. 2.4. The area of the minimal concentrations, *en grand ligne,* follow that of the maximum of the alumina as shown in Figs. 2.2 and 2.4.

An example for the horizontal distribution of the silica to compare with that of the alumina is shown in Figs. 2.3 and 2.5.

C. Iron

The iron occurs in two modifications namely: ferric and ferrous iron. The ferric iron is one of the main elements of the bauxites while ferrous iron is rather an accessory element, even if in certain levels of some deposits the ferrous may be the dominant form of the iron such as in certain sedimentary/paralic bauxites in China. The ferrous iron is a component of pyrite and in less extent of siderite and chamosite (see details in Sect. 2.3.1). In the routine analyses, where both modifications are present, the iron content is given in form of $\sum Fe_2O_3$. The ferrous iron content can be calculated on the basis of the mineralogical assays.

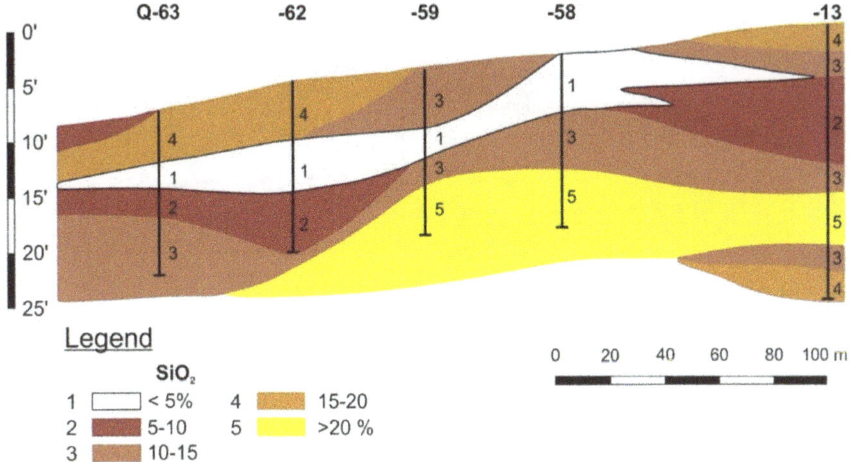

Fig. 2.4 Distribution of silica in section—Western Ghats, Goa, India [11]

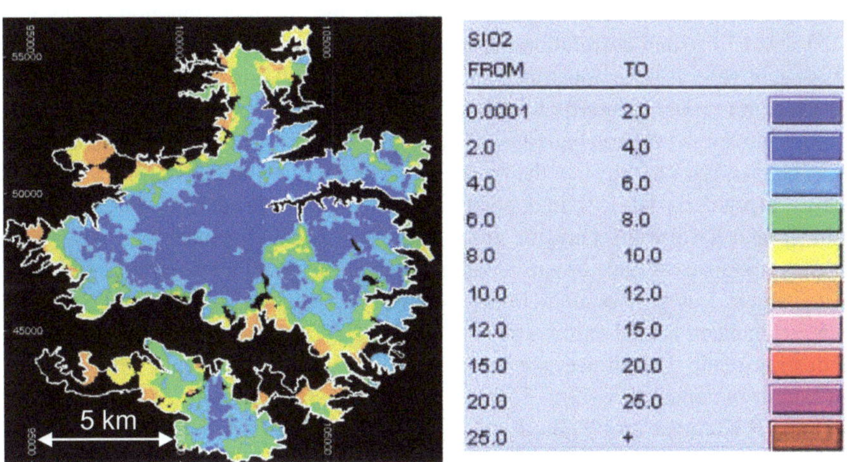

Fig. 2.5 Horizontal distribution of silica in % at the Gove deposit—Australia (Rio Tinto Expl. Ltd. [12]) i(8.)

The ferrous iron may be both primary element (syngenetic, formed at the same time as the bauxite) in sedimentary/paralic bauxites and secondary one in karst bauxites. In the latter case the ferrous iron is the product of supergene processes due to swampy environment, developed on its surface, resulting in complete or partial reduction of ferric minerals and pyrite (FeS_2).is formed. Later, because of the decomposition of the pyrite, acidic media formed (H_2SO_4) mobilising the ferric iron and results in iron-poor type of bauxites such as in cases of Kwakwani (Guyana) or many Mediterranean karst bauxites.

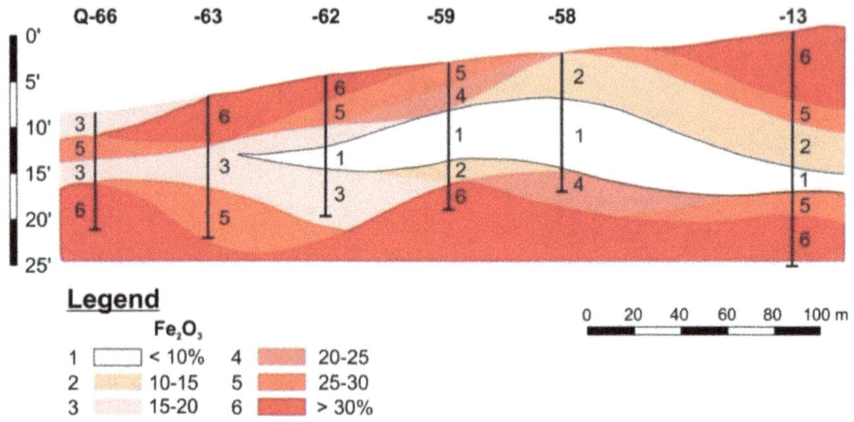

Fig. 2.6 Distribution of iron in section—Western Ghats, Goa, India [11]

The ferric iron content is in goethite, hematite and to a small extent in chamosite. As a general rule the ferric oxide concentration changes in the function of alumina.

There is a strong correlation between the alumina and iron: in laterite bauxites it is negative as it is demonstrated when Figs. 2.2 and 2.6 is compared. Correlation coefficient $r^2 > 0.8$ in laterite bauxites,

In contrary, in the karst bauxites the correlation is positive, however, the coefficient may be less. It is obvious that the premium grade laterite bauxites ($Al_2O_3 > 50\%$) the Fe_2O_3 is regularly 10–15% in deposit averages. e.g., Los Pijiguaos, Venezuela 13%, Linden and Kwakwani, Guyana, and a number of Indian bauxites, etc. Several ore bodies contain extremely low iron content type of ores used for different purposes as listed above. Low iron content, in deposit average, occur e.g. in Guinea at Sangaredi 6.5% calculated for the entire deposit [13]. The Fe_2O_3 content of laterite bauxites in industrial grade deposit average most frequently is between 18 and 25%. Bauxites of high iron content ($Fe_2O_3 > 25\%$) are known in some deposits of the East Coast in India at Karlapat 28.4%, Baplimali: 26.0%, Sijimali 29.4% [14], but also occur at Pakaraima Mt: in Guyana: 25.8%, [15] and at many ore bodies in West Africa.

In karst bauxites the iron content is typically between 20 and 24%. In Greece, e.g., Delphi—Distomon, or in Turkey Milas—Mugla it attains as much as 26%.

D. Titanium

In the Bayer process itself view titania content does not play a specific role in the alumina manufacturing, however the TiO_2 content may be harmful for the equipment due to scaling that form in the course of high temperature digestion. Its content, in overwhelming majority of bauxites, is between 2 and 3% in deposit averages both in lateritic and karst bauxites. As a general rule titanium is lower in granite, gneiss or alkaline rocks derived bauxites (1–2%) such as for example at Los Pijiguaos, in Venezuela, at Pocos de Caldas in Brazil and higher in basalt derived ones (4–6%) at the West Coast bauxites, in India. In extreme cases it may surpass 10%, e.g.

at Mainpat, Chattisgarh State, India [1]. In karst bauxites, in individual samples, the highest concentrations were recorded at Ishlahiye deposit (SE Turkey) attaining almost 8%. (East–West Mining. Int. Ply. Ltd. 2005).

E. Combined water

Instead of combined water (bound in alumina, silica and iron minerals) the loss on ignition (LoI) at 1050^0 C is determined which, involve the volatiles of carbonates (CO_2) as well. High combined water values are in the try-hydrate (gibbsitic) bauxites. In this type of ores, the goethite is the dominant iron mineral therefore contributes to the high combined water content, so that, LOI content attains 28–30% such as at Sangaredi in Guinea, at Onverdacht in Suriname. The LOI in the mono-hydrate type karst bauxites, if not contaminated by sulphur and/or carbonates ranks typically between 11 and 14%.

2.2.1.2 Accessory Elements

Those elements are ranked into this group which occur in order of magnitude of 0.0x% up to 1.0.x % in practically all the deposits. They are more or less contaminant components depending on their concentration. There is no data for all of deposits as their analyses are occasional. Regular analyses are carried out in such cases when their presence have already be revealed and extend over only those section of the profiles (usually the upper levels) where they may concentrate. Available data restricted mainly to those occurrences which are being mined or for which Feasibility Studies were prepared and made including detailed technological tests were carried out.

A. Carbonates

Carbonates are bound in Ca, Ca-Mg and Fe (details in Sect. 2.3.1). In analyses they are given in forms of CaO% or $\sum CO_2$. These constituents occur mainly in karst and sedimentary/paralic bauxites and less extent in laterite bauxites. In karst and sedimentary bauxites, they are primary (syngenetic) associations while in laterite bauxites it is a secondary one and found only in buried or at once buried deposits. In this respect the laterite bauxites of Gujarat State (India) shows the maximum where the CaO content reaches in deposit average as much as 2% at Jamnagar and 1,5%, at deposits of Kutch area, in contrary its content in bauxite from Maharastra State was detected in traces only [16].

CaO content both in karst and laterite bauxites is usually between 0,x and 0,0x % in order of magnitude. As much as 0.1% can be estimated to be typical worldwide. In karst and sedimentary/paralic bauxites, in certain levels, it may attain 1% or even more e.g., Timan. Russia (personal data collection 2003). At Severuralsk, (Russia) in the "porphyry and discoloured type of bauxites the $\sum CO_2$ attain 4–8% bound dominantly to Ca (personal data collection at site visit 2006). In case of re-deposited type of karst bauxites, because of the limestone or dolomite environment, in certain levels, the $CaCO_3$ or $CaMgCO_3$ occur 1–3%.e.g. at Halimba and Nagyegyháza in

Hungary. Moreover, the bauxite deposits may be intercalated by fragments carbonates in the form layers or lenses due to rapid accumulation and/or re-deposition of the bauxite. During the mine operation—because of dilution—the carbonate contamination increases. In these cases, checking of each shipment is important. When the carbonates may exceed 1% the routine analyses (applied during the exploration) must be completed with the determination of the CaO or $\sum CO_2$ content. In the reserve calculation barrier should be given for its maximum value making possible to satisfy the demand of the consumer. For Halimba, in Hungary the cut off for $\sum CO_2$ was1.5%. Niksic mine (Montenegro) applied cut off for CaO max 1% [17].

Carbonate is also bound in siderite as a primary constituent. Siderite, as primary mineral, is common in sedimentary/paralic bauxites identified at Lang Son deposits in Vietnam [18] and also at Severuralsk, Russia (Beneslavskij 1965) and many deposits in China. The carbonate content in siderite is estimated to be negligible in most of the industrial grade ores.

B. Sulphur

Sulphurs are expressed in forms of $\sum S$ or $\sum SO_3$ in chemical analyses. They occur dominantly in form of sulphides (pyrite FeS_2 and its dimorph: marcasite) and in form of sulphates (most commonly alunite: (KNa) $Al_3(OH)_6$ $(SO_4)_2$, melanterite $FeSO_4 7H_2O$ and gypsum: $CaSO_4 2H_2O$). Number of other sulphates in several bauxites are rather curiosities. The sulphide/sulphate rate in bauxites is practically unknown. In karst bauxites as a general rule the grey bauxites are rather pyritic while reddish are sulphatic. Sulphur, similarly to the carbonates, occur in primary form in sedimentary/paralic type bauxites and in secondary form dominantly in karst bauxites, and exceptionally in laterite bauxites (e.g., Kutch, in Gujarat State, India). In sedimentary/paralic bauxites, such as Severuralsk (Russia), the industrial grade ore in averages of ore bodies contains 0.1–1.3% of $\sum SO_3$ (VAMI document 2003). In the upper levels, in grey pyritic type of ores ($Al_2O_{3:}$ 44–45%, SiO_2 1.0–1.5%) the concentration of the $\sum SO_3$ is 12–15%. In China, in the same type of deposits the high sulphur content bauxites its concentration attains as much as 2%. By desulphurisation treatment (flotation) it is reduced to 0.2% [5].

In the karst bauxites in the industrial grade sections the $\sum SO_3$% ranks between 0x and 0.0x%, however, in those deposits which are covered by lignite (indicating H_2S producing swampy environment) its concentration may increase up to 1–2%. In Hungary, at her historical deposits, in the grey bauxites the sulphur content exceeded 10%, e.g. Iszkaszentgyörgy deposits (Komlóssy 1968) and Szőc deposits in Hungary [19]. Such high values may occur at other Mediterranean deposits, too. The grey piritic bauxites, occurring in the upper level of the ore bodies, rarely satisfies the specification of the industrial grade ore. In such cases the regular five component analyses should be completed for the whole profile with the component of $\sum SO_3$ in order to avoid the excess caustic soda consumption in the refinery. In Hungary e.g. a constraint was applied for sulphur and no bauxite of $>$ 0.6% $\sum SO_3$ content was qualified as industrial grade ore.

C. Organic Carbon

In chemical analyses it is given as C^{org} %. It is one of the critical contaminants in the Bayer process. Its average concentration, in karst bauxites is usually 0.0X % and in laterite bauxite 0.X% in order of magnitude [20]. Based on the available data the highest C^{org} content, in deposit average, was revealed at Gove with value of 0.25% (Hill et al. 1983). Of course, at that occurrences where the industrial grade ore crops out the surface the remnants of the roots of the vegetation increases, the C^{org} content may be above 1%. A common practice to decrease the organic input of the alumina refineries is to strip the topsoil at least one wet season before mining of the bauxite ore. This allows for natural leaching of the organic material by the ample precipitation of the wet season(s).

D. Phosphate

Its concentration is given as P_2O_5 determined by chemical analyses, as P determined by spectrometer and also as crandallite detected by mineralogical tests P_2O_5 content, as a general rule, varies between 0.X and 0.0X% in order of magnitude both in karst and laterite bauxites. In the Jamaican karst bauxite at certain levels it may attain almost 4% (Porter and Anderson 1982), though in deposit average it should not be more than 0.X% in order of magnitude.

2.2.1.3 Trace Elements

Comprehensive investigation of trace elements became widespread, comprising whole sections of bauxites. In the early sixties when Q spectrometers applied for their detections, later on the introduction of the electron microprobe analyser supported the analyses of the microstructures including trace element containing detritic minerals e.g. xenotime (YPO_4), monazite (Ce,La,Nd,Th)PO_4, zircon ($ZrSiO_4$), etc.

About nearly 60 elements were detected in bauxites [1], from which about 40–50 elements can be classified into trace elements. In the followings those elements are grouped into this class which are dominantly in the order of magnitude of g/t or ppm (0.000X%). There are some elements, such as P, Mn, Ca, Mg, C^{org}, etc., which may occur in order of magnitude of both accessory- and trace elements. Analysis of trace elements do not comprise all constituents, at each investigated deposit. There are no more than 8–10 elements which are regularly investigated: such as V, Cr, Mn, Co, Ni, Cu, Zn, Ga, Pb.

The concentration of the trace elements depends on two factors: namely: (a) their concentration in the source rocks and rather (b) the mobility (solubility) of the elements under the physical–chemical conditions of bauxite formation (temperature, annual precipitation, drainage, pH and Eh). The less mobile elements the higher the relative enrichment [21].

In the followings the concentrations of the most prominent elements are summarized, those ones which have been analysed in most bauxites. Compilation is basically based on the comprehensive review of Bronevoi et al. [22]. His figures are in accordance with the results of later publications.

In practical, economic respects Bárdossy and Aleva [1] divided them into three groups (1–3). Nowadays, the rare earth elements (REE) in bauxite residue, as one of their possible source, have been investigated more thoroughly than earlier, the bauxite and bauxite residue became the subject matter of several researches. For that reason, the above groups an independent one is distinguished (4).

(1) Elements that can be extracted economically as by-products of the Bayer process: Ga and V.

They are homogenously disseminated in bauxites, thus large-scale concentration changes do not occur [1]. The Ga concentration is in between 40 and 110 ppm in deposit averages. The V concentration is between 77 and 1370 ppm. Both elements can be extracted in the course of the Bayer process e.g., the cooling a part of the spent liquor.

(2) Contaminant elements, that increase the costs of the alumina processing (apart from those ones which are enumerated among the accessory elements)

Pb, Cu, Zn. The enrichment of these elements are thought to depend mainly on the source rock, thus, the magmatic rocks (granites, granodiorites, basalts, etc.) should contribute to their higher concentrations. In deposit averages their concentrations are as follows (Table 2.1).

(3) Majority of the trace elements are indifferent in the alumina processing; their concentration is shown in Table 2.2

Mn and Li may turn into solution in different measure in the function of process parameters.

Ni and Cr concentrations in the Mediterranean karst bauxites, in the Dinarids are in between 50–250 ppm and 130–840 ppm resp. (Crnicki and Jurkovic 1989). These ranges are valid for the Hungarian bauxites, as well. (Maksimovic et al. 1991).

The highest Ni and Cr contents have been revealed in basic rocks (basalt, gabbros) derived laterite bauxites attaining 1000 ppm. Moreover, nickeliferous laterites might have formed with 1–2% of Ni (e.g., Minas Gerais, Brazil). The highest Mn concentrations (above 1000 ppm are typical in karst bauxites, while there are others, like e.g. the Mn of which highest concentration (above 1000 ppm) is typical also in karst bauxites e.g., ex-Yugoslavia, Hungary (Crnicki and Jurkovic 1989; Maksimovic et al. 1991).

Table 2.1 Trace elements contaminant in the alumina processing (Komlóssy, G)

Element	Range ppm	Average
Pb	20–185	48
Cu	10–133	28
As	5–136	48
Zn	28–180	46

Table 2.2 Trace elements indifferent in the alumina processing (Komlóssy, G.)

Element	Range ppm	Average
Li	06–320	12
Be	0.7–7.4	0.9
Sc	8–180	25
Cr	85–5100	419
Mn	87–1250	188
Co	7–70	18
Ni	7–320	29
Sr	35–2010	298
Zr	174–675	380
Nb	14–93	45
Th	2–73	30
U	2–22	6.4

(4) Rare earth elements (REE)

The first comprehensive investigation of the rare earth elements was made on the ex-Yugoslav karst bauxites by Crnicki and Jurkovic (1989). The following components have been investigated: Y, La, Ce, Nd, Sm, Eu, Gd, Tb, Dy, Ho, Er, Tm, Yb, and Lu. As a general information the following data in ppm have been collected from different published sources (Table 2.3).

The Y, La, Ce and Nd are believed to be the most prominent elements. The La, and Ce, are bound in phosphate mineral: monazite (Ce, La, Nd, Th)PO_4 and carbonates e.g. bastnasite: (Ce, La)F(CO_3).

The ΣREE ranks between 60 and 800 ppm, in deposit averages (Bronevoi et al.). That of in the ex-Yugoslavian bauxites is between 600 and 900 ppm (Crnicki and Jurkovic 1989) and in Hungarian ones: 1200 ppm [23]. Nevertheless, in strict sense the Sc is not a REE element, due to similarities in the chemical behaviour and the unique economic importance, Sc is usually discussed and investigated with the REE elements.

Table 2.3 Rare earth element concentration (Komlóssy, G)

Elem	Y	La	Ce	Pr	Nd	Sm	Eu	Gd	Tb	Dy	Ho	Er	Tm	Yb	Lu
Min	40	9	0,4	23	56	8	2	9	3	9	2	7	1	6	1
Max	170	340	504	55	180	52	5	100	15	31	17	22	2,5	13	3

2.3 Mineralogy and Condition of Mineral Forming

2.3.1 Mineralogy

The mineralogy of the raw material determines the technology applied in the refineries. In the followings the description of bauxite mineralogy will be restricted to that part of the bauxite deposits (geological term) which is classified as industrial grade ore, that is, suitable for the processing in the plants. Consequently, no curiosities or extreme phenomena will be considered.

The mineralogical association of any given bauxite deposit may change both vertically and horizontally. To clarify the mineralogy of the bauxite is an integrated part of geological exploration, more over indispensable information in characteristic/representative bulk sample compilation made for alumina technological tests [24]. See Appendix 2.3 Representative sampling for elaboration of the alumina manufacturing process (Theory and practice).

Alumina minerals

Based on technological considerations bauxites are divided into the following three groups by their dominant alumina bearing minerals (a) tri-hydrate-bearing (gibbsitic), (b) monohydrate bearing (boehmitic-diasporic), (c) mono-hydrate-bearing (diasporic-chamositic) bauxites.

Gibbsite $Al(OH)_3$. Its concentration varies between the extremes of 0 (in older karst and sedimentary (paralic) bauxites, (see details in Sect. 2.1.3.2) and 95% in nodules or boulders of some laterite bauxites, such as, in Gujarat and Goa in India or in Cataguaes in Brazil, Chiapas in Mexico, etc. The Manantenina (Madagascar) deposits contain 1–20 cm size gibbsitic nodules with 70–90% gibbsite embedded in a kaolinite-quartz rich matrix [25]. Most laterite bauxites are predominantly gibbsitic. On the average, its concentration varies between 60 and 75%. Caribbean gibbsite is the only alumina-mineral. Gibbsite content of the Jamaican karst bauxite varies between 48% at Smithfield and 72% at Farenough [8, 26].

As to the mineralogy of the textural element gibbsite is concentrated in pseudomorphs after feldspars or it may occur as cavity- or fissure-fills [27]. The crystal size of gibbsite ranges from ten to some hundreds of μm, however, sometimes it may also attain the mm-size (e.g., Guyana, India, Mexico). It has been observed that in the alumina-plants, at 105 °C, under conditions of atmospheric digestion, well crystallized gibbsite lowers the efficiency of the processing [27, 28].

Boehmite γ-AlO(OH) It is one of the main alumina bearing minerals in almost all karst bauxites of Italy, Hungary, Croatia, Bosnia, Montenegro, Romania and in Russia (Timan). Boehmite content of these deposits ranks usually between 55 and 60%. It also occurs in laterite bauxites most frequently in 1–3% on the average. It is usually concentrated at the upper level of the ore body, attaining 7–15% locally. At Weipa (Australia) in the upper "boehmite" zone it may reach even 30%. (Komlóssy

1993) The Central Indian bauxites (Chattisgarh State—Maikala Range) contain 14–16% of boehmite (deposit average). There are dominantly boehmitic laterite deposits, as well. Typically, they are buried ones such as Az Zabirah (Saudi Arabia) where the average gibbsite to boehmite ratio is 1:2 [29, 30].

As to texture, boehmite is usually disseminated in the matrix and may concentrate in textural elements like in pisolites (Weipa—Australia) (Morgan 1991). In the Arkansas bauxite boehmite occurs exclusively in the pisolites (Gordon et al. 1958). Its crystal size is generally less than 10 μm [27].

Diaspore α-AlOOH It is the characteristic alumina-mineral of sedimentary (paralic) bauxites e.g., Russia (Severuralsk), Chinese bauxites and elder karst bauxites in Romania (Padurea Crailiui), Turkey (D'Akseki-Seidisehir), Greece (Parnassos – Kiona) and Iran (Jajarm). The diaspore content of these deposits varies between 45 and 55% usually. The highest values were detected in the Uralian and the Chinese bauxites, where the most important ones are located in Lianoning, Shanxi, Shandong, Guizhu and Guanxi provinces. In the high-grade Chinese bauxites (TAA > 60%) the diaspore content may reach 70–80%. (Wangsing Li 2013). Diaspore is associated with chamosite in sedimentary (paralic) bauxites.

Diaspore may occur also in some laterite bauxites but always in minor quantities (~1%) In Kutch (Gujarat—India) as much as 4% was detected in the bauxites [1].

It is a general rule that the younger bauxites, in age Mio-Pliocene (<25 Ma), are much more gibbsitic with 64–83% of gibbsite than the elder ones, in age Mesozoic (250–65 Ma), with 4–11% of gibbsite content. In contrary the highest diaspore concentrations are found in the eldest deposits, in age Palaeozoic (500–250 Ma) with diaspore concentration of 42–54% (Bárdossy 1973), more over it may attain 80% as mentioned above.

Iron minerals

Goethite (FeOOH) is the most important iron mineral in laterite bauxites. There is a strong negative correlation (typically $r^2 = 0,9$) between the gibbsite(boehmite) and goethite(hematite) contents. Goethite concentration typically varies between 5 and 20% on the average. When the highest goethite content ranks between 30 and 40%, (Bougainville- Australia) then the bauxite is called ferriferous bauxite, however, bauxites of such high goethite content, occurring on the top of the deposits, hardly satisfies the requirements of industrial grade ore, therefore it is called iron-rich laterite.

In karst bauxites, as a general rule, goethite plays a subordinate role compared with hematite. Goethite is associated with gibbsite, so that, gibbsite-bearing bauxites are always more goethitic than their boehmitic counterparts. Goethite content is—as a rule—greater in younger deposits (Paleocene and younger < 65 Ma) while it is missing in old deposits (Paleozoic > 250 Ma). Normally it does not occur as a primary mineral phase in diasporic ores (see Fig. 2.12 in Sect. 2.3.3).

Because of the structural isomorphy of goethite and boehmite often there may be AlOOH isomorphic substitution in the crystal lattice of goethite. Above 10 mol% of substitution the mineral is called alumo-goethite. At low digestion temperatures

(120–140 °C) alumina cannot be extracted from the goethite In the Jamaican karst bauxite the substitution varies between 14 and 24 mol% [12]. In the Hungarian karst bauxites, the substitution varies between even more extreme values (from 5 to as much as 30 mol% [27] these extreme values are deemed to be typical for the Mediterranean Bauxite Province. As a general rule higher goethite contents mean also greater percentages of alumina substitution in karst bauxites.

Alumina substitution in goethites of laterite bauxites occurs mainly in the younger ores (eg. Mio-Pliocene or younger (<25 Ma) bauxites of Suriname the Paranam. It contains 9–13 mol% of alumina, while that in the Moengo district is characterized by 20–25 mol% Al-substitution (Grubbs et al. 1982.)

Hematite Fe_2O_3 is usually the main iron mineral of karst bauxite. Its quantity, in the industrial grade ore, is changing between 5 and 25% typically. The differences are composed of goethite. Goethite to hematite ratio is changing both vertically (see Fig. 2.8 in Sect. 2.3.2) and horizontally. In the Dinaric Alps (ex-Yugoslavia) and Greece, and in Turkey the boehmitic/diasporic bauxites are usually purely hematitic with 15–25% of hematite. There may occur extreme values—iron rich and iron poor bauxites types—within one and the same ore body, but they are not believed to play any role in the alumina processing. Hematite in sedimentary(paralic)bauxites is a secondary mineral, due to the supergene oxidation of chamosites [11]. In these cases, the percentage of hematite depends on the degree of oxidation of the bauxite. In Vietnam (Lang Son) the oxidation was apparently highly efficient: the hematite to chamosite ratio is > 1, whereas in Russia (Severuralsk) < 1) in China it is almost nil (showing that those bauxites were not subject to oxidation).

In the ore grade laterite bauxites hematite plays subordinate role compared with goethite. Locally it concentrates at the top level of the surface called duricrust (or cuirasse) where its concentration may surpass even 60%. There are exceptions: e.g., Aurukum (Australia) hematite is the sole iron mineral. At Onverdacht (Suriname) and at Los Pijiguaos (Venezuela) hematite concentration surpasses the goethite content [1].

Siderite $FeCO_3$ as primary mineral it occurs in association with chamosite. In sedimentary (paralic) bauxites the carbonate content of 1 to 3% is mainly bound in siderite. It also may occur typically in minor quantity in karst bauxite due to the epigene secondary processes. There is no siderite in laterite bauxite apart from some exceptional cases such as in Gujarat (India) where its content may vary between 0.4% and 6% due to inundation of bauxite by reducing swamp [1].

Pyrite and its dimorph modification marcasite FeS_2. Grey pyritic-marcasitic karst bauxites—developed by reducing conditions provoked by swampy environment on the surface. It requires also high-enough S-activity, (e.g., the presence of sulphate-rich marine or brackish pore-waters). Pyritic bauxites are common in the Mediterranean region. They are situated at the upper level of the deposits. This type of bauxite, because of its high sulphur content, is not processed in the refineries, even if its alumina and silica contents satisfy the requirements of industrial grade ore. Moreover, epigenetically, because of the gradual oxidative decomposition of the sulphides, the descending migration of the solution may contaminate the lower (non-pyritic) levels, as well, forming principally alunite (K.Na.) Al_3 $(OH)_6$ $(SO_4)_2$. In such cases the regular five component analyses are to be completed with the component of $\sum SO_3$ in order to avoid the high caustic soda consumption in the refinery. In Hungary, e.g. a barrier was applied for sulphur and no bauxite > 0.6% $\sum SO_3$ content was qualified as industrial grade ore. Sedimentary (paralic) bauxites usually contain more or less syngenetic pyrite (Beneslavkij 1963; Komlóssy [11]).

There are several laterite bauxites buried by lignite formations (such as Gujarat—India, or Linden-Mackenzy in Guyana). Decomposing sulphides resulted in acidic media by which the ferric iron has been mobilized and bauxite of low iron content ($Fe_2O_3 < 5\%$) was formed. They are utilized as abrasive-, refractory-, or even chemical grade types of ore.

Silica minerals

Kaolinite: $Al_4(OH)_8$ (Si_4O_{10}) is by far the most important silica mineral both in karst and laterite bauxites. significant impurity of the ore. The grade of bauxite, beside the quantity of the free alumina bearing minerals, depends on the quantity of kaolinite expressed in the qualification as reactive silica. High grade bauxite contains less than 10% kaolinite (<5% $rSiO_2$). In bauxite profiles the minimum values are found in the upper-middle levels, downwards its quantity is gradually increasing and the ore body, after a zone of transition, turns into kaolinite rich formation, called saprolite in lateritic areas. At Kutch (Gujarat State, India) pure china clay formed between the basalt and bauxitic horizon (see Fig. 2.15 in Sect. 2.4.3).

Kaolinite is more or less evenly disseminated in the bauxite, but by secondary processes the kaolinite precipitates along the fissures and joints or fills the pores. When it is dominant, in mature laterite bauxites the appropriate drilling method applied in the exploration phase should be carefully chosen. It may happen that using the technique of air flush drilling, kaolinite is blown out to the surface from among the bauxite fragments resulting in an artificial "beneficiation" of the samples during drilling. This way the reserve estimate will indicate significantly better-quality bauxite than later on the quality of the run off mine ore. The differences in reactive silica content may reach up to 1–3 abs. % (Suriname—Onverdacht, Accaribo deposit or Saudi Arabia—at Az Zabirah deposit—Komlóssy experiences in 1993 and 2003 resp.). Similar apparent up-grading may happen when laterite bauxites are stored on the stockpile for a long time. During the heavy monsoon periods kaolinite is washed down from the surface of the bauxite fragments or boulders so that on the surface of the stockpile the ore will show better quality, than the average of the stock. As a consequence, when the quality check is made sampling should be extended to the whole section of the stored bauxite (Gujarat—India,—Komlóssy. experience 1999).

Halloysite $Al_4(OH)_4Si_4O_{10}$ and **metahalloysite $Al_4(OH)_4SiO_{10}$** may occur in several bauxites (Tamil Nadu and Maharastra States—India) mainly as secondary minerals forming fissure fillings. Halloysite is the main silica mineral in the Central

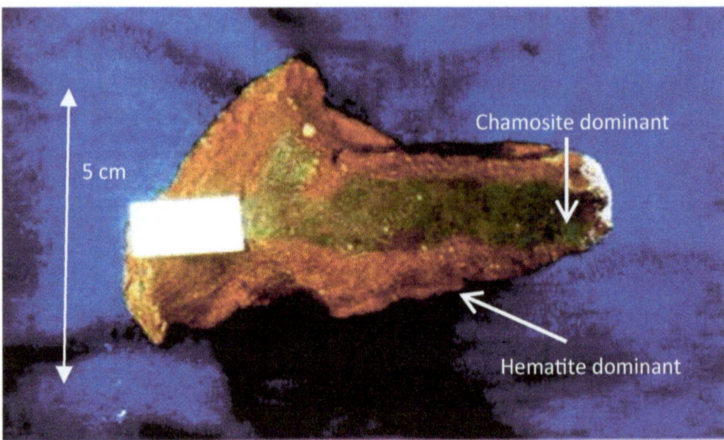

Fig. 2.7 Re-oxidized fragment of diasporic-chamositic bauxite—Ban Long, Lang Son Province, Vietnam [31]

Highlands—Vietnam (Novikov et al. 1985). The silica in halloysite and meta-halloysite is also reactive. The behaviour of these minerals during the alumina processing, is similar to that of kaolinite (Fig. 2.7).

Chamosite $(Fe^{2+}\ Fe^{3+},Mg)_6(Al_2Si_2\ O_{10})(OH)_8$ is the characteristic iron bearing clay mineral in some karst, but mainly in sedimentary—paralic bauxites accumulated primarily in a reducing environment (see Sect. 2.5.2.2) Chamosite occurs in Iranian (Jajarm), Russian (Uralian) and most Chinese bauxites and in North Vietnam (Cao Bang – Lang Son). Its quantity in greenish-grey reductive types of ores ranks typically between 25 and 40%. In extreme cases its concentration in selected specimens may attain even 80% [11]. In deposit averages in Lang Son bauxites (partly oxidised type ore) it is between 13 and 16% [18, 26]. By secondary oxidation manifested along fissures chamosite decomposes to hematite, diaspore and kaolinite. Its silica content is reactive, the surpass in alumina can be extracted at high temperature digestion (280 °C).

Quartz SiO_2 is a relict mineral in bauxites derived from quartz-bearing parent rocks (granitoid rocks gneisses, charnokites, khondalites etc.). It attains 10 to 25% in bauxites of Manantenina (Malgas Republic), Mulanje—Malawi, Darling Range (Australia), Los Pijiguaos (Venezuela), etc. At low temperature digestion, applied for gibbsitic ore types, it is a non-reactive component. Almost all the laterite bauxites contain 0.5–1.0% quartz, (it was detected even in basalt-derived ores, too (e.g.Western Ghats—India).

Titanium minerals

Anatase TiO_2, is the most- and **rutile Rutil TiO_2,** is the less frequent titanium minerals in the bauxites. Their concentration in laterite bauxites is typically around 3% and in karst bauxites that of is 2%. Rutile is mainly of detrital origin, while anatase is considered to be syngenetic formed on ferralliltic alteration of Ti-containing primary silicates. The maximum values of Ti were found in basalt derived deposits where total TiO_2 percentages may reach up to 10% (e.g., Kuch—India). In the South Vietnamese laterite bauxites, the major titanium mineral is **Ilmenite $FeTiO_3$.**

Titania concentration does not influence the efficiency of the alumina extraction, however, the anatase may take part in the process when perovskite ($CaTiO_3$) forms at lower temperatures in the presence of carbonates (Banvölgyi oral comm. 2017).

Carbonate minerals

Carbonate content of bauxite increases the caustic soda consumption. Above a certain level the refinery has to be prepared for modification of the routine process. Difficulties are increased by the fact that the carbonate concentration varies between extreme values and its distribution is uneven within the ore body. Therefore, the refinery may be supplied with bauxites of fluctuating carbonate concentration and, for that reason, in due cases, the regular checking both during the exploration and mining operations is important. Please refer to Chap. 8: Bayer Process Impurities and Their Management for a more detailed discussion of the carbonate issue.

Calcite $CaCO_3$. Bauxites are contaminated by carbonate minerals of which the most frequent is usually calcite. It may occur in between 0.x and up to x,0% both in karst, laterite, but mainly in sedimentary (paralic) bauxites. Calcites are partly epigenetic (karst and laterite bauxites) partly syngenetic in the sedimentary (paralic) bauxites forming carbonate rich layers in their upper and lower horizons where their concentration may rise to > 10%. $\sum CO_2$ in the Severuralsk (Russia) bauxite varies between 1.9 and 4.0% in average it is 2.9%. For that reason, it is blended with bauxite of low carbonate content in order to the contamination be dropped to an acceptable level of 1%. Calcite occurs in fine distribution in the matrix, but in majority as fissure fillings. Crystal size attains 1–6 mm [32].

Dolomite $MgCa(CO_3)_2$ in bauxite is usually a physical contaminant in the run off mine ore. Karst bauxites having been accumulated on dolomite surfaces are regularly contain smaller or larger fragments of this mineral. The contamination is higher close to the bedrock or to fault planes. When bauxite was re-sedimented during its final accumulation, dolomitic debris are common in the ore body. In extreme cases the ore body may be divided by several m thick layer(s) of dolomite debris or dolomite silt distributed in the bauxitic matrix (see Fig. 2.27 in Sect. 2.4.4). In such cases because of dilution the extracted bauxite may be extremely rich in dolomite, blending is needed.

2.3.2 *Mineral Distribution in the Ore Bodies*

The mineralogy of bauxite deposits is heterogeneous in space—there may be quite a variation of the mineralogical composition both vertically and horizontally.

In this respect—apart from the reactive SiO_2 content (kaolinite)—the alteration of the gibbsite/boehmite goethite/hematite and goethite/alumo-goethite ratios are the most significant.

Secondary alterations play, at places, also important role such as pyritisation. The possible alterations are to be taken into consideration by the mining companies and refineries as the efficiency of the alumina extraction might be, more or less, influenced by them. Examples for vertical change in mineralogy are shown in Figs. 2.8 and 2.9.

At CBG mine Guinea in the gibbsite rich Sangaredi deposit inside the ore body there is an interbedded boehmite rich layer as it is shown in Fig. 2.9. Bauxite mining is going on with selection of the tri-hydrate (TAA 143 °C) and monohydrate (TAA 243 °C–TAA143 °C) type ores for processing at different refineries. For that reason, the five component routine analyses are completed regularly with *bomb digestion tests.*

To be aware of the horizontal variation of the mineralogy of the deposit, or of the group of deposits, within a given mining unit is important in order to avoid the feeding of refinery with bauxite of various mineralogical make up. Unfortunately, these kinds of investigations are sometimes neglected. Apart from the increasing kaolinite content towards the edge of the karst bauxite deposits, the most important element of the inhomogeneity is the change-in-space of the gibbsite/boehmite and goethite/alumo-goethite ratios.

In plateau type laterite bauxites, developed on quasi flat surfaces (at same relative and absolute elevations) the horizontal change in these ratios is deemed to be negligible. (e.g., West Africa—India). On the contrary, in cases of those deposits which formed on hill tops and re-accumulated on slopes the gibbsite/boehmite ratio may

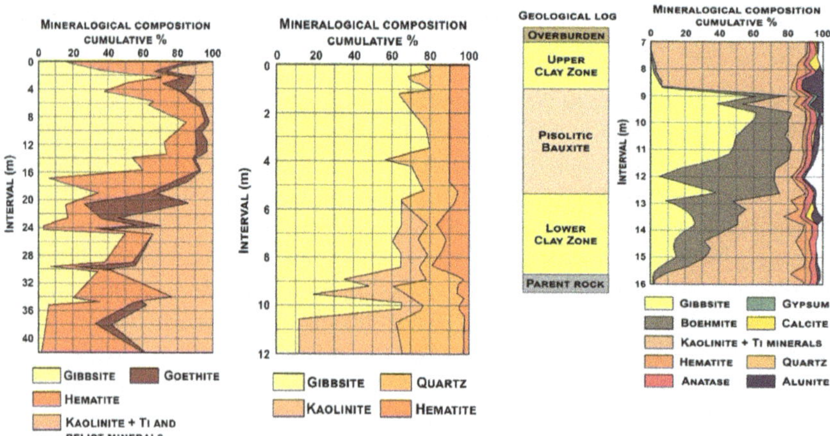

Fig. 2.8 Vertical change in mineralogy at Los Pijiguaos—Venezuela, at Panjpatmali—Orissa, India and at Az Zabirah—Saudi Arabia (after in rank Alusuisse - CVG [33], Nalco document 1985 and Black et al. [30])

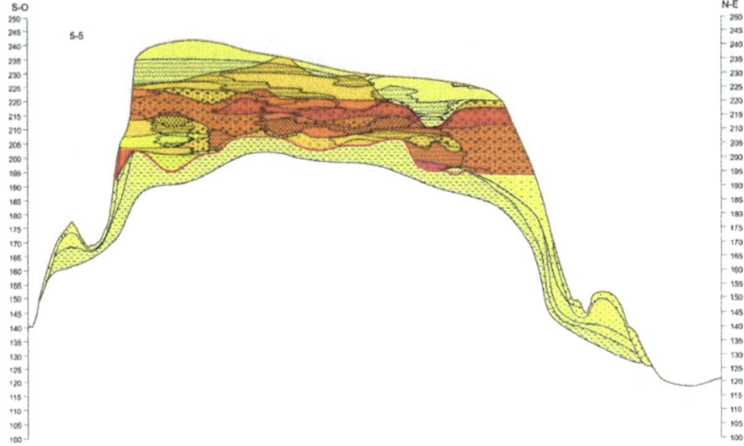

Fig. 2.9 Tri-hydrate—mono-hydrate variation in the Sangaredi—Guinea, CBG mine [13]. Explanation: yellow: mono content < 3%, red: mono content > 10%

drop. The measure of the change is supposed to be more when the difference in relative elevation of the levels is greater. Gibbsite-rich bauxite may turn into boehmite -rich bauxite (e.g. Bao Loc—Vietnam), In case of the Zabirah (Saudi Arabia) buried deposit, the gibbsite/boehmite ratio is considerable lower in the deeper situated parts of the ore body Ross, Y et al. [34].

The same phenomenon has been revealed in the Iszkaszentgyörgy (Hungary) karst bauxite deposits. The gibbsite/boehmite ratio at Rakhegy deposit, along 2 km of length, towards the deeper levels, changed from 3:2 to 1:3. The same alteration at the Bito deposit shows values from 1:5 to 1:10 [21]. The similar phenomenon is supposed to apply to other karst deposits and also to the case of boehmite/diaspore ratio.

2.3.3 General Genetic Aspects, and Mineral Forming

Conditions of bauxite forming have been investigated by a number of professionals from the early twentieth century. Establishments are summarized as follows:

(1) Climatic conditions: Recent bauxite forming takes place in the tropical belt where the mean annual temperature is around 22 °C and the rainfall minimum 1200 mm/annum, distributed over 9–11 rainy months and 1–3 months of fair seasons. The rate of rainy and dry seasons is changing. Less time is compensated by more rain. Position of the continents have changed during the geologic times. The bauxite is a reliable climatic indicator.
(2) Vegetation plays important role in bauxitisation. Bauxite deposits are found below tropical rain forests and grass lands. The root activity plays a role in

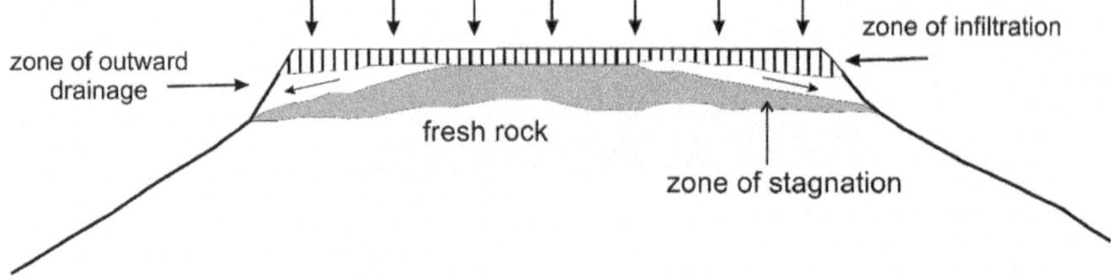

Fig. 2.10 Outward lateral drainage on dissected peneplain [35]

fragmentation of the parent rocks provoking decomposition of alumo-silicates on one hand and on the other one silica is extracted and bound by the plants (Mikhailov 1964).

(3) Geomorphology: There is a strict relation between the geomorphology and bauxitisation. Bauxitisation is a land-forming process. Deposits, in the followings, are classified by geomorphological characteristics. Laterite bauxites associate to positive—while karst bauxite to negative landforms. In both cases the fundamental condition is the good drainage, free flow of the solvents, that is, a suitable hydrological system. On higher levels the relative elevation is determinant, on lower levels the hydrological pattern, the density of networks of rivers and vados waters.

(4) Hydrogeology: In the laterite forming there are three zones: zone of infiltration, zone of outward drainage and zone of stagnation as it is shown in Fig. 2.10.

During the lateritisation, the mobile elements, such as alkalis and silica bound to alumo-silicates turn into solution in the fluctuation zone of the ground water table where the oxidisation and solution are the most effective, resulting in relative concentration of alumina, iron and titania. Silica bound to quartz is resistant, less soluble than the bound one.

The solved elements are leached out by drainage (alkalis and part of the silica). At the same time begins separation of the alumina from the iron. The removal of silica and Al/Fe separation is the **process of bauxitisation** in strict term. See Fig. 2.11.

In the zone of stagnation there is a direct argillation from which the iron is partly or completely depleted mainly at the beginning of the process of alteration.

(5) Protolith (parent rock). **Laterite bauxite** formed by the weathering of alkali- and calc-alacali alumo-silicates of different origin and chemical, mineralogical composition e.g. granites with its metamorphosed variants like gniesses, charnockite, along with diorites, dolerites, syenites, anorthosites, etc.) and rocks of volcanic origin (from which the basalt is the dominant) and marine sediments such as sandy or silty clays and their metamorphosed formations like, claystones, schists, gneisses, etc. are the most frequent. The role of the chemistry of the parent rocks (protolith) is in general subordinate. In the formation of the alumina minerals the protolith has no qualitative effect. Instead, in case of the iron bearing minerals the relation is rather quantitative since significantly more iron minerals were formed from basic rocks (e.g., basalts with

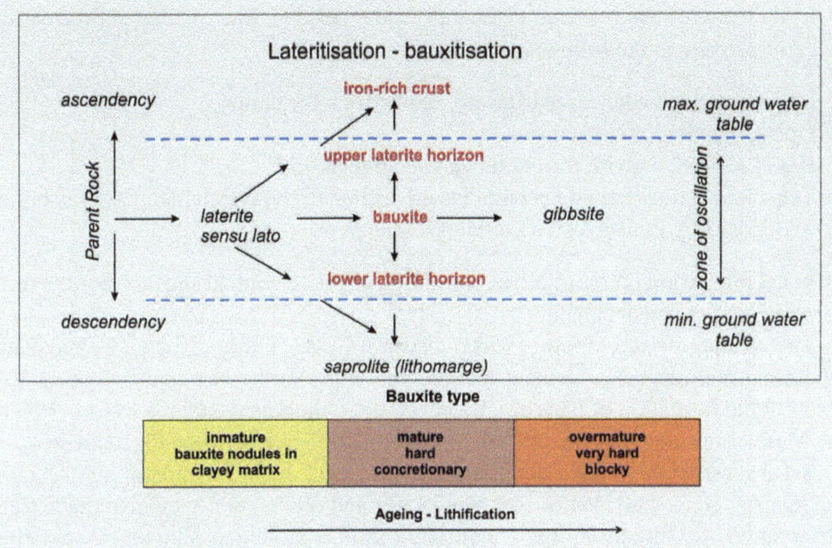

Fig. 2.11 In situ laterite bauxite forming [36]

their 12–16% of $\sum Fe_2O_3$) than from acidic ones (e.g., granites and their metamorphic variants—with their 2–5% of $\sum Fe_2O_3$ content). The quartz content of the rocks of granitoids is inherited by the bauxites, so that, non-reactive silica concentrates in weathering profiles developed over families of granitic (igneous and metamorphic) rocks, along with q-rich (meta)sediments. It has been concluded that there are two determinant conditions of bauxite forming rocks. They are the water conductivity depending on the permeability and, as far the chemical composition is concerned, it is no more than the alumina/iron ratio. The rule of thumb is that rocks in which this ratio is less than one no bauxite, but ferruginous laterite may form. The relative enrichment of the elements, compared with the protoliths, depends on their solubility under given Eh and pH conditions.

The origin of the karst bauxite was heavily disputed during the last century. Neglecting the overview of historical scientific details the theories can be classified into two main groups: *(para)auchtochtonous* theories regarding the source rock of the carbonate formations and *allochtonous* theories regarding the source lateritic weathering product of alumo-silicates locating out of the carbonates. The last one was not possible to maintain for all karst bauxites in frame of the general development of geo-sciences.

(6) Tectonic position: Chemical weathering manifests in tectonically calm periods. Laterite deposits are found on tectonically stabile continental plates (South America, Africa, India, Australia). Karst bauxites occur in orogenic belts, but

they are always relating tectonically calm periods. Bauxite deposit formation took place in the following steps [37]:

- Large-scale subsidence: marine carbonatic rock formation
- Uplift, denudation, planation
- Karstification, bauxite formation and accumulation
- Subsidence: final bauxite accumulation, filling up the karstic depressions, burial of deposits by transgressive marine sediments.

Mineral formation is controlled by prevailing physic-chemical and thermo-dynamic conditions of the environment.

The weathering of carbonate rocks—even if it was highly disputed by the overwhelming majority of prominent bauxite geologists in the last century—may also result in the formation of bauxites, however, their alumina content is low (1–3%) in the Mediterranean Province and 0.X % in the Caribbean Province [5, 21] resp.

Several researchers have dealt so far with clarification of mineral synthesis in laboratories (e.g., [38]; Pedro and Melfi [39]; and others) and they also made field observations for measuring pH conditions within bauxite deposits which are being formed recently (Belinga [40]; Valeton [41]; Jepsen and Schellmann [42]). Summarizing the results completed with field experience and bauxite genetic considerations an Eh–pH diagram was constructed showing the stability fields of the main Al and Fe minerals as shown in Fig. 2.12.

The bauxite mineralogy in karst bauxites depends on the paleo relief and relation to fluctuating karst water table as described in Sect. 2.5 and shown in Figs. 2.50 and 2.54.

2.4 Bauxite Lithology

2.4.1 Generalities

The lithological characteristics of the bauxite fundamentally determines the exploration technics pitting its sampling and drilling technics. The reliability of the applied exploration technics is the bases of the reliability of the reserve/resource estimation in terms both quantity and quantity. It is the bases of the whole industries from the economic decision making at the beginnings, through the mine planning up to the establishing of the most economic procedure used in the refinery. Selection of the most proper exploration technics is the responsibility of the explorer geologist.

As far as the mining is concerned the lithological properties of the bauxites determines the following characteristics:

(a) Hardness: blasting is needed or not.
(b) Heterogeneity in grade: Selective mining is needed or not. When the bauxitic complex is divided by barren (non-industrial grade) large layer in min. 1 m thickness and it can be visually differentiated, as it is usually possible in laterite

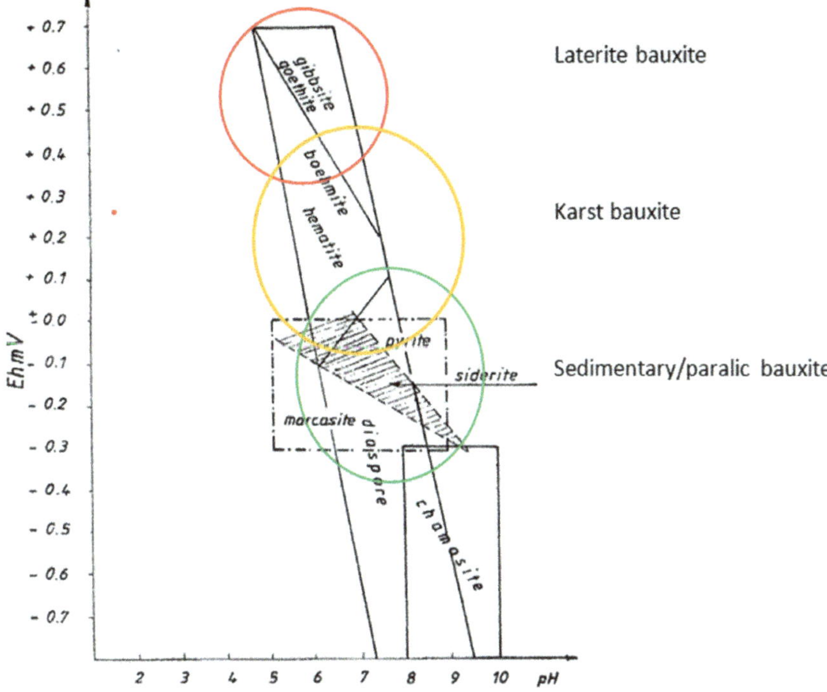

Fig. 2.12 Stability fields of the main Al and Fe minerals in the function of Eh–pH conditions in different deposit types (Komlóssy [3], 1997)

deposits, selective mining is possible provided the company is prepared technically. In case of karst bauxites visual differentiation is exceptional. As a general rule when the grade of the whole section of the operating mining level or bench in case of open pit, including the barren layer, does not drop below the industrial grade ore it dilutes the run of mine ore, if it drops the quality below the industrial grade the whole section of the mining block is mining loss. Delineation of barren intercalation(s) is the task of the mine development exploration.

(c) Heterogeneity both in grade and hardness. When the industrial grade particles (nodules or fragments) are in soft clayey matrix the bauxitic complex can be beneficiated by physical methods: by dry- or wet screening. See details in Sect. 2.7. Principles of Physical Bauxite beneficiation.

2.4.2 Physical Properties

The **wet bulk density** of the mine products is a determinant factor both in reserve estimate and in mining operation. Wet bulk density, in case of laterite bauxites, varies most frequently between 1.8 (e.g., Los Pijiguaos, Venezuela) and 2.3 t/m^3 (e.g., West Coast, India). In extreme cases, such as at Mokanji Hill (Sierra Leone) for the light porous type ore it is 1.1 t/m^3. In case of beneficiation the bulk density of the washed product is between 1.1 and 1.4 t/m^3 (e.g. Trombetas, Brazil, Manantenina, Madagascar).

In karst bauxites the wet bulk density of gibbsitic, gibbsitic-boehmitic type deposits is typically 2.0–2.4 t/m^3 depending on the porosity of the rock. In the diasporic bauxite such as at Parnassos (Greece) it may reach 3.0–3.2 t/m^3 (Nia 1969). It is typically 2.6–2.8 t/m^3 for the old (Palaeozoic > 250 Ma) diasporic—and diasporic-chamosite bauxites, e.g. at Severuralsk, Russia and at the Chinese sedimentary/paralc deposits.

The **moisture content** is between 6 and 11% in the laterite bauxites which alternates seasonally. For that reason, the figure is to be given in yearly average which is roughly around 8%. That of the karst bauxites is 12–18% in the gibbsite and boehmitic type of ores, and 8–10% in the diasporic ores.

The **grindability** of the bauxite is expressed by the **Work (Bond) Index** defined by F. Bond as the specific energy (kWh/ton) required to reduce a particulate material from infinite grain size to 100 μm. The highest Wi values have been found in diasporic-chamositic karst bauxites, e.g., bauxite of Jajarm (Iran) this index is 25 kWh/ton. That of Chinese sedimentary/paralic bauxites (in mineralogy similar to Jajarm) is estimated to be even higher (25–27 kWh/ton). The Wi of Greek diasporic ore is 17–20 kWh/ton. For gibbsitic-boehmitic karst bauxite this it is around 12–14 kWh/ton. This figure in laterite bauxites varies between 9 kWh/ton (e.g., Eastern Ghats India) and 17 kWh/ton (e.g. Fria, Guinea). Wi of bauxite nodules extracted by beneficiation, in Trombetas and Paragominas (Brazil) is estimated to be less than 9 kWh/t.

2.4.3 Lithology: Structure and Texture of Laterite Deposits

In lithology two types of structure are distinguished:

- Large scale structure: expressing the structure of the bauxite bearing laterite sequence and ore body.
- Small scale structure expressing the feature of the rock in hand spacemen size: such as e.g., banded, fragmented, etc.

In lithology of the laterite bauxites reflects to the subsequent stages of the development of the *in-situ* deposits. The process is the lithification itself (Fig. 2.13).

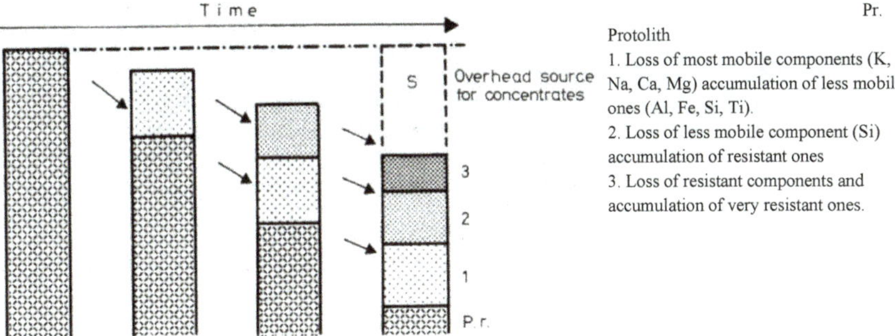

Fig. 2.13 Stages of lateritic weathering

Remark The Fe is a very mobile element in short distance and a very resistant element is long distance forming the whole lateritic profile and inner structure of it. The Al is less mobile, but its rhythmic reorganization also contributes to structure of the bauxite confined by the oscillation zone of ground water table [11, 43].

Bauxites in stage (1) are *concretionary or nodular* bauxites. It corresponds with bauxites in *immature stage,* see Sect. 2.3.3 (Fig. 2.11). They consist of soft, friable clay and harder, gibbsitic bauxite nodules. The nodules are the product of continuous accretion increasing by time, by the progressing lithification itself. Their size in dimension (depending on their age) varies most frequently between 0.01 m and 0.1 m in order of magnitude, however, they may attain 1 m e.g., at Bao Loc (Vietnam). Their shape is irregular. In extreme case (e.g., Ostuacan, Chiapas State Mexico) the gibbsite content of the particles attains 80%. See Fig. 2.14.

Most significant deposits, in this type, are found in the Lower Amazon Basin, Brazil (e.g. Trombetas, Juruti, Paragominas, etc.), at Manantenina, Madagascar, at Tayan in Kalimantan Island (see Fig. 2.42 in Sect. 2.5.1.3), Indonesia, at Jahore in Malaysia, at Bao Loc in South Vietnam. The crude ore is beneficiated by washing and screening. The yield of concentrate (above 1 mm) varies between 30% (e.g.,

Fig. 2.14 Gibbsitic nodules in Ostuacan bauxitic laterite

Fig. 2.15 Bauxite bearing laterite profile at Kutch in Gujarat State, India

Manantenina and 70%, at Trombetas where hydrocyclone is used for extraction of finals and super-finals (+400Mesh that is 0.037 mm). At Trombetas the bauxite is covered by clayey sequence celled Balttera clay which is up to 11 m in thickness. Downwards it turns into nodular bauxitic clay layer in 0–2.5 m thickness. It contains knobby aggregates in 25%. The ore bodies themselves are 5–6 m in average. They are also nodular in structure. In the upper part of them the nodules are in iron-oxide rich matrix. This zone is very hard, blasting is needed. Downward the iron-oxide diminishes, and the ore can be excavated by front end loader [44].

Structures and textures of bauxites in stage (2) and (3) is more complex, by aging more and more heterogeneous. The typical complete laterite profile is made up by the following units as shown in Fig. 2.15.

At several places the continuous bauxite forming has interrupted indicated by interbedded nonindustrial grade (ferruginous) zones resulting in heterogeneous profile called deposit of *poly-phase origin* such as at Nyinahin and Kibi in Ghana. Where the bauxitic sequence of 10–12 m is divided by 1–2 m iron rich laterite formation ($Fe_2O_3 > 30\%$, $Al_2O_3 < 35\%$). The structures of the bauxites are classified into (1) relic and (2) neo-conform structures:

(1) **Relict structures**: The bauxites preserved the structure of the parent rocks: e.g. its *bedded* structure as e.g. at several Guinean bauxites and at the deposits of the Eastern Ghats (India) derived from meta-sediments (khondalithes).

Other deposits ranked into bedded structure type of ores are at Mokanji Hill in Sierra Leone preserving the stratification of the granulite protolith, and at Kibi and Nyinahin in Ghana, where the stratified large-scale structure of the parent rocks: slates, schists and metavolcanics can be recognized in the lithology of the bauxite profiles (Fig. 2.16).

In special cases, not only the structure, but the texture of the parent rock is also inherited, e.g., at Az Zabirah (Saudi-Arabia) where the bauxite formed on siltstone

Fig. 2.16 The bauxite (**a**) developed on well stratified Ordovician sandy slate (**b**) Kindia (Guinea)

and fine-grained sandstone (Lower Cretaceous Biyadh Formation - Black et al 1985)). This type of ore is medium hard, very compact, grain size of the particles of the matrix is < 0.05 mm, in which small (~ 1 mm) bauxite "fragments" or pisoids are found. It is the aphanitic *texture* occurring as a layer in a structurally complex bauxitic profile. See Fig. 2.17. This type of texture is believed to be common elsewhere where the ore bodies developed on (silty) claystone—siltstone protoliths (Figs. 2.18 and 2.19).

The boulder (rounded blocky) structure is also a typical relict structure. Its formation is derived from the spheroidal weathering of the basalt e.g., at Kutch deposits in Gujarat and at Mainpat in Chattisgarh States, India. The space among the hard boulders (up to 1–1.5 m in width) is filled up by clayey matrix and smaller bauxite

Fig. 2.17 Az Zabirah—trial mine and core sample from bore hole AZ-1919 Bauxite of aphanitic texture

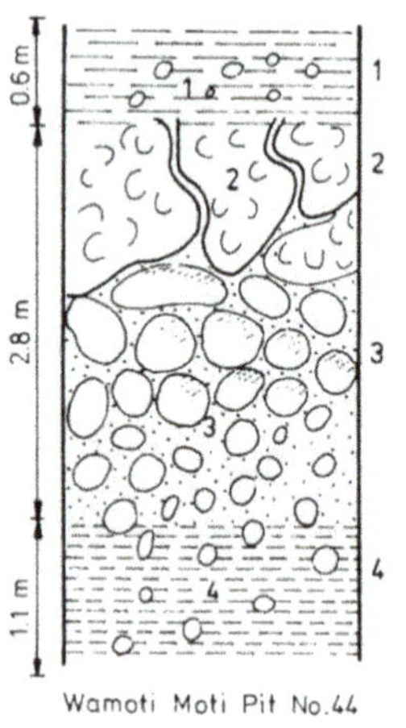

Legend:
1 soil.
2 bouldery bauxite.
3 boulders with fragments.
4.saprolite with bauxite nodules

Fig. 2.18 Bouldery bauxite at Kutch Wamoti Moti deposit Pit No 44 (Komlóssy 1978 in Bárdossy and Aleva [1])

Fig. 2.19 Maininpat BALCO mine—Chhattisgarh State, India

fragments. This kind of bauxites were manually selected, crushed, and stored at the BALCO owned Mainpat mine.

It is important to emphasise that this kind of deposits needs a careful selection of the reliable drilling technics and the results is recommended to check by pitting.

(2) Neoconform structures and textures: the term was introduced by: Bárdossy and Aleva [1]

They developed in course of continuous or, at places, interrupted bauxitisation. All the structures and textures in this group correspond to the reorganisation of the original structure. These types of ores are generally *massive* in structure (blasting is needed), however, at lower zones *friable* structure may also occur. Three main sub-types are distinguished in the followings:

Fig. 2.20 Blocky bauxite: Udgiri mine—Maharastra State—India

- *Simple large structure*: relatively homogeneous: the upper part is hard massive in structure with different kind zones in textures such as: aphanitic, gravelly, brecciated, concretionary, cellular, etc. turning downwards less and less compact or even friable type of ores e.g., at Moengo (Guyana) at Kindia, Aye-Koye, Pita Labé (Guinea), etc.
- *Blocky large structure.* The bauxitic assemblage is composed of big blocks, up to several metres in size, and the fissures filled in ferruginous clay, the small-scale structure and texture of the blocks are different at different occurrences. Among them the *breccia* small scale structures are typical at many deposits. As the lithification is progressed, the evolution of the bauxitic sequence results in compacting. Gradually opening fissures and cracks are filled in by rhythmically precipitated iron-rich or silica rich (or both) minerals, while the fractural elements become more and porous with relative increasing of alumina and decreasing iron. By that way in situ *big blocks and breccia* are forming. See Fig. 2.20.

Similar structure and texture were also observed at Nassau Mts. in Suriname.

Complex large structure with facies of laterally arranged small scale structures and textures. It is the most common lithological profile of those deposits where the oscillation zone of the ground water table has changed during the geological time. Deposit of Los Pijiguaos (Venezuela) is believed to be one of the most complex in small scale structure as shown in Fig. 2.21.

Professional papers distinguish number of **textures** such as: aphanitic, breccia, gravely (conglomerate) colloform, concretionary, gravelly, tabular, spongy, vuggy, etc. Some of them, for sake of a general view are described in the followings, however, the texture itself of has not influence neither in mining nor in processing.

Peculiar texture in the neo-conform structures is the *pisoidal* one occurring at number of deposits occur locally in lenses passing laterally into other textures, e.g., Digo Mokouedou, Ivory Coast, Kindia, Guinea, West Coasts bauxites, India, etc.

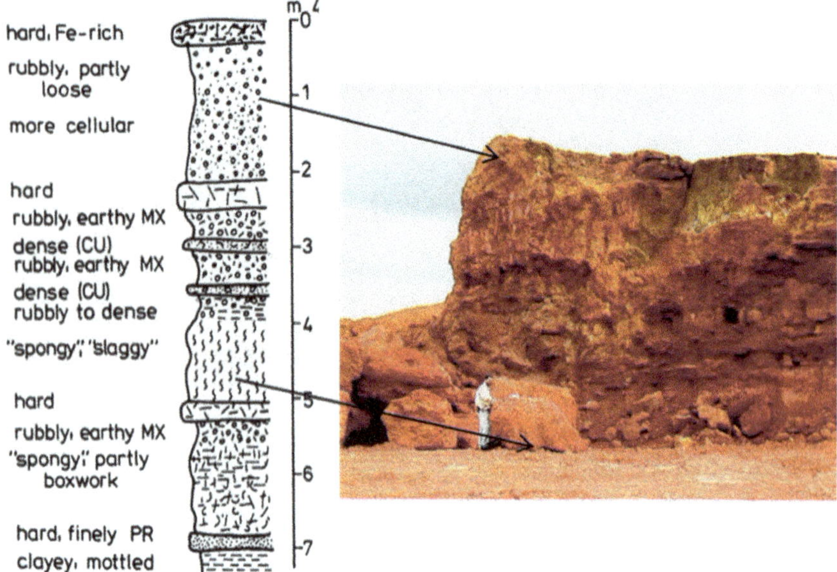

Fig. 2.21 Complex large structure—Los Pijiguaos, Venezuela (ALUSUISSE-CVG [33] in Bárdossy and Aleva [1]. *Photo*

Pisoids are round grains of 1 mm to 1–2 cm in size which are usually harder than the matrix. See Fig. 2.22.

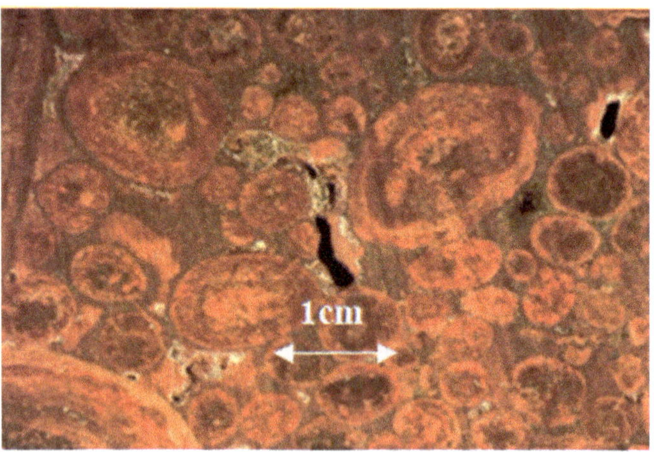

Fig. 2.22 Pisiodal bauxite—Az Zabirah, Saudi Arabia (polished surface)

By their inner structure 5–7 different types of pisoids were distinguished (e.g., Boulangé, B. for Ivory Coast 1984 and Taylor, G., Eggleton, T for Weipa, Australia [45]).

The upper part of the large bauxite deposits at Weipa and Gove in Australia consists of loose unconsolidated pisolites. At Gove, the lower part is cemented in *aphanitic* bauxites. Their uniform diameter is 3–6 mm in both occurrences (Fig. 2.23).

Another common texture is the *breccia*, composed of angular bauxite fragments cemented by ferriferous (bauxitic) laterite. They formed by in situ physical fragmentation and cementing. The pseudo *breccia* formed by inner chemical processes: iron mobilization and re-precipitation). See Fig. 2.24.

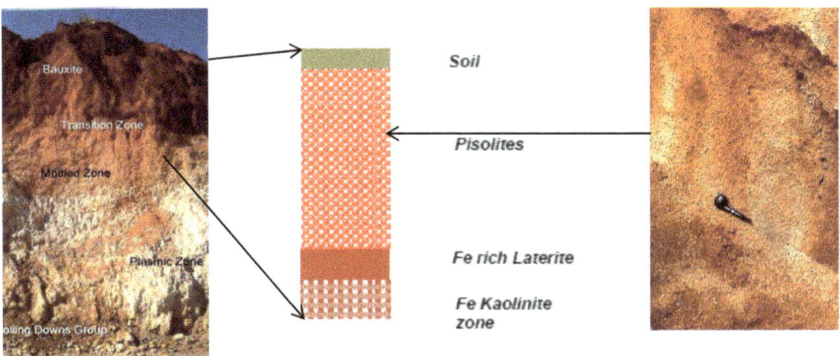

Fig. 2.23 Lithology of Weipa-Aurukum deposits – Australia [45]. Photos: Parcel, E and personal observation (2008)

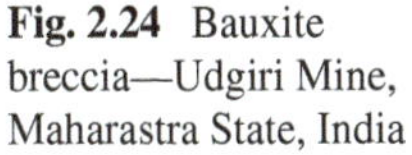

Fig. 2.24 Bauxite breccia—Udgiri Mine, Maharastra State, India

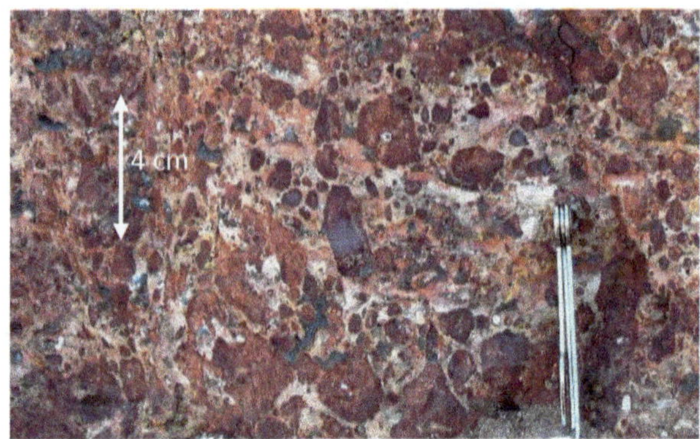

Fig. 2.25 Bauxite conglomerate—Az Zabirah trial mine, Saudi Arabia

Bauxite *conglomerates* formed also by physical fragmentation made up by physically eroded (rounded) particles and cemented by bauxites, see Fig. 2.25.

Local re-worked bauxites, formed during the bauxitisation process, are massive, compact in lithology and variegated in texture: for which Sangaredi (Guinea) is a world class example comprising textures of relict, arenitic, gravelous, conglomeratic, pisoidal, aphanitic type of ores [13].

There are local re-worked bauxites along most of the bauxite bearing plateaus in West Africa and India. In the Russian (Soviet) terminology they are called as sedimentary bauxites. There are re-worked bauxites along most of the bauxite bearing plateaus in West Africa and India. In the Russian (Soviet) terminology they are called as sedimentary bauxites.

(3) Re-accumulated (transported)—slope bauxites.

Where the bauxite deposits formed on the top of dome-shaped hills, because of the permanent physical erosion, beginning with in situ fragmentation, the debris, by gravitation, are transported towards the slopes forming new secondary deposits with special lithological characteristics. The new formation consists of bauxite fragments and clayey matrix. See Fig. 2.26. Both the size of the fragments and their quantity are diminishing towards the depth. Further details are presented in Sect. 2.5.1.3.

It is to be mentioned that in the professional literature sometimes it is also named as *nodular* or *nodular-gravely* bauxite. The difference between the nodes and fragment is that the first one is a product of continuous accretion (precipitation), their size increasing while the fragments are products of physical disintegration of the formerly lithified bauxite. Why the difference is important in practice? In the first case the auger or bucket auger drillings may serve reliable samples. In contrary in the second case, it is inconvenient because the larger particles (sometimes boulder in size) instead of cutting them they are pushed into the soft clay and just the bauxite itself is not taken out from the hole. Applied drilling technics is to be checked by pitting. These type of bauxites needs beneficiation just like the nodular ores and the procedure is to be fitted to the lithology of the rock.

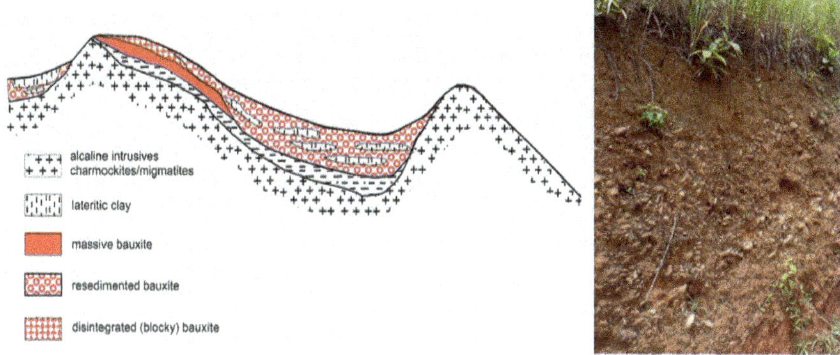

Fig. 2.26 Conceptual section showing the in situ and associated re-worked (transported) bauxite Atlantic Shield, Brazil [46]. Photo: personal observation

For typical examples Barro Alto (SE Brazil) and Amargosa (E Brazil) deposits are detailed in Sect. 2.5.1.3

2.4.4 *Lithology of Karst Bauxites*

Lithology is much simple in karst bauxites. The deposit assemblages, in petrographic term, are built up, from bauxite and associated formations such as bauxitic clay, clayey bauxite, clay, rarely ferruginous clay, see Fig. 2.27. Its texture consists of the base material (matrix) and more or less different kind of textural elements. They formed either as in situ (authigenous) elements or by re-accumulation (allothigenous) as elements.

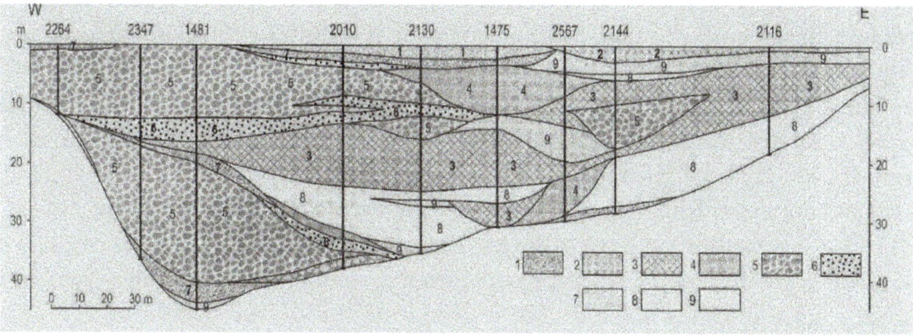

Fig. 2.27 Genetic profile—SW Halimba, Hungary (Bárdossy [47]). Legend: 1—Senonian conglomerate, 2—Grey pyritic-marcasitic clayey bauxite, 3—Highy grade bauxite ($SiO_2 > 4\%$), 4—Low grade bauxite (SiO_2 4–10%), 5—Carbonate conglobreccia, 6—Carbonatic silt with bauxitic cement, 7—Conglobreccia cemented by clayey bauxite, 8—Clayey bauxite, 9—bauxitic clay

The base material is pelitomorph (grain size is less than 1 μm) which by aging or tectonic effects (compression) may turn into *fine-grained-* (1–5 μm), to *medium-* and *coarse grained* (more than 100 micron) textures resulting in increasing hardness from about 3 up to 6 in the Mohs scale.

Neglecting the review on number of theories, in which principally the source material and way of transport were disputed it can be concluded that the material of bauxite deposits, in contrary of the in-situ laterites, was collected by vadose waters, and as seasonal stream loads accumulated in depressions. The younger karst bauxites (Cretaceous and younger, less than 140 Ma) are dominantly *homogenous* in lithology (e.g., Jamaican deposits, Hungarian- and some of ex-Yugoslavian bauxites) while the elder ones may be *bedded*, as well. (e.g., some ex-Yugoslavian or Greek ones). Cross-bedding structure also occur, indicating fluviatile sedimentation.

A part of the textural elements might be formed during the short distance transport, and a part when after the final accumulation. The bauxitisation is going on by which different kind of textural elements form in place such as: e.g. pisolithic (spherical grains), fluidal-, pseudo-brecciated-, nodular-, etc. textures [27]. These textural elements formed mainly by the partial solving and re-precipitation of the iron. Which is intensive when grey pyritic bauxite occurs on the top level and the decomposing FeS_2 produced H_2SO_4 mobilizes the iron which at neutralised lower-level precipitate.

Some bauxite deposits, partly or completely, have been re-accumulated (quasi in situ) or re-deposited in a shorter distance. In such cases bauxite silts, fragments or gravels or concretions are found in the pelitomorph matrix (see Fig. 2.27).

2.4.5 Lithology of Sedimentary/paralic Bauxites

The lithology, just like at the karst bauxites, reflects to the circumstances of their accumulation. Bauxites were formed on the carbonate surface along the sea shore and accumulated in the zone of oscillation of the see and inner basin of the more ore less closed golfs. So that, these bauxites are typically *stratified and bedded*. See Fig. 2.28.

The rock itself is very compact pelitomorph in texture, however pisolitic or detrital layers may also occur. The bauxite sequences are often interrupted by clayey or lignitic clay layers, indicating reduction medium. This type of ores are extremely hard and compact. Bulk density may attain 2.8 t/m^3 (see Sect. 2.4.2).

When the overburden eroded, the bauxite primary formed in reducing media will be more or less oxidised resulting in significant mineralogical alteration as presented in Sect. 2.3.1 and shown in Fig. 2.7.

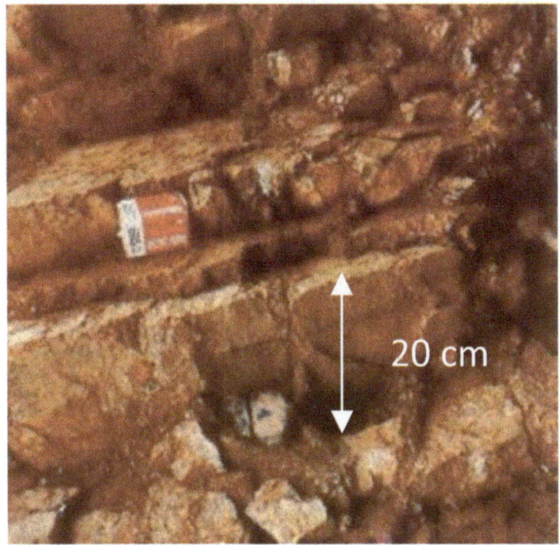

Fig. 2.28 Bedded sedimentary/paralic bauxite Ba Xa, Lang Son County, Vietnam

2.5 Deposit Types (Features Determinant in Mining and Processing)

Bauxite deposits have been classified from the early twentieth Century by a number of scientists (e.g., [5, 48–52]; Harrassowitz 1926; Fox 1927, 1932; Patterson 1967; Grubb 1973; Grubic 1975; Bárdossy and Aleva [1], etc.). Grouping of these deposit types have been made using different criteria, such as parent rock, shape and size of the ore bodies, relief of bauxite formation and accumulation and media of accumulation or by their lithological megastructures. See details in Sect. 2.4.3.

From a genetic aspect, based on the groupings of Fox (1927) and Vadász [50], the deposits have been classified into two main types: (1) laterite bauxites and (2) karst bauxites.

2.5.1 Laterite Bauxites

This type of deposits is generally found as surficial deposits in Australia, India, West Africa, Atlantic Shield in Brazil, Venezuela and at the high level of Guianas). There are also buried deposits (overburden > 10 m) covered by Cretaceous marine sediments at Az Zabirah (Saudi Arabia), Eocene-Miocene marine sediments in Kutch—Gujarat India, and various sediments of Miocene-Quarternary age, like at the Coastal Plain in Guyana and Surinam. They are exploited by open pit mining..

There are also deposits of lateritic origin covered by thicker marine sediments e.g., Belgorod in Russia and Vysokopolye in Ukraine [53]. They are insignificant in the global bauxite-alumina industry context and have not been reviewed in this review.

The basis of the current classification where the geomorphological features are the determinant factors in the formation, the shape, size, thickness and lithology and the preservation of the ore bodies.

The dominant laterite deposit types and subtypes are distinguished as:

2.5.1.1 Plateau Type Bauxite: ith lat (errain nclination ess han 5°) or ndulating, hale ack urfaces (errain nclination ess han 15°)

From a genetic aspect, based on the groupings of Fox (1927) and Vadász [50], the deposits have been classified into two main types: (1) laterite bauxites and (2) karst bauxites.

(1) Plateau bauxites on morphological terraces: bauxite developed on erosional surfaces at altitude between 40 and 2000 m MSL.
(2) Plateau bauxites developed on low land interfluves:

    (2.1) Coastal plain interfluves (altitude 10–100 m MSL and
    (2.2) Inner basin interfluves (altitude: 100–150 m MSL).

2.5.1.2 Whale Back Topped Plateaus

Dome shaped hills and associated slope bauxites formed on rolling hilly terrain at elevation of 200–1600 m MSL

(1) Bauxites confined to the hill tops
(2) Bauxite formed on the hill tops with re-accumulation on slopes
(3) In situ slope bauxites.

This classification based on published and non-published papers along with field observations made on five continents. It is believed to be applicable world-wide, at least for those deposits which play important role in the current alumina industry. The laterite bauxite types discussed are characterised by the most prominent economic examples. Each deposit has its own depositional character in detail; however, consistent dominant themes can be recognised.

Plateau type bauxites

(1) Plateau bauxites developed on morphological terraces: deposits on (1.1) flat planation surfaces and (1.2) whale back topped plateaus.

(1.1.) **Terraced flat plateau bauxites** on erosional surfaces as shown in Fig. 2.29 are the dominant in India, West-Africa and in the Guiana Shield (South-America) providing for significant part of the global bauxite production.

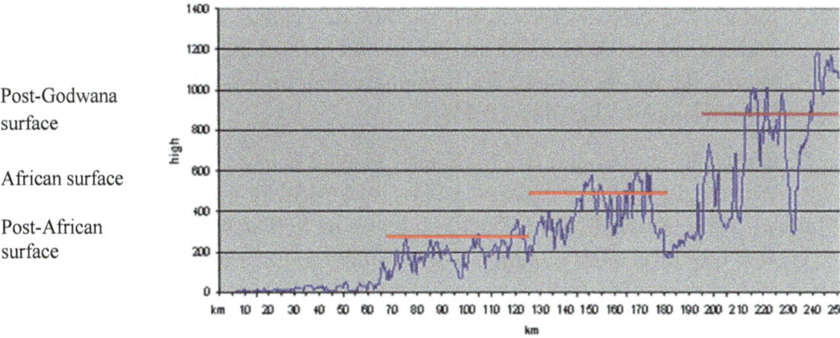

Fig. 2.29 Erosion surfaces between Fouta Djalon Mts. and Boké (Guinea). Planation levels as recognised by King [54] (Courtesy of Rio Tinto Expl. Ltd. 2013)

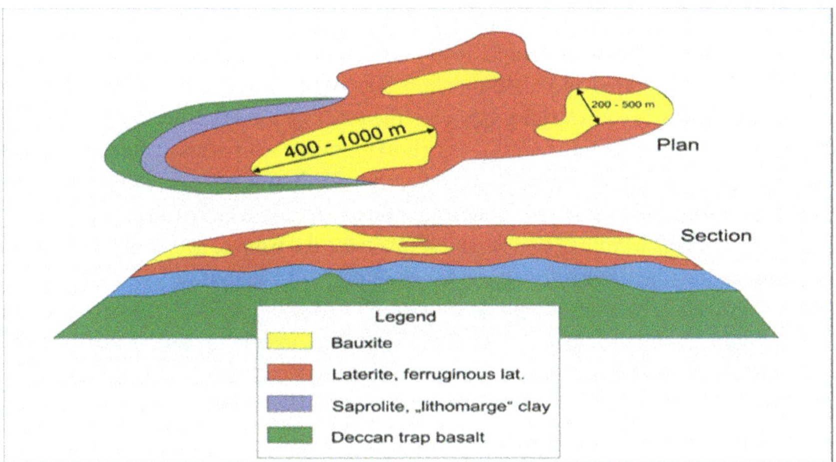

Fig. 2.30 Conceptual plan and section Maharastra State, Kolhapur District, India (Courtesy of Rio Tinto Expl. Ltd. 2012)

Similar terraces are found at the Western rim of the Guiana Shield (Venezuela) called Kamarata (1000 m), Nuria(600 m), Intermediate (400 m) and Los Llanos (100 m) planation surfaces (Menendez and Sarmento 1984).

Defined by their elevation these types of deposits, high level and low-level deposits are distinguished:

(a) **High level deposits** developed above 400–500 m MSL at the levels of Gondwana Post-Gondwana and African with slightly undulating surfaces. Relative elevation is several 100 m. Massive structure and pseudomorph textures with high porosity is common. They are typically Pre-Miocene: more than 25 Ma.

(b) **Low level deposits** developed on the Post-African and youngest surfaces are quasi flat at 40–500 m MSL Relative elevation is lower, usually from 10 to

100 m. The bauxites are clastic, pisolitic or nodular. It is in the zone of the alternating iron mobilisation and precipitation, resulting in pallid and cemented bauxite. In age they are typically younger than high level deposits (Miocene, Plio-Pleistocene: less than 25 Ma).

Economically most important high level plateau bauxites are found e.g., in the Western and Eastern Ghats (India) at 900–1450 MSL in the Fouta—Djalon Mts, (Guinea), in the Atewa Range (Ghana), in Sierra de Serbatana—Venezuela, etc.

Typical low level plateau deposits are found e.g. at Fria and Boké in Guinea, Port Loco, Mokanji Hills in Sierra Leone, Onverdacht in Suriname Linden and Ituni, Kwakwani in Guyana.

At a high level where the relative elevations are higher, the drainage conditions are more effective. These deposits are typically greater in thickness (industrial grade ore often exceed 10 m) than that developed at the low level, however, both the thickness and quality is highly dependent on the parent rock as well, e.g. at East Coast, India. There are several ore bodies where the maximum thickness of the ore grade bauxite surpasses 30 m (IBM Monograph 1993) due to the excellent drainage conditions provoked by the bedded and tilted metamorphic rocks.

Bauxite deposit—industrial grade ore (the term of the later one varies between different deposits and mining units)—as a horizontally continuous mantle is regularly interlayered between the upper and lower laterite horizons (non-industrial grade bauxites) resulting in *sandwich* structure. Transition between the layers is gradual. The boundary is determined by the cut off given for the bauxite ore. It also may occur that that the bauxite ore forms lenses: for example, in Kolhapur district (West Coast, India) there are lenses of sub grade laterite of 1–2 km long 300–400 m wide and 2–3 m thick. This type of bauxite ore is quasi floating inside the laterite capping. See Fig. 2.30. Industrial grade ore lenses in the laterite capping—Hedavde Kalamwadi deposit.

Further details on lithology are given in Sect. 2.4.3.

Plateau deposits can be grouped by their size, arrangement, and shape, as well. There are (a) Single elongated ridges, (b) Combination of elongated ridges and (c) Spherical (roundish) surfaces.

(a) **Single elongated ridges**: they may surpass 10 km in length and they are 1–3 km in width. An example for this type of deposits is at Panjpatmali—East Coast in India as shown in Fig. 2.31.

The world class Panjpatmali deposit has formed on tilted konadalites (metasediments) and is about 22 km in length and 1.5 km in width on average of the total area of 33 km^2 almost 15 k m^2 have been bauxitised. The maximum bauxite thickness is 54 m Comprising a number of well-defined individual plateaus the Atewa Range (Ghana) is at an elevation of around 700 m MSL. It is dissected by erosional valleys forming a series of smaller plateau units within its 60 km length and 1–10 km in width. See Fig. 2.33. The total area of planation surface is 280 km^2. The bauxite coverage on the explored central area is estimated to be 80–90% (Fig. 2.32).

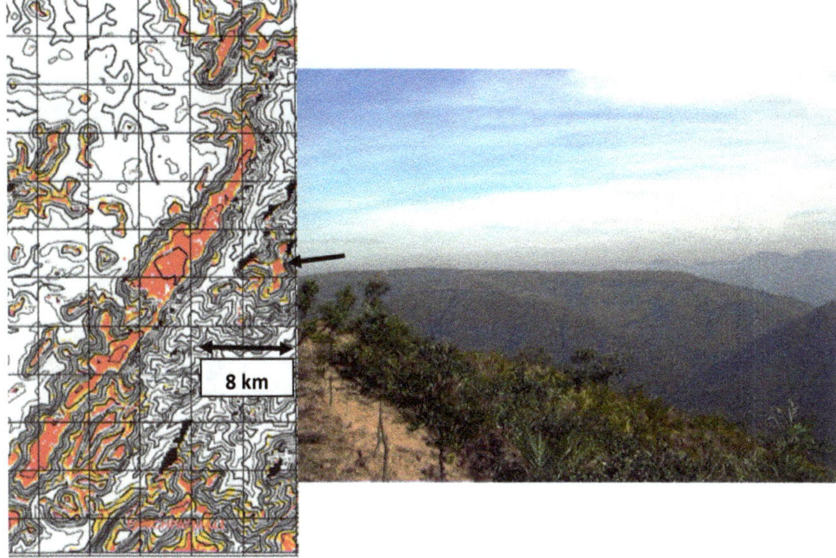

Fig. 2.31 Panjpatmali Plateau—East Coast, Orissa Stat (Map: courtesy of Rio Tinto Expl. Ltd (2006). Photo: personal observation

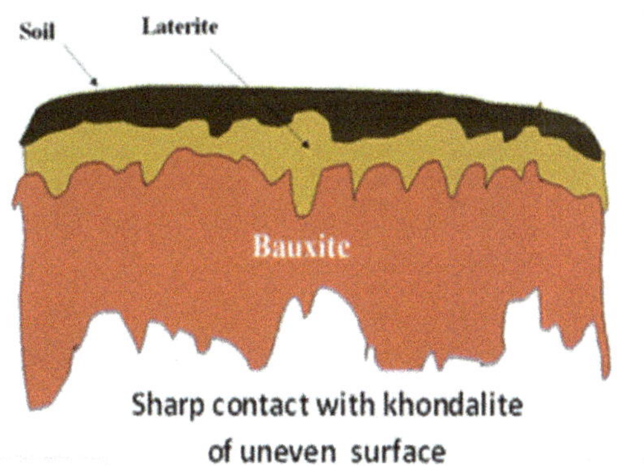

Fig. 2.32 Cross section of the Panjpatmali deposit, Orissa State, India

(b) **Combination of elongated ridges**: arranged in two or more dominant orientations reflecting the structure within the basement. Example for two dominant orientation is e.g. at Atewa Range in Ghana, and Divo Mokueodou plateaux in Ivory Coast See Figs. 2.33 and 2.34.

At Kibi, the bauxite thickness in the mineralised block averages varies between 6 m (Asokawa N) and 12.6 m (Atiwiredu) [26, 55].

The common plateaux systems, identified in West Africa, are also ranked into this type of ore bodies forming an *amoeba-like surface*. They occur as a typically a single

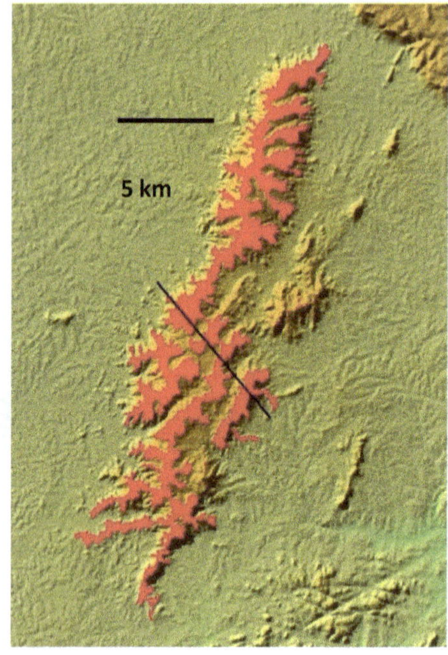

Fig. 2.33 Atewa Range with Kibi deposits—Ghana (Courtesy of Rio Tinto Expl. Ltd. 2014)

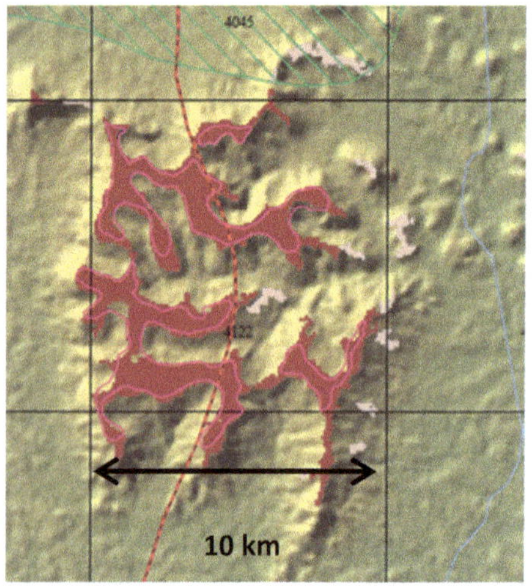

Fig. 2.34 Digo-Mokouedou plateau—Ivory Coast (Courtesy of Rio Tinto Expl. Ltd. 2014)

cluster emerging from their surroundings, e.g. Digo Mokouedou (Ivory Coast) shown in Fig. 2.34 or appear in groups, accompanied by plateaus of a single elongated type occupying large areas as in the districts of Boké, Kogon Touminé Interfluve and Fatala, etc. in Guinea (Fig. 2.35).

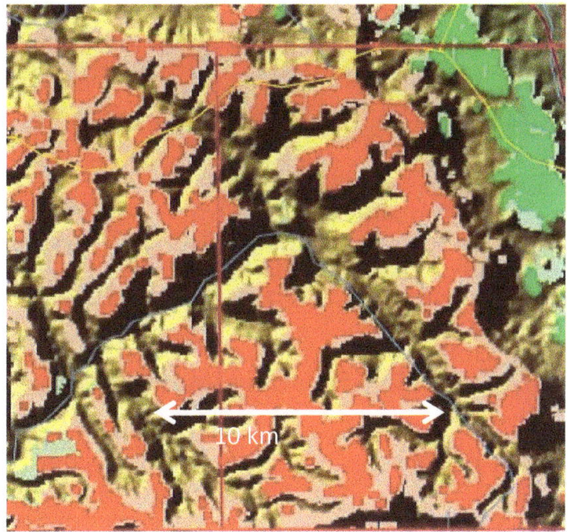

Fig. 2.35 Series of bauxite bearing complex plateaux in Kogon Toumińé Interfluve district, Guinea (Courtesy of Rio Tinto Expl. Ltd. 2014)

Individual plateaus are between 1 and 20 km^2, most commonly 5–10 km^2 in extent. These plateaus are only partly bauxitised. Industrial grade ore (TAl_2O_3 > 40%) cover these plateaus about in 40–90%. Industrial grade bauxite thickness, for this type of deposits, is estimated to be 6–8 m in average on a global scale.

Guinea has by far the largest high grade bauxite resources in the world. Identified high grade bauxite reserves + resources are estimated at 29 Bt. Additional further prospects are estimated to be 11 Bt. (Diallo 2007). In Guinea the total identified bauxitic prospective surfaces cover almost 5000 km^2, comprising 865 identified deposits. (Boufeev et al. 2003). These figures have been increased during the past decade by new exploration. The bauxite horizon is typically 5–8 m thick in certain domains exceeds 20 m.

Bauxite thickness typically increases at the plateau margins. Hard, massive bauxites require blasting in mining.

Not all the laterite cappings have been bauxitised (TAl_2O_3 > 40%). In the bauxite horizon quality suddenly drops where the Devonian - Silurian sediments have been intruded by dolerite dykes. See Fig. 2.36.

In Guinea the bauxite deposits are typically restricted to almost flat plateaus with the basement mainly composed of quasi horizontal marine sediments.

Now the lowermost step of the terraced plateaus is present on the Coastal Plain of South America—Suriname and Guyana. Deposits formed before at a higher level, are of Paleocene-Eocene age [56] which equates to the African surface (at elevation above 500 m) Because of the gradual subsidence the planation surfaces dropped to 40 m MSL. Bauxite is of massive type ore. They are covered by sandy and clayey unconsolidated sediments of 5–40 m thickness.

The Mananatenina (Madagascar) deposit is interpreted to be a special case from the aspect of point of planation formation. It is related to marine abrasions of the Indian Ocean. Two distinct levels, at 60 m (Pleistocene 2 Ma) and at 30 m (Holocene),

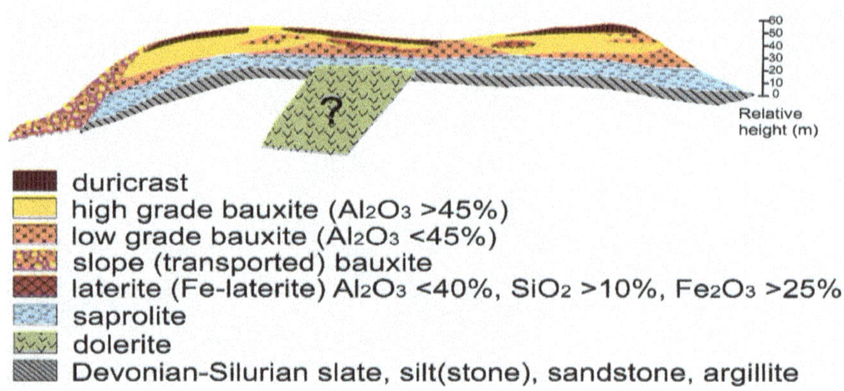

Fig. 2.36 Typical bauxite bearing laterite profile for Boké and Kogon Touminé Interfluve bauxite district—Guinea. Schematic section (Courtesy of Rio Tinto Expl. Ltd. 2013)

have been identified. (Balkay, B. from Bárdossy and Aleva [1]). These deposits are believed to most recent level. The gibbsitic bauxite profile consists of two horizons: (a) *Upper level* is a reworked layer consisting of nodules and concretions cemented by loose matrix (1–3 m in thickness, (b) *Lower level* is in situ, earthy, concretionary (1–10 m in thickness) Beneficiation is needed to create an export grade product. Tests show at a 30% of recovery 40–48% TAA and 1–2% $RSiO_2$ can be achieved [25].

(c) **Spherical (roundish) forms** e.g., at Sangaredi in Guinea, at Baphlimali in East Coast, India, see Fig. 2.37. The best-known typical plateau is the Sangaredi deposit (Guinea) covering $4km^2$. It is almost entirely a locally reworked deposit with complex structure. Both from lithological and mineralogical aspects. Its maximum thickness surpasses 50 m. Baphlimali tri-hydrate type bauxite (East Coast, India) is formed within an area of almost $10km^2$ with thickness varying between 2 and 28 m [57].

The bauxite extended over the whole plateau reflects an extremely efficient drainage system (Sect. 2.5.1.2). They are related to granitic intrusions on the Western rim of the Guiana Shield (Venezuela). The planation surfaces show banded forms at surface following basement structure. They are dissected by deep valleys. The relative elevation is more than 300 m. The terrain makes the mining and transport difficult, with rare large continuous deposits (Fig. 2.38).

The whale back plateau bauxite has formed on the top (slope < 5°—red domain in the map cut) and slopes (<15° yellow domain). The later is a transitional type toward the in situ slope bauxites (Sect. 2.4.3).

Bauxite thickness is typically 6–8 m, but locally 15 m. On the Western and Northern rims of the Guiana Shileld there are hudge prospect. The lithological profile is much more complex than that of at plateau bauxites in West Africa or India. (see Fig. 2.8. in Sect. 2.4.3) This compexity is also expressed in the mineralogical make

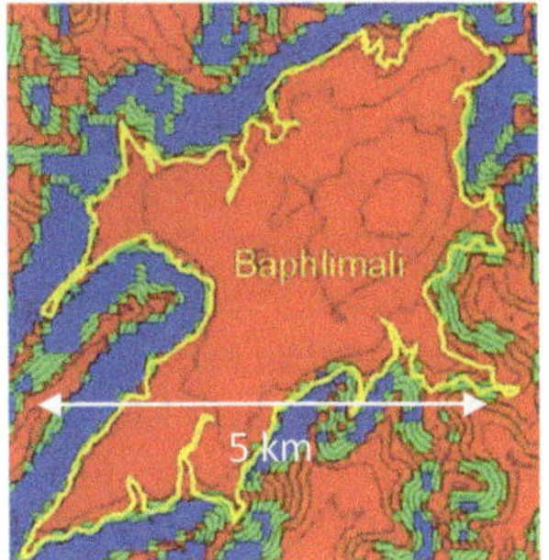

Legend:

Red domain: is the plateau established by remote sensing technics.
Yellow contour: identified bauxite within the plateau.

Fig. 2.37 Baphlimali plateau at East Coast, Orissa State—India (Courtesy of Rio Tinto Expl. Ltd. 2006)

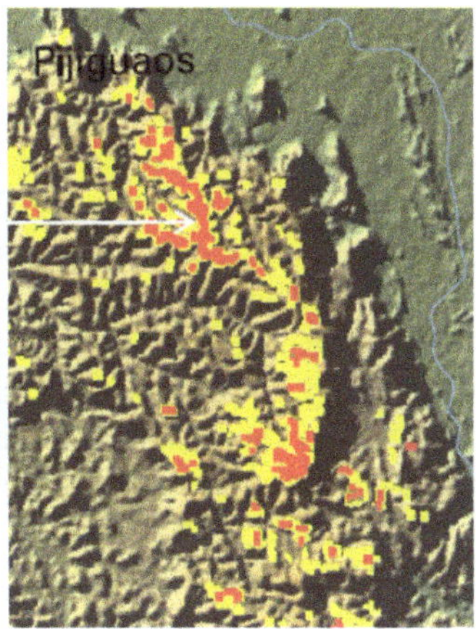

Fig. 2.38 Whale back topography at Sierra de Cerbatana—Venezuela (Courtesy of Rio Tinto Expl. Ltd. 2007)

up (see Fig. 2.8 in Sect. 2.3.2). The phenomenon refers to a multiple bauxitisation (Bárdossy and Aleva [1]).

Mulanje—Lichenia deposit (in Malawi) belongs also to this type, however it is simple in lithology. The Lichenia bauxite developed on a quartz syenite intrusion elevated abbruply from the surroundings by more than 1000 m.

2.5.1.3 Plateau Bauxites Developed on Low Land Interfluves Terrains

Interfluve's terrain reflects a less dissected upland between adjacent streams flowing in the same direction. Their relative elevation is typically less than 20 m.

There are two subtypes:

(1) Deposits of coastal interfluves.
(2) Deposits of inner basin's interfluves.

(1) Deposits of coastal interfluves

The coastal plateaus, in contrary of the terraced ones, formed by the emergence of flat laying marine sediments (e.g., silty clay) and flat basaltic lava flows. The planation surfaces formed without significant physical erosion and occupy large areas which have been incised by the rivers in depths of 20–80 m. Their geomorphology, shape and size and also genesis differ, at least in part, from terraced plateau bauxites.

At Cape York, Northern Australia, the laterite on the Weipa plateau is 180 km in length and occupies an area of 11,000km^2 of which 520 km^2 is potentially productive (Shaap, A.D. 1990). The bauxite thickness is between 1 and 9 m, in average 2.5–3 m. The bauxite body consists, in most cases, of loose free running pisolites although in places they are cemented. The crude ore is beneficiated by wet screening. The deposits are homogenous, both in thickness and grade, compared with the bauxites of terraced plateaus.. The saprolite horizon with its Fe rich laterite are believed to be of in situ formation while the pisoidal bauxite layer on the top of the ferric laterite is interpreted by some academic studies as being derived from reworked material transported by simple gravitational gliding or by seasonal water-courses from higher topographical levels where it has been completely eroded [45]. Lithology of this type of bauxite is shown in Fig. 2.23 in Sect. 2.4.3 The mineralisation is dominantly gibbsitic boehmitic although zones amenable to low temperature processing have been defined and exploited.

The Gove deposit is also ranked as a world class occurrence. The deposit morphology, and characteristics are in part analogous to that of the Weipa ore bodies. The laterite blanket is significantly smaller in area, 120 km^2 of which 63 km^2 contains commercial grade ore (defined by TAA > 44% and reactive SiO_2 < 5%).

The bauxite thickness on average is around 6 m. Compared with Weipa, the thickness and lateral grade variance is greater.

Mitchell Plateau and Cape Bougainville bauxite, in North Australia, can also be ranked into this group of deposits, however, the elevation at Mitchell Plateau. is greater (200–400 m MSL) developed on a large area, l on flat basaltic lava flows and pyroclastic rocks. The grade is lower than that of discussed above. The bauxite of Mitchel Plateau requires beneficiation based on current economic modelling. At Cape Bougainville there is ferriferous bauxite with extremely high iron content: Fe_2O_3:35–40%. (AMAX Annual Report 1970).

(2) Deposits of inner basin's interfluves

Bauxites of the Lower Amazon Basin (Brazil) are of this type of deposit. There are multiple occurrences (group of deposits) such as: Trombetas, Jurity, Jurupari, Almeirim, and Paragominas, etc. Deposits are located at low levels: 200 m MSL in the West (Trombetas) and around 100–150 m MSL in the East (Paragominas), The relative elevation is 10–20 m. The plateaus form irregular, elongated bodies showing distinct orientations in harmony with the network pattern of rivers. See Fig. 2.29.

At Paragominas the Miltonia3 plateau is the largest covering 90 km^2. Average bauxite thickness is typically 4–6 m, but locally exceeds up to 10 m. In the laterite profile typically two distinct bauxite horizons are distinguished (e.g., Trombetas, Jabuti-Ipixuana) separated by Fe-rich laterite indicating a polycyclic origin. The upper bauxite horizon is nodular, whilst the lower is blocky. Both are friable, o blasting is needed. The deposits are covered by the Belterra Clay which is 10 m thick on average. The parent rocks are uncertain , they are supposed to be sedimentary rocks (mainly sandy clays) of terrestrial or marine origin. These bauxites are young (Plio-Pleistocene, or pre-Pleistocene, less than 5 Ma). The bauxites are exclusively gibbsitic. The crude ores are beneficiated applying (crushing) washing and screening. At Trombetas the bauxite is crushed, wet scrubbed and passed through hydrocyclones to recover the final fine fraction (between 150 and 400 mesh). The recovery varies between 50 and 70%. Product's TAA% varies between 46 and 48% (Juruti, Paragominas) and > 53% (Almeirim). Reactive silica ranks between 4and 5% (Bauxite Training—Rio Tinto Expl. Ltd. [58]). Typical plateau morphology is shown in Fig. 2.39.

In the Amazon Basin the most important district is Paragominas with its 2,5Bt of crude ore, which comprises about the 2/3 part of the total reserves/resources of the inner basin type of deposits. Bauxite of the Trombetas deposits is of the highest quality, as far as its extremely good technological properties are concerned. The main stages of bauxite formation ofTrombetas are identical with that of Jabuti-Ipixuana, as shown in Fig. 2.40.

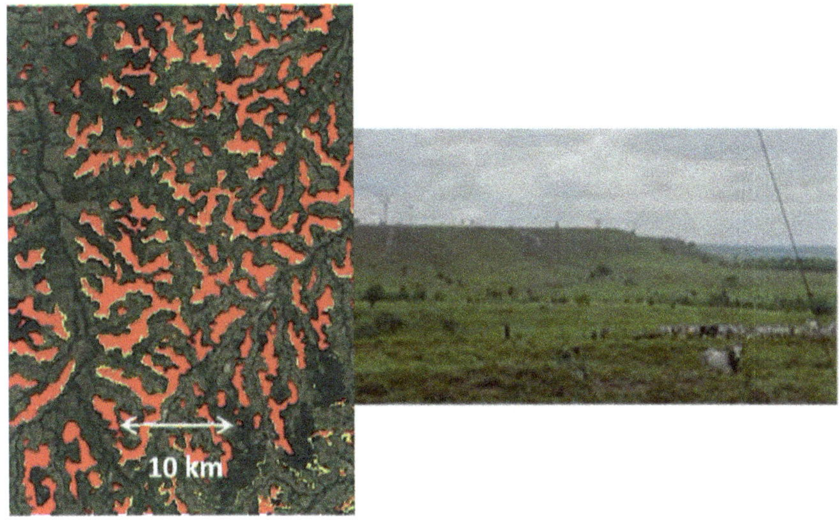

Fig. 2.39 Bauxitic plateaus in Paragominas bauxite district—Lower Amazon Basin, Brazil (Photo: Rio Tinto Expl. Ltd. Plt 2008, Map: courtesy of Rio Tinto Expl Ltd. 2012)

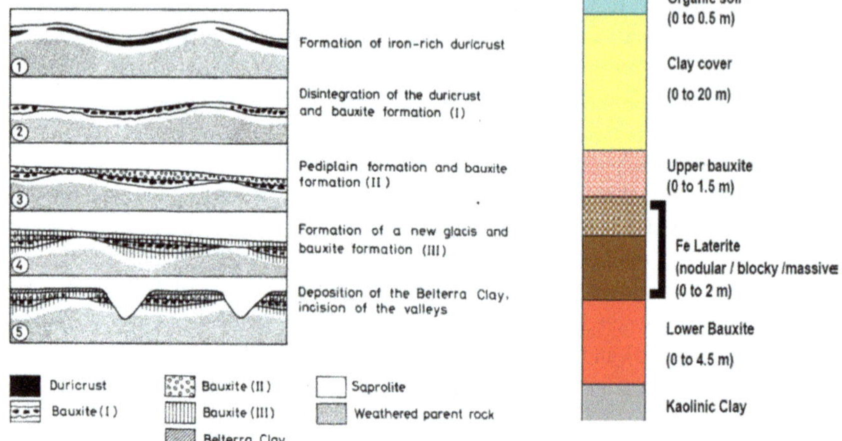

Fig. 2.40 Bauxite formation at Jabuti-Ipixuana and lithological profile of deposit. Lower Amazon Basin—Brazil (Kotschoubey et al. [59] from Bárdossy and Aleva and courtesy of Rio Tinto Expl. Ltd. 2008)

2.5.1.4 Dome Shaped Hills

For this type of deposits, the following sub-types are distinguished based on their depositional characteristics.

(1) Bauxites confined to hilltops

Low relief (20–200 m MSL) dome shaped hilltop deposits examples are found.in West Kalimantan—Indonesia, Sarawak and Malacca—Malaysia, at Mokanji Hills and Port Loco in Sierra Leone, Bauxites are developed on gently undulating small hills with a relative elevation of 20–100 m. The slope inclination varies between 5° and 15° as shown in Fig. 2.41 with one of the typical occurrences found in West Kalimantan.

The deposit is typically 200–400 m × 200–400 m in order, that is, 4–16 ha. Those on the Malacca Peninsula (Malaysia) are smaller, most frequently around 2–3 ha. The lithological structure of the profile is simple. Bauxite thickness is typically around 4 m only, covered by topsoil, or by silty clay of 3–4 m. The bauxite lithology is shown in Fig. 2.42.

Bauxite is beneficiated by washing and screening. Yield on beneficiation is 50–65% with Av. Al_2O_3: 38–45%, reactive SiO_2: 2–5% *(Antam Annual Report 2011)*, the bauxite is of gibbsitic type.

On **high relief** (700–1000 m MSL) deposits e.g., at Dak Nong district—Vietnam, there are both plateau and dome shaped deposits developed on Neogene- Paleocene basalt (1–5 Ma) where the industrial grade ore is confined to the top of the domes as marked on the Fig. 2.43. No industrial grade ore is accumulated on the slopes.

Fig. 2.41 Legend: Red domain on the map indicates the identified deposits and further prospective targets (Photo: Komlóssy 2007 and Map: courtesy of Rio Tinto Expl. Ltd. 2013)

Fig. 2.42 Typical two zoned bauxite profile of Tayan—West profile W Kalimantan, Indonesia [60]

Fig. 2.43 Bauxite bearing plateaux and dome shaped hills at Dak Nong—Vietnam [60].

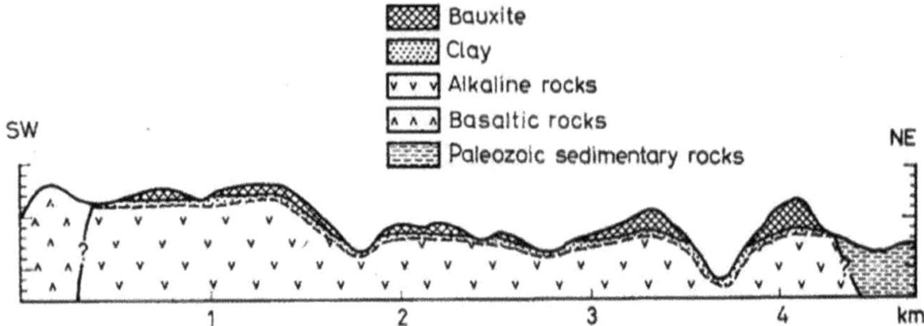

Fig. 2.44 Geologic cross section at Lages—Brazil [61]

The bauxite is blocky, nodular in structure requiring beneficiation. With a 45–47% recovery (at > 1 mm), produced bauxite containing Av Al_2O_3:45–48% and reactive SiO_2:2.5%. The ore is gibbsitic in mineralogy. The deposits of Cambodia and Laos, derived from the same basalt, are deemed to be analogous with the Vietnamese deposits.

(2) Bauxites developed on hill tops with associated slope bauxite

These deposits are restricted to high relief dome shaped hills with slope associated bauxite have typically been developed on intrusive magmatic and associated rocks often of circular structure. They have formed on a dissected, undulating terrain at an elevation between 700 and 1400 m MSL. The terrain inclination varies between 15° and 45°.

Typical high level dome shaped bauxite deposits are found, for example, in Central Brazil and the Atlantic Shield district, such as deposits of Sierra dos Carajas, Baro Alto and Catagues, Pocos de Caldas, Lages, Amargosa, etc. The ore has been partly eroded from the hill tops and accumulated on the slopes forming significant secondary deposits. Lithological characteristics of Atlantic Shield are detailed in Sect. 2.4.3 (Fig. 2.44).

The re-accumulated type of ore comprises hard bauxite fragments from fines (silts) (<150 mesh) up to boulders (up to 40–60 cm) cemented by soft lateritic clays. The crude or is beneficiated by washing and screening. Recovery of the concentrate is typically 60–70% on average (Fig. 2.45).

The massive bauxite thickness on the hill top at Pocos de Caldas, is between 1 and 11 m, the average 3–4 m (Bárdossy and Aleva [1]). The thickness of the re-accumulated bauxitic increases downslope. The size of fragments and recovery (yield) decreases downslope. At other deposits of this type of the maximum productive thickness may attain almost 30 m, on average it is about 8 m. South American hilltop—slope bauxites are typically gibbsitic. Porous isalteritic bauxites predominate, marked by light and dark bands relict after the original layering. Preservation of structures and textures of the primary rocks is evidence for an in-situ origin [62].

Fig. 2.45 Dome shaped bauxite of Barro Alto—Brazil [62]

Re-accumulated (secondary) slope bauxite can be found flanking most of plateau deposits in Guinea and India. They are simple talus. They do not represent always industrial quantity of bauxite.

(3) Bauxite formed on slopes

Primary, in situ slope bauxites are formed on gentle hillsides (slope less than 20^0) They developed either around an intrusive block, forming a skirt-like mantel e.g., Los Islands (Guinea) as shown on Fig. 2.46 or on the flat, slightly inclined top of ridges around the eruptive centre of lava flows, e.g. at Fongo Tongo (Cameroon) (Fig. 2.47).

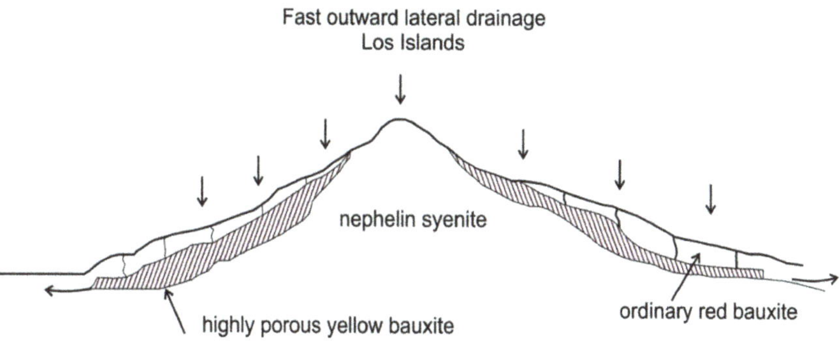

Fig. 2.46 Formation of slope bauxite at Los Islands—Guinea [35]

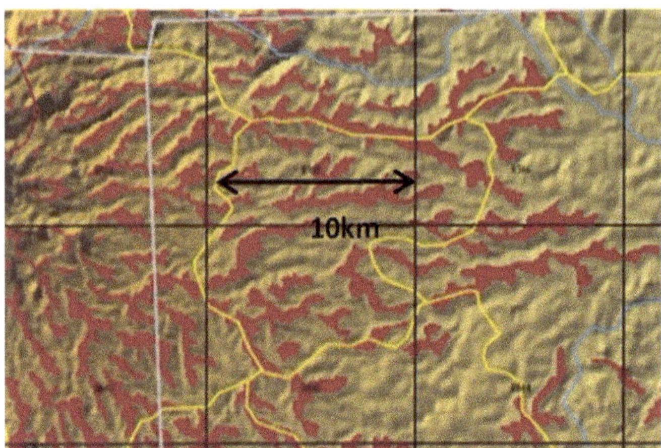

Fig. 2.47 Eastern flank of the eruptive centre at Fongo Tongo—Cameroon (Courtesy of Rio Tinto Expl. Ltd. 2014)

The lithological profile is identical with, or at least remarkably similar to plateau bauxites. On the top the deposits a ferruginous duricrusthas developed, in contrast to the reworked (secondary) slope bauxites.

The Fongo Tongo deposit (Cameroon) is another example of in situ slope bauxites. The mineralisation is developed on the crests of the basaltic lava flows which often exceed 10 km in length and are 400–700 m in width, as shown on Fig. 2.47.

Geomorphology established with the aid of digital elevation model. The red domain shows the bauxites lava flows of the basalt at elevation above 1000 m MSL (Fig. 2.48).

Planation surfaces of plateaus of Kontum Pleicu (Vietnam) at elevation of > 600 m seem to be identical with that of Fongo Tongo. With the exception of the northern tenements, owned by the Bauxite Resources Limited, the Darling Range (Australia) has a complex geomorphology, where the in-situ slope bauxite is typical. The extent of the lateralized capping, developed on granitic rocks, is 350 km in length (N-S) and 40–60 km in width. The elevation is between around 200–600 m MSL. where several Bt bauxitic laterite formed on slopes. The alumina concentration increases down slope. See Fig. 2.49.

The bauxitic laterite is partly cemented, partly loose, friable, so that, it either needs to be blasted or ripped with large machinery. Its reactive silica content is extremely low (1–2%), even if its available alumina content is low (30–35%), it is one of the best raw materials in the global alumina industries as it can be easily extracted, treated, and processed in the proximal refineries. The annual production was almost 50 Mt in 2008 (ALCOR data) and represented as much as 20–21% of the world's production at that time.

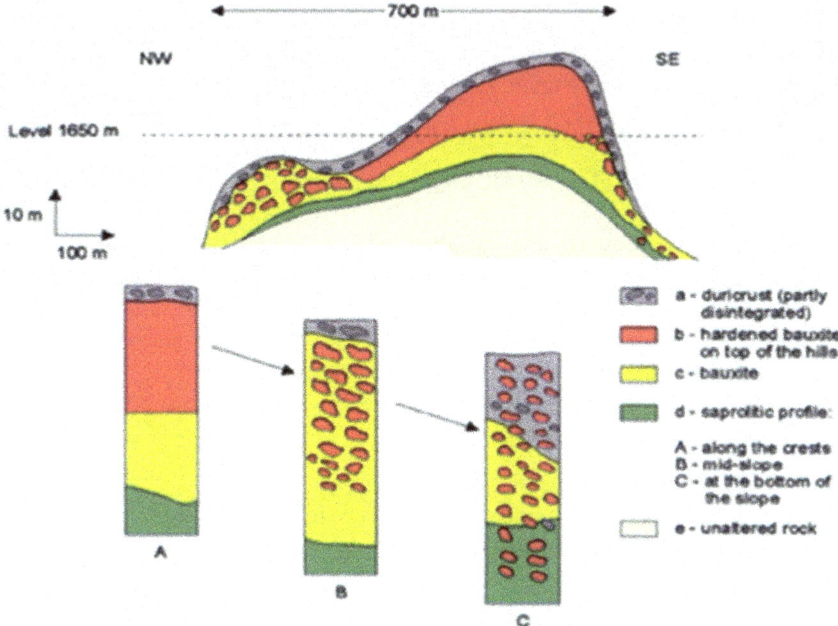

Fig. 2.48 Cross section of the in-situ slope bauxite at Fongo Tongo—Cameroon (Hieronymus [63] from Carboni [64])

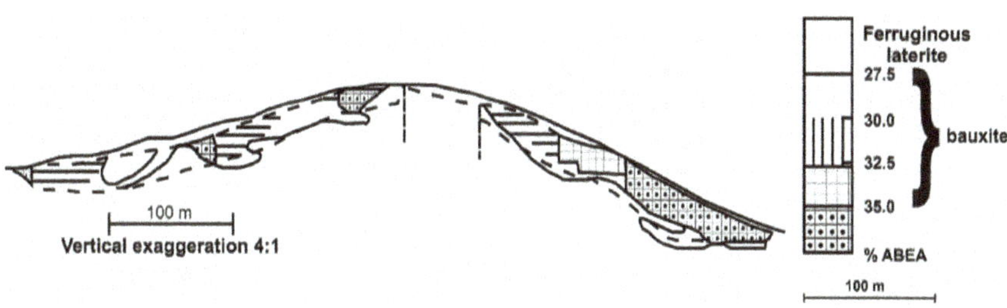

Fig. 2.49 Cross section showing the grade distribution within the Darling Range bauxite ore bodies—Australia (Kirke and Murray from Bárdossy and Aleva [1])

2.5.2 *Karst Bauxite*

Classification of karst bauxite deposits, in harmony with the above-described laterite bauxites, is based on the paleogeography in the time of bauxite formation and accumulation. Two main karst bauxite types are distinguished:

- Continental (terrestrial) deposits
- Sedimentary (paralic) deposits.

The typing is derived from the classifications of Grubic (1975) and Sapoznyikov (1975) and supported by studies made on the Mediterranean Region (Hungary, ex-Yugoslavia, Russia, and Vietnam: (Komlóssy [3, 9, 11]).

2.5.2.1 Continental (Terrestrial) Deposits

The bauxites were formed over large areas and were accumulated by vadose waters percolating through depressions, resulting in their being classified as para-autochton. In lithology they can be stratified, or in some cases, cross-bedded indicated intermittent fluvial sedimentation. While the laterite bauxites are associated with positive landforms, the karst bauxites are restricted to negative landforms.

Continental (terrestrial) deposits can be ranked into three subtypes: bauxite deposits formed and accumulated on.

- karst plateaus in deep sinkholes and canyons,
- morphological terraces: in undulating surfaces: lenticular deposits,
- inner basins: blanket-like stratiform deposits.

Karst depressions deepened to the minimum level of the fluctuating water table. control the form, shape, size and thickness, with the mineralogy of the deposit's dependent on the relation to the karst water table that existed during the bauxite forming and accumulation. as shown in Fig. 2.50. Insert the figure, deleted from the previous page here

- *Deposits within sinkholes* may attain up to100 m in thickness and are.

10–20 m in diameter at Iharkút, Hungary, and 70 m at Niksic, Montenegro, in the Dinaric Alps. They are typically 10–40 m deep and several 100 m in diameter. Complex sinkholes have formed. at Tavernes in France and Ghiona Mts.in Greece [27].

- *Canyon-like deposits* are found in several 100 m long and 10–30 m deep trenches.

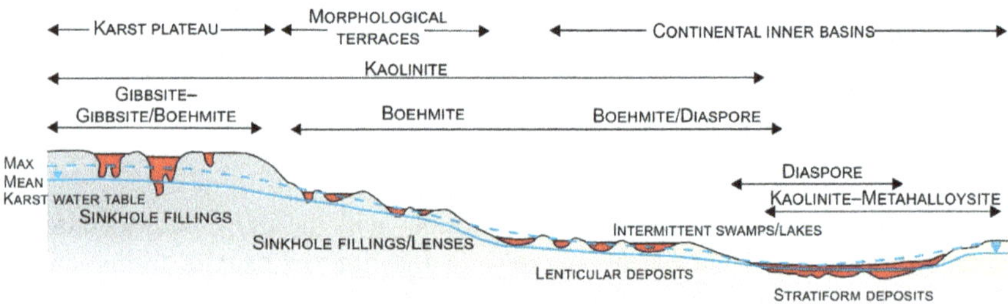

Fig. 2.50 Sketch profile of the palaeogeographic morphology of the continental bauxite deposits [3]

Fig. 2.51 Sinkhole deposits at Iharkút (Hungary) (**a**) and at Niksic (Montenegro) (**b**). Red bauxites are covered by yellowish-white marine limestones. **a** Pataki, personal contribution (2017) and **b** [9, 17]

At the time of bauxite formation, due to instability of the terrain the relief (hydrogeological) conditions, might vary, resulting in complex deposits formed post-initial formation. Where the buried deposits were tectonically disturbed tilted, faulted and folded types of deposits formed, e.g., Dinaric Alps, Greece.

The world class Jamaican bauxites, locating at a high level, are also classified as this type of karst deposits. They developed on Miocene (25 Ma) limestone formations. In contrast to the European bauxites, they are situated on sub-surface, and not buried by marine sediments (Fig. 2.51).

On karst terraces bauxite deposits of lenticular type geometry can be formed. They are widespread in the Mediterranean region. Their shape follows the main directions of fault systems within the bedrock. In section they show dinner plate-like forms pinching out gradually towards the rims (see Fig. 2.52) or bordered by tectonic faults.

Lenticular deposits vary in shape and size. Typical dimensions are most frequently between 1 and 2 km^2. Bauxite thickness, as a function of the undulating bedrock surface, varies between 1 and 20 m, on average 4–6 m.

- *Blanket type stratiform deposits* can be large in their areal extent and can cover up to 10 km^2 (e.g. Halimba—Hungary) (Fig. 2.53).

As a general rule the bauxite grade decreases inwards into the basin. In some cases, the bauxite sequence is interbedded by carbonate fragments forming layers or lenses within the ore body increasing the carbonate content of the bauxite for

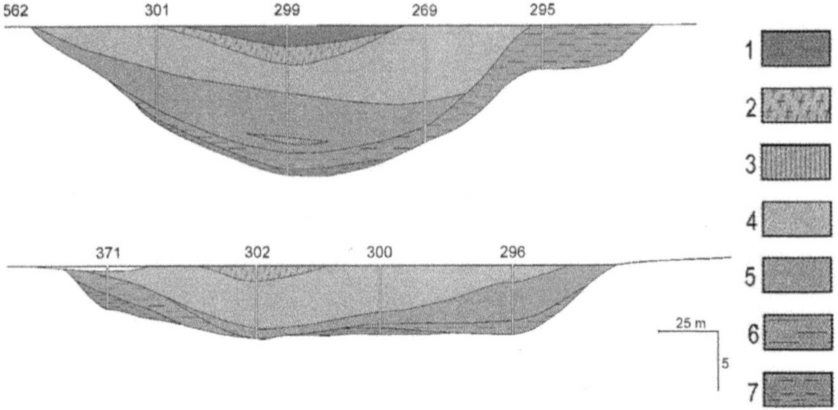

Legend: 1. Pyritic, lignitic clay, 2. Grey pyritic clayey bauxite,
3- Partly re-silicified upper zone.4. Bauxite, 5. Clayey bauxite, 6. Bauxitic clay, 7. Red kaolinitic clay.

Fig. 2.52 Lithologic cross section across the Felix II lens—Transdanubia—Hungary [19]

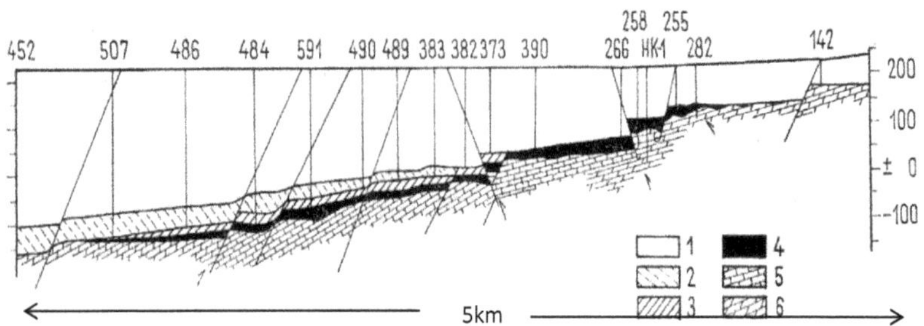

Lgend: 1. Eocene sediments, 2. Cretaceous lignite, 3. Cretaceous mottled marl, 4. Bauxite, 5. Triassic Dachstein limestone, 6. Triassic dolomite.

Fig. 2.53 N-S cross section of the Halimba (Hungary) stratiform deposit (Erdélyi 1965 from Bárdossy [27]). Lgend: 1—Eocene sediments, 2—Cretaceous lignite, 3—Cretaceous mottled marl, 4—Bauxite, 5—Triassic Dachstein limestone, 6—Triassic dolomite

that reason regular checking of the carbonate content is necessary in order to avoid impaired in the refinery performance. In addition to the alumina and silica cut offs, a cut-off for $\sum CO_3$ is also necessary to apply in the reserve calculation.

Karst bauxite deposits are covered, in many cases, by lignitic clay, indicating swampy conditions post formation. These conditions can produce H_2S which reacts with ferric minerals forming pyrite and marcasite (FeS_2) and grey bauxite forming in the upper levels. The decomposing pyrite results in significant epigenetic alteration in the ore body. The industrial grade ores can be contaminated by sulphate minerals. It is another deleterious component reducing the industrial value of the run of mine ore. The usual five component routine analyses need to be extended to include the determination of the $\sum SO_3$ content, and a cut-off applied for sulphate content in the reserve estimate. (In Hungary it was 0.6% $\sum SO_3$).

2.5.2.2 Sedimentary/paralic Deposits

The term of sedimentary bauxites was applied to transported bauxites deposited on unconsolidated sediments (Groeneveld 1963). and for slope bauxites (as talus) re-deposited marginal to the laterite plateaus (Mamedov 2007). Sapozhnikov (1975), in his karst bauxite classification, distinguished "**polygenetic sedimentary-karstic deposits**". According to the Chinese professional literature, the bauxite, accumulated in marine, lagoonal facies is classified as **sedimentary bauxite.** This type of bauxite in some cases may be interbedded with marine sediments, commonly composed of siltstone, silty clay and clay and covered by lignite (Yang 1989).

Bauxite explorations carried out at Tam Lung, Lang Son county, Vietnam revealed bauxite deposits supposed to be accumulated in marine lagoons. Primary bauxites were formed under terrestrial conditions in oxidizing media and deposited in the lagoon in a mainly reducing environment. In such examples the bauxite layers are found in continuous marine sediments or in the oscillation zone of the marine environment where the sedimentation could be interrupted resulting in a narrow stratigraphic gap between the bedrock or overlying cover. Towards the inner basin the gap diminishes the bauxitic bed terminates with interfingering with the marine sediments See Fig. 2.54. This type of bauxite has been classified as ***paralic bauxite*** [11]. As the "polygenetic sedimentary karst deposit", the "sedimentary" and "paralic" ones are the same formations an unambiguous term: **sedimentary (paralic) deposits** have been applied in this description.

Sedimentary (paralic) deposits are found in China in Shanxi, Shandong, Henan, Guizhou, Guangxi, Sichuan, and Yunan Provinces where 98% of the proven resources has been ranked into this type of ore (Yang 1989). Annual production in 2007 was as much as 40 Mt. Bauxite deposits of the historical mines at Severuralsk are believed

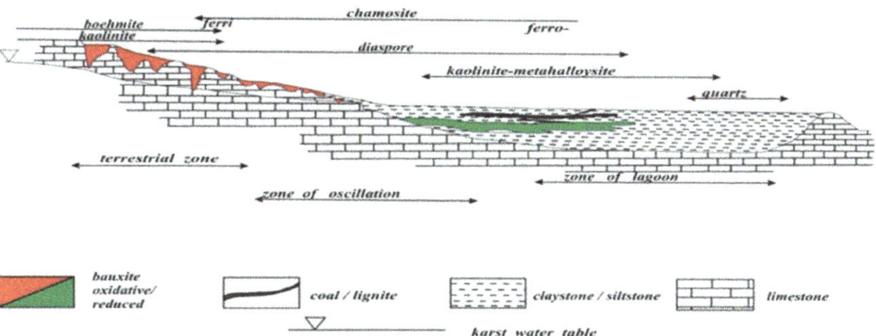

Fig. 2.54 Conceptual geologic cross section on sedimentary (paralic) bauxite forming

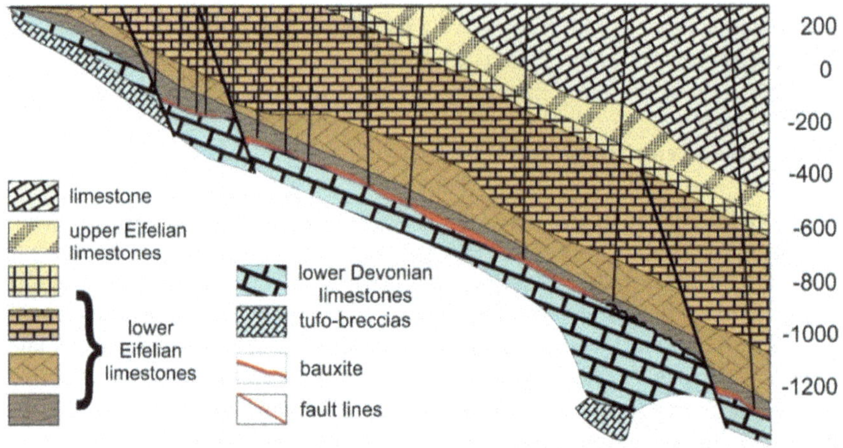

Fig. 2.55 Geologic cross section of a bauxite deposit in the North Ural Mts. (Severuralsk)—Russia (after: Kirpal and Teniakov [53])

to be the member of this type of ore, too, even if there is gap between the karstified bedrock and the ore. See Fig. 2.55. The lower level is oxidative red in colour, where the bauxite accumulated in a terrestrial environment, while the upper zone is reductive grey in colour indicating a marine transgression and simultaneous bauxite accumulation.

These types of deposits are Middle-Early Palaeozoic in age (400–250 Ma). The extent of multiple series of these deposits in strike may attain 100 km, as has been identified in Vietnam in Lang Son and Cao Bang Counties. The bauxite is generally stratified and homogenous in texture. The industrial grade ore (Al_2O_3/SiO_2 ratio $\geq$ 2.6, Al_2O_3 min. 40%) is not always a continuous layer in the bauxitic horizon.

The free alumina is bound in diaspore and the silica in chamosite. In several intervals the diaspore content may attain 80–90%. Impurities are in the form of carbonates (mainly in siderite: $FeCO_3$) and sulphide (pyrite FeS_2).

The sulphides are concentrated in the upper zone. In the Uralian reserves these contaminants are 4% and 1.5%, respectively. In China the low-grade high silica and high sulphur content bauxites are beneficiated where 15% of silica is reduced to 6.6% and $\sum SO_3$ from 1.9 to 0.2% by flotation developed for this purpose [65].

In Vietnam, at Ma Meo, the sedimentary (paralic) deposits where the covering carbonate rocks have been eroded, the primary bauxite largely oxidised, resulting in mineralogical alteration (see details in Sect. 2.3.1) The outcrops at surface fragmented and re-accumulated on the slopes (Fig. 2.56).

This type of bauxitic assemblage, made up by bauxite fragments and clay. Beneficiation by simple wet screening is needed to gain industrial grade ore.

Fig. 2.56 Sedimentary (paralic) bauxite talus at Ma Meo, Lang Son county, Vietnam [11]

2.6 Global Bauxite Reserves and Resources by Provinces, Countries, and Group of Deposits

2.6.1 Terminology: Reserves and Resources Categories

Inconsistencies exist in the way in which the resource content of a deposit is defined. Often statistics on a given deposit report "reserve" as the only category without further explanation. From the 1970s the concept of resources was introduced and since refined. The two concepts are clearly worded in the JORC code 1992 as follows:

- "Resource": is quantified on the basis of geological data and an assumed cut-off grade only. There is an implication that there is a reasonable prospect of eventual economic exploitation.
- "Reserve": it is that part of the resource which, after application of mining factors, results in an estimated tonnage and grade which can be the basis of a viable project after taking account of all relevant metallurgical, marketing, environmental, legal, social and governmental factors.

There are different classification systems for the reporting feasibility of the exploration results. The JORC, US, UN and Soviet systems are compared in the Table 2.4.

In the international practice the application of the JORC code or its equivalent is normally required in general. It is worth emphasising that reserve/resource categories are to be completed with calculated numeric data of possible errors separately for tonnage and components of grade (Al_2O_3 and SiO_2). As a general rule for tonnage the following limit of errors are expected in different reserve categories:

Table 2.4 Increasing level of knowledge

Stagesof knowledge	Reconnaissance		Prospecting	General exploration	Detailed exploration
US before 1976	Possible			Probable	Proven
US after 1976	Undiscovered		Inferred	Demonstrated	
				Indicated	Measured
JORC 2004			Inferred	Indicated	Measured
UN 1996	Undiscovered		Identified		
			Inferred	Indicated	Measured
	Speculat	Hypoth	Inferred	Indicated	Measured
Soviet	D_3 D_2	D_1	C_2	C_1	A + B

Fodor [66]

$$\text{Measured}: \ \text{max.} +/-10\%$$
$$\text{Indicated max} +/-20\%$$
$$\text{Inferred max} +/-30\%$$

For grade in measured category the Al_2O_3 (total or available) should be kept below $\pm$ 1 absolute % and the SiO_2 (total or reactive) below $\pm$ 0.5 absolute%.

2.6.2 Principle of Grouping of the Global Deposits

The pattern of grouping of the global bauxite resources is shown in the follows (Fig. 2.57).

It is difficult to apply a prescriptive system for subdividing the bauxite provinces because the due to extremely variable conditions/controls in different regions. Moreover, the higher unit does not mean necessarily bigger resources, in some cases a

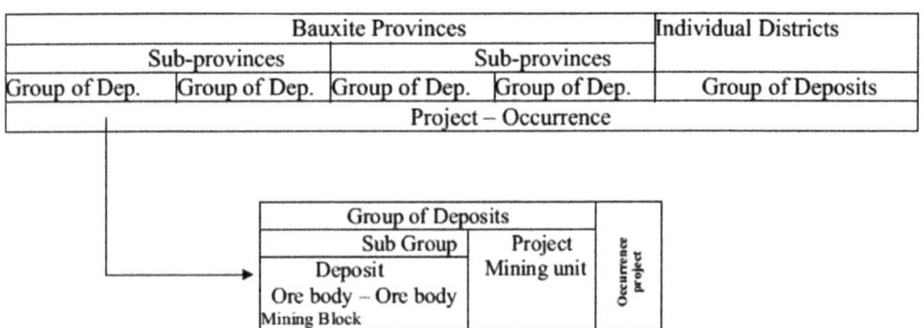

Fig. 2.57 Komlóssy [67]

bauxite province may be considerably less significant than a sub-province or even a group of deposit. The term "project" is also widely used. It is an area delineated for industrial utilisation. Generally, it means an area where geological activities are carried out by a company in order to delineate mining unit(s). In volume it may comprise anything from a Group of Deposits (sometimes 100 Mt of ore in order) to a single deposit/ore body with 0.1–10 Mt tons of bauxite.

2.6.3 Bauxite Provinces Determinant in the Global Aluminium Industries

A grouping of the world bauxite deposits, on global scale, both for karstic and laterite, was established by Bárdossy [32] and Bárdossy and Aleva [1]. It is now time, after 28 years, to issue an up-dated version for the global bauxite deposits, as new world class deposits have been discovered and the minimum demand for a viable deposit considerably increased. Resources increased predominantly in the laterite realm (e.g., Guinea, Brazil and Southeast Asia). Based on global experiences of the Author, the delineation of bauxite provinces and individual districts was reconsidered, taking into account important related publications (Mamedov [68], Bogatirev and Zhukov [69]). The data are inconsistent, hence the criteria for reserve/resource calculation and estimation are different in different countries and bauxite districts. The date of reference is also different. Some of the data are reliable, some of them are at an acceptable estimation. Even if the data are not consistent, they may give a general view as to which territories would be the basis for establishing modern green field refineries (Fig. 2.58; Tables 2.5 and 2.6).

Prominent global bauxite resources by provinces, districts, and groups

Laterite bauxites

2.6.3.1 South American Bauxite Province

A. Coastal Plain sub-province

Total bauxite output of **Suriname** dropped from 5.5 Mt (2009) to 2.7 Mt (2017). In 2017 Meonego (Coermotibo mine) produced 2.6 and the remaining part came from Paramaribo-Onverdaht deposits (Kaaimangrassie and Klaverblad mines). The Coastal Plain deposits are expected to be depleted in the near future.

Guyana produced 1.8 Mt of bauxite (2017). Infrastructure is developed. Bauxite is high grade gibbsitic ore, with commonly more than 50% of available alumina. The reactive silica is less than 3.5%. Overburden thickness is 20–50 m. At the Mackenzy-Linden occurrence, lignite was formed on the top of bauxite. The swampy medium resulted in secondary deferrization; abrasive, chemical and refractory grade bauxite was also formed. In **French Guyana** the identified ore bodies of the Kaw Mts,

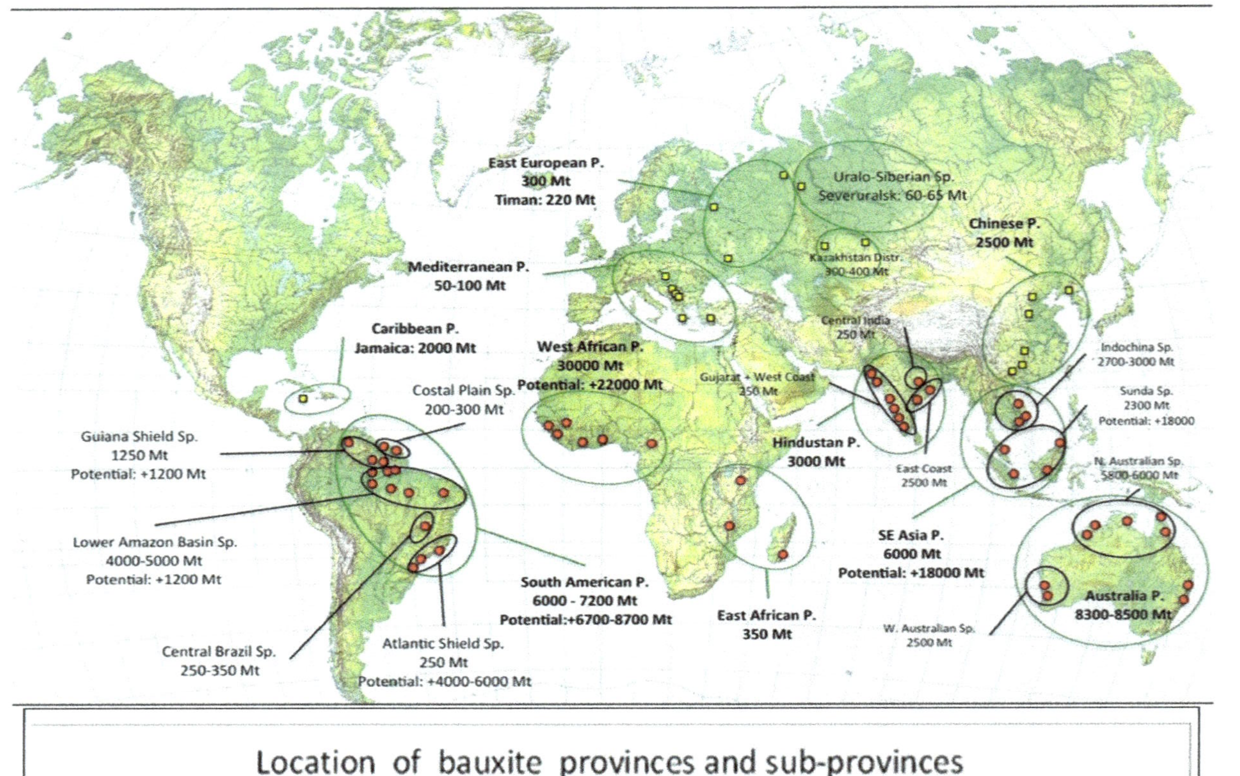

Fig. 2.58 Komlóssy [67]

Table 2.5 Laterite bauxites

Prov	Sub province	District	Group of dep.	+ Resources and potential Mt
South America	Coastal Plain	Suriname	Paranam, Onverdaht	(being) depleted
			Moengo Ricanao	(being) depleted
		Guyana	Pomeroon	200–300?
			Mackenzy - Linden	
			Ituni	
			Kwakwani	
		Fr. Guyana	Kaw Mts	60
	Guyana Shield	Suriname	Bakhuis	200
		Guyana	Pakaraima	450–(1000?)
		Venezuela	Los Pijiguaos	320–350
			Serrania de Cerbatana	250
			Further potential	1200
	Lower Amazon Basin		Pitinga	?
			Trombetas	550 cc
			Juriti	700 crude
			Jauarí	?
			Jurupari	210 crude
			Almeírim	400 crude
			Paragominas	2500 crude
			Further Potential	> 1500 crude
	Central Brazil		Serra dos Carajás	40
			Barro Alto	220–350
	Atlantic Shield		Ouro Preto,Cataguaes	250
			Serra de Mantiquerra	
			Pocos de Caldas	
			Lages	
			Further potential	4000–6000 crude
West Africa	Guinea	Boké	84 deposits	5900
		Kogon – Toumine Interfluve	144 deposits	10,000
		Fatala	115 deposits	1700
		Debele - Kindia	49 deposits	250
		Mali	67 deposits	500
		Labe	85 deposits	2550

(continued)

Table 2.5 (continued)

Prov	Sub province	District	Group of dep.		+ Resources and potential Mt
		Dalaba Mamou	57 deposits		465
		Dongol Sigon	49 deposits		1700
		Balinko	26 deposits		(930)
		Tougoue	111 deposits		3500
		Dabola	64 deposits		750
		Bafing Tinkisso	39 deposits		90
			Further potential		10–20,000
		Boé (Guinea Bissau)	5 deposits		160
		Mali	Falea		170
			Beléa		180
			Bamako West		185
			Further prospect		700
		Sierra Leone	Mokanji Hills/Gondama		24
			Port Loko		100 cc
		Ivory Coast	Digo Mokoudeu + Orumba		20–40
			Inferred		300–400
		Ghana	Ninyahin		330
			Atewa R. - Kibi		120
		Cameroon	Minim Martap		550
			Ngaoundal		120
			Zamba inc. Fong-Tongo		564
East Africa		Malagasy	Manantenina		> 230 cc
		Mulanje (Malawi)			26
		Tanzania	Usambara Mts.(Leshoto)		37
			Pare Mts		50
Hindustan		Gujarat	Ahmedabad		120
			Jamnagar - Junagadh		
			Kutch		
		West Coast	Low level	Thane	~ 20
				Raigadh	
				Goa	35
			High level	Sartara	80
				Kolhapur	
				Belgaum	

(continued)

Table 2.5 (continued)

Prov	Sub province	District		Group of dep.	+ Resources and potential Mt
		Southern India		N. and S. Kanara	Minor-scattered
				Kerala Coastal Plain	
				Tamil Nadu	
		Central and Eastern India		Maihal/Rewal/Katni	30–40
				Maikala – Mainpat Hazaribagh Range	+ 200
				Balaghat	~ 20
				Ranchi	15–20
		East Coast -Orissa	Bolangir	Gandamardan	213
			Kalandhari	Karlapat	300 (+100)
				Lanjigarh	
			Rayagada	Sijimali	570
				Kutrumali	
				Baphmali	
			Koraput	Sasbahumali	> 600
				Kondingamali	
				Panchpatmali	
				Maliparbath	
				Karnapanikonda	
				Pottangi	
				Ballada	
		East Coast Andra Pradesh		Anantagiri Chintapalle and Gurtedu	550
South-East Asia	Indochina	Bao Loc – Di Linh		Than Rai	460 cc
				Bao Loc	
		Dak Nong (Mnong Plateau East)		Dak Song	1400 cc
				Bu Dong	
				Dak Rung	
				Quang Tan	
				Dak Ton Nhan Co	
				Gian Nghia	

(continued)

Table 2.5 (continued)

Prov	Sub province	District	Group of dep.		+ Resources and potential Mt
			Dak Tik		
			Bac Gian Nghia		
			"1–5′ 1st of May		
			Quang Son		
		Phuoc Long			110 cc
		Cambodia	Mnong Plateau W		350–400 cc
		Laos	Bolawen Plateau		200–400 cc
			Xanxai Dakchung		200–400 cc
	Sunda Islands	Riau island	/Potential		1250 cc
		W Kalimantan (Borneo)	Ssanggau Reg	Tayan	700–800 cc
				Simpang Dua	
				Munggu Pasir	
				Pantas	
				Belai Berkuak	
				Sandai	
			Keta-pang R	Kendawangan	150 cc
				Air Upas	
				Pening Kinyit	
				Riam Kusik	
			Sing-kawan	Karagan	250 cc
				Toho	
				Landak	
			Further potential		6700 cc
		Central – E-Kalim	Further potential		7400 cc
		Malaysia	Further potential		4200 cc
Australia	Northern Australia	York Peninsula	Weiapa&Aurukum		1200
			North Wenlock R		1100
			Adoom/Ely		300
			South Embey R		1900
		Gove			200–400
		North Kimberly	Mitchell Plateau		340
			Cape Bougenville		800
	Western Australia	Darling range	Jarradahle		No data
			Huntly-Willowdale		> 2000
			Boddington		300–400

(continued)

Table 2.5 (continued)

Prov	Sub province	District	Group of dep.	+ Resources and potential Mt
	Eastern Australia		Tambory Mt–Hampton	26
			Inverell	
			Moss Vale	
			South Gibbsland	
			Tasmania	

Komlóssy [67]

representing as much as 60 Mt of bauxite. TAl_2O_3 varies between 40 and 42%, and $TSiO_2$ is around 2%). Tonnage is insufficient for establishing greenfield refinery, the low alumina content does not satisfy the demand of export.

B. Guiana Shield sub-province

The most important producer is Venezuela. The annual output in Los Pijiguaos was 7.3 Mt of bauxite in 2009 which dropped to 2.2 Mt by 2017. The TAA is > 46% and reactive SiO_2 is 1.2% in average. Bauxite to alumina ratio is 2.4 t/t. There are additional resources in the Sierra de Cerbatana area but for their exploitation an expensive haulage system will be necessary to develop in a difficult terrain. The bauxite of Venezuela is purely gibbsitic.

The Bakhuis deposits of Suriname were proposed for development, however, development plans have stalled. Other deposits, located on the top of inselbergs (island hills), like Nassau, Lely and Browsberg because of their small size the development of required infrastructure is not viable.

The northern margin of the Guyana Shield (in Venezuela, Guyana and Suriname) seems to be prospective, but for a lock of infrastructure they are not likely to be utilized in the next future.

C. Lower Amazon Basin

This sub-province is one of the most important bauxite sources in the world. Bauxite is of a fragmented nodular type ore which requires beneficiation. The biggest occurence is at Paragominas (Para State) with its 2,5 Bt of crude ore from which about 1,5 Bt can be extracted by beneficiation. The total output is estimated to be around 10 Mt/a. The concentrate is supplied to Belém in pipe line as slurry. At Trombetas hydrocycloning is applied to extract the fine and superfine fractions (<14 Tyler mesh). The total recovery was about 70% in the 1990s which is not thought to have changed significantly. Total bauxite output in 2009 was as much as 21 Mt of crude ore, which represents about 15 Mt of concentrate. The Trombetas bauxite is amongst the best bauxite ore for its extremely good refining properties. The TAA concentration of Lower Amazon bauxite concentrate varies between 46.5% (Juruti) and >53% (Almeirim). Reactive silica ranks commonly between 4 and 5%. This sub-province

Table 2.6 Karst bauxite

Prov	Subprovince	Districts	Deposit Groups	Identified Mt
Caribbeans		Jamaica	Manchester	~ 2000
			St.Ann (Discovery	
			St Elisabeth	
			Trelawny Clarendon	
			St Chaterine	
Mediterranean		Dinarids	Jajce	< 10
			Mostar	< 10
			Niksic	15
		Hellenids Central Greece	Fokis	40 (100)
			Ftiothis	
			Boetia	
		Anatolia	Menderes Massiv	~ 80
East Eur		Russia	Belgorod	50
			Tikhvin	20
			Timan	220
Central Asia		Kazakhstan	Turghay and Pavlodar	300–400 (+1000)
East Asia	Uralo – Siberian	Ural	Severuralsk	60–65
	Chinese Subprovince	Chinese–Korean platform	Lianoning Shandong Shanxi–Henan	1,460
		Fold belts		330
		South Chinese Yangze platform	Guizhou-Gunaxi	575
			Vietnam Cao Bang	~ 100
			Lang Son	25

Komlóssy [67]

is typically gibbsitic. The main transport route is the Amazon and its tributaries i.e., Trombetas and Rio Negro. Thus, the accessibility of these resources is relatively easy making the resources one of the most competitive suppliers in the world. Identification of new world class groups of deposits is predicted for this region. A further 1.5 Bt. of bauxite is estimated in the hypothetical category in this sub-province.

D. Atlantic Shield Sub-province—District

Total annual bauxite production was about 1.7–1.8 Mt in 2009. On the hilltops the upper part of the bauxitic horizon typically is made up by fragmented, nodular type of ore while the lower parts consist of massive bauxite. Along the slopes, a deluvial product (glacis), detrital type of ore accumulated. The best quality ore is in the Pocos de Caldas and Mantequirra districts/groups with TAl_2O_3: 50–60% and $TSiO_2$ 1–5%. Bauxites are purely gibbsitic. Along the eastern margin of the Atlantic Shield substantial potential exists within a belt several hundred km in length and in 20–30 km width. The basement is comprises Proterozoic migmatite, granulites, charnockite and alkaline intrusives. The total belt may contain up to 4–6 Bt of crude bauxite.

E. Central Brazil Sub-province – District

From the bauxite deposits, identified so far in in this sub-province, the Barro Alto deposits, which are developed on anorthosite, is the most significant. The bauxitic blanket is typically made up both by nodular/detrital (upper level) and massive type ores (lower level), analogous with the Atlantic Shield ores. Its TAl_2O_3 surpasses 60% and the $TSiO_2$ is around 1%. It represents the best-known quality of ore all over the world! Its, exploitation has only recently commenced. Serra dos Carajás deposit with its limited quantity and 44% of TAA and 1.5% of reactive silica is much less important.

2.6.3.2 West African Bauxite Province

A. A. Guinea Sub-province—Guinea District

The annual production of Guinea attained 45 Mt by 2017. Production was initially confined to the Sangaredi, Fria and Kindia mines. Post 2010 this production has been supplemented by multiple new mine development utilizing river transport and offshore transshipment. Total identified reserves attain as much as 2.8 Bt of bauxite 'with further proven and probable resources estimated at 26 Bt, prognostic resources may attain additional 11 Bt. Bauxite reserves + resources with their prognostic ore total as much as 40 Bt tons. From this perspective Guinea is the number one bauxite bearing country, especially if its concentration (number of ore bodies or tonnage/sqkm) is taken into account. This fantastic bauxite richness is has historically been challenged by an underdeveloped infrastructure.

Remote bauxite districts such as Mali, Labé, Dabola, Tougoué, Dongol Sigon, Kogon-Toumine Interfluve districts, etc. are not expected to be developed in the foreseeable future. It is obvious that only those deposits will be developed which need less investment in infrastructure, which are located close to the existing railways between Sangaredi and Kamsar port, between, Kindia and Conakry and Fria and Conakry port or have access to river transport corridors. installation of new transport routes will be needed in case of further increasing of the mine production. It is the same for the Russal owned Dian Dian project, located West of Sangaredi. It needsan independent railway line and moreover a new port, near to Kamsar both of which are nearing completion. If we take into account, the aspects of above the total reserves and resources will drop below 6 Bt which are likely to be extracted with moderate development. Guinean bauxite, in general, is a medium–high grade ore. The bauxite is typically gibbsitic (boehmite is < 3%), however, in several lenses the mono content may attain 10% or even more. This type of bauxite is suitable for export. The range of TAl_2O_3 is between 45 and 48%, TAA is typically around 42%, the $TSiO_2$ is around 2%. At Sangaredi and its vicinity (inside the Boké district) there are higher grade ore, TAl_2O_3 is > 52%, and TAA 49%.

B. Mali District

Mali's deposits are not competitive with compared to those of the already developed Guinean resources. There is no infrastructure to access the ore bodies. They are located about 40–100 km from the Bamako—Dakar railway, with the quality of lower grade (TAl_2O_3: dominantly is below 47%) the mono content is typically higher varying between 4 and 25%. A large-scale refinery project, based on the Falea group of deposits, was proposed in 2009, by the Central African Mining and Exploration Co.- London but not developed.

C. Sierra Leone District

After the end of the civil war (2002) bauxite mining resumed operations at Gondama. In 2011 1,3 Mt crude ore was extracted from which 0,9 Mt of high-grade concentrate gained by washing and screening with 53–57% of Al_2O_3 (TAA = > 50%). The $TSiO_2$% is between 2.5 and 4.5%. Bauxite is beneficiated, with a recovery around 70%. Existing railway and road systems are positive factors. A belt of the coastal low-level hills (Post African Surface) in a length of about 250 km seems to be the most promising area for discovering and developing new deposits. The other prospects on the North are located in remote areas.

D. Ivory Coast District

As many as nine groups of deposits have been identified in South Eastern part of the country representing 300-400Mt of bauxite resources of which the Divo Mokouedou is the most important with 20-40Mt of proven reserves. They are tri-hydrate type of ores in which 1–10% of boehmite may occur. Available grade data is inconsistent. It is estimated that at a cut off at 40% TAl_2O_3 the average product should be around 43–45% and the $TSiO_2$% is 1–4%.

E. Ghana District

The Awaso deposit has largely been depleted, with the opening of new mines at Ninahin and/or Kibi a high priority to the government. Investors have been reluctant participate to date because of the expensive development of access roads (distance is 250 km between Ninahin and Tokaredi port and Kibi is 100 km from Tema port and significant environmental constraints and concerns. The bauxite quality is comparable with other West African ores. TAA is 45–50% for Ninahin and 41% for Kibi. The estimated bauxite tonnage and grade potentially could satisfy the demand of a new green field alumina refinery.

F. Cameroon District

Utilization of the Cameroon bauxite has also been hindered by the underdeveloped infrastructure. Minim Martam and Ngaoundal are world class occurrences with their 670 Mt of identified ore and further 200Mt of potential resources at Minim Martap, its quality is comparable to the better remaining the Guinean resources. The bauxite is almost pure gibbsitic with 45–46% of TAA and 1.8% of $TSiO_2$. The deposit is located about 40 km from the narrow-gauge railway track leading to Daoula port. Access to a newly constructed Panamax port at Kribe would require an additional more than 200 km spur line. Compared to many Guinean deposits (e.g., Labe, Dalaba Mamou, Dongol, Sigon Dabola, etc.) it is in a more favourable geographical position. No recent assessment of the smaller Fongo Tongo resource also logistically challenged. Explored resource amounts as much as 50Mt only, but its surroundings have not been explored for further prospects. It is located 250 km from the port but no more than 50 km (as the crow flies) from the coast. It is a potentially a premium grade ore with its almost 50% of TAA and low silica:2.1%. .

2.6.3.3 East African Bauxite Province

The most valuable known bauxite resource of East Africa is Manantenina—**Malagasy.** Bauxite deposits of 550 Mt of crude ore have been revealed by exploration from which about 230 Mt of concentrate can be extracted at grain size of > 2 mm, with 42% of recovery. The ore is gibbsitic, TAA is about 41% accompanied by 2% of reactive silica.

The terrain along the seashore is difficult (swampy) but the deposits are situated relatively close to harbour Port Dophin. There are further prospects to the West of the identified deposits.

Malawi. The Linje-Lichenya plateau emerges by about 1400 m from its surroundings, attaining 2800–3000 m in above MSL. As it is an absolutely isolated occurrence with only 26Mt of bauxite as a metallurgical grade bauxite it is not of interest. The TAl_2O_3 content is 43% (from it about 40% is estimated to be extractable) and $RSiO_2$% is about 1.5%. The utilisation of the deposit is not considered to be feasible.

In **Tanzania** on the Usambara Mts and Pare Mts. the deposits are located in 1600–1750 m MSL, with 1200–1350 m relative elevation from the main transport route (narrow gauge railway) leading to Tanga port. The bauxite deposits are scattered and defined by steep and deep valleys making the access difficult. Individual bauxite bearing hill tops may represent, in the best case, some ten million tons of bauxite. Most of them are smaller.

2.6.3.4 Hindustan Bauxite Province

A. Gujarat and West Coast Districts

It is a 1000 km long bauxite belt where deposits are largely parallel the Arabian Sea coast. They are valuable deposits due to their close location to the sea, however, the individual ore bodies are scattered, representing typically 8-12Mt of bauxite or even less. The bauxite is typically gibbsitic. In Gujarat the mono content surpasses 5%, at places it is contaminated by CaO in 1–3.0%. Thus, blending is needed to decrease the mono- and carbonate contents. No mining unit can be developed for an independent green field refinery but for export purposes (Ukraine, China, Japan), these bauxites have further potential. Total output from Gujarat, in 2015, was as much as 1.2 Mt. Its TAl_2O_3 is ~ 50%, TAA is some 46–47%, $TSiO_2$ is 2–4%. High grade bauxites ($Al_2O_3 > 58\%$ and $Fe_2O_3 < 5\%$ are also available.

In the Western Ghats the Al_2O_3 varies between 46 and 52% and the SiO_2 2–4%, CaO content, at places, is significant surpassing 1%. Raigadh mines produced 3.5 Mt of bauxite in 2009. Low grade ores are utilized by the local cement industries.

B. Central and Eastern India District

The bauxite mines satisfy the demand of the local aluminium industries (Hindalco, Balco). In 2009 the total output of the two companies was 3.2Mt from Chandgad, Jharkhand and Lohardaga mines. Typical bauxite quality is: Al_2O_3 52–54% and SiO_2: < 3.0%. The bauxites are composed of gibbsitic, and gibbsitic-boehmitic type of ores. Korba plant, running at high temperature, is fed by Bodai Daidali and Mainpat mines with high mono content (>10%) bauxite.

C. East Coast District

The most important Indian deposits are located in this bauxite district. Single plateaux contain large ore bodies, such as Ghandamardan 210 Mt, Panchpatmali 300 Mt, Sijimali 270 Mt, Karlapat 210 Mt, Baphlimali 200 Mt, etc.). This phenomenon makes the mine development of large capacity mines as Pancpatmali (6,4Mt/a 2015) where bauxite thickness (average about 15 m) makes the operating costs of mine low. The railway line between harbor Visakapatnam and Raipur township passes through the bauxite district in Orissa. Bauxite export is not permitted, but the establishment of refineries in joint venture with local companies is supported by the Indian Government. Total alumina content, in the main East Coast deposits, is between 46 and 47% and total silica is less than 4%. The available alumina is typically around 38–40%. Production rate of Panchpatmali was as much as 6.4 Mt in 2014. Latest opened mine is Baphlimali with annual capacity of 4.5 Mt (2014). There are further prospects in Orissa mainly in the so called "satellite plateaus" in Andhra Pradesh there are potential for the identification of larger bauxite capped plateaus as well. Development of these resources has been locked to date by environmental and indigenous owner objections.

2.6.3.5 South-East Asian Bauxite Province

A. Indochina Sub-province

Vietnam is one of the countries with an interest in establishing bauxite—alumina industries. Its laterite bauxite deposits have been intensively explored since the seventies. The resources, identified up to now, appear to be suitable (both in terms of quantity and quality) to establish infrastructure for mine development. In this respect the Bao-Lok—Tan Rai deposits are in the best position. In the next step, when haulage system will be introduced for these deposits, further extension towards the Dok Nang district is believed to be also feasible. Two small alumina refineries have recently been established based on Chinese design. Bauxite is of a nodular type ore. Beneficiation is needed. Recovery is 45–47% (at > 1 mm grain size). Average grade of the concentrate for Bao Lok—Di Linh distric is 44.7% of total alumina and 2.4% of total silica. For Dak Nong it is 48.4% and 2.5%, respectively. Extractable alumina is 40% for Bao Lok—Di Linh and 43.4% for Dak Nong. Bauxite is dominantly gibbsitic.

In Cambodia and Laos the similar grade bauxites have been identified as in Vietnam. The industrial value of these deposits is questionable due to the lack of transport routes for material of big volume. Mekong is not navigable. Terrestrial transport towards the South Chinese Sea seems to be expensive, even if the concentrate in form of slurry may be transported in pipeline through Vietnam.

B. Sunda Islands Sub-province

Until 2014 when the Indonesian Government **stopped** its export Indonesia was the largest bauxite exporter in the world attaining 40 Mt. In 2012 the Chinese bauxite imports, originated from the Riau Islands (**Banka** and **Beilitung**) and from **Mempawah** mines (NW Kalimantan, Singkawang Regency), reached as much as 37 Mt. In the West Kalimantan district, the total reserves + resources may attain 1.1–1.2 Bt expressed in washed product. Tayan is in the best geographical position locating nearby the Kapuas River which serves as a natural transport route towards Pontianak port The other deposits in Sanggau, Ketapang and Singkawan Regencies locating far from the River and from the Karimata Strait need development of transport routs. **Tayan** is adequately explored, whilst the other resources probably need further investigation. Indonesian bauxite is of a loose, nodular type (similar to Vietnam). Average recovery of bauxites of Sanggau Regency is above 60%, TAA: 40% and reactive SiO_2 1.2%. Compared this bauxite to the Vietnamese ones its recovery is significantly higher, but the ore bodies are smaller in size (mining coasts will be higher). on the other hand the terrain is the difficult because of marshy lands separating access to the deposits. Identified **Malaysian** deposits, including Sarawak (North Borneo) and Malacca Peninsula, are similar in lithology to the bauxites of Riau Islands and. Kalimantan. Identified crude ore in Sarawak is about 10–30 Mt, that of In Malacca Peninsula 70 Mt. Beneficiation product represents about 60% of the crude ore. With aid of remote sensing techniques significant prospects are supposed to be formed attaining 4.2 Bt of washed product.

2.6.3.6 Australian Bauxite Province

Australian bauxite operation commenced in the late 1950s and reached first in rank in world production in 1971. It has kept its position since that time. In 2017 the total output was 83 Mt, representing 27.7% of the total world production of 300 Mt. This position i due to the following favorable conditions:

(1) Deposits are located at favorable geographical positions (close to ports).
(2) Deposits are large, concentrated in relatively small area permitting large scale mining operations of easily extractable type of ore and with low the mining costs.
(3) Well developed, integrated infrastructure.
(4) Reliable and stable legal and economic background.
(5) Developed economy, with responsible commercial and environmental regulations.

A. Northern Australian Sub-province

The Gove bauxite is gibbsite-, whereas Weipa and its surroundings are boehmite dominant ores. The TAA at Gove is 44% and that of at Weipa is around 50%, reactive silica is some 3% and 4% respectively. Weipa bauxites are beneficiated with greater than 60% recovery. In North Kimberly District the deposits have not been yet developed for mining purposes. The deposits are located in remote areas, Bougainville has low quality bauxite: TAl_2O_3 36%, SiO_2 1.9%, that of for Mitchell Plateau are: 44.6% and 6.6% respectively. It is considered that it is unlikely that these resources would be developed because of its high environmental value.

B. Western Australian Sub-province

It is the most significant sub-province in the world. Its production was almost 50 Mt in 2008 representing 20–21% of the world production. These bauxites are of low-medium quality regarding their TAA content which varies between 30 and 33%. In spite of the low alumina concentration, because of its many other advantageous properties, it is one of the most valuable raw material of the aluminium industries. Bauxite is gibbsitic, with low reactive silica of 1–2%. It is soft, earthy type, only the hard cap needs blasting, thus easy to extract in large volumes. Mining operations extends over a large area with aggregations of small ore bodies. The sub-province is located in excellent geographical position.

Karst bauxites

Karst bauxite are no more attractive for establishing new refineries, but they may feed established and continuously modernized refineries. Karst bauxite resources of Jamaica, Russia and China will play relatively less and less role in the world alumina industries even if the bauxite production has been increased in from 46 Mt (2013) up to 68 Mt (2017) in China. That in Jamaica dropped from 14.6 Mt (2008) to 9.8 Mt (2017). In Russia the production is between 5 and 6 Mt/a in the last ten years.

2.6.3.7 Caribbean Bauxite Province

In the Caribbean island's bauxite occurs in Jamaica, Haiti, Dominican Republic, Costa Rica and Cuba. Jamaica is far the most important bauxite producer in this Province.

Discovery of the Jamaican bauxite after the 2nd World War revolutionized the bauxite alumina industries. It was the 1st in world rank between 1957 and 1970, but when Australia and Brazil ramped up production its significance began to decline and in 2017 its position was the 6th in the world rank with its output less than 10Mt. Deposits have been explored in detail, The infrastructure is established and reserves/resources are still significant even in a global scale of about 2000 Mt of ore although a signification proportion of these are sterilized by competing land use and reserves further inhibiting exploitation is that the ore bodies are scattered and small in size. Ore bodies containing of 50–100 kt of bauxite are common. The Al_2O_3 usually varies between 47 and 50%. From the ores of best quality only ~ 2% Al_2O_3 cannot be extracted in the refinery. The reactive SiO_2 content of this type of ore is 1–2%. The bauxite is gibbsitic and gibbsitic-boehmitic type. Jamaica is expected to maintain its position among the prominent bauxite producers for a long time.

2.6.3.8 Mediterranean Bauxite Province

The historical and significant bauxite producers like France, ex-Yugoslavia, Hungary, Turkey are not even the shadow of their former selves. Bauxite deposits have been depleted or became uneconomic, occurring as small pockets in faulted and folded zones. Sometimes situated 50–400 m below the surface. From the beginning of the twentieth century, generations of the professionals have grown up discovering, exploring, evaluating, material testing, mining, processing of these bauxites and contributed significantly to the global bauxite knowledge of our time. The total production of this province recently amounted to 3–4Mt of which the bauxite mines of Itea and Distomon (Greece) are dominant with their almost 2 Mt/a production. The Mediterranean bauxites are typically boehmitic or boehmitic-diasporic. The Greek bauxites are diasporic. The total alumina content, in this Province, was regularly between 52 and 57%, the silica content varied between 5 and 7%. The Greek bauxites were and still are the best in quality in this province, with their relatively higher alumina and lower silica concentration.

2.6.3.9 East European Bauxite Province

The Russian Belgorod and Tikhvin are out of the scope of this review because of their minor economic importance. The only bauxite district in this scope is Timan with its 200 Mt of bauxite. It is a medium grade mono-hydrate ore with its 50% of Al_2O_3 and ~ 7% of SiO_2. Reserves may be increased by further detailed exploration. Bauxite production comes from both open pit and underground operations, Timan's output was 3 Mt in 2016 and supplied to the Bogoslovsk refinery (Ural) in a distance of about 2000 km. Refinery and smelter were proposed to be established in 2004 based on Timan's resources but have not been established.

Kazakhstan Bauxite District

Kazakhstan is not grouped to any province as it is an independent district under the proposed classification. The total reserves are < 360 Mt but resources may attain 1000 Mt. Groups of deposits are located at West and Central Turghay and Pavlodar. Lishakovsk and Arkalyk (Turghay Group) mines produced about 5 Mt of bauxite in 2017. Its quality is low grade ore with 45% of total alumina and 8–10% of total silica, with carbonate as a significant impurity. Bauxite is processed in a domestic plant at Pavlodar.

East Asian Bauxite Province

A. Uralo-Siberian Sub-province

The Severuralsk deposits remained in the scope of the global bauxite-alumina industries. It is a large blanket type diasporic-chamositic ore. Geological resources are estimated as 60–65Mt from which 70% may be extracted. Deposits outcropped along a strike of 30 km long with a dip of 25°–35° East. It may extend over 2000 below the surface. Explorations identified the bauxite to minus 1000 m from and mining operation to around minus 800 m which is a world record in bauxite mining. Its average alumina content is between 53 and 57% whereas the silica is 1.5 and 4.5%. Bauxite is contaminated with carbonate and sulphur. Formerly it was processed in the Bogoslovsk refinery. Since it was closed the bauxite production of 3.25 Mt has been utilised elsewhere.

B. Chinese Sub-province

In the past ten years China dramatically increased its bauxite production from 20 Mt (2007) to 68 Mt (2017). Such an increase in bauxite production in this century has only been rivalled by Guinea. Deposits are concentrated in Lianoning, Shanxi, Shandong, Henan, Guizhou, and Guangxi, provinces to which the Vietnamese karst deposits of Cao Bang and Lang Son provinces are also included this Sub-province. The majority of Vietnamese bauxites are slope type of ore. Beneficiation is needed, washing and screening applied.

In China there are some 140 identified deposits from which the most important (containing >50 Mt of bauxite) are as follows: Fushun-Fuxian- Benyi-Fuzhou (Lianoning) Goucun, Goupiang, Xiaping (Shanxi), Zhengzhou, Xianguan, Gun (Henan). Xinwei, Maling, Nadou, Taiping, Pingguo (Guanxi), Gansu (Gansu) Kunming Yunnan), Guiyang and Zuny (Gouizhou). Deposits are classified into karstic and "sedimentary"(paralic) bauxites. The estimated 2400–2500 Mt of total reserves/resources are scattered over a huge area. Because of the high silica and in some cases high sulfur content the bauxite needs beneficiation or selective mining in many cases. Application of flotation techniques have been developed for these types of bauxites on a pilot scale where 15% of average SiO_2 concentration in crude bauxite is reduced to 6,6% and alumina increased from 56.8% up to 66.9%.

2.7 Principles of Physical Bauxite Beneficiation

Run of mine (ROM) ore consists of a mixture of valuable and gangue minerals. Ore dressing are the operations that prepare this material for further processing or use, by producing a somewhat standardized material with well-defined physical and chemical characteristics. This material, after dressing, might be processed into metals or used as aggregate or fuel, depending on the mineral.

Bauxite ROM may vary in a broad range of compositions, textures, particle breaking resistance and other since many rocks—or mineral aggregates—might be weathered into bauxite. Bauxite is a broad definition for an aggregate of minerals from which aluminum might be obtained. With that, it is relevant for the process engineer to know the specificities of the bauxite deposit to develop a suitable process flow diagram. Bauxite dressing should access relevant matters to the alumina refinery, and its handling operations and must be adapted to each bauxite ore. These operations aim at separating the valuable mineral grains from the gangue minerals producing concentrate and tailings. They might include crushing, disaggregating, screening, and classifying, concentrate and tailing dewatering operations.

Ore dressing aims at developing two fundamental operations, minerals liberation and subsequently separating those of interest from the gangue. Liberation is achieved by reducing the particles up to a size where the individual minerals are sufficiently free from each other. Specifically, for bauxite, the liberation concept should also include a large-scale liberation where the clay agglomerates, that are not part of the aggregate but present in small pockets inside the ore lumps, are exposed so they might be mechanically removed by attrition or scrubbing operations.

Concentration or separating minerals of interest is dependent of the previous operation. If the particles are not well liberated separation processes might not be able to achieve the required grades. On the other side if the process over-reduces the particles energy is wasted. In addition, recovering smaller particles is increasingly difficult since more reagents might be needed, and the differentiation forces are smaller (Fig. 2.59).

The separation is achieved by utilizing physical or chemical specific differences between the mineral of interest and the gangue. Clay minerals, that are naturally fine, or artificially fine particles created by the comminution process also interfere in the separation process by adhering on the surface of coarser particles. This adhesion prevents the bond between a surface characteristic reagent and the mineral particle. This is the reason that between liberation and concentration operations the mobility between particles should be accessed.

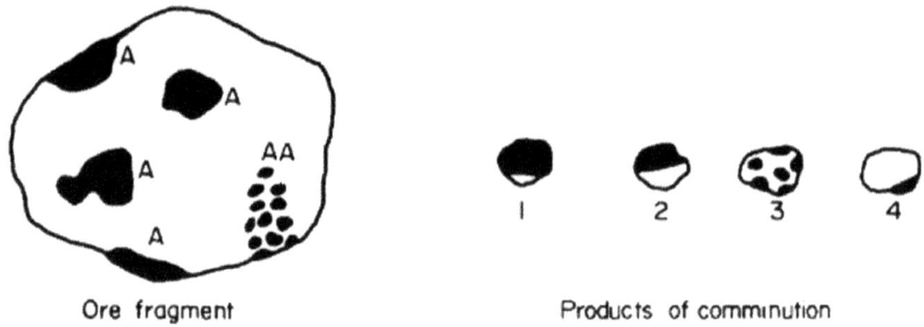

Fig. 2.59 Liberation of minerals from ROM particles

2.7.1 Why Beneficiate Bauxite?

To obtain smelting grade alumina, it is necessary to refine the alumina contained in the bauxite. The Bayer process relies on the fact that the alumina present in bauxites is soluble in hot solutions of sodium hydroxide, while iron and titanium minerals are insoluble. Thus, leaching of the alumina bearing minerals and precipitating it from the filtered solution is a way to obtain purified aluminum hydroxide that might be calcined to alumina and subsequently reduced to aluminum.

The dissolution of these minerals can be done at two temperatures depending on the ratio of gibbsite (alumina trihydrate), boehmite and diaspore (both alumina monohydrates) present in bauxite. The gibbsite is solubilized at about 150 °C and the Bayer process done at this temperature is said to be low temperature. The dissolution of boehmite and diaspore is done at about 250 °C and this Bayer process is termed high temperature. If there is a mixture of these minerals, the process can be done at low temperature if the amount of bauxite monohydrate does not exceed 5% of the total ore mass.

In the process at 250 °C, both quartz and kaolinite (a hydrated aluminum silicate) are solubilized. Once sufficient silica has entered into solution, the precipitation of artificial Bayer Sodalite, a hydrated sodium aluminum silicate, occurs. Although bearing the name of the mineral of the same chemical composition this is an artificial material and, thus, not a mineral. The precipitation of this compound sets sodium ions in the solid phase, which will be discarded from the process. This loss of sodium must be compensated by NaOH solution make up. In the process at 150 °C, the quartz reaction rate is low enough so it might be considered inert, only kaolinite reacts [20]. This means that, depending on the aluminum bearing mineral, quartz might or not be a contaminant in the process while kaolinite always is.

At this point it is relevant to make the distinction between minerals, a natural occurring material, and artificial materials such as the aluminum hydroxide that is precipitated in the form of artificial Bayerite in the refining process, which is a gibbsite polymorph [70].

The cost of alumina refining is heavily influenced by the required amount of NaOH make up per ton of alumina produced. In a simplified way, it can be affirmed that,

Table 2.7 Chemical analysis per size fraction

Material	Fraction from ROM (%)	Available alumina (AA) (%)	Reactive silica (RS) (%)	AA/RS
ROM	100	32	10	3.2
+ 37 μm	70	40	3.0	13.3
− 37 μm	30	7	26	0.3

excluding other factors of the analysis, bauxites with a ratio of Available Alumina (AA)/Reactive Silica (RS) greater than 10 are economically attractive because the production of alumina would compensate the soda's cost.

Bauxites that have gibbsite ($Al_2O_3 \cdot 3H_2O$) as the main alumina bearing mineral have this alumina refined a 150 °C. At this temperature, only the kaolinite present in the material reacts with the sodium hydroxide solution. Clay minerals, such as kaolinite, are naturally fine. For this reason, they can be separated by beneficiation processes such as screening and classification once they have been disaggregated from coarser particles.

Mono hydrated bauxites which have boehmite and diaspore ($Al_2O_3 \cdot H_2O$) as the alumina bearing minerals must be processed at around 250 °C. At this temperature both kaolinite and quartz react. Thus, the bauxite must have naturally low content of these minerals or they must be reduced to an acceptable level during beneficiation.

As an example, a sample of Brazilian gibbsitic bauxite chemical analysis per size fraction is shown in Table 2.7.

For this sample, it is easy to conclude that, if the −37 μm fraction is removed from the ROM, the AA grade is increased while reducing the RS grade. This increases the ratio between them from 3.2 to more than 13, while reducing the mass delivered to the refinery with the bonus of reducing the amount of residue generated by the metallurgical process.

Bauxite dressing activities aim to improve the ore quality to ease the next processing steps. With that, each activity done has to be justified by the improvement it delivers to the ore.

The first step in a bauxite processing plant is particle size reduction, and, at some cases, this is the only alteration promoted to the ore besides lot homogenization. The crushing might be done in one or more steps depending on the feed top size and the desired crushed product size. Typically, each crushing stage has a reduction ratio of up to five, meaning that if the larger dimension of a block is 1 m, the product should have a top size of 0.2 m. Naturally, this depends on the crusher type selected and the desired effect on the material. Other characteristics of the material, such as stickiness and abrasiveness, might impose different equipment selection with a diverse reduction ratio. The size reduction improves the material handling activities and, by liberating the particles and clay, allow the separation process to happen.

The separation process relies on the mobility between the mineral particles. Since bauxite is the product of rock weathering, the presence of clay and moisture makes the particles bind together with each other and with the clay itself. To overcome this issue, a unit operation to disassemble these structures is needed. Water is the media used to suspend the particles and gentle attrition gives the energy needed to dilute the clay and free its particles from the coarser ones. This disaggregation might be done in a dedicated equipment such a drum scrubber or combined with the size separation operation.

The separation of the particles of interest from the gangue depends on their characteristics, specifically the differentiating ones. For instance, coarser particles might be separated from finer particles using a screen. Screening is a unit operation that aims to reduce the number of finer particles from the retained fraction. For this to be done at an industrial scale the material must be characterized. If the particles, after breaking, have similar main dimensions any screen aperture might be used, but if the particle has two large dimensions and one small, elongated aperture in the screen might allow larger than desired particles pass through implicating in inefficiencies.

Other characteristics might influence the screening process, such as the presence of a high proportion of near size particles and fines. The near size particles, that are just larger than the screen aperture, are prone to obstruct the openings. Particles that are just smaller need several opportunities to fit and pass through the screen. This means that probability of such fragments to be directed to the desired flow is low when compared to particles that are smaller than half of the aperture. The fines influence the wet screening process by increasing the apparent density and viscosity of the fluid. With higher density and viscosity, the fluid hinders the mobility of the particles, reducing the process efficiency. As discussed previously, the correlation between size and quality of the particles implies that, by screening, the product might have its available alumina grades incremented while reducing the reactive silica.

Other separation processes rely on diverse differentiating characteristics. If the particles have similar sizes and shapes but differ in specific gravity, this might be used to promote their separation. Depending on the particle size distribution, several gravimetric techniques might be applied.

As shown in Fig. 2.60 the optimum particle size distribution (PSD) for two gravimetric methods—jigs and spirals—differ. This is a consequence of the method of separation imposed by the equipment geometry and flow patterns.

The hydraulic diameter is a way to compare particles combing the effects of several aspects of the particle. To elaborate on this concept, suppose two perfect spheres with the same diameter, surface roughness, chemical composition, same crystalline structure, and any other characteristic that might influence its movement in a fluid, but they differ only on total weight, due to a difference in density. If those two particles are let to settle in an undisturbed fluid, the terminal velocity of the denser will be higher. This happens on the account that the movement resistant forces are the same. The drag perceived by each particle is a function of the apparent area perpendicular to the movement direction, the fluids viscosity, and the particle velocity. The accelerating force, product of the particle mass and the gravity field within it is in, promotes a variation in velocity until it is balanced by the drag. Once the forces balance each other, there is no more acceleration, and the terminal velocity is achieved. Particles with more mass have higher terminal velocities and given enough residence time, they will be in a different region of the separating device. After the

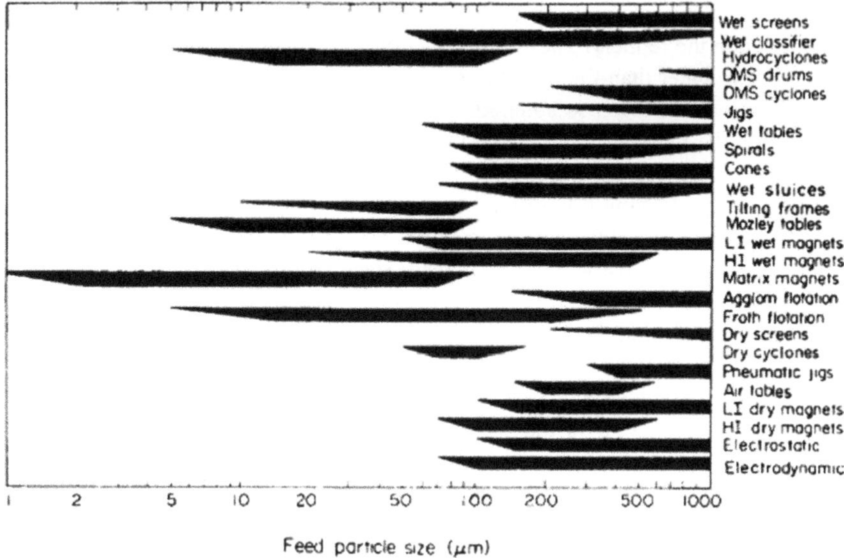

Fig. 2.60 Separation techniques [71]

separation has occurred a splitter of some sort diverges the flows, and the feed is divided into two streams: one with higher density spheres and another with lower.

Another equipment that uses the hydraulic diameter to classify particles is the hydro cyclone. Slurry cycloning relies on the fact that particles with larger hydraulic diameter have more inertia. The flow is fed tangentially into a cylinder. The speed of the flow makes the slurry spin inside the equipment and particles with larger hydraulic diameter accumulate closer to the cylinder wall. One end of the equipment is conical in shape and forces the coarser particles to converge in a stream that exits through an aperture in the apex. The central portion of the flow, with particles having a smaller hydraulic diameter, is forced to the opposing orientation from the apex and exits through a central discharge named vortex finder.

Similarly, to screening, concentration techniques have the ultimate goal of increasing the value of the product by adjusting its quality to what is perceived as relevant by the refinery. At this stage, reducing the quantity of the contaminants—reactive silica—and diluents—iron and titanium minerals and quartz, for low temperature refineries—influence the cost of refining and residue disposal. The soda make-up is directly influenced by the reactive silica amount, but the physical loss in the residue is also significant. Quartz and iron minerals increase the amount of generated residue, thus also increasing the soda loss. Along with that, those minerals increase wear in the refinery pipes, pumps, and other equipment. The reasoning of including a unit operation in the beneficiation stage of processing is that its cost will be offset by the savings in freight costs or in the next processing stages.

After crushing, screening, and concentrating, another operation might take place. Dewatering the material is a final processing step prior shipping or feeding into the refinery. The rationale is straightforward: less mass to be transported or water to be evaporated in the refinery. The coarser fractions, after having the clay removed, tend to carry less water than finer fractions. The moisture is influenced by the particle's size since that the water is adhered at its surface. Having higher specific surface means that the material will be able to hold more water. For coarser fractions, the dewatering might be done alongside the shipping process. Just by stockpiling the concentrate and leaving it to drain for a period its water content might be reduced to acceptable levels. This dewatering might be aided with the use of additives that reduce the water viscosity or the particles surface affinity to water.

Finer fractions might need to be dewatered with the use of a dedicated equipment. Both filters and dewatering screens have been successfully used. Nevertheless, each equipment is better suited for a specific particle size distribution. Dewatering screens are used with particles smaller than 1 cm and up to 200 μm. Larger particles might promote excessive wear in the screening media. These screens must have small apertures so only water passes through. If larger particles are fed, their own weight associated with the equipment's motion might promote premature failures. Filters are the recommended equipment when smaller fractions are still of interest and must be dewatered prior being exported. The slurry is fed directly over the filtering cloth or the filtering element is submerged in the suspension. In both options the vacuum promotes the movement of the slurry and the solids are deposited over the cloth, while the water passes through. The size range for this application might include particles as fine as 30 μm. When compared with screens, filters are more sensitive to the presence of clay particles. These will wear and clog the filtering cloth reducing its ability to efficiently separate water from the finer particles up to a point where it will have to be replaced. Regarding this matter, desliming operations will typically precede.

Bauxite grinding is done to increase the particles surface area and, thus, increase the chemical kinetics of the leaching process at the refinery. This operation might be done on the mine, using water as medium, or at the refinery using spent liquor. If it is done at the mine, the resulting slurry must be dewatered prior being fed into the refinery. The water addition in the refinery must be done in a controlled manner, and preferably, this water is used to reduce the amount of soda loss by washing the residue or to improve the precipitation process. This means that reducing the bauxite moisture before it is fed into the refinery is necessary. The particles, after being ground, should have an ideal top size for the extraction process. If particles are excessively coarse, the extraction might not be fully accomplished, and valuable alumina is lost to the residue. Also, coarser particles increase wear in the equipment. If the material is overground, there is energy and grinding media waste.

Processing the ROM to produce the concentrate imply in generating a tailings flow. Due to the processing characteristics of bauxite, its tailing is a diluted slurry containing no more than 30% solids. These solids are the minerals that were rejected during the concentration stages and include kaolinite or the clay as the finer fraction. A less fine fraction contains particles of iron bearing minerals such as hematite, magnetite and goethite, titanium bearing minerals such as rutile and anatase, and quartz. Some gibbsites or other aluminum bearing mineral is also present due to imperfections in the process.

The processing tailings might be deposited hydraulically inside a reservoir. This option does not require any previous treatment but is capital intensive due the reservoir building and the least efficient option in space occupation. Alternatives to this method are thickened options. Thickening might be done inside a pond, with or without the aid of flocculants, or in a dedicated equipment or in a decanter centrifuge. All these options will produce a solids flow with increasingly lower water concentration. The ultimate alternative for water recovery is the press filter. The latter options that recover more water are becoming increasingly more common in mining operations and new projects.

This trend arrives as an answer to the public demand of more rational use of resources and safer operations. Reducing the water locked in with the slurry signifies lower raw water catchment. New legislation in different countries is introducing charges over the water that is pumped from the environment to drive reduction. In addition, safety issues must be taken in account. The risk of having an embankment full of slurry fail might be minimum if proper attention is given to the system. Adequate project and management should be enough to guarantee the physical stability of the slopes.

Nevertheless, the danger these structures impose are massive. This means that, even when professionally managed, the risk perception of the society might prohibitive. Recent dam failures in Hungary, Canada, Brazil and many other, are pushing public demand towards slurry free deposits. This means that drier deposits might be, soon, mandatory for a new operation to receive its operating license. On the financial spectrum of the analysis, not having an unconsolidated solid accumulation to handle during mining closure is a benefit.

With that, Bauxite beneficiation might include one or more of the following activities:

1. Comminution toe ease handling and to exposing and liberating the minerals.
2. Disaggregation of the minerals.
3. Separation of the minerals.
4. Dewatering the concentrate; and
5. Dewatering and disposing the tailings.

2.7.2 *Technological Characterization*

To elaborate a process flow diagram or to improve an existing beneficiation plant, a technological characterization program must be developed. These activities should be done each time a new ore body is set to be developed. Technological characterization differs from the geological characterization of an ore occurrence. The geological characterization, aside from the structural aspects of the deposit, aims at evaluating the characteristics needed to estimate the value of the material once it is processed into concentrate. As illustration, bauxite ores are evaluated for their chemical composition, mainly aluminum, silicon, iron, titanium and phosphorus and available alumina and reactive silica. The AA and RS are to be evaluated initially at low and high digestion temperatures to evaluate whether the deposit is suited for low or high temperature Bayer process. Once it is decided, the AA and RS grades are to be evaluated only at the assigned temperature.

A technological characterization program aims at understanding how the ore will behave in many other aspects regarding its handling and processing conditions. Such a program should evaluate all the ore dressing options for the ore and ancillary activities.

The samples that are used for this program must, as accurately as possible, represent not only the mean aspects of the deposit but also the extremes. These characteristics will eventually be used as definition criterion for the flowchart of the beneficiation plant, and, as such, they must evaluate the behavior of the material in every considered unit operation. A tentative list follows:

- Non-specific tests
  - Particle size distribution (PSD) with chemical analysis for several different top sizes, with screens and laser diffraction
  - Stickiness
  - Abrasion
  - Particle top size versus alumina extraction
  - Particle specific gravity versus PSD
  - Dense media separation; and
  - Particle liberation
- Crushing
  - Compressive strength; and
  - Breakage matrix
- Disaggregation
  - Optimization of solids concentration, energy applied and residence time; and
  - Particle autogenous degradation (or grinding)
- Screening
- Particle shape analysis
- Concentrate dewatering
  - Leaf (or filtering) test; and
  - Natural non saturated drainage
- Concentrate grinding.
  - Specific energy consumption for product top sizes (Bond index)
- Investigative tests
  - Magnetic concentration versus PSD; and
  - Electrostatic concentration versus PSD

- For the tailings:
  - Geotechnical characterization
  - Thickening with and without polymers
  - Filtering (including press filtering)
  - Centrifuge settling; and
  - Rheology tests
- Pilot equipment tests
  - Hydro-cyclone
  - Humphrey spiral; and
  - Decanter centrifuge.

Another specific set of tests should be developed to evaluate the behavior of the bauxite and its products in the refinery. E.g., the settling rate of the residue and its rheology.

2.7.3 Unit Operations

The following text is a short introduction to each individual operation. The goal is to bring to the reader's attention key aspects of each technique and equipment. The intention is not to dive into each subject, but to open the doors and incite curiosity in the reader's mind.

Crushing

A bauxite crushing equipment selection must be done regarding the next processing operations.

If it is the only operation needed, i.e., the bauxite does not need to be further improved prior being sent to feed the refinery, it should be done having the handling and grinding phases as drivers. Bauxite grinding demands particles with a definite top size depending on the grinding concept. Refineries might have autogenous or semi autogenous grinders, rod-ball, and ball mills. Each one accepts different top sizes, from coarser to finer, respectively. Typical figures are up to 20 cm for the first, 10 cm to the second and 5 cm to the latter. Still, if there is no size or gravity separation, the crushing operation does not need to be done in such a way that the fine particle generation is reduced. However, the equipment selection still has to consider other material characteristics, such as its stickiness. If the clay and moisture content is low, ordinary jaw crushers might be used, yet their reduction ratio is low, and many stages might be needed. For higher reduction ratios and fine particles generation, hammermills are well suited.

On the other side of the spectrum, when further operations rely on particle size to separate the minerals of interest and the ore has higher clay and moisture content, sizer crushers are indicated. This kind of crusher has various desirable features for bauxite processing. The bauxite crushing is done generating lesser fines and all the particles three dimensions are controlled. In addition, the rotating movement, instead of a reciprocating one, pulls the material in the desired direction reducing clogging.

The crusher selection has to be done respecting the ore characteristics. Besides that, the volumetric output required and the ROM top size must likewise be taken into consideration.

Scrubbing

The disaggregation of bauxite, when needed, has be done minimizing the grinding of particles that should be recovered afterwards. This means that proper sizing of the equipment must be done, as well as the installation of parts that allow better control of the process. Once the residence time, energy input and water ratio ranges are defined in the laboratory a proper drum must be designed. These variables are to be controlled by altering the discharge outlet diameter, rotation speed and by using a proportional water valve, respectively. Typical residence times range around 1.5 min, rotation speed at 60% of the critical speed and the water ratio is 1:1.

A pulp discharger, like a pebble port, is an appropriate feature in view of that it removes larger or denser particles that might otherwise accumulate inside the equipment promoting undesired particle grinding. The length to diameter ratio of the drum will affect the energy input and residence time, commonly it varies between 1 and 2.5 and is selected using a benchmark operation or with discreet element analysis combined with computational fluid dynamics.

An undesired event is the presence of clay agglomerates in the scrubber's discharge. These clay balls are the result of low dilution and energy in the scrubbing process. This might happen due to a change in the operating conditions. If, for any reason, the water ratio is reduced clay balls might present themselves over the screens. Other trigger is a change in the ROM. If the clay content is increased the water to clay fed might not be enough resulting on this phenomenon. To prevent this from happening, proper monitoring of the quality of the material fed into the plant must be done. Small process adjustments will always be deemed necessary once the ore varies according to its position in the mine.

Screening

The screening operation of coarser fractions is straightforward. Typical screen sizing might be used, evaluating the material characteristics—PSD, near size particles, particles shape and moisture—screen characteristics—deck position, open area, opening format and type of movement—and process conditions—loading and desired efficiency. Once the screening area is defined, the geometry of the screen might be evaluated. The length and width of a screen must be defined as a function of the material layer thickness—no more than three times the top size—over the screen, the speed of the particles over the screen and the defined unitary volumetric capacity.

Finer fractions screenings are peculiar. The low solids content in the slurry and the high liquid phase viscosity are specificities that make this application stand out. The main goal, as already described, is to remove contaminant and reduce the diluent content. The contaminant for low temperature bauxites is the clay, which is an order of magnitude smaller than the particles of interest. This means that this screening process might be observed as a dewatering unit operation. Thicker material layers are acceptable; these might reach up to 20 times the particle top size. Since the main goal is to remove the finest of the particles and these move alongside with the water, substantial amounts of process water are added on the screen using spray nozzles. The length of the screen determines the residence time of the material over the equipment, and, for this application, longer screens are the appropriate option. Longer residence time means that the clay slurry will have enough time be drained and washed. The total water addition in ratio in such operations, including on the scrubber vary, but ranges around 1:1 solid to water up to 1:5.

Hydro-cyclones

If the ore being processed has recoverable material at sizes that prevent industrial screening from performing satisfactorily other separation method has to be utilized. Hydro-cycloning is an operation that can deal with large slurry volumes and to classify particles according to their hydraulic diameter, particularly the mass. A 15-inch hydro-cyclone can process an excess of 150 m^3 per hour while separating particles around 100 μm. The main difference from a screen to other separation methods is that a screen has a physical barrier that prevents coarser particles to be found in the undersize. In a different manner, hydro-cyclones do not have physical barriers, they rely solely on the separation phenomenon. Cyclones are said to be classification equipment because they separate particles probabilistically. There will always be a possibility to find a coarser particle on the overflow slurry, even though the bulk of them will exit trough the apex. The classification of hydro-cyclones is characterized by the particle size that has the same probability of exiting trough the apex and vortex finder.

Hydro-cyclone sizing might be done adjusting mathematical models available; Nageswararao [29] and Plitt [72] are examples. These models require some knowledge about the operation and a pilot unit may be of great value once it permits proper calibration. The main variable information set to be collected at tests is, for the feed and both discharges, the PSD with chemical analysis and specific gravity, the volumetric flow, feed pressure and the solids content. In addition, the equipment must be characterized knowing its geometric configuration: main body diameter, inlet, apex and vortex finder diameters, length and cone angle.

Filtering

Bauxite fines, after being classified in hydro-cyclones, are still in a slurry that has to be dewatered. Filtration equipment vary in configuration; however, the working principle remains the same. A flow of slurry is forced through a cloth by pressure difference. This pressure difference might be achieved with vacuum, positive pressure or by mechanically squeezing the slurry against the cloth. The solids, or cake, that remain over the cloth have their moisture reduced and must be discharged so that a new cycle may occur.

To size a filter, the filtration rate and desired/possible moisture must be evaluated. A laboratory test where a suction device with a cloth is submerged in the solid's suspension is used to determine the better suited cloth, suction and drying times, as well as the cake thickness. With this information, the sizing of a continuous filter is possible.

Spiral concentration

Once the minerals are liberated and disaggregated into a suspension, the particles of interest might be concentrated. The concentration in Humphrey spirals allow particles of similar shape and size to be separated by specific gravity. The flow after having the excess clay removed, is fed into spiral shaped troughs. These troughs are inclined, and gravity forces the slurry to descend. The spiral shape forces the flow into a constant turn. The acceleration towards the outer edge of the trough has greater influence on less dense particles. Particles with higher specific gravity stay closer to the axle where the spiral is attached. In the lower end of the spiral, splitters divide the flow. The flow with less dense particles also carries the greater part of the water since it tends to behave as if it was a fine particle. The flow with denser particles has higher solids concentration. Eventually, a third flow might be separated at the end of the trough. This flow is the portion of material that is in between the other two. It might be recirculated or fed into a secondary spiral for further processing.

This unit operation separates iron and titanium minerals from gibbsite. Quartz is just slightly denser than gibbsite, hence, the differentiation factor of both is small. Small differentiation factors imply in low selectivity, to improve the separation results the residence time must be increased. Alternatively, to this technique, elutriation might be used. Elutriation relies in different terminal velocities in a fluid and tends to be more selective while having less throughput per occupied area.

Centrifugation

Tailings dewatering conditions the disposal method to be utilized. After centrifugal dewatering, the solids are no longer in a fluid suspension, but in a paste like condition. This means that they can be hauled and dumped being disposed in exhausted pits or, after some sun drying, stockpiled. Intense dewatering avoids the need for hydraulic deposits, increased the occupation rate per unit area and reduces the water makeup necessities.

The centrifuge dewatering comprises in feeding the slurry flow, eventually mixed with flocculants, into a decanter centrifuge. The equipment's bowl spins rapidly and creates an artificial gravity field with thousand times the acceleration over earth's surface. The clay particles have higher densities than water and so accumulate closer to the bowls perimeter. This accumulation forces the water out of that region. It is essentially a forced sedimentation phenomenon. The central shaft, trough witch the slurry is fed, spins in a different speed than the bowl. By also having a helical structure attached to it, the differential speed makes the shaft perform as a rake pushing the settled solids out of the bowl.

Press filtering

This technique delivers the driest possible solids after mechanical dewatering. The only way to drier material is by applying thermal energy. The filter plates are brought together creating between them voids that will be filled with tailings slurry. These chambers are the space between plates and are lined with filtering cloth and, if the case requires, with a pressing membrane. The slurry is fed through openings on the plates and, after filling the chambers and reaching a predetermined pressure, they are pressed together, and the membrane is filled with pressing fluid to dewater the cake. Pressurized air might be forced to further improve the drying process. Once filtering is complete, the plate's pack is opened, and the cakes fall freely. After opening all chambers, the filter might be closed again, and the operation can start over.

Filter pressed material is stackable and can be loaded into trucks just as it exits the equipment. The advantage of this method is that it recovers all waters that could be recovered mechanically. When the liquid phase is comprised by not only water the increased amount of recovered fluid might justify the effort. The choice of dewatering method to be used, if any, must include proper assessment of the geotechnical characteristics of the solids and how those may vary according to different levels of moisture.

2.7.4 Process Flowchart—Example

One possible combination of theses unit operations is shown in Fig. 2.61. There, number 1 through 3 represent crushers. These are sizer crushers since this is a

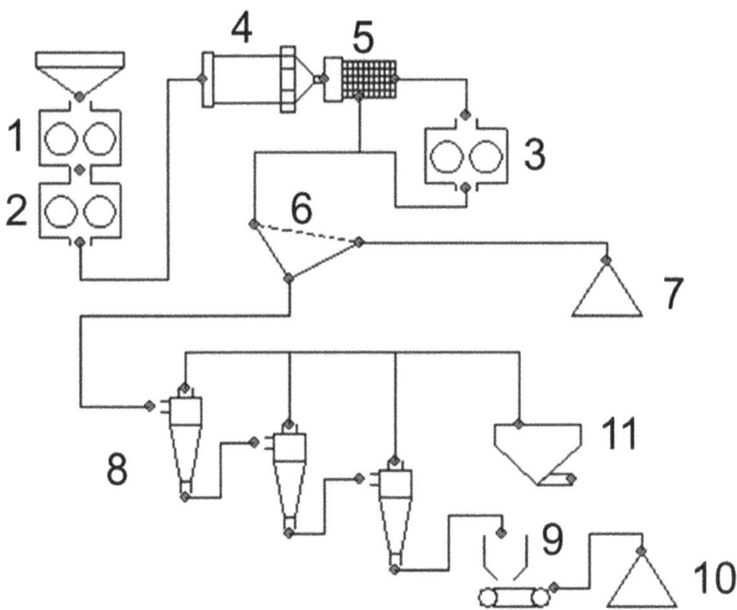

Fig. 2.61 Bauxite beneficiation flow sheet [73]

flowchart for a wet processing route. Crusher 1 has the largest aperture and is responsible for eliminating the biggest blocks that are fed into the plant. The primary crusher is sometimes into the responsibility of the mining operation and might be located adjacent to the mining site, whereas the secondary closer to the ore dressing plant.

To reduce over crushing, a tertiary crushing operation is put after the scrubbing (4) and is to receive only coarser particles that are the oversizer of the trommel screen (5). The screening operation (6) might be comprised of many aperture screens with one or more decks. This operation is also responsible for finishing the disaggregation of the clay with abundant water spraying. The coarser washed bauxite is represented in number 7.

By number 8 is represented the hydro-cycloning operation. There, the slurry, that contains most of the clay already in suspension with finer aluminum rich particles, is processed in subsequent classification stages to separate the gangue (finer clay particles) from the mineral of interest (coarser particles rich in gibbsite). Number 9 is the filtering stage where the coarser fraction from the classification is dewatered to be stockpiled (10). The slurry with the clay (11) is sent for further processing for water recovery and tailing deposition.

2.7.5 *Modifying Factors*

Non-operational modifying factors might include Economic, Marketing, Legal, Environmental, Social and Governmental aspects that might influence the access to the deposit. Once those are considered, two other operational modifying factors should be accessed: mining and metallurgy.

Modifying factors include alterations promoted by the mining and beneficiation operations that alter the characteristics of the material described by the geological sample. One obvious factor e.g., is a surface restriction that obligates the modeling professional to delete a certain area of mineralized material. Such a restriction might be of any nature, one example might be some settlement that cannot be removed, a heritage site or a wildlife sanctuary. These are all examples of resource to reserve modifying factors.

The first one must deal with the ability of the mining operation. Some areas might not be accessible due to topographical deterrent, presence of water etc. Once it is decided that an area is to be mined, the mining operation itself modifies the resource by not perfectly executing the modeled volume. The sheer size and precision of the excavator bucket implies that some desired material will be left behind while some sterile material will be dug unintentionally. These factors are called loss and dilution. To properly evaluate those, the exploration phase of the deposit must sample and characterize it, so its PSD, AA and RS and other grades are known.

The beneficiation modifying factors are the alterations imposed by the reality of the operation. The geological sample is the ideal standard to pursue. This sample is ideally crushed, disaggregated, screened, and classified. Exceptional care must be taken to minimize uncontrolled alterations. Nevertheless, during the industrial operations some deviation from the ideal condition is expected. The main alterations are coarser fraction loss, undesired finer fraction recovery and recovery at a different particle size from the geological samples.

The sample from the exploration drill hole is carefully reduced in size, while the ROM is loaded into trucks, dumped into hoppers or piles than pushed with dozers, crushed, disaggregated, and screened. All the operations done with the ROM will induce abrasion between particles and between particles and equipment. This abrasion will reduce naturally coarser fractions in such a way that they will not be recoverable in the process. This will reduce the quantity of recoverable material, thus reducing the resource available quantity.

On the other hand, those samples are thoroughly washed, meaning that no further clay could be removed. At industrial scales, some inefficiencies are expected, and a certain amount of undesired fraction is to be left with the product. Benchmarks indicate that less than 5% of the product mass is comprised of it.

At last, the geological model might have been developed by using a certain particle size cut off. This cut off on to the beneficiation operation might differ. Proper consideration is to be taken while developing the mass balance to consider such alterations. Eventually, it might be discovered, at a latter phase of the project, that finer fractions are recoverable. The amount and characteristics of the recoverable material should be known so it might be included in the reserve. This should be accomplished carefully so it does not materially change the resource with a less known material decreasing the confidence in the reserve.

2.8 Principles of Thermo-Chemical Bauxite Beneficiation/Activation

In the previous section bauxite beneficiation by physical methods was presented. The objective was maximizing the content of alumina and minimizing the content of solid impurities, like silica and iron oxides in the bauxite feed to the refinery.

As seen from Table 2.8, some commercial bauxites contain significant amounts of silica and iron oxides that has a major impact on the "Red Mud to Alumina Ratio", which can vary from, say 0.39 for Boke´ to about 1.25 for Darling Range bauxite [74].

But, as indicated in Table 2.1, there are other components of ROM bauxite like organic carbon and monohydrate content that need chemical/thermal bauxite beneficiation or activation technologies to optimize the extraction of alumina from the bauxite in the Bayer process.

A brief discussion of chemical/thermal bauxite beneficiation or activation work reported in the literature will be presented in this section covering ROM bauxite with high organic carbon content and Monohydrates, Boehmite and Diaspore.

Table 2.8 Typical composition of bauxites (dry basis, com. digestion) [75, 76]

Composition (wt%)	Trombetas	Boke	Darling range	Gove	Weipa	Henan
Al_2O_3	55–56	58–59	35–40	50–51	53–56.3	64.35
Fe_2O_3	9–10	1–5	15–25	16–17	6.2–15.5	5.93
Total SiO_2	4.5–5.5	0.5–2.5	6–15	3–3.5	2.9–7.1	9.94
Quartz	0.1–0.5	< 0.5	5–13	0.5–1.0	0.6–2.3	7.17
LOI	28–29	30–32	20–25	25–26	23.1–26.8	–
Organic carbon	0.02	0.12	0.30	0.20	0.23	–
Ratio mono-/tri-hydrate	0.01	0.14	0.02	0.02	0.14–0.32	Diaspore

2.8.1 *Organic Carbon in ROM Bauxite.*

The content of organic material in ROM bauxite may vary from 0.05 to 0.40 wt% carbon according to Pohland and Tielens. During digestion part of the carbon will be dissolved in the liquor depending on the digestion temperature, while the undissolved part will be discharged with the red mud. The organic material in the ROM consists mainly of cellulose, lignin and Ca/Mg pectates. Cellulose and lignin are converted into soluble long chain compounds of sodium humates. The degree of cracking of the long and intermediate chains of dissolved organic material in the liquor strongly accelerate with higher digestion temperature and concentration of the caustic liquor. The long chain organics are subsequently broken down to two stable products, sodium oxalate and sodium carbonate, which will reach their equilibrium concentration in the circulating Bayer liquor.

Sodium oxalate and sodium humates are the two compounds most harmful to the Bayer process as the sodium humates appears to have a stabilizing effect on the dissolved sodium oxalate explaining the supersaturated concentration of sodium oxalate in the green/pregnant liquors.

When the sodium oxalate concentration reaches a certain level of supersaturation, the sodium oxalate precipitates together with the trihydrate and appears as needle shaped crystals in the precipitated trihydrate [75].

When the concentration of the sodium humate compounds increases about a certain level, the below effects on the Bayer Process have been experienced [77]:

- Reduced digestion yields—up to several percentage points.
- Reduced red mud settling rates and capacity.
- Increased foam and scaling tendency of precipitator tanks reducing capacity.
- Decreased hydrate yield in precipitation.
- Reduced settling rates in hydrate classifiers.
- Decreased efficiency of hydrate washing.
- Foam formation in spent liquor evaporators causing operating problems.

From the above experience it is easy to understand that Alumina Refineries strive to minimize the impact of organic carbon on the Bayer process.

In doing that there are two fundamental options:

(1) Eliminate/minimize the content of organic carbon with the ROM bauxite by chemical/thermal bauxite beneficiation before it becomes feed to the refinery.
(2) Minimize the content of dissolved organic carbon in the circulating caustic liquor in the Bayer process to a level where the above experience is acceptable or even eliminated.

Option 2 is dealt with in Chap. 8 Bayer Process Impurities and Their Management, while a brief discussion of option 1 will be presented here.

2.8.2 Thermal Treatment of ROM Bauxite to Reduce Organic Carbon in the Bayer Process Liquor

Partial calcination or roasting of bauxite have been tried many times in the past [77, 78] with the prime objective to remove the organic carbon from the ROM bauxite. In their review of the literature Rijkeboer and van der Meer point out that roasting/calcination of bauxite to burn off the organic matter was mentioned in one of the patents by Charles Martin Hall from 1900.

To be a viable option, the basic process requirement for roasting/calcination of bauxite, are destruction of Total Organic Carbon (TOC) in the bauxite without loss of alumina in digestion of the bauxite.

In the Giulini refinery at Ludwigshafen, Germany, partial calcination of a trihydrate bauxite at 300–400 °C in a rotary kiln was tried to burn the organics [77]. But a good deal of the trihydrate was converted into monohydrate, which together with high investment and operating cost was thought to be economically unattractive.

The result shown in Fig. 2.62, of the bench scale roasting of the Darling Range bauxite was obtained from two-stages of fixed-bed reactors in the work by Rijkeboer and van der Meer.

Testing in a two-stage continues fluid-bed reactor set-up with 10 kg/h bauxite feed rate, confirmed the operating conditions of 350 °C at 10 min retention time, and 450–500 °C at 15 min retention time, respectively.

The first stage is designed for removal of crystal water and the second stage is designed for reducing TOC and good alumina extraction. For the second stage operation, the partial water vapor pressure should be below 2 kPa, with superficial gas velocity above 0.1 m/s and bauxite feed ground to below 8 mm.

Under the above conditions, above 90% TOC removable was achieved making bauxite roasting a potential technology at the time for elimination of the major problems associated with co-crystallization of oxalate and gibbsite.

Other potential benefits emerged from the bauxite roasting process [78]:

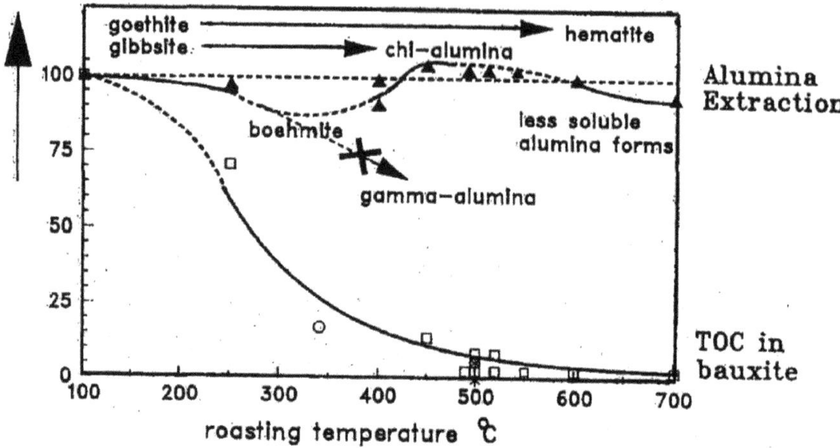

Fig. 2.62 Effect of roasting temperature on TOC content and alumina extraction efficiency [78]

- Improvement in mud settling, filtration and pregnant liquor stability.
- Reduction in mud load per ton alumina produced.
- About 8-ton water/ton alumina reduction in water input with the roasted bauxite.
- Reduction in consumption of chemicals (caustic soda and lime).

Rijkeboer and van der Meer made a dynamic mass-balance model to simulate the build-up and subsequent reduction in liquor TOC after introduction of the Darling Range bauxite roasting, Fig. 2.63.

The reduction in liquor TOC about 4 years after initiation of the bauxite roasting process, will allow the plant to be operated a higher precipitation yield.

The specific energy consumption of the roasting process was estimated to 4.3 GJ/ton alumina, with estimated savings of 2.8 GJ/ton alumina from recovery of latent heat, decrease in energy of dissolution, and capitalizing on operation at higher yield in precipitation.

Despite the potential benefits resulting from the work of Rijkeboer and van der Meer, no industrial bauxite roasting facility has yet been built, most likely due to un-attractive economics.

Furthermore, it is an open question what the environmental impact would be from such a unit with respect to odor emissions? The relatively low roasting temperatures suggested indicating that odor emissions could be an issue that require costly end-of-pipe treatment of the flue gases to burn away odors.

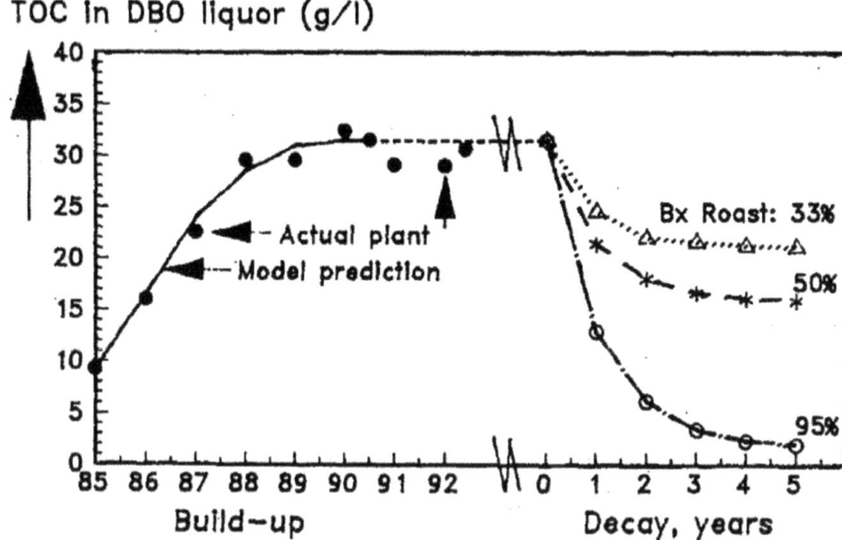

Fig. 2.63 Build-up of liquor TOC in the plant and decay after initiation of bauxite roasting [78]

2.8.3 *Thermal Treatment of Commercial Bauxite to Activate Boehmite for Low Temperature Digestion*

The objective of the work presented in the previous section was primarily focused on thermally removing organic matter, and secondly minimizing loss of bauxite extraction efficiency, before feeding the thermal beneficiated bauxite to the refinery.

The objective of the Comalco Bauxite Activation Process [76] presented in this section, primarily focus on activating monohydrate in the form of Boehmite to allow low temperature digestion, and secondly minimizing organic matter in the thermal activated bauxite before feeding to the refinery.

Weipa bauxite can be seen in Table 2.1 to have a mono-/tri-hydrate ratio ranging from 0.14 to 0.32.

The work reported by Hollit et al. has surprisingly shown that thermal activation of Boehmitic bauxite can simultaneously achieve (1) the extinction of crystalline alumina bearing phases, and (2) the destruction of extractable organic carbon and (3) the avoidance of deactivation of the activated alumina by water vapour.

The elimination of the previously reported impact from a high water vapor pressure [79] was explained by conducting the activation process at relativey high temperature and very short retention time as seen in Table 2.9.

For fast decomposition rates of Weipa bauxite, the thermal dehydration products of Gibbsite and Boehmite is X-ray amorphous; simultaneously, Kaolin and Goethite is dehydrated as well. No crystalline form of alumina or hydrated alumina was found by X-ray diffraction analysis of digestion residues [76].

Table 2.9 Laboratory processing condition and extraction results [76]

Apparatus	Fluid-bed	Fluid-bed	Fluid-bed	Fluid-bed	Gas supension
Bauxite passing size, microns	1000	1000	1000	1000	106
Temperature, Celsius[a]	510	510	540	540	750
Residence Time, minutes	5	5	11	11	0,025
Extraction, wt%[b]	83,7	83,5	85,6	81,2	90,4
Final A/C	0,685	0,684	0,691	0,676	0,707
Water vapor pressure, kPa	8	18	< 5	18	30

[a] Gas temperature in case of Gas suspension
[b] Net extraction of available alumina to liquor, rather than conversion

Based on the results of the laboratory test with gas suspension activation, the process has been piloted at a scale of one (1) tonne per hour at FL Smidth Test Center in Allentown, Pensylvania, US.

Here a two-step approach was applied. First, the commercial bauxite feed was dried and milled to a d_{50} of 106 μm as seen from the red curve in Fig. 2.64.

Then, the dried and milled bauxite was fed to a gas suspension calciner through a three-cyclone-stage gas suspension pre-heater. The temperature profile was optimized to maximize the extractability of alumina from the activated bauxite including dust. The particle size distriibution of dust is shown as the green curve in Fig. 2.64.

Digestion in a single stage fixed time batch test showed that activated gibbsite was more quickly digested than the activated boehmite, explaining

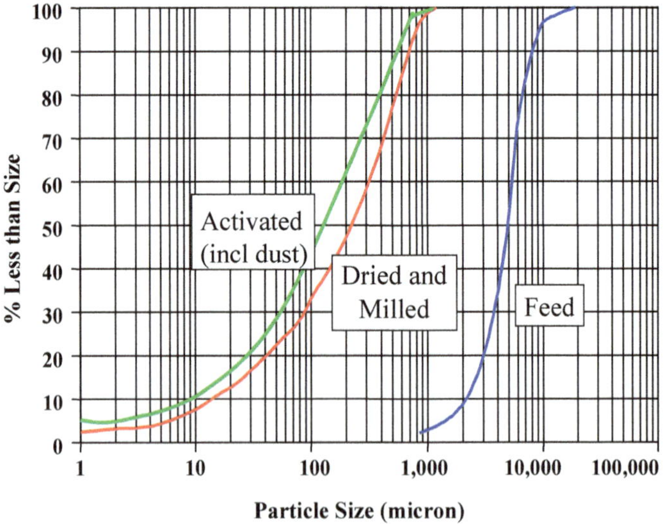

Fig. 2.64 Typical particle size distribution in one (1) tonne per hour activation test [76]

the extraction dependence on the boehmite content of the original bauxite as seen from Fig. 2.65 [76].

The activated bauxite was digested at 175 °C in a single stage digestion test conducted at elevated caustic strength resulting in 90.3% extraction of "available alumina" that transfer to the liquor to 0.71 A/C ratio. The result of the optimized activation of the dried and milled bauxite is shown in Table 2.10.

The optimized thermal processing conditions simulated in the 1.5 Mtpa flowsheet in Fig. 2.66, are such that the process can be conducted with a low energy consumption of approximately 2 GJ/ton original bauxite using fossil fuel. A gain of approximately 1.2 GJ per tonne original bauxite feed to activation is predicted from operating the

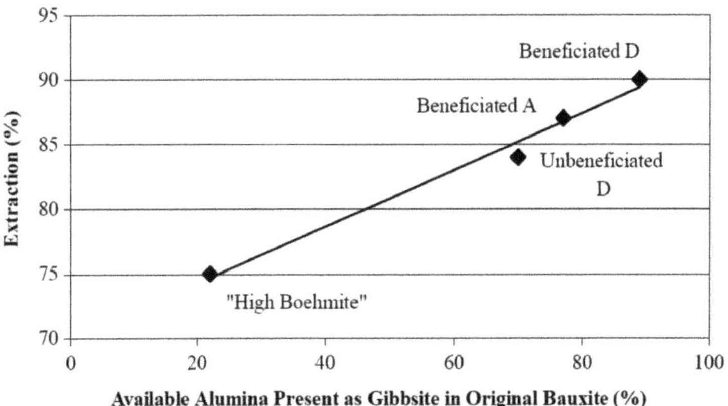

Fig. 2.65 Extraction depends on the original gibbsite/boehmite content [76]

Table 2.10 Composition and characterization of commercially versus optimally activated bauxite [76]

Composition	Unit	Weipa A Com	Weipa A Activated	Weipa D Com	Weipa D Activated
Al_2O_3	wt%	53.0	67.8	56.4	72.2
Fe_2O_3	wt%	15.5	19.8	6.2	8.2
Total SiO_2	wt%	2.9	4.2	7.1	9.4
Quartz	wt%	0.6	0.9	2.3	3
LOI	wt%	23.1	5.5	26.8	5.6
Organic carbon * commercial Bx basis	wt%	0.23	0.06*	0.23	0.06*
Organic carbon rate (excl. oxalate)	Kg/t A	1.9	0.28	1.9	0.28
Sodium oxalate rate	Kg/t A	3.5	0.095	3.5	0.049
Sodium carbonate	Kg/t A	19	2.2	19	1
Ratio mono-/tri-hydrate	–	0.32	–	0.14	–

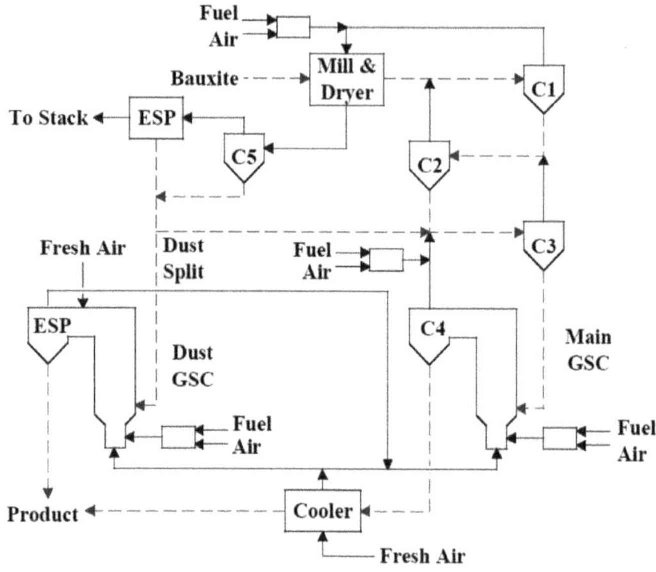

Fig. 2.66 Flowsheet of 1.5 Mtpa plant showing gas and solids (dashed lines) streams [76]

precipitation circuit at 94 gpl yield, as a result of of reduced energy losses per unit hydrate hydrate produced in a higher productivity circuit [76].

Despite many indicative and simulated benefits derived from the testwork made during development of the Comalco Bauxite Activation Process it has never been applied in commercial scale, either as a retrofit to an existing refinery or a part of a green field refinery.

Odor emissions from the Bayer process using activated bauxite according to the Comalco Bauxite Activation Process may be significantly reduced, or even eliminated, but it is not known if the activation process by itself will emit odor with the stack gases?

2.8.4 Sintering or Pyro-Genic Attack on Bauxite—Process Integrated with the Bayer Process

Mono hydrate bauxites (Boehmite or Diaspore) with low to high silica content mixed with limestone and/or sodium carbonate and sintered at 1300–1400 °C in direct fired rotary kilns or on sinter pan/belt has been used in various combinations with the Bayer process since the second World War [80, 81]. One example is described in Chap. 1, Sect. 1.5.1.

The main sintering chemistry can be summarized as follows [80]:

With Na_2CO_3:

$$Al_2O_3 + Na_2CO_3 \rightarrow 2NaAlO_2 + CO_2 \tag{2.1}$$

$$Fe_2O_3 + Na_2CO_3 \rightarrow 2NaFeO_2 + CO_2 \tag{2.2}$$

With CaO:

$$2CaO + SiO_2 \rightarrow 2CaO \cdot SiO_2 \tag{2.3}$$

$$CaO + Al_2O_3 \rightarrow CaO \cdot Al_2O_3 \tag{2.4}$$

$$5CaO + 3Al_2O_3 \rightarrow 5CaO \cdot 3Al_2O_3 \tag{2.5}$$

$$2CaO + Al_2O_3 + SiO_2 \rightarrow 2CaO \cdot Al_2O_3 \cdot SiO_2 \tag{2.6}$$

The above solid state reacton products has been verifiied by X-Ray examination of the clinker as well as an examination of leaching behavior.

The leaching behavior of the above compounds depends on several factors. On the one hand, temperature, and length of time of the effective sintering process, and on the other hand, composition of the causticizing solution and temperature of the leaching process.

The useful leaching reactions are:

$$NaAlO_2(s) + 2H_2O \rightarrow Na^+(l) + Al(OH)_4^- = NaAlO_2 \tag{2.7}$$

$$2NaFeO_2(s) + 2H_2O = 2NaOH + Fe_2O_3 \cdot H_2O \downarrow \tag{2.8}$$

$$CaO \cdot Al_2O_3(s) + Na_2CO_3(s) = 2NaAlO_2 + CaCO_3 \downarrow \tag{2.9}$$

Causticizing reactions:

$$5CaO \cdot 3Al_2O_3(s) + 5Na_2CO_3(s) + 2H_2O = 6NaAlO_2 + 4NaOH + 5CaCO_3 \downarrow \tag{2.10}$$

$$2CaO \cdot SiO_2(s) + 2H_2O \rightarrow CaO \cdot SiO_2 \downarrow + Ca(OH)_2 \tag{2.11}$$

$$Ca(OH)_2 + Na_2CO_3(s) \rightarrow 2Na(OH)(l) + CaCO_3 \downarrow \tag{2.12}$$

In the presence of excess Na_2CO_3, dilution and high temperature the above reactions (2.7–2.12) is shifted to the right [80].

If the sintering temperature is too low, and the concentration of the solution and temperature of leaching is too high $2CaO{\cdot}SiO_2$ tends to react with $NaAlO_2$

detrimental to the process:

$$3(2CaO \cdot SiO_2)(s) + 6NaAlO_2(s) + 15H_2O = 3Na_2SiO_3 + 2(3CaO \cdot Al_2O_3 \cdot 6H_2O) \downarrow + 2Al(OH)_3 \quad (2.13)$$

$$2Na_2SiO_3 + 2NaAlO_2(s) + 4H_2O = Na_2O \cdot Al_2O_3 \cdot 2SiO_2 \cdot 2H_2O \downarrow + 4NaOH \quad (2.14)$$

In the overall reaction of (13) and (14) three molecules of Al_2O_3 are lost for removing two molecules of SiO_2, not a desirable outcome.

A 12–15 tpy Al_2O_3 pilot plant of the above-described sinter process was operated successfully at Montecatini's Porto Maghera Works, Italy [80] since 1957, but an industrial facility was never built.

In China, the Lime-Soda-Sinter Process is used in various flowsheet combinations with the Bayer Process using low grade bauxite (see Table 2.8, Henan) and/or red mud as feed materials [81].

Appendix: Representative sampling for elaboration of the alumina manufacturing process (Theory and practice)

Generalities

All kinds of geological sampling for analyses need comprehensive geological knowledge about the rock (geological formation) to be sampled. Representative sampling for tests serving the elaboration of technology of processing a raw material is one of the most important tasks of a geologist.

As much as 50–200 kg of representative sample for laboratory tests is needed when an operating refinery needs new feedstock to process. In case of establishing a new refinery, depending on the technology envisaged to be applied 1–10 tons, or even more may be needed when test is carried out on pilot plant scale. To take out hundred kilograms or even tons of bauxite from a deposit consisting of ten-, or even hundred million tons of raw material the geologist must be aware of its local (specific) conditions of genesis, accumulation, chemical and mineralogical make up and their variances both vertical and horizontal directions, secondary alterations, along with the shape and size of the deposits.

Representative samples are composed based on the exploration data. Bulk sample has to involve individual samples of extreme concentrations of each determinant element in proper ratio. How far a carefully taken sample is really representative in practice, that is, how far the chemical and mineralogical composition of the sample and the mine product are uniform (deviations are in acceptable ranges) it highly depends on how reliable the exploration and reserve calculation data are, and how far the mining method fits to the deposit characteristics. Reliability of the representative sample comprises all the errors of the exploration method, techniques, sampling, analyses and reserve calculation. Alumina refinery pays bonus after shipments of ore when better than agreed in contract. Mining company pays penalty after lower quality. Contract between the producer and consumer, among the others, is based on the results of the technological tests carried out on representative sample.

When no adequate data are available additional mineralogical reconnaissance is to be made for setting up a sampling plan taking into account the variances in mineralogical composition of the deposit.

What is Representative Sample?

Representative samples are taken either for:

(a) establishing the technology in a new plant, or
(b) satisfying the demand of an operating refinery.

In the second case the technology may need to be modified according to the new feedstock.

For satisfying the demand of the representation a composite sample, both in its chemical and mineralogical composition, should be fitted, to that of the run-off mine ore's characteristics planned to be extracted in a long term. Quality, mineralogy of mine product changes during the mine operation period. For that reason, depending on the measure of expected changes, collecting additional subsamples may also be needed.

From the point of **chemical view**, the bauxite sample can be regarded as a representative one provided the differences satisfy the following deviation conditions:

i. $TAl_2O_3 \pm 1.5$ abs.%, when $TAl_2O_3 \geq 45\%$ and $\pm$ 1,0 abs. % when $TAl_2O_3 <$ 45% and,
ii. $TSiO_2 \pm 0.5 - 0.8$ abs.%, when $TSiO_2 \leq 5\%$ and $\pm$ abs.%1,0 abs.% when $TSiO_2 > 1\%$.

It is better to apply these conditions for available alumina and reactive silica, but no sufficient amount of data are always available for the deposits in this respect.

Conditions for acceptable differences must be clarified with the responsible process engineer before the sample is composed. When the deviations are somewhat looser the sample is regarded as **characteristic sample**.

In case of laterite bauxites, the difference for Fe_2O_3 is not so important because there is a very close negative correlation between the $TAl_2O_3\%$ and $Fe_2O_3\%$. The correlation coefficient (R) is around 0.90–0.95. Therefore, if the bauxite is representative for alumina it must also be representative in its iron oxide content within an acceptable range ($\pm$ 1–2%). In case of karst bauxites, the correlation between the alumina and iron, in contrary, is a positive one and it is not so tight. The representative sample should be taken by that way that the difference in iron oxide be less than $\pm$ 2 abs. %.

The contaminants, like C^{org}, ΣSO_3, MnO_2, etc. are related typically to the epigenetic (secondary) processes in karst bauxites. $\Sigma CO_2 > 1\%$, in karst bauxite, due to a secondary re-accummulation. C^{org}, $\sum CO_2$ and P_2O_5 occur in laterite bauxites. Their variances are extremely high both vertically and horizontally. See details in Sect. 2.2.1.2. These elements are not involved in the routine analyses of the exploration, but there are, in most cases, data enough to estimate them in the deposit average. Based on geostatistical calculation extreme and most frequent values may also be known. Process engineers must be informed about these data and it is also necessary to take them into account when the bulk sample is composed, in case additional subsamples may be needed.

The sample must be representative, or at least characteristic, in its mineralogical make up, as well. Systematic quantitative mineralogical analyses usually are not available. Representative samples are composed based on the available information on mineralogy. If adequate data are not available, additional mineralogy tests are needed to make on individual samples before the bulk samples blended.

Classification of the Bauxites by Their Mineralogical Composition

Based on the fundamental processing properties of the bauxites they are classified into the following main types:

(a) *Gibbsitic bauxite: at least 95% of its free Al_2O_3 content, (aluminum oxihydroxides) must be in gibbsite.*[1]
(b) *Gibbsitic/boehmitic and gibbsitic/boehmitic + diasporic bauxite: at least 90% of free Al_2O_3 content must be in gibbsite.*[2]
(c) *Boehmitic bauxite: $\geq$ 10% boehmite, at least 95% boehmite + gibbsite.*
(d) *Boehmitic/diasporic bauxite 5–10% free alumina content is in boehmite and/or diaspore.*
(e) *Diasporic bauxite more than 10% of its free Al_2O_3 content is in diaspore.*
(f) *Chamositic/diasporic bauxite: chamosite content is > 5%.*

---

[1] It is also qualified as "low mono content" ore (Sangaredi – Guinea).

[2] It is also qualified as "high mono content" ore (Sangaredi – Guinea).

Grouping of Bauxites on Their Mineralogy and Processing Properties

Homogenous Bauxite Deposits

The total amount of the ore can be processed by either low (140–150 °C) or high (240–280 °C) temperature digestion, that is, the whole resource can be ranked into one of the mineralogical type of ore, as listed above, the alterations can be followed on the basis of their chemical data, so that reliable sample can be taken following their chemical makeup. Furthermore, the kaolinite/quartz ratio is, more or less, constant. Variance in alumina substitution in iron minerals is tolerable. The loss in total alumina is expressed in the value of extractable alumina. Contaminants, such as concentrations of $\sum CO_2$, $\sum SO_3$, MnO_2, P_2O_5 and C^{org} , in general, can be kept at an acceptable level. What is acceptable? What is tolerable? These criteria should be discussed between the geologist and process engineer.

Heterogeneous Bauxite Deposits

Laterite Bauxites

Based on their **aluminium bearing minerals** different types of ores may occur in one deposit: so called low mono content type (a) and high mono content type (b, d) (most frequently in laterite bauxites type (a) and (b), that is, gibbsite/boehmite ratio changes horizontally and/or vertically. There are two possibilities:

(1) High mono content type of ore cannot be extracted separately double digestion method is needed to apply when the boehmite content exceeds 8–10% in the tri-hydrate type bauxites. One representative sample may be enough + 2 separate samples, representing 10% of the bulk sample in weight are recommended to collect from the extreme parts of the deposit. (e.g., Az Zabirah—Saudi Arabia).
(2) High mono content bauxite can be separately extracted. The run-off mine ore is stocked and supplied separately to different plants (e.g., Sangaredi—Guinea). Two representative samples are needed for plants of different types.

Based on the **iron concentration** some deposits can be divided into an iron rich (Fe_2O_3 typically $\geq$ 24–26%) and an iron poor (Fe_2O_3 typically: 12–15%) (e.g., Aya Ninahin, Ashanti Region—Ghana—Kesse [82]). In such cases process engineers are satisfied by three samples: One main bulk sample representing the average values and two (smaller) additional samples representing the two extremes. In case of higher iron concentration, the probability for higher alumina substitution in iron minerals (dominantly in goethite) is bigger.

Karst Bauxites

Karst bauxite mining is usually going ahead toward the depth. It may occur that the bauxite grade and mineralogy gradually change toward the depth. No average bauxite can be shipped to the plant but only an ore of continuously changing composition. When the difference does not result in necessity of modification of the technology, as it is most frequently, the consumer pays for bauxite according to its value, following the costs changed in processing. In such a situation one representative sample should be composed on the average values and two additional samples from the extremes are recommended to take in order to learn whether modifications in technology is needed and how far the processing costs may change.

When a deposit, regarding its alumina bearing minerals, is built up by two types of ore and they cannot be selectively extracted the bauxites are processed on the highest required digestion temperature, e.g., Kethro deposit where in boehmitic bauxite a diasporic ore body can be found (Nia 1971). In such cases, one representative sample may satisfy the demand with average boehmite/diaspore concentration, in case of need the bulk sample may be supported by two sub samples, as well.

At Manchester Plateau (Jamaica), in horizontal term the alumina substitution on goethite changes between 5 mol% and 18 mol% [8]. This is also the case when three samples are recommended to be collected.

Compilation of a Sampling Plan

Data Required

(1) *Capacity of the refinery, that is, bauxite demand (quality and quantity).*
(2) *Concept of mine planning for the next 10–30 years in order to foresee the bauxites characteristics.*
(3) *Run-off mine ore reserves (tonnage and grade by mining units; ore bodies).*
(4) *Chemical and mineralogical analyses of the bauxite, accessory (contaminant) elements.*
(5) *Quantity of the sample.*

Overview of the geochemical, hydrological (drainage conditions) and geomorphological features which are determinant in deposit geology: establishing the variance of each main element and mineral both vertically and horizontally.

Marking of the Sampling Sites on Map.

Number of the sampling points depends on:

(1) Tonnage of the reserves
(2) Number of the ore bodies taking also into account their size and shape.
(3) Heterogeneity of the reserves to be processed (details in Appendix Sect. 2.4).

ad (1) As a general rule, provided the deposit represents >100 million tons of bauxite, number of sampling points should be 1 whole section (bore hole, pit, channel sampling from mine face) for every 5–6 million tons of bauxite. In case of a smaller deposit (ore body) relatively more sampling points are recommended to be marked (e.g., in case of a deposit of about 10 million tons one sampling points should represent 2.5–3 million tons.
ad (2) When a deposit consists of more ore bodies all of them are recommended to be sampled, provided its reserve reaches the tonnage of minimum annual production of the mine. Each ore body is to be sampled proportionally to their tonnage.
ad (3). If the bauxite to be sampled is heterogeneous (see Appendix Sect. 2.4) it is not always necessary to mark more sampling sites, as given above in point ad (1), but in their location its heterogeneity is to be taken into account and composition of three sample(s) are advised such as: one big **bulk sample** which represents the average values and **two smaller additional samples** for the extremes.

Location of Sampling Sites on Laterite Deposits

(1) Sampling points (bore holes, pits), having been marked on a map, must be checked on site to clarify the access possibilities.
(2) In case of surface laterite bauxite, the shape and the morphology of the ore body give the guideline for sampling points location, as the mineralogy changing in the function of drainage conditions, which are determined, beside the bedrock permeability, and the surface morphology.

- Based on the surface morphology there are three main types of deposits such as:
  - flat plateau,
  - hill tops on undulating surfaces,
  - slopes as shown in Figs. 2.67 and 2.68.
- In all the three the points must be located in such a way that the possible horizontal variation of mineralogy be taken into account.
- The deposit consists of number of ore bodies developed on an undulating surface and the commercial grade ore restricts to the positive morphological forms (singe or complex heaps (e.g., Fria—Guinea, Mulanje (Malawi), Bao Loc (Vietnam, etc.). The points are to be located at the **weight point(s),** close to the centre and at the slopes at different topographic levels (Fig. 2.67).
- The bauxite deposits developed on an elongated, amoeba like, quasi flat plateaux such as the bauxites developed on Deccan basalt, or khondalite in India. Sampling

points are proposed to be located at the "reference points" of the deposits as it is shown in Fig. 2.67.

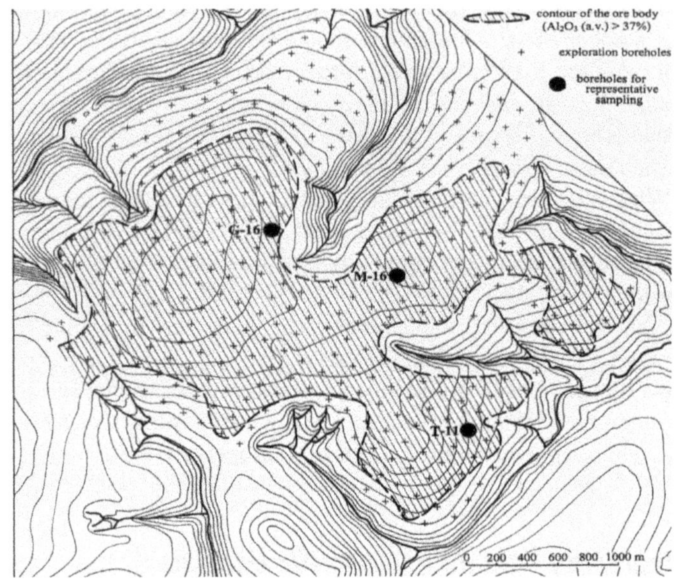

Fig. 2.67 Bore holes showing the representative sampling sites on hill top and slopes of the Kondekoure (SW) ore body Fria—Guinea—Komlóssy [9]

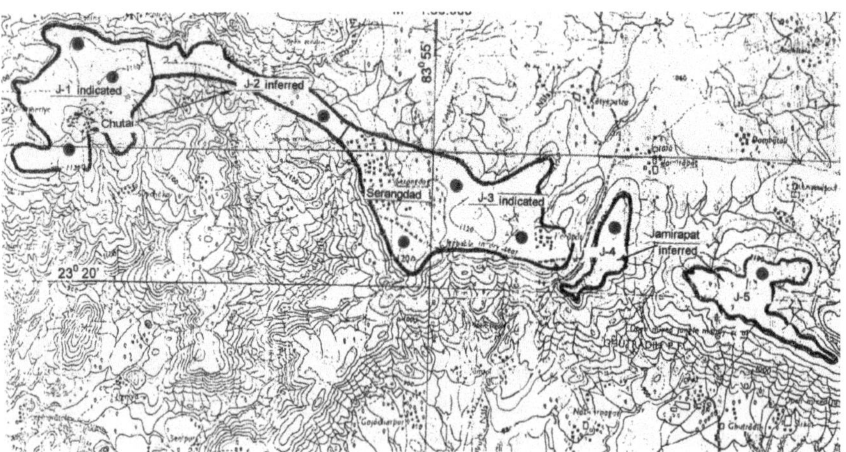

Fig. 2.68 Bore holes marked for representative sampling on flat surface, Jamiraplat Plateau,—Chattisgarh State—India. Komlóssy [83]

Sampling Method and Analyses

Sample Taking and Preparation for Analyses

- Sampling can be made from bore holes, from pits or mining face (channelling).
- Samples must not be taken from old pits (along the wall of old pits the bauxite quality has considerably changed (kaolinite is washed out, iron precipitated, even within a half meter width even in a very short time). As it is shown in Figs. 2.8 and 2.9 in Sect. 2.3.2 the whole extractable section must be sampled in order to cover the vertical changes in mineralogy.
- Samples are advised to take in one meter intervals and analysed for TAl_2O_3, $TSiO_2$ and Fe_2O_3. It is useful if:

(a) Average $AvAl_2O_3$ and $RSiO_2$ are also should be determined, if not in every interval, but at least for a composite sample of one section,
(b) The accessory contaminants (C^{org}, ΣSO_3 ΣCO_2, MnO_2, etc.) are also advised to be analysed in each section in order to learn how far the expected average values of these elements fit to the reserve averages.

- Mineralogical tests of every sampled section are highly recommended to be made before the sample compilation.
- It is advantageous if the analyses are made at the existing plant's (if any) which utilises the bauxite or at the mine's laboratory with the same procedure as it was done during the exploration. When the analytical methods are different the correspondence between the methods should be clarified.
- Depending on the required total weight of the representative sample from one sampling interval 1–10 kg of wet sample is needed + two additional spare bulk samples representing about 2 × 20% in weight of the total representative sample. One spare sample may be needed for ***upgrading*** and another one for ***downgrading*** purposes.
- For preparation of bauxite for analyses it must be ground to:
  - In case of 1 kg: grain size $\leq$ 4.5 mm
  - In case of 10 kg grain size $\leq$ 14 mm (Richards-Cherchette's formula)
- Before the bulk sample prepared as described above a separate bulk sample is also needed to compose **for crushing and grinding test**.

Sample Compilation

- Weighted average chemical composition of the individual samples is to be calculated.
- These results must be compared with that of the run-off mine ore planned to be extracted.

- Such a calculated (theoretical) blending is to be made which satisfies the demand by adding proportional quantities from the individual samples. It is a mathematical iteration, and the result is still a *theoretical composition.*
- In case of heterogeneous bauxites subsamples are also recommended to be composed (from each section) for making analyses for accessory elements (contaminants) and mineralogy. In this case the representative sample should be combined on the basis of the average values of the additional samples.
- The well blended bulk sample must be sampled by forming a cylinder which not higher than 20 cm in case of 100 kg sample and 50 cm in case of a ton. At least three channel samplings are to be made by mixing the bauxite prior to each sampling. Sample taken out in this way must be reduced by quartering and prepared for analyses. The analytical results may be still different from the required composition.
- Reaching of the desired composition can be made by proportional addition of the spare samples (up-grader or down-grader).

If we make the procedure correctly, we will have a bulk and reliable sample which is representative or at least very close to it.

References

1. Bárdossy, G., - Aleva, G.J.J. (1990) Laterite bauxite. - Akad. kiadó Budapest - Hungary
2. G. Komlóssy, *Assessment of the West African Bauxite Deposits. Identified and Potential* (Rio Tinto doc. Melbourne, Australia, 2014)
3. Komlóssy, G. (1985) Paleogeographic implication of karst bauxite genesis - ICSOBA Travaux Vol 14-15 Zagreb Yugoslavia
4. Rio Tinto Ltd. Plt. (2008) Bauxite Training, Brazil. - Lecture
5. J.G. Weisse, Bauxite latéritique et bauxite karstique. (in French Laterite bauxites and karst bauxites), in *Symposium Bauxites et hydroxides d'aluminium. – Zagreb – Yugoslavia* (1964)
6. Huashu Yang (1990) Bauxite deposits of China. - Chinese Journal of Geochemistry. Vol.8. No 4.
7. Djokic, V., Kelezic, M.,Voros, I. (1978): New aspects of bauxite geology in Montenegro, Yugoslavia. - 4th Int. Congr. of ICSOBA Vol. I. Athens, Greece
8. ALUTERV-FKI, *Report on the Feasibility Study of the JAVEMEX Alumina Plant* (Budapest, Hungary, 1976)
9. G.Komlóssy, *Representative sampling of the Fria—Guinea Deposit* (Manuscript, Budapest, Hungary, 2004)
10. Gubbs, D.K., Rodenburg, J.K., Wefers,K.A. (1982) Origin of aluminous goethite in Suriname bauxite - in Bell, G.M. ed. Light Metal Comm. 110th A.I.M.E. Ann. Meeting Feb. 1981 Chicago USA
11. Komlóssy, G. (1976) Formation of laterite bauxite and iron ore deposit in Goa - India. - Ált. Földt. Szemle (General Geol. Review) Budapest - Hungary
12. ALUTERV-FKI (1978) Report on the feasibility of the Kibi (Ghana) bauxite and alumina project Budapest, Hungary
13. Mamedov, V.I., Anufriev, A.A., Jakubivits, I.K., Suma, N.L.(1985) Particularities of the Sangaredi bauxite, Guinea (in Russian). Vyssh. Utaheb Zavod. Geol. Razvdka, Moskow, No. 4.

14. Rao, M.G., Raman, P.K. (1979) The East Coast bauxite deposits of India. - Bull. Sc. Eco. Geol. GSI. Vol. 46.
15. Patterson, S. (1986) World bauxite resources. - US Geol. Surv. - Washington, US
16. Panov, A., Suss, A. (2002). World bauxite market scenario and new raw material sources for alumina production. - Travaux ICSOBA. Vol. 29 No 33 Vienna, Austria
17. G. Komlóssy, *Bauxite Reserves and Resources and Further Prospects at Niksic Bauxite Mine (Montenegro)* (Glencor Archive, Baar Switzerland, 1999)
18. ALUTERV-FKI. (1975). *Report on the Geology and Reserves of Bauxite Deposits of Lang Son*, Vietnam
19. Bárdossy, G. (2010) The Szőc bauxite deposit. – Occasional Papers of the Geological Inst. of Hung. – Budapest, Hungary
20. Ostap, S. (1984) Effect on bauxite mineralogy on its processing characteristics. In Jacob, Jr. El. Ed., Bauxite Proc. , 1984 Bauxite Symposium, Los Angeles USA., AIME., New York, pp 651–671
21. Komlóssy, G. (1969) The Iszkaszentgyörgy bauxite (SE Bakony Mts, Hungary). Problem of genesis and mineral formation. – Annales Instituti Geologici Publici Hungarici – Vol. LIV. Fasc.3. – Budapest
22. Bronevoi, V.A., Zhilberminc, and Teniakov, V.A. 1985: Average of chemical composition of bauxites and their evolution in time (in Russian). – Geokhimia, Moscow, 4.
23. G. Pantó, A ritja földfémek geokémiája és hasznosításuk (in Hungarian), in *Geochemistry of the Rare Earth Elements and Their Utilization* (Acad. Doctoral Theses. Manuscript. Budapest, Hungary, 1980)
24. S.Beaulieu, Impurities Control from Mine to Plant. in *35th International ICSOBA Conference*, Hamburg, Germany, 2–5 Oct 2017
25. Chemokomplex – Aluterv (1974): Broullon de rapport definitive sur une érude préliminaire de factibilité sur la conversion locale des bauxites Malgaches (in French Draft report on the feasibility of local processing of bauxites of Madagascar) – Onudi contract 72/53
26. ALUTERV-FKI, Rapport on the feasibility of bauxite mines and processing alumina refinery at Lang Son, Vietnam – in Frech., - Budapest, Hungary (1977)
27. G. Bárdossy, Karsztbauxitok. Bauxittelepek karbonátos kőzeteken (in Hungarian) - Karst bauxites. Bauxite deposits on carbonate rocks. – Akad. Kiadó. Budapest – Hungary (1977)
28. Nandi, A.K. (2004):Bauxite alumina industry of India. – Present status and future prospects. – Travaux of ICSOBA Vol.31 No 35. St. Petersburg,Russia.
29. K. Nageswararao, *Further Developments in the Modelling and Scale-Up of Industrial Hydrocyclones* (Ph. D. thesis, The University of Queensland, 1978)
30. R.Y. Black, G.-P. Lozej, S.S. Maddah (1984) The geology of bauxite deposits of Ghana, in *Bauxite Proceedings 1984 Bauxite Symposium Los Angeles USA*, ed. by Jacobs Jr. (A.I.M.E., New York)
31. G. Komlóssy (1976a) Minérealogy, géochemie et génétique des bauxites Vietnam du Nord (in French Mineralogy, geochemistry and genetics of North Vietnamese bauxite). Acta Geol. Acad. kiadó . Sci. Hung. Tom
32. Bárdossy, G., Csanádi, A., Csordás-Tóth, A. (1978) Scanning electron microscope study of bauxites of different ages and origins. – Clays. Clay Miner No 26 pp. 245–262
33. ALUSUISSE-CVG (1980) Field work and evaluation of ore reserves.
34. Ross, Y., Black, R. Y., Lozej, G. P. and Maddah, S. S. (1984), Geology and mineralogy of the Az Zabirah Bauxite, Northern Saudi Arabia, Proceedings of the 1984 Bauxite Symposium, Los Angeles, California, USA
35. B. Balkay (1973) Bauxitization and underground drainage. Travaux, ICSOBA No. 9.Part B. Pp 151–161. Zagerb, Yugoslavia
36. Komlóssy, G. (1976b) Formation of lateritic bauxite and iron ore deposits in Goa, India. – Ált. Földt. Szemle. (General Geol. Review). Budapest
37. E. Dudich, G. Komlóssy (1969) Paleogeographical and structural considerations to the question of the age of the Hungarian bauxite deposits (In Hungarian). Földt. Közl. 99. Budapest. – Hungary

38. G. Pedro, J. Berrier, Sur altération expérimentale de la kaolinite et sa transformation en boehmite par lessivage a l'eau. (in French) – Experimental alteration of the kaolinite and its transformation to boehmite by dissolution in water. C. R. Acad. Sci Paris **226**. Série D, 551–554 (1963)
39. G. Pedro, A. J. Melfi (1983): The superficial alteration in tropical region and the lateritisation phenomena. – In Melfi, A.J. and Carvalho eds. Proc. "n dint. Seminar on Lateritisation Process, 1982. – Sao Paulo, Brazil
40. S. M. Belinga (1971) Étude géochemique des eaux de circulation recueillis dans la zone des plateau bauxitiques du sedcteur Haléo: Minim Martap (Adamaoua), Cameroun. Ann. Fac. Sci Cameroun 5. Pp 33-37.
41. Valeton, I. (1972) Bauxites. – Elsev ier Publ. co. Amsterdam.
42. K. Jespsen, W. Shellmann (1974): Über den Stoffbestand und die Bildungsbedingungen der bauxite lagerstatte Weipa, Astralien. Geol. Jahrb. Hannover, D.
43. A. Mindszenty (1978) Tantitive interpretation of the micromorphology of bauxitic laterites. – In Proc. 4th Int. ICSOBA Congress, Athens, Greece
44. G. Komlóssy (1993a) Worsley Alumina Pty. Ltd. (Australia) – Project Kudu. Final Report. Part Geology. HATCH and Ass. London
45. G. Taylor, A.R. Eggleton (2004) "The little balls" The origin of the Weipa bauxite. - Roach I.C. ed. 2004. Regolith 2004. CRC LEME
46. G. Komlóssy, et al., *Report on the Reconnaissance Field Visit to Gujarat and Maharastra* (Rt doc. Melbourne, Australia, 2012)
47. G. Bárdosy, The Halimba bauxite deposit. Occasional Paper Geol. Inst. Hungary **2008**, 87 (2007)
48. D.V. Sapozhnyikov, Geneticeskaya klassifikacia boksitovih mestorozhgenyja USSR (in Russian)—Genetic classification of Soviet Union's bauxite occurrences), in *Problems of Bauxite Genesis* (Ikzd. Nauka, Moscow, 1975)
49. D.V. Sapozhnyikov, Geneticeskaya klassifikacia boksitovih mestorozhgenyja USSR (in Russian)—Genetic classification of Soviet Union's bauxite occurrences), (Problemy I genezhisa boksitov, Moscow, 1975)
50. E. Vadász, Bauxitföldtan (in Hungarian)—*Bauxite Geology* (Akad. Kiadó Budapest, Hungary, 1951)
51. J.G. Weisse, Bauxite karstique sur clacaire recent (In French)—*Karst Bauxites on Recent Limestones*. (Travaux de l'ICSOBA. Zagreb, Yugoslavia, 1976)
52. V.A. Zans, *Recent Views on Origin of Bauxites* (Geonotes 1. 5 Kingston—Jamaica, 1959)
53. G.R. Kirpal, V.A. Teniakov, (1978). The deposits of aluminium (in Russian) – Ore deposits of the USSR Niedra, Moscow
54. L.C. King,. (1962): Morphology of the Earth. – Oliver and Boyd Publ. House, London UK
55. ALUTERV-FKI, Rapport sur une etude de factibilité des mines de bauxite et d"une usine d'alumine a Lang Son – Vietnam (in French) Report on the feasibility bauxite mines and processing in alumina refinery at Lang Son, - Vietnam (1977). G.Bárdossy, K.Jónás, A.Imre, K. Solymár, Interrelation of bauxite texture, micromorphology, mineral individualism and heteromorphism. – Econ. Geol. No. 72 (1977)
56. T.E. Wong, Outline of the stratigraphy and the geological history of Suriname coastal plain. Geol. Mijnbouw **65** (1986)
57. A.K. Nandi and L. Mueller (2001): The Indian bauxite scenario and prospects for export. – Metal Bulletin No 7. Bauxite and Alumina Seminar, Miami, Florida, USA
58. Rio Tinto Ltd. Plt.(2008) Bauxite seminar - Brazil
59. Kotschoubey, B.L., de Queiros Menezes, Truckenbrodt, W. (1984) Nature et evolution des bauxites de secteur de Jabuti- Ipixuana (District de Paragominas – État du Pará, Brésil. 109e Congr. Nat. Soc. Savantes, Dijon Sciences 1 pp. 335–345
60. G.Komlóssy (2008), *Explanation Notes to SRTM Maps of Indochina Showing Its Bauxite Potential* (RT Archive, Melbourne, Australia, 2008)
61. A.J. Melfi and Carvalho (1983) Bauxitization on alkaline rocks in Southern Brazil. – Sci. Géol. Mém. Strassbourg. France

62. A.T.C.Veiga, L.A.Vessani, G.A. Guerra, J.L.G. Costa, The Barro Alto bauxite deposit. A large and rich bauxite deposit recently discovered in the Goias State, Central Brazil. ICSOBA Travaux **37**(41) (2012)
63. Hieronymus, B (1985) Étude de l'altération des roches éruptives de l'oues de Cameroun. – Latéritisation, bauxitisation et évolution bauxitique. (in French Study on alteration of eruptive rocks in West Cameroon) These de doctorat d'état Université Pierre et Marie Curie. – Paris – France
64. V. Carboni (2005): Bauxite potential of Cameroon, West Africa. – AlCAN - Monteral. RT archive Melbourne- Australia.
65. W. Li, Alumina industry in China, in *19th International Bauxite and Alumina Conference* (Miami, US. Metal Bulletin, 2013)
66. B. Fodor (1994) UN Economic Commnission for Europe: Assessment of mineral resources and reserves under market conditions. – New York/Geneva.
67. G. Komlóssy (2010) Review on global bauxites, origin and types. Travaux ICSOBA Vol 35.No. 39. Zhengzhou, China.
68. V.I. Mamedov (2005) Bauxite Provinces in Guinea (2005) in Syposium on Mines Guineé. Ministerium of Mines and Geology.2011 Lecture, Conakry, Guinea
69. B.A. Bogatirev, V.V. Zhukov (2009) Bauxite provinces of the World. Geology of the Ore Deposits Vol.51 Pleaides Publishing Ltd
70. I. Valenton, New York: Elsevier Publishing Company (1972)
71. B. Wills, T. Napier-Munn, *Mineral Processing Technology* (Elsevier Science & Technology Books, New York, 2006)
72. I.R. Plitt, A mathematical model of the hydrocyclone classifier. CIM Bull. 114–122 (1976)
73. C.M. van Deursen, Métodos de desaguamento e disposição de rejeito da bauxita: estudo de caso e avaliação econômica. São Paulo. 2016, 127 p. (Mestrado) Escola Politécnica, Universidade de São Paulo, São Paulo (2016)
74. D.J. Milne, The influence of Mineralogy on alumina processing. Chem. Eng. Australia **ChE7**(2) (1982)
75. N. Brown, T.J. Cole, The behavior of sodium oxalate in Bayer Alumina Plants. Light Metals 105–118 (1980)
76. M. Hollitt et al., The Comalco Bauxite activation process, in *International Alumina Quality Workshop* (2002), pp. 115–122
77. H.H. Pohland, A.J. Tielens, A new Bayer liquor purification process. Light Metals 211–222 (1983)
78. A. Rijkeboer, A.P.van der Meer, Bauxite roasting—an option to reduce organics input to Bayer Plant Liquor, in *International Alumina Quality Workshop* (1993), pp. 254–269
79. G.J.J. Aleva, Lateritization, bauxitization and the cyclic landscape development in the Guiana Shield, in *Bauxite Proceedings 1984 Bauxite Symposium, Los Angeles USA*, ed. by L. Jacob Jr. (AIME., New York, 1984)
80. P. Lecis, A.. Guidi, Pyrogenic attack of bauxite, in *Extractive Metallurgy of Aluminum*, Vol. 1 (1962), pp. 231–249.
81. G. Songqing, Chinese Bauxite and Its Influence on Alumina Production in China. Light Metals 44–47 (2008)
82. G.O. Kesse, The geology of bauxite deposits of Ghana, in *Bauxite Proceedings of 1984 Bauxite Symposium*, Los Angeles USA., ed. by Jacobs Jr. (A.I.M.E., New York) (1984)
83. G. Komlóssy, Report on the tonnage vs. grade reserve calculation of the BALCO captive bauxites and the Khuria Highland occurrence. Chattisgarh State, India (Manuscript, - BALCO doc. Bhubaneshwar – India Sept. 2002 (2002)
84. G.J.J. Aleva, *Laterites. Concepts, Geology, Morphology and Chemistry* (ISRIC, Wageningen, The Netherlands, 1994)
85. G.Komlóssy, P.Siklósi, G.Tóth, Zs. Csillag, *Techno Economic Study for the Commercial Extraction of Alumina from Bauxitic Clay*. Draft final report (UNIDO, Mexico City, Mexico, 1988)

86. G. Komlóssy, *Report on the East Coast Bauxite Deposits and Their Further Prospects Established by Land Sat Interpretation* (RT Archive, Melbourne, Australia, 2006)
87. G.Komlóssy, J. Heim, *Assessment of Guinea's Bauxite Resources and Potential* (RT Archive, Melbourne Australia, 2013)
88. F.Lelong, Y.Tardy, G.Grandin, J.J.Trescases, B. Boulange, *Pedogenesis, Chemical Weathering and Processes of Some Supergene Ore Deposits* (Elsevier, Amsterdam, 1976) (Reprinted from strata-bound and stratiform ore deposits edited by K.H.Wolf) (1976)
89. G. Millot, Géologie des argiles. (in French) *Geology of Clays* (Masson Ed., Paris, 1964)
90. G. Komlóssy, Representative sampling for elaboration of the alumina manufactoring process (Theory and practice), in *Proceedings of ICSOBA XVI International Symposium*, Nagpur, India (2005)
91. G. Komlóssy (1993b) Mineracaoin Rio de Norte S.A. (Brazil). – Project Kudu Final Report, Part Geology. HATCH and Ass. London. 1993

Dr. György (George) Komlóssy Bauxite Consulting Geologist, Retired—Rio Tinto Expl. Pty. Ltd.—Australia. Email: geokom 38@gmail.com

Born in Budapest, Hungary in 1938, graduated as geologist at the Eötvös Loránd University Budapest (1962) Ph.D. degree received in 1969, thesis: "Iszkaszentgyörgy Bauxite Deposits: Geology, Mineral Formation and Genetics". Employed by the Bauxite Prospecting Co (1962–1970) and ALUTERV-FKI Engineering Centre and Research, Engineering and Prime Contracting Centre of the Hungarian Aluminium Corp. (1970–1990). Leader of bauxite explorations in North Vietnam (1972–1975) reconnaissance activities in Algeria (1976), in Mexico (1981–87), in Cuba (1987), in Zimbabwe, in Malawi, in Tanzania (1989). Co-author of feasibility studies made for Indian and Iranian projects. During the change in political regime, as the president of the Central Geological Bureau, managed the reorganisation of two scientific entities: the Hungarian Geological Institute and the Eötvös Loránd Geophysical Institute (1990–1993). In the followings became independent consultant in bauxite geology working for number of companies, such as HATCH and Ass. (Montreal, Canada), Kaiser Engineers Twickenham, UK), Hydro Aluminium (Oslo, Norway), BALCO (Nagpur, India), MA'ADEN (Riyadh, Saudi Arabia), SRK Consulting (Johannesburg, South Africa), etc. Between 2004 and 2017 served the Rio Tinto (Melbourne, Australia) as retainer geologist. Experiences in exploration, auditing of reserves, representative sampling for industrial tests extended over five continents, in 22 countries. 1992–1996 president of the ICSOBA (International Committee for Studying Bauxite, Alumina and Aluminium).

Author or co-author of about 200 studies and reports on bauxite geology with special regards to the reliability of exploration data and their industrial values. Scientific and technical publications on deposit geology, genesis, mineral formation and sampling methods, exploration technics. In 2006 introduced a method for identification of laterite bauxite deposits with the aid of remote sensing technics, making possible estimating the bauxite potential, in cases resources, of areas of interest.

Caio van Deursen Innovation Manager—Nexa Resources. Email: caio.deursen@nexaresources.com

Caio van Deursen graduated as a mining engineer from the Escola Politécnica da Universidade de São Paulo. After a brief period in the financial market, he started his career at Votorantim Metais in 2011 participating in the development of an aluminum mine and refinery project. He took part and executed the technological characterization program of the bauxite in order to define processing routes and flowcharts. He developed FEL1, 2 and 3 activities for this project, being responsible for the beneficiation plant, grinding, filtration and waste disposal systems. He defended his master's thesis entitled "Dewatering methods and disposal of bauxite tailings" by the same school where he graduated. After some years of exclusive dedication to the project, in 2014 and with CBA, he also took the lead in other projects in the bauxite mining units and in the aluminum plant, where he had the opportunity to test processes on an industrial scale. He currently holds the Innovation management role at Nexa. In this position, his scope includes leadership in the development of circular economy projects using mineral-metallurgical waste as inputs and the development of innovation management facilitating the cultural transformation desired by the company.

Benny E. Raahauge Deceased—Owner—Director Raahauge-SGA ApS, Denmark

Benny received his M.Sc. Chemical Engineering from the Danish Technical University in November 1972. Start working with programming minicomputers controlling the raw material mixing for production of Cement Clinker, December 1972 in Process Technical Department, FLSmidth, Cement Division, Copenhagen, Denmark.

From 1975–76, Benny worked as process and plant engineer in a Danish Sugar Factory before re-joining FLSmidth as R&D engineer and later R&D Manager in the Mining Division in Copenhagen. Benny was assigned the task to develop the stationary Gas Suspension Technology for Smelter Grade Alumina (SGA) to replace the rotary kilns.

After successful commissioning of the first 1000 tpd Gas Suspension Calciner (GSC) unit for alumina at Hindalco Industries, India, in 1986, Benny worked in the role as General Manager-Pyro & Alumina Technology based in Copenhagen. This assignment included the responsibility for design, marketing, sales and commissioning of GSC units for Alumina, culminating with commissioning of 3 x 4500 tpd GSC units at Queensland Alumina 2004–05, the world's largest stationary calciners.

During +44 years employment with FLSmidth & Co. contributed with many papers on calcination to international ICSOBA, Alumina Quality Workshop and TMS meetings until retiring in 2018.

Together with Don Donaldson, Benny was Lead-Editor to "Essential Readings in Light Metals—Volume 1—Alumina & Bauxite" with Fred Williams as Co-Editor, Published by

Wieley & Sons. Inc. Copyright © 2013 by The Minerals, Metals & Materials Society (TMS).

In 2018 Benny was editor and contributor to Chapter 12.1 Alumina, in the "SME Mineral Processing & Extractive Metallurgy Handbook", Published by Society for Mining, Metallurgy & Exploration (SME), Copyright © 2019.

Chapter 3
Physical Bauxite Processing: Crushing and Grinding of Bauxite

Michael Wanyo, Anthony T. Filidore, and Benny E. Raahauge

Abstract The most common process step to feed an alumina refinery with bauxite is sizing of the raw bauxite material that is extracted from the mine. The first crushing of the bauxite may often take place at the mine before transport to the alumina refinery if located nearby, or to the shipping port for export. At the refinery further crushing and/or washing may take place before the crushed bauxite is subjected to a grinding operation for final sizing before digestion. This chapter present some of the various crushing, washing, and grinding equipment available in the market as well as the various options for designing a grinding flowsheet.

3.1 Introduction to Crushing and Grinding of Bauxite

The most common initial process step to feed an alumina refinery with bauxite is the crushing or sizing of the raw bauxite material that is extracted from the mine. The feed material is crushed or sized so that it is conveyable, as well as correctly dimensioned, for the next step in the process. After crushing, the downstream circuit consists of grinding mills located in the refinery. In most cases, the crushed bauxite will proceed to a shipping/transport terminal for final processing at a sometime remote alumina refinery.

Benny E. Raahauge is deceased.

M. Wanyo (✉) · A. T. Filidore
FLSmidth, Bethlehem, PA, USA
e-mail: mike.wanyo@flsmidth.com

B. E. Raahauge
FLSmidth A/S, Midvale, UT, USA

B. E. Raahauge and F. S. Williams (eds.), *Smelter Grade Alumina from Bauxite*, Springer Series in Materials Science 320,
https://doi.org/10.1007/978-3-030-88586-1_3

3.2 Crushing of Bauxite

The equipment selected for the crushing area is chosen based on the physical characteristics of the raw bauxite and the capacity rates required. Typical crusher types evaluated for a specific project include Jaw Crushers, Cone Crushers, high-speed Impactors or Hammermills, and low-speed Sizers. The crushing area may also include equipment for separation by particle size such as vibrating screens or roller-screens; both having openings of a known size for proper size separation.

Preparing the Run-of-Mine (ROM) bauxite for the grinding mill circuit must be completed in order to efficiently size the grinding mill circuit. The focus of the Sect. 4.1.2 discussion is about choosing the correct crusher for the bauxite application. Including the considerations for using a drum scrubber or not.

Bauxite mines are in many places. In tropical or sub-tropical climates like Australia, West Africa, the West Indies, and South America mining of laterite or Gibbsitic type bauxite, sometimes with minor content of Boehmitic bauxite takes place. In China, Greece and countries on the Balkan peninsula, mining of Diasporic type bauxite is common.

For selecting of the best crushing solution, the characteristic bauxite parameters in Table 3.1 [1] are important.

Limestone and quartz have been included in Table 3.1 as these components are frequently found in Diasporic type Bauxites. For selecting the best grinding circuit solution, the characteristic bauxite parameters in Table 3.2 [1] are important.

The work indices are important for determination of crushing energy requirements and the abrasion index is important for estimating equipment metal wear and grinding media consumption. The hardness according to Mohs scale of Gibbsitic

Table 3.1 Characteristic bauxite parameters for crusher selection [1]

Run of mine material	Average impact work index (kWh/ton)			Average abrasion indices [2]		
	No. test	Average	Range	No. test	Average	Range
Bauxite	8	5.3	2.5–12.2	11	0.02	0.003–0.12
Limestone	178	11.1	3.3–27.6	52	0.05	0–0.66
Quartzite	11	12.8	6.8–22.1	7	0.69	0.19–0.99

Table 3.2 Characteristic bauxite parameters for grinding circuit selection [1]

Run of mine material	Average rod mill work index (kWh/ton)			Average ball mill work index (kWh/ton)		
	No. test	Average	Range	No. test	Average	Range
Bauxite	33	10.8	2–20	29	14.5	1–31
Limestone	84	13.7	7–50	177	9.9	4–36
Quartz	1	14.4	–	13	14.4	11–21

and Boehmitic bauxites are typically 1–3, while it ranges from 6½–7 for Diasporic bauxites [3].

Needless to say, the parameters corresponding to the bauxite being considered as feed to a new alumina refinery needs to be determined by laboratory testing based on samples obtained during exploration of the bauxite deposit.

To process the bauxite, the bauxite must be mined, crushed, and delivered to the refinery. The first step to reaching the refinery after mining the bauxite is crushing.

The crushing circuit is defined by a series of variables and how those variables come together to affect each other.

Those variables include the following:

- Feed Size maximum to the crusher
- Feed Size estimated distribution range
- Product Size required
- Feed rate to crushing circuit
- Delivery system to crushing circuit
  - Haul trucks
  - Railcars
  - Front End Loaders
  - Shovel
- Feed method to the primary crusher
  - Direct-dump
  - Metered feed via apron feeder, vibrating grizzly feeder, etc.
- Hardness of the feed
  - Unconfined Compressive Strength (UCS)
  - Crusher Work Index
- Percentage of moisture in the feed
- Specific Gravity (SG) of the ore in situ
- Bulk Density of the blasted or excavated ore
- Abrasion Index.

Bauxite ore typically falls into the lower UCS ranges and therefore lends itself to various types of crushing technologies. Of all the variables listed, the two main variables that drive the crusher selection are the feed rate and the percentage of moisture in the feed.

Crusher Types Considered

Compression crushers (Fig. 3.1a, b), which include primary gyratory, primary jaw, and secondary/tertiary cone crushers are normally not required to crush Gibbsitic and Boehmitic type bauxite. The compression-type machines are designed for the hardest, most competent, and abrasive feed materials that are typical in hard-rock mining like some Diasporic bauxite.

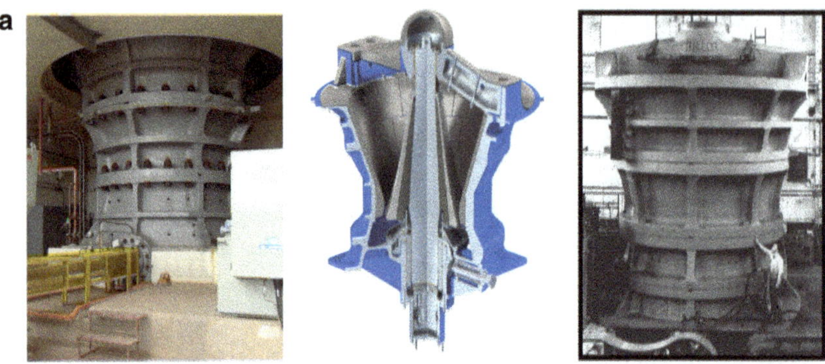
Primary Gyratory Crusher Gyratory Crusher Cutaway Primary Gyratory Crusher

Primary Jaw Crusher Secondary/Tertiary Cone Crusher

Fig. 3.1 **a** Compression—type crushers. **b** Compression—type crushers

Compression crushers work best with dry, friable feed. The general acceptable moisture limit for a compression crusher is 5%. The risk to "pack" the crusher increases with moisture levels greater than 5%, and especially when processing feed materials that become sticky with moisture. Gibbsitic and Boehmitic type Bauxite is one of those feed materials that is ever more challenging to crush as the moisture level increases.

This is not to say though that compression machines are never used. There are examples of sites utilizing primary jaw crushers due to the very high compressive strength of the intrusion rock mixed in with the bauxite. In a jaw crusher circuit, the feeder can be a vibrating grizzly feeder, which removes the wet and sticky fines. With the fines removed, the jaw crusher does not "pack", and therefore able to reliably crush the more competent portion of the feed stream.

Historically, Gibbsitic and Boehmitic type bauxite applications utilized high-speed impact type crushers (Fig. 3.2). Important to note about these historical applications is the fact that they were normally processing relatively dry material and at rates up to approximately 1500 mtph. Dry, friable feed is ideal for high-speed impact crushers. Additionally, when the downstream process involved grinding mills, the

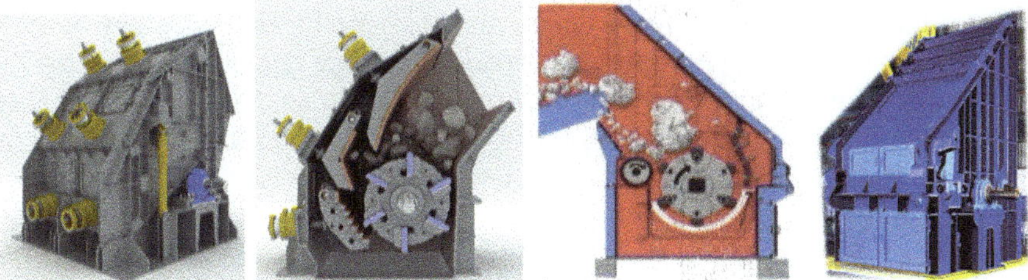

Blow-Bar Style High-Speed Impactor **Hammermill Style High-Speed Impactor**

Fig. 3.2 High speed impact-type crushers

fines created from high-speed crushing circuit were a benefit to the comminution process. In general, if the feed material, application parameters, and downstream process match with the capabilities of a high-speed impact style crusher, this is the crusher type to use. The high-speed impact machines inherently have a high size reduction-ratio capability. This means that the high-speed impact crushers can reach the final product size in less crushing stages than a compression crushing circuit, or a low-speed sizer circuit, given that all the feed parameters are the same, and when those feed parameters fit with the capabilities of the high-speed impact machines.

Later, as bauxite exploration continued, new suitable deposits were identified. Some of those deposits were in locations with higher annual rainfall rates and therefore elevated moisture levels in the ore itself. In order to successfully mine these wetter deposits, other crusher types were needed. High moisture feed into an impactor can cause build-up problems. The same is true for high moisture feed into a compression type crusher. The material sticks to the wear surfaces and jams the machine. The bauxite industry needed crushers that could be self-cleaning. Operators turned to machines with rotating shafts and teeth; toothed twin-roll crushers and Low-Speed Sizers (Fig. 3.3). Adding scrapers to twin-roll machines and sizers was easy. The scrapers worked reliably to keep the shafts free from excessive build-up of the wet & sticky feed material, and essentially made the sizers self-cleaning.

Another factor driving the need for using twin-roll crushing technology was the increasing feed rates of some of these new mines. Hourly feed rates in the 2,500

Primary Low-Speed Sizer

Secondary Low-Speed Sizer

Tertiary Low-Speed Sizer

Fig. 3.3 Low-speed sizers

to 5000 metric tph range were becoming more typical. In addition, many of these high-capacity mines happened to be in the high-moisture regions. Lengthening the crushing zone on the sizers and twin-roll crushers gained more volumetric capacity. In effect, the twin-roll crushers and low-speed sizers easily addressed both recent processes demand, higher capacities, and elevated moisture levels, with their standard models and designs.

One of the major benefits of the twin-roll crushers and low-speed sizers is their low relative head-height. Then if multiple machines or crushing stages were required for a project, the crushing station was still of an acceptable overall height. Another advantage of the roll crusher/sizer technology is the low dynamic forces when compared to a compression-type crusher or a high-speed impact crusher. This too helps keep the capital cost of the crushing station down.

Another major step in the crushing circuit, for bauxite ore applications and nearly all crushing applications, is the screening stage. The screening stage separates particles in the material flow by their particle size. The purpose of the screening stage is to remove material from the circuit that is already correctly dimensioned and does not require further size reduction in the crushing circuit. Removal of correctly sized material decreases the volume requirements of the downstream crushers when we consider a multi-stage crushing circuit.

Depending upon the actual characteristics of the feed, two main types of screens are utilized: traditional vibrating screens or Roller-screens.

Vibrating screens (Fig. 3.4) process dry material or material that is water-sprayed while it is on the screen. If the material is sticky with elevated moisture, and due to the process, using water spray is not possible, vibrating screens can blind-off. This means that the sized openings in the screen deck get blocked by material and do not allow the small particles to pass through the mesh as was intended.

The option then for high-moisture, sticky feed like bauxite is the Roller-screen. Roller-screens have rotating shafts with overlapping disks (Fig. 3.5). Like the rotating-shaft crushers and sizers, roller-screens can be fitted with a scraper system that prevents material build up and blinding of the discs. The self-cleaning nature

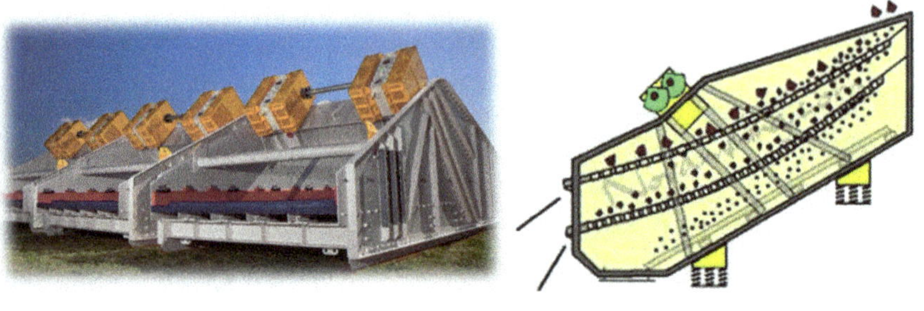

Fig. 3.4 Vibrating screens

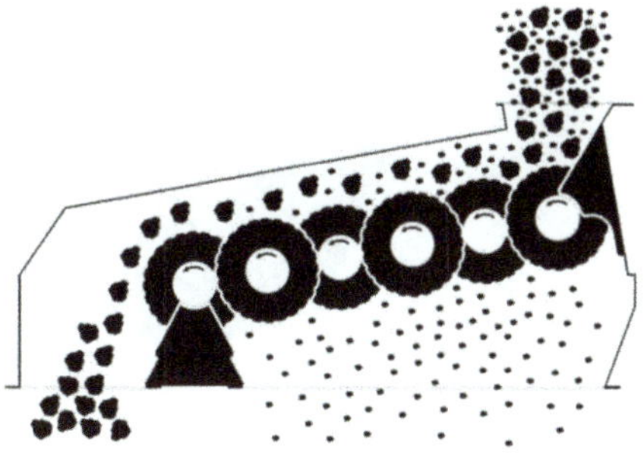

Roller-screen Cutaway

Roller-screen

Roller-screen

Fig. 3.5 Roller-screens

of the roller-screen allows higher screen efficiencies to be achieved especially when considering challenging wet, sticky feed characteristics.

In general, the crushing circuit equipment selection for a bauxite mine is based on the site-specific variables related to the application, and not based solely on the fact that this is a bauxite mine versus a mine for some other type of ore. A crusher does not know what type of material is passing through it. The crusher does however "know" the feed and product sizes, the compressibility, the work index, the level of stickiness, and the abrasion level. With all the application variables defined, the correct type and model of crushing circuit equipment can be properly specified.

3.3 Rotary Scrubber for Bauxite

In addition to the use of grinding mills in Bauxite preparation circuits, the use of rotary drum scrubbers is finding increasing application. Figures 3.6 and 3.7 illustrate the use of a scrubber in an application where the bauxite is very sticky and contains rock intrusions that are of no economic value. The scrubber pulps the bauxite into a slurry which then passes over a trommel screen at the scrubber discharge. The screen separates the fine bauxite from the coarse intrusions and the trommel underflow is

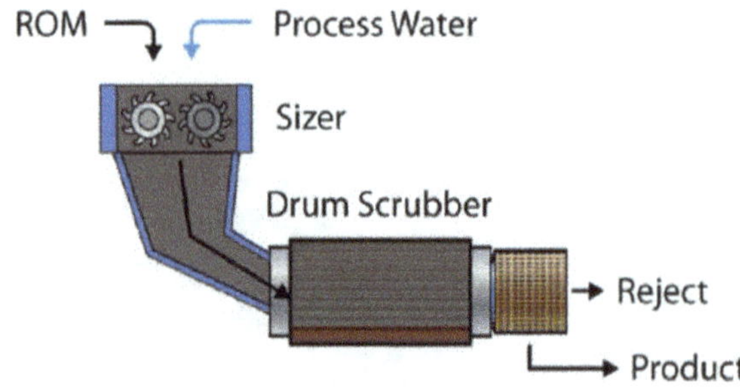

Fig. 3.6 Flow sheet of drum scrubber fed by sizer

Fig. 3.7 Typical 5 m × 10.4 m, 1300 kW drum scrubber application

then pumped to the downstream process plant. The intrusions are rejected as trommel overflow and conveyed to a waste dump.

It is apparent from Fig. 3.6 that the scrubber is preceded by a low-speed sizer. The Run of Mine (ROM) feed is fed to the sizer, which is mounted directly above the scrubber feed chute, thus minimizing mechanical handing of the sticky feed material. Process water is introduced with the feed and acts to flush the material thought the sizer and directly to the scrubbing chamber.

Feed to the scrubber is nominally 250 mm top size and very little breakage occurs in the scrubber. It is thus necessary to install "rock lifters" at the discharge end of the scrubbing chamber to remove the rock from the scrubber and assist with the feeding the trommel screen. Build up of the lump material in the scrubbing chamber is thus prevented.

Fig. 3.8 Typical 3 m × 8 m, 200 kW drum scrubber application

Scrubbers have also been proposed to slurry the feed material prior to the grinding circuit. In such applications this allows the pre-classification and removal of fine material that is naturally occurring in the feed and is of a partial size suitable for subsequent processing without grinding. This option reduces the occurrence of very fine particle in the plant feed that occur due to over-grinding in the grinding circuit. Such options have been considered for SAG and SAG/Ball circuits (Fig. 3.8).

3.4 Grinding of Bauxite—Design Requirements and Variables [4]

Bauxite feed preparation to digestion by grinding in the alumina refineries is a critical first step in the alumina production process. The grindability of the bauxite type in question can vary a lot as indicated in Table 3.2.

At the very low end with respect to Work Index ranging from 3 to 6 we find Caribbean Gibbitic and Boehmitic type bauxite [5, 6] which hardly needs grinding before digestion, but to ensure a maximum particle size for the downstream processing.

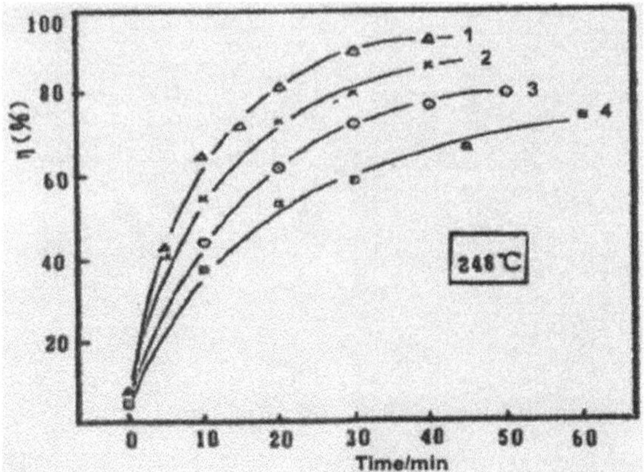

Fig. 3.9 Digestion curves of different size fractions of Pingguo Bauxite [7]. 1: 100% < 45 μm; 2: 50% < 45 μm + 5 0% 56–75 μm; 3: 100% 56–75 μm; 4: 100% 100–150 μm

At the middle range with respect to Work Index ranging from 12 to 16 we find Gibbsitic and Boehmitic type bauxite from Suriname and Indonesia [5]. These types of bauxite need about 20 min digestion time [7] at low temperature, 143 °C, to approach 98% extraction efficiency, but with no clear dependence on particle size.

At the high end with respect to Work & Abrasion Index and digestion temperature we find Diasporic type bauxite from i.e. China [7], where the extraction efficiency depends on particle size of the bauxite feed, see Fig. 3.9.

By careful grinding of bauxite with relatively high content of quartz that do not dissolve at low temperature digestion conditions [8], these particles may be removed after digestion as sand size particles, see Sect. 5.1.

There are a full range of bauxite grinding options available an each has its advantages and disadvantages. The final choice of which option best suits the application is influenced by several variables. These can be broadly categorized into the feed characteristics and the target product requirements.

Some of the feed variables are

- Feed size distribution and feed top size.
- Grinding and breakage characteristics of the feed i.e., Bond Work Index.

Some of the common target product requirement that are considered in the final selection are:

- Product particle size distribution. In this respect the main objective is to limit both the presence of course material and very fine material. Typically, specifications will place limits on the percentage of particles
  - > 1.0 to 1.7 mm
  - < 45 μm

- Overall grind size is typically expressed as the size at which 80% of particles will pass a sieve aperture of the nominated size (P80).

These target product requirements are generally common to all applications; however, it is also common that one or two of the criteria will take precedence.

Which criteria take precedence will have a major influence in the final grinding circuit choice.

In addition to the above criteria there are number other criteria that will influence the final circuit design such as.

- Specific Power consumption expressed in kWh/t
- Circuit simplicity
- Grinding plant footprint
- Ease of operation
- Health and safety considerations
- Availability
- Capital cost
- Operating cost.

In this section we will review the various grinding circuit options and consider the strengths and weakness of those circuits against the criteria outlined above.

We do not expect to come to definitive conclusions here, but the objective of the discussion is to explore the options and highlight where each circuit is most appropriate. Inevitably the final choice will be a compromise between competing criteria and the best outcome is an optimum solution that meets best the primary criteria specified at the project inception while not building in excessive on-going problems due to other criteria that are less fulfilled.

In the experience of the authors' the following product particle size criteria are most common.

- % passing 1.7 to 1.3 mm, 98% or greater
- % passing 45 μm
- P80, 250 to 700 μm.

In the experience of the authors' typical feed size distribution specifications are

- Maximum particle size, 20–40 mm
- 80% passing, 8–20 mm

These feed and product size parameters are generally the primary drivers in the selection of the grinding circuit, followed closely by Specific Power Consumption, capital cost and operating cost.

It is very important that the key criteria are clearly defined at the outset of the grinding circuit selection and design process.

3.5 Grinding Circuit Options [4]

Grinding of bauxite as feed to alumina refineries is undertaken in a variety of grinding circuit types utilizing a range of mill types such as Rod Mills, Rod/Ball Mills, Ball Mills, Rod Mill & Ball Mill circuits, AG/SAG Circuits and SAG/Ball Mill circuits both in open circuit and closed circuit.

This section considers the selection criteria in terms of the objectives of the grinding process and reviews the range of options available particularly those used in recent installations. The advantages and disadvantages of these options are discussed. Important evaluation criteria such as product size distribution, specific power consumption and cost are considered, summarized and compared with the design objectives. The following bauxite grinding circuits are the main circuits employed in the alumina industry today.

Rod/Ball Mill—Open Circuit

This is probably the most common grinding mill type and has been used for over 40 years. It is often the base case considered for bauxite grinding. The mill is operated in open circuit and comprises two compartments. The first compartment is charged with rods to handle the course grinding of the primary feed while the second compartment is charged with steel balls to perform the finer, finish grinding of the final product. Figure 3.10 is a diagrammatic illustration of this option.

This option has the following characteristics:

The rod compartment ensures that all the course particles are ground to a size suitable for the subsequent grinding in the ball compartment. This helps to minimize undesirable oversize particles while the second, ball charged compartment produces a reliable finished product with the desired P80. Ball-milling however will produce a higher proportion of finer particles due to the high surface area of the grinding media. It is also difficult to control the shape of the product particle size distribution in open circuit, i.e., without some form of classification.

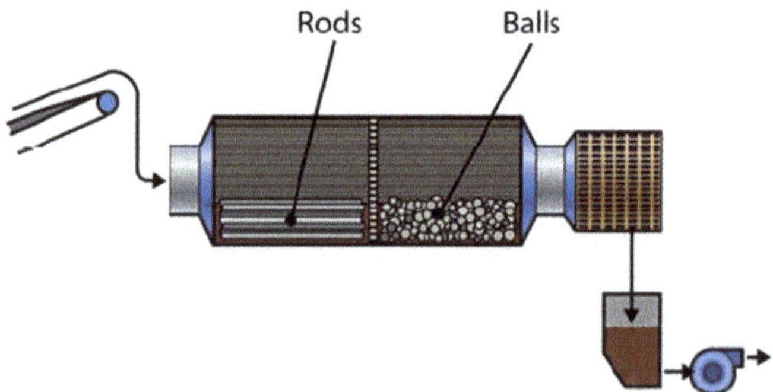

Fig. 3.10 Rod-ball mill

While this solution has been successfully employed in many bauxite grinding circuits it has a number disadvantages that are considered important, particularly in modern plants. The mill incorporates a central diaphragm to separate the two grinding compartments and allow the material to be moved from the first to the second compartment. This requires ongoing and relatively regular maintenance which requires personnel to access the interior of the mill and given that bauxite grinding is undertaken in hot, caustic liquor, it is desirable to limit personnel access to the interior for safety and health reasons. Modern safety procedures make this a lengthy exercise. Regular entry into the mill will therefore affect the operating availability of the mill.

This circuit also requires a relatively large footprint when compared to other solutions. Not only docs the space occupied by the mill need to be considered but access at the feed end for rod charging equipment is required. This can be space hungry and difficult to arrange at the mill feed platform; particularly where there are multiple mills. In contrast, single compartment rod mills are usually charged from the discharge end where more space is available.

Open Circuit Rod Mill followed by Closed Circuit Ball Mill

In this case the grinding circuit comprises an open circuit rod mill followed by a closed-circuit ball mill. The ball mill is most often closed over a classifier like a hydro cyclone (Fig. 3.11) or DSM style screen with a fine cut of 1–1.5 mm. The course material is returned to the ball mill feed and the fine screen underflow reports to the downstream process. Figure 3.11 is a diagrammatic illustration of this option.

The primary advantages of this circuit are the ability of the rods mill to handle course and variable feed sizes while the closed circuit secondary grinding helps to minimize the presence of undesirably course particles in the final product. By minimizing the residence time in the secondary circuit, it is also possible to reduce the proportion of very fine particles in the product.

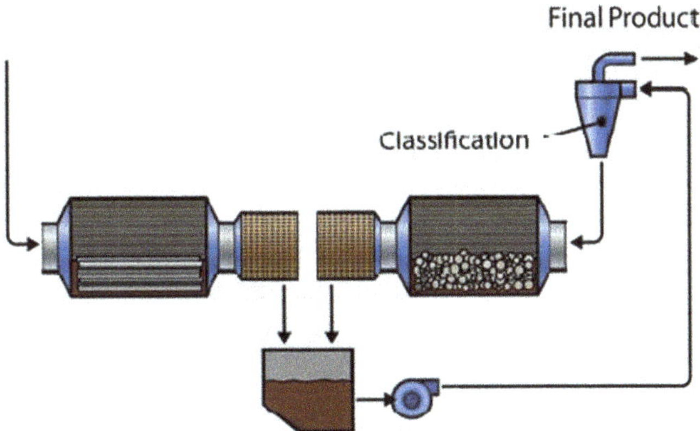

Fig. 3.11 Open-circuit rod mill followed by closed-circuit ball mill

Some other operating and maintainability advantages with this option are that both mills are overflow discharge and as such there are no diaphragms or discharge grates to maintain. By careful layout of the circuit, it is possible to arrange the rod charge at the discharge end of the mill thus improving the layout at the feed end.

This option does however require a significant footprint comprises more mechanical components, particularly in the mill drives, bearings and lubrications systems which leads to higher maintenance requirements. The classification screen also requires significant maintenance and can be problematic from an operational point of view. Overall availability can be impacted by the availability of the screen and as such it may be necessary to add redundant capacity in the form of standby screens; thus, adding to the complexity of this option.

Single Stage Closed Circuit Ball Mill

This circuit has been recently applied in Australia and India. It is essentially a development of the rod/ball mill circuit described earlier except that the particle size reduction occurs in one compartment. The rod mill is eliminated, and the fresh feed is fed directly to the ball mill. Because ball mills have limited ability to handle course feed it is necessary to ensure that the feed is not too course and the mill is required to operate in closed circuit, usually with fine aperture screens in order to control the proportion course particles in the final product.

This option has the advantage that the handling and management of rods is eliminated along with the attendant availability and safety issues. The need to enter the mill for maintenance of the charge is limited too however the screens require regular maintenance and can limit the overall availability of the grinding circuit. As such redundant screen capacity is often employed. Figure 3.12 is a diagrammatic illustration of this option.

This option offers a relatively small footprint and medium capital cost (Fig. 3.13).

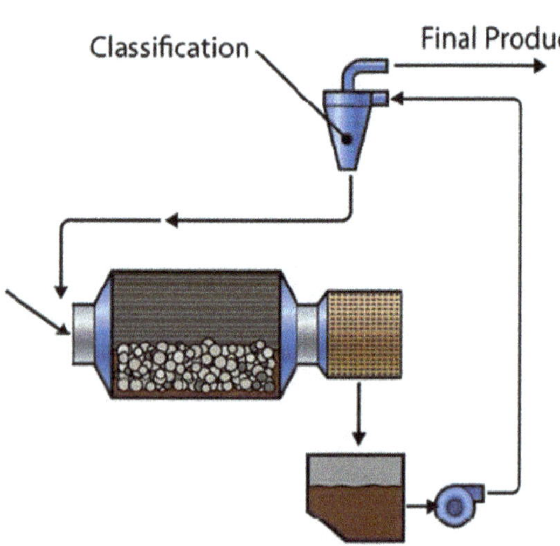

Fig. 3.12 Closed-circuit ball mill

Fig. 3.13 3500 kW closed-circuit ball mill

Single Stage Open Circuit Rod Mill

This circuit comprises a single stage rod mill operating in open circuit. Of the options under consideration, it is the simplest and has been employed successfully on recent alumina refinery expansion projects. Careful and comprehensive test work and modeling was undertaken before the decision to use this option was finalized.

The primary advantage of this circuit is its ability to handle course and variable feed to the grinding circuit while at the same time limiting the proportion of course particles and very fine particles in the final product. It employs large rods (~100 m diameter) in order to ensure that there is enough impact energy available to break to course particles. Characteristically, rod mills produce less fines than other mill types.

Figure 3.14 is a diagrammatic illustration of this option.

The footprint for this option is small. In a recent project in Australia three (3) 15′ × 21.5′ rod mills mill replaced three (3) rod/ball mills and were incorporated into the same footprint while simultaneously increasing production by a factor of two. The foundations of the original mills were partially reused and, by innovative design of the modified foundations disruption to the existing operation was kept to an absolute minimum during the decommissioning of the old mills and installation and commissioning of the new rod mills (Fig. 3.15).

While simple, economical, and compact, this option has several characteristics that should be recognized, understood, and considered at the selection stage.

While being very effective at controlling the size and proportion of top size particles in the product it must be recognized that this option produces a generally courser

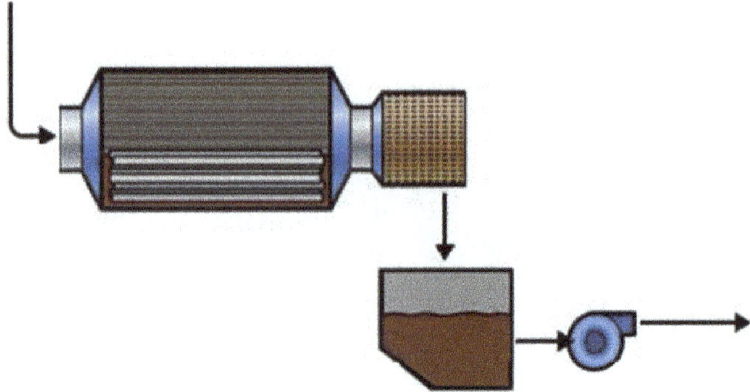

Fig. 3.14 Open-circuit rod mill

Fig. 3.15 1800 kW open-circuit rod mill

grind in terms of the product P80. Typically grinding circuits that employ ball mills will produce a P80 in the range 200–400 μm while the rod mill typically produces a P80 in the range 500 to 800 microns. At the top end of the product size distribution this rod mill circuit produces a P98/99 of approximately 1,700 microns.

Comparative size distributions are discussed later.

This circuit is also demanding less energy than ball mill circuits and typical specific energy consumption is in the range 4.0–6.0 kWh/t. The primary reason for the lower energy requirement is the generally courser grind. It is therefore very important to verify that this courser size distribution is suitable to the downstream processes in the alumina refinery.

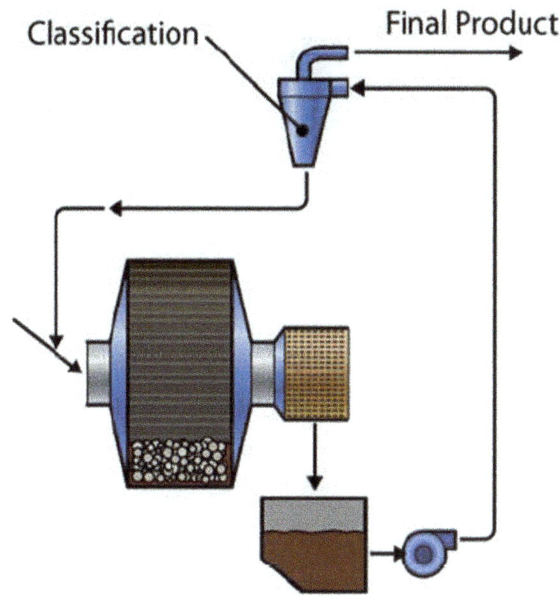

Fig. 3.16 SAG mill in closed-circuit

This circuit offers a very simple and economical plant in the right application. With the proper management of the rod charge and liners, high availability is achievable with limited need for personnel to enter the grinding chamber for maintenance purposes.

Closed circuit single stage SAG mills

The advantage of the SAG mill is that it can handle very course primary crusher feed and achieve very high reduction ratios. As such secondary crushing is not required in this option and the SAG mill can produce final product in a single stage. High throughputs are also achievable in single line plants.

Figure 3.16 is a diagrammatic illustration of this option.

This option requires the use of classification devices in the circuit in order to control the top size. It is capable of high availability provided the classification stage is adequately sized and incorporates necessary redundancy (Fig. 3.17).

In terms of capital and operating cost this option tends towards the high end due to the relatively higher cost of SAG mills in comparison to other grinding mill types. Due to high reduction ratios involved and the predominance of attrition grinding of small particles in this circuit the circulating loads may be relatively high. As such the SAG mill will produce a higher proportion of fine particles which may be undesirable in subsequent processes in the refinery.

The primary advantage of this circuit is the ability the handle high throughputs in single line plants and the simplification of the feed preparation area.

Open-Circuit SAG Mill following by Closed-Circuit Ball Mill

Fig. 3.17 4850 kW SAG mill in bauxite grinding

An extension of the previous circuit is the SAG/Ball Mill circuit. This option comprises an open circuit SAG mill followed by a closed-circuit ball mill. Due to the high throughputs associated with this circuit, the ball mill is most often closed with a hydro-cyclone classifier. The course material is returned to the ball mill feed and the fine overflow product reports to the downstream process. The primary advantage of this circuit is the ability to handle high throughputs in a single line plant. The SAG mill can handle very course and variable feed sizes while the closed-circuit ball mill helps to minimize the presence of undesirable course particles in the final product.

The capital cost is high with this option as is the maintenance and footprint requirements. It is slightly more energy efficient due to the use of a ball mill as the secondary grinding machine. Figure 3.18 is a diagrammatic illustration of this option.

While the type of grinding circuit selected for a given application will be based on a wide range of criteria, the primary criteria for an application are generally the process parameters that the grinding circuit must achieve. The various mill types are best suited to different ranges of feed size and product size distributions.

3.6 Comparison of Grinding Circuit Options [4]

The following is intended as a general comparison of the various circuits with respect to their optimum ranges of application. The final selection will always depend upon a

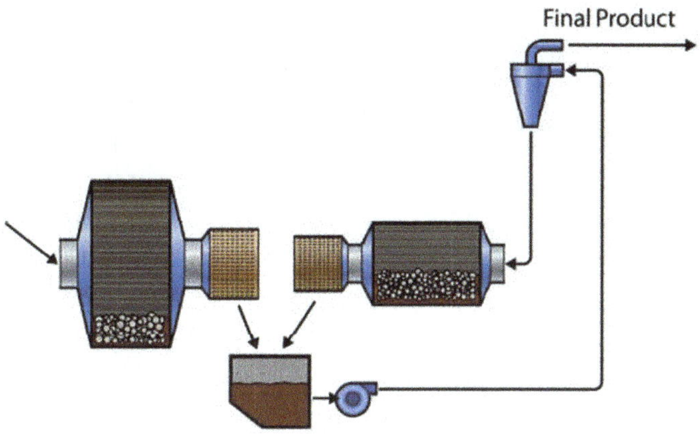

Fig. 3.18 Open-circuit SAG mill following by closed-circuit ball mill

thorough characterization of the Bauxite feed to the grinding circuit. Key parameters to be established are.

- Bond Work Index
- General breakage characteristics parameters.

Comparison of Feed Size Distributions.

Table 3.3 shows the range of application for each circuit type described above.

Comparison of Product Size Distributions

Table 3.4 shows the range of application for each circuit type described above.

Comparison of fine particles in the final product

Fine particles in the final product (i.e., % < 45 μm) is an important consideration for the down stream processes in an alumina refinery. It is difficult to quantify this criterion other than to make some general remarks about the relative performance of each circuit under discussion here.

Fines generation is largely a function of the breakage characteristics of the given feed material, but the selection of mill type and circuit configuration will influence

Table 3.3 Feed size comparison

Circuit	F80 (mm)	F98 (mm)
Rod/ball	10 to 25	25–40
Rod and ball mill in closed circuit	10 to 25	25–40
Single stage closed circuit ball mill	4 to 10	8–16
Open circuit rod mill	10 to 25	25–40
SAG mill	100 to 175	150–300
SAG/ball circuit	100 to 175	150–300

Table 3.4 Product Size Comparisons

Circuit	P80 (μm)	P98 (μm)
Rod/ Ball in open circuit	200–400	1000–1500
Rod and ball mill in closed circuit	200–400	1000–1300
Single stage closed circuit ball mill	200–400	1000–1300
Open Circuit rod mill	500–800	1600–2000
SAG	250–500	1000–1500
SAG/Ball Circuit	200–400	1000–1300

Table 3.5 Comparison of fine particle generation

Circuit	Contribution to fines generation
Rod/ball in open circuit	Neutral
Rod and ball mill in closed circuit	More
Single stage closed circuit ball mill	More
Open circuit rod mill	Less
SAG	More
SAG/ball circuit	More

the final product in terms of fines generation. The following comments are intended as a general guide only as to the influence that the mill/circuit will have. We have simply ascribed the descriptions, "more", "less" or "neutral" to describe the effect on the generation of fines in the grinding process.

Table 3.5 given a guide to relative propensity fines generation for each circuit.

Comparison of Specific Power Consumption

The final specific power consumption of a given circuit will depend largely on the grindability (i.e., Work Index) of the Bauxite and the feed and product size distributions. Each grinding option will, however, influence the power required to meet the required throughput under a given set of feed and product conditions. For example, if the mill is operating in open circuit the main criteria is the minimization of the course material in the product, then the mill power will be de-rated to ensure a longer residence time to reducer the probability of course particles "short circuiting" the grinding chamber. Table 3.6 shows the range of application for each circuit type described above.

Comparison of Other Criteria

In the final selection of the most appropriate grinding circuit for a given application there are many criteria to be considered apart from the process parameters. The most important of these are as follows.

- Capital Cost
- Plant Footprint
- Availability

Table 3.6 Specific power comparison

Circuit	Specific power consumption (kWh/t)
Rod/ball in open circuit	8.0–12.0
Rod and ball mill in closed circuit	6.0–9.0
Single stage closed circuit ball mill	5.5–8.5
Open circuit rod mill	4.0–6.0
SAG	8.5–13.0
SAG/ball circuit	8.0–12.0

- Maintenance and other operating costs
- Health and safety of personnel.

The priority given to each criterion will vary from project to project and it is not the intension here to deal with these in any ranking but simply to give a general idea of the relative merits of each option.

In order to do this the simple notation is ascribed to each option; "High", "Medium" or "Low". Table 3.7 is a summary of the comparative strengths of each option with respect to the other criteria.

The above assessment is, of course, subjective and open to debate however this is considered a reasonable guide to the relative merits of the various options.

It is clear from the above that the final selection of grinding mill type and circuit is a complex balance of many competing criteria. While the process criteria will always take precedence there are many other important issues to consider. There is no universal solution.

Careful definition of the most important criteria is essential. Thorough characterization of the Bauxite from a breakage perspective must be completed. Finally,

Table 3.7 Comparison of other criteria

	Capital cost	Foot print	Availability	Ease of maintenance	Risk to personnel
Rod/ball in open circuit	Med	Med	Low	Low	High
Rod and ball mill in closed circuit	High	High	Low	Low	Med
Single stage closed circuit ball mill	Med	Low	Med	Med	Low
Open circuit rod mill	Low	Low	High	Med	Med
SAG	High	Med	High	Med	Low
SAG/ball circuit	High	High	High	High	Low

each circuit option must be modeled and evaluated against the key criteria and other important criteria that will affect the long-term success of the final installation.

References

1. M.D. Flavel, C.A. Rowland, Jr., Selecting Circuits to Prepare Beneficiation Circuit Feed from Primary Crusher. SME of AIME, 18–20 Nov 1981
2. N.L. Weiss, ed., Testing and calculations, in *SME Mineral Processing Handbook* , Vol. 1 (1985), pp. 3A-25–3A-27
3. C. Pellant, Rocks and minerals, in *Eyewitness Handbooks* (1992), p. 96
4. J. Hadaway, A. Filidore, Bauxite grinding practices and options. Light Metals (2009)
5. T.L. Capron, An evaluation of alternative bauxites for Kaiser's Bayer plant in Gramercy Luisiana. Light Metals 11–14 (1998)
6. L.N. Schooler, Comminution of Caribean bauxites in rod mils. Light Metals 19–37 (1980)
7. G. Songqing, Y. Zhonglin, Preheaters and digesters in the Bayer Digestion Process. Light Metals 35–40 (2004)
8. D.J. Milne, The influence of mineralogy on alumina processing. Chem. Eng. Australia **ChE7**(2) (1982)
9. W.E. Wahnsiedler, The kinetics of bauxite digestion in the Bayer Process. Light Metals 145–182 (1984)

Michael Wanyo FLSmidth. Email: mike.wanyo@flsmidth.com

Michael Wanyo is an Applications Engineer with FLSmidth, the globally known supplier of cement and mining process equipment. He is an engineering graduate from the Pennsylvania State University, and he has twenty-five years of global hands-on cement and mining experience. The majority of Michael's time in industry was devoted to crushing applications and related technology including compression-crushing equipment and low-speed sizer technology. His ongoing career stretches across a wide variety of mineral types and incorporates in-depth process/application reviews, equipment installation and commissioning, as well as troubleshooting assignments.

Anthony T. Filidore FLSmidth, Director—Comminution Technologies. Email: anthony.filidore@flsmidth.com

Anthony Filidore joined FLSmidth in 1995 and is currently Director of Comminution Products. He received a BS in Mechanical Engineering from the University of Virginia and an MBA from Pennsylvania State University. Anthony has spent 25 years modeling and sizing grinding circuits for all types of ores. His experience includes sizing over 12 bauxite grinding circuits that are currently in operation.

Benny E. Raahauge Deceased—Owner—Director Raahauge-SGA ApS, Denmark

Benny received his M.Sc. Chemical Engineering from the Danish Technical University in November 1972. Start working with programming minicomputers controlling the raw material mixing for production of Cement Clinker, December 1972 in Process Technical Department, FLSmidth, Cement Division, Copenhagen, Denmark.

From 1975–76, Benny worked as process and plant engineer in a Danish Sugar Factory before re-joining FLSmidth as R&D engineer and later R&D Manager in the Mining Division in Copenhagen. Benny was assigned the task to develop the stationary Gas Suspension Technology for Smelter Grade Alumina (SGA) to replace the rotary kilns.

After successful commissioning of the first 1000 tpd Gas Suspension Calciner (GSC) unit for alumina at Hindalco Industries, India, in 1986, Benny worked in the role as General Manager-Pyro & Alumina Technology based in Copenhagen. This assignment included the responsibility for design, marketing, sales and commissioning of GSC units for Alumina, culminating with commissioning of 3 x 4500 tpd GSC units at Queensland Alumina 2004–05, the world's largest stationary calciners.

During +44 years employment with FLSmidth & Co. contributed with many papers on calcination to international ICSOBA, Alumina Quality Workshop and TMS meetings until retiring in 2018.

Together with Don Donaldson, Benny was Lead-Editor to "Essential Readings in Light Metals—Volume 1—Alumina & Bauxite" with Fred Williams as Co-Editor, Published by Wieley & Sons. Inc. Copyright © 2013 by The Minerals, Metals & Materials Society (TMS).

In 2018 Benny was editor and contributor to Chapter 12.1 Alumina, in the "SME Mineral Processing & Extractive Metallurgy Handbook", Published by Society for Mining, Metallurgy & Exploration (SME), Copyright © 2019.

Chapter 4
Chemical Processing of Bauxite: Alumina and Silica Minerals—Chemistry, Kinetics and Reactor Design

György (George) Bánvölgyi and Brady Haneman

Abstract This Chapter provides an introduction to the Bayer process and how the mineral composition of bauxites affects the process variants and the principal parameters to be selected. The Chapter covers the rational of sizing of principal equipment (such as heat recovery system and reactors) and also of calculating the energy (heat) requirement of the digestion unit operation. Beside the processing behaviour of the main constituents of bauxite such as hydrated alumina and silica minerals, the Chapter covers the impact of other constituents, such as titania, iron compounds, organics, and other impurities. The use of lime in the Bayer process and the chemistry behind that is also discussed.

4.1 The Bayer Process and Low Temperature Digestion

The Bayer process is a chemical process for refining aluminium hydroxide, $Al(OH)_3$ from bauxite; this aluminium hydroxide is subsequently calcined to produce alumina, Al_2O_3. The basis of the Bayer process is an understanding of the characteristics of the sodium-hydroxide—sodium-aluminate solution relationship, namely its ability to keep sodium-aluminate in a dissolved state over a wide range of conditions, especially on the temperature and caustic soda concentrations.

G. Bánvölgyi (✉)
Bán-Völgy Ltd, 49 Dessewffy st, 1066 Budapest, Hungary

B. Haneman
Hatch, Brisbane, Australia

B. E. Raahauge and F. S. Williams (eds.), *Smelter Grade Alumina from Bauxite*, Springer Series in Materials Science 320,
https://doi.org/10.1007/978-3-030-88586-1_4

4.1.1 Chemistry of Dissolution of Hydrated Alumina Minerals and Silica Minerals, Low Temperature Digestion

Initially, let us consider a hypothetical bauxite which consist of pure aluminium hydroxide $Al(OH)_3$. (In fact, there are few bauxites which consist of almost pure, 90–95 phase% gibbsite [1]). When this "bauxite" is treated with NaOH solution, the following reversible reaction takes place (4.1).

$$\underset{\text{(gibbsite)}}{Al(OH)_3} + OH^- \underset{\text{cooling}}{\overset{\text{heating}}{\rightleftarrows}} \underset{\text{(aluminate anion)}}{[Al(OH)_4]^-} \tag{4.1}$$

Having reviewed the results of several researchers, Glastonbury concludes [2], that the most probable structure of the aluminate ion is $Al(OH)_4^-$ up to 25% Na_2O. In the liquid phase $[Al(OH)_4]^-$ complex anion and its derivatives such as $[Al(OH)_4]^- \cdot 2H_2O$ most probably exist [3]. Higher temperatures and higher caustic concentrations are favourable for the dissolution (so-called digestion), the lower temperatures and lower liquor caustic concentrations are favourable for the crystallisation of the aluminium hydroxide (so-called precipitation). The equilibrium solubilities of gibbsite, boehmite and diaspore as a function of the temperature and caustic soda concentrations are shown on Fig. 4.1.

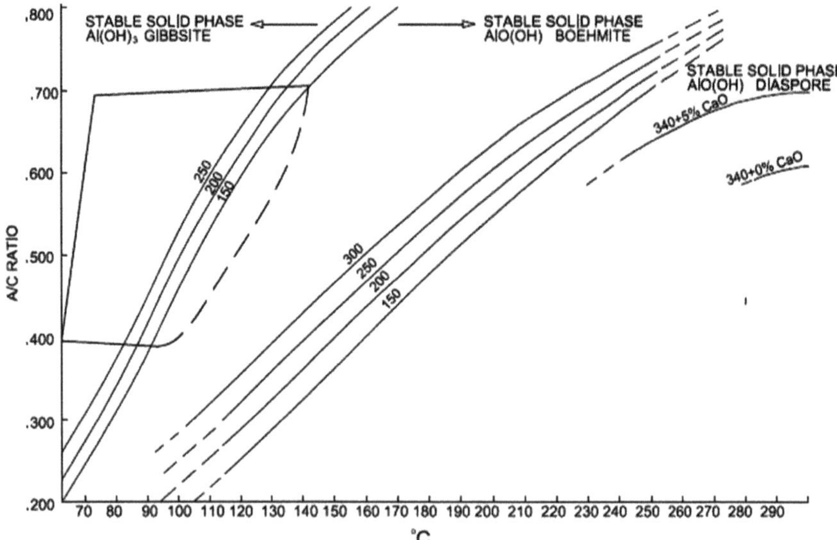

Fig. 4.1 Equilibrium solubilites of dissolved alumina for gibbsite, boehmite and diaspore [4]. Copyright © 1981 by The Minerals, Metals & Materials Society. Used with permission

In Fig. 4.1, the numbers 150-200-250 on the graphs of equilibrium relate to caustic soda concentrations in Na_2CO_3. A/C denotes aluminate ion concentration as Al_2O_3 (A)/caustic soda concentration as Na_2CO_3 (C). A typical Bayer process cycle with low temperature digestion is also shown.

Virtually each bauxite contains more or less amount of clay minerals, mainly kaolinite ($Al_2(OH)_4(Si_2O_5)$ or $Al_2O_3{\cdot}2SiO_2{\cdot}2H_2O$). The kaolinite content in bauxites ranges from 0.1 to 15% with an average [5] of 4–8%; kaolinite is the principal silicate containing mineral. In addition to kaolinite, the other main silicon containing component of bauxite is quartz (SiO_2); to differentiate the sparingly soluble crystalline silica, quartz, from kaolinite, the latter is often referred to as "reactive silica".

In the course of the digestion, kaolinite reacts with sodium hydroxide as a side reaction. The dissolution reaction of kaolinite can be described by (4.2). For the composition of silicate anion Pevzner et al. is referred [6].

$$\underset{\text{(kaolinite)}}{Al_2(OH)_4(Si_2O_5)} + 6OH^- + H_2O \rightleftharpoons \underset{\text{(silicate anion)}}{2[SiO_2(OH)_2]^{2-}} + \underset{\text{(aluminate anion)}}{2[Al(OH)_4]^-} \quad (4.2)$$

Subsequent to dissolution, sodium aluminium hydrosilicate compounds forms in the liquid phase as per the following desilication reaction (4.3).

$$\begin{aligned} &6[SiO_2(OH)_2]^{2-} + 6[Al(OH)_4]^- + 20Na^+ + X^{2-} \\ &\rightleftharpoons [Na_2O \cdot Al_2O_3 \cdot 2SiO_2 \cdot (1 + k)H_2O]_3 \cdot Na_2X \\ &+ 12OH^- + 12Na^+ + (9 - 3k)H_2O \end{aligned} \quad (4.3)$$

where X: $2OH^-$, $CO_3{}^{2-}$, $SO_4{}^{2-}$, $2Cl^-$, $2[Al(OH)_4]^-$, etc., $k = 0 - 1$

at temperatures of about 140–160 °C. The sodium aluminium hydrosilicate formed at this temperature range is mostly sodalite.

It can be seen that the Al_2O_3/SiO_2 ratio (A/S) in kaolinite and desilication product (DSP) is practically identical. In other words, formation of DSP causes sodium hydroxide losses but not losses of soluble alumina since the alumina in kaolinite is left out when the available alumina content is determined.

In the gibbsitic bauxites quartz (SiO_2), iron mineral(s), typically hematite and/or goethite (Fe_2O_3, α-FeO(OH), respectively) and titania (TiO_2, anatase and/or rutile) can also be found. In the case of goethite and hematite, aluminium may replace up to 30 molar % of the iron in the lattice. Some boehmite (γ-AlOOH) can usually also be found even in the **gibbsitic type** bauxites, and occasionally some diaspore (α-AlOOH) as well. None of these constituents is reactive at the typical low temperature digestion temperatures (140–150 °C). Bauxites are considered to be of gibbsitic type when the boehmite content is less than 5% [7].

The quantitive mineralogical composition derived from the XRD measurement by the XDB™ program package [8] of a good quality bauxite from Trombetas, Brazil is shown in Table 4.1 and Fig. 4.2.

Table 4.1 Quantitative mineralogical composition of a bauxite from Brazil (used with permission of I. Sajó)

[BXT-03A] Trombetas std										
	SUM	Gibs	Hema	Kaol	Goet	Anat	Quar	Ruti	Chem	
Phase%	99.10	78.00	6.00	7.50	6.50	0.70	0.20	0.20	Anal	Diff
Fe_2O_3%	11.70		6.00		5.70				11.70	0.00
TiO_2%	0.90					0.70		0.20	1.05	0.15
SiO_2%	3.69			3.49			0.20		3.79	0.10
Al_2O_3%	54.08	50.98		2.96	0.14				54.10	0.02
H_2O%	28.74	27.02		1.05	0.67					
LOI%	28.73	27.02	0.00	1.05	0.67	0.00	0.00	0.00	28.90	0.17

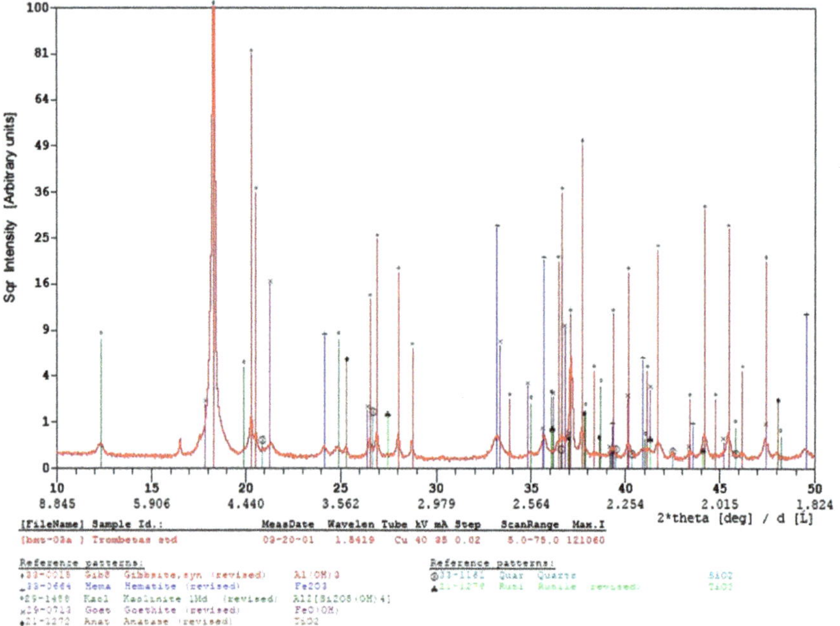

Fig. 4.2 XRD pattern of a bauxite from Brazil (used with permission of I. Sajó)

In the case of the digestion of gibbsitic bauxites at low temperature (LT, about 140–150 °C), gibbsite dissolves in about 1–3 min; the kaolinite also dissolves but at a somewhat slower rate. About 30–45 min retention time is required so that the dissolved silica content falls fairly closely to its equilibrium solubility.

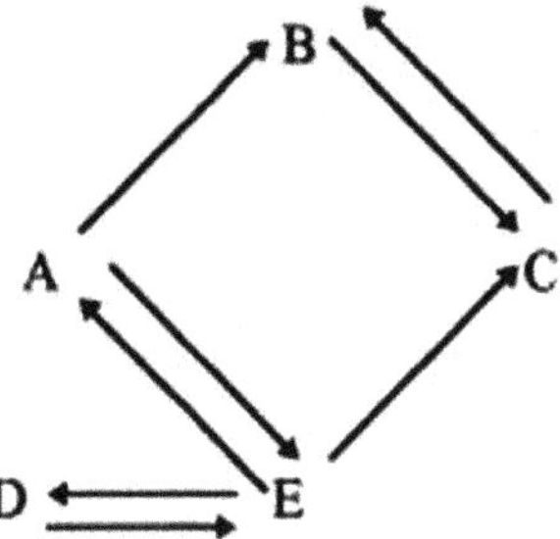

Fig. 4.3 Reaction mechanism for the dissolution of gibbsite and transformation of kaolinite to DSP

The reaction mechanism of the dissolution of gibbsite and transformation of kaolinite to sodalite

The basic reaction Eqs. (4.1)–(4.3) depict the dissolution of gibbsite and the transformation of kaolinite to sodalite. The schematic of the reaction mechanism of the dissolution of gibbsite and transformation of kaolinite to DSP is shown in Fig. 4.3.

In Fig. 4.3 A denotes kaolinite, B dissolved silica, C sodium aluminium hydrosilicate (sodalite, DSP), D gibbsite and E dissolved alumina.

Reaction rate equations of dissolution of kaolinite and gibbsite, formation of desilication product

The reaction rate equations which best describe the substantial reactions are the following [9, 10]. The rate equation for dissolution of kaolinite that was developed originally for predesilication (see later) has been adopted here to the conditions of LT digestion:

$$r_1 = dc_A/dt = k_1\, AH\, c_A \quad \text{dissolution of kaolinite} \tag{4.4}$$

$$r_2 = dc_B/dt = k_2\left(c_B - c_B^{eq}\right) \quad \text{formation of DSP} \tag{4.5}$$

$$r_3 = dc_D/dt = k_3\, AH\, c_D \quad \text{dissolution of gibbsite} \tag{4.6}$$

where

$$AH = 2 * 17\left[\left(c_E^{eq} - c_E\right)/102 - c_B/60\right] \quad \text{reactive OH concentration,} \tag{4.7}$$

c_A, ..., c_E concentrations of the constituents as denoted in the schematic of the reaction mechanism (Fig. 4.3), AH is the reactive OH^- concentration, 17, 102 and 60 are the molar masses of OH, Al_2O_3 and SiO_2, respectively. k_1, k_2 and k_3 are the kinetic rate constants, to be determined experimentally.

It should be pointed out that the OH^- ions are the principal reactants which attack both gibbsite $Al(OH)_3$ and kaolinite $(Al_2(OH)_4(Si_2O_5)$ as depicted in the reaction Eqs. (4.1) and (4.2). The reactive OH^- concentration is believed to be proportional to the difference between the actual and the equilibrium alumina concentration against gibbsite as denoted in (4.7). The dissolved silica content plays some, but only a minor role in the reactive OH^- concentration as shown in (4.7). In other words, the reactive OH^- ion concentration comprises the common driving force of the dissolution reactions of gibbsite and kaolinite. The dissolution of gibbsite and kaolinite can be explained by their competition for the reactive OH^- ions in the liquid phase.

Similar rate equations for the gibbsite extraction and kaolinite dissolution/desilication were developed by Raghavan and Fulford [11]. The basic difference in the two explanations is that Raghavan and Fulford claim the "free OH" (the difference between the total caustic soda and the caustic soda which is present as sodium aluminate) is the reactant that attacks gibbsite and kaolinite.

LaMacchia has suggested [12] a kinetic rate equation for the desilication of the liquor at the predesilication where the rate is proportional with the multiplication of the amount of DSP (g/L) and the relative supersaturation ratio of dissolved silica to the power of 1.81 ± 0.63. The relative supersaturation ratio: $(c_B - c_B{}^{eq})/c_B{}^{eq}$.

Desilication, predesilication and post-desilication

In the literature there are various terms such as desilication, predesilication and post-desilication; several professionals are confused about their meaning. Predesilication aims to convert the kaolinite into DSP as soon as possible before the soluble alumina content of bauxite is digested at higher temperature. Desilication in general, is the reduction of the dissolved silica content in the liquid phase by the formation of DSP. Post-desilication is the same, but it is carried out after the digester stage, either out of the liquid phase of the digester slurry effluent such as in the case of atmospheric digestion, or, from the pregnant liquor obtained from the separation of bauxite residue (BR), such as in the case of ILTD (Improved Low Temperature Digestion) Process [10] or Sumitomo New Bayer Process (SNBP) [13].

Addai-Mensah et al. summarized the formation mechanism of DSP (**sodalite** and **cancrinite**) [14]:

$$[\text{Na-aluminosilicate ions}] \rightarrow \text{amorphous} \rightarrow \text{zeolite A} \rightarrow \text{sodalite} \rightarrow \text{cancrinite} \tag{4.8}$$

The formation of an amorphous phase and its transformation to zeolite A (LTA) were shown to occur at temperatures lower than 85 °C by Barnes et al. [15, 16]. Previously, formation of zeolite A (LTA) as an intermediate in the course of the formation β-hydroxy-sodalite at temperatures of 50–90 °C, and at higher temperatures, conversion of sodalite to cancrinite were revealed by Nemecz et al. [17]. Peng et al. investigated the kinetics of desilication [18] at 75 and 90 °C. Four distinct phases were observed: (1) formation of amorphous material shortly after the start of the reaction; (2) dissolution of some of the amorphous solid; (3) growth of the crystalline phases; and (4) maturation of the crystalline phases. At 75 °C, LTA appears as a crystalline phase, at 90 °C, LTA and sodalite simultaneously appear early in the process.

A thin amorphous layer and zeolite were detected during formation of sodalite and cancrinite at 140 °C by Shi et al. [19]; the reaction time and the temperature favour the formation of sodalite and then cancrinite.

In the conditions of the low temperature (LT) digestion, sodalite is the typical DSP product in BR. However, in a scaling sample taken from a LT digester (operating as 145 °C) after several months of operation, 46% of the scaling was found to be cancrinite [20].

The rationale of the predesilication, is to shorten the retention time in the digestion stage to attain the same dissolved silica concentration as in conventional digestion. The composition of the DSP that forms during predesilication [10] can be represented with the following equation:

$$2[SiO_2(OH)_2]^{2-} + 2[Al(OH)_4]^- + 2Na^+ \rightleftharpoons Na_2O \cdot Al_2O_3 \cdot 2SiO_2 \cdot (2+k)H_2O + 4OH^- + (2-k)H_2O \quad (4.9)$$

where $k = 0 - 2$ in the predesilication rather close to 0.

Batch predesilication test results of a high reactive silica boehmitic bauxite from Nyírád (Hungary) are shown on Figs. 4.4, 4.5, 4.6, 4.7, 4.8 and 4.9 [9]. The bauxite contained 48.8% Al_2O_3, of which 37% was present as boehmite, 9.1% SiO_2, all in kaolinite. The kinetic test was carried out at 100 °C in a liquor that contained 149.5 g/L Na_2O_{caust} (257 g/L C as Na_2CO_3), 73.1 g/L Al_2O_3. The calculated values with the kinetic model are also shown. The following rate equation proved the best fit for the dissolution of kaolinite:

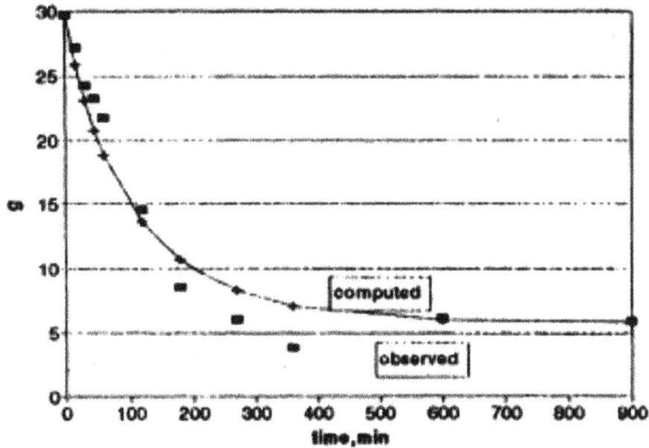

Fig. 4.4 SiO_2 in kaolinite, g/L slurry

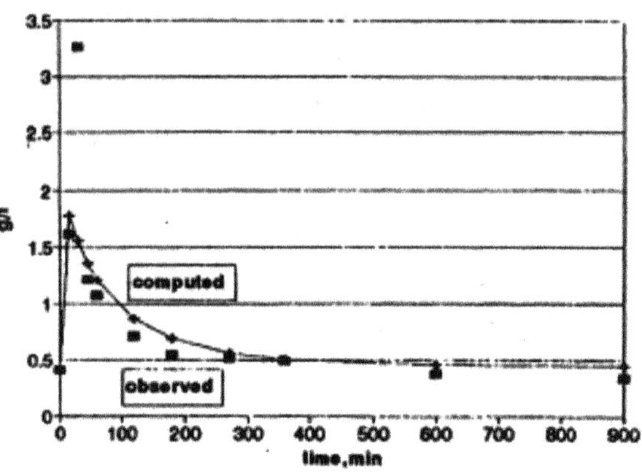

Fig. 4.5 SiO_2 concentration, g/L

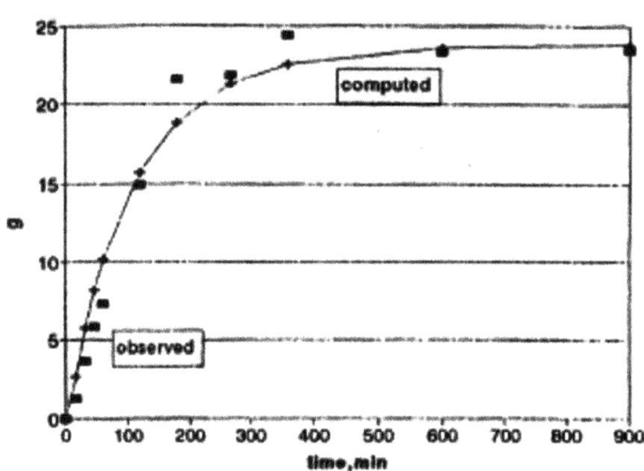

Fig. 4.6 SiO_2 in DSP, g/L slurry

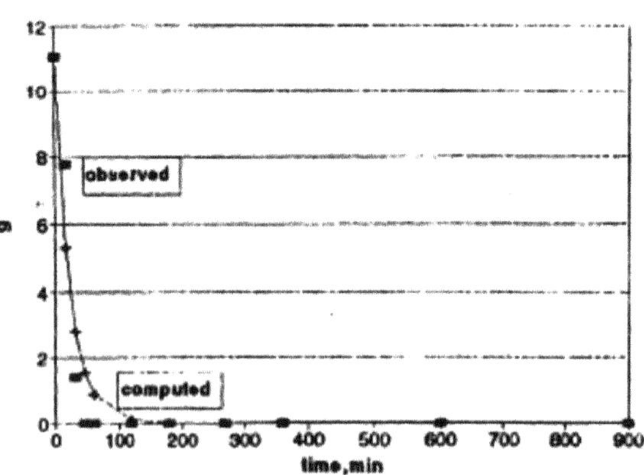

Fig. 4.7 Al_2O_3 in gibbsite, g/L slurry

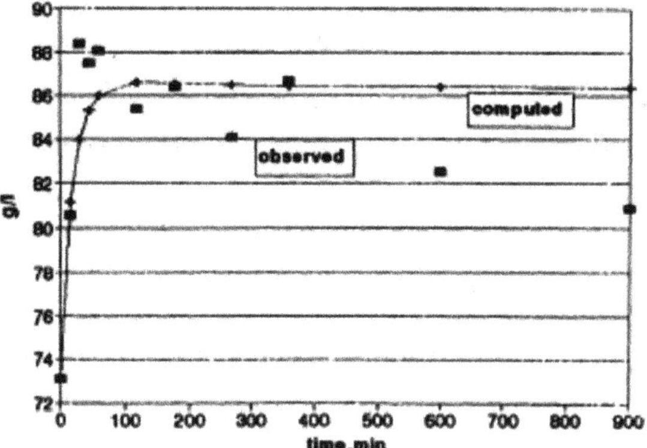

Fig. 4.8 Al_2O_3 concentration, g/L

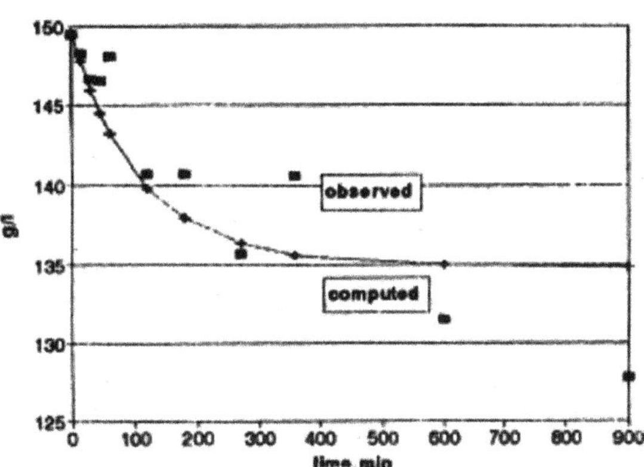

Fig. 4.9 Na_2O_{caust} concentration, g/L

$$r_1 = k_1\,AH\left(c_A - k_4 c_C / c_F^2\right) \quad \text{dissolution of kaolinite} \qquad (4.10)$$

where

c_C amount of DSP, g/L slurry

c_F Na_2O_{caust} caustic concentration, g/L.

The c_C/c_F^2 part of the rate equation suggest that, at least in certain conditions, the formation of DSP slows down, or even stops the further dissolution of kaolinite. This is in line with the observation of Roach on the role of DSP [21].

The rest of the kinetic model, i.e. the rate equations are (4.5)–(4.7).

A particular feature of the predesilication stage, i.e. the conversion of kaolinite to DSP, may take place in-situ, without dissolution of kaolinite into the bulk of the liquor has been shown by Bánvölgyi et al. [22].

The DSPs that form in the course of predesilication (sodalite and probably zeolite) can dissolve in the LT digester autoclaves at a rate of more than 50% [23], and then re-precipitates as DSP as per the reaction Eq. (4.3).

The effect of silica in the Bayer Process

A separate book would be required to cover the subject in depth; a vast number of papers have been published which discuss the different aspects of the reactions of the silica minerals in the Bayer process. The principal consequences of the reaction of silica minerals are:

1. NaOH losses which are largely proportional to the amount of reactive silica entering the Bayer process;
2. Contamination of the product with silica;
3. Formation of DSP-containing scalings in the digester trains depending on the liquor compositions (caustic, alumina and silica contentrations) and temperature;
4. Formation of DSP scalings in the evaporators;
5. DSP acts as a sink for certain anions, thereby removes them from the Bayer liquor circuit.

Equations, graphs for the equilibrium concentration of soluble alumina against gibbsite, boehmite and DSP

The dissolved alumina equilibrium solubility against gibbsite has been defined by Siklósi [24]:

$$c^{eq}{}_{alumina} = t'\left(78c'_{Na2Oc} + 50c'^{2}_{Na2Oc} - 20c'^{3}_{Na2Oc}\right) + \left(8.9c'_{Na2Oc} - 62.8c'^{2}_{Na2Oc} + 32.8c'^{3}_{Na2Oc}\right) \quad (4.11)$$

where t' is the temperature in °C/100, c'_{Na2Oc} is the caustic concentration Na_2O_{caust} in g/L/100. This formula is based on the graphs of Adamson et al. [25] and can be used in the range 80–180 g/L Na_2O_{caust} and 100–140 °C.

The dissolved alumina equilibrium formula for gibbsite based on thermodynamic considerations, which covered a wide range of caustic concentration (4–400 g/L caustic soda expressed as Na_2CO_3), temperature ranges (55–175 °C), and the liquor contaminants were also taken into consideration, was developed by Rosenberg and Healy [26]. The formula calls for some improvement at the temperatures of the Low Temperature digestion, since higher A/C ratios have been measured in the course of kinetic tests with Trombetas bauxite [10] then the formula of [26] suggests as equilibrium solubility for gibbsite.

A formula that was developed for the equilibrium solubility of dissolved silica against sodalite by Bánvölgyi [27] is as follows

$$c^{eq}{}_{SiO2} = (3.392 - 4.97t' + 1.96t'^{2}) * C'_{Na2Oc*}c'_{Al2O3} \quad (4.12)$$

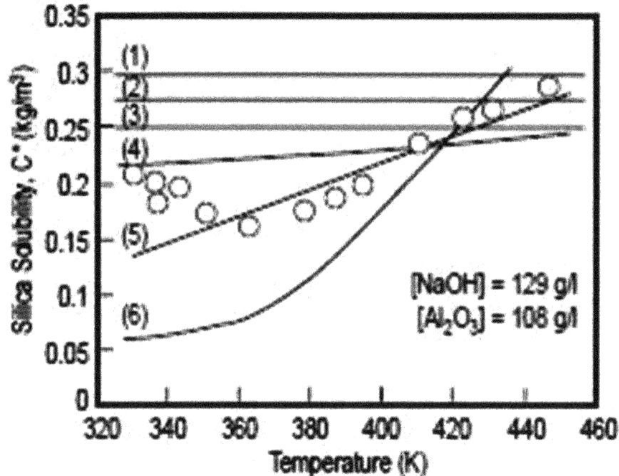

Fig. 4.10 Silica solubilities (C°) as measured by various authors [28]. Copyright © 1998 by The Minerals, Metals & Materials Society. Used with permission

where t′ is the temperature in °C/100, C'_{Na2Oc} is the caustic concentration in Na_2O_{caust}/100 and c'_{Al2O3} alumina concentration in Al_2O_3 g/L/100. The formula is valid in the range of 100–220 g/L Na_2O_{caust}, 60–120 g/L Al_2O_3 and 70–150 °C.

A summary of previous attempts for the measurement of the silica solubility is given by Jamialahmadi and Müller-Steinhagen [28]. A comparison of the silica solubilities as measured by various authors is shown in Fig. 4.10. At low temperatures, most probably Zeolite A was formed and the equilibrium solubility of silica in this case is higher than that of with sodalite at higher temperature. At the highest temperatures, most probably cancrinite is formed. This may explain the non-linear shape of the experimental points.

Curves in Fig. 4.10 were taken from work of (1) Oku and Yamada (1971), (2) Adamson et al. (1964), (3) Leiteizen (1972), (4) Cresswell (1984), (5) Hewett et al. (1987), (6) Jamialahmadi and Müller-Steinhagen. The test results of Jamialahmadi and Müller-Steinhagen (1998) are shown with small circles [28].

The regression equation for the equilibrium A/C ratios of alumina against boehmite [29] based on the Kotte diagrams [4] (Fig. 4.1) is as follows:

$$A/C^{eq} = -0.3733 + 0.44911\,t' - 0.02194t'^2 + 0.00095\,c_{caust} - 3.0855e^{-7}\,c^2{}_{caust} - 0.0001\,t'\,c_{caust} \quad (4.13)$$

where t′ is the temperature in °C/100, c_{caust} is the caustic soda concentration in Na_2CO_3. The formula is valid at temperatures between 100 and 200 °C and 150 and 300 g/L caustic soda as Na_2CO_3.

Conversion of gibbsite to boehmite

The dissolution of boehmite can be described as a reversible reaction as well.

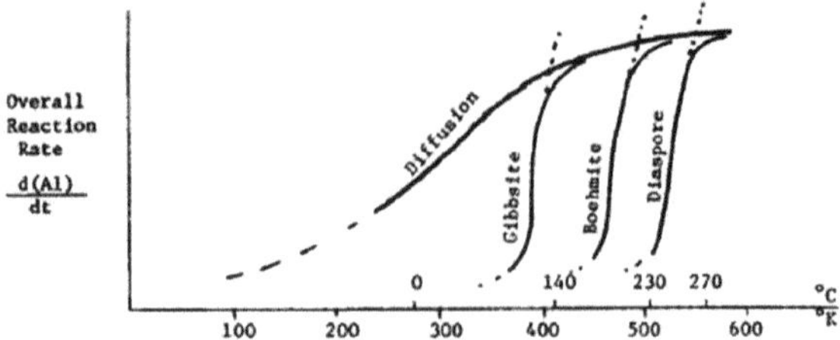

Fig. 4.11 Schematic representation of dissolution reaction rates versus temperature [4]. Copyright © 1981 by The Minerals, Metals & Materials Society. Used with permission

$$AlO(OH) + H_2O + OH^- \rightleftharpoons [Al(OH)_4]^- \qquad (4.14)$$

A review of the equilibrium alumina concentrations for gibbsite and boehmite suggest that solutions which are saturated for alumina against gibbsite, at the same time are supersaturated for dissolved alumina against boehmite. Therefore a driving force exists to precipitate boehmite from the dissolved alumina. However, as for all the reactions, the actual reaction rate is crucial; this is confirmed by the fundamental paper of Kotte [4] which demonstrates the dissolution reaction rate of boehmite rapidly increases in the temperature range of 150–200 °C (see Fig. 4.11).

It has been demonstrated that the amount of boehmite in scale taken from the LT digester autoclave (at 145 °C) and the flash tanks (last flash tank at 115 °C) increased from 25 to 50% respectively. In settler overflow pipe scales, after 5 years, boehmite was detected [20].

There have been several attempts to investigate the formation of boehmite [30–37]. However, no comprehensive explanation and/or kinetic rate equations have been developed as yet.

Reaction of organic matter

In the course of the LT digestion, the organic matter content of bauxite degrades, and a series of organic compounds form. This metabolic reaction scheme is described in a more detailed way in the Sect. 4.1.5.

Reactor design principles for low temperature digestion

Optimally, the target A/C ratio should be as close to the equilibrium as possible, taking into consideration the variance of the A/C ratio due to the imperfection of the A/C ratio control. If bauxite is overcharged, a part of the gibbsite stays undissolved and facilitates the formation of gibbsite as seed during the settling and washing of bauxite residue.

When a cascade of continously stirred tankreactors are used, the design, the number and volume of each reactor has to be determined. The authors prefer agitated tanks in these situations. The actual residence time distribution must be taken into consideration. Earlier measurements by Bujdosó et al. with vertical autoclaves suggest [38] that about 80% of the nominal volume is effective.

In the case of a tube reactor, it is easy to translate the batch kinetic test results to sizing of the reactor since the residence time distribution [39] closely corresponds to the batch kinetics.

Conventional process

The low temperature digester reactor should be designed to provide sufficient time so that an adequate low dissolved silica concentration can be attained at the effluent. As a rule of thumb, this is about 30 min in the case of predesilication. Without pre-desilication the residence times in the digester autoclaves is usually about 45–60 min. As mentioned, gibbsite is dissolved in the first 1–3 min.

The longer the retention time and the closer the target A/C ratio to the equilibrium, the higher is the risk of the hydrothermal formation of boehmite in the course of digestion.

Improved/modified low temperature digestion process

The objective of the Sumitomo New Bayer Process (SNBP) and also the Improved Low Temperature Digestion (ILTD) Process concept is to apply a short digestion residence time, which is just sufficient for the dissolution of gibbsite as well as separation of the bauxite residue as quickly as possible, preferably at the temperature of digestion, with the aim of reducing the reaction of kaolinite to an extent of less than about 50%. In the gibbsitic bauxites, gibbsite is present in a greater amount than kaolinite by at least an order of magnitude and has a significantly higher dissolution reaction rate [10, 13]. The Sumitomo process uses digestion temperature of 135 °C and aimed at maximum caustic savings during the Bayer process, while the ILTD process concept uses about 150 °C digestion temperature and aims at the maximum liquor productivity. The silica concentration in the overflow of a pressure decanter or in the filtrate of a hyper-baric (Hi-Bar) filter is in the range of 3–5 g/L and calls for a desilication of the liquor preferably in a crystallizer designed specifically for this purpose [10, 13]. The very low chemically combined soda and high Fe_2O_3 content of bauxite residue are further potentially significant benefits [40]. Sintering of the DSP by-product of the ILTD process makes possible to further reduce the soda losses at great extent and make the otherwise unsoluble alumina being originally in kaolinite soluble [41].

4.1.2 High Temperature Digestion

During the early decades of the history of Bayer process, those bauxites which contained siginificant amount of boehmite were digested at about 180–200 °C operating temperature, usually with batch autoclaves.

The chemistry of the High Temperature (HT) digestion is significantly more complex that that of the LT digestion, because in addition to the dissolution of boehmite and diaspore, several other reactions take place.

Dissolution of boehmite and diaspore

The prevention and reduction of scales have evolved as our understanding of the chemistry, reaction kinetics, heat recovery and materials for equipment have developed, firstly the autoclave trains, later the tube digester systems were introduced. Currently, the typical digestion temperature of the autoclave trains is about 240–260 °C, for the tube digester systems up to about 270 °C.

Diaspore is digested at higher temperatures usually with more concentrated Test Tank Liquor in presence of lime or its derivative compounds.

The temperatures at which the dissolution reactions of gibbsite, bohemite and diaspore start to rise rapidly, and also where the kinetic controlled reactions become diffusion controlled ones are shown in Fig. 4.11.

In the course of the LT digestion process, both the principal reactive mineral and the product of the process are the same compound, gibbsite ($Al(OH)_3$). When boehmite/diaspore having one mole of crystal water to one mole of Al_2O_3, is digested and gibbsite having three moles of water to one mole of Al_2O_3 is produced, it is associated with taking two moles of water from the Bayer circuit for each mole of Al_2O_3 crystallized.

Looking at the equilibrium solubilities for boehmite and diaspore (Fig. 4.1) the rationale of using higher temperature should be evident.

Quartz dissolves much less readily under high temperature conditions than kaolinite does under low temperature conditions. However, the precipitation rate of the desilication product is increased under high temperature conditions. In fact, a choice between boehmite and quartz extractions needs to be made under high temperature conditions in choosing a digestion time to minimize caustic and alumina losses related to the quartz attack while maximizing boehmite extraction. Bayer sodalite/cancrinite formation from quartz attack represents both an alumina and a caustic loss in the HT digestion process.

A kinetic rate equation for the dissolution of boehmite was proposed by Korcsmáros [42] as it follows:

$$dc_{A}/dt = D_b/\delta_b * s_b * (c_{Atb} - c_A)\left(c_{Abeq} - c_A\right) \quad (4.15)$$

where

c_A	dissolved alumina concentration, kmol/m^3
c_{Atb}	alumina concentration if all the soluble alumina (at the given temperature) were dissolved, kmol/m^3
c_{Abeq}	equilibirium alumina solubility for boehmite, kmol/m^3
D_b	diffusion coefficient, m^2/s
δ_b	thickness of diffusion layer, m
s_b	specific surface of material transfer (in the given case for boehmite), m^2/kmol.

Pei et al. found the following core shrinking rate equation to be the best fit for the digestion of boehmite in Weipa bauxite [43]

$$dD/dt = -2kM^n (D > 0) \tag{4.16}$$

where

D the particle diameter
M the driving force, equal to $A/C_{eq} - A/C_{(t)}$. The A/C_{eq} is the equilibrium solubility for boehmite
k rate constant
n order of reaction.

Reactions of silica, iron oxides, titania

The stoichiometry of the dissolution reaction of kaolinite and the formation of DSP are very similar as has been described in Sect. 4.1.1. The higher the reaction temperature, the higher the reaction rates. During HT digestion, formation of cancrinite is preferred. At a temperature of 180 °C (or, possibly somewhat less) quartz tends to react [44] as well, resulting in alumina and soda losses.

Sodalite and cancrinite are usually described with the same stoichiometry [45], $Na_6[AlSiO_4]_6 \cdot 2NaX \cdot nH_2O$, where X is $½CO_3^{2-}$, $½SO_4^{2-}$, Cl^-, AlO_2^- and OH^-. Sodalite has a cubic space group crystal structure, while cancrinite's is hexagonal. The chemically combined water content of DSP as per Whittington is $n = 0–2$ [46].

At temperatures of about 140–150 °C, certain iron compounds, especially goethite, tend to react. Investigation of the scalings from HT digester trains where boehmitic bauxites were processed showed that the transformation of goethite to hematite (via sodium ferrite in the liquid phase) started at 140–150 °C and became more dominant with the increase of the operating temperature [20].

The reactions of **titania** minerals in caustic solutions were published by Wefers [47], Schultze-Rhonhof and Winkhaus [48, 49], Hazainé Borsiczky and Solymár [50], Malts [51], Croker et al. [52] and others. Verghese claimed the following reaction at high temperature [53]

$$3TiO_2 + 2NaOH \rightleftharpoons Na_2Ti_3O_7 + H_2O \tag{4.17}$$

The work by Verghese showed that the sodium titanate almost completely hydrolizes during the washing of the bauxite residue.

Lime is widely used in the HT digestion process step, either as quicklime (CaO), or in the hydrated form ($Ca(OH)_2$) for various purposes such as reduction of soda losses, to facilitate the transformation of goethite into hematite, or to facilitate the dissolution of diaspore, etc. Karst bauxites contain some calcite and/or dolomite. Therefore the behaviour of CaO, $Ca(OH)_2$ and other calcium and magnesium compounds is an important aspect of the Bayer process. Smith [54] recently published a comprehensive review of the state of the art, the equations and reaction paths of lime with alumina, silica, titania, iron oxides and other species, with a special emphasis on HT digestion. There are several papers which deal with various aspects of the chemistry of calcium

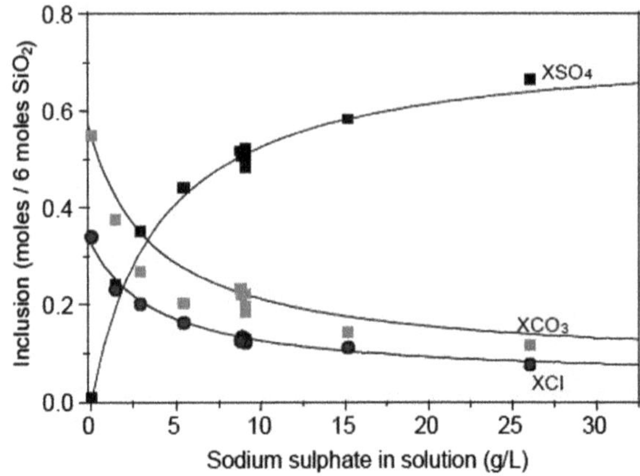

Fig. 4.12 Effect of SO_4^{-2} on inclusion of CO_3^{-2} and Cl^- into DSP formed [56]. Copyright © 1999 by AQW Inc. Used with permission

species in the Bayer liquors; see also, papers by Malts and Rosenberg et al. [51, 55]. Section 4.5 provides a comprehensive summary on the Lime Chemistry.

Composition of various DSPs, their structure

A peculiar aspect of the desilication reaction is that DSP comprizes a sink of certain liquor inorganic contaminant anions, especially for sulfate, carbonate and chloride ions [56–58]. Figure 4.12 demonstrates that sulfate anions show the highest affinity to be included into the sodalite cage at the expense of CO_3^{2-} and Cl^- ions.

In the alumina refineries where gibbsitic bauxites with high reactive silica content are processed, the process liquor is almost free from contaminants such as sulfate, carbonate and chloride. In these alumina refineries, the caustic soda per total soda level is 96–98%, high liquor productivity at precipitation (nearly 100 kg Al_2O_3/m^3 pregnant liquor) can be attained without any removal of liquor contaminants except for their spontaneous inclusion into the sodalite lattice. This effect may offset to a greater or lesser extent, the higher caustic soda costs due to the high reactive silica content in bauxite. Accurate techno-economic assessment, probably using mathematical models are required to find the optimum.

A list of DSPs commonly found under different conditions is shown in Table 4.2 as Ken Evans has collected [59].

Figure 4.13 depicts the cubic structure of Bayer sodalite with a monovalent ion in each cage or a divalent ion in every other cage to give full occupancy. The aluminosilicate cage structure has a net three negative charge which is balanced by three sodium ions. At full occupancy, Bayer sodalite has a monovalent anion (and associated sodium) in each cage, or a divalent anion in every other cage. Soda can be recovered from this structure only if the sodalite cage structure is broken down, or if the included ions can be removed through the openings of the sodalite cage structure [59, 60].

Table 4.2 List of DSPs formed at different conditions in the Bayer Process

Name	Formula	Structure
Zeolite X (ZEO-X)	$32Na_2O{\cdot}32Al_2O_3{\cdot}64SiO_2{\cdot}256H_2O$	Cubic
Zeolite A (ZEO-A)	$Na_2O{\cdot}Al_2O_3{\cdot}1.85SiO_2{\cdot}5.1H_2O$	Cubic
Zeolite (ZEO)	$1.08Na_2O{\cdot}Al_2O_3{\cdot}1.68SiO_2{\cdot}1.8H_2O$	Cubic
Sodalite (SOD)	$3Na_2O{\cdot}3Al_2O_3{\cdot}6SiO_2{\cdot}4H_2O$	Cubic
Hydroxy-sodalite (OH-SOD)	$3Na_2O{\cdot}3Al_2O_3{\cdot}6SiO_2{\cdot}2NaOH{\cdot}2H_2O$	Cubic
Aluminate-sodalite (A-SOD)	$3Na_2O{\cdot}3Al_2O_3{\cdot}6SiO_2{\cdot}NaAl(OH)_4{\cdot}4H_2O$	Cubic
Cancrinite (CAN)	$3Na_2O{\cdot}3Al_2O_3{\cdot}6SiO_2{\cdot}24H_2O$	Hexagonal
Ca containing cancrinite	$3(Na_2O{\cdot}Al_2O_3{\cdot}1.8Si0_2){\cdot}1.42CaC0_3{\cdot}2.4H_2O$	Hexagonal
Tri-calcium aluminate (TCA)	$3CaO{\cdot}Al_2O_3{\cdot}6H_2O$	Cubic
Hydrogarnet (HG)	$3CaO{\cdot}Al_2O_3{\cdot}mSiO_2{\cdot}nH_2O$	Cubic
Andradite hydrogarnet	$3CaO{\cdot}Fe_2O_3{\cdot}mSiO_2{\cdot}nH_2O$	Cubic
Andradite-grossular hydrogarnet	$Ca_3[Al,Fe]_2(SiO_4)_n(OH)_{(12-4n)}$	Cubic

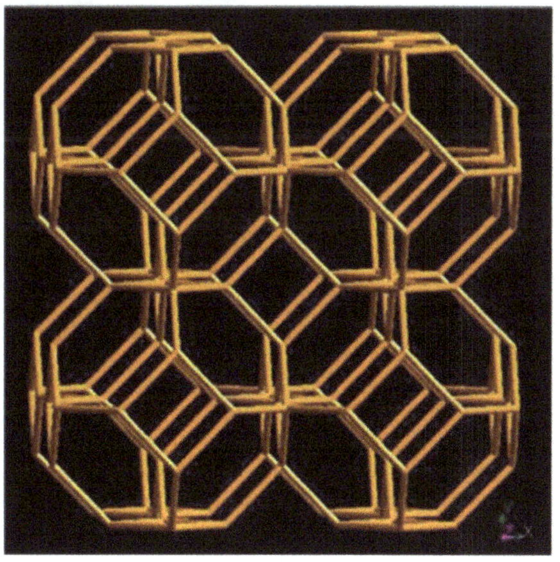

Fig. 4.13 Bayer sodalite–cubic structure

At high digestion temperatures cancrinite, another cage-structured aluminosilicate can form from sodalite. Cancrinite has the same cage stoichiometry as sodalite, but has a different structure, see Fig. 4.14. Cancrinite has two possible locations for included ions, either in the cage structures (which are smaller than those in sodalite) or the larger linear channels that form part of the hexagonal structure. The size and location of the channels mean that the preferred inclusion for cancrinite is calcium carbonate ($Na_6[AlSiO_4]_6{\cdot}2CaCO_3$). Thus cancrinite contains 25% less sodium than sodalite and the benefit of favouring a non-useful anion.

Other compounds that can be considered desilication products and important in processing high silica bauxites, are the hydrogarnet series of minerals; these are hydroxy substituted grossulars of the general formula $3Al_2(SiO_4)n(OH)_{12-4n}$.

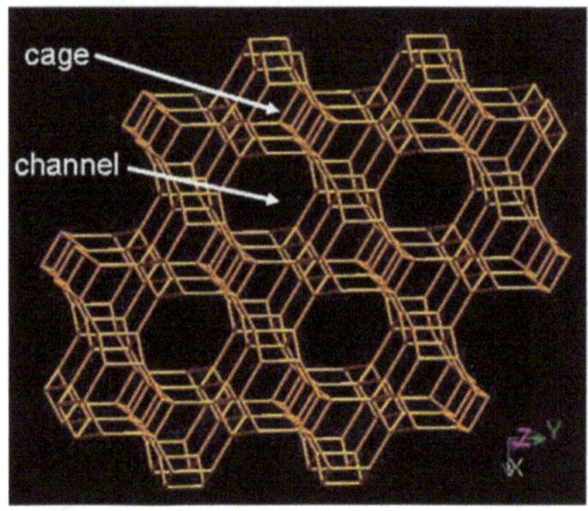

Fig. 4.14 Structure of cancrinite (hexagonal)

Natural minerals exist for integer n values, 0 (hydrogrossular or tricalcium aluminate, TCA), 1 (katoite), 2 (hibbschite) and 3 (grossular). Isomorphous substitution of Si_4^+ for $4H^+$ gives rise to a range of solid solution compounds under Bayer conditions, ranging from $0.6 > n > 0$.

4.1.3 Available Alumina, Reactive Silica, and Breakpoint

When bauxites are characterized, usually the most important apects that account for its value in the Bayer process are the "available alumina" and "reactive silica" contents. When considering these values, it is important to take into account the reaction temperature. The available alumina is traditionally determined by "bomb digestions". The major alumina producers have developed their own standardized determination methods. The "available alumina" determined at atmospheric digestion (at 108 °C) or, at about 143 °C represent well the gibbsite content of bauxite. The standard low temperature bomb digestion of Alcoa is carried out in 100 ml liquor having NaOH concentration of 102 g/L NaOH at 143 °C for 30 min [61]. The „available alumina" content measured at 225 or 240 °C is close to the gibbsite + boehmite content. When reactive silica is discussed/measured, the temperature of the determination such as for the available alumina must be indicated. This issue and its complexity of the available alumina (eg. aluminium substitution in alumo-goethite) is extensively discusssed in the paper of Authier-Martin et al. [62].

Bárdossy defines the most important terms (ABEA, MEA, R.SiO_2, T.SiO_2) in his monograph on the lateritic bauxite [1]:

Al_2O_3, SiO_2	the total values of these oxides, if not otherwise indicated.
T.Al_2O_3, T.SiO_2	the total values of these oxides in places where without this additional T the text could be ambigous.
ABEA	the amount of alumina extractable when using the common, low temperature (ca. 145 °C) process conditions (ABEA—American Bayer Extractable Alumina, this is equivavent to the THA (tri-hydrated alumina), or gibbsite content).
MEA	the amount of alumina extractable when using maximum temperature (up to 250 °C) process conditions (Maximum Extractabe Alumina).
R.SiO_2	reactive silica, i.e. silica present as silicate, generally in the mineral kaolinite.
F.SiO_2	free or non-reactive silica (at low temperature digestion) generally present as crystalline silica (quartz), it is frequently calculated as T.SiO_2–R.SiO_2.

Breakpoint

Laboratory tests are carried out in order to measure the actual amount of the available alumina and the A/C ratio (or caust. molar ratio of Na_2O_{caust}/Al_2O_3) in the digestion effluent, where the digestion extraction yield sharply falls with gradually increasing bauxite dosages. The results are called "breakpoint measurement" (Fig. 4.15) or "characteristic digestion curve" by various authors.

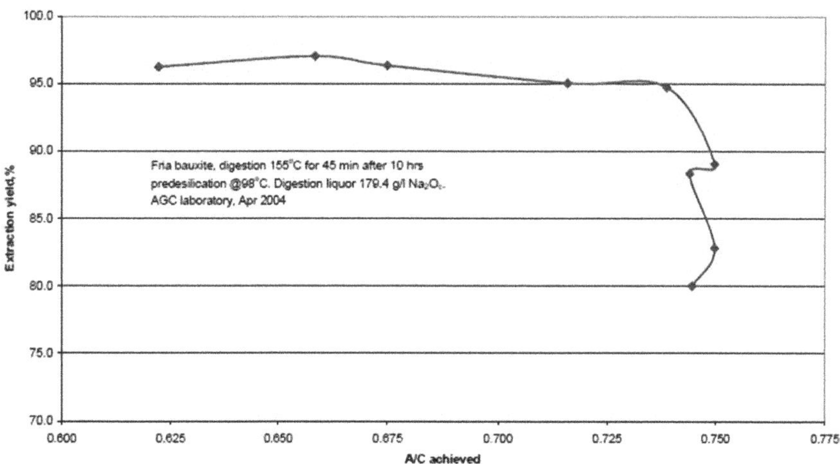

Fig. 4.15 Laboratory test results to find the breakpoint for a gibbsitic bauxite (Fria, Guinea)

4.1.4 Reactor Design

There are many excellent books which review the design of chemical reactors. Thomas and Pei quote [63] fundamental book of Levenspiel [64] “Chemical Reaction Engineering (CRE) uses information, knowledge and experience from a variety of areas—thermodynamics, chemical kinetics, fluid mechanics, heat transfer, mass transfer and economics [...] with the aim of properly designing a chemical reactor”. The key elements of CRE are

- Interpretation of laboratory or pilot test data (most often “batch” data). This includes understanding of impact of process parameters (temperature, pressure, concentration, agitation rate, catalyst concentration, etc.) on reaction rate and selectivity.
- Using this understanding to select appropriate reactor system (plug flow, Continuously Strirred Tank Reactor (CSTR), recycle reactor, etc.) and conditions.
- Optimisation of the reactor design. This includes trade-offs between reactor costs, process performance, impacts on upstream and downstream equipment costs.”

As the book of Denbigh and Turner [65] depicts, the mathematical models of the chemical reactors as based on the mass balance for any component in a small volume in the reactor as follows:

$$\underset{1}{\left\{\begin{matrix}\textit{number of moles}\\ \textit{entering the volume}\end{matrix}\right\}} = \underset{2}{\left\{\begin{matrix}\textit{number of moles}\\ \textit{leaving the volume}\end{matrix}\right\}} + \underset{3}{\left\{\begin{matrix}\textit{number of moles}\\ \textit{reacted}\end{matrix}\right\}} + \underset{4}{\left\{\begin{matrix}\textit{change of moles}\\ \textit{in the volume}\end{matrix}\right\}}$$

Any of these four members can be negligible or even zero. In batch reactor the members 1 and 2 are zero. In a unit where physical operation separation, or, mixing of streams take place the third member is zero. In the steady state the value of the fourth member is zero.

Similar balances can be written to the heat (energy) and impulse [66].

A general trend can be observed in the chemical process industries, which also tends to apply to the Bayer process as well. Initially, batch unit operations are implemented, then as the industry becomes more mature (process technology, equipment design and manufacturing, automatization, health and safety, environmental protection), the unit operations become continuous. The digestion and precipitation unit operations were implemented as batch reactors at first then as CSTRs. The next stage, at least in digestion, is the use of plug flow (tube) reactors.

Subsequent to the idea of the tube reactor mentioned in a patent of Müller and Hiller [67] early bench scale trials of Lányi [68] were carried out in 1953 in Hungary. These trials reached digestion temperatures of 270–365 °C. At the Nabwerk plant in Schwandorf [69] tube reactor technology was implemented at a 20 kt/year commercial scale in 1966. Plant scale tube digesters were implemented in West Germany in 1973 in Stade [70]. A pilot tube digester was commissioned in Hungary in 1973 and a plant scale unit in MOTIM [71] in 1982. The developments in (West)Germany and in Hungary, later in Czechoslovakia made it possible to achieve digestion temperatures of up to about 270 °C, resulting in short retention times as well as advanced techniques for the prevention and removal of scalings.

Low temperature digestion systems were also implemented with tube digester in Japan [72] and later in China and Vietnam.

4.1.5 Degradation of Organics

All bauxite contains some organic material with the organic content of bauxite normally ranging from 0.03 to 0.3%, and in some cases up to about 0.5% C_{org} [73, 74]. In general, the lateritic bauxites contain a higher level of organics than the karst type boehmitic or diasporic bauxites.

The surface deposits covered by lush vegetation,—particularly with deep roots—have the highest organic content. E.g. the Darling Range bauxites (Australia) contain 7.5–8.2 g/kg fulvic acids and about 1 g/kg humic acids. The fulvic and humic acid content in Boké bauxites (Guinea) are 3–4 g/kg and 0.5 g/kg, respectively [75]. The vegetation in the topsoil is certainly associated with carbohydrates and lignin. In general, the deeper the bauxite layers, the lower their organic content. A common practice to decrease the organic input of the alumina refineries is to strip the topsoil at least one wet season before mining of the bauxite ore. This allows for natural leaching of the organic material by the ample precipitation of the wet season(s).

There are several other methods to reduce the organic content of bauxites, e.g., roasting bauxite at about 300 °C, but these methods are generally prohibitive for cost reasons.

Bauxite is the principal source of organics input of the alumina plants. Formerly, the use of starch (flour) as a flocculant was an additional source of organic carbon, but this is now an outdated practice and the input from modern synthetic additives is negligible. Depending on the digestion conditions, up to 90% of the organic carbon in bauxites may be extracted into the Bayer liquor where it gradually degrades and re-arranges to a whole spectrum of organic compounds. Figure 4.16 shows the breakdown of the carbon content of a Gove bauxite to oxalate, non-oxalate organic carbon (NOOC) and carbonate in the course of its low temperature digestion as reported by Clegg and Armstrong [76].

Verghese reported that Alcan had found that most of the degradation of the humic matter of bauxite occurs during the first pass of (low temperature) digestion [53]. The oxalate formed in one hour is usually more than 90% of that formed in 24 h. The number of the identified organic compounds was between fifty and one hundred, their concentrations ranging from a few mg/kg to several g/kg [53]. Lever (Alcan) claimed that during the digestion process the organic material in bauxite breaks down to sodium carbonate, sodium oxalate, sodium formate, sodium acetate, sodium succinate and a host of complex heterocyclic, carboxylic and phenolic acids, which are the building blocks of the original long-chained humic acids [77].

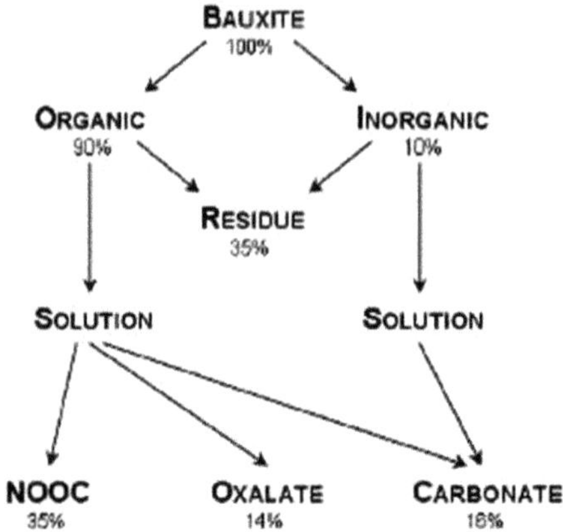

Fig. 4.16 Breakdown of Gove bauxite carbon during low temperature digestion [76]. Copyright © 2005 by AQW Inc. Used with permission

A dissolution reactions study of carbohydrate and lignin of Wilson et al. at low temperature Bayer digestion conditions shows that low molecular weight compounds such as C_2–C_6 mono, di- and tri-aliphatic carboxylic acids, which have hydroxy substituents are formed from humic material. Some 56 organic species were identified in this study as organics degradation products [78].

Another study by Armstrong and Healy compared the degradation kinetics of liquor organics under anaerobic conditions through digestion at 145 and 220 °C. Formation of sodium oxalate was found to be about 0.8 g/L at 145 °C over the 24-h digestion treatment. No obvious change in sodium carbonate was detected within the instrumental precision (±0.3 g/L). In the course of 13–42 h digestion treatment at 220 °C the oxalate generation averaged 3.0 g/L, while 42 h of treatment caused an increase of 9.5 g/L sodium carbonate. Half of the formation of oxalate and carbonate occurred within the first 2 h of treatment [79].

The organic compounds in the Bayer process liquor have a lower average molecular weight in those refineries where HT digestion (240–255 °C) is applied than in those where LT digestion (140–150 °C) is practised.

Except for oxalate, the solubility of organics in the Bayer liquor is relatively high. They affect the physical properties of the liquor (density, viscosity, boiling point rise (BPR)) in a way that is similar to the highly soluble inorganic impurities. The sodium salts of different organic acids decrease the rate of alumina precipitation from the liquor since they tie up the hydroxide ion and make it unavailable for extraction purposes.

The organic compounds in Bayer liquor have been classified into three main groups by Lever [77]:

- Low molecular weight acid salts such as formate, acetate, oxalate and succinate and lesser amount of salts of hydroxy acids, such as lactic, glycolic and glyceric acid

- Sodium salts of intermediate molecular weight acids, including hydroxy aliphatic acids, benzene carboxylic acids, and phenolic acids with molecular weights of less than 500
- Humate compounds of high molecular weights, over 500.

Yamada et al. reported, in the precipitation circuits the actual concentration of sodium oxalate may exceed the true equilibrium solubility by a factor of 2–2.5 due to the stabilizing effect of other organics, mainly humates [80]. When oxalate precipitates, some of the stabilizing organics co-precipitate with it. Reyhani et al. observed that the nature of the stabilizing organics determines the size of the oxalate needles. The greater the level of highly coloured (dark) humic compounds present, the finer the oxalate needles tend to be. When fine oxalate crystals precipitate they may act as nuclei for the precipitation of fine hydrate [81].

It has been the experience e.g., of Alcan refineries, that whenever the oxalate problem occurs, the product hydrate becomes more fragile and more prone to breakdown during calcination. The occluded soda content in alumina also tends to increase [53].

It is worthwhile to look at the total organic content and the breakdown of the organic material of some refineries [82], see Table 4.3. Whilst some of the refineries have now closed, the data shows the variation of the organics of different refineries that can occur.

Inorganic sodium carbonate is the ultimate product of the organic's degradation. Desilication Product (DSP) is a sink for carbonate; the soda causticisation and the liquid phase of the bauxite residue (especially in the case where bauxite residue is disposed of in ponds) are also principal routes whereby carbonate may be removed from the process. The more effective the dewatering of bauxite residue is, the higher the risk of having difficulties with maintaining the carbonate and/or oxalate balance in the Bayer liquor circuit.

Table 4.3 Total, oxalate and humate organics of alumina refineries [82]. Copyright © 1996 by The Minerals, Metals & Materials Society. Used with permission

Alumina plant	C_{total}	$C_{oxalate}$	$C_{humates}$
	g/l	g/l	g/l
Kwinana (Australia)	37.8	0.91	1.13
Pinjarra (Australia)	27.6	1.13	0.20
PortoScuso (Italy)	22.5	0.72	0.35
Gramercy (United States)	15.2	0.26	0.41
Renukoot (India)	11.6	0.80	0.73
Almásfüzitő (Hungary)	6.1	0.20	0.08
Zaporoshye (Ukraine)	5.2	0.36	0.15
Ajka (Hungary)	4.1	0.49	0.05

Data refer to 256 g/l caustic as Na_2CO_3

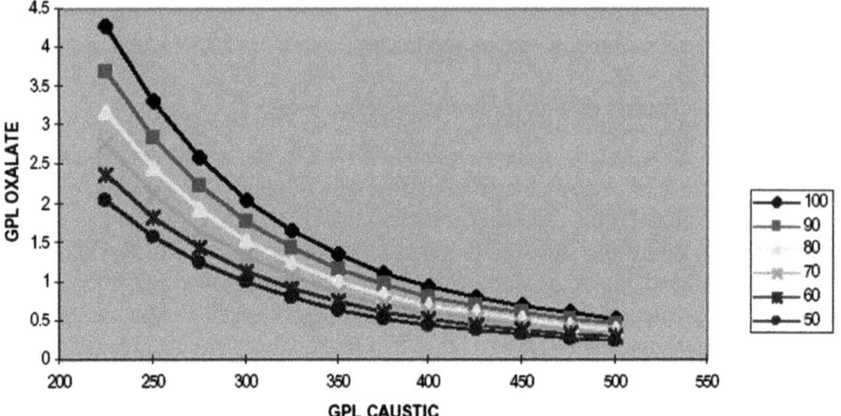

Fig. 4.17 Oxalate solubility versus temperature and caustic concentration [83]. Copyright © 2002 by AQW Inc. Used with permission

The high molecular weight humic compounds are primarily responsible for the dark brown colour of the Bayer process liquor, and consequently an off-white colour of the alumina hydrate.

The oxalate equilibrium solubility has been shown to be a function of caustic concentration and temperature, as in Fig. 4.17 [83]. The equilibrium solubility of sodium oxalate decreases with the liquor caustic concentration (common ion effect) and increases with temperature. A thermodynamically based model for the oxalate solubility was developed by Beckham and Grocott [84].

When sodium oxalate crystallizes, the stabilizing organics also precipitate, thereby reducing the apparent solubility of oxalate, leading to even further oxalate precipitation. This phenomena is referred to as an "oxalate shower", and reduces the actual oxalate concentration towards the equilibrium solubility. Fine oxalate needles form mats which carry fine hydrate into and over the tertiary thickeners of the hydrate classification, thereby causing a severe disturbance to hydrate agglomeration and a recycle of precipitated hydrate to digestion.

4.2 Digestion Conditions, Bauxite Mineralogy and Yield

4.2.1 Bauxite Mineralogy

Bauxite is the primary mineral for alumina and subsequently aluminium production. It is named after the town of Les Baux in southern France, where the chemist Berthier first recognized it in 1821. World reserves of readily extractable bauxite total approximately 29 billion tonnes, with Guinea, Australia, Brazil, Vietnam and Jamaica accounting for over 70%.

Table 4.4 Mineralogical composition of some Bauxites

Minerals	Weipa	Greece	Jamaica	Guinea Boké	Brazil Trombetas
Gibbsite ϒ-$Al(OH)_3$	59.6	0	67.3	78	83
Boehmite ϒ-$AlOOH$	13.3	35	2.0	8	0
Diaspore α-$AlOOH$	0.2	25	0.4	0	0
Kaolinite $Al_2O_3{\cdot}2SiO_2{\cdot}2H_2O$	8.8	8	15	3	3
Quartz SiO_2	1.2	0	0.1	0.3	0.5
Hematite Fe_2O_3	9.2	20	10.2	1.2	10
Goethite $Fe_2O_3{\cdot}H_2O$	3.3	5	10.4	6	2
Anatase TiO_2	2.5	2	2.5	3.5	1.5

The main alumina containing minerals that occur in bauxites and their associated digestion temperatures are:

1. **Gibbsite**
   - Chemical Formula: $Al(OH)_3$, or in hydrated form $Al_2O_3{\cdot}3H_2O$
   - Digestion temperature, typically 150 °C
2. **Boehmite**
   - Chemical Formula: $AlOOH$, or in hydrated form $Al_2O_3{\cdot}H_2O$
   - Requires higher digestion temperatures, typically 240–270 °C
3. **Diaspore**
   - Chemical Formula: $AlOOH$, or in hydrated form $Al_2O_3{\cdot}H_2O$
   - Has the same composition as boehmite but a higher density crystal structure. It is mainly found in deposits throughout China.
   - Requires the highest digestion temperatures to process, typically >260 °C (Table 4.4).

The Bayer process is the dominant hydrometallurgical process used for alumina production. This utilizes a recirculating sodium hydroxide (caustic soda) stream to dissolve the alumina bearing minerals in the bauxite. The bauxite mineralogy dictates the refining conditions that must be used and ultimately the economics of processing a particular bauxite. The basic equilibrium chemistry for dissolution of the alumina bearing minerals in the sodium hydroxide solution is per below:

$$[\text{Mineral}]_{(s)} + (OH)^-{}_{(aq)} \rightleftharpoons Al(OH)^-_{4(aq)} \quad (4.18)$$

Gibbsitic bauxites

Bauxites of primarily gibbsitic mineralogy are digested at 145–150 °C. The Gibbsite dissolution kinetics are described by

$$d(A)/dt = f\{(A)_\infty - (A), SSA_{gib}, e^{-k/T}\} \quad (4.19)$$

where

SSA_{gib}	the specific surface area of the gibbsite particles.
k	equal to E/R with E being activation energy and R the ideal gas constant.
T	absolute Temperature, K.
(A)	alumina concentration in solution (mol/m^3).
$(A)_\infty$	alumina equilibrium concentration (mol/m^3).

For Bayer digestion of Gibbsitic bauxites, typical ratios of alumina and caustic (A/C ratio) throughout the process cycle are depicted by Fig. 4.18.

The reaction kinetics are typically very fast with the extraction occurring within minutes. Hence reactor sizing for Gibbsitic bauxites is not determined by the kinetics of the bauxite extraction but by liquor desilication kinetics. The time for liquor desilication is typically 10–30 min. In this process, bauxite is added to achieve a target solution A/C ratio with a set margin to the solubility A/C ratio.

Boehmitic bauxites

Bauxites of primarily boehmitic mineralogy are digested at typically 250–280 °C. The Boehmite dissolution kinetics are described by:

$$d(A)/dt = f\{(A)_\infty - (A), SSA_b, e^{-k/T}\} \quad (4.20)$$

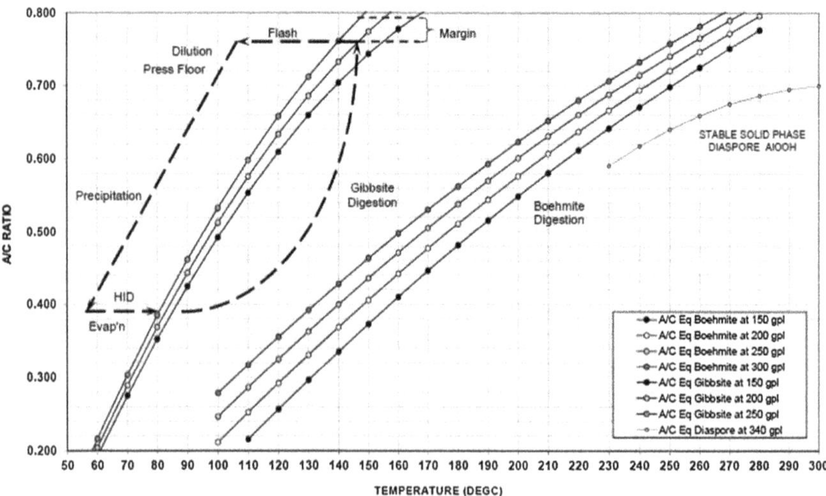

Fig. 4.18 Gibbsite bauxite digestion conditions (indicated by dotted line). Solubility data from Kotte [4]

where

SSA_b	the specific surface area of the boehmite particles
k	equal to E/R with E being activation energy and R the ideal gas constant
T	absolute Temperature, K
(A)	alumina concentration in solution
$(A)_\infty$	alumina equilibrium concentration.

The reaction kinetics are typically much slower than for Gibbsite at equivalent caustic soda concentrations, however at elevated temperatures of up to 280 °C, the kinetics for dissolution may also approach 4–5 min. Hence reactor sizing for Boehmitic bauxites is determined by the kinetics of the bauxite extraction. Deposits with boehmitic bauxite grades of 3–5% or more are typically processed using an intermediate or high temperature digestion process at ≥250 °C (refer Fig. 4.29) or a double digestion process at ≥210 °C.

Although dependent on the individual bauxite mineralogy and caustic concentrations, typical laboratory test batch reaction times of 10–30 min are required at elevated temperatures for economic extractions.

As evident from Fig. 4.19, Boehmite has approximately half the solubility of Gibbsite at the same temperature in caustic soda solution. For the intermediate-high temperature digestion flowsheet at 250 °C, the initial bauxite charge ratio is limited by the equilibrium solubility of the boehmite. To enhance yield of the circuit, a sweetening (gibbsitic) bauxite is injected into the flash vessel train to increase the alumina to caustic ratio to that limited by the gibbsite equilibrium solubility.

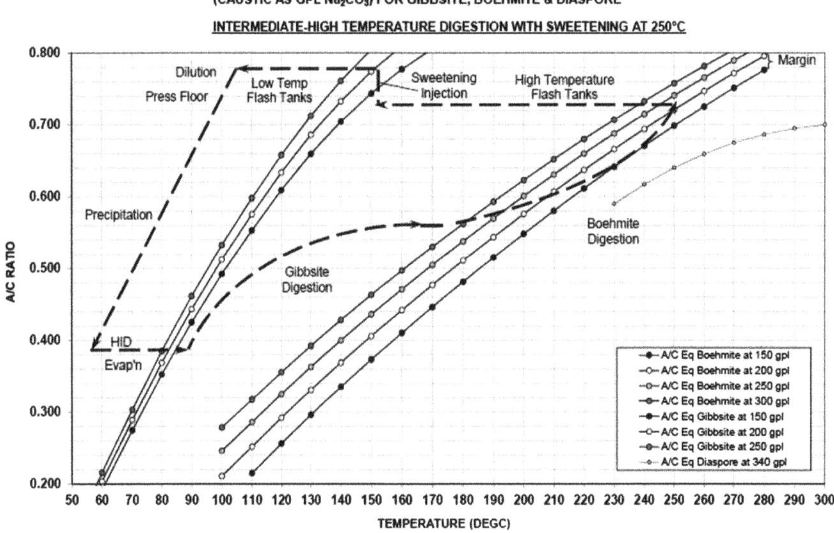

Fig. 4.19 Boehmitic bauxite high temperature digestion conditions with sweetening (indicated by dotted line). Solubility data from Kotte [4]

Double digestion (refer Sect. 4.3.4) minimizes quartz attack and therefore caustic consumption by extraction of the Boehmitic alumina at reduced temperatures.

4.2.2 *Effect of Lime Addition*

Lime or 'Quick lime' as Calcium Oxide (CaO) is generally prepared as a 'milk of lime' reagent of Calcium Hydroxide, by reacting calcium oxide with water. The reaction is exothermic.

$$CaO_{(s)} + H_2O_{(aq)} \rightarrow Ca(OH)_2 \quad (4.21)$$

Lime is added to the Bayer process for a number of reasons including to assist with digestion extraction efficiency, filter aid preparation in security filtration, causticisation and oxalate control. These are further discussed below.

Predesilication

Lime may be added prior to the digestion stage to form Hydrogarnet ($Ca_3Al_2(SiO_4)_n(OH)_{(12\text{-}4n)}$). Hydrogarnet is a series of compounds containing CaO, SiO_2 and the oxides of other elements such as Al_2O_3, MgO, Fe_2O_3 and TiO_2.

Hydrogarnets have a low equilibrium solubility in caustic liquor and can therefore reduce the silica content of the precipitated gibbsite [85]. Dependent on the pre-existing levels of calcium present in the bauxite, the relative merits of lime addition versus a traditional predesilication stage (to convert kaolin to DSP (refer Sect. 4.2.3) using temperature and residence time) need to be determined via chemical analysis [86].

Extraction Efficiency

The addition of lime to digestion can improve extraction efficiencies for certain types of bauxite, especially if diasporic bauxites are digested. As an example, Chinese alumina refineries add at least between 2% and up to 5% wt CaO to the bauxite feed to achieve satisfactory extraction efficiencies.

Enhancing Transformation of Aluminogoethite to Hematite

Aluminogoethite transformation (FeAlOOH) to Hematite (Fe_2O_3) has been found to be beneficial in the presence of lime for low temperature Gibbsitic Jamaican bauxite and high temperature digestion of Weipa bauxite [87, 88].

Impeding Sodium Titanate Formation

Reaction between sodium hydroxide and titanium compounds in bauxite can lead to the formation of sodium titanates resulting in formation of a film [51] over the aluminium compounds. This in turn lowers the aluminium extraction efficiency. The addition of $Ca(OH)_2$ to high titanium bauxites at digestion conditions may favour substitution of sodium titanate with insoluble calcium titanate.

Impurities Control

Lime is added to control impurities present in Bayer liquors including carbonate, and phosphorous. Phosphorus is removed by the formation of calcium carbonate phosphate (or apatite), $Ca_5(PO_4,CO_3)_3{\cdot}(F,OH,Cl)$ and, as indicated by the formula, CO_3^{2-} can substitute PO_4^{3-}, and OH^- and Cl^- can substitute F^- in the apatite structure. Exact substitutions can only be determined by mineralogical analysis. Stoichiometrically, each ton of P_2O_5 consumes about 1.3 tonnes of CaO based on the formula for apatite. In practice however, more than stoichiometric amounts are required to completely convert phosphorous in liquor to apatite. The excess amounts of lime are due to substitution of PO_4^{3-} by CO_3^{2-} in the structure and largely due to the formation of Tri Calcium Aluminate in the Bayer liquor.

Filter Aid Preparation

Filter aid known as Tri-calcium aluminate hexahydrate ($Ca_3Al_2(OH)_{12}$ or in oxide form, $3CaOAl_2O_3{\cdot}6H_2O$) enhances the filtration rate of the liquor purification process through security filtration by minimising the filter cake resistance. Blinding of the filter cake results in a reduction to the liquor flow and signifies that a cleaning cycle of the filter is required. Filter aid can be recycled during causticisation or desilication. It may be prepared via:

- CaO and spent liquor
- $Ca(OH)_2$ slurry and thickener overflow liquor or
- CaO and pregnant liquor.

$Ca(OH)_2$ has also been used as a filter aid using a higher lime addition than was required to achieve a similar filtration efficiency as that of the filter aid formed from CaO and spent liquor.

Filter aid is reported to lower the concentration of calcium in the liquor and reduce the concentration of soluble iron in liquors, while the calcium precoat used in the filter beds can remove phosphorous.

Oxalate Destruction

Sodium oxalate ($Na_2C_2O_4$) is an organic decomposition product in the Bayer liquor. If not controlled, oxalate accumulation interferes with the alumina crystallization process, impedes yield and generates fines.

To control the oxalate concentration in a co-precipitation circuit, fine hydrate slurry is filtered and washed with cold water to precipitate solid phase oxalate. The filter cake is then washed with hot water at above 55 °C which dissolves the oxalate. This low caustic soluble oxalate rich liquor stream is reacted with slaked lime to form insoluble calcium oxalate and sodium hydroxide. This is an old technology and rarely practised currently.

Causticisation

Lime is used for the causticisation of liquors by removal of CO_3^{2-} as $CaCO_3$. High lime efficiencies are obtained with increasing temperature, decreasing caustic concentration and longer reaction times. The basic reaction chemistry is:

$$Na_2CO_3 + Ca(OH)_2 \rightarrow CaCO_3 + 2NaOH \quad (4.22)$$

In Bayer liquors, the removal of carbonate is more accurately defined by an equilibrium causticisation reaction [89]:

$$Ca(OH)_{2.}2Al_2O_3 + 3Na_2CO_3 = 3CaCO_{3(s)} + 2NaAl(OH)_4 + 4NaOH \quad (4.23)$$

Prior to the formation of calcium carbonate, $3CaOAl_2O_3.CaCO_3.11H_2O$, an intermediate monocarbonate is formed.

Conversion of the monocarbonate is accelerated at temperatures up to 95 °C. Formation of the monocarbonate results in the removal of carbonate species and transforms to calcium carbonate or tricalcium aluminate, dependent on the total caustic and total alkalinity of the liquor. An increase in causticity will favour the formation of tricalcium aluminate over calcium carbonate.

4.2.3 Impact of Impurities

Impurities typically found in bauxites include:

1. Silica (Kaolin and Quartz)
2. Phosphorous
3. Iron oxide and iron hydroxy-oxide (Hematite and Goethite)
4. Titanium dioxide in the form of anatase or rutile
5. Organic Carbon
6. Sulphur.

The impact of these on alumina production is discussed below.

Silica

Silica minerals in bauxite are mainly kaolin and quartz. Kaolin is highly reactive, while quartz requires more severe conditions to dissolve. Under standard digestion conditions all kaolin dissolves and re-precipitates as Desilication Product (DSP).

At higher temperatures (>180 °C) quartz attack becomes significant and increases the total amount of silica dissolved during digestion. Reactive silica (Rs) can be defined by:

Rs = Kaolin × Kaolin Dissolution + Quartz × Quartz Attack

Rs = [Total Silica − Quartz] × Kaolin Dissolution + Quartz × Quartz Attack

Kaolin: $Al_2Si_2O_5(OH)_4$ or $Al_2O_3 \cdot 2SiO_2 \cdot 2H_2O$

Quartz: SiO_2

Caustic Soda consumption is directly proportional to the amount of reactive silica, as all silica dissolved initially will re-precipitate as Desilication Product (DSP) which has a clearly defined soda to silica ratio. The rule of thumb is that, on a weight basis, any mass of silica that reacts in the refining process will cause an almost equal loss of caustic soda, and an equal loss of alumina to the red mud in the De-silication Product (DSP). Thus silica is a key determinant of any bauxite's economic value.

The Kaolin dissolution reaction is:

$$Al_2O_3 \cdot 2SiO_2 \cdot 2H_2O \xrightarrow{6NaOH} 2\{Na^+Al(OH)_4^-\} + 2\{(Na^+)_2(SiO_3^{2-})\} + H_2O \tag{4.24}$$

Quartz attack:

$$SiO_2 \xrightarrow{2NaOH} (Na^+)_2(SiO_3^{2-}) + H_2O \tag{4.25}$$

DSP Precipitation:

$$\{Na_2O \cdot Al_2O_3 \cdot 2SiO_2\} \cdot 1/3Na_2O(CO_3^{2-}, SO_4^{2-}, Cl^-, Al(OH)_4^- \ldots) \cdot 2H_2O \tag{4.26}$$

The DSP stoichiometry is 1.333 Na_2O: $2SiO_2$ = 0.667 [$^{mol}/_{mol}$], or 1.18 g Na_2CO_3/g SiO_2. Hence for every ton of reactive silica dissolving in the process, 1.18 tons of sodium carbonate (or 0.89 tons of NaOH) are lost.

Silica related caustic soda consumption typically represents between 80 and 90% of total plant caustic soda consumption, depending on bauxite grade and soluble soda losses. One advantage of a high reactive silica level in bauxite when combined with a high digestion temperature will be that the precipitation of high amounts of DSP assist to purify the liquor and strip it of most impurities such as carbonates, sulphates and chlorides, as shown in the DSP formation reaction.

Phosphorous

Most bauxites contain between 0.05 and 0.20% phosphate, of which most is soluble, except for Jamaican bauxites which contain between 0.8 and 1.0% of caustic soluble phosphate.

The adverse effects of excessive phosphate in the liquor on the process are:

- scaling problems as a result of double salt formation with other anions
- blinding of filter cloth in security filtration using lime generated filter aids
- settling problems for goethite rich muds due to reduction of flocculants activity
- contamination of product alumina and subsequent reduction of current efficiency in smelting process.

Iron Oxides (Hematite and Goethite)

Iron minerals normally are found in bauxite as:

- Hematite (Fe_2O_3)
- Goethite ($FeO(OH)$ or $Fe_2O_3{\cdot}H_2O$).

The solubility of iron oxides increases with increase in temperature, and free caustic. Free caustic is the caustic concentration not bound with dissolved alumina. The nature of iron contained in bauxites also has significant influence on its solubility. The settling and filtering properties of red mud are largely influenced by the concentration, fineness, degree of dispersion and the nature of the iron minerals in the bauxite.

Hematite has no significant adverse impact on mud settling and largely passes through the refining process unchanged.

For bauxites with high goethite content, incomplete conversion of goethite to hematite will cause significant problems in mud settling and washing. Due to aluminium ion substitution, it can be a minor contributor of alumina; accessible in high temperature refining.

Iron contamination of product alumina has been a problem for many high Temperature alumina refineries. There are two main sources of iron in product alumina, (1) particulate iron, and (2) precipitated iron from solution.

Particulate iron can be minimised by good operating performance of the settlers and security filtration, minimising mud passing through to the precipitation circuit.

Any increase in iron in product alumina due to particulate iron is usually matched by increases in other impurities like, calcium, silica and titanium.

While particulate iron can be minimised by good plant operation, there is however a "base level" of iron in product related to the precipitation of iron from the liquor phase, which is more dependent on dissolved iron, related to bauxite grades and certain operating conditions, rather than security filtration performance.

The two main factors impacting on iron in solution for high temperature alumina refineries are:

- Free Caustic in Digestion and
- Iron Content of Bauxite.

Seeding Effect

High iron content of gibbsitic bauxite, used for sweetening, has been found to reduce dissolved iron at some refineries. It is thought that high iron bauxites generally provide more seed for iron re-precipitation from solution. It has been suggested to recycle red mud to increase iron oxide seed surface area and hence lower iron in solution and improve product quality.

Iron substituted in Gibbsite

Soluble iron can originate from iron substituted in gibbsite. Many low iron bauxites are more likely to produce high iron in liquor because iron substituted in gibbsite is released during digestion.

Titanium Dioxide

Chemical Formula: TiO_2

Associated with "rock hard" scales inside refining equipment. These reduce refinery efficiency and increase maintenance costs.

Organic Carbon

Bauxites typically contain 0.1–0.4% organic carbon that emanate from entrained and decomposed vegetative materials. When processed in caustic soda, these organics degrade to form carboxylic acids. These can interfere with the alumina production process in the following ways:

1. reduces alumina yield
2. generation of excessively fine aluminium hydroxide particles
3. reduced red mud settling rates
4. increase in product impurity in the calcined alumina
5. increased caustic soda consumption due to formation of sodium organic salts
6. co-precipitation of sodium oxalate in the product hydrate
7. increased scale formation and cleaning costs.

Various treatment methods are used in refineries to reduce organic carbon in liquor streams including:

- Wet oxidation. Wet oxidation has been practiced at Aluminium Oxid Stade GmbH since 1980 [90, 91]. This involves injecting oxygen at ~270 °C in the tube reactors to decompose the organic material. The main organic compounds (approx. 50%) of the organic carbon in AOS liquors are sodium acetate, sodium succinate, sodium formate and sodium oxalate.
- Oxalate removal. Oxalate removal is typically via precipitation of calcium oxalate using slaked lime (Sect. 4.5.10) or oxalate calcining [92]. In oxalate calcining, oxalate crystals are processed in a kiln to produce sodium carbonate and oxide for regeneration of caustic.
- Liquor calcining. Liquor calcining [92] takes a side-stream of liquor which is first dried to a paste then calcined in a kiln.
- Continuous biological oxalate destruction. From ~1999 to 2010, a continuous biological oxalate destruction technology was developed, trialed and implemented by Alcoa using specific alkali tolerant bacteria that could efficiently break down the oxalate. The first full scale plant utilizing this technology was installed and commissioned at Alcoa's Kwinana refinery in Western Australia. The process is non thermal (maximum temperature of 40 °C) with very low emission levels. Although the biological breakdown of the organics generates carbon dioxide, this is quickly mineralized to inorganic carbonates and bicarbonates within the liquor

stream passing through the process. Alcoa has since rolled out this technology to its Pinjarra refinery and at the Point Comfort Refinery in Texas.

Sulphur

Causes iron contamination and poor mechanical properties in the product alumina, elevated equipment corrosion, reduced liquor caustic to soda ratio. Costly and difficult to remove; originates from sulphur minerals such as pyrite or organic materials in the bauxite.

4.3 Applied Digestion Technology and Heat Consumption

The modern Bayer cycle for alumina production is represented by Fig. 4.20. The four main processing facilities are generally referred to as:

1. Digestion—where extraction of the alumina from bauxite is undertaken
2. Clarification—where solids separation and caustic recovery occur
3. Precipitation—nucleation and growth of alumina hydrate
4. Evaporation—to remove water and attain target caustic concentrations.

Note that Calcination (removal of chemically bound water from the hydrate to produce Al_2O_3) is not considered part of the Bayer cycle.

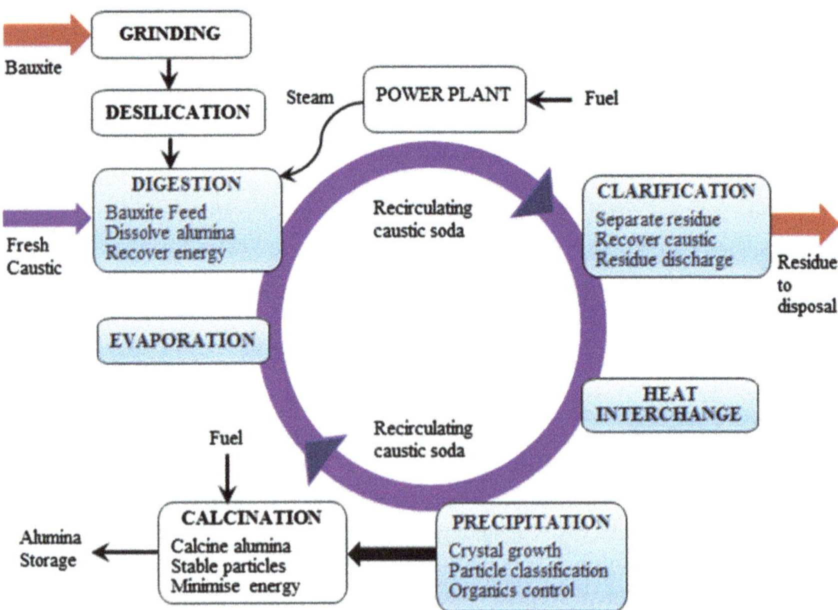

Fig. 4.20 The Bayer cycle for alumina production

Within the Bayer process, digestion may be undertaken under atmospheric conditions for gibbsitic bauxites but is typically undertaken at elevated temperature and pressure of 150 °C and 4–5 atm. For boehmitic and diasporic bauxites, digestion temperatures of 250–270 °C are used.

The digestion facility has the combined aims of:

(i) maximizing extraction of the alumina content of the bauxite
(ii) optimizing energy recovery
(iii) maintaining the clarifier feed at less than or equal to atmospheric boiling point
(iv) evaporating water from the digestion slurry to:
   - maintain the alumina to caustic ratio at appropriate molar concentrations
   - avoid auto-precipitation of the dissolved alumina in downstream pressure filtration while maximizing yield in precipitation,
(v) production of a clean vapour for condensate recovery for re-use for washing and/or boiler feedwater
(vi) minimising release of volatile organic odours to the environment [93–95].

4.3.1 Atmospheric Digestion

This digestion technology is in operation in a plant, Nalco in India; nameplate capacity 2.27 MTpa. The atmospheric digestion process uses low digestion temperatures at ~105°–108 °C. This ensures a low reactivity of silica. The extraction yield with atmospheric digestion is ~70–77% available alumina. In Fria, Guinea, there is another refinery with atmospheric digestion, with a nameplate capacity of 700 ktpa.

Atmospheric pressure digestion presents some advantages such as:

- higher plant operating factor,
- low reactivity of silica,
- less digestion maintenance and cleaning costs,
- no pressure safety specific problems,
- no wear problems during digestion and flashing etc.

However, it has the following disadvantages:

- during atmospheric digestion, gibbsite extraction is lower in efficiency causing higher bauxite consumption,
- the undigested gibbsite can lead to elevated alumina losses post digestion in the residue,
- due to atmospheric digestion, the supersaturation of the digested liquor cannot be increased to a desired level causing lower liquor productivity,
- raw material and energy consumption is generally higher resulting in higher costs of production.

4.3.2 Autoclave Digestion

The Autoclave

Other than the two refineries listed above, all other refineries utilize 'pressure leaching' to extract alumina from the bauxite. The use of pressure leaching to extract alumina from caustic liquors dates back to 1888, where German Patent No. 43977, entitled A Process for the Production of Aluminum Hydroxide, was issued. The discovery which led to the patent was made by the Austrian chemist Karl Josef Bayer. The process he invented involved the pressure leaching of bauxite with NaOH solution to obtain sodium aluminate from which aluminum hydroxide was precipitated by seeding [96].

The pressure autoclave itself incorporates the coincident hydrometallurgical unit operations of:

- chemical reaction engineering and
- heat transfer.

Over the decades since Bayer's initial patent, pressure autoclaves have grown in scale with an approximate two hundred and 70-fold increase in reactor volumetric capacity and over a one thousand-fold increase in mass throughput.

Figure 4.21 illustrates an autoclave in use at the time of Bayer, with size of ~1 m diameter and 2.5 m length. The evolutionary growth in autoclave capacity since the 1890s is illustrated in Fig. 4.22.

Throughout the rapid expansionary decades of the 1960s–1980s, two primary types of autoclave design were implemented (Fig. 4.23). These were autoclaves employing indirect steam heat transfer (eg Pechiney designs) and those with direct steam injection (eg Kaiser designs). The agitated indirect steam heated autoclaves incorporate internal steam distribution bundles and condensate collection piping and have the advantage of not diluting the process stream. This in turn eliminates

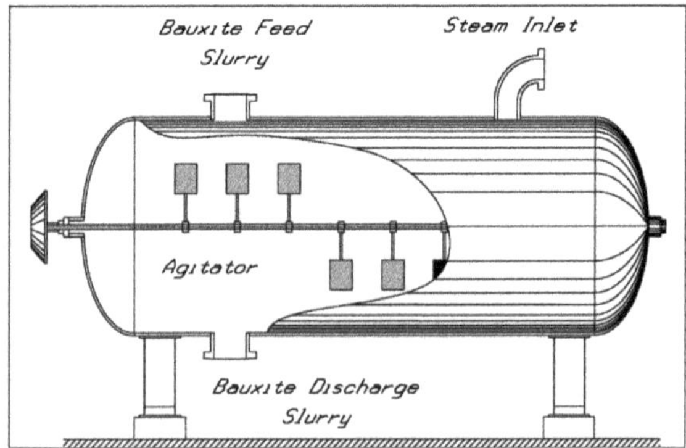

Fig. 4.21 Bayer autoclave, pressure 300–400 kPa, 160–170 °C. Bauxite rate 4–5 tonnes per day [96]

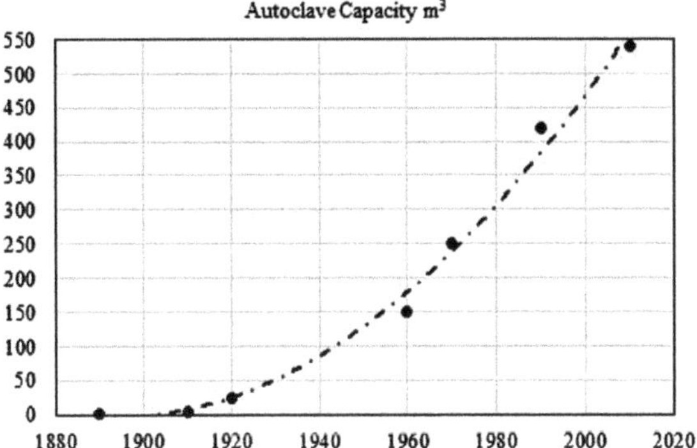

Fig. 4.22 Growth in autoclave capacity

additional liquor circuit evaporation requirements, maximizes the liquor productivity and reduces the peak caustic soda concentration required upstream of the autoclave.

The limitation for the indirect heat transfer autoclave becomes the practical size of the vessel and internal steam bundles such that cleaning and descaling operations remain manageable. The indirect heated autoclave of Fig. 4.23 is ~3 m in diameter by 10 m in height.

For the direct contact heater autoclave, steam is sparged directly into the bauxite slurry stream to achieve the requisite temperature for alumina extraction. These autoclave sizes have increased in capacity to ~5 m in diameter × 30 m in height (~600m^3). For this autoclave design, the limiting solubility ratio for extraction of alumina is dictated by the resultant liquor concentration after steam injection.

Following alumina extraction in the autoclave and the evaporative effect of the pressure let down circuit to ambient boiling point, the caustic concentration must be maintained in the optimal range for crystallization of alumina in the supersaturated liquor. If the liquor is too highly supersaturated, there is the risk of auto-precipitation through the security filtration facility within Clarification. This auto-precipitation can rapidly 'blind' filter cloth leaves reducing filter capacity with subsequent interruptions to process flows throughout the refinery circuit.

Autoclave Sizing Principles

Direct Contact Heater Autoclave.

For the direct steam injection autoclave, the heat transfer coefficient (HTC) is typically an order of magnitude higher than its indirect equivalent. Steam sparge nozzle design is subject to mass flux and plume length considerations with the mass flux a key determinant in the HTC. The number of injection nozzles needs to be cognizant of both design flow and refinery turndown requirements to avoid unstable flow regimes. Dependent on injection nozzle design and mass flux, HTC's ranging from 8000 to 30,000 W/(m^2 K) can be expected. As a result, the heat transfer coefficient is not limiting and reaction kinetic criteria drive the autoclave sizing.

The direct contact heater autoclave of Fig. 4.23 incorporates an internal agitator, longitudinal wall baffles and circumferential baffles in what is essentially a series of

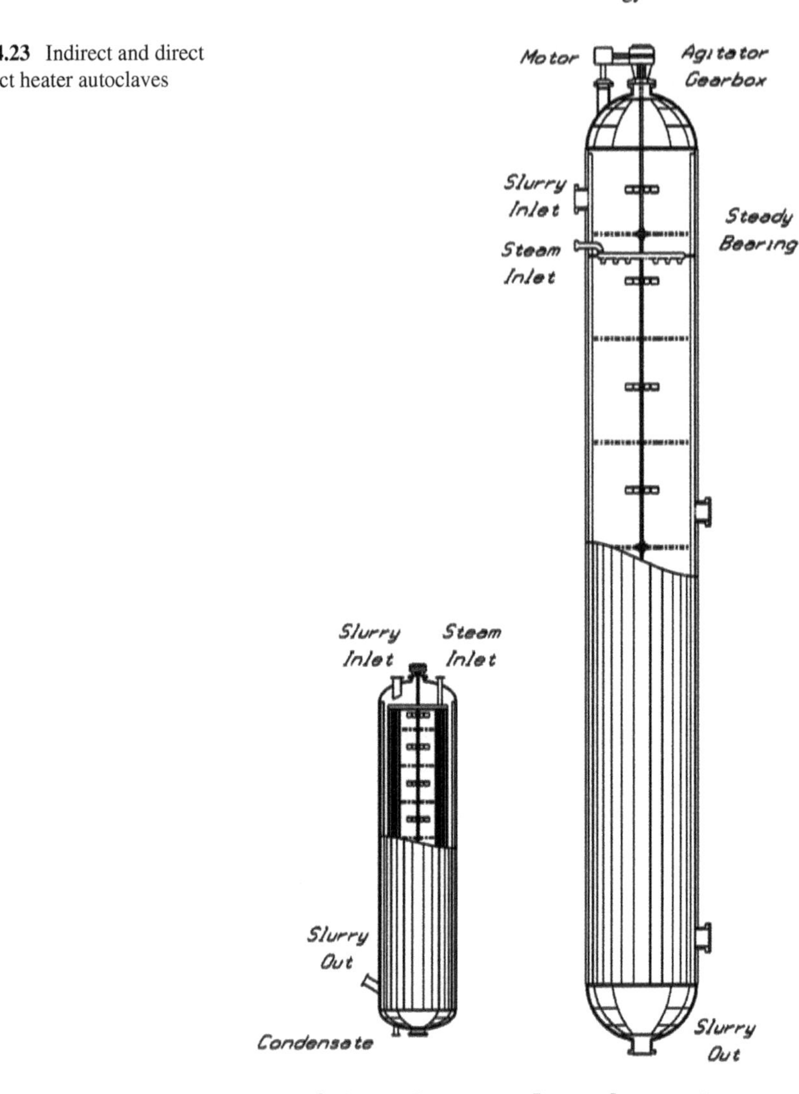

Fig. 4.23 Indirect and direct contact heater autoclaves

continuous stirred tank reactors (CSTR's) in series. As discussed in Sect. 4.2.1. For Gibbsitic bauxites, the kinetics of liquor desilication will drive the autoclave size. Alternatively, for Boehmitic bauxites, the kinetics of boehmite dissolution drive the autoclave size.

For a typical Boehmitic ore exhibiting 2nd order reaction kinetics, the target residence time from a 'bomb' batch autoclave test must then be converted to a continuous reactor flow equivalent. Figure 4.24 illustrates reaction kinetics approach to idealized plug flow for 'N' CSTR's in series for a 2nd order single reaction [64].

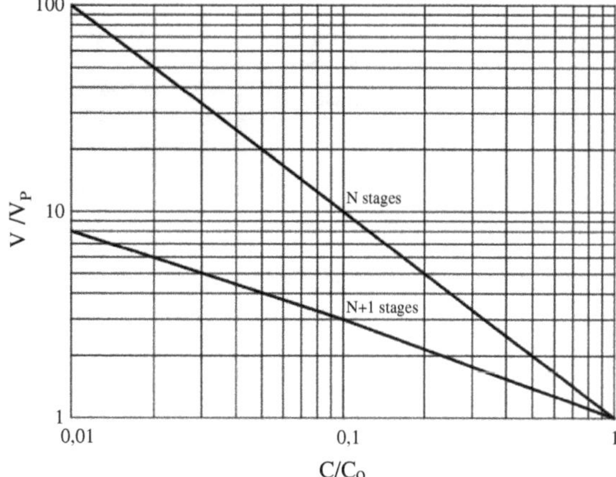

Fig. 4.24 Plug flow equivalence to 'N' CSTRs in series—2nd order [64]

In this figure the ordinate represents (1 − X) with X being the conversion (ie. 0.1 represents 90% conversion). The coordinate represents the ratio of the required CSTR reactor volume to that of a plug flow reactor achieving the same conversion. As the number of CSTRs increases this ratio approaches unity. For the autoclaves of Fig. 4.23, the aspect ratio and compartments are designed to provide 'N' stages of CSTR's in series such that the target alumina extraction is attained. For autoclaves designed for single order liquor desilication reactions and high temperature boehmite dissolution, aspect ratios ranging from 3.5:1 to 7.5:1 have typically been used.

Number of Recuperative Stages

The digestion facility is essentially a multiple effect slurry evaporator. Each effect or stage (comprising a 'flash' tank and associated condenser) will provide a finite degree of heat transfer. The more stages incorporated, the greater the amount of heat recovery. Consequently, the lower will be the energy consumption required to heat the fluid from the final stage of recuperative heat transfer to the target reaction temperature in the autoclave. The incorporation of multiple stages also allows a more controlled depressurization of the slurry stream throughout the flash tank train.

The condenser is a heat exchanger designed to transfer the latent heat of vapourisation of the flash steam to the liquor or slurry being preheated. The heat exchanger or condenser area required for a stage will be primarily a function of the energy to be transferred and the heat transfer coefficient determined for the particular duty. This is turn will be driven by the thermal design of the exchanger and fluid physical property data. An assessment of fouling conditions on heat transfer coefficient will be needed to determine changes in heat transfer from clean to scaling conditions and consequent cleaning cycles for the equipment. Sodalite scaling rates tend to increase with temperature. Under clean caustic liquor heat transfer conditions, heat transfer coefficients of up to ~2000–2400 W/m^2 K are typical. Under fouling conditions at the end of the heater cycle, heat transfer coefficients can reduce to ~700–1000 W/m^2 K.

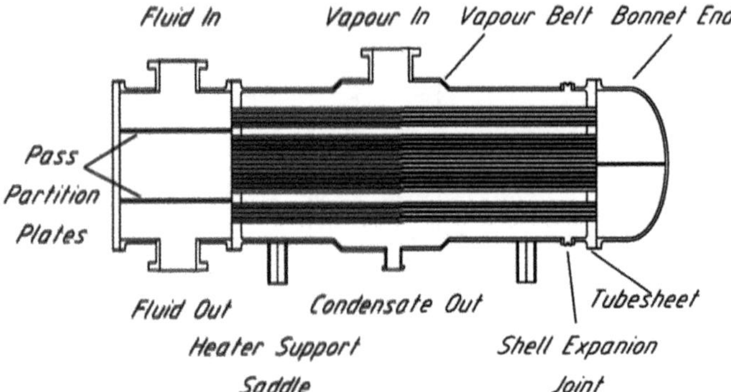

Fig. 4.25 Shell and tube heater/condenser

For the shell and tube type condenser (Fig. 4.25), there will be a difference between the actual area installed and the effective heat transfer area. The effective heat transfer area must consider the tube area 'eroded' by the tube-sheets, baffles and, for a vertically oriented condenser, the stagnation zone of condensate to the condensate outlet nozzle. For condensers operating with differential design temperatures between the shell and tubes above ~60–80 °C, shell expansion joints are often incorporated into the heater shell to limit mechanical stresses arising from differential thermal expansion.

The'flash' tank (Fig. 4.26) is a pressure vessel designed to allow depressurization of the incoming slurry from a pressure normally corresponding to its vapour pressure, to a lower pressure. In this depressurization process, steam is liberated from the incoming slurry as it boils (or flashes) with the remnant slurry now at a lower pressure and temperature. The flashed steam is then used to pre-heat the incoming liquor or slurry stream in heat exchangers (condensers), in counter-current flow to the flash tank slurry stream. The condenser heat transfer area and heat transfer coefficient determines the net evaporative load from the flash tank. In the absence of any control devices in the vapour system, this net evaporative load across the condenser then defines the operating pressure in the flash tank stage. The condenser and flash tank therefore operate in a thermodynamic equilibrium.

As part of the depressurization process, the flash tank must perform several coincident objectives:

- de-energize the slurry in a controlled manner so as to avoid excessive erosion of associated piping and control elements
- produce a relatively clean steam to avoid contamination of the condensate produced in the condenser
- maintain a hydraulic seal between stages such that vapour does not bypass the condenser and short-circuit to the underflow piping, reducing the thermal efficiency of the digestion facility.

In sizing the flash tank to avoid excessive vapour contamination, upward vapour rise rates are restricted to less than the estimated terminal settling velocity of entrained

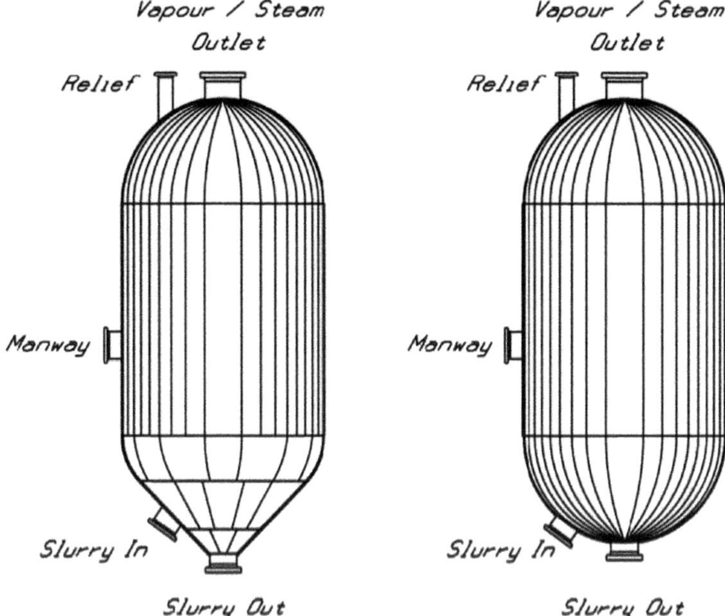

Fig. 4.26 Flash vessel–conical and hemispherical head designs

caustic soda of a maximum drop size. In practice, condensate quality factors developed through industrial scale experience are often employed in proprietary algorithms to optimally size this equipment. Further integrity in condensate quality may be provided with the incorporation of entrainment separators. However, as the vapour is superheated by the boiling point elevation of the fluid (Sect. 4.4.2), internal separators such as Chevron style corrugated plates, tend to suffer from scale accretion and blockage and are generally avoided in this duty. External centrifugal separators are favoured when incorporating additional vapour cleaning systems.

Several flash tank designs are in use throughout industry to accomplish the coincident objectives listed above. The most common are referred to as side entry and bottom entry.

The side entry design discharges the incoming boiling slurry downward into the slurry pool and incorporates a minimum of piping components within the flash vessel. Unless significant staging of the upstream connected flash tanks is employed, slurry cavitation will commence early in the external piping.

The bottom entry design discharges the incoming boiling slurry upward into a deflection plate and into the slurry pool. With an inventory of slurry retained in the vessel, the slurry pool assists the dis-entrainment duty by acting as a 'wet scrubber' eliminating a large fraction of the population of liquid droplets and particulate that would otherwise be carried upward with the disengaging vapour. For the same upward vapour rise rate and vessel size, the bottom entry configuration has been shown to

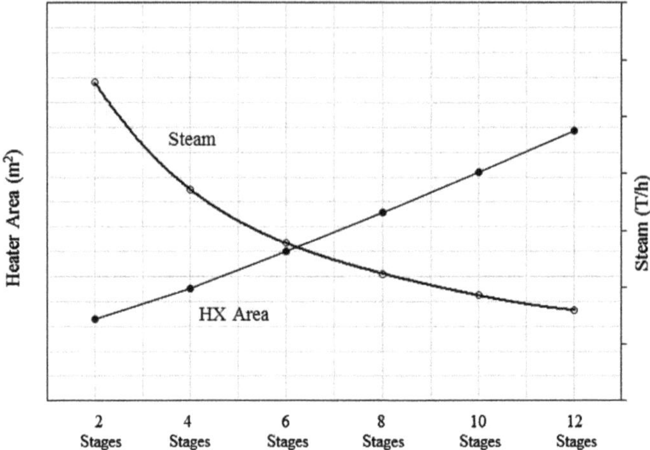

Fig. 4.27 Stage optimization

achieve an approximate five-fold reduction in vapour contamination when compared to the side entry configuration.

In modern Bayer plants, temperature pickups of 10–15 °C per flash tank and condenser stage are provided. The ideal number of heat transfer stages required is best determined by an economic analysis as the incremental benefit in lowering energy consumption generally reduces with each stage.

Figure 4.27 illustrates the relationship between increasing installed heat transfer area and reducing steam consumption (digestion energy) for a high temperature digestion facility processing a Boehmitic bauxite. The installed heat transfer area is the summation between the stage recuperative area from the vessel train and additional heat transfer area required to attain the reaction temperature in the autoclave. Each stage addition will require an assessment of the following installed costs:

- mechanical equipment (heat exchanger/s, flash tank and condensate receiver),
- pressure relief stations,
- interconnecting slurry, vapour and condensate piping and valving,
- control elements and associated instrumentation,
- instrument and electrical cabling and ladder,
- structural access platforms and stairways for access/egress,
- structural supporting steel and pipe-racks,
- concrete foundations,
- earthworks and underground drainage/piping etc.

The economic evaluation may take the form of a Net Present Value (NPV) assessment using the following as inputs:

- a selected discounted cash flow rate,
- the life of the facility,
- the installed cost of a stage of heat recuperation,

- digestion energy usage,
- energy cost.

For a low temperature digestion refinery processing a Gibbsitic bauxite, 3–4 stages of pressure let down and recuperative heat recovery are found economic. For a high temperature digestion facility processing a Boehmitic bauxite, 8–12 stages of pressure let down and recuperative heat recovery are typical.

Pressure leach digestion circuits

As outlined in Sect. 4.2.1, the bauxite mineralogy will define the type of hydrometallurgical pressure leach flowsheet used to extract the alumina, namely:

1. low temperature digestion for Gibbsitic bauxites
2. high temperature digestion for Boehmitic and Diasporic bauxites
3. double digestion or high temperature digestion for mixed grade Gibbsitic/low grade Boehmitic ores.

Each of the above pressure leach flowsheets may be further classified as:

1. dual stream digestion
2. single stream digestion
3. hybrid circuits using indirect or direct contact autoclaves.

Each processing route utilizes different equipment to achieve its process targets. These dual stream, single stream and hybrid flowsheets will have different impacts on salient features of the refinery, including:

- materials of construction
- the digestion facility capital cost
- the liquor circuit evaporative requirements and cost
- the boiler plant capacity
- equipment maintenance cycles and associated costs for on-site/off-site labour
- operational labour demand
- operating procedures for startup, shutdown, equipment isolation and cleaning
- digestion energy consumption
- refinery energy consumption
- refinery operating costs.

Dual Stream Digestion

In a dual stream process, the caustic liquor and bauxite slurry are fed to the digestion unit in separate streams as shown in Fig. 4.28. For a high temperature digestion flowsheet treating a boehmitic bauxite, the caustic liquor is typically heated in a series of heat exchangers to temperatures of ~210 °C to limit the deposition rate of sodalite (a silica based scale) in the heat transfer equipment. If the autoclave design incorporates direct contact heat transfer via steam injection, free caustics will be elevated throughout the heating cycle.

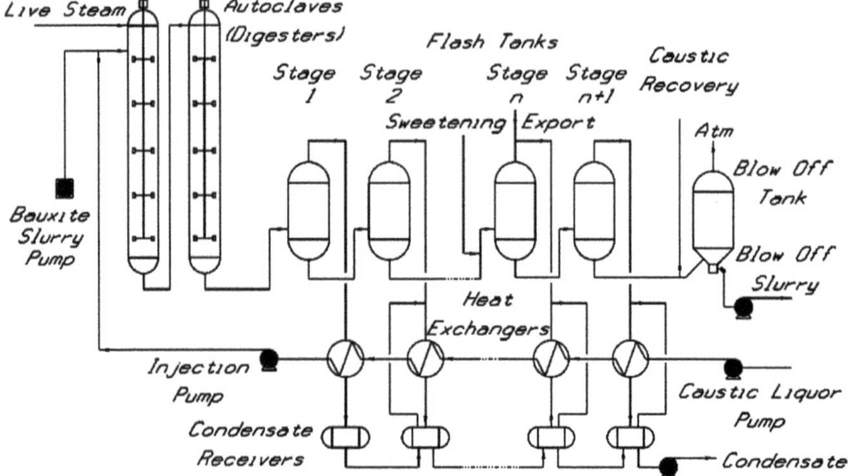

Fig. 4.28 Dual stream flowsheet for a Boehmitic Bauxite incorporating Sweetening Bauxite

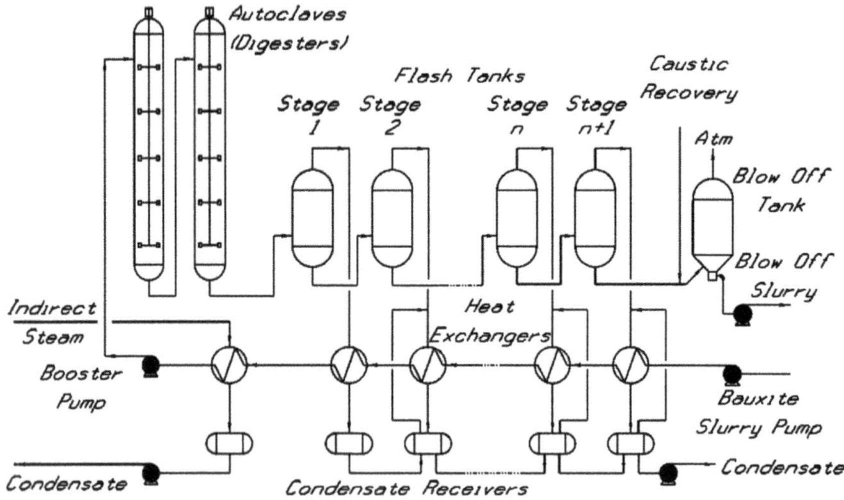

Fig. 4.29 Single stream digestion for a boehmitic bauxite

The bauxite slurry is generally preheated to temperatures of ~100 °C via either direct steam injection or indirect heaters to accelerate slurry predesilication. The predesilicated slurry and the heated caustic liquor are then fed to a series of digestion autoclaves (or digesters). For the flowsheet of Fig. 4.28, steam is injected into the digesters directly to achieve and maintain the required digestion temperature and to allow for the endothermic digestion reactions to occur. Following alumina extraction, the slurry is cooled to atmospheric conditions by depressurising it in a series of flash vessels. The steam generated by the depressurization process is used in the recuperative heaters to heat the incoming spent caustic liquor.

To improve the efficiency of boehmitic bauxite processing, sweetening bauxite (gibbsitic) is often added to the vessel train at temperature of 180–150 °C. This effectively utilizes the low temperature end of the high temperature digestion circuit as a parallel low temperature digestion train. The sweetening bauxite injection raises the molar concentration of alumina to a ratio that is normal for gibbsitic bauxite digestion (refer Fig. 4.19). The addition of a sweetening bauxite to a high temperature boehmitic digestion facility can contribute an additional ~10–15% in equivalent alumina yield (6–10 gpl) and refinery production.

For the dual stream circuit, there is always a thermal mismatch between the heat 'sink' available in the incoming liquor stream and the slurry being depressurized. This thermal mismatch results in additional energy needing to be 'exported' to other refinery heat sinks, to mitigate excessive atmospheric vapour emission and therefore energy loss.

Commonly used heat transfer equipment includes shell and tube heat exchangers (Fig. 4.25). These incorporate several hundred relatively small diameter special thin walled tubes in bundles held in line with horizontally or vertically aligned pressure plates. Several standard arrangements are available in accordance with the Tubular Exchanger Manufacturers Association (TEMA). If standard grades of carbon steel are used, shell and tube heaters may be used up to temperatures of 150 °C. Above this temperature, life cycle cost analyses for materials selection tend to favour the use of more expensive materials of construction for the tubeside wetted surfaces, such as nickel and duplex stainless steels.

Despite an approximate 20-fold increase in cost, nickel heater tubes are not uncommon and have survived for many years the arduous exposure to frequent acid cleaning cycles. However, maximum temperatures are still limited to 210–230 °C due to the elevated free caustic of the spent liquor and the increased risk of erosion-corrosion mechanisms.

A key advantage of the shell and tube heater is the large heating area per unit volume and hence high heat transfer rates per unit cost. The heaters may be oriented horizontally on elevated concrete floors or steel support frames or vertically in structural buildings. A secondary advantage is that the equipment is universally used throughout the process industry with many suppliers of "off-the-shelf" heater designs.

For reactor residence times in excess of 15 min, the autoclaves represent a practical option due to their large specific volumes. One of the disadvantages of the autoclave however, is the potential for short circuiting which can negatively impact on alumina extraction. Hence the use of high aspect ratios for this equipment item. High temperature digestion autoclaves operating at $\geq$250 °C incorporate aspect ratios in the range 5–7.5:1.

To achieve target residence times, multiple autoclaves in series may be installed. To allow for periodic cleaning and maintenance activities, the autoclaves and associated pressure equipment are typically interconnected by piping and valving such that any vessel (or pressure stage) may be isolated.

Single Stream Digestion

A typical schematic for a high temperature single stream digestion plant is shown in Fig. 4.29. In this flow-sheet, predesilicated bauxite slurry and evaporated spent caustic liquor are mixed prior to being fed to the recuperative heaters at temperatures below 100 °C. The slurry heaters typically experience less scaling than their dual stream counterparts due to the presence of desilication product seed in the bauxite slurry stream. Following the recuperative heaters, the slurry is heated to the digestion temperature via indirect heating using live steam heaters.

As the slurry is being heated, the endothermic dissolution of gibbsite takes place, providing an additional heat sink and hence reducing the requirement for heat exchange area in the lower temperature range between 100 and 150 °C. As with the dual stream process, following digestion, the slurry is cooled to atmospheric conditions by depressurising in a series of flash vessels. The steam generated is used in the recuperative heaters to heat the incoming slurry. One of the advantages of the single stream process is that it has the benefit of achieving a tighter balance between the heat source (energy generated by the flash vapour) and the heat 'sink' (energy required for slurry heating).

The heat transfer equipment used in single stream flowsheets has included agitated autoclave designs, the shell and tube heater and the jacketed pipe heater. For high temperature digestion facilities, disadvantages of the shell and tube heater relate to thermo-mechanical aspects. These are:

- the compounding effects of higher scaling rates at elevated temperatures above 200 °C, particularly titanate scale
- tube-end erosion
- tube wall erosion leading to condensate contamination
- blockage of shell-side expansion joints which can lead to excessive stress in the tube to tubesheet and shell to tubesheet joints.

As outlined above, there are several advantages associated with using a single stream digestion flow-sheet relative to the dual stream alternative. These are particularly relevant for the high temperature digestion circuits. A summary of these is provided below.

1. there is an optimal balance between the heat supply and heat sink removing the requirement for export of energy from the digestion facility whilst minimising energy consumption
2. it significantly reduces liquor circuit dilution in digestion and hence minimizes requirements for spent liquor evaporation and boiler plant load
3. for mixed grade boehmitic and gibbsitic ores, dissolution of the gibbsite in the low temperature recuperative heaters suppresses the free caustic allowing the use of standard grades of carbon steel
4. lower scaling rates through the low and mid temperature heat transfer equipment are generally observed reducing chemical and/or mechanical cleaning
5. it reduces digestion and refinery complexity and therefore installed capital cost.

4.3.3 Tube Digester

The high temperature tube digester is a particular application of single stream digestion technology that presents strategic advantages for processing mixed grade Gibbsitic-Boehmitic, Boehmitic or Diasporic bauxites. The bauxite charge ratio remains constrained by the same equilibrium solubility limits set by the Boehmitic or Diasporic bauxite, however the chemical reactions occur at elevated temperatures such that the alumina to caustic ratio is equivalent (or higher) to those of the intermediate temperature sweetening flowsheet of Fig. 4.19.

For the tube Digester, typical ratios of alumina and caustic throughout the process cycle are depicted by Fig. 4.30.

In addition to the single stream advantages outlined above, the supplemental benefits of this technology are:

1. it further maximises reaction temperature and therefore bauxite charge ratio; as a result, flow productivity is maximised (alumina production per unit plant flow) and bauxite consumption minimised.
2. at the elevated reaction temperatures, reaction kinetics are accelerated reducing required reaction residence time; this eliminates the need for the use of the autoclave and plug flow (pipe) reactors are used reducing mechanical complexity.

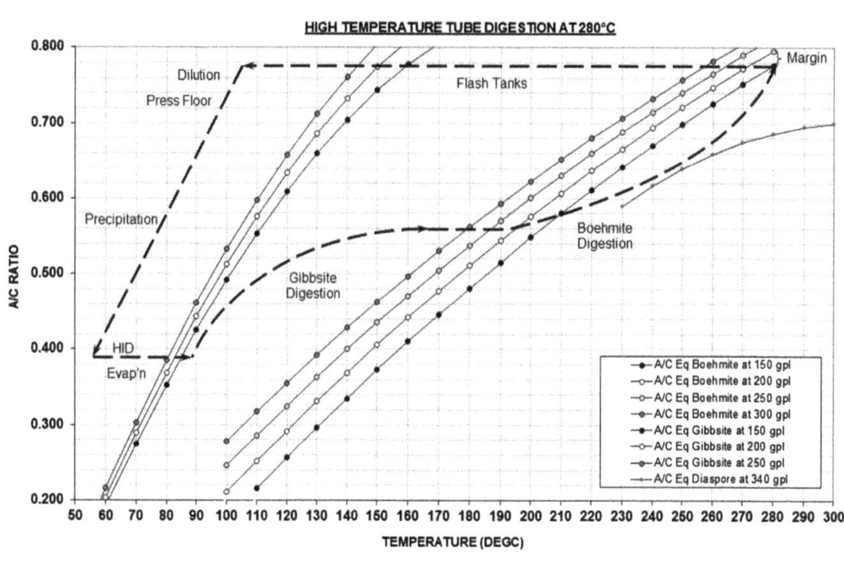

Fig. 4.30 Tube digester conditions

3. it incorporates jacketed pipe heaters that mitigates thermo-mechanical challenges associated with other types of condensers exposed to high temperature indirect heat transfer slurry duties.
4. it allows the safe use of wet oxidation technology to reduce the build-up of long chain organics in the liquor thereby improving liquor productivity, reducing liquor 'dead' volume and stabilising alumina particle size.

For the tube digester flowsheet (Fig. 4.31), the shell and tube or autoclave condensers are replaced by jacketed pipe heaters incorporating tubes of larger bore diameter than utilised in the shell and tube heater. Autoclaves are replaced by holding tubes providing plug flow reactor kinetics. Multiple centrifugal liquor pump-sets are replaced by single stage positive displacement slurry pump-sets which drive the slurry through the regenerative heaters to a temperature of ~230–235 °C. The final heating to the target reaction temperature is accomplished via molten salt or indirect steam heating using high pressure boiler plant steam at up to ~300 °C and 100 Bar pressure.

Molten salt heating utilises an increase in log mean temperature difference to maintain heat transfer under fouling conditions in the tube reactor (from sodium or calcium titanate precipitation). This has the advantage over indirect steam of avoiding the exponential rise in pressure associated with saturated steam temperatures.

Molten salt at up to 400 °C has been used in heaters [90] to maximise operational life of the units. For high-capacity tube digesters however, the utilisation of high-pressure steam has had the advantage of improved integration into the refinery steam and power grid avoiding the need for installing multiple molten salt heater trains of lower energy delivery capacity, based on commercially available salt systems.

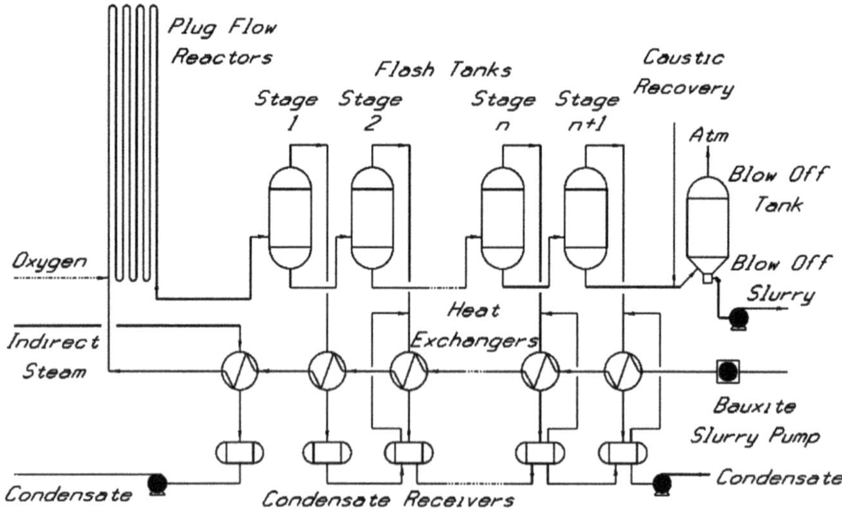

Fig. 4.31 Tube digester flowsheet

The jacketed pipe heater utilizes standard schedule piping rather than thin walled heater tubes. As such, it is less prone to slurry flow mal-distribution and therefore tube blockages, tube ruptures, high erosive localized wear, free caustic related corrosion or other issues associated with traditional shell and tube heat exchangers. They are typically 50–70 m in length and assembled from standard piping components making them simple and cost effective to fabricate.

Tube velocity is selected to avoid erosive slurry mechanisms and to provide an optimal compromise between heat transfer coefficient and hydraulic delivery pressures. Chemical cleaning frequency of the jacketed pipe heaters is reduced to every 600–2000 h. This is possible due to significantly reduced silica scaling of heater tubes in the presence of desilication product seed in the single stream slurry. As a result, a relatively slow and a steady degradation of heat transfer coefficient over time is observed.

High capacity tube digester units incorporate multiple tiers of jacketed pipe heater trains operating in parallel to maximize unit capacity. This has accommodated unit capacity growth from ~0.25 MTpa smelter grade alumina to ~1.0 MTpa per unit.

4.3.4 *Double Digestion*

For Double Digestion, ratios of alumina and caustic throughout the process cycle are depicted by Fig. 4.32. Double Digestion was developed by Alcan (~1993)

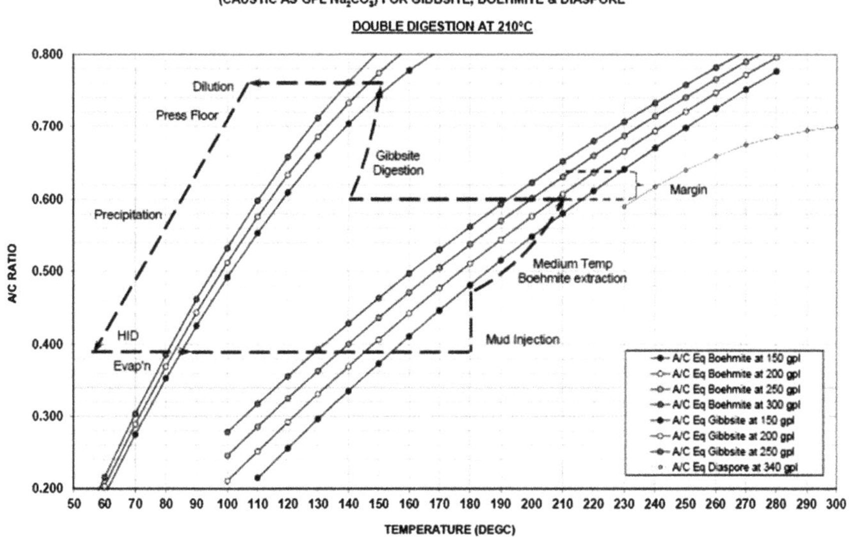

Fig. 4.32 Low grade Boehmitic bauxite double digestion conditions

and incorporates solid/liquid separation under pressure into the flowsheet. This is accomplished with the use of a Pressure Decanter [97–99].

The incorporation of pressure decanters into the double digest flowsheet also assists with energy recovery from the low temperature digestion circuit. Several flowsheet variations have been developed and are in use. These may incorporate the pressure decanter as part of both the low and medium temperature digestion circuits, after the low temperature digester or as part of the medium temperature digestion circuit.

The Pressure Decanter is essentially a high aspect ratio thickener operating under elevated temperature and pressure. This method of solid–liquid separation takes advantage of both the static pressure from the column of fluid to increase mud compaction densities and reduced liquor viscosities at elevated temperature to accelerate particle settling rates. Internal rake mechanisms are utilized and synthetic flocculants are generally added into both the feedline and feedwell to assist particle agglomeration and settling. The high aspect ratio and high settling rates achievable in the Pressure Decanter may assist with brownfield retrofits into existing operations where real estate constraints prevent implementation of other digestion facility cost reduction alternatives.

Some of the advantages of the double digestion process are:

1. it presents a reduced caustic soda loss for treatment of mixed grade Gibbsitic and Boehmitic bauxites as a result of the reduced temperature for Boehmite extraction. This lower reaction temperature (~210–220 °C) reduces quartz attack (Sect. 4.2.3) from silica in the bauxite and associated alumina losses in the desilication product
2. for certain red muds, e.g. Jamaican bauxites, it avoids solid–liquid separation issues following depressurization of the slurry through flash cooling
3. the elevated settling rates in the pressure decanter allows for a reduced residence time and therefore a decreased loss of alumina due to boehmite reversion.

An example flowsheet is illustrated in Fig. 4.33.

Fresh bauxite is contacted with liquor that has already been through a high temperature boehmite digestion. The gibbsite from the fresh bauxite dissolves in the low temperature digestion facility producing a liquor close to equilibrium solubility. The slurry is then separated under pressure (~6–7 Barg) and temperature using a pressure decanter. The clarified liquor overflow from the pressure decanter is flashed, releasing energy to pre-heat the incoming slurry stream, prior to a final stage of solids separation in the Gibbsite thickener. The underflow solids from the Pressure Decanter are next contacted with spent liquor and directed to the medium temperature Boehmite digester (at ~210–220 °C) to dissolve the boehmite. After this second digestion the slurry is again separated, with the solids sent to caustic recovery and residue and the liquor directed to the low temperature Gibbsitic digestion.

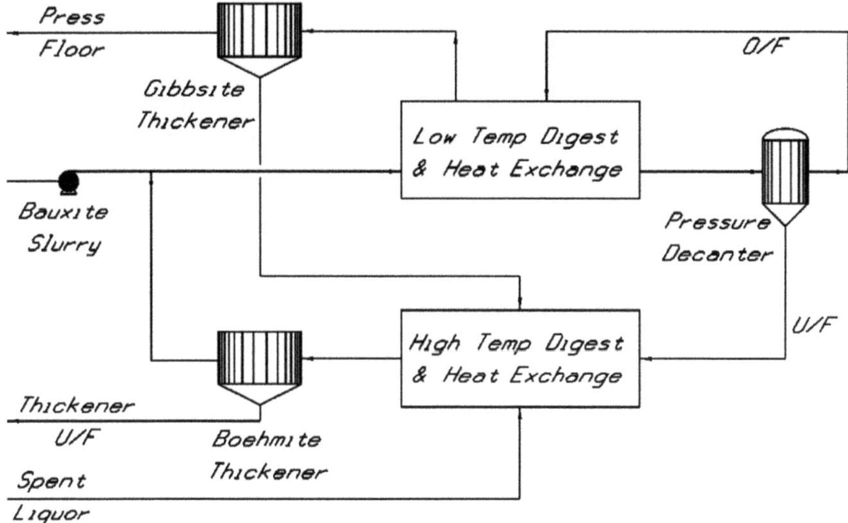

Fig. 4.33 Double digest flowsheet

4.3.5 *Energy Intensity*

Representing 30–40% of the operating costs (Table 4.5), energy usage is one of the main components determining the profitability of a refinery. Within the period 2000–2017, operating refineries in Australia, the United States, Jamaica and Europe have closed resulting from operating cost pressures.

Several of these installations have historically relied on heavy fuel oil to sustain operations with conversion to natural gas increasingly incorporated into modernization projects to reduce energy costs. Global trends in refinery energy (2013) are presented in Table 4.6.

Table 4.5 Bayer refinery operating cost breakdown

Operating cost	% total
Energy	30–40
Bauxite	20–25
Caustic	10–16
Labour	7–10
Maintenance	7–10
Lime	2.5–4
Water	2.5–3
Flocculants and reagents	2.5–3

Table 4.6 Energy costs by region (USD, 2013)

Refinery energy	USD (2013)/t
Australia	53–137
Brazil	68–161
Europe	74–244
Nth America	40–78
India	61–142
Russia	119–272
MENA (Middle East/Nth Africa)	188–213

Table 4.7 Refinery energy cost breakdown

Facility	Refinery energy % total
Digestion/evaporation	52
Calcination	36
Electricity	6
Other	6

The Digestion and Evaporation facilities typically constitute in the vicinity of 50–55% of a refinery's energy costs (Table 4.7). This energy consumption is equivalent to a steam consumption of ~1.35–2.6 t/t alumina. Hence, minimizing energy consumption in digestion has a significant effect on refinery economics.

As outlined in Fig. 4.28, irrespective of digestion flowsheet selection, increasing the number of recuperative stages will have a positive impact on reducing digestion energy, however this is dependent on its own economic assessment.

The dual stream flowsheets of Fig. 4.28 utilizing direct contact heat transfer in the autoclave, incur additional energy consumption based on the necessary inclusion of the steam enthalpy (as opposed to the latent heat of condensation), albeit some of this inefficiency is recovered by exporting recuperative vapour to other refinery users. In addition, the dilution effect of the steam injected must be removed by evaporation from the liquor circuit. The energy requirement of this additional evaporation may be reduced by using export from the digestion train and/or utilizing increased steam economies in the evaporator design, albeit with increased installed capital cost.

For the dual stream digestion facility, the thermal decline in heater transfer results from sodalite formation in heater tubes. The use of chemical additives to the liquor has been effective at inhibiting the silica scale formation and maintaining higher average heat transfer coefficients.

For the tube digester, the prominent source of heat transfer decay stems from both sodalite deposits through the low temperature heaters and sodium or calcium titanate formation in the high temperature heaters operating above ~200 °C.

As for the dual stream flowsheet, chemical cleaning using dilute sulphuric acid is effective at removing the silica scale. For removal of the titanate scale, mechanical cleaning is required. The use of mechanical cleaning methods such as 'pigging' of heater tubes assists to reduce downtime and thereby maximize heater circuit availability.

As a result of their improved thermal efficiency, tube digester flowsheets place these single stream refinery operations in the lowest quartile of refinery energy users (Fig. 4.34). Relative to a global average of ~14.5 GJ/t, high temperature refineries incorporating tube digestion operate at 7.5–9.5 GJ/t. By comparison with the historical dual stream flowsheet used for high pressure Bayer refinery designs of the 1960's (refinery energy of ~11.6 GJ/t), the tube digester flowsheet provides an approximate 15–20% reduction in energy use.

Other avenues to reduce refinery energy consumption include the use of Co-generation or Combined Heat and Power generators (CHP). Over the last 15yrs,

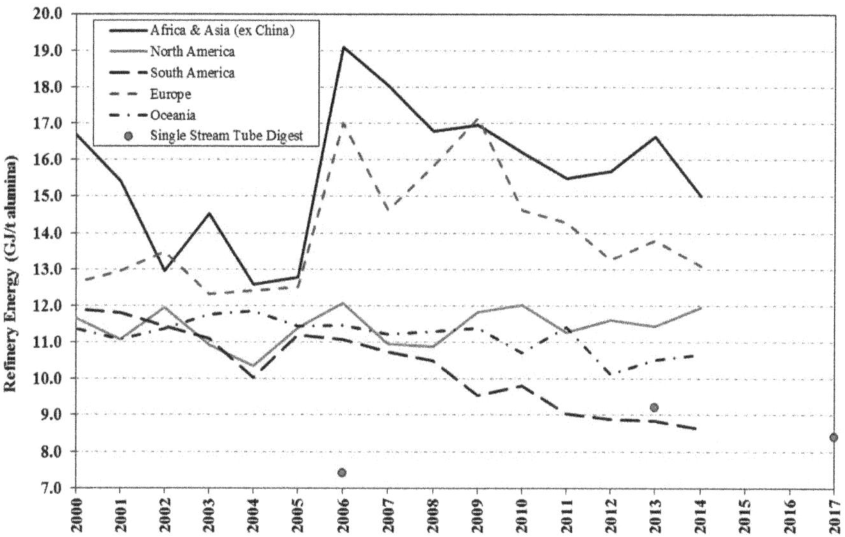

Fig. 4.34 World metallurgical refinery energy intensity

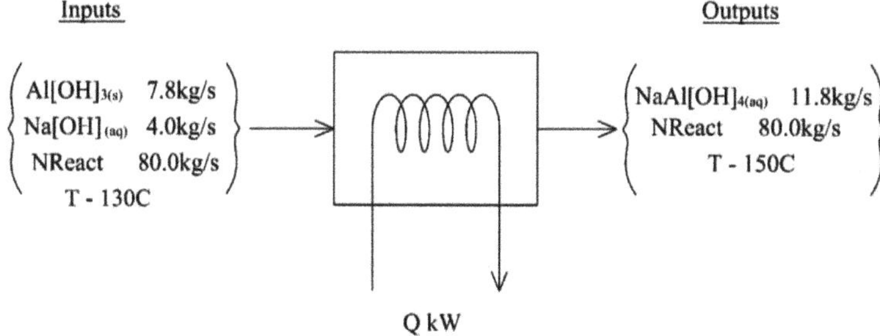

Fig. 4.35 Simple gibbsite reactor

CHP units have been incorporated into several refineries including operations in Australia, Brazil and Ireland.

The CHP may eliminate the need to import electrical energy from a regional electricity grid. A gas-fired cogeneration facility, comprising of an electrical generator, gas turbine and heat recovery steam generator combusts natural gas in the gas turbine. The gas turbine is linked to an electrical generator which produces electrical energy. Waste heat from the gas turbine exhaust is captured and used to produce steam for the refinery with surplus electricity being exported to the electricity grid. The electricity from a CHP plant using natural gas has about one-third the emissions intensity of coal-fired electricity.

Calcination is the other main energy user in the Bayer refinery. Modern calciners (including Circulating Fluid Bed, Gas Suspension and Fluid Flash) have largely replaced rotary kilns of the 1960's driven by improved energy efficiency and increased capacity.

4.4 Digester Heat and Mass Balances

The digestion area heat and mass balance is the critical pre-cursor to the facility mechanical design, its installed capital cost, energy usage and annual operating cost. As part of the Bayer flowsheet, it will impact the refinery evaporative load and water balance, residue treatment requirements (via the bauxite to alumina ratio in digestion) and capacity of the boiler/power plant needed. For any selected bauxite, an understanding of the bauxite characterization, reaction chemistry and fluid physical property data is needed, to compare alternate flow-sheet options or optimize a selected flowsheet. Key elements of the heat and mass balance development process are outlined below.

1. *Bauxite Characterization*
   - Chemical analysis (XRF/LOI/Impurities/Quartz), Mineralogy
   - Prediction of plant performance, raw material consumption / operating costs
   - Initial selection of digestion technology.
2. *Test work—Quantification of Key Process Design Parameters*
   - Digestion temperature, caustic concentration and alumina to caustic ratio
   - Liquor desilication, residence time and impurity balances
   - Heats of dissolution, and physical property data
   - Optimize/define salient process design parameters
   - Feedback to improve mineralogy calculation and models.
3. *Process Modelling*
   - Define the process design criteria
   - Develop mass & energy balances, raw material & energy consumption rates
   - Equipment sizing and selection

- Determine the plant availability analysis to determine actual refinery stream flows, pressures, temperatures, sparing of equipment, operations and maintenance personnel, etc.
- Define operating costs (energy, raw materials, labour, consumables etc).

4.4.1 Theoretical Heat of Digestion

Heat of Reaction

The heat of reaction for alumina dissolution in caustic is based on the enthalpy difference between the products and reactants under the conditions that:

- stoichiometric quantities of the reactants are fed and the reaction proceeds to completion,
- the products of a reaction emerge at the same temperature and pressure as the reactants.

For both Gibbsitic and Boehmite, the heat of reaction, ΔH_r is positive, i.e. endothermic, consuming energy. Heats of reaction may be determined from the difference between the heats of formation of the products and the heats of formation of the reactants.

For Gibbsite dissolution, the heat of reaction developed from the heats of formation (at a reference temperature of 298°K, or 25 °C) is as follows:

Gibbsite dissolution: $Al[OH]_{3(s)} + NaOH_{(aq)} \rightarrow NaAl[OH]_{4(aq)}$

Heats of formation [100]:

$$
\begin{aligned}
&\Delta H^{\circ}_{f}\,(298\,K) : Al[OH]_{3(s)} = -1293.1\,kJ/mol \\
&\Delta H^{\circ}_{f}\,(298\,K) : Na[OH]_{(eq)} = -470.3\,kJ/mol \\
&\Delta H^{\circ}_{f}\,(298\,K) : NaAl[OH]_{4(eq)} = -1743.1\,kJ/mol \\
&\Delta H^{\circ}_{f}\,(298\,K) : Al[OH]_{3(s)} = -20.37\,kJ/mol, \\
&\quad \text{or} \sim 261.2\,kJ/kg\,Gibbsite
\end{aligned}
$$

For Boehmite dissolution, the heat of reaction developed from the heats of formation (at a reference temperature of 298°K, or 25 °C) is as follows:

Boehmite dissolution: $AlO.OH_{(s)} + NaOH_{(aq)} + H_2O_{(l)} \rightarrow NaAl[OH]_{4(aq)}$

Heats of formation [101]:

$$
\begin{aligned}
&\Delta H^{\circ}_{f}\,(298\,K) : AlO \cdot OH_{(s)} = -998.5\,kJ/mol \\
&\Delta H^{\circ}_{f}\,(298\,K) : Na[OH]_{(eq)} = -470.09\,kJ/mol \\
&\Delta H^{\circ}_{f}\,(298\,K) : H_2O_{(1)} = -258\,kJ/mol
\end{aligned}
$$

$$\Delta H^{\circ}_{f}\,(298\,K) : NaAl[OH]_{4(eq)} = -1743.1\,kJ/mol,$$

$$\Delta \widehat{H}^{\circ}_{r}\,(298\,K) : AlO \cdot OH_{(s)} = 10.5\,kJ/mol$$
$$\text{or} \sim 174.5\,kJ/kg\,Boehmite$$

Energy balance over a simple reactor

For an adiabatic system where we may assume heat is not exchanged with the surroundings, our energy equation may be simplified to:

$$E_{(Accumulation)} = 0 = E_{(Input)} - E_{(Output)} + E_{(Generation)} - E_{(Consumption)},$$

where $E_{(Generation)}$ and $E_{(Consumption)}$ refer to heats of reaction for exothermic and endothermic systems respectively. Expressing the above energy equation algebraically:

$$\Delta H = 0 = \frac{n.\Delta \widehat{H}^{\circ} r}{v} + \sum_{out} ni.\widehat{H}i - \sum_{in} ni.\widehat{H}i$$

where:

n	the moles of a reactant or product consumed or produced
v	the stoichiometric coefficient of the reactant or product
$\Delta \widehat{H}^{\circ} r$	the heat of reaction at the reference temperature
$\widehat{H}i$	the enthalpy of component i with respect to the reference temperature.

For the simple Gibbsite reactor below, having determined the heat of dissolution above, we may use the energy equation to determine the heat input Q, required to supply both sufficient heat for complete conversion of the available Gibbsite as well as sensible heat to reach the product temperature.

Our reference temperature will be the condition for which the heat of reaction is known, i.e. 298 K.

Estimated Heat Capacities for Species (at Tin/Tout):

$$Al[OH]_{3(s)} : 1.48\,kJ/kg\,^{\circ}C$$
$$NaOH_{(aq)} : -1.97\,kJ/kg\,^{\circ}C$$
$$Na[OH]_{4(aq)} : 1.2\,kJ/kg\,^{\circ}C$$

$Al[OH]_{3(s)}$: 1.48 kJ/kg °C

Heat Inputs:

1. $Al[OH]_{3(s)}$ at 130 °C: 7.8 × 1.48 × (130–25) = 1214.9 kW
2. $NaOH_{(aq)}$ at 130 °C: 4.0 × −1.97 × (130–25) = −828.9 kW

3. N React at 130 °C: 80.0 × 3.2 × (130–25) = 26,880 kW
4. Q—To be determined.

$\Delta\widehat{H}^{\circ}r$:

1. $Al[OH]_{3(s)}$ at 25 °C: 7.8 × 261.2 = 2038 kW

Heat Outputs:

1. $NaAl[OH]_{4(aq)}$ at 150 °C: 11.8 × 1.2 × (150–25) = 1763.8 kW
2. N React at 150 °C: 80.0 × 3.2 × (150–25) = 32,000 kW

Solving for Q: 8535 kW.

Therefore 8535 kW must be transferred to the reactor. Typically, this energy transfer may be via indirect steam and comprise the mass flow of steam multiplied by its latent heat at a given temperature and pressure.

4.4.2 Boling Point Elevation

Boiling point elevation relates to the increase of a solution's boiling point as a result of dissolved chemical species in a solvent. In effect, boiling point elevation suppresses a fluid's vapour pressure (or pressure at which boiling occurs) at a given temperature.

Water at atmospheric pressure (101.325 kPa abs) has a boiling point of 100 °C. Expressed alternately, the vapour pressure of water at 100 °C is 101.325 kPa abs. If, as a result of dissolved species in water, a solution has a boiling point elevation of 10°, then at atmospheric pressure, the solution will not boil until the solution temperature equates to 110 °C. This is illustrated in Fig. 4.36.

For Bayer liquors, various correlations have been developed to estimate the boiling point of solutions based on the molar concentration of dissolved species. The main inputs to these correlations generally incorporate concentrations of:

- sodium carbonate (Na_2CO_3)
- sodium hydroxide (NaOH)
- sodium chloride (NaCl)
- sodium sulphate (Na_2SO_4)
- alumina (Al_2O_3)
- organics (including oxalate $Na_2C_2O_4$).

Boiling point elevations of 5–15 °C are typical throughout the refinery circuit, including the digestion facility. Further, as liquors mature, varying concentrations of organics may alter the liquor physical properties including boiling point. If allowed to accumulate in the liquor circuit, boiling point elevations will also increase.

The dominant impact that boiling point rise has within the digestion facility is to reduce the efficiency of heat transfer. The most efficient form of heat transfer is that resulting from condensation. The transfer of the latent heat of condensation occurs at the saturated steam temperature of the vapour that, if superheated, must

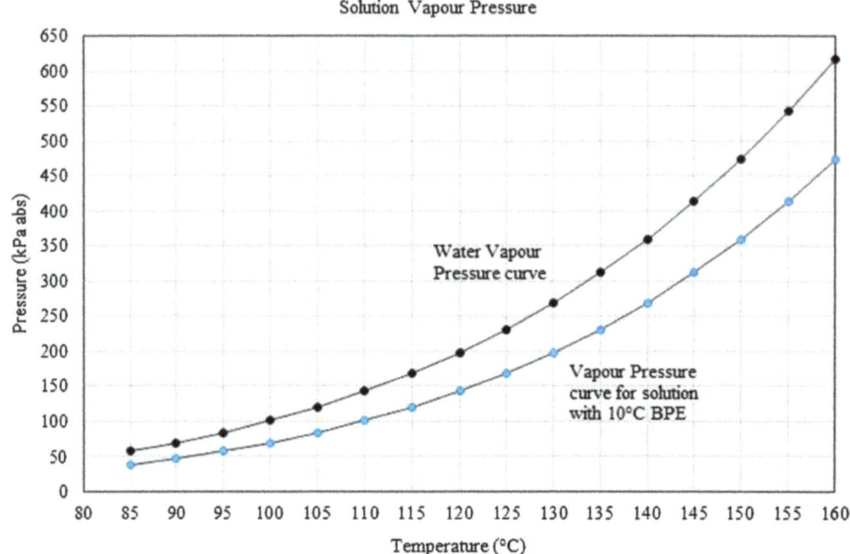

Fig. 4.36 Effect of boiling point elevation (BPE)

first de-superheat to the condensing temperature. If there is a large superheat, this detracts from the saturated steam temperature which defines the driving force for heat transfer.

The governing rate equation for heat transfer is: $Q = U_o \cdot A \cdot \Delta T_{lm}$,

where:

Q	heat transfer across the heat exchanger/condenser
U_o	overall heat transfer coefficient
A	heat transfer area
ΔT_{lm}	log mean temperature differential (LMTD).

For the condenser, the log mean temperature differential may be illustrated graphically by Fig. 4.37. As indicated below, for two vapours at the same temperature, the vapour with a lower boiling point rise provides a higher LMTD. This consequentially provides a higher rate of heat transfer for the same heat transfer area. To offset the effect of elevated boiling points for the same heat transfer, increased condenser areas are required. As example, for an increase in boiling point elevation from ~5 to 10 °C, the heat transfer area indicated in Fig. 4.27 would increase from 2% for 2 stages to 17% for 12 stages.

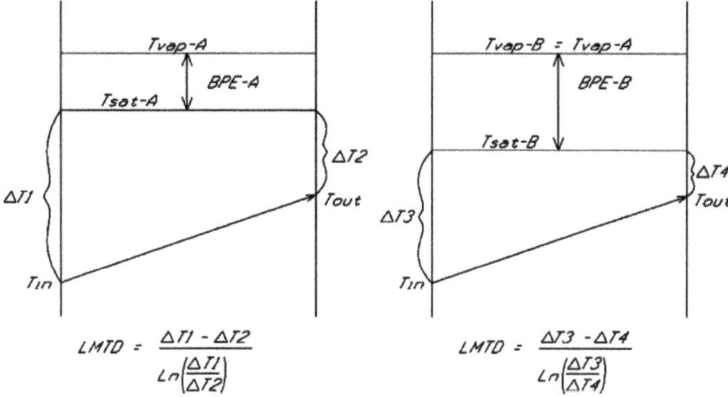

Fig. 4.37 Condenser log mean temperature difference

4.4.3 Steam Requirements

As outlined in the preceding sections, the energy required in the digestion facility is driven by a multitude of parameters, many requiring an iterative process of analysis to achieve an optimal condition.

These input parameters are summarized generally as:

- the chemical species present
  - the bauxite characterization (Gibbsitic, Boehmitic, Diasporic, mixed grade)
  - molar ratios of reactants and products
  - molar ratios of impurities
  - heats of reaction
- the thermodynamic properties of the fluids
  - fluid enthalpies
  - fluid heat capacities
  - fluid boiling point elevations
  - conditions of pressure and coincident temperature of reactants and products
  - other fluid-dynamic properties such as viscosity, density, etc.
  - heat transfer coefficient (for selected equipment)
- aspects of the process and mechanical facility design
  - fluid flows
  - flowsheet selection (low temperature, high temperature, dual stream, single stream, double digest, etc.)
  - autoclave selection (direct injection, indirect steam heated, plug flow)
  - number of effects or heat transfer stages
  - heat transfer equipment selection
  - plant availability analysis (sparing of equipment and cleaning cycles)

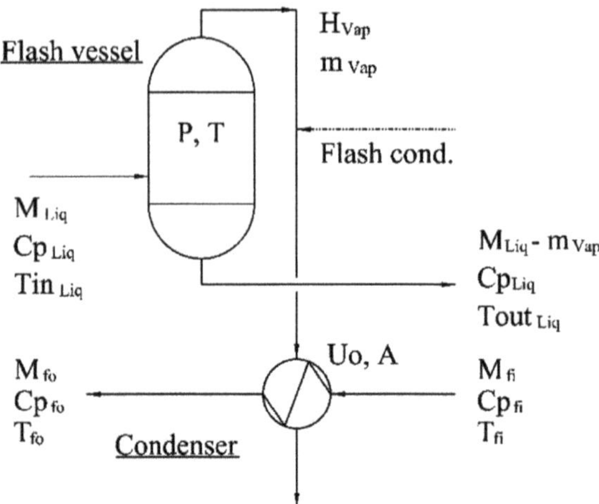

Fig. 4.38 Single effect—flash tank and condenser

- the cost of energy.

Once a bauxite resource is selected for processing, many of the chemical species (and impurities) will be defined. Similarly, the alumina content of the bauxite, selected A/C ratio and digestion reaction temperature, will provide a finite set of boundary conditions for the thermodynamic properties to be used. To define the digestion energy demand however, the ideal number of stages must still be determined.

In Sect. 4.3.2, the generic impact of the number of stages on digestion energy was introduced. Increasing the number of stages reduces digestion energy requirements, however, generally at diminishing increments.

To assess the ideal number of stages, the relevant thermodynamic relationships for the flash tank and condenser must be established.

For the single effect flash tank-condenser system indicated in Fig. 4.38, the relevant thermodynamic relationships are presented below under the simplifying assumptions of:

- no heat loss to the surroundings
- no heats of reaction are applicable through the condenser
- only the latent heat of condensation is transferred across the condenser
- the vapour stream is at its saturation temperature for the given stream pressure
- no solids are present in the stream entering or exiting the flash tank
- liquor heat capacities entering and exiting the flash tank are equal
- fluid heat capacities entering and exiting the heat exchanger are equal
- no re-flash condensate is present as an additional energy stream.

1. Energy transferred from the flash vapour (at selected P, T):

$$M_{Liq}.Cp_{Liq}.Ti = m_{Vap}.H_{Vap} + \left(M_{Liq} - m_{Vap}\right).Cp_{Liq}.To$$

$$m_{Vap} = \frac{M_{Liq}.Cp_{Liq}.(Ti - To)}{\left(H_{Vap} - Cp_{Liq}.To\right)}$$

Energy transferred to exchanger: $m_{Vap}.\Delta\widehat{H}v$

where:

$\Delta\widehat{H}v$ is $H_{Vap} - H_{Cond}$

H_{cond} is the enthalpy of the condensate exiting the condenser.

2. Energy transferred to condenser fluid: $M_{fo}.\ Cp_{fo}.(T_{fo} - T_{fi})$

where:

$Mf = M_{fo} = M_{fi}$

$Cpf = Cp_{fi} = Cp_{fo}$

3. Energy transferred across exchanger: $Uo.A.\Delta Tlm$

where:

$$\Delta Tlm = \frac{\Delta T1 - \Delta T2}{Ln\left(\frac{\Delta T1}{\Delta T2}\right)}$$

$\Delta T1 = T_s - T_{fo}$

$\Delta T2 = T_s - T_{fi}$

Uo overall condenser heat transfer coefficient (W/m^2 K)

A condenser heat transfer area (m^2)

T_s vapour saturation temperature at inlet pressure

T_{fi} fluid temperature at inlet to the condenser

T_{fo} fluid temperature at outlet of the condenser

As the energy (4.1, 4.2 and 4.3) above must equate, rearranging gives the fluid temperature exiting the condenser as:

$$T_{fo} = T_s - \frac{(T_s - T_{fi})}{e^{\left(\frac{Uo.A}{Mf.Cpf}\right)}}$$

These energy equations may be applied to a multiple effect system to define the fluid temperature exiting the recuperative condenser circuit. From this maximum recuperative temperature, the resultant energy to be transferred to reach the autoclave reaction temperature may be determined.

4.5 Lime Chemistry

Lime is widely considered throughout the alumina industry to be a panacea to prevent or cure various difficulties that may happen in the course of processing various types of bauxites and production of various types of alumina having different characteristics, such as particle size distribution, impurities, attrition indices etc.

As long as the use of lime is looked at somewhat closely, it can be seen that there are several points in the Bayer process, where some sort of lime is used, and the effectiveness of the use varies in different refineries.

Lime is used first of all in a slaked form $Ca(OH)_2$, in some cases as quicklime (pebble lime, burnt lime, mostly CaO). If CaO is added to water or a processs liquor, it quickly reacts with water and $Ca(OH)_2$ forms. The reaction equation of lime slaking is as follows:

$$CaO + H_2O \rightarrow Ca(OH)_2 \quad (4.27)$$

The reaction is exothermic, the heat of formation ($\Delta\ H_f{}^o$) is—65.3 kJ/mol [102]. Here is a list where lime is being used in the Bayer process:

- in the digestion for different purposes, such as to
  - convert the caustic soluble P-compounds to insoluble Ca-phosphates,
  - convert sodium titanates to possibly complex Ca-titanates,
  - convert sodium-aluminium-hydrosilicates (DSP) to Ca-aluminium-hydrosilicates,
  - enhance the conversion of the (alumo)-goethite into hematite,
  - enhance the digestion (solubilisation) of diaspore,
- in the settling, counter-current washing, processing of washed bauxite residue (BR), security (control) filtration
  - preparation of tricalcium aluminate hexahydrate (TCA) filtration aid,
  - causticisation of sodium carbonate content of a washer overflow,
  - recovery of soda and possibly alumina content of DSP content of BR,
- in the precipitation, where hydrate separation and oxalate removal included
  - causticisation of sodium oxalate,
- in the evaporation, removal of sodium carbonate included
  - causticisation of sodium carbonate.

4.5.1 Phosphorus Control

Usually 0.5–1% CaO for the bauxite weight is added to the low temperature (LT) digestion to convert P-content to insoluble compounds. Ostap claimed [73] that the formula of the insoluble carbonate apatite that is formed in the course of P-control with lime is the following:

$$(3.3 + 0.76x)CaO \cdot 0.33x \cdot Na_2O \cdot xCO_2 \cdot P_2O_5. \quad (4.28)$$

Kirwan et al. studied [103] the "Bayer" carbonate-apatite and cited the stoichiometry proposed by Bonel [104] et al. as follows:

$$Ca_{10-x}Na_{2x/3}(PO_4)_{6-x}(CO_3)_x(H_2O)_x(OH)_{2-x/3} \quad (4.29)$$

where $x = 2.5$ or 2.7 depending of the lime source (calcite or slaked lime respectively).

4.5.2 *Reaction of Sodium Titanates with Lime*

Wefers synthetised crystalline sodium titatane in pure caustic soda solution [47]. At concentrations of higher than about 100 g/L Na_2O (171 g/L caustic soda as Na_2CO_3), and temperatures higher than 150 °C the product was $Na_2Ti_3O_7.2H_2O$. At temperatures of higher than about 260 °C, $Na_2Ti_3O_7$ product was formed.

Bánvölgyi studied scales [20] on the walls of the slurry preheaters of a high temperature (HT) digestion. The preheaters operated at 145–155 °C. Anatase and rutile (both TiO_2) were not found, however, Ca(Mg)-titanate were observed instead. More detailed description will be given later in the Sect. 4.5.7.

4.5.3 *Reaction of Sodium-Aluminium-Hydrosilicates with Lime, Conversion of (Alumo)-Goethite into Hematite*

Zöldi et al. claimed [105] that $3CaO{\cdot}(Fe_xAl_{1-x})_2O_3{\cdot}kSiO_2{\cdot}(6\text{-}2\,k)H_2O$ is the composition of the hydrogarnet type silicates which are formed from a considerable proportion of silica by the conversion of iron minerals. During the low temperature digestion, where the liquid phase is nearly saturated for dissolved alumina $k \leq 0.5$. In the course of causticisation of bauxite residue (BR), the liquid phase has a low dissolved alumina content, $k = 1.0–1.2$. During high temperature digestion of karst bauxites, when the calcite ($CaCO_3$) and possibly dolomite ($CaCO_3{\cdot}MgCO_3$) content react with the caustic liquor and lime may also be added to bauxite, $k = 0.5–0.8$, only a few percent of alumina is substituted with iron. When goethite is converted into hematite in presence of Ca-containing catalytic additive, the SiO_2 content increases, and may attain $k = 1$, and the iron substitution of Al^{3+} may exceed 10%. The overall formula of hydrogarnet is

$$A_3B_2(SiO_4)_{3-x}\left[(OH)_4\right]_x \tag{4.30}$$

where

A denotes divalent cation, such as Ca^{2+}, Mg^{2+}, Fe^{2+}, Mn^{2+}, Co^{2+}

B denotes trivalent cation, such as Al^{3+}, Fe^{3+}, Cr^{3+}.

Fe^{2+}, Mn^{2+}, Fe^{3+} in the hydrogarnet structure, and presence of some anions as $SO_4{}^{2-}$ facilitate the conversion of (alumo-)goethite into hematite. During the high temperature digestion, some 20% savings in the chemically combined soda can be attained [106] due to partial replacement of Na_2O of desilication product (DSP) with CaO.

4.5.4 Lime Addition to Enhance Digestion of Diaspore

Addition of lime to solubilise diaspore is a precondition to make diaspore in bauxite soluble. A minimum amount of 2% CaO was needed in the fundamental tests of Mercier and Magrone [107]. As long as the lime addition increased, exctraction yield of diasporic alumina sharply increased and at CaO dosages higher than 3–4% it slowly decreased (Fig. 4.39), evidently due to formation of some sort of Ca-aluminate. When increasing amount of lime was added to digestion, partial replacement of Na_2O of the DSP to CaO took place, the Na_2O/SiO_2 ratio continuously decreased from 0.65 to 0.3, the CaO dosage reached a ratio of 2 mol CaO/SiO_2 (Fig. 4.40).

Maltz explained [51] the catalytic effect of CaO in such a way that Ca-titanates and hydrogarnets formed instead of sodium-titanates and DSP which impede the dissolution of diaspore.

Gu Songquing reported [108] a Lime-Bayer process. Here DSP is converted to hydrogarnet by extra lime addition in the course of the digestion of diaspore; the hydrogarnet, therefore the BR, does not contain chemically combined soda. However, this process route is associated with lower alumina extraction yields in comparison with the conventional Bayer process.

When boehmitic bauxite is digested, CaO addition is not a precondition to make boehmite soluble. The replacement of Na_2O in DSP is similar to Fig. 4.40, probably to a lesser extent.

4.5.5 Hydrothermal Treatment of Bauxite Residue in Presence of Lime

The Bayer process per se is a hydrothermal processing of bauxites in caustic liquor; the presence of lime makes the reactions more complex.

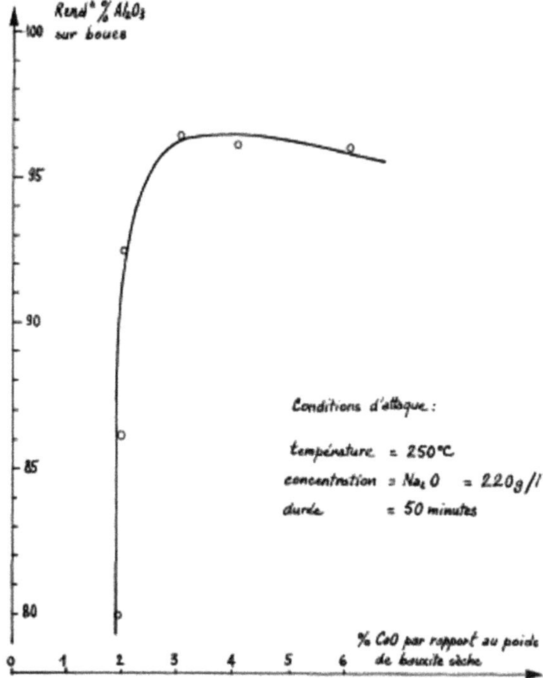

Fig. 4.39 Extraction yield of a diasporic bauxite in the function of CaO dosage

Causticisation of bauxite residue

The idea of causticisation has been known for several decades and it was applied in the Hungarian alumina refineries from mid 1950' years for about 25 years.

Nemecz et al. treated [17] hydroxy-sodalite with $Ca(OH)_2$ at 90 °C, the product had the composition of $3.1CaO{\cdot}Al_2O_3{\cdot}1.65SiO_2$. Juhász et al. reacted [109] sodium-aluminium-hydrosilicates with $Ca(OH)_2$ and $CaCl_2$ and the product had the composition of $3CaO{\cdot}Al_2O_3{\cdot}kSiO_2{\cdot}H_2O$, where k might be even 3 and the calcium aluminium silicate had a cubic structure similar to $3CaO{\cdot}Al_2O_3{\cdot}6H_2O$ (TCA).

Vörös et al. studied [110] the causticisation of sodalites, cancrinites and sodium titanates as model substances and those of in the BR. It was confirmed that the CaO/Al_2O_3 ratio in the product was approximately 3, the SiO_2/Al_2O_3 ratio varied between 0.8 and 2.2. The causticisation of cancrinite proceeded rapidly. The hydrogarnet type Ca-aluminium–silicate product had the composition of $3CaO{\cdot}Al_2O_3{\cdot}kSiO_2{\cdot}(6\text{-}2\ k)H_2O$, no matter if the starting material was sodalite or cancrinite. Solymár et al. claimed [111] k value to be of about 1.5.

Causticisation theoretically requires a lime consumption of 2 kg CaO per kg recovered NaOH. In practice 3–4 kg CaO/kg NaOH is needed [111]. This is probably due to the side reaction of formation of TCA. Causticisation works fairly well at low

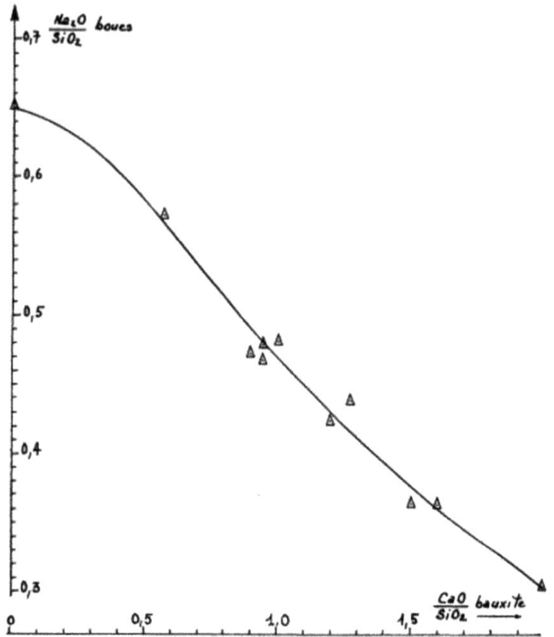

Fig. 4.40 Na_2O/SiO_2 in bauxite residue in the function of CaO dosage (CaO/SiO_2)

caustic concentrations and the spike in the concentration of the liquid phase is to be balanced with additional washing and/or filtration of BR.

A novel process route has been proposed by Zhao Qiuyie et al. [112]. The DSP content of BR is causticised, washed and separated. The $3.1CaO{\cdot}Al_2O_3{\cdot}1.7SiO_2{\cdot}2.6H_2O$ product is converted to $Al(OH)_3$ and insoluble $CaSiO_3$ and $CaCO_3$ by carbonization. The $Al(OH)_3$ formed is recovered by caustic leaching, $CaSiO_3$ and $CaCO_3$ can comprise feedstocks for manufacturing of cement.

Hydrothermal treatment of bauxite residue at elevated temperatures

The recovery of caustic soda from BR by lime addition significantly increases at elevated temperatures. The results of Solymár et al. are shown [111] in Fig. 4.41.

Creswell and Milne carried out laboratory scale hydrothermal treatment experiments in a high temperature reactor of continuous operation [113]. The feedstock material was washed BR probably from a high temperature digestion of Weipa bauxite in presence of negligible CaO addition. 2 mol CaO/mol SiO_2 was found to be the optimum lime dosage. Soda recovery of about 90% and an alumina recovery of about 70% could be achieved at temperatures of at least 260 °C and residence times of 10–15 min. The A/C ratio of about 0.12 proved to be a strict limit, above that both the soda and alumina recoveries fell. The alumina recoveries reduced above 270 °C. The main reaction product was $3CaO{\cdot}Fe_2O_3{\cdot}2SiO_2$, i.e. no chemically combined soda or alumina losses connected with SiO_2 occurred.

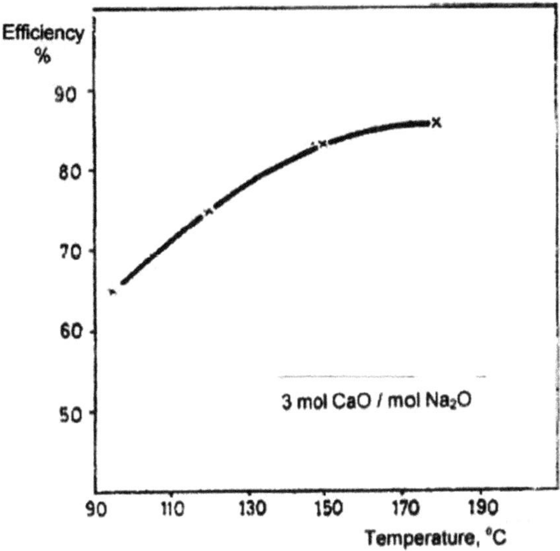

Fig. 4.41 Caustic soda recovery from bauxite residue [111], with the permission of G. Baksa

The results of a pilot scale hydrothermal treatment of BR with CaO are presented in a paper of Solymár et al. [114]. The bauxite was submitted to a high temperature digestion (240 °C) in the presence of lime in order to recover some caustic soda and facilitate the conversion of goethite into hematite. The Na_2O recovery from BR reached 90–94% at 265–270 °C treatment, at a CaO dosage of 2.7–3.1 mol CaO/mol SiO_2 in the course of the hydrothermal treatment. However, the Al_2O_3 recovery was only 7–13%, probably due to the liquor A/C ratio, and a relative CaO overdosage. The feedstock material of different character from that of Creswell and Milne might have been the reason of the lower alumina recovery.

The low A/C ratio of the hydrothermal treatment effluent could not be harmonised with the use of **all** the bauxite residue for neither development. This was probably the reason why none of these processes were implemented in a commercial scale.

4.5.6 Theoretical Background of Reactions of Lime in Sodium Aluminate Solutions

The principal reactions of lime in sodium hydroxide–sodium aluminate solutions are described by Rosenberg [55] et al. as follows.

Calcium hydroxide is very unstable in sodium aluminate solutions and the following reaction takes place in a sodium hydroxide–sodium aluminate solution.

$$4Ca(OH)_2 + 2Al(OH)_4^- + 2X^- + nH_2O$$

$$\rightarrow \left[Ca_2Al(OH)_6\right]_2X_2 \cdot nH_2O + 4OH^- \quad (4.31)$$

As a result of the reaction depicted by the reaction (4.31), member(s) of the family of the metastable layered double hydroxide (LDH) compound is formed consisting of repeated units of two $[Ca_2Al(OH)_6]^+$ cations (C_4AH). The cavities between the layers are occupied by water and charge balancing monovalent, divalent, or, a mixture of the two type of anions. X-represents the charge balancing anion that is intercalated between the layers.

The charge balancing anion depends on the composition of the solution. In pure sodium aluminate–sodium hydroxide solution the following reaction takes place:

$$\begin{aligned} &4Ca(OH)_2 + 2Al(OH)_4^- + 6H_2O \\ &\rightarrow \left[Ca_2Al(OH)_6\right]_2(OH)_2.6H_2O + 2OH^- \end{aligned} \quad (4.32)$$

In real Bayer liquors, which contain other anions as well, some will be intercalated into the compound. The sort and the amount of anions that is intercalated depends on their concentration, though the most preferred anion is carbonate. The amount of carbonate is variable; two distinct forms are known, i.e. hemicarbonate $[Ca_2Al(OH)_6]_2 \cdot {}^1/_2CO_3 \cdot nH_2O$ and monocarbonate $[Ca_2Al(OH)_6]_2 \cdot CO_3 \cdot nH_2O$.

At first, hemicarbonate forms:

$$\begin{aligned} &4Ca(OH)_2 + 2Al(OH)_4^- + {}^1/_2CO_3^{2-} + 5{}^1/_2H_2O \\ &\rightarrow \left[Ca_2Al(OH)_6\right]_2{}^1/_2CO_3 \cdot OH \cdot 5{}^1/_2H_2O + 3OH^- \end{aligned} \quad (4.33)$$

As it is evident from the reaction (4.33), aluminate anion is removed from the liquid phase and for each formula unit half mole of carbonate anion is removed from solution as well. Formation of LDH is favoured by high aluminate and carbonate concentrations and inhibited by high OH^- ion concentration. Roach identified [89] monocarbonate as an intermediate compound. Rosenberg claims it is characteristic to the synthetic liquors, but rarely found in plant liquors.

Above 80 °C the quaternary calcium aluminates become progressively more unstable, react with the carbonate ions, causing formation of calcium carbonate.

$$\begin{aligned} &\left[Ca_2Al(OH)_6\right]_2{}^1/_2CO_3.OH.5{}^1/_2H_2O + 3{}^1/_2CO_3^{2-} \\ &\rightleftharpoons 4CaCO_3 + 2Al(OH)_4^- + 5OH^- + 5{}^1/_2H_2O \end{aligned} \quad (4.34)$$

This reaction (4.34) is deemed to be the main causticisation reaction, its rate rising rapidly with temperature. At temperatures at 80 °C or at lower temperatures, the reversion reaction is favoured.

Formation of TCA (tricalcium-alumino-hexahydrate) also occurs via hemicarbonate species:

$$3\left[Ca_2Al(OH)_6\right]_2.{}^1/_2CO_3.OH.5{}^1/_2H_2O + 2Al(OH)_4^- + OH^-$$

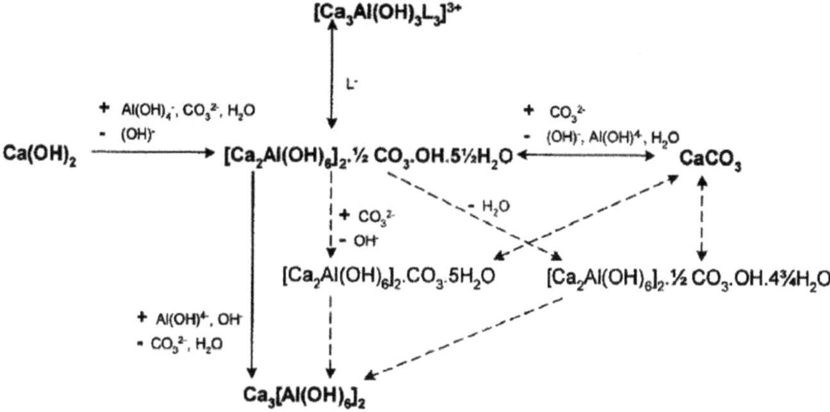

Fig. 4.42 Map of reactions of calcium species in Bayer liquors [55]. Copyright © 2001 by The Minerals, Metals & Materials Society. Used with permission

$$\rightarrow 4\left[\mathrm{Ca_3Al(OH)_6}\right]_2 + 1/2\mathrm{CO_3^{2-}} + 16^1/_2\mathrm{H_2O} \quad (4.35)$$

The rate of formation of C_3AH_6 is controlled by diffusion, it increases with intensity of agitation, also with increasing concentrations of aluminate and hydroxide, though it is inhibited by high carbonate concentrations.

Based on the reaction map of Fig. 4.42 and its background, improvements in the causticisation of washer overflow liquor was invented by Rosenberg et al. [115]. The C/S ratio was improved from 0.882 to 0.906 and the lime efficiency from 55 to 95%, by inhibiting the formation of TCA meanwhile it is competing with formation of $CaCO_3$.

Solubility of calcium

There are some papers in the literature on this subject, however, there are still some aspects to be clarified. The and Sivakumar [116] revealed that some organics, such as humic acid and sodium gluconate appreciably increased the solubility of calcium. It seems to be a reasonable explanation that these organic compounds formed soluble complexes with calcium.

It can be deducted from the reactions that there are some consecutive reactions, where $CaCO_3$ and TCA, or complexes (with the formula of $[Ca_3Al(OH)_3L_3]^{3+}$, where L is a complexing ion (ligand), possibly gluconate or other complexing anion, or, OH^- ion) are the final products. It was demonstrated that there are fairly siginificant variations in the calcium solubility in the liquid phase.

4.5.7 Theoretical Considerations of the Reactions of Lime Under High Temperature Bayer Digestion Conditions

This part of the Chapter is largely based on the paper of Smith [54].

Reactions of lime with individual liquor components

When quicklime or slaked lime contacts Bayer liquor the slaking process involves reaction with both aluminate and carbonate ions. Either calcite or TCA forms through the formation of an intermediate product, which is a layered double double hydroxide (LDH) of the hydrotalcite with a carbonate ion inclusion. Rosenberg et al. describes [55] LDH as hemicarbonate $[Ca_2Al(OH)_6]_2{\cdot}\frac{1}{2}CO_3{\cdot}OH{\cdot}5\frac{1}{2}H_2O$ that is also called hydrocalumite. The decomposition of the LDH to either TCA ($Ca_3Al_2(OH)_{12}$ or $3CaO{\cdot}Al_2O_3{\cdot}6H_2O$) or calcite ($CaCO_3$) defines the causticisation equilibrium.

$$Ca_3Al_2(OH)_{12} + 3Na_2CO_3 \rightleftharpoons 3CaCO_3 + 2NaAl(OH)_4 + 4NaOH \quad (4.36)$$

TCA, once formed, can incorporate other anions by substitution for some of its hydroxyl groups. Wilson et al. revealed [117] that fluorine can be incorporated into TCA and this compound comprises one of the sinks for fluorine in liquor. They reject that fluoride exists as CaF_2 co-precipitated with TCA. Some other anions such as V is likely to be scavenged in a similar way as fluorine. Titanium and phosphorus are other elements that can also be substituted into TCA, but there is no sufficient amount of experimental evidence for this.

Bayer sodalite is the compound that forms out of reactive silica during predesilication, digestion and the flash train. Lime can act to reduce soda losses associated with sodalite by including silica into the structure of TCA. There is a continuous solid solution series from TCA (mineral name hydrogrossular—$Ca_3Al_2(OH)_{12}$) to grossular ($Ca_3Al_2[(SiO_4)]_3$). The silica substituted TCA is generally known as, "hydrogarnet—HG" ($Ca_3Al_2(SiO_4)(OH)_{12\text{-}4n}$). The Bayer hydrogarnets have relatively low silica substitution in predesilication (n ~ 0.1) rising to n ~ 0.6 during digestion to 250 °C and possibly as high as 1.0 at 280 °C. Figure 4.43 depicts a summary of these compounds.

In the case of lime added to predesilication under conditions where TCA forms and then digested at 250 °C in presence of silica, it is likely that TCA transforms into

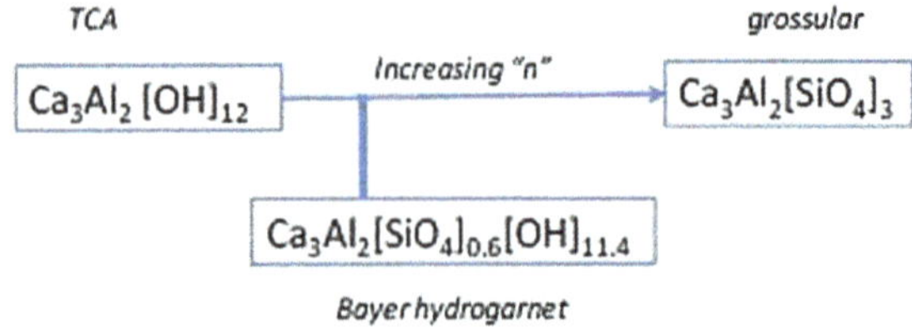

Fig. 4.43 Silica substitution into TCA [54]. Copyright © 2015 by AQW Inc. Used with permission

HG. Nevertheless, if lime is added directly to the digestion reactor, its dissolution releases calcium ions that may be consumed in a variety of reactions and formation of HG is less likely. Addition of lime to the digestion reactor is the preferred way in case of a HT digestion.

At HT digestion conditions cancrinite (an alternative zeolite type DSP) forms. Some sodalite re-dissolves and precipitates as cancrinite. Bayer cancrinite can form at HT digestion conditions, but its transformation from sodalite is very slow. The presence of calcium and carbonate in liquor dramatically enhances formation rate of cancrinite.

It has been known for decades that anatase reacts in Bayer liquor more readily than rutile [118]. The amorphous reaction product having a stoichiometry $Na_2O{\cdot}3TiO_2{\cdot}2.5H_2O$ can coat the surface of dissolving boehmite inhibiting its extraction [51, 119]. Gu et al. [120] identified reaction products of lime with titanium in the liquid phase. Calcium hydroxy-titanate (CTH), its relationship to perovskite ($CaTiO_3$) and another compound named titanium substituted TCA were described. It was found that CTH is an intermediate product during digestion of diasporic bauxite that finally reverts to perovskite; high level of lime increases the proportion of Ca reporting to titanium substituted TCA. A reasonable reaction equation for the formation of CTH from anatase, based on kassite:

$$Ca(OH)_2 + 2TiO_2 \rightarrow CaTi_2O_4(OH)_2 \tag{4.37}$$

Smith [54] claims that the structure of CTH may be closer to caffetite ($CaFe_2Ti_2O_{12}{\cdot}4H_2O$) than kassite.

Croker [121] showed that lime can concurrently react with silica from sodalite to form cancrinite and with titania to form perovskite, no CTH intermediate was observed. Smith reported similar results.

As far as the chemical formula of titanium substituted TCA is concerned, Suss et al. [122] suggest the following general formula: $Ca_3(Al,Fe)_2[Si,TiO_4]_n(OH)_{12\text{-}4n}$, based on analyses of scales.

As per Malts et al. [119] the transformation of goethite to hematite is significantly accelerated by the catalytic action of lime by the temporary formation of iron hydrogarnets (IHG) which subsequently decomposes to hematite:

$$\begin{aligned} &3Ca(OH)_2 + 2FeOOH \rightarrow 3[CaO \cdot Fe_2O_3 \cdot 2H_2O] + 2H_2O \\ &\quad \rightarrow 3Ca(OH)_2 + Fe_2O_3 \end{aligned} \tag{4.38}$$

The ideal stoichiometry of the formation of IHG from hydroxy-sodalite is as follows:

$$\begin{aligned} &Na_6(AlSiO_4)_6.2NaOH + 9Ca(OH)_2 + 3Na_2Fe_2O_4 + 12H_2O \\ &\quad \rightarrow 3Ca_3Fe_2(SiO_4)_2(OH)_4 + 6NaAl(OH)_4 + 8NaOH \end{aligned} \tag{4.39}$$

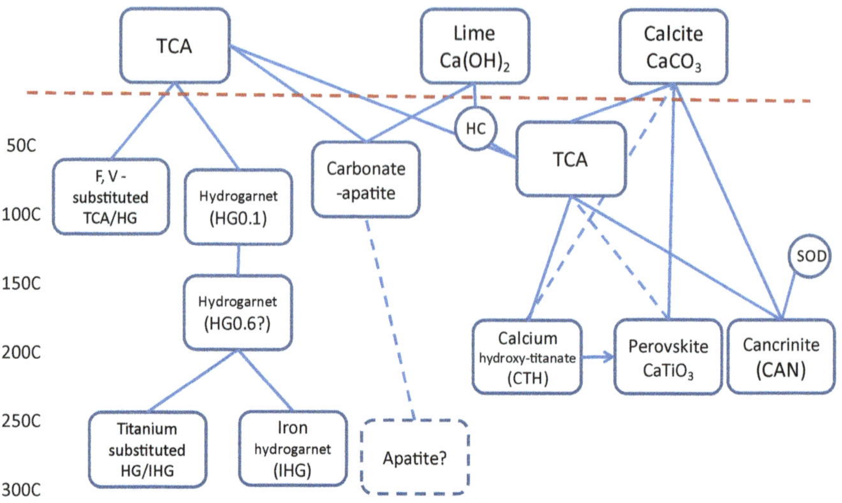

Fig. 4.44 Map of possible products of the reaction of lime showing the approximate temperatures at which they start to form. HC = hydrocalumite, SOD = sodalite. Pathways indicated with dotted lines are uncertain [54]. Copyright © 2015 by AQW Inc. Used with permission

A method for scavenging colloidal iron from solution is suggested by Fulford [123], i.e. addition of lime to digestion could result in the formation of IHG. The competing lime reactions with titanium species remove sodium titanates that otherwise coat iron particles, leaving them more likely to act as scavengers.

The map of possible reaction paths of various lime sources, the products and the temperatures at which they start to form are shown in Fig. 4.44, that is taken from Smith [54].

4.5.8 Preparation of Tricalcium Aluminate Hexahydrate

TCA is the abbreviation of the tricalcium-aluminate-hexahydrate compound, commonly described as $3CaO{\cdot}Al_2O_3{\cdot}6H_2O$. The abbreviations C_3A or C_3AH_6 are also being used, these originate from the cement industry. TCA has been found to be a fairly cheap and inert material that has very low solubility in the Bayer liquors. Its principal use is as a filter aid for the security (control) filtration of the settler overflow pregnant liquor. Continuous addition of the filter aid to the liquor to precipitation (LPT) to be filtered is important.

The high throughput and complete removal of the fine BR particles are two, largely contradictory requirements. One of the drawbacks of the conventional process to make TCA filter aid is that the particle size distribution is „wide", contains large amount of fine particles, $\leq 10\ \mu m$. In several alumina refineries the amount of fine BR in the settler overflow should be ≤300 mg/L, in extreme cases might be siginificantly more. The expected amount of BR in the filtrate is ≤5 mg/L.

There are several publications discussing the various aspects of the security (control) filtration, preparation of TCA. Some are in the references [124–133].

Rosenberg et al. have proposed [134] the following procedure to prepare a good quality TCA filter aid (with minimum amount of ≤10 μm fraction):

- slaking lime (preferably quicklime) in a solution which has an alkalinity between 5 and 30 g/L expressed as equivalent g/L of sodium carbonate,
- dosing the slaking solution or slaked lime slurry (milk of lime) with a suitable surface active agent
- reacting the dosed slaked lime slurry with a Bayer process liquor
- ageing the reaction products for a sufficiently long reaction time to permit substantially complete conversion of reaction products to TCA.

The novelty of the improved process was the use of a surface active agent, which can be glucose or sucrose, polysaccharides, such as starch, more preferably an anionic organic surfactant, e.g. polyacrylic acids or its co-polymers with acrylamide, or polymers having hydroxamate functional groups, etc. The Bayer liquor should have preferably a high A/C ratio. The flows of slaked lime slurry and the Bayer liquor are controlled so that the molar ratio of Al_2O_3 and CaO exceeds 0.33:1 during the ageing step. More preferably the molar ratio of Al_2O_3 and CaO is within 1:1 and 1.5:1. The temperature of this ageing step is preferably between 100 and 105 °C, the residence time is between 2 and 4 h. The median of the particle size distribution of the TCA by this improved method is about 20–30 μm. The throughput of the filters increased by 35%, the calcia content of the product alumina decreased by almost 30%, when the improved process to prepare the TCA filter aid was applied.

4.5.9 *Causticisation of the Sodium Carbonate Content of a Washer Overflow*

Recovery of caustic soda from dissolved sodium carbonate in an overflow of a red mud washer is one of the most traditional causticisation process. The reaction equation that has been widely used is as follows:

$$CO_3^{2-} + Ca(OH)_2 \rightleftharpoons CaCO_3 + 2OH^- \tag{4.40}$$

The most probable reaction route is that calcium hemicarbonate forms at first, which is an intermediate compound [55] and it converts to calcium carbonate (see the map of reactions on Fig. 4.42).

It is important to select the washer overflow having an appropriate concentration [106]. If the total alkali concentration is too high, due to the decrease of the solubility of $Ca(OH)_2$, the efficiency of causticisation is low. If the total alkali concentration is low, the amount of sodium carbonate is low to maintain a reasonable carbonate balance. The concentration range that was found to best satisfy the contradictory considerations is about 70–85 g/L total alkali as Na_2CO_3. Dosage of $Ca(OH)_2$ in excess should be avoided due to the formation of TCA. The preferred temperature of this causticisation reaction is about 100 °C. In case of the carbonate level of the causticiser effluent is less than 5–6%, the causticisation process is associated with high alumina losses.

It has been known for quite some time that there are competitive reactions as well. The calcium hydroxide readily reacts with aluminate ion, seemingly to form TCA. TCA forms most probably by two reactions, i.e. calcium ions react with aluminate anions and the „reversion reaction" when calcium carbonate reacts with aluminate and TCA forms. Some authors suggested that in Bayer liquor caustisation takes place via a "hydrated tricalcium aluminate" intermediate [135] or carboaluminate phase and TCA forms as this material ages.

A comprehensive paper on the side stream causticisation of Chaplin is referred [136]. Some patents that describe the development and the state of the art of the liquor causticisation are listed in the reference [137].

4.5.10 Causticisation of Sodium Oxalate

A widely used way for the removal of oxalate is carried out with lime from the solution obtained from washing of the deliquored fine seed with hot water. This method is sensitive to the remaining aluminate and carbonate content of the oxalate solution due to imperfect deliquoring of the fine seed. The aluminate and carbonate content cause low lime efficiency in regard to oxalate removal since formation of TCA and other compounds. The caustic soda and sodium carbonate concentration of the oxalate liquor must be kept low, the temperature of the reaction is usually not warmer than 60 °C.

An improvement was proposed by Rosenberg et al. [138] where lime is added to the wash liquor, the hemicarbonate (C_4A) that readily forms and strips the aluminate and carbonate ions is separated and used efficiently in the refinery causticisation circuit with recovery of alumina. The clarified liquor being essentially free of aluminate and carbonate with addition of further amount of lime results in an efficient precipitation of Ca oxalate.

4.5.11 Causticisation of Sodium Carbonate

In some alumina refineries the so-called sodium carbonate salt is removed from the main process liquor by the so-called salting out method. A fraction of the concentrated spent liquor is further concentrated to a total alkali concentration of about 450–500 g/L (expressed in Na_2CO_3) and from this superconcentrated liquor, the sodium carbonate salt (mainly $Na_2CO_3·H_2O$) crystallizes. The sodium carbonate salt contains several other compounds, such as organics, sodium sulfate, etc. as well.

In order to recover the caustic soda content of the sodium carbonate salt, the simplest way is the causticisation reaction usually with lime milk. The total alkali concentration target is usually about 70–85 g/L as Na_2CO_3; the reaction map of the causticisation reaction is shown in Fig. 4.42 and was discussed earlier.

There are several embodiments of the causticisation of the sodium carbonate salt, e.g. if it is dissolved in a washer overflow of appropriate concentration, the effectiveness of lime can be increased. It should be pointed out that the causticisation of the sodium carbonate by the conventional ways does not reduce the organic compounds present.

Acknowledgements The Authors are grateful for Robert LaMacchia for his active involvement during the outlining of the Chapter. Special thanks should be emphasized to Steve P. Rosenberg and Peter Smith for their contributions to the Lime Chemistry.

References

1. G. Bárdossy, G.J.J. Aleva, Lateritic bauxites, in *Akadémiai Kiadó*, Budapest (1990)
2. J.R. Glastonbury, Nature of sodium aluminate solutions, in *Chemistry and Industry* (1969), pp. 121–125
3. J. Zámbó, Structure of sodium aluminate liquors: molecular model of the mechanism of their decomposition, in *Proceedings of TMS 115th Annual Meeting*, ed. By R.E. Miller, New Orleans, LA, 2–6 Mar. 1986 Light Metals 1986, pp. 199–215
4. J.J. Kotte, Bayer digestion and predigestion desilication reactor design, in *Proceedings of 110th AIME Annual Meeting*, ed. by G.M. Bell, Chicago, Illinois, 22–26 Feb. 1981 Light Metals 1981, pp. 45–79
5. A Atasoy, An investigation on characterization and thermal analysis J. Therm. Anal. Calor. 81 (2005) 2 pp. 357-361
6. I.Z. Pevzner, A.S. Dvorkin, M.J. Fiterman, N.I. Eremin, J.B. Rozen, Matematicheskoe opisanie processa obeskremnivania aluminatnih rastvorov (Mathematic description of desilication process of aluminate liquors). Tsvetnie Metalli **7 (1975)**, pp. 49–52
7. P.A. Lyew-Ayee, Bauxite reserves for the world aluminium industry, in *Proceedings of 8th International Congress of ICSOBA, "Energy and Environment in Aluminium Industry"*, Milan, Italy, 16–18 Apr 1997. Travaux ICSOBA, vol. 26, No. 29 (1999) pp. 9-31
8. I.E. Sajó, X-ray diffraction quantitative phase analysis of Bayer process solids, in *Proceedings of Xth Conference of ICSOBA*, held in Bhubaneswar, India, 28–30 Nov 2008. Travaux ICSOBA, vol. 34, No. 38 (2009), pp. 46–51
9. G. Bánvölgyi, Reactions of gibbsite and kaolinite in the Bayer liquor: a comprehensive kinetic model and an improvement of the low temperature digestion, in *Proceedings of the 7th International Congress of ICSOBA*, Balatonalmádi, Hungary, 22–26 June 1992. Travaux ICSOBA, vol. 23, No 25 (1996), pp. 155–171
10. G. Bánvölgyi, P. Siklósi, The Improved Low Temperature Digestion (ILTD) process: an economic and environmentally sustainable way of processing gibbsitic bauxites, in *Proceedings of 127th TMS Annual Meeting*, ed. by B. Welch, San Antonio, TX, 15–19 Feb 1998. Light Metals 1998, pp. 45–53. (The paper was reprinted in the *Essential Readings in Light Metals, Vol. 1, Alumina and Bauxite*, TMS-Wiley, 2013, pp. 362–370)

11. N.S. Raghavan, G.D. Fulford, Mathematical modelling of the kinetics of gibbsite extraction and kaolinite dissolution/desilication in the Bayer process, in *Proceedings of 127th TMS Annual Meeting*, ed. by B. Welch, San Antonio, TX, 15–19 Feb. 1998, Light Metals 1998, pp. 29–36
12. R. LaMacchia, Toward a better understanding of desilication product (DSP) precipitation kinetics, in *Proceedings of the 9th International Alumina Quality Workshop*, Perth, WA, 18–22 Mar 2012, pp. 214–218
13. T. Harato et al., The development of a modified Bayer process that reduces the desilication loss on soda by 50% compared to the conventional process, in *Proceedings of 8th International ICSOBA Conference*, Milan, Italy, 16–18 Apr. 1997 Travaux ICSOBA, vol. 24, No 28 (1997), pp. 205–213
14. J. Addai-Mensah et al., Sodium aluminosilicate solid phase specific fouling behaviour, in *ECI Conference of Heat Exchanger Fouling and Cleaning. Fundamentals and Applications*, Santa Fe, New Mexico, USA, 18–22 May 2003, Year 2004, Paper 11 (2004)
15. M.C. Barnes, J. Addai-Mensah, A.R. Gerson, A methodology for quantifying sodalite and cancrinite phase mixtures and the kinetics of the sodalite to cancrinite phase transformation *Microporous Mesoporous Mater.* 31 (1999) pp. 303-319
16. M.C. Barnes, J. Addai-Mensah, A.R. Gerson, The mechanism of the sodalite-to-cancrinite phase transformation in synthetic spent liquors *Microporous Mesoporous Mater.* 31 (1999) pp. 287-302
17. E. Nemecz, K. Solymár, S. Elek, A Bayer-rendszerű timföldgyártás során keletkező nátrium- és kalcium-alumínium-szilikátokról. (On the sodium and calcium-aluminium-silicates formed in the Bayer process.) *Bányászati és Kohászati Lapok – Kohászat*, 101, 3 (1968), pp. 89–97
18. H. Peng, D. Seneviratne, J. Vaughan, Role of the amorphous phase during sodium aluminosilicate precipitation Ind. Eng. Chem. Res. 57 (2018) pp. 1408–1416
19. L. Shi, J. Li, A.R. Gerson, The kinetics of desilication of synthetic Bayer spent liquor seeded with synthetic sodalite and cancrinite, in *Proceedings of the 10th International Alumina Quality Workshop*, Perth, WA, 19–23 Apr 2015, pp. 155–163
20. G. Bánvölgyi, Scale formation in alumina refineries, in *Proceedings of the 34th International Conference and Exhibition of ICSOBA*, 3–6 Oct. 2016, Quebec City, Canada. No 45 (2016), pp. 101–114
21. G.I.D. Roach, A.J. White, Dissolution of kaolinite in caustic liquors, in *Proceedings of 117th TMS Annual Meeting,* in ed. by L.G. Boxall, Phoenix, Arizona, 25–28 Jan. 1988, Light Metals 1988, pp. 41–47
22. G. Bánvölgyi, A. Csordás-Tóth, I. Tassy, In situ formation of sodium aluminium hydrosilicate from kaolinite, in *Proceedings of the 120th TMS Annual Meeting*, ed. by E. Rooy, New Orleans, LA, 17–21 Feb. 1991, Light Metals 1991, pp. 5–15
23. G. Bánvölgyi, Reducing of the silica problem in the alumina refinery of C.V.G. BAUXILUM conclusive report of ICF Kaiser-Aluterv Ltd., Budapest, Hungary, Manuscript. Apr. 1998, Attachment 6. Detailed Material, Heat and Silica Balances
24. P.P. Siklósi, Mathematical model of the technology of the Bayer alumina manufacturing process, in *Aluterv-FKI - UNIDO*, Budapest, Manuscript (1984), p. 40
25. A.N. Adamson, E.J. Bloore, A.R. Carr, Basic principles of Bayer process design, in *Extractive Metallurgy of Aluminium*, ed. by G. Gerard, P.T. Stroup, vol. 1, Alumina, Interscience Publishers, New York, London, Sydney, 1963
26. S.P. Rosenberg, S. Healy, A thermodyanamic model for gibbsite solubility in Bayer liquors, in *Proceedings of the 4th International Alumina Quality Workshop, Darwin, Northern Territory*, 2–7 June 1996, pp. 301–310
27. G. Bánvölgyi, Zárójelentés. Gibbsites bauxitok kovasavtalanításának és feltárásának vizsgálata. (Final Report on the Investigation of the Desilication and Digestion of Gibbsitic Bauxites). Manuscript, Aluterv-FKI, Budapest (1990), p. 23

28. M. Jamialahmadi, H. Müller–Steinhagen, Determining silica solubility in Bayer process liquor. *JOM* **50**(11) (1998), pp. 44–49
29. I.G. Zsélyi, Improved Low Temperature Digestion (140–150°C), modellezés, érzékenységanalízis and paraméterbecslés. (Modeling of the ILTD Process (at about 140–150°C), Sensitivity Analysis, Parameter Estimation) *Interim Report* as of 23 Nov 2008
30. K. Yamada, M. Yoshihara, S. Tasaka, Properties of scale in Bayer process, in *Proceedings of the 114th Annual Meeting of TMS*, ed. by H.O. Bohner, New York, 24–28 Feb. 1985, Light Metals 1985, pp. 223–235
31. T. Harato, T. Ishida, K. Yamada, Autoprecipitation of gibbsite and boehmite, in *Proceedings of 111th AIME Annual Meeting*, ed. by J.E. Andersen, Dallas TX, 14–18 Feb. 1982, Light Metals 1982, pp. 37–48
32. J.G. Lepetit, Autoprecipitation of alumina in the Bayer process, in *Proceedings of TMS 115th Annual Meeting*, ed. by R.E. Miller, New Orleans, LA, 2–6 Mar. 1986, Light Metals 1986, pp. 225–230
33. M. Authier-Martin et al., Boehmite reversion: predictive test and critical parameters for bauxites from different geographical origins, in *Proceedings of the 6th International Alumina Quality Workshop*, 8–13 Sept. 2002, Brisbane, Queensland (2002), pp. 109–114
34. J. Loh et al., Boehmite vs. gibbsite precipitation, in *Proceedings of 134th TMS Annual Meeting,* San Francisco, California, 13–17 Feb. 2005, Light Metals 2005, pp. 203–208
35. F. Picard, V. Ouellet, G. Forté, Towards a precise and accurate method to determine bohemite reversion, in *Proceedings of 7th International Alumina Quality Workshop*, Perth, WA, 16–21 Oct 2005, pp. 193–198.
36. A.G. Suss et al., Geological mineral and process features of Fria bauxites (Guinea), in *Proceedings of 7th International Alumina Quality Workshop*, Perth, WA, 16–21 Oct 2005, pp. 193–198
37. V. Ouellet, G. Forté, F. Picard, Utilization of thermogravimetric analysis for quantification of aluminium phases in red mud effects of Bayer liquor on boehmite reversion, in *Proceedings of 17th International Symposium of ICSOBA*, Montreal, Canada, 1–4 Oct. 2006, Travaux, vol. 33, No. 37 (2006), pp. 166–177
38. E. Bujdosó, A. Imre, L. Tóth, Autoklávkeverők keverési hatásfokának vizsgálata radioaktív izotóppal (Investigation of the mixing effectiveness of agitators of autoclaves by radioactive isotopes) *Fémipari Kutató Intézet Közleményei V*. Budapest (1961), pp. 99–105
39. G. Winkhaus, Measurements by means of radioisotopes in industrial tube digester and autoclave lines of bauxite plants, in *Proceedings of 2nd International Symposium of ICSOBA*, Budapest, 6–10 Oct 1969, vol. III (1971), pp. 147–157
40. G. Bánvölgyi, Production of bauxite residue with low soda and high iron content: the ILTD process option, in *Proceedings of the 2nd Bauxite Residue Valorisation and Best Practices Conference*, ed. by Y. Pontikes, Athens, Greece, 7–10 May 2018, pp. 67–76
41. G. Bánvölgyi, How to increase the economic viability of Bayer alumina by the ILTD process? Lecture at the *IBAAS—JNARDDC Webinar 2020*, 4–6 Nov 2020
42. I. Korcsmáros, Computerized evaluation of the operation mechanism of digester-lines in alumina plants by mathematical modeling, in *Proceedings of 4th International Congress for the Study of Bauxites Alumina and Aluminium*, Athens, Greece, 9–12 Oct. 1978, vol. 3 (1978), pp. 189–205
43. B. Pei, D. Taylor, D. Thomas, Extraction kinetics of Weipa monohydrate bauxite, in *Proceedings of the 6th International Alumina Quality Workshop*, Brisbane, Queensland, 8–13 Sept 2002, pp. 123–127
44. T. Oku, K. Yamada, The dissolution rate of quartz and the rate of desilication in the Bayer liquor, in *Proceedings of Symposia 100th Annual Meeting*, ed. by T.G. Edgeworth, New York, 1–4 Mar. 1971, Light Metals 1971, pp. 31–45
45. K. Zheng, The influence of sodium carbonate on sodium aluminosilicate crystallisation and solubility in sodium aluminate solutions *J. Crystal Growth* 171 (1997) 1–2 pp. 197-208 https://doi.org/10.1016/S0022-0248(96)00480-0

46. B.I. Whittington, Quantification and characterisation of hydrogarnet and cancrinite present in desilication product (DSP) by powder X-ray diffraction, in *Proceedings of 4th International Alumina Quality Workshop*, Darwin, 2–7 June 1996, pp. 413–422
47. K. Wefers, Zur chemischen Technologie des Bauxitaufschlusses IV. Untersuchungen im system Na_2O-Al_2O_3-TiO_2-H_2O. (To the chemical process of bauxite digestion. Studies in system of Na_2O-Al_2O_3-TiO_2-H_2O) *Metall* **25**(3), (1971) pp. 239–250
48. E. Schultze-Rhonhof, G. Winkhaus, Untersuchungen im system Na_2O-CaO-Al_2O_3-TiO_2-H_2O bei 100 °C unter Normaldruck, (Studies in system of Na_2O-CaO-Al_2O_3-TiO_2-H_2O at 100 °C under normal pressure) *Zeitschrift für anorganische und allgemeine Chemie* **390**(2), (1972), pp. 97–103
49. E. Schultze-Rhonhof, Untersuchungen im system Na_2O-CaO-Al_2O_3-TiO_2-H_2O im Temperaturbereich 100–275°C (Studies in system of Na_2O-CaO-Al_2O_3-TiO_2-H_2O in 100–275 °C temperature range) *Zeitschrift für anorganische und allgemeine Chemie* **396**, (1973) pp. 303–307
50. V. Hazainé Borsiczky, K. Solymár, A bauxitok TiO_2 tartalmának szerepe a Bayer-rendszerű timföldgyártásban. (The role of the TiO_2 content of bauxites in the Bayer process.) *Bányászati és Kohászati Lapok – Kohászat* **101**(5), (1968) pp. 187–193
51. N.S. Malts, Efficiency of lime use in Bayer alumina production, in *Proceedings of the 120th TMS Annual Meeting*, ed. by E. Rooy, New Orleans, LA, 17–21 Feb. 1991, Light Metals 1991, pp. 257–262
52. D. Croker, M. Loan, K. Hodnett, Sodium titanate formation in high temperature Bayer digestion, in *Proceedings of 17th International Symposium of ICSOBA Combined with Light Metals 2006 and of the Metallurgical Society of CIM*, Montréal, Canada, 1–4 Oct. 2006, Travaux ICSOBA, vol. 33, No. 37 (2006), pp. 154–165
53. K.I. Verghese, Impact of impurities on the Bayer process, in *Proceedings of 8th International Light Metals Congress*. Leoben-Vienna (1987), pp. 42–46
54. P. Smith, Reactions of lime under high temperature Bayer digestion conditions, in *Proceedings of 10th International Alumina Quality Workshop*, Perth, WA, 19–23 Apr 2015, pp. 373–381
55. S.P. Rosenberg, D.J. Wilson, C.A. Heath, Some aspects of calcium chemistry in the Bayer process, in *Proceedings of 130th TMS Annual Meeting*, ed. by J.L. Anjier, New Orleans, Louisiana, 11–15 Feb. 2001, Light Metals 2001, pp. 19–26
56. G. Riley et al., Plant impurity balances and impurity inclusion in DSP, in *Proceedings of the 5th International Alumina Quality Workshop*, Bunbury, WA, 21–26 Mar. 1999, Report 41
57. G.W. Riley et al., A generic system of plant impurity balance models, in *Proceedings of the 6th International Alumina Quality Workshop*, 8–13 Sept 2002, Brisbane, Queensland, pp. 144–148
58. A. Kuvyrkina et al., The effect of DSP structure on removal of impurities from refinery liquor and efficiency of alkali recovery process, in *Proceedings of the TMS 137th Annual Meeting*, New Orleans, LA, 9–13 Mar. 2008, Light Metals 2008, pp. 127–132
59. Personal communication with Ken Evans. Nov 2018
60. P Smith The processing of high silica bauxites—review of existing and potential processes *Hydrometallurgy* 98 (2009) pp.162 176
61. W.E. Wahnsiedler, The kinetics of bauxite digestion in the Bayer process, in *Proceedings of the TMS 113th—AIME Annual Meeting*, ed. by J.P. McGeer, Los Angeles, CA, Feb 27–Mar 1. 1984, Light Metals 1984, pp. 146–182
62. M. Authier-Martin, G.D. Fulford, F. Feret, Bauxite extractable phases in the Bayer high temperature process: Re-assessment of boehmite content aluminium substitution in alumino-goethite, in *Proceedings of 5th International Alumina Quality Workshop*, Bunbury WA, 21–26 Mar 1999, Report 37
63. D. Thomas, B. Pei, Chemical reaction engineering in the Bayer process, in *Proceedings of the TMS 136th Annual Meeting*, ed. by M. Sorlie, Orlando, Florida, USA, Feb 25–Mar 1. 2007, Light Metals 2007, pp. 49–54
64. O. Levenspiel Chemical Reaction Engineering, *Wiley* USA (1972)

65. K.G. Denbigh, J.C.R. Turner, Kémiai reaktorok (Chemical Reactors). *Műszaki Könyvkiadó*, Budapest (1971), p. 22
66. P. Benedek, A. László, A Vegyészmérnöki Tudomány Alapjai. (The Basics of the Chemical Engineering Science) *Műszaki Könyvkiadó*, Budapest (1964), p. 54
67. W.J. Müller, H. Hiller, Verfahren zur Herstellung der Tonerde. (Process for production of alumina) *German Patent* DE579114, Issued 1933
68. B. Lányi, Des Schnelle, Kontinuerliche Bauxitaufschluss (The fast, continuous digestion of bauxite), in *Proceedings of Symposium sur les bauxites, oxydes et hydroxides d'aluminium*, Zagreb, 1–3 Octobre 1963, Zagreb, vol. III (1964), pp. 105–117
69. H. Peters, Tube digestion—late success of a VAW technology, in *Proceedings of 35th International Conference of ICSOBA*, Hamburg, Germany, 2–5 Oct. 2017, Travaux ICSOBA, No. 46 (2017), pp. 145–153
70. K. Bielfeld, W. Arnswald, Aluminium, 43, 6 (1967), p. 355
71. F. Orbán et al., Processing of monohydrate bauxites with tube digestion, in *Proceedings of the TMS 118th Annual Meeting*, ed. by P.G. Campbell, Las Vegas, Nevada, Feb 27–Mar 3. 1989, Light Metals 1989, pp. 1027–1034
72. T. Ogava et al., Effect of acid cleaning using inhibitor in a tubular digestion and in a vapour recompressor evaporator at Sumitomo's Alumina Refinery, in *Proceedings of the 7th International Alumina Quality Workshop, Perth, WA,* 16–21 Oct 2005, pp. 41–45
73. S. Ostap, Effect of bauxite mineralogy on its processing characteristics, in BAUXITE *Proceedings of 1984 Bauxite Symposium*, ed. by L. Jr. Jacob, Los Angeles, California, Feb 27–Mar 1 1984, pp. 651–670
74. J.G. Pulpeiro et al., Sizing an organic control system for the Bayer process, in *Proceedings of TMS 127th Annual Meeting*, ed. by B. Welch, San Antonio, TX, 15–19 Feb. 1998 Light Metals 1998, pp. 89–95
75. G. Bárdossy, G.J.J. Aleva, Lateritic bauxites, in *Akadémiai Kiadó*, Budapest, pp. 147–148. (1990) (In the monograph referred the contents of organic compounds are given as g/L. This g/L unit is deemed to be a mistype. GB)
76. R.L. Clegg, L.G. Armstrong, Development of liquor purification at Alcan Gove, in *Proceedings of 7th International Alumina Quality Workshop*, Perth, WA, 16–21 Oct 2005, pp. 137–141
77. G. Lever, Identification of organics in Bayer liquor, in *Proceedings of 107th AIME Annual Meeting*, ed. by J.J. Miller, Denver, Colorado, Feb 26–Mar 2. 1978, Light Metals 1978, pp. 71–83
78. M.A. Wilson et al., Transformation of organics input to alumina refineries, in *Proceedings of the 6th International Alumina Quality Workshop*, Brisbane, Queensland, 8–13 Sept 2002, pp 67–72
79. L.G. Armstrong, S.J. Healy, Degradation of liquor organics during Bayer digestion cycles, in *Proceedings of 17th International Symposium of ICSOBA Combined with Light Metals 2006 and of the Metallurgical Society of CIM.*, Montréal, Canada, 1–4 Oct 2006. Travaux ICSOBA, vol. 33, No. 37 (2006), pp. 291–301
80. K. Yamada, T. Hashimoto, K. Nakano, Behaviour of organic substances in the Bayer process, in *Proceedings of the 102nd AIME Annual Meeting*, ed. by A.V. Clack, Chicago, Illinois. Light Metals 1973, pp. 745–753
81. M.M. Reyhani et al., Gibbsite nucleation at sodium oxalate surfaces, in *Proceedings of 5th International Alumina Quality Workshop*, Bunbury, WA, 21–26 Mar 1999, Report No 19
82. K. Solymár, M. Gimpel-Kazár, E. Molnár, Determination and evaluation of organic balances of alumina refineries, in *Proceedings of TMS 125th Annual Meeting,* Anaheim, California, USA, 4–8 Feb. 1996, Light Metals 1996, pp. 29–35
83. O'Connell et al., Maintaining an oxalate free precipitation circuit at Aughinish alumina, in *Proceedings of the 6th International Alumina Quality Workshop*, Brisbane, Queensland, 8–13 Sept 2002, pp. 270–273
84. K.R. Beckham, S.C. Grocott, A thermodynamically based model for the oxalate solubility in Bayer liquor, in *Proceedings of TMS 122nd Annual Meeting*, ed. by S.K. Das, Denver, Colorado, 21–25 Feb. 1993, Light Metals 1993, pp. 167–172

85. B. Whittington, The chemistry of CaO and $Ca(OH)_2$ relating to the Bayer process *Hydrometallurgy* 43 (1996) pp. 13-35
86. X. Pan et al., Effect of lime addition on the predesilication and digestion properties of a gibbsitic bauxite, in *Proceedings of TMS 141st Annual Meeting*, ed. by C. Suarez, Orlando, FL, 11–15 Mar. 2012, Light Metals 2012, pp. 39–42
87. G. Roach, E. Jamieson, Effect of bauxite and digestion conditions on iron in SGA, in *Proceedings of 6th International Alumina Quality Workshop*, Brisbane, Queensland, 8–13 Sept, 2002, p. 340, Paper 56
88. P. Basu, Reactions of iron minerals in sodium aluminate solutions, in *Proceedings of 112th AIME Annual Meeting*, ed. by E.M. Adkins, Atlanta, Georgia, Mar. Light Metals 1983, pp. 83–97
89. G. Roach, The equilibrium approach to causticisation for optimising liquor causticity, in *Proceedings of TMS 129th Annual Meeting*, ed. by E.D. Peterson, Nashwille, TN, 12–16 Mar. 2002, Light Metals 2002, pp. 228–234
90. A. Lalla, R. Arpe, 30 years of experience with tube digestion, in *Proceedings of 17th International Symposium of ICSOBA Combined with Light Metals 2006 and of the Metallurgical Society of CIM*, Montréal, Canada, 1–4 Oct 2006, pp. 105–113
91. A. Lalla, R. Arpe, 12 years experience with wet oxidation, in *Proceedings of TMS 131st Annual Meeting, Seattle*, ed. by W.A. Schneider, Washington, 17–21 Feb. 2002, Light Metals 2002, pp. 177–180
92. D. Mulligan, Environmental management in the Australian minerals and energy industries, principles and practices (1996)
93. G. Graham, R. Capil, R. Davies, Odour destruction for digestion vent gases, in *Proceedings of the 6th International Alumina Quality Workshop*, 8–13 Sept 2002, Brisbane, Queensland, pp. 316–319
94. S.J. Cox, Odour emissions reduction case study at Alcoa's Wagerup alumina refinery, in *Proceedings of 6th International Alumina Quality Workshop*, Brisbane, Queensland, 8–13 Sep, 2002, pp. 309–315
95. C. Dobbs et al., Mercury emissions in the Bayer refinery—an overview, in *Proceedings of 7th International Alumina Quality Workshop*, Perth, WA, 16–21 Oct 2005, pp. 199–20
96. F. Habashi, A hundred years of the Bayer process for alumina production, in *Proceedings of TMS 117th Annual Meeting*, ed. by L.G. Boxall, Phoenix, Arizona, 25–28 Jan. 1988, Light Metals 1988, pp. 85–93
97. P. Landry, H. Edwards, Pressure decantation at Gramercy Alumina, in *Proceedings of the TMS 136th Annual Meeting*, ed. by M. Sorlie, Orlando, FL, Feb 25–Mar 1. 2007, Light Metals 2007, pp. 470–475
98. J. Doucet et al., Pressure decantation technology: the Kaiser Gramercy experience, *6th International Alumina Quality Workshop,* Brisbane, Queensland, 8–13 Sept 2002, pp. 94–99
99. J.M. Lamerant, Y. Perret, Boehmitic reversion in a double digestion process on a bauxite containing trihydrate and monohydrate, in *Proceedings of TMS 131st Annual Meeting*, ed. by W.A. Schneider, Seattle, Washington, 17–21 Feb. 2002, Light Metals 2002, pp. 377–380
100. Q. Chen, X. Yuming, L. Hepler, Calorimetric study of the digestion of gibbsite, $A1(OH)_{3(cr)}$, and thermodynamics of aqueous aluminate ion, $A1(OH)_4{}^-{}_{(aq)}$. *Canadian Journal of Chemistry* **69**, (1991) pp. 1685–1690
101. B. Hemingway, R, Robie, J. Apps, Revised values for the thermodynamic properties of boehmite, AlO(OH) and related species and phases in the system Al-H-O *Am. Miner.* 76 (1991) pp. 445-457
102. https://chemistry.stackexchange.com/questions/47742/theoretical-temperature-change-in-lime-hydration . Accessed on 25 Dec 2018
103. L. Kirwan et al., Characterisation of iron mineralogy in Jamaican bauxite and associated aspects of alumina and soda losses, in *Proceedings of TMS 138th Annual Meeting*, ed. by G. Barne, San Francisco, California, 15–19 Feb. 2009, Light Metals 2009, pp 133–138
104. G. Bonel, J.C. Labarthe, C. Vignoles, Contribution á l'etude structurale des apatites carbonatées de type B, (Contribution to the structural studies of carbonate apatites type B) *Colloques Internationaux C.N.R.S.*, 230 (1975), pp. 117–125

105. J. Zöldi et al., Iron hydrogarnets in the Bayer process, in *Proceedings of the TMS 116th Annual Meeting*, ed. by R.D. Zabreznik, Denver, Colorado, 24–26 Feb. 1987, Light Metals 1987, pp. 105–111
106. G. Pásztor, P. Siklósi et al., Könnyűfémek metallurgiája (Metallurgy of Light Metals) *Tankönyvkiadó*, Budapest (1991), p. 57
107. H. Mercier, R. Magrone, Influence de l'addition de chaux a l'attaque alcaline en voie humide des bauxites a diaspore. (Influence of the addition of lime to the wet alkaline attack of diaspore in bauxites.) *Sommaire de ICSOBA 3e Congres International,* Nice, France (1973), pp. 513–522
108. S. Gu, Key technology development of Bayer process in China, in *Proceeding Book of 32th International Conference of ICSOBA-2014* in Zhengzhou, China, 12–15 Oct 2014, *No 43* pp. 154–162
109. Á. Juhász, Orbán F-née, M. Matula, A Bayer-eljárás során keletkező nátrium-alumínium-szilikátok kémiai összetételének és szerkezetének vizsgálata. (Investigation of chemical composition and structure of the sodium-aluminium-silicates that formed in the Bayer process) *Bányászati és Kohászati Lapok – KOHÁSZAT* **11**, (1965), pp. 513–521
110. I. Vörös et al., Causticization of red mud, in *Int. Colloqium on Alumina Production from Low Grade Bauxites, Banska Bystrica – Ziar nad Hronom (Czechoslovakia)*, 6–8 June 1972. Travaux ICSOBA No 12 (1974), pp. 121–134
111. K. Solymár, M. Orbán, J. Zöldi, G. Baksa, Methods for reducing NaOH losses in the Hungarian alumina plants, in *Proceedings of the 5th International Congress of ICSOBA*, 26–28 Sept 1983. Zagreb, Yugoslavija No 18 (1983), pp. 377–390
112. Q. Zhao et al., Economic analysis of producing alumina with low-grade bauxite (red mud) by calcification-carbonization method, in *Proceedings of TMS 143rd Annual Meeting*, ed. by J. Grandfield, San Diego, CA, 14–16 Feb. 2014, Light Metals 2014, pp. 165–168
113. P.J. Cresswell, D.J. Milne, Hydrothermal Recovery of Soda and Alumina from Red Mud: Test in a Continuous Flow Reactor. In *Proceedings of the TMS-AIME 113th Annual Meeting*, ed. by J.P. McGeer Los Angeles, CA, February 27 – March 1 1984, *Light Metals 1984*, pp. 211–221.
114. K. Solymár, J. Steiner, J. Zöldi, Technical pecularities and viability of hydrothermal treatment of red mud, in *Proceedings of the TMS 126th Annual Meeting*, ed. by R. Huglen, Orlando, FL, 9–13 Feb. 1997, Light Metals 1997, pp. 49–54
115. S.P. Rosenberg, D.M. Wilson, C.A. Heath, Improved Bayer causticisation. *Australian Patent*, No. 769770, Granted: 05.02.2004.
116. P.J. The, T.J. Sivakumar, The effect of impurities on calcium in Bayer liquor, in *Proceedings of the TMS 114th Annual Meeting*, ed. by H.O. Bohner, New York, 24–28 Feb. 1985, Light Metals 1985, pp. 209–222
117. D. Wilson et al., Fluoride chemistry in the Bayer process, in *Proceedings of the Sixth International Alumina Quality Workshop*, Brisbane, Australia, 8–13 Sept 2002, pp. 281– 287
118. J. Zámbó, M. Orbán-Kelemen, Calcium oxide and magnesium oxide compound formation in processing calcite-dolomite bearing bauxites by the Bayer method *Acta Tech. Acad. Sci. Hung.* 82 (1976) 3–4 pp. 333-352
119. N.S. Malts, The intensifying action of upon the kinetics of bauxite leaching *Sov. J. Non-Ferrous Met.* 26 (1985) 11 pp. 38-40
120. S Gu, R Cao, X Chen, Calcium hydroxytitanate in the Bayer digestion process *Rare Met. (China)* 8 (1989) 1 pp. 14-18
121. D. Croker, M. Loan, B.K. Hodnett, Kinetics and mechanisms of the hydrothermal crystallization of calcium titanate species *Cryst. Growth Des.* 9 (2009) 5 pp. 2207-2213
122. A. Suss et al., High titanium bauxites: specific features of mineral composition and behaviour in Bayer cycle (by the example of Indian bauxites), in *Proceedings of the Sixth International Alumina Quality Workshop*, Brisbane, Australia, 8–13 Sept 2002, pp. 180–184

123. G.D. Fulford, Scavenging non-filterable iron from Bayer liquors using activated digester sands, in *Proceedings of the TMS 118th Annual Meeting*, ed. by P.D. Campbell, Las Vegas, Nevada, Feb 27–Mar 3. 1989, Light Metals 1989, pp. 77–89
124. A. Suss et al., The impact of magnesium on tricalcium aluminate (TCA) filter aid properties, in *Proceedings of the 8th International Alumina Quality Workshop* in Darwin, Northern Territorry, 7–12 Sept 2008, pp. 37–43
125. L.J. Andermann, G.J. Pollet, The manufacture of tricalcium aluminate, in *Proceedings of the TMS 132nd Annual Meeting*, ed. by B. Sadler, San Diego, California, USA, 2–6 Mar. 2003, Light Metals 2003, pp. 11–17
126. S.C.A. Franca et al., Some aspects of tricalcium aluminate hexahydrate formation on the Bayer process, in *Proceedings of the TMS 139th Annual Meeting*, ed. by J.A. Johnson, Seattle, Washington, USA, 14–18 Feb. 2010, Light Metals 2010, pp. 63–66
127. V.C. HuiLing, Optimal concentration of filter aid for Bayer alumina precoat filters: optimisation of filtration rate through an effective filter-aid particle size distribution. *Ph.D. thesis, Unversity of Queensland* (2002)
128. M. Mohapatra, S. Acharya, Tricalcium aluminate hexahydrate (TCA) synthesis and characterization. In: *Proceedings of the TMS 139th Annual Meeting ed. by J.A. Johnson,* Seattle, Washington, USA, February 14–18, 2010, *Light Metals 2010*, pp. 119–124
129. N. Mugnier, P. Clérin, J. Sinquin, Industrial experience of polishing filtration performance. Improvement and interpretation, in *Proceedings of TMS 130th Annual Meeting*, ed. by J.L. Anjier, New Orleans, Louisiana, 11–15 Feb. 2001, Light Metals 2001, pp. 33–39
130. J.T. Malito, Improving the operation of red mud pressure filters, in *Proceedings of TMS 125th Annual Meeting*, ed. by W. Hale, Anaheim, California, USA, 4–8 Feb. 1996, Light Metals 1996, pp. 81–86
131. J.M. Rousseaux, P. Ferland, Impact of excess synthetic flocculant on security filtration, in *Proceedings of TMS 138th Annual Meeting*, ed. by G. Bearne, San Francisco, California, USA, 15–19 Feb. 2009, Light Metals 2009, pp. 157–161
132. J. Romero, F. Gauthier, Improvements in overall performance of security filtration area, in *Proceedings of the TMS 121st Annual Meeting*, ed. by E. Cutshall, San Diego, California, 1–5 Mar. 1992, Light Metals 1992, pp. 97–98
133. B.I. Whittington, T.M. Fallows, M.J. Willing, Tricalcium aluminate hexahydrate (TCA) filter aid in the Bayer industry: factors affecting TCA preparation and morphology Int. *J. Miner. Process.* 49 (1997) 1 29
134. S.P. Rosenberg, D.J. Wilson, Process for filter aid production in alumina refineries, *US Patent* 7,192,568 (Date of patent: 20 Mar 2007)
135. R.C. Young, Chemistry of Bayer liquor causticisation, in *Proceedings of TMS 111th AIME Annual Meeting*, ed. by J.E. Andersen, Dallas TX, 14–18 Feb. 1982, Light Metals 1982, pp. 97–117
136. M.T. Chaplin, Reaction of lime in sodium aluminate liquors, in *Proceedings of Symposia 100th Annual Meeting,* ed. by T.G. Edgeworth, New York, 1–4 Mar 1971, *Light Metals 1971*, (Reprinted in the "*Essential Readings in Light Metals, vol. 1, Alumina and Bauxite*", TMS-Wiley, 2013, pp. 202–207)
137. US 2,992,893 (Patent Date: July 18, 1961), US 3,120,996 (Date of Patent: Feb. 11, 1964), US 3,210,155 (Date of Patent: 5 Oct 1965), US 4,486,393 (Date of Patent: 4 Dec 1984) AU 2004201629 (Date of Patent: 13 July 2006)
138. S.P. Rosenberg, D.J. Wilson, C.A. Heath, W. Tichbon, Process for the removal of oxalate and/or sulphate from Bayer liquors, *Australian Patent*, No. 768730 (Date of Patent: 8 Jan 2004)

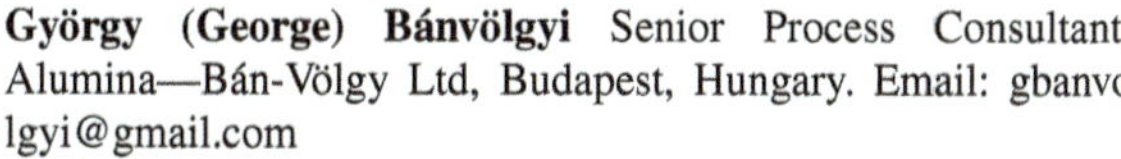

György (George) Bánvölgyi Senior Process Consultant, Alumina—Bán-Völgy Ltd, Budapest, Hungary. Email: gbanvolgyi@gmail.com

Senior Process Consultant with more than 49 years of experience in the alumina industry. Graduated in 1972 in Veszprém University of Chemical Engineering, Hungary (M.Sc.) and as a Patent Attorney, at the Budapest University of Economics, in 1992. His expertise covers processing of bauxites of different types and grades using the low or high temperature digestion processes, process modelling, material and energy conservation of the alumina production. Specialises in on-site techno-economic evaluations leading to process concepts, Feasibility Studies and de-bottlenecking proposals. Further specialties are the environmental aspects, valorisation of bauxite residue, prevention and removal of scales, control of the impurities, also co-ordination and conducting of research and development projects. Travelled extensively in Brazil, Canada, China, France, Guinea, India, Kazakhstan, Russia, Spain, UK, USA, Venezuela and Vietnam. Published several papers, gives lectures at various symposia and workshops. Principal inventor and driver of the Improved Low Temperature Digestion (ILTD) Process aiming at manufacturing alumina in an environmentally sustainable manner and significantly improving its profitability. Author or co-author of 41 publications. Awarded with the Commemorative Medal of ICSOBA.

Brady Haneman Principal Metallurgist (Digestion Technology), High Pressure Metals, Hatch, Canada. Email: brady.haneman@hatch.com

Brady has a Hons Degree in Chemical Engineering from Sydney University Australia and has over 25 years experience in industry. He has extensive experience in conducting feasibility studies, plant and equipment design, engineering management, development of new process technologies and project execution having spent 18 years on feasibility and subsequent detailed engineering phases of major projects. Brady has designed high temperature and low temperature digestion units totaling 9.0 million tpa of alumina refining capacity and these refinery projects have incorporated Tube Digestion, Double Digestion and Hybrid Tube Digestion flow sheets. He has also acted as a Consultant to numerous refineries globally, in particular in Australia, Ireland, Italy, Ukraine, the United States and Saudi Arabia on design improvements for digestion facilities. Currently, Brady is the Principal Metallurgist for Hatch's High Pressure Metallurgy practice specialising in Digestion technologies. Previously as the Global Design Lead—Alumina, Brady led major efforts to improve process plant safety by applying a fundamental understanding of refinery dynamic behaviour during process upsets to the design of overpressure protection systems. He has also developed mitigation strategies for complex pressure transients in piping systems and led the detailed design of Digestion Units and their safety relief systems for over 15 years.

Chapter 5
Liquid–Solid Separation: De-Sanding—Flocculation, Sedimentation and Liquor Filtration

Manfred Bach and Fred Schoenbrunn

Abstract Post-digestion, the solids and aluminum bearing liquor need to be separated, and the solids washed to remove and recover the bulk of the liquor. Sedimentation and filtration equipment are commonly used for this. High temperatures, high scaling rates, and very fine particle size solids make these applications challenging. De-sanding equipment technologies and combinations thereof are described which frequently have been installed to protect flat bottom thickeners from high loads of coarse solid material. Today's thickener technology providing relatively small diameters, steep bottom cones and a high mud bed mostly are operating without sand separation units. Flocculation of the fine solids into larger agglomerates is a key aspect of thickening, along with diluting the feed to an optimal concentration for the flocculation reaction. Good flocculation allows both the production of both clean overflows and high underflow densities that result in better recoveries and washing efficiencies in the CCD circuit. Scale growth is a major operating issue and eventually requires equipment shut down and descaling, so all thickeners are designed to be bypassed during the descaling operation. Historically, descaling was done largely by hand, but newer equipment is frequently designed for use with mechanical descaling arms that improve operator safety. Filtration technologies are described that have been developed by various companies for pregnant/green liquor filtration along with their specific features. A novel filter design is described that is operated completely without the addition of tri-calcium aluminate (TCA) as a filter aid. That specific characteristic is achieved due to a unique filter media cleaning technology comprising an automated hydro jet system.

M. Bach (✉) · F. Schoenbrunn
FL Smidth, Copenhagen, Denmark
e-mail: manfred.bach@flsmidth.com

B. E. Raahauge and F. S. Williams (eds.), *Smelter Grade Alumina from Bauxite*, Springer Series in Materials Science 320,
https://doi.org/10.1007/978-3-030-88586-1_5

5.1 De-Sanding Equipment

In the Bayer process bauxite is mixed with caustic soda during grinding and fed to the digestion area where most of the alumina dissolves by forming soda aluminate. The insoluble residue, the "red mud" must however be separated from the solution and washed in order to recover soluble soda and soda aluminate. This is done by counter-current decantation in raked clarifiers and washers.

The coarse fraction of the red mud, the "sands", can cause mechanical problems in thickeners, mainly in thickeners with flat or slightly sloped bottom due to their limited sand raking capacity, resulting in accumulation of sands and finally in an overload situation of the thickener rake drive. Therefore, the sand fraction must be removed before reaching the clarifiers[1] and washers whereby the relevant particle size is the fraction above 100–150 μm. This can effectively be achieved by the installation of de-sanding equipment.

Typically, de-sanding equipment is installed ahead of the clarifiers. However, there are also examples with de-sanding equipment installed at the rear end of a washer train mainly for producing sands for roads or dam's construction.

Today in modern Bayer process flowsheets de-sanding equipment is not considered as a standard unit operation. Modern clarifier and washer designs are more tolerant to elevated sand loads and do not require upstream de-sanding equipment any longer. However, de-sanding equipment may be used to remove sand for road construction or for building red mud dams.

De-sanding equipment used in the Bayer process mainly is represented by the following list:

- Hydro Cyclones
- Rake Classifiers
- Screw Classifiers
- Bended Screens

A typically de-sanding installation comprises an initial de-liquoring unit with the overflow pumped to the clarifiers and the sand fraction discharged to the sand washing unit. The sand washing unit comprises a certain number of de-sanding stages operated in a countercurrent installation to remove caustic liquor from the sand fraction at minimum wash media consumption. Condensate or plant water with a low caustic concentration are thereby added to the final sand washing stage and either mixed to the slurry before the final separation step takes place, or, in case of the rake classifier, added directly onto the sand on the dry deck.

Most de-sanding flowsheets are considering hydro-cyclones in the feed to the de-liquoring stage for reduction of the hydraulic load.

As an option to rake and screw classifiers also bended screens can be used resulting in a very efficient installation at minimum footprint.

---

[1] The term "Clarifier "is used as a synonym to the term "Settler".

Bended Screens, Cone Classifiers, Hydro separators and Bowl Classifiers have been used in the few flowsheets but not reached significant footprint in the industry. Since most of these classifier designs are not used today any longer or have not reached significant installation base, they are not described in further details hereunder.

5.1.1 De-Sanding by Hydro-Cyclones

In a hydro-cyclone a gravimetric separation takes place due to the buoyancy effect of the media forcing the lighter solids to the center of the cyclone where they are transported upward and through the vortex finder. The dense mineral matter spirals toward the apex and exits through that orifice.

In the Bayer flowsheet sand separation hydro-cyclones are installed for reducing the hydraulic flow rate to the rake or screw classifiers. A typical installation is shown in Fig. 5.1. The hydro-cyclones in these cases typically are installed on top of the de-liquoring classifier with the underflow directly discharging into the feed trough. The overflow of the hydro-cyclones is transferred to the clarifier area.

Wear resistant steel is used as construction material of the hydro cyclones. Depending on the flowrate a single hydro-cyclone or several hydro-cyclones may be installed in a parallel feed arrangement as shown in Fig. 5.1.

Due to rather high flow rates and coarse separation cut, large diameter hydro-cyclones are used, depending on flow rate 24″, 36″ or 48″ diameter. Due to

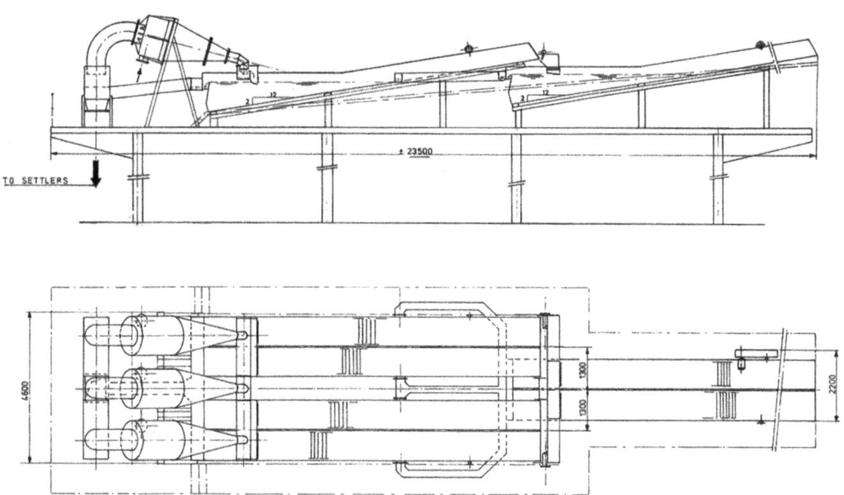

Fig. 5.1 Three hydro cyclones connected to a single feed header with the underflow discharging to a series of rake classifiers

the abrasive character of the solids, abrasion resistant steel hydro-cyclones are recommended.

5.1.2 De-Sanding by Rake Classifiers

The rake classifier is a combination of a classification trough for gravity separation of the coarse sand fraction in the feed slurry and a reciprocating rake for continuously discharging the settled sand from the unit. The feed slurry enters the classification trough through a feed device which provides an equal distribution of the solids over the complete width of the trough. Any coarse sand that will settle in the classification trough will be removed by the action of the rake. The remaining solids fraction will leave the rake classifier with the liquor through the classifier overflow outlet nozzles.

The reciprocating rake is pushing the sand on an inclined deck towards the discharge by a series of single strokes. The rake thereby comprises a motor operated cranking device that pushes forward, lifts up—pushes back and lowers down the rake in order to transport the settled sand towards the discharge end of the rake classifier.

After leaving the submerged part of the classifier the sand on the inclined deck can be washed with a set of spray nozzles and thereafter be drained by gravity to a certain minimum moisture which is depending on the nature of the sand and the liquor.

The classifier size is selected upon the required capacity whereby for large rake classifiers two rakes can be installed in one unit. The two reciprocating rakes of the DSF type Rake Classifier are installed in parallel but operate in a directly opposed raking mode: When one rake is pushing the sand towards the discharge the other rake is in the return stroke.

A lifting device allows adjusting the gap between the rake blades and the classifier trough manually.

The DSF type Rake Classifier (*trademark FLSmidth*) shown in Fig. 5.2 comprises two classifying rakes in one classifier unit for raking the settled sand to the discharge end of the machine. For reduced sand load or smaller feed capacities single rake classifiers are used.

In order to improve the classification efficiency and reduce the caustic concentration in the sand discharged, classifiers are installed in a countercurrent washing scheme comprising an initial de-liquoring classifier followed by sand washing classifiers, typically two or three units (Fig. 5.3).

In a typical countercurrent washing scheme, the overflow from the de-liquoring classifier is fed to the clarifiers.

Rake Classifiers are reliable units, typically operated with a low number of instruments and control loops and are low in power consumption. One distinct advantage is that none of the bearings required to move the rake is in direct contact with the slurry or the sand.

Fig. 5.2 Rake classifier unit type DSF (courtesy of FLSmidth)

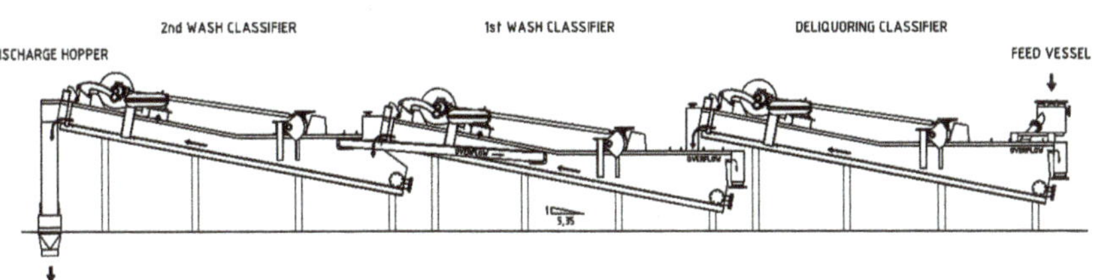

Fig. 5.3 Typical rake classifier installation in a countercurrent washing arrangement comprising one de-liquoring classifier and two wash classifiers

5.1.3 De-Sanding by Screw Classifier

Screw classifiers are very similar in design in comparison to rake classifiers. In both cases an open feed tank is provided allowing the coarse sand to settle. In a screw classifier a large diameter spiral is used to remove the settled sand while in a rake classifier a reciprocating rake is installed for the same reason.

The screw in a screw classifier is a large diameter single pitch steel spiral, mounted on the shaft by spiral arms with spiral flights made of high wear resistant steel. Screw classifiers are providing a submerged lower anti-friction bearing and a lifting device to adjust the spiral slope (Fig. 5.4).

Fig. 5.4 Workshop picture of typical screw classifier (courtesy of FLSmidth)

5.1.4 *Sizing Principles*

Important sizing parameters for rake and screw classifiers are.

- Pool area,
- Sand raking capacity,
- Length of the drying deck.

The pool area is defined by the settling area of the feed tank. The ratio of the available pool area versus flow rate defines the separation cut of the classifier whereby correction factors need to be taken into account related to the number of rake strokes per minute respectively the rotational speed of the spiral. A low number of rake strokes is also advantageous for a sharp separation cut. A minimum dry deck length is required to reach a low moisture in the discharged sand.

Pool area, sand raking capacity and dry deck length are parameters by the specific rake or screw classifier model selected.

Single, double and triple pitch screws are possible, depending on the solid's capacity. With double pitch a nearly twice as high solids capacity is reacted compared to single pitch.

The inclination of the deck also has an influence on the capacity of the classifier: a steeper deck is resulting in a lower capacity compared to a deck with a reduced slope.

The rate of fines removal can also be improved by reducing the rotational speed of the screw and by selecting a screw with a reduced pitch.

5.1.5 DSM Screen with Tilting and Rotating Device

DSM Screens (bended screens) in Alumina Refineries are typically characterized by a 45-degree section and a large radius of over 2 m. DSM screens are used in the bauxite grinding circuit and for removal of sand ahead of the red mud clarifier/washer train (Fig. 5.5).

Feed slurry is introduced, at a slight overpressure, into the feed box distribution device on top of the screen. An adjustable deflection plate provides even distribution over the screen width. The slurry flows by gravity onto the top of the screen deck thereby allowing most of the liquor and the fines to pass the screen slots on the passage towards the lower end of the screen. The coarse particles are falling off the screen by gravity at the lower end into the discharge trough. The liquor with the fine fraction of the solids is collected in the screen box and discharged at the lower end of the box by a pipe.

Fig. 5.5 Workshop picture of bended screen with feed box, tilting device and hood (courtesy of FLSmidth)

A spray nozzle arrangement can be provided on top of the screen deck in order to remove fines from the coarse fraction and. A vapor hood can be arranged on top of the screen deck for vapor removal.

When handling abrasive slurry, the leading edge of the profile bars are subject to higher wear rates and in time become slightly rounded. To maintain constant screening efficiency and capacity, it is necessary to periodically reverse the direction of flow across the screen surface so that the wear pattern on the profile bars may be equalized.

The tilting device of the bended screen shown above provides a manual tilting and rotating device to reverse the screen towards the direction of flow.

5.1.6 T-Type Screen with Automated Tilting Device

The T-type screen is a fully enclosed design with the provision of reversing the screen position automatically. Due to the design with two collecting funnels for the coarse fraction as shown in (Fig. 5.6), two alternate directions of flow (A and B) are possible just by tilting the screen. For tilting the screen, the feed flow needs to be stopped for a short moment only until the change in screen position is completed (Fig. 5.7).

5.2 Flocculation and Sedimentation Principles

5.2.1 Introduction

Effective separation of red mud from the digested bauxite stream is crucial for the operation of a Bayer alumina refinery. The separation is carried out in a series of thickeners whereby the initial thickener is called a decanter, settler, or clarifier designed to produce a clear overflow stream, followed by a series of thickeners called red mud washers which are operated to separate most of the remaining liquor from the red mud. The decanters produce a clear overflow of high strength liquor which is filtered and sent to alumina precipitation. Two decanters are often operated in parallel for the total stream capacity with the underflow of both decanters collectively fed to a single line of washer tanks operated in counter-current decantation (CCD) mode with condensate or wash water added to the last stage washer for mud washing. Last stage washer overflow goes to the 2nd to last stage, and the overflows cascade upstream to wash the mud from the previous washer. Figure 5.8 shows a typical alumina plant decanter/washing circuit arrangement.

The design and performance of the separation circuit is directly coupled to the performance of flocculants. Flocculants agglomerate micron size particles into millimeter size clumps, increasing the settling velocity by orders of magnitude. Over the last half of the twentieth century, flocculant technology changed dramatically

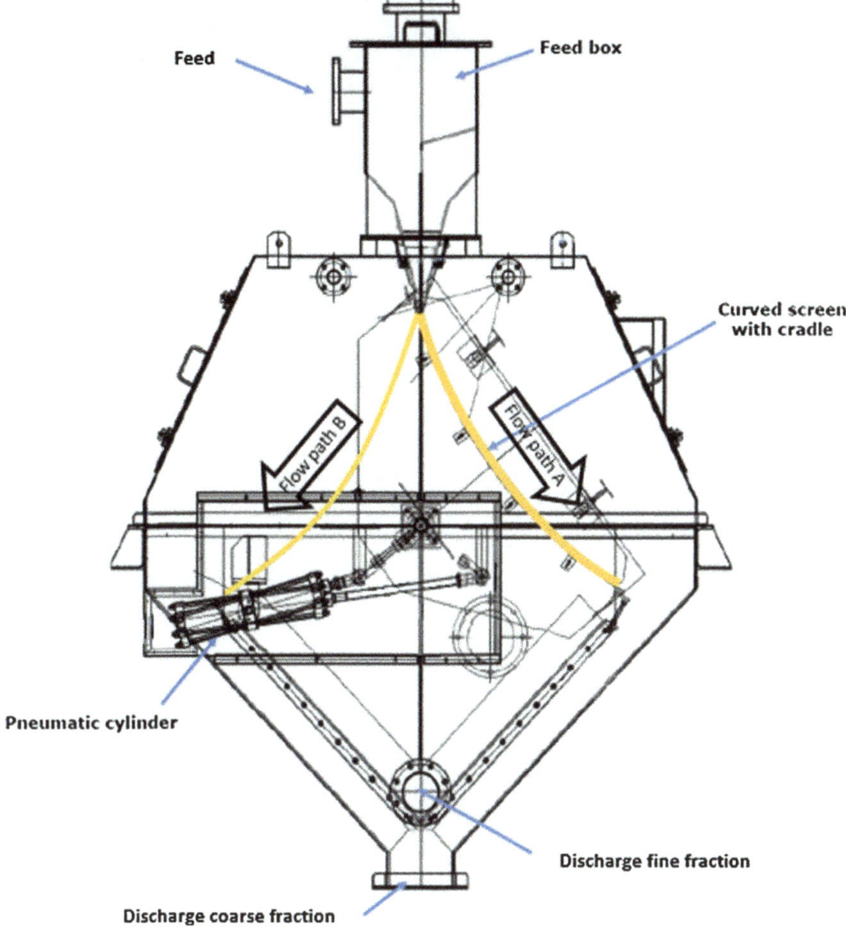

Fig. 5.6 T-type screen general arrangement (courtesy of FLSmidth)

from using natural starches, guar gum, wheat bran etc. to highly specialized synthetic polymers. The synthetic polymers are typically millions of molecular units long with charged sites to aid in attaching to the solids. These polymers, in combination with advanced auto-dilution feed well systems resulted in much higher settling rates allowing significantly reduced diameter of settler and washer tanks [1].

Flocculant suppliers have developed flocculants specialized for decanter applications and for washer applications. Decanter flocculants, in most cases emulsions or liquid in nature, are designed for capturing the fine red mud particles thereby efficiently reducing the overflow turbidity. Often these flocculants are used in combination with other polymers which are promoting underflow density ("co-dosing"). In many cases it is also advantageous to split the flocculant addition into two or more dosing points ("split dosing") and control the split ratio.

Fig. 5.7 Workshop picture of T-type screen with pneumatic cylinder for reversing the screen position (courtesy of FLSmidth)

5.2.2 *Red Mud*

The term "red mud" is widely used in the industry and is a synonym for the fraction of the bauxite that has not been dissolved during digestion and that after passing the decanters and the washers, is finally disposed either as a slurry, a paste or a filter cake. The high concentration of iron compounds in the bauxite gives the by-product its characteristic red color, and hence its common name "red mud". A typical chemical composition is shown in Table 5.1. The main constituents of the red mud after the extraction of the aluminum component are mainly iron oxides, titanium oxide, silicon oxide and un-dissolved alumina together with a wide range of other unreacted metallic oxides which will vary according to the country of origin of the bauxite.

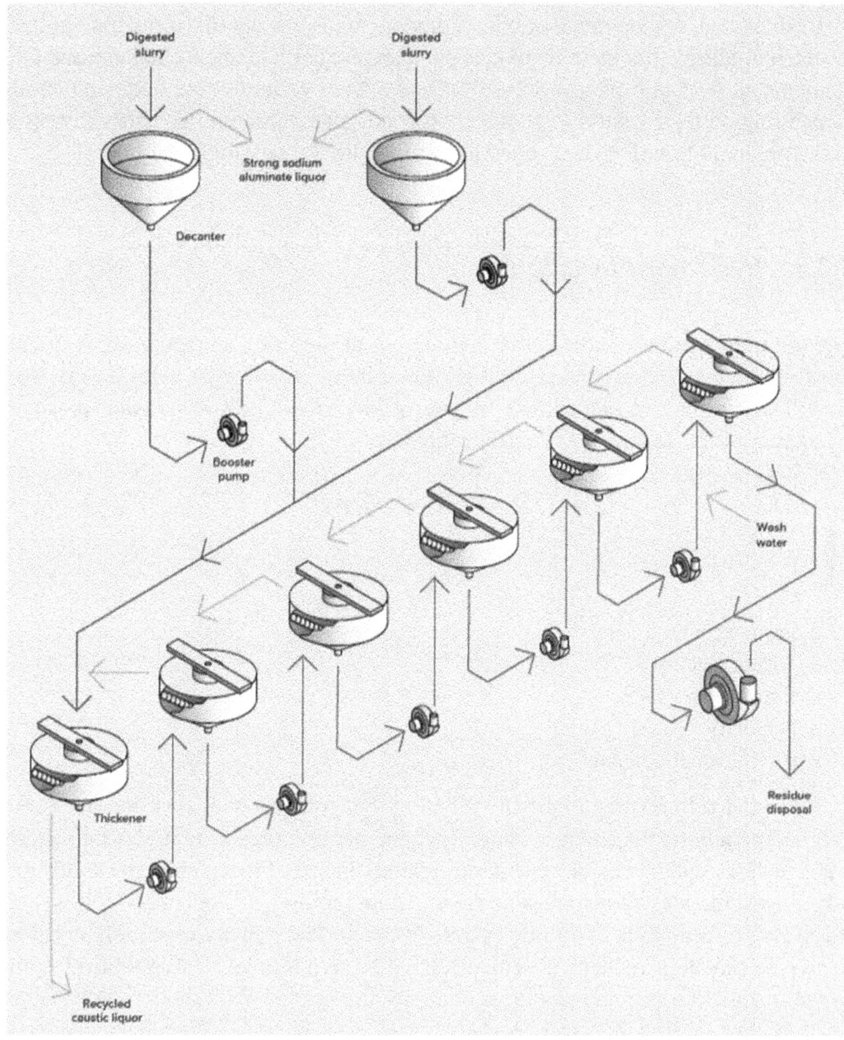

Fig. 5.8 Typical alumina plant mud separation and washing arrangement

Table 5.1 Chemical composition range (%) for bauxite residue for main components [2]

Component	Typical range (%)
Fe_2O_3	20–45
Al_2O_3	10–22
TiO_2	4–20
CaO	0–14
SiO_2	5–30
Na_2O	2–8

Besides that, a wide range of other components is present in the red mud at trace levels, including other metallic oxides but also rare earth elements and non-metallic elements such as phosphorus and sulfur and a wide variety of organic compounds. Depending on their solubility at digestion conditions, these will be partly dissolved in the pregnant liquor or discharged as a solid with the red mud.

5.2.3 *Red Mud Flocculants*

The red mud particles in the slurry coming out of digestion are generally very fine; mostly less than 10 microns, and the settling rate of very fine particles is very, very slow. The settling rate of a particle is governed by Stokes Law, where the velocity is a function of the square of the particle diameter.

$$V = d^2(\rho_s - \rho_l)/\mu g$$

where

d diameter
ρ_s solids density
ρ_l liquid density
μ viscosity
g gravitational constant

Red mud settling was originally aided by coagulating with starches and guars, and later synthetic flocculants. These agglomerate micron size particles into aggregates or flocs, which can be several millimeters in size. Flocculating the solids into larger agglomerates allows the solids to settle out rapidly; the flocculated solids settle much faster, commonly in the range of 50–100 m/h. The starches used initially before the development of modern flocculants, resulted in rather low settling rates leading to the installation of relatively large diameter thickeners. With modern flocculants, much higher settling rates can be easily achieved, leading to a significant change in decanter and washer thickener technology and improved productivity. Polyacrylate and polyacrylate-polyacrylamide co-polymers are one family of commonly used flocculants for settler conditions. Another family are the hydroxamate polymers containing salicylic acid groups, siloxane groups or polymethacrylate based polymers. “Synthetic polymers can be made from a variety of feed-stocks and the nature of the polymer can be tailored in the manufacturing process to deliver products with different performance characteristics” [3]. To achieve required plant capacities, flocculants need to be added to the digested bauxite slurry in order to settle the solids at reasonably high settling rates, thereby achieving target overflow turbidities and underflow solids concentrations. Unit area requirements for thickening have been reduced from 1 to 5 m^2/tpd to less than 0.1 m^2/tpd (Figs. 5.9 and 5.10).

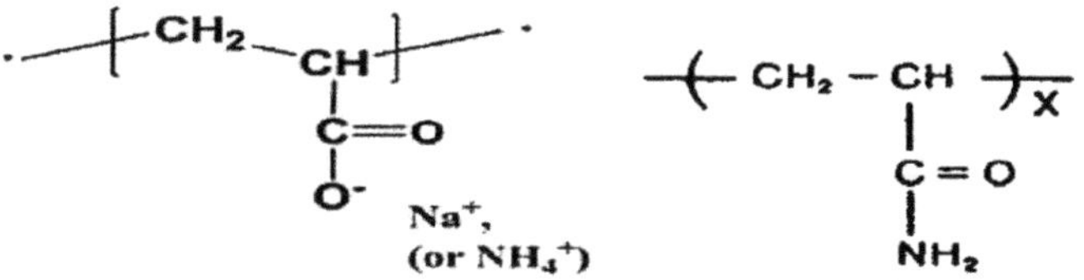

Fig. 5.9 Structure of acrylate (left) and acrylamide homopolymers. *Source* Nalco Ecolab Company

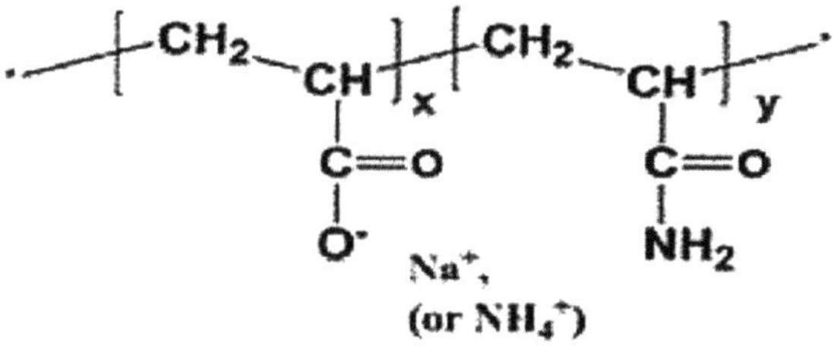

Fig. 5.10 Structure of sodium acrylate/acrylamide copolymer. *Source* Nalco Ecolab Company

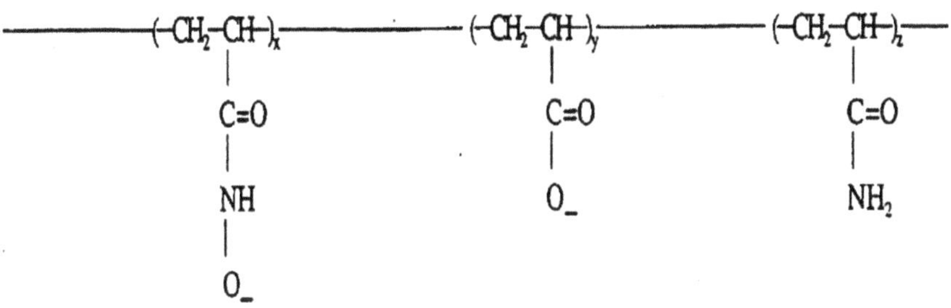

Fig. 5.11 Structure of hydroxamate polymers. *Source* [6]

Anionic charges of the polyacrylate/acrylamide copolymer can be tailored whereby the anionic charge is increasing with increasing acrylate quantities. Higher anionic flocculants are more effective under settler conditions with high caustic concentration. Lower anionic flocculants are most effective under last washer conditions with low caustic concentration [4, 5] (Fig. 5.11).

5.3 Mud Rheology, Rake Drives and Mud Pumping

The development of flocculants as well as feed dilution technology led to much higher capacities on a unit area basis. Accordingly, throughputs have increased, and new equipment designs are based on much higher throughputs and generally smaller diameter equipment.

Where historically large (typically 30–40 m diameter) flat or shallow slope tanks were used, they are now being replaced with small diameter (commonly around

20 m), tall, steep bottom tanks which can process 4–5 times the amount of red mud [7].

In the 1980s and 90s, Alcan developed deep thickener technologies and were able to dramatically increase the density of underflow that could be consistently achieved. These machines evolved to use comparatively high torque drives, streamlined rakes, very deep mud beds and steep floors to produce thick muds from generally tall but small diameter machines. Traditional settlers used in the alumina industry have a relatively low aspect ratio, or height to diameter. Traditional settlers are able to achieve about 30 wt.% (380 gpl) solids in the underflow. Deep cone style thickeners with relatively high aspect ratio are commonly able to produce concentrations in the range of 50 wt.% (750 gpl).

The feed solids concentration has a big impact on the flocculation reaction. There is an optimal concentration for flocculation that is a function of particle size, where finer particle sizes generally flocculate better at lower concentrations. Flocculating at high solids concentrations generally requires high flocculant dosages and results in slow settling and poor overflow clarity. The optimal feed concentration depends on a number of factors including specific gravity and particle size, but by diluting the feed it may be possible to increase the flocculated settling rate by an order of magnitude. Red mud being quite fine, the optimal feed concentration is generally much lower than the concentrations in the red mud washing circuit. Both internal and external dilution can be used, with internal dilution utilizing clarified liquor from within the thickeners to dilute the feed. Internal dilution does not increase the overflow rate and can use either eduction or density differentials to drive the dilution flow. The optimal feed concentration can be determined by measuring the settling velocity as a function of feed concentration at a constant dosage. The product of the settling velocity and the concentration is a settling flux, or solids rate per unit area. There is generally a maximum in the settling flux curve, which starts at the origin, with zero settling flux at zero concentration, and hits a maximum at a low concentration before declining again at higher concentrations. This can be seen in Fig. 5.12 below. Maximizing the settling flux maximizes the capacity of the thickener.

Developments in thickening technology over the past 50 years have enabled significantly higher underflow densities to be consistently produced. In the 1960s and 70s

Fig. 5.12 Settling flux curve

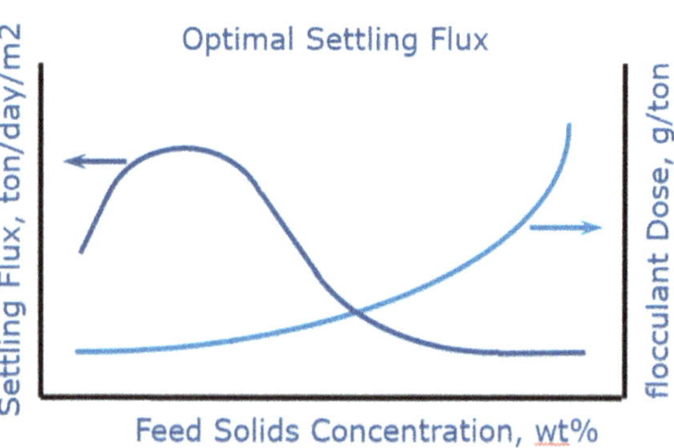

the British National Coal Board (NCB) developed and installed a number of Deep Cone Thickeners (DCTs) to process coal flotation refuse. They were designed to take advantage of modern flocculants that were newly developed at that time and discharged the underflow directly onto a conveyor belt. These early designs were relatively small and utilized a steep bottom cone and a rotating picket assembly to produce high underflow solids concentrations. There were operational difficulties, but these units opened the door to high density underflows. Alcan developed the technology for red mud washers, gradually increasing underflow densities from roughly the 30 wt.% (380 gpl) range up to the 50 wt.% (750 gpl) range. These DCTs were used in the red mud washing circuits and markedly improved the CCD circuit efficiency. The thicker mud also enabled them to use dry tailings disposal methods at the end of the circuit. Very deep mud beds and steep floors allow them to be able to produce and discharge muds near the limit of pumpability, while operating at very high rates in terms of throughput. High throughput rates also help with heat retention, which can impact scaling rates.

Alumina remains the market with the highest level of penetration for DCT technology. This technology started as an extension of high-rate thickening, utilizing a deeper mud bed to augment the thickening capacity. High density or high compression thickeners usually add depth to a high-rate design to aid in increasing the underflow density. Deeper mud beds increase the mud compressive force, reducing the time required for thickening and increasing the underflow density. There are significant modifications necessary to be able to produce and handle the higher viscosity materials, including significantly higher torque capacity than high-rate machines, low drag rake arms, dewatering pickets, higher raking capacity, and improved underflow discharge arrangement.

Deep cone thickeners are currently the best technology for achieving maximum underflow densities utilizing sedimentation equipment alone. These units typically utilize feed dilution with high settling rates to minimize diameter and very deep mud beds to take advantage of mud compressive forces for dewatering and provide long retention time for the mud to dewater to a thick consistency. Mud retention time in a high-rate thickener is typically around one hour whereas mud retention time in a DCT could be 6–12 h. The tank height to diameter ratio is frequently 1:1 or higher. Due to the high underflow viscosities, mechanism torques are generally 5–10 times higher than high-rate machines in order to be able to handle production of very high viscosity material. The underflow pumping systems also need to be designed for high viscosities, as the pipeline pressure losses and power requirements are substantially higher than with Newtonian slurries (Fig. 5.13).

At the densities produced from DCTs, the red mud slurry is generally outside the realm of Newtonian behavior. The thickened slurry begins to exhibit a yield stress, meaning it will pile up to a degree instead of flowing like water. The yield stress is a function of the solid's concentration, and generally increases asymptotically with increasing solids concentration up to a point where the slurry behaves like a solid. The yield stress translates directly into torque requirement; a thickener rake is not too different from a large rheometer and higher rheology requires higher torque. Producing a high yield stress mud requires a high torque thickener drive. It also

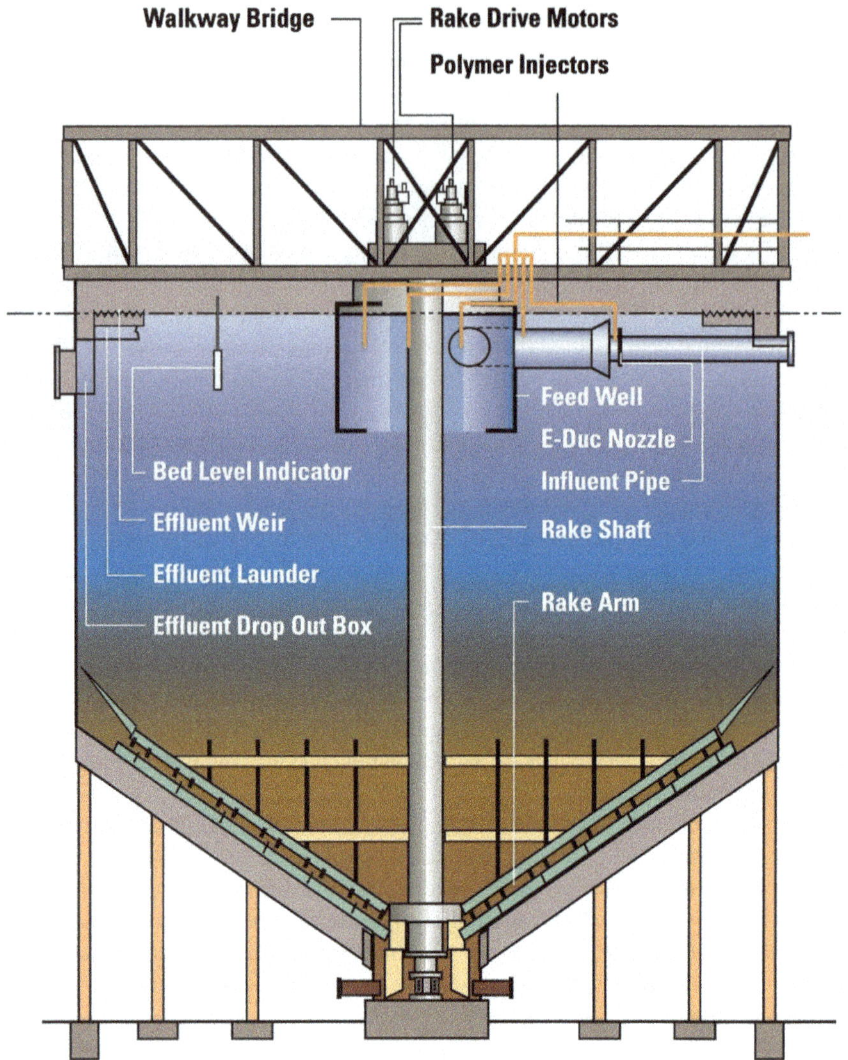

Fig. 5.13 Deep cone thickener arrangement

translates into different pumping requirements as the flow behavior changes. What would be a turbulent flow regime with a Newtonian fluid may be a plug flow regime for a high yield stress material (Fig. 5.14).

The gel point is a solids concentration where the solids flocs are in contact with one another, creating a network or lattice. This is generally the point at which a yield stress begins. Beyond this point, compression of the floc networks allows for further water release. This occurs by self-weight consolidation as well as the mechanical action of thickener rakes and pickets and shearing from mud flow (Fig. 5.15).

Fig. 5.14 Rio-tinto alcan Style DCTs in a 6-stage washer circuit (courtesy of FLSmidth)

Fig. 5.15 Yield stress curve for red mud, slump photos

Fig. 5.16 Conventional style washer circuit

Typical yield stresses result from the particle interactions, which are affected by flocculants. Subjecting the material to shear or energy input can substantially reduce the yield stress; this reduction can be about 60% after pumping a short distance. When reporting yield stresses, it's important to know if it refers to a sheared or unsheared condition. Different technology suppliers may use test methodologies specialized for their particular requirements. For example, thickener rakes are subjected to unsheared conditions, whereas pumping systems see some degree of sheared behavior. So, thickener suppliers are most concerned with unsheared yield stresses while pump manufacturers are concerned with behavior at higher shear rates.

Underflow pumping systems can be configured to allow for the shearing behavior of the mud and recycling sheared mud to help move unsheared mud out of the thickener. Alcan pioneered this technology and, in some situations, it can reduce overall required pumping power. Centrifugal pumps are used for most duties with the exception of pumping thickened red muds some distance to a disposal area, for which positive displacement pumps may be indicated (Fig. 5.16).

Rake drives for thickeners can be either electric or hydraulic, hydraulic drives using a separate hydraulic power pack. Advantages of hydraulics include easy and accurate torque indication by hydraulic differential pressure, elimination of the high-speed reducer (hydraulic motors run about 300 rpm versus electric at 1500 rpm), torque limiting by pressure release, variable speed, and soft starts, although they do cost somewhat more. Electric drives typically use load cells for torque indication, and mechanical springs and switches for overload protection. It is possible to add VFDs to electric drives for variable speed as well as load sharing and soft start capabilities. Both electric or hydraulic drives can provide long life.

5.4 Decantation and Clarification Technology

The strong liquor coming out of digestion commonly contains about 10 wt.% insoluble residue. Flocculating and settling the solids can reduce the liquor suspended solids content to roughly 100 ppm or less. At that point the liquor can be cleaned up using clarifying filters. The thickeners used for the digested slurry are generally called decanters or settlers. The strong liquor overflow from these gets filtered and precipitated, while the underflow goes to the red mud washing circuit.

The primary objective in the decanters is producing a good overflow quality. Strong liquor production is also maximized by maximizing the underflow density, so that is generally the secondary objective. To get the best clarification, a different flocculant or combination of flocculants may be used than in the washer circuit.

Due to the high scaling rates in this duty, plants frequently have three 100% flow decanters, with one online, one in standby and one in descale mode. A hard scale builds up on all submerged surfaces, and in a period of several months can reach thicknesses of 300 mm or more. The scale generally builds until it interferes with the rakes and increases the rake torque to the point of tripping the drive. It's important to maintain temperature as much as possible to minimize scaling and maximize operating time between descaling, so these tanks are commonly covered and insulated. Putting a cover on the thickeners also minimizes environmental exposure and emissions. Rake designs for decanters are generally very simple to minimize surface for scale buildup. Robust construction is needed to allow for the mechanical loads and abuse of descaling. Scale loads need to be planned for in the design of the decanters and settlers, including tanks, rakes, feed wells and feed pipes, and the associated loads applied to the drives and bridge structures.

5.5 Thickener and Washer Tank Design

With decanters and settlers frequently being offline for descaling, piping to bypass any thickener in the circuit is needed. Scale traps are used to screen out large chunks to protect pumps. Angle slurry valves are typically used because of their suitability in scaling service. Common construction practice uses elevated steel tanks for both decanters and washers, typically using DCT configurations. Allowance for access for descale drives a variety of features for access and observation as well as scale removal from the tank. Access for vehicles and equipment laydown may be needed as well. Particularly with DCT style tanks, remote access mechanical descaling tools are commonly used to limit or avoid personnel exposure to both chemical and mechanical hazards. Steep floor slopes and deep tanks with scale on all surfaces makes for a hazardous environment for both slip and falling scale hazards. The use of mechanical descaling tools can greatly reduce exposure and personnel risks.

Mechanical descaling typically involves everything from jackhammers to high pressure washers mounted on the end of hydraulic driven robotic arms. These can be configured to access most surfaces from one or several access locations built into the tank. In addition, viewing, lighting and ventilation access is needed, and arranging safe platforming and observation as well as solids removal is an important aspect of the design.

Tank covers are frequently used throughout the decanter and washer circuits, although some end-of-circuit tanks are left uncovered, as temperatures decrease through the washer circuit. Early-stage decanters and washers are at relatively high temperatures with significant caustic contents in the vapors, so covering those is also a personnel safety issue. As well, minimizing temperature loss help minimize solubility changes and scaling, which helps prolong tank operating life.

Concrete construction is an option for the tank but depends on local concrete access and relatively low labor cost. Benefits include high compressive strength for descaling, no corrosion issues, and a degree of insulating value compared with steel. Most new thickeners are built using elevated steel tanks, with insulation and covers as needed.

The scaling rate is impacted by temperature and concentration. Tank life is typically a few months in early stage washers and a few years in late stage washers. In early-stage washers, the rake design needs to be simple to minimize surface area for scaling. That usually results in using pipe or tubular steel members, instead of trusses, for example. With the decreasing scaling tendency in later stage washers, it's possible to add features such as pickets to help improve underflow densities.

5.6 Pregnant/Green Liquor Filtration Options and Trends

Filtration of the clarifier overflow liquor, also called security filtration, is an essential unit operation of the Bayer Process flowsheet to control the quality of the final hydrate. The application is mainly characterized by the high temperature of the feed liquor around 100 °C and the low feed solids concentration in the range of typically 50–150 mg/l. Also, the system shall be able to operate under off-set conditions when the feed solids concentration might increase due to upstream process fluctuations such as bauxite quality changes or issues related to flocculant dosing, including overdosing of flocculant.

Due to the process requirements in combination with high volumetric flow rates, pressure leaf filters have been most successful for this application. Horizontal Pressure Leaf Filters and Vertical Pressure Leaf Filters are providing the required large filtration areas and the possibility to clean out the filter cake after each filtration cycle. To achieve the desired filtrate clarity a suitable filter aid such as Tricalciumaluminate (TCA) is added.

5.6.1 Horizontal Pressure Leaf Filter

For several decades the Horizontal Pressure Leaf Filter, also known as "Kelly Filter" has been the state of the art pressure filter design for security filtration in the Bayer Process. The filter consists of a retractable pressure vessel, stationary head, quick opening "umbrella" type closure, heavy duty filter leaves and a structural steel support

Fig. 5.17 Horizontal pressure leaf filter ("Kelly Filter") in closed condition (courtesy FLSmidth)

frame with tracks, to move the pressure vessel. A hydraulic power pack is provided to operate the "umbrella" type quick opening/closing system and to open and close the pressure vessel. The filter leaves are arranged in a common cage in a parallel arrangement. The filter media cleaning initially had to be done manually by sluicing the cake from the filter leaf with a water lance operated from a platform on top of the filter. Meanwhile also automatic cleaning systems have been developed allowing a complete automated filter operation. Single feed and individual discharge nozzles for each filter leaf are provided, allowing to turn off each single filter leaf separately in case high turbidity is encountered. The largest Horizontal Pressure Leaf Filters available are comprising a filtration area of up to 440 m^2. The achievable solids concentration in the filtrate is well below 10 mg/l. Horizontal Pressure Leaf Filters for a long period have been the state-of-the-art type of filter for security filtration and are still in operation in large numbers in Alumina Refineries worldwide (Figs. 5.17 and 5.18).

5.6.2 *Vertical Pressure Leaf Filter*

The Vertical Pressure Leaf Filter is comprising a vertically arranged pressure vessel with rectangular filter frames which are covered with a textile filter bag. By changing the orientation of the filter vessel into a fixed vertical position it was possible to install a cloth wash header for cleaning the filter media inside the pressure tank after the tank was drained. The cloth wash header comprises a system of pipes with spray

Fig. 5.18 Horizontal pressure leaf filter ("Kelly Filter") with pressure vessel opened (courtesy FLSmidth)

nozzles which are installed on top of the filter leaves. The nozzles are spraying along the surface of the filter media, thereby removing the filter cake that has been formed on the surface of it. By introduction of this principle the downtime of the filter could greatly be reduced compared to the Horizontal Pressure Leaf Filter, also because opening the filter after each filtration cycle for cleaning was not necessary any longer. Opening of the vessel now only was required after around 100 filter cycles, depending on the operating and process parameters. Vertical Pressure Leaf Filters are available up to a filtration area of 440 m^2 (Fig. 5.19).

5.6.3 Vertical Pressure Leaf Filter—Backflushing Type

Backflushing type vertical pressure leaf filters are comprising a vertically arranged pressure vessel with a small overhead vessel for backflushing (Fig. 5.20).

Many small filter leaves are installed whereby several leaves are always grouped together on a single filtrate collecting pipe on top. These pipes are arranged in a star type arrangement with single filtrate discharge nozzles through the tank wall and discharging into a circumferential receiving pipe. These filters are using filtrate to backflush the filter elements and remove the TCA/red mud layer formed on top of the filter media surface. For backflushing the static pressure of the filtrate flowing back from a buffer tank installed right on top of the filter is used. The backflush is affected

Fig. 5.19 Vertical pressure leaf filter (VPF) (courtesy FLSmidth)

Fig. 5.20 Vertical pressure leaf filter with overhead tank for backflushing

following each filtration cycle followed by a mud settling cycle in the bottom cone to achieve a concentrated slurry which then is drained. The backflush type filter cycle time typically is shorter compared to the cycle of a Vertical Pressure Leaf Filter or a Horizontal Pressure Leaf Filter.

5.6.4 ClariTube Pressure Tube Filter With Automated Hydro Jet Cleaning Device

The main advantages of the ClariTube filter are the operation without addition of filter aid, such as tricalciumaluminate (TCA), and the automated filter media regeneration with a hydro jet cleaning device. Another unique design feature of the filter is the use of wire mesh as filter media instead of textile filter bags.

The filter provides a series of vertically installed pressure rated filter tubes. The filter tubes are installed in individual, non-pressure rated filtrate vessels for collecting the clear filtrate. All filter tubes are fed from a single feed pipe. Filtrate from the filtrate vessels is collected in filtrate launders discharging to a filtrate tank.

When the filtration rate drops due to the build-up of filter cake on the inside of the filter tubes, the hydro-jet cleaning process is affected. For hydro-jet cleaning typically one or two filter tubes are isolated from the others which remain in filtration mode. The hydro-jet cleaning is carried out with a small amount of wash media only whereby plant water or washer overflow can be used as wash media. Criteria for automatic filter media regeneration can by a fixed cycle time, maximum feed pressure, minimum flow rate or a combination of two of them. A positioning device carrying typically one or two cleaning lances with cleaning heads is installed to perform the hydro-jet cleaning. The mud discharged can be added to the washer circuit or directly filtered on a small filter press.

ClariTube filters can be operated without interruption of the filtration process when operating in media regeneration mode. Filter media maintenance is greatly reduced by the installation of wire mesh or other long-lasting filter media. Due to the very effective filter media cleaning system the ClariTube filter can be operated without adding TCA as filter aid as it is required by other filter designs which are typically using a low pressure backflush effect.

The ClariTube filter design is a modular concept allowing to increase the filtration capacity by adding further filter elements. The filter tubes can be installed in one or two parallel lines. When filter tubes are installed in a parallel configuration always two tubes are cleaned at the same time (Figs. 5.21 and 5.22).

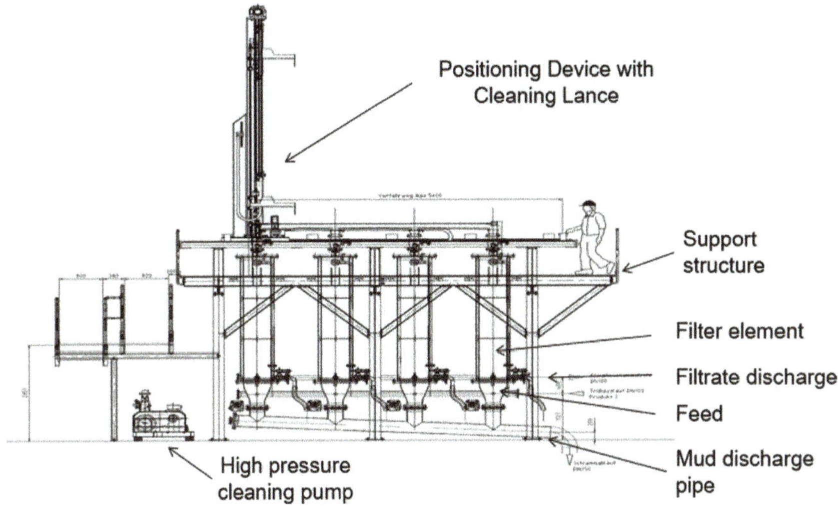

Fig. 5.21 ClariTube filter design comprising four filter elements and positioning device (courtesy FLSmidth)

Fig. 5.22 ClariTube filter installation (courtesy FLSmidth)

References

1. F. Ballentine, M.E. Lewellyn, S.A. Moffatt, *Red Mud Flocculants Used in the Bayer Process* (Cytec Industries Inc, USA, 1937)
2. Bauxite Residue Management: Best Practice; World Aluminium (2015)
3. J. Kildea, *New Developments in Red Mud Flocculantion and Control of Solid/Liquid Separation* (Nalco—An Ecolab Company, Presented at Icsoba 2015, Dubai)
4. L.J. Connelly, D.O. Owen, P.F. Richardson, Synthetic flocculant technology in the Bayer process. *Light Metals* 61–68 (1986)
5. E.C. Phillips, New water-continuous red mud flocculants for the Bayer process. *Light Metals* 39–43 (1999)
6. R.G. Ryles, P.V. Avotins, *Superfloc® lx, a New Technology for the Alumina Industry*. 4th International alumina quality workshop, Darwin June 1996.
7. T.J. Laros, F.A. Baczek, *Selection of Sedimentation Equipment for the Bayer Process: An Overview of Past and Present Technology* (TMS, 2013)

Manfred Bach Sales Manager Alumina and Pyromet Technology, ENAR. Email: Manfred_bach1@gmx.net

After passing the bachelor's degree in environmental protection, Manfred Bach has started his professional career in the field of solid liquid separation as application engineer with company Pannevis B.V., now part of Outotec. After a short employment at Westinghouse Environmental Services, Germany, he joined the solid liquid separation division of Orenstein and Koppel, Germany, now part of Thyssen-Krupp. In 1998 Manfred joined Dorr-Oliver GmbH in Germany, now part of FLSmidth A.S., Denmark. Based on his previous experience Manfred soon started working on alumina refinery related projects, including filtration, sedimentation and classification applications. Manfred developed the ClariTube polishing filter technology as an alternative to current technologies for security filtration. Manfred today is in the position of a sales manager and responsible for FLSmidth's alumina refinery related activities covering Europe, Russia and Northern Africa.

Fred Schoenbrunn, PE, BS ChE Director for Thickeners, FL Smidth. Email: Fred.Schoenbrunn@flsmidth.com

Fred has been with FL Smidth for 30 years, starting with Eimco Process Equipment in the Technology department and working the last 25 years in sizing and design of thickeners and clarifiers as a product manager and group manager, and currently as the company's Director for Thickeners. Prior to coming to FL Smidth, he spent 5 years as the supervising engineer in a chemical plant. He has been instrumental in the development of the deep cone paste thickener market over the past 20 years and helped develop the standards for paste thickeners. Over the same span he has worked in a broad variety of minerals and industrial applications and had a hand in the design of many of the world's largest and highest capacity thickeners. Among other achievements, he has been involved in the design of tailings thickeners accounting for about 25% of global copper concentrate production.

Fred has authored or co-authored 24 books and papers including articles on thickener design and control for the SME Plant Design Handbook. He is also a contributing author in the latest editions of Perry's Chemical Engineer's Handbook, and has several patents related to thickener design and operation. He is a certified Professional Engineer with a bachelor's degree in Chemical Engineering from Lehigh University.

Chapter 6
Bauxite Residue/Red Mud

Quentin D. Avery, Manfred Bach, Stephan Beaulieu, David Cooling, Ken Evans, and Fred Schoenbrunn, PE

Abstract The chapter on bauxite residue (red mud) comprehensively reviews all aspects of bauxite residue from separation, management and disposal through to the ever-increasing use of the Material. Mud washing and the optimum recovery of soda and alumina values is presented. Dewatering options, filtration theory is described in detail and the benefits that can be achieved using various filter types. The management and storage of bauxite residue is then reviewed from a historical perspective together with the trends and changes in disposal methods. Topics covered include sea water neutralisation, costs of disposal, dry mud stacking, filtration and as well as recent developments in the remediation of closed bauxite residue areas. Some current best practices in Europe and Australia are discussed with a detailed review of mud farming methods and rehabilitation trials that have been carried out at Aughinish Alumina in Ireland. Alcoa's vision for more sustainable management of bauxite residue in their Australian operations is described together with a historical overview of best practices covering their move into thickened slurry disposal, solar drying through to filtration. The mineralogy and chemical composition data for bauxite residue are discussed prior to discussing possible uses. The very wide spectrum of uses is reviewed from laboratory studies that have been undertaken through to current significant industrial uses. Areas covered include use of bauxite residue in cement, iron and steel extraction, materials recovery (particularly rare earths), landfill capping, heavy metal and phosphate removal, manufacture of construction

Bauxite residue and red mud is the same material and the names are used interchangeably as it fits the subject.

Q. D. Avery · M. Bach · F. Schoenbrunn, PE
FLSmidth, Westwood, NJ, USA

S. Beaulieu
ENAR, Tauansstein, Germany

D. Cooling
Alcoa World Alumina - Retired, Dawesville, Western Australia, Australia

K. Evans (✉)
FLSmidth, Cottonwood Heights, UT, USA
e-mail: Ken.evans111@btinternet.com

B. E. Raahauge and F. S. Williams (eds.), *Smelter Grade Alumina from Bauxite*, Springer Series in Materials Science 320,
https://doi.org/10.1007/978-3-030-88586-1_6

products, heat storage medium, and fire retardants. Some of the recent extensive work programmes being undertaken under EU HORIZON 2020 funding are also reviewed.

6.1 Mud Washing—Recovery of Soda and Alumina

Mud washing for recovery of soda and alumina is carried out in a series of thickeners, called washers, Fig. 6.1. Typically, the combined underflow of two clarifiers is fed to the front-end washer of a single washer chain for counter-current decantation (CCD) with condensate which is added to the feed of the rear-end washer. The number of washers in the washer train greatly depends on the settling characteristics of the red mud, the tank design, the flocculation technology used and the amount of plant condensate available for mud washing and inter-stage mixing efficiency. Achieving recovery efficiencies of around 99% is common for modern washer train designs.

Plants with additional red mud vacuum or pressure filters installed can be operated with a lower number of washers and/or reduced condensate consumption since the separation efficiency of both filtering systems is higher compared to a thickener. Modern plants today using deep cone thickener (DCT) technology for clarifiers and washers are typically operating washer trains with not more than four washers. In

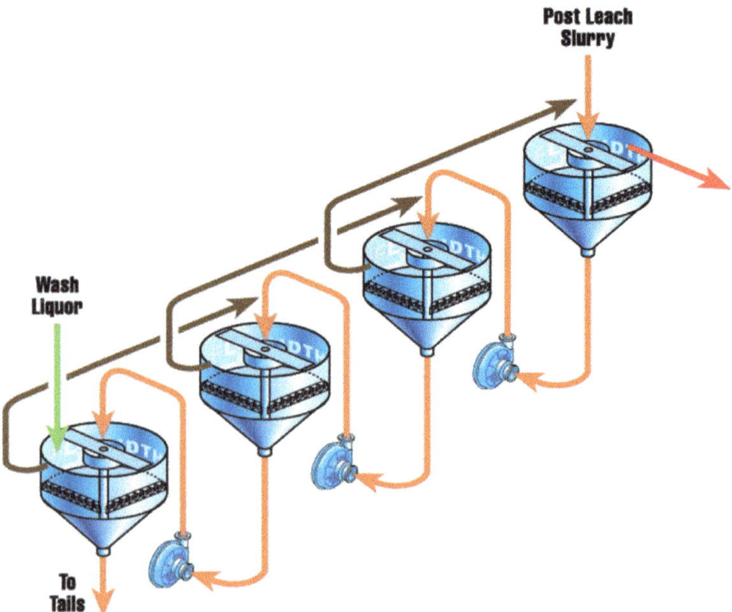

Fig. 6.1 Mud washing for recovery of soda and alumina is carried out in a series of thickeners or washers

the contrary plants designed based on flat bottom thickener technology and starch as a flocculant may operate eight to ten washers and sometimes even more.

Since the settling velocity of the red mud particles in washer service is higher compared to the clarifier, the tanks can be designed for higher specific throughput per square meter settling area. This allows to combine the underflow of two clarifiers and feed it into a single washer tank.

In the washer counter current decantation circuit comprising four washer tanks the mud from the clarifier underflow is pumped to the feed system of the 1st washer (front end washer) whereby the plant condensate is added to the thickener feed of the 4th washer (rear-end washer) after proper mixing with the underflow of the 3rd washer. In a washer CCD the thickener underflows and the overflows are pumped in opposite direction—the overflow from the rear-end washer to the 3rd washer, 2nd washer and finally to the 1st washer, whereby the mud is pumped from the front-end washer to the 2nd washer, the 3rd washer until it reaches the rear-end washer. By mixing the overflow and the underflow of the respective washer stage a caustic concentration profile is achieved that is depending on the mud solids concentration in the washer underflow and the amount of plant condensate provided. Proper mixing of the underflow and the overflow in the thickener feed is important. Also, it is essential to have a proper feed dilution system in place since the feed solids concentration that is achieved after mixing the underflow and the overflow stream together is above the level most efficient for achieving a high settling rate and good overflow quality.

Descaling frequency depends on the location in the circuit, with decanters frequently needing descaling after operation for about 3 months, at the end of a circuit, tanks generally last for years without needing to be shut down.

Thickener underflow densities generally increase through the washer circuit as the dissolved solids content of the liquor decreases. In addition, some of the flocculant retains effectiveness from stage to stage and it's common the reduce floc dosage by about 20% with each washer stage. DCTs, last stage underflow densities over 50 wt% are typical.

6.2 Dewatering Options

Red mud waste residue management has been a continuing issue for review within the alumina refinery operations. With anywhere from 1 to 2.5 tonnes of red mud residue per tonne of Al_2O_3 produced, the alumina producer must carefully consider the various options available to handle this material. The most common practice 75 years ago was to dump residue slurry directly to rivers, swamps, oceans, etc. This is no longer an acceptable and environmentally-sound solution for most refineries so alternative approaches have been developed worldwide including mud lakes. These are wet impoundment areas engineered to hold the large quantities of slurry. Red mud underflow slurry from the final stage washer can be 'dry-stacked' for evaporative dewatering in a residue disposal area. Slurries for dry stacking can be pumped directly from washer underflow *or* dewatered first, typically utilizing rotary drum vacuum filters, thereby increasing solids concentration in the material to be stacked. An alternative approach is to even further dewater red mud slurries with pressure filters for "dry-storage" of the resulting high-solids filter cakes. A key to success in such applications is to produce a red mud filter cake that is dry enough to have the required geotechnical properties for stability of the storage area. Automated conveying and stacking equipment become integral components in system designs for dry storage. In this chapter we will discuss the options for red mud dewatering with a focus on the current trend towards pressure filtration.

6.2.1 Background

The production of alumina from bauxite results in a tremendous amount of residue referred to as red mud. With a potential more than 2.5 tons of red mud residue for every single ton of Al_2O_3 produced, how to deal with this waste material in a safe and environmentally friendly way is of extremely high importance to the plant operators. The amount of residue produced will depend upon the bauxite ore grade made available to the refinery at a particular plant site. However, much is produced, and regardless of location, a plan must be in place to handle red mud.

When the retaining wall holding on the red mud impoundment broke at the Ajka alumina refinery in October 2010, thousands of cubic meters of caustic red mud slurry flowed down the Hungarian countryside and into the neighboring village, with devastating results [1]. The worldwide alumina industry responded quickly by closely evaluating the options and investing heavily in upgrades to their existing sites. Minerals producers, regulatory bodies and engineering companies alike are heavily investigating alternate ways to manage tailings, rather than utilizing wet tailings impoundments [2].

Each alumina refinery is unique, and the amount of residue produced will depend upon the bauxite ore grade made available to the refinery at a particular plant site.

The evolution of disposal strategies for red mud residue has taken us from low density waste slurries to techniques for disposal of high-density slurries and now, high-solids filter press cakes.

- RIVER SEA DUMPING/MUD LAKES In the earliest cases, river and sea dumping was the normal practice. These waste slurries had low densities (25–30 wt%). The practice is largely abandoned today. Mud lakes or "impoundments" were created instead to contain large volumes of low-density slurries. Dyke heights have to be above slurry level to contain the large volumes of material in the waste disposal area.
- DRY STACKING. Thickeners are now routinely employed in alumina refineries to produce high density underflows. The thickeners are also used in counter current decantation (CCD) settler-washer trains to recover caustic for the plant.

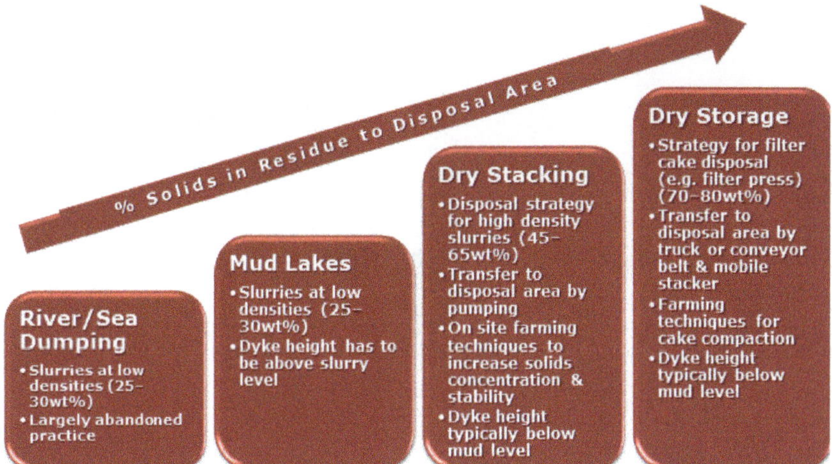

Fig. 6.2 Disposal strategies have evolved from low to high density techniques

- DRY STORAGE of even higher solids concentration filter press cakes (70–80 wt%) is now the technology of keen interest for the refineries with an eye on extending the useful life of a disposal area (Fig. 6.2).

Early on in the development of the alumina refineries, an acceptable method of disposing of red mud slurry was to pump it directly to a river, lake or into the ocean. This approach has been largely abandoned due to environmental regulations and poor community acceptance. Purpose built impoundments or mud lakes have been used to collect and store the residue waste.

Red mud underflow slurry from the final stage washer can be 'dry-stacked' for evaporative dewatering in a residue disposal area. Slurries for dry stacking can be pumped directly from washer underflow or deep cone type thickeners *or* dewatered first, typically utilizing rotary drum vacuum filters, thereby increasing solids concentration in the material to be stacked.

6.2.2 *Red Mud Filtration*

Red mud filtration options of thickened and washed red mud from the final washer include rotary vacuum drum filters and filter presses. On vacuum drum filters in most cases cake wash is part of the filter cycle whereas standard filter press cycles are without. Filter cake from filter presses is well suitable for transportation by conveyor belts and mobile stacking equipment can be utilized (Fig. 6.3).

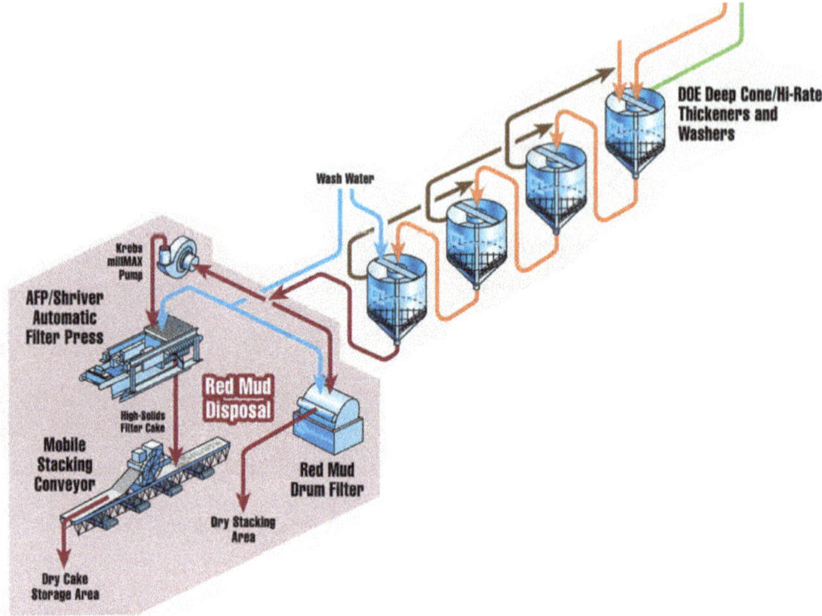

Fig. 6.3 Counter-current decantation (CCD) circuit followed by red mud filtration options

6.2.3 Filtration Theory

To understand the process of selecting a liquid solids separation device for red mud filtration we look at the basic principles of filtration that impact our choices.

The feed pressure affects the rate at which the liquid passes through the cake. When the filtration pressure is greater, a higher filtration rate and shorter filtration time occurs. The relationship is nearly inversely proportional to the filtration pressure, only subject to differences in the cake resistance being formed at the different pressures. Basic relationships may be understood by reviewing 6.1, which is the basis for the filtration theory.

$$\frac{1}{A}\frac{dV}{d\theta} = \frac{\Delta P}{\mu\alpha\omega V/A + R_{\mathrm{m}}} \tag{6.1}$$

where

V = filtrate volume.

θ = cake formation time.

Ω = weight of solids per unit volume of filtrate.

α = average specific cake resistance.

μ = liquid viscosity.

R_{m} = resistance of media or other system resistance.

Vacuum drum filters are in practice limited to a pressure difference of approx. 0.8 bar. Due to the elevated temperature of the filtrate and the liquid ring vacuum pump operating curve it is not possible to reach higher vacuum, even when the red mud filter cake layer is very low in porosity. The filter press recessed chamber plates can be fed at pressures up to 15 bar. The effect of increasing the feed pressure, and the associated feed rate, is to increase the cake solids concentration and to reduce the filtration time to maximize the throughput.

6.2.4 Red Mud Vacuum Drum Filters

The vacuum drum type red mud filters are in use for red mud filtration for more than 50 years. They are of special design tailored for this demanding application, including internal piping, heavy duty agitator, cake wash spray pipes and a unique roller discharge device.

The red mud slurry coming off the final stage of the CCD settler-washer circuit is introduced to the filter vat and a semi-dry filter cake is continuously discharged off the red mud drum filter by using the roller discharge device. For additional caustic recovery, cake washing is accomplished on the red mud drum filter by applying wash liquor via precisely positioned spray headers (Fig. 6.4).

A typical red mud vacuum drum filter will produce a filter cake for discharge with a dry solids content of approximately 50–65%. The cake may be thixotropic in nature but generally solid by appearance. Through shear-thinning in an agitated vessel the filter cake being discharged from the red mud drum filter will become liquefied and then may be pumped to the disposal site (Figs. 6.5 and 6.6).

Fig. 6.4 Rotary drum vacuum filter

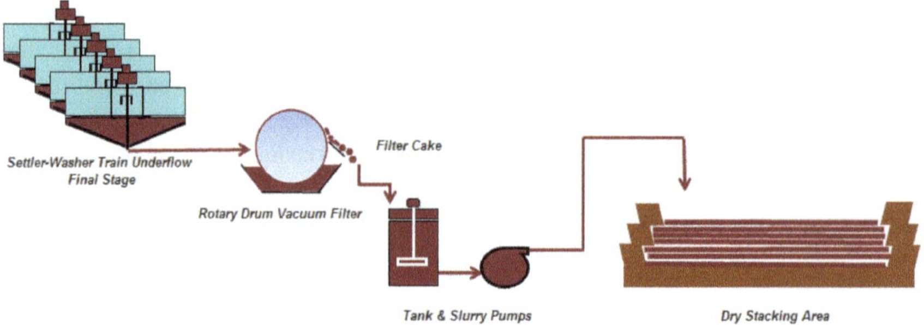

Fig. 6.5 Thickener underflow to filtration with washing, re-slurry, and pumping

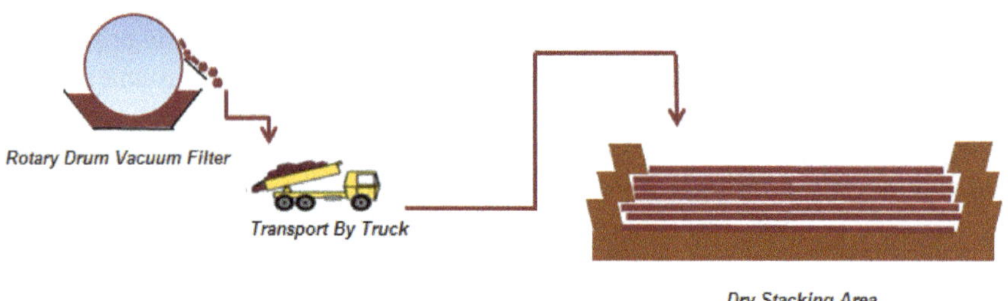

Fig. 6.6 Rotary drum vacuum filter cake transported by truck to dry stacking area

The maximum solids content of a rotary drum vacuum filter is limited due the limited dewatering forces available by vacuum alone. An alternative approach to vacuum filtration is to even further dewater red mud slurries with pressure filters for "dry-storage" of the resulting high-solids filter cakes.

6.2.5 Pressure Filters

Many alumina refineries are seeking an alternative approach for the transportation and distribution of the thickened red mud slurry to a dry stack area. Solutions for lower moisture content filter cakes to reduce volumes of residue in red mud disposal sites are very desirable. Pressure filters can produce these very dry cakes. On the red mud drum filter, the driving force for filtration is generated by vacuum source, meaning the difference between atmospheric pressure and absolute 0 pressure is the maximum driving force available for filtration. Unlike a vacuum filter where the filtration pressure is limited to amount of vacuum available, the pressure filter can be fed with high pressure pumps. The 16-bar feed pumping pressures typical of filter presses results in cakes that are extremely high solids content and have ideal geotechnical qualities for placing the material safely and securely in the storage area. Filter presses offer efficient cake washing for caustic recovery. With modern filter presses, fully automatic operation is offered (Fig. 6.7) and an example of the filter cake produced is shown in Fig. 6.8).

When considering the handling of dewatered red mud filter cakes discharged from pressure filters, automated conveying and stacking equipment become key components in system designs for the dry storage area.

The filter press is one of the oldest and most venerable filtration technologies in use for liquid solid separations. The filter press design, with plate stacks numbering 150 chambers or more, allows a large amount of filtration area to be packed into a

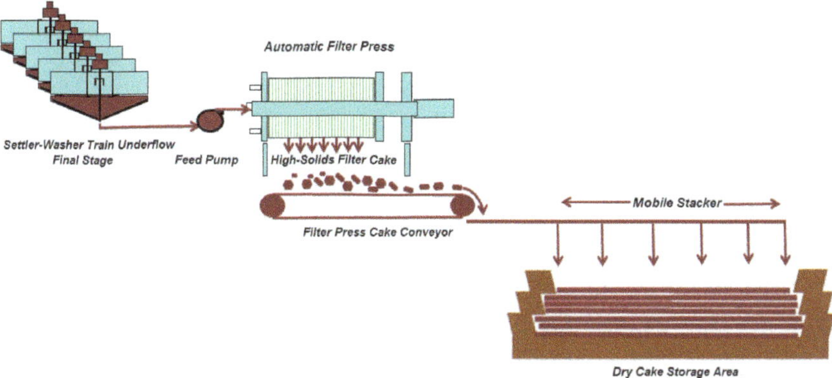

Fig. 6.7 Automatic filter presses producing high solids filter cake, transported by conveyors (red mud cake may be distributed with automated mobile stackers)

Fig. 6.8 High-solids red mud filter press cake suitable for conveyor transport

Fig. 6.9 Picture of 2 m × 2 m AFP type automatic filter press (courtesy FLSmidth)

relatively small footprint. With the high design operating pressures for cake formation on a filter press, extremely high solids filter cakes can be produced. Consequently, the range of applications for filter presses is immense. However, the design and scope of the filter system can be different from case to case. The mining industry presents challenging and demanding applications that are being handled with newly developed filter press systems.

Today's large-scale fully automatic filter presses for minerals processing are staggering in size with single units being capable of handling 900 tons per hour of solids. These filters run in a fully unattended mode and, while fundamentally similar to basic filters, engineering of the complete filter package for the minerals processing plant can be far more complex and extends well beyond the filter press itself.

Automatic filter presses with plates that measure 2 m × 2 m are quite common today but with the demand for larger production, plate sizes are becoming even larger. One such filter, as an example, is the FLSmidth Colossal™ filter press with plates that measure 2 m × 4 m. Fitted with 152 chambers, these filters provide a filtration area approximately 2080 m^2 (Figs. 6.9 and 6.10).

6.3 Bauxite Residue Generation and Management

As described in previous chapters, bauxite residue is an inevitable by-product from the Bayer Process. This chapter reviews the composition and characteristics of bauxite residue, the factors that influence the amount generated, the history and trends in the management and storage of bauxite residue disposal, some specific examples

Fig. 6.10 Picture of 2 m × 4 m Colossal™ filter press (courtesy of FLSmidth)

from different regions, the critical environmental issues, an overview of uses, a review of large scale/industrial scale uses and some recent innovative approaches including major EU HORIZON 2020 initiatives.

6.3.1 Quantities of Bauxite Residue Generated

Bauxite residue, sometimes called red mud, or bauxite tailings, is the solid material remaining after digestion in the Bayer process and includes the compounds not dissolved by the sodium hydroxide and the compounds formed during the desilication, autoclaving, and other stages of the process (see Chaps. 4 and 5). The main constituents are iron oxides (haematite and goethite), aluminous compounds (gibbsite, boehmite, diaspore), silicon compounds (quartz, sodalite, kaolinite), calcium compounds (calcite, tri-calcium aluminate), and titanium compounds (rutile, anatase, perovskite)—the composition is discussed in detail later in the chapter.

Annual production of smelter grade and chemical grade alumina in 2020 was approximately 134 million tonnes [3] which, except for some plants in Russia, Iran and China, is all produced using the Bayer process. The quantity of bauxite residue generated using the Bayer process at a particular refinery is primarily governed by the bauxite quality and the processing conditions: the key bauxite characteristics being the alumina content; the nature of the aluminium hydrous-oxide component present (whether as gibbsite, boehmite or diaspore); and the amount and nature of the silica present, either as reactive (kaolin) or crystalline silica (quartz). The bauxite processing conditions used are fundamental to the plant design but are driven by the nature of the bauxite, local energy and transport costs; whilst not always the case, generally when long bauxite transport distances are involved, a higher quality of bauxite is used together with more aggressive extraction conditions which leads to lower residue production per tonne of alumina produced. The global average of bauxite residue generated per tonne of alumina is between 1 and 1.5 tonnes [4], though the amounts from different alumina plants is much broader. LCA data covering a large percentage of the global alumina plants indicates a global average of approximately 1.3 tonne per tonne of alumina whilst the EA (European Aluminium) [5] indicates that the average quantity of bauxite residue produced per tonne of alumina is currently at approximately 0.8 t in Europe. Whilst a number of European plants have good local bauxite deposits, e.g. diasporic bauxite from Greece, Turkey or the Balkans; most of the alumina refineries in Western Europe tend to use bauxite from Guinea and to a lesser extent Brazil. Refineries in Australia are focussed in two locations: in Western Australia, where bauxite is especially plentiful, there are four large refineries which utilise bauxite from the Darling ranges. The bauxites used are relatively low in alumina and have a high coarse silica component; the bauxite residue generated per tonne of alumina is therefore higher than the global average. Meanwhile in Queensland, Australia there are two alumina refineries which use Weipa bauxite, this is predominantly boehmitic in nature and has a high alumina content so consequently produces a significantly lower level of bauxite residue per tonne of alumina.

Currently there are about 80 active Bayer plants in the world of which approximately 30 are in China, eight in Europe, six in Australia, six in South America, seven in India, three in Jamaica, two in North America, four in Russia plus others in Azerbaijan, Iran, Kazakhstan, Malaysia, the Philippines, Saudi Arabia, South Korea, Taiwan, Turkey, Ukraine, United Arab Emirates and Vietnam. Based on 2020 production data [3] it is estimated that approximately 170 million tonnes of dry bauxite residue are being generated annually, and ever-increasing level which is predicted to increase as primary aluminium demand continues to grow.

In addition to bauxite residue sites at these active locations, there are at least another 50 closed legacy sites so the combined stockpile of bauxite residue at active and legacy sites is estimated at over four thousand million tonnes [6].

6.3.2 Methods of Disposal/Storage

Management and methods of storage of bauxite residue has evolved markedly since the first plants built in the 1890s [7]. In the early Bayer alumina plants, the residue generated was normally stored on site or an area adjoining the alumina plant. As these areas filled, exhausted bauxite mine or quarry sites were used to store the bauxite residue. In other situations, residue was disposed of in nearby estuaries or sea lagoons and then later as the closest sites were filled, valleys were dammed to contain the growing volumes of residue. These methods of storage were especially true of the early European sites. Examples include Bergheim (Germany), on site storage then in former lignite mines adjacent to the plant. In Burntisland (UK) disposal initially was into an estuary, when this was full behind a sea wall, see Fig. 6.11a, which shows the

Fig. 6.11 **a** Bauxite residue disposal into an estuary to recover land, Burntisland, Scotland, **b** aerial view of refinery and residual disposal area; the closed residue disposal area is located between the plant and the sea wall, **c** aerial view of refinery after remediation showing housing estate and residual disposal area

Fig. 6.12 **a** Bauxite residue disposal in a former oil shale mine, Whinnyhall, Scotland, **b** bauxite residue disposal area at Whinnyhall, Scotland after partial restoration

alumina refinery and residue disposal area. Figure 6.11b shows the bauxite residue disposal area after remediation and Fig. 6.11c shows the alumina plant site after the refinery had been demolished, the land had been remediated and a housing estate built on the site; the remediated residue disposal area is shown in the foreground. When the area behind the sea wall was full, the bauxite residue was disposed of in an old shale quarry nearby, see Fig. 6.12a, b shows the residue disposal area during the restoration period. In Gardanne (France) the residue was stored in a nearby dammed valley then disposal by pipeline to the sea. In La Barasse (France) storage on site, then a nearby dammed valley, then disposal to sea via a pipeline. In Larne (UK) disposal was carried out was putting it into sea water lagoons whilst in Newport (UK) disposal was initially into the River Severn estuary from barges then lagoons. In Salindres (France), initially on-site storage then in a dammed valley. Ludwigshafen (Germany) and Schwandorf (Germany) both used on site storage [6].

Prior to 1980, most of the inventory of bauxite residue was stored in lagoon-type impoundments and the practice is still carried out at a few refineries. In this method, the bauxite residue slurry from the mud washing circuit is pumped, with a solids content of 18–30%, into storage areas created by dams and other earthworks for secure containment.

In many instances valleys were dammed, for example: Ewarton (Jamaica), Gardanne, Salindres, Saint Cyr (France), Ouro Preto (Brazil). In addition to Bergheim and Burntisland, other examples of old mine storage were bauxite mines at Kirkvine (Jamaica) and the former refineries at Bauxite, Arkansas (USA). In the case of sites constructed in the past three or four decades, the storage areas have normally been sealed to minimise leakage to the underlying ground and ground-water, however, this tended not to be the practice in earlier years. Sealing approaches cover a range of materials including compacted or impermeable clays and/or the use of plastic and other membrane materials.

Wherever possible, the supernatant liquor above the residue was normally returned to the plant for reuse thereby recovering most of the caustic soda value and avoiding contaminating the environment. Various drainage and seepage collection systems have been incorporated into the design and construction of the facilities. The construction of the storage area was often dictated by the type of bauxite residue and differed for clay like muds compared to more sandy residues. At Gramercy (USA), sand-bed filtration was used and "French Drains" were used with drainage pipes and covering layers of sand of different size and gravel to give permeability through the base of the lake. This was termed the DREW (Decantation, Drainage and Evaporation of Water) system.

In some cases, low dykes or levees were built to the final expected height at the start and a new area constructed when more volume was required whilst in others the dam walls were successively increased in height when the space was filled. The area of suitable land readily available dictated the approach. There are many examples of this storage method and they include Stade (Germany), see Fig. 6.13, Burnside (USA), Aughinish (Ireland) and Vaudreuil (Canada).

If the residue material is not neutralised before discharge to the storage lagoon, it becomes a highly alkaline, poorly compacted mud area covered by a highly alkaline

Fig. 6.13 Examples of dyked area remote for the plant, Stade, Germany

Fig. 6.14 Bauxite residue disposal into a dammed valley to create a lagoon, early stages, Mount Rosser, Jamaica

lake, see for example the lake in Fig. 6.14 which had a pH > 12 and a soda level in excess of 2000 mg/L many years after bauxite residue slurry pumping had stopped; this pond at a later stage is shown in Fig. 6.15. This creates safety and environmental hazards including the potential for contact of humans and wildlife with alkaline liquor and mud, and contamination of surface and ground waters by leaching of caustic liquor and other contaminants. Regrettably, in some cases the integrity of these early ponds has proved of limited effectiveness and caustic liquor, plus other contaminants, has subsequently seeped into the surrounding environment. Ongoing remediation of these situations can prove to be a costly exercise. In certain cases, dam

Fig. 6.15 Bauxite residue disposal into dammed valley, later stages, Mount Rosser, Jamaica

failure has occurred, Ajka 2010, resulting in loss of life and widespread contamination of land [1]. Addressing the risk and to eliminate the potential for catastrophic failure of the dam/impoundment and consequent environmental hazard to the surrounding area/communities introduces high monitoring, maintenance and remediation costs. This has led the alumina producers to constantly look at opportunities for improving the management and storage of bauxite residue in a secure, safe and cost-effective manner.

6.3.3 Sea, River, Estuary Disposal

Sea, river or estuary disposal was another technique adopted by some plants, particularly in the period 1940–1970. In at least three plants, two located in France (Gardanne and La Barasse) and one in Greece (Distomon) discharge was carried out by pipeline to the sea. In the case of Gardanne and La Barasse, the pipeline extended seven kilometres into the sea; extensive investigations were carried over a prolonged period and no irreversible effect was noted. In Japan, three refineries, Shimizu, Ehime and Yokohama, discharged bauxite residue into deep trenches from ocean going vessels. The discharge occurred between a depth of 30–50 m in an area where the ocean depth was between 2500 and 4500 m. Extensive investigations were undertaken to assess any adverse effect on marine life; no impact of bauxite residue on water quality and marine organisms was observed in the water column throughout the process of its sedimentation to the sea bottom. However, some differences were observed in particle size of sediment and the number of benthic organisms between the reference point and the area close to the centre point where measurable amount of bauxite residue had settled. It was concluded that there seem to be certain changes caused by the ocean disposal of bauxite residue, but its impact was expected to be rather small and limited to the area smaller than 20 km radius from the centre point of the disposal [8].

Other alumina plants disposed of the residue into rivers or estuaries, for example into the Mississippi River (Gramercy), and Severn Estuary (Newport). In other cases, in Northern Ireland (Larne), Wales (Newport) and Scotland (Burntisland), land was reclaimed from the sea by disposing of the residue in tidal lagoons or behind sea walls. Bauxite residue is no longer discharged into rivers or estuaries at any alumina refining facilities and all sea discharge ceased at the end of 2015 [6].

6.3.4 Dry Mud Stacking

As land for lagoon storage became scarce for many plants, “Dry stacking” methods were progressively adopted to minimise the land required for disposal. A dry stacking regime was adopted nearly 75 years ago in the UK [9] and about 50 years ago in Germany [10]. Since the 1980s the trend has been increasingly towards dry stacking to reduce the potential for leakage of caustic liquor to the surrounding environment, reduce the land area required, and maximise the recoveries of soda and alumina. Considerable work was undertaken in Jamaica using the Robinsky [11] sloped thickened tailing disposal system and Ewarton adopted the practice in the mid-1980s [12]. The current practices in Western Australia and in Aughinish (Ireland) are described in more detail later in the chapter. As discussed earlier in the chapter, improved methods for thickening, washing of the residues prior to storage and recovery of decant water during storage, have been developed to further increase the recovery of valuable caustic soda and alumina to the Bayer process plants and to minimise the potential for offsite impact to the environment.

The current trend in residue storage practice is towards increasing use of dry stacking as the preferred technology and further research to optimise this technology is appropriate. Many plants now use equipment such as Amphirols to aid dewatering of the mud in order to compact and consolidate the residue, see Fig. 6.16. Partial neutralisation using seawater is practiced at a number of Australian plants close to the sea (Yarwun and QAL); carbonation by using waste carbon dioxide from ammonia production has been used—Kwinana (Australia); and accelerated carbonation using intensive farming methods, for example at Aughinish, Kwinana, Worsley (Australia), is showing considerable benefits.

Fig. 6.16 Farming with an Amphirol at Aughinish Alumina Limited

6.3.5 Filtration

Filtration, using drum filters and plate and frame filter presses to recover caustic soda, produces a lower moisture and more handleable bauxite residue; this method has been employed for about 80 years but is now growing in usage, partly to minimise disposal volumes. Plate and frame filter presses were adopted in Vaudreuil in 1936 when the plant was constructed and in Burntisland in 1941 as the areas adjacent to the site were full. In the case of Burntisland, the bauxite residue needed to be transported on public roads through the nearby town to a nearby old oil shale mine, so a high solids content mud which was more readily shipped in a truck was a key requirement. In the mid-1960s at Ludwigshafen, vacuum drum filters were adopted [6]. Figure 6.17 shows a typical press filter now adopted for bauxite residue prior to disposal.

In addition to helping recover more caustic, this trend opens up considerable benefits in terms of reuse as the material is normally produced as a friable cake, with typically less than 28% moisture. Additionally, the higher solids content and lower soda level, dramatically reduce transport issues and costs which is important for reuse. Alunorte (Brazil), Distomon (Greece), Gardanne, Kwinana, Seydisehir (Turkey), Gramercy USA), several Indian refineries including Utkal and very many plants in China have already adopted or plan to adopt plate and frame filter presses. Some of the recent installations are reportedly producing a bauxite residue with a moisture content of 21%. Hyper Baric filters are reported to achieve particularly low moisture content material with the performance of the press being enhanced when

Fig. 6.17 Plate and frame filter press, Gardanne, France

steam is used; moisture contents lower than 25% have been reported [13]. The use of plate and frame filter presses prior to bauxite residue disposal as practiced at the Kwinana refinery is described in more detail later.

The nature of the industry has also markedly changed in scale, in a number of cases, relatively small alumina plants has been converted to a specialty alumina plants which uses aluminium hydroxide as feed instead of bauxite; this trend has been particularly apparent in Europe and includes Ajka, Bergheim, Ludwigshafen, Salindres and Schwandorf. In the USA the two former Bayer plants at Bauxite have closed but the specialty alumina operation at the former Alcoa plant continues in operation. The three alumina refineries in Japan, Shimizu, Ehime and Yokohama all closed their Bayer operations after restrictions on the disposal of bauxite residue into the sea and imported aluminium hydroxide as feed to make speciality aluminium oxides and hydroxides; however, the Yokohama plant has since closed.

6.3.6 Neutralisation

Various liquid acids and acidic gases have been evaluated and employed for neutralising or partially neutralising bauxite residue: these include waste acids, pickling liquors, iron sulfate/sulfuric acid solutions, acidic fly ash, carbon dioxide [14, 15] and sulfur dioxide. Sea water has been used for partial neutralisation [16, 17]. The addition of gypsum prior to remediation and vegetation when the bauxite residue areas is to be closed has been employed in a number of cases and has the effect of reducing pH and the sodium content of the residue making it more amenable to plant growth; this is discussed later.

Large scale neutralisation of bauxite residue with carbon dioxide at the Alcoa Kwinana refinery started in 2002 and reached full capacity to treat all Kwinana residue by 2007 with the construction of a pipeline to a nearby ammonia plant to convey the carbon dioxide.

At the alumina refinery at Shimizu in Japan, sulfuric acid was used to neutralise bauxite residue before transport by ship for disposal into the sea.

A desulfurisation unit was built by Sumitomo in 1977 which used bauxite residue to remove sulfur dioxide from flue gases; this unit was termed the Sumitomo Bauxite Residue Desulphurisation System was effective at removing up to 96% of the SO_2 in the flue gas. It was employed for a time at the Eurallumina refinery in Sardinia and it is understood that sulfuric/sulfurous acid gases from boiler off-gas was used to neutralise bauxite residue at the alumina refinery in Surinam.

In Australia at Gove, QAL and Yarwun, sea water has been used to neutralise bauxite residue whilst retaining storage of the neutralised residue on land.

In the Saguenay region, mass dosing of one of the bauxite residue disposal ponds has been undertaken with sulfuric acid addition in order to neutralise the overlying high pH water.

6.4 Residue Disposal Cost

The bauxite residue disposal costs for a plant are obviously very dependent on the availability of a safe suitable disposal site, the distance from the plant to the disposal area and the method on conveying used, normally by pumping, but occasionally by conveyor belt or truck. It should be noted that pumping over long distances can be achieved, even in excess of 50 km, but such long distances are rare.

There is relatively little published data on the cost of disposal of bauxite residue, but it is generally estimated to be between 1 and 3% of the total production cost, equating to an operating cost of perhaps US$5 to 15/t of alumina produced although it can be significantly higher for some plants. As well as the distancc to the residual disposal area and the storage method used, the overall cost is very dependent on the site's final end use requirements and legislation in place in the region where the site is located.

6.5 Remediation

The costs for the eventual closure and remediation of bauxite residue disposal site vary widely depending on their age, area, method of construction and location. In some situations, these sites have been left fallow and over time have become colonised by wind and animal dispersed plant species, meanwhile in other cases, capping layers of various types and thickness have been used, whilst in other cases efforts have been made to achieve rehabilitation by the addition of minerals such as gypsum to convert the residue into a medium suitable for sustaining plant growth. If the site is simply fenced and made secure, the costs can be extremely low, and this has been allowed to occur with some older legacy sites. Whilst at the other extreme, costs can be very high if substantial regrading, and capping is employed which might require both an engineered or plastic membrane (typically HDPE) impermeable layer and importation of topsoil prior to seeding. Additionally, closure can be particularly onerous if there are complex ongoing water management issues and the site is in an area of high rainfall or if the residue disposal pond has not been lined prior to the commencement of disposal. Following closure, ongoing collection of any rainwater run off must be controlled and may or may not be need to be treated prior to discharge. The collection of any leachate is likely to be required, this will need to be monitored and normally treatment to remove any unacceptable species; furthermore, these operations may need to be carried on for decades after the site is closed. If the refinery remains in operation post closure of a residual disposal site, any caustic containing leachate can be returned to the plant for reuse or treatment. However, if the refinery closes at the same time, management of any caustic leachate becomes more troublesome to deal with. Approaches to solving the issue have included mass acid dosing of residue lakes or acid neutralisation of the collected leachate to a pH of 9 or lower; flocculation of any solids may also sometimes be required to meet discharge requirements. An alternative option to deal with leachate that has shown considerable promise is the use of constructed wetlands which has been trialled by the University of Limerick at Aughinish in Ireland [18].

Whilst passive treatment systems such as constructed wetlands have been used extensively over the past 30 years to treat mine waste, little work has been done on alkaline discharges until recently. The University of Limerick/Aughinish study evaluated influent and effluent water qualities of a constructed wetland in order to assess treatment performance on bauxite residue leachate representative of a closed residue disposal area. Residual risk associated with metal(loid) uptake by biota was also assessed. Efficiency of the system to buffer alkalinity, trace element loading and possible mechanisms for alkalinity removal were examined; soils also exhibited increases in soluble and exchangeable sodium. The results demonstrated that a reduction in residue leachate pH and element loading is possible through passive treatment systems and that there were no significant increases in bioavailable levels of trace elements and adverse effects to plant growth.

Rehabilitation of former residual disposal sites by capping has been very effectively carried out at closed residue sites in Scotland, France and Germany; the type and thickness of the caps used has been driven by national legislation, the availability of suitable capping materials and final intended use. Examples of final uses include local recreation areas, the siting of wind turbines, hunting, reforestation, dairy farms, cinema complexes, shopping malls, warehouse areas and motoring events.

Whilst remediation of residue disposal sites can be accomplished by capping with topsoil, frequently it is scarce in the areas close to alumina refineries, additionally it is wasteful of resources to use good quality topsoil, so a highly desirable target is to aim for direct remediation of the bauxite residue by treatment or adding amendment agents to the bauxite residue to convert it to a suitable material for growth. In practice, this means converting a highly alkaline (pH > 12), highly saline (Electrical Conductivity >8 mS/cm), highly sodic (exchangeable sodium percentage > 75%), high bulk density (>2.5 g/cm^3) material to a 'soil' like material with a pH between 6 and 9, Electrical Conductivity < 4 mS/cm, exchangeable sodium percentage < 9.5% and bulk density of < 1.0 g/cm^3. Revegetation of bauxite residue areas without capping layers has been successfully carried out in Jamaica at the now closed Kirkvine refinery using gypsum, manure and fertilisers by Rio Tinto Alcan [6, 19, 20] and is almost complete at Mount Rosser, Ewarton; numerous trials have been undertaken elsewhere, in particular by the University of Limerick and Aughinish Alumina Limited, Ireland [21, 22].

The residue disposal areas at Kirkvine had been closed for over 15 years and the top 300 mm of bauxite residue was conditioned with agricultural equipment prior to gypsum addition. Consequently, by the time planting commenced the pH of the residue and the Electrical Conductivity values had fallen substantially as the solid phase alkaline compounds present had slowly converted to soluble species which could be washed away by rain [6, 19, 21, 22]. The Aughinish studies reviewed the changes in sodicity, exchangeable sodium, salinity, phospholipid fatty acid analysis, organic carbon, total carbon, total nitrogen, magnesium and microbial community composition and are discussed in more detail later in the chapter.

Gypsum additions to bauxite residue lowers the pH by precipitating OH^-, $Al(OH)^{4-}$, and CO_3^{2-} as calcium hydroxide ($Ca(OH)_2$), tri-calcium aluminate (TCA, $Ca_3Al_2(OH)_{12}$), hydrocalumite ($Ca_2Al(OH)_7{\cdot}2H_2O$) and calcite ($CaCO_3$) [21–24]. The efficacy of gypsum in transforming the alkalinity is limited to gypsum's ability to readily dissolve and is affected by the particle size and the nature of the gypsum (anhydrite—$CaSO_4$, bassanite—$CaSO_4{\cdot}0.5H_2O$ or gypsum—$CaSO_4{\cdot}2H_2O$).

A substantial programme of work has been undertaken by Santini and her group working at the Universities of Queensland, Western Australia and McMaster examining the characteristics of bauxite residue on a number of sites which has been left fallow [25, 26]. Later this has been extended to ways of accelerating in situ rehabilitation of the sites by encouraging the multiplication of micro-organisms which favour pH reduction. The work has shown that neutralisation of pH through organic acid production during fermentation and CO_2 production are particularly promising pathways for bioremediation of bauxite residue, requiring potentially no or little addition of substrates, and being carried out by micro-organisms known to tolerate physico-chemical conditions similar to those present in un-amended bauxite residue. Organic acids which are commonly produced during the decomposition of organic matter that might be added as manure or plant residues are acetic, citric, lactic, propionic acids. Some micro-organisms can support the production of sulfuric acid by chemo-autotropic or chemo-heterotropic metabolism which could aid neutralisation of the alkalinity of the bauxite residue.

Carbon dioxide may be generated either by respiration of micro-organisms under aerobic conditions or fermentation under aerobic conditions, this will dissolve in water to form carbonic acid. The carbonic acid will readily react with the readily soluble alkaline species and will slowly react with the tri-calcium aluminate according to the equation:

$$Ca_3Al_2(OH)_{12} + 3CO_2 \leftrightarrow 3CaCO_3 + Al(OH)_3 + 3H_2O$$

The kinetic of this reaction are slow and depend on the partial pressure of the carbon dioxide. Respiration processes generate carbon dioxide rapidly and whilst this will aid neutralisation of readily soluble alkaline species, fermentation processes are generally slower and may be better for neutralising slowly released alkalinity stored in desilication products and other solid phase alkaline compounds.

Following closure and remediation, ongoing land management must be considered. In some instances the land has been returned to the national government, in other cases given to the local community for leisure activities whilst in most cases it remains managed by the original plant owner. Site security, dam wall integrity and management represent a very long-term commitment and expenditure.

The earlier closure options are considered in a plant's history, the better the closure can be planned with normally a lower cost. The ongoing management and remediation costs are obviously an important factor when potential bauxite residue reuse options are considered. Corporate attitudes have changed markedly over the past decade, reflecting growing community awareness and the desire to meet the demands of concerned shareholders, stakeholders and NGOs; producers now have a more holistic attitude to resolving the problem, which includes reducing the area given over to residual disposal/storage and increased effort in investigating uses for bauxite residue.

6.6 Current Best Practice Europe—Residue Processing at Aughinish Alumina Limited

At the RUSAL owned Aughinish Alumina Limited refinery on the west coast of Ireland which produces some 1.9 million tonnes per year of alumina, the bauxite residue is dewatered by vacuum filtration to a solids concentration of 58% before being slightly diluted and transported, by a 2 km pipeline, to the Bauxite Residual Disposal Area (BRDA) where it is discharged, spread and allowed to consolidate and dry in layers. Two-metre high rockfill embankments form a stable boundary to stack the layers and increase the BRDA in height.

There are several stages to post deposition treatment. As mentioned after vacuum filtration, the residue is diluted with water, sheared, thinned in an agitated tank and then pumped as a 58% solids paste to the BRDA. In this state, the deposited residue cannot yet be traversed by conventional machinery and first must be dewatered and compacted. An amphibious vehicle called an Amphirol is employed to carry out this de-watering and compaction process known as farming—see Fig. 6.16.

The Amphirol travels using scrolls, to allow the vehicle to move through the residue. As the Amphirol travels, it compresses the residue and creates tracks or furrows. These furrows allow the water, which has been "squeezed" from the residue to drain along the sloping stack towards the perimeter wall of the cell and into the perimeter channel.

Once the residue has compacted to >70% solids by multiple passes of the Amphirol, the surface is then graded by a bulldozer to level the surface and generate a constant gradient from the discharge (high point) to the perimeter wall (low point). This makes the residue suitable for conventional agricultural machinery to travel and operate on its surface. Atmospheric carbonation of the residue by the Amphirol and agricultural machinery allows for exposure of the residue to CO_2 in the air. Sufficient exposure and carbonation reduces the causticity below 30% and reduces the residue pH below 11.5. This is the mechanism by which the residue is exposed to atmospheric CO_2. Once carbonation is completed as evidenced by pH measurements of samples from the cell, the area is then re-graded using a bulldozer to remove any depressions. The cell is then ready for the subsequent layer of bauxite residue.

Although earlier planning permissions granted to the refinery contain requirements with respect to landscaping and restoration of the BRDA, the Integrated Pollution Prevention Control Licence (IPPCL) introduced in 2008 issued by the Irish Environment Protection Agency contained many stringent conditions for BRDA restoration and aftercare. Since its issue in 2008, the licence has been updated in 2012 and most recently in 2014 to the Industrial Emissions License. Over the years, the major change has been the introduction of partial neutralisation of the residue surface by farming which is a well-established practice at Aughinish.

Conditions stipulated in the Industrial Emissions Licence issued to Aughinish state that partial neutralisation of the residue using mud farming is to be used and that revegetation work be continued on the BRDA. Provision of a dedicated research trial cell for demonstration of the proposed closure technique for the residue is also a licence condition.

History of Rehabilitation work at Aughinish Alumina Limited

Aughinish in collaboration with the University of Limerick in 1996 implemented a series of revegetation trials on the BRDA to develop a restoration technique that can be established on the residue and to demonstrate the effectiveness and sustainability of the closure technique (Fig. 6.18; Table 6.1).

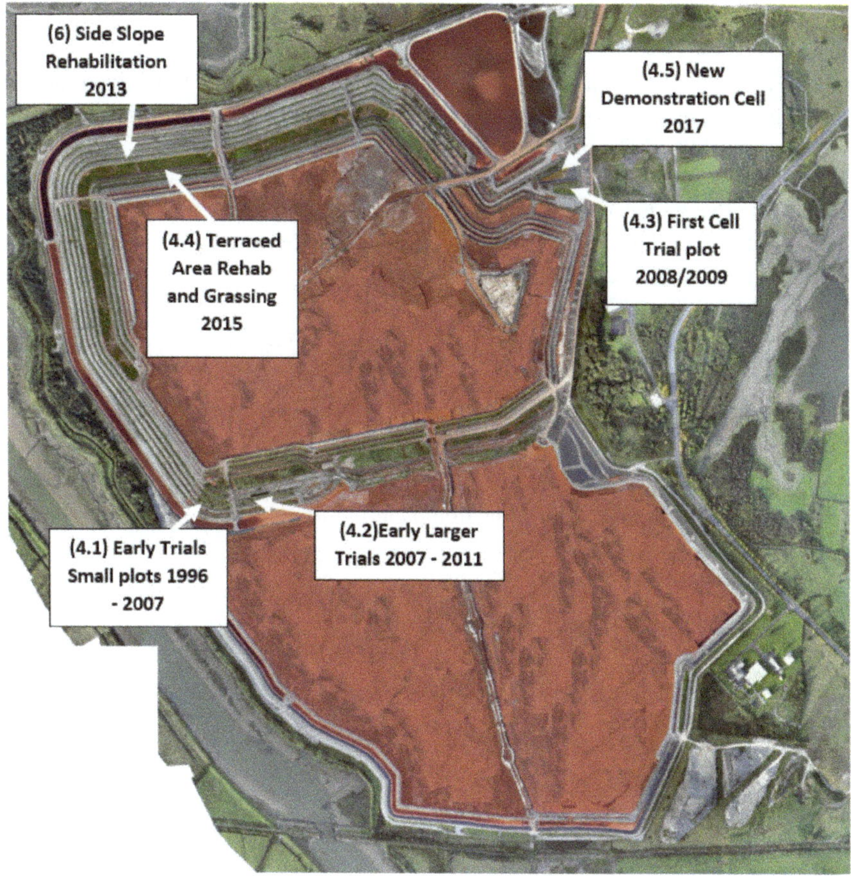

Fig. 6.18 Overview of BRDA and area where trials have been carried out

Table 6.1 Legend to Fig. 6.18

Reference in Fig. 6.18	Description
4.1	Early trial plots 1996–2007
4.2	Larger trials 2007–2011
4.3	First cell trial plot 2008/2009
4.4	Terraced area rehab and grassing 2015
4.5	New demonstration cell
6	Side slope rehabilitation 2013

Early Revegetation Trials at Aughinish 1996–2007

The BRDA is designed using an upstream system using 2 m high rock embankments. The rock lifts are terraced in an upstream fashion to increase the height of the BRDA and manage the disposal of residue. Consequently, available space for the initial revegetation work was restricted to terraced areas between the raises (see Fig. 6.18). The restoration strategy adopted on the BRDA was to seed temperate grassland species on amended residue with a view to establishing amenity type grassland that is sustainable. A series of revegetated areas were implemented on these terraced areas during 1996–2007 as outlined below (Table 6.2).

Different scenarios were experimented with during the early trials. Plots were fully treated amended with gypsum, process sand, spent mushroom compost and seeded with a grassland mix. Partially treated plots were amended only with process sand, spent mushroom compost and then seeded. Other plots were left untreated, see Fig. 6.18.

The revegetation recipe now used at Aughinish is based on the success of these trials. Residue that was treated and revegetated between 1996 and 1999 was surveyed in 2005. Species diversity was recorded and compared to the initial seed mixture of six species:

- There were 50 species belonging to 40 genera and 16 families.
- *Asteraceae* and *Poaceae* were the dominant families.

Table 6.2 Experimental history

Year	1996–1999	1997–1999	2000–2005	2006	2007
Experiment type	Laboratory characterisation and greenhouse trials on bauxite residue and amended bauxite residue	Field experiment testing effect of amendments on residue properties and plant uptake in revegetated residue	Continued field experiments testing efficiency of revegetation methods employed in 1997–1999	Field experiments investigating variations of the procedure to optimise conditions for preparing residue prior to seeding	Larger scale field experiments on a 0.6 ha site demonstrating revegetation prescription is effective on residue typical of a closure scenario

- Seven leguminous species were recorded growing.
- Dominant grass species were *Holcus lanatus* with *Festuca rubra* and *Agrostis stolonifera.*
- Woody species *Betula, Salix* and *Alnus* established on the revegetated areas.
- Patches of hay, spread on residue surface acted as a seed source.

The trials also have demonstrated that addition of process sand and gypsum is effective in lowering uptake of sodium, aluminium and iron in plants.

Large Scale Field Trial Implementation 2007–2011

Previous field trial work conducted on the bauxite residue at Aughinish focused on small-scale (2 m^2) plots; see Fig. 6.19. A series of large-scale trials were implemented in 2007 to develop practices for residue amendment and seedbed preparation at large-scale level. Findings from this work shows that the key stages in the revegetation programme can be achieved at a large-scale level. A range of grassland species can be used in the seeding once the inhibitory properties of the residue are overcome and a seedbed with adequate nutrients and organic matter is established.

Due to the success of the trials and experiments onsite, methodologies for optimising plant establishment on the bauxite residue are well developed:

Fig. 6.19 Early small plot trials experimenting different amendments

- Addition of process sand to improve texture and structure of the residue substrate, gypsum for reducing pH and Exchangeable Sodium Percentage (ESP) and organic matter for nutrients are essential components of the revegetation prescription;
- Several indigenous species are capable of growing in amended bauxite residue;
- Effective amendment of the residue results in lower plant content of Na, Fe and Al;
- Nutrient cycling in the residue is seen as a critical parameter to demonstrate that the vegetation cover is self-sustaining cover.

Lack of organic matter and nutrient deficiency is recognised as a limiting factor in establishing vegetation on any soil including residue. Incorporation of organic matter into the rooting medium is a critical component of the revegetation prescription.

First Cell Trial Revegetation of Residue

A large-scale (0.6 ha) dedicated research trial cell was constructed in 2008 to test rehabilitating unfarmed residue. The lined cell was filled with fresh bauxite residue and underwent amendment as per the developed restoration technique (i.e. process sand, gypsum, organic matter, and organic matter). Revegetation was undertaken in September 2009; the amended area was seeded with species that had previously been trialled at small plot and large plot scale.

Key research areas within this trial area were:

- Vegetation establishment, survival and succession;
- Vegetation productivity, sustained growth and structure development;
- Fauna colonisation and habitat development;
- Ecosystem processes such as soil development and nutrient cycling;
- Colonisation of specific fauna groups that are involved in these processes;
- Microbiological studies e.g., colonisation by mycorrhizal fungi and microbial biomass.

The cell area is sampled bi-annually to monitor the emerging plant/residue soil system and assess functioning ability of the system, see Fig. 6.20; it is expected that this system can be proven sustainable/self-regulating. See Fig. 6.21, five months after seeding; Fig. 6.22, ten months after seeding and Fig. 6.23, soil profile five years after planting.

Based on results and observations from residue research field trials the surface residue (0–15 cm) pH and exchangeable sodium (ESP) indicates improvement in residue properties following amendment, seeding and subsequent soil development. As previously found, amendment procedures resulted in sufficient macro nutrient supply and there was no evidence of excessive uptake of elements associated with bauxite residue (e.g. aluminium or sodium).

Grassing of Terraced Area on BRDA

As part of the continuous improvement programme, a new area of the BRDA was rehabilitated. The area chosen was approximately 30 m wide at an elevated location

Fig. 6.20 Selection of species growing on revegetated residue

Fig. 6.21 Cell five months after seeding

Fig. 6.22 Cell 10 months after seeding

on the BRDA stack, in total 4 ha of residue were seeded. Since its seeding, the area has been continually monitoring for grass growth.

The aims of the work undertaken in 2015 were:

- To establish grass on the wide terraced area of the BRDA; and
- Monitoring and maintaining the grass.

Analysis was undertaken in 2015 to characterise the bauxite residue. From the results of the characterisation, an inclusion rate of customised organic matter along with an incorporation depth and application method was devised. The organic matter provision was to provide the nutrients for grass establishment.

In 2015 the following was conducted:

- Addition of gypsum to the surface at a level of 3%.
- Customised organic matter was incorporated at a rate of 30% and sown with the following nine salt tolerant grass species:
  - Creeping Bent
  - Common Bent
  - Crested Dog's Tail

Fig. 6.23 Soil profile of amended and seeded residue after 3 years in trial cell

- Sheep's Fescue
- Red Fescue
- Perennial Rye grass
- Salt Marsh Grass
- Red Clover
- White Clover.

A drainage plan was designed and a management plan for the established grass area was implemented. This involved mowing the grass 2–3 times during the year to return the nutrients to the ground; the ground was aerated as required.

Nine salt tolerant grass species were sown and grew well, the grass continues to thrive and currently presents a green belt along the north and west of the BRDA perimeter. This result shows that it is possible to establish grass on the BRDA. Aeration is required to ensure that the activate root zone remains aerated, due to compaction a hard pan can develop resulting in restricted movement of the roots limiting the plants access to vital nutrients. See Figs. 6.24 and 6.25.

Fig. 6.24 Steps in rehabilitation and revegetation of deposited residue in terraced area

Fig. 6.25 Terraced area demonstrating healthy grass growth

Rehabilitation of Farmed Residue

During 2017 Aughinish constructed a new closure demonstration area. The amended layer will be constructed from carbonated residue, neutralised process sand 25%,

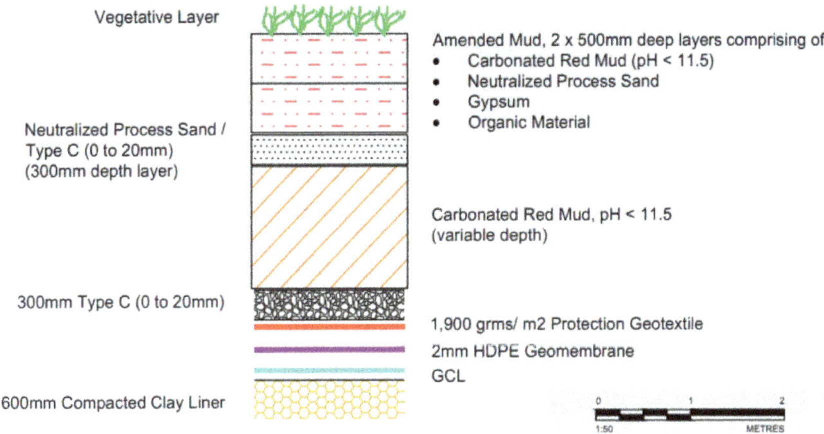

Fig. 6.26 Section through new closure demonstration cell

gypsum 3% and organic material 20%; the trial area will also be used to test the effectiveness of the amended layer without gypsum.

Once completed, this cell will reflect a typical closure scenario of the BRDA. To demonstrate its sustainability and comply with the license conditions it will be intensively monitored (Fig. 6.26).

Sustainability of Remediated Bauxite Residue

In September 2015, in collaboration with the University of Hull and Leeds, trial pits were dug at 5 cm intervals to a depth of 50 cm in the revegetated plots seeded in 1999 on the BRDA, sampling was undertaken 20 years after deposition and 16 years after treatment. Three plots within the BRDA were investigated. The fully treated plot which was amended with gypsum (3% w/w rotavated-in to a depth 30 cm), process sand (10% w/w rotavated-in to a depth of 30 cm), spent mushroom compost (80 t/ha rotavated-in to a depth of 20 cm), and seeded with a grassland mix (100 kg/ha). The partially treated plot which was amended only with process sand, spent mushroom compost, and then seeded. The third plot was left untreated. Samples of bauxite residue were collected to a depth of 50 cm from the trial pits in each of three different treatment zones.

Both the fully treated and partially treated sites were vegetated with a variety of perennial grasses (*Holcus lanatus*), trifoliate clovers *(Trifolium pratense),* and occasional small shrubs. The untreated plot was largely unvegetated with one or two areas of stunted grasses. The root zone of the fully treated and partially treated sites extended approximately 15 cm beneath the surface, and below 20 cm the substrate had the appearance of dewatered bauxite residue with little change in appearance to 50 cm depth. The untreated profile had no root zone and at all depths had a very similar appearance to the residue in the other profiles at depths below 20 cm.

The pH of the treated plots was notably lower. The plots exhibited a decrease of 2.5 pH units at a given depth, compared to the untreated plot, giving the surface zone of the treated plots pH < 8. The treated plots also displayed a 3–4-fold decrease in aqueous available sodium at all depths. Treatment also decreased the overall availability of trace metals aluminium, chromium and vanadium to 50 cm, compared to the untreated plot. These tests proved that the positive effects of treatment extend well beyond the 20 cm deep treatment zone and are evident by the lush vegetation growing on the surface.

Figure 6.27 shows profiles after various treatments: untreated bauxite residue 20 years after deposition (A, B); bauxite residue 16 years after full treatment with gypsum, organic matter, and process sand (C, D); and bauxite residue 16 years after partial treatment with organic matter and process sand (E, F). The tape measure in B, D, and F measures 50 cm depth in each trial pit.

Side Slope Rehabilitation

A side slope rehabilitation programme of the BRDA was undertaken in 2013 to improve visual impact of the area (see Fig. 6.28 overview). The objective was to establish a relatively dense green vegetative cover and to establish a dispersed and relatively low-level scrub style planting. The planting has a random and naturalistic appearance when viewed from areas surrounding the structure. The total area for revegetation was 36,870 m^2 in which over 4000 trees and scrubs of varying species were planted. Many plant species have also naturally developed on the side slopes of the BRDA improving the naturalistic appearance. Side slope rehabilitation differs from dome rehabilitation, as the residue is not directly amended, see Fig. 6.28. A layer of drainage stone is first placed on the residue surface followed by subsoil and topsoil. Aughinish Alumina Limited have a proactive landscaping programme that targets the appearance of the BRDA and each year invest heavily on planting to improve the visual impact.

Final Land Use

The long-term sustainable land-use of the BRDA at Aughinish is the goal for the rehabilitated facility. In deciding the most suitable end use for the BRDA, it has been determined that activities which may lead to over-grazing, poaching, cultivation, uprooting of trees by wind-blow and other surface disturbance will be avoided. The preferred land-use option, based on current knowledge of the chemistry and biology of the sown grassland cover, is to develop the area for nature conservation, see Fig. 6.29.

6.7 Current Best Practice Australia—Background

Alcoa of Australia Limited (Alcoa) has three refineries in Western Australia, at Kwinana, Pinjarra and Wagerup, with a combined capacity of approximately nine million tonnes of alumina per year. As early as the mid-1970s, Alcoa recognised the

Fig. 6.27 Slices taken through trial plots

Fig. 6.28 Side slope rehabilitation

Fig. 6.29 Evidence of thriving ecosystem on revegetated residue (trial 2007–2011)

need to investigate alternative residue management strategies. The Company made a significant commitment to continuous improvement in bauxite residue management with the formation of a research and development group charged with investigating alternate storage practices. The primary focus for this effort came from the discovery of ground water contamination below the Kwinana storage areas. This led to research and field trials, from which a range of significant improvements to residue storage methods have been developed and implemented over the past two decades. However, through recent long-term planning for residue storage, it has become increasingly apparent that community and government expectations are changing, and that further improvements to the way residue is managed into the future will be needed.

Alcoa's vision for residue management is an overall transition to a more sustainable method of residue storage, where the impact on the environment and the surrounding community is being progressively reduced. The move to dry stacking was a significant step toward a more sustainable method of storage, allowing a greater volume of residue to be stored within a given footprint, and providing a stable storage area which could be used for a range of land uses once closed. Residue filtration, a process where the slurry is filtered to produce a filter cake which can be mechanically stacked is a further step in this direction.

Neutralisation of the residue was also investigated as a potential step along this pathway, as the more significant hazard associated with the residue (its high pH) can be reduced. Work on residue carbonation commenced in the early 1990s with preliminary laboratory testing. This work demonstrated the viability of carbonation and was followed in the mid to late 1990s with small scale pilot testing. Encouraged by the success of these trials, a full-scale prototype of residue carbonation was installed and commissioned mid-2000 at the Kwinana refinery. This prototype allowed detailed evaluation of both the initial treatment of a thickened residue slurry with a high concentration CO_2 gas, and the performance of this thickened slurry in a full scale drying bed.

However, the ultimate aim in terms of sustainability of residue management is to progressively reduce the volume of residue stored. Dry stacking, and more recently residue filtration, have been important steps toward re-use of the residue, as it has made access and reclamation of the residue more cost effective. Residue neutralisation may also be a key step toward re-use of the residue, as it will remove the hazardous nature of the residue, removing constraints on its handling and use. With an increased focus on re-use over recent years, a number of re-use opportunities have been developed and progressed toward possible commercialisation.

Early Developments

The original containment areas were constructed on the sandy coastal plain and relied upon a 380 mm thick clay blanket to prevent contamination of the underlying aquifer. Whilst the clay seal prevented general seepage, there were a number of defects in the blanket placed on the embankment. These were thought to be the result of either cracking due to desiccation or erosion caused by rainfall; because of the ground water contamination, the early emphasis of research was placed on improving the design of the base seal.

In 1980, a trial underdrain storage area was constructed to evaluate the concept of base drainage for improved storage of the fine fraction of the residue. The design was identical to the conventional storage area except that a one-meter deep sand layer was installed above the base clay seal. A network of perforated pipes was installed in the sand layer and this drained via gravity to a sump. The incorporation of an underdrainage system offered a number of advantages. Pumping of the sump maintained a low hydrostatic pressure on the base seal reducing the potential for seepage. Base drainage also improved consolidation of the fine-grained material and proved an efficient means of recovering alkali for re-use by the refinery. Monitoring of the trial deposit showed an improvement in the density from 55% solids (0.9 t/m^3 dry density) achieved in the conventional storage area to 62% solids (1.08 t/m^3 dry density) thereby providing a 20% improvement in storage efficiency during the operating life of the area.

This improved density increased the strength of the fine-grained material providing a more stable deposit which in turn aided surface reclamation. It was thought that long term rehabilitation would also be enhanced by gradual downward leaching of the deposit. A decision was also made that new containment areas would be constructed with a composite clay/synthetic membrane seal which, when installed with a drainage layer placed above this composite seal, provided a very high factor of safety against any seepage.

The Move to Thickened Slurry and Solar Drying

Early development work was also focussed on improving the overall efficiency of the residue storage areas. Traditional "wet" storage areas, as discussed earlier, required full height containment embankments. Economics and the availability of construction materials at some locations dictated a limit on the height to which these embankments could be constructed, typically around 20 m. The alternative to high embankments was that large amounts of land were going to be required for the ongoing construction of these wet storage areas. This led to the development and adoption of the thickened slurry deposition and solar drying process.

Efficient solar drying requires initial dewatering of the residue to produce a thickened slurry, but only thickened to the point where the slurry can still be pumped and hydraulically spread over the drying area. Through the early 1980s, Alcoa investigated a wide range of thickening technologies before adopting a large diameter gravity thickener, called a super thickener, to de-water the fine tailings and produce a thickened slurry.

The residue from processed Darling Range bauxite is characterised by a high coarse fraction (nominal particle size > 150 μm). This coarse fraction can be considered as a fine to medium grained sand. This sand fraction is removed prior to the thickening of the finer red mud and is used for construction of drainage layers and upstream perimeter embankments. The overall storage area is constructed as a progressive stack, avoiding the need for full height perimeter embankments. It has also facilitated continued deposition on areas which were previously "wet" impoundments.

Rainfall run-off from the storage area is dilute alkaline water. It is stored during winter, and then added to the super thickener during summer, along with fresh water,

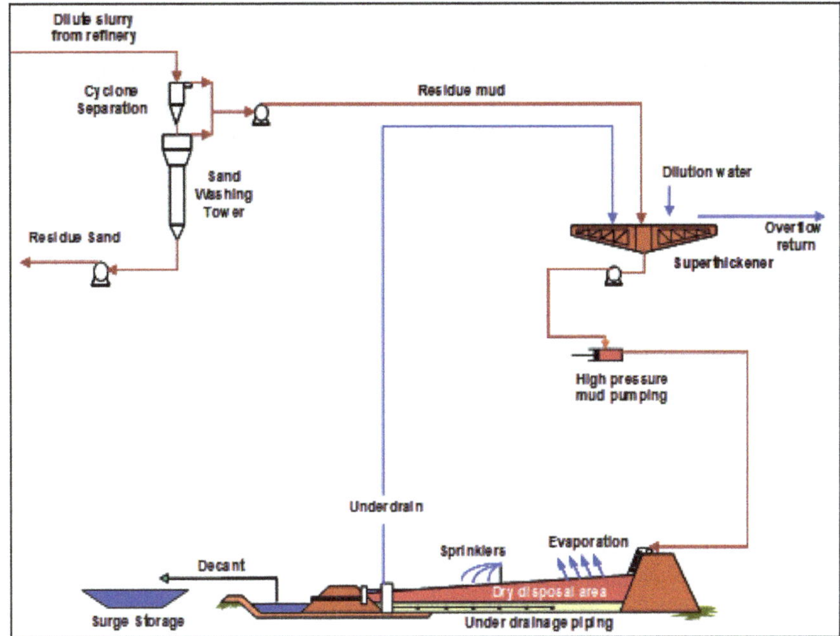

Fig. 6.30 Schematic of the dry stacking process

as make-up water to compensate for refinery and residue area evaporative losses. Because this water is dilute, it acts as a final stage of dilution washing of the mud. The overflow from the super thickener is returned to the refinery. Underdrain from the residue storage areas is a combination of water released from the initial slurry via self-weight consolidation and rainfall which infiltrates through the drying beds and is either added to the super thickener feed to aid dilution washing or returned directly to the refinery, depending on the time of year and hence alkalinity.

A schematic diagram of the Dry Stacking Process is shown in Fig. 6.30; the initial costs of establishing dry stacking at Alcoa's three Western Australian refineries exceeded AUD$150 million. However, there were many benefits, which helped justify the transition to dry stacking including:

- a higher density deposit can be achieved reducing the overall volume of stored tailings;
- the progressive stacking allows the deposit to be taken to a height which would not be economic with conventional wet impoundments;
- the higher density and increased deposit height means less land is used;
- the exposure of less land area to residue;
- the drained condition of the dry stack and the smaller footprint significantly reduce the risk of ground water contamination;
- improved surface stability and drainage mean that completed areas can be reclaimed and re-vegetated quickly; and

- safety hazards to people and wildlife are reduced.

Optimising the coverage on the drying beds has become one of the key operational objectives in recent times. As the height of the stack grows, drying area is lost due to the encroachment of the embankments as they are progressively raised via upstream lifts. This means that additional area needs to be provided and this is at a significant capital cost. If drying of the slurry can be sustained on a smaller area, significant capital for expansion of the drying areas can be deferred. Alcoa measures drying rate in terms of the overall (or average) annual storage rate expressed as the amount of residue that can be stored per unit area each year. To achieve the required moisture loss, through the winter drying cycles, Alcoa routinely ploughs the mud with machinery. This is generally done using a low ground pressure swamp dozer or an Archimedean screw vehicle, called an Amphirol—see Figs. 6.16 and 6.31. This allows a storage rate of between 13,000 and 14,500 t/yr/ha to be sustained.

Routine ploughing of the mud with mechanical equipment has been termed "mud farming". Mud farming helps achieve a maximum density which allows the dry stack to be developed with maximum outer slopes (a minimum strength of 25 kPa is achieved allowing an outer slope of 6:1 to be maintained), and maximises the storage efficiency of the stack. Mud farming also minimises the potential for dust generation, which is important given the location of the refineries close to residential areas. Ploughing the surface prevents a wet surface, buries fine particulates that can

Fig. 6.31 Typical drying bed, with amphibious machines "ploughing" the freshly deposited residue slurry to aid drying

accumulate at the surface, and provides a surface roughness that prevents dust lift-off once the tailings have dried.

Introduction of Residue Filtration within Alcoa

To sustain this method of "dry stacking", a large active surface area is required. As mentioned above, the available area for stacking is continually reducing as the stack height increases, necessitating the construction of new drying beds to maintain the required minimum drying area. This has led to a detailed study into the potential use of residue filtration. The concept for such a process involves:

- Coarse and fine residue separation as per the current system;
- Thickening of the fine fraction, again as per the current system;
- Filtration of the fine fraction to a low moisture content cake (moisture below the point at which the cake will re-slurry);
- Conveying of the cake to a localised stacking area;
- Mechanical stacking of the cake from a mobile "spreader";
- Machinery utilised to aid the stacking and to compact the cake to a minimum volume.

This removes the requirement for residue drying and therefore reducing the surface area required to store residue. Because a minimum drying area is no longer needs to be sustained, the existing footprint can be used to store the filtered residue for many years, delaying the ongoing expansion required.

Over the past few years, an assessment of filtration technologies available has been undertaken by Alcoa with plate and frame pressure filters being selected as the preferred equipment. The overall function of plate and frame filters from different suppliers is similar, however these filters can have a number of associated features and ancillary functions which may be specific to the filter type or supplier. The specific features of potential filters were evaluated in relation to the preferred features for the red mud application.

Following extensive trial and pilot testing, the decision was made in 2015 to install Alcoa's first full scale filtration system at its Kwinana refinery; this system was commissioned mid-2016—see Figs. 6.32, 6.33, 6.34 and 6.35 showing the filter, the mobile spreader being used to distribute the filtered mud and a dozer compacting the filtered mud. Based on the success of the plant, the first stage of residue filtration is now being installed at Alcoa's Pinjarra refinery with commissioning expected early 2019.

Closure and Rehabilitation

While the active residue storage areas are not generally visible outside of the operations, Alcoa is committed to rehabilitating the outer embankments of the storage areas to as natural an appearance as possible. The aim of embankment rehabilitation is to generate a self-sustaining ecosystem, but this can take years to establish. As the external embankments are lifted to contain the residue, the new embankment sections

Fig. 6.32 Kwinana filtration building

are progressively rehabilitated. In effect, this allows a large proportion of the overall residue footprint to be rehabilitated before the storage areas close (Fig. 6.36).

Field trials and fundamental research on residue rehabilitation as a means of optimising the vegetation performance is undertaken at all three Alcoa of Australia refinery residue areas. This work varies from large scale field trials such as demonstration areas at Pinjarra, to bench scale laboratory experiments by local, interstate and international universities. This research aims to better understand water-nutrient-plant-residue sand dynamics as a means of optimising the rehabilitation prescription and identifying potential residue area closure strategies.

The current closure strategy has three main objectives:

- The decommissioned residue areas should have the capability to be used for productive community benefit.
- Be a safe and self-sustaining structure in the long term; and
- Allow future access to residue for alternate uses.

An example of returning residue storage areas to “productive land use” is the current use of Alcoa’s early residue storage areas at the Kwinana refinery. The land on which these storage areas are located is leased from the Western Australia State Government under the Alumina Refinery Agreement Act. The act requires this land to be returned to the government when Alcoa has finished using it for residue storage.

Fig. 6.33 Inside the Kwinana filtration building—filter floor level

Parts of the area were handed back to Landcorp, (the Western Australian Government's land and development agency) in 2000 and have since been incorporated into the Perth Motorplex, which was opened in December 2000, see Fig. 6.37. Since opening, there have not been any issues related to managing the site in the context of it being a former residue storage area.

6.8 Uses of Bauxite Residue

Reuse of bauxite residue has been a goal of the alumina industry for over a century but despite thousands of trials, the identification of dozens of technically feasible uses, only some four million tonnes a year of the 170 million tonnes of bauxite residue produced annually is used in a productive way. The challenge remains to find good economically viable uses for the amount generated every year in addition to utilising some of the material already stockpiled. The most successful large-scale uses include cement production, raw material for iron and steel manufacture, manufacture of building materials, landfill capping, road construction and soil amelioration.

The current focus on sustainability and the desire to utilise our global resources more efficiently is leading industry, universities and entrepreneurs, often aided by government support, to seek new/improved ways of using bauxite residue. From

Fig. 6.34 Mobile spreader used to distribute the filter cake at Kwinana

Fig. 6.35 Dozer spreading and compaction of the filter cake deposited by the mobile spreader

Fig. 6.36 Vegetation established on the outer slopes of a residue dry stacking area

Fig. 6.37 Motor sports complex built partially over the original residue storage areas

the process side, improvements include: the increasing use of press filters will give residues with lower moisture levels, lower soda levels, lower contaminants, lower pH levels; higher efficiency electro-magnets are now available which allows for more effective iron recovery from bauxite residue. Meanwhile the growing demand for scandium in aluminium alloys or the demand for specific rare earth elements also present new opportunities. The increased funding for research, greater industry/university co-operation, improvements in process technology, public, corporate and government attitudes have never presented such an encouraging environment for developing and implementing bauxite residue uses.

Residues from different bauxites vary very much in terms of composition, mineralogy and particle size so behave quite differently. They will have different handling and pumping characteristics and display widely different settling and particle packing characteristics thereby influencing moisture contents after treatment and handling costs. The variation in composition naturally has an overriding effect on potential applications so any practical work looking at applications must take into account the specific chemical composition, mineralogy, pH, particle size distribution, morphology and nature of the residue emanating from a particular plant.

6.8.1 Composition of Bauxite Residue

The key first steps in discussing uses is a consideration of the chemical compounds present in the bauxite residue, the levels present and the physical characteristics of the material. Bauxite is formed by the weathering of various aluminosilicate-based rocks and are frequently divided into lateritic bauxites (approximately 88% of the total reserves) and karstic bauxites (about 12% of the reserves). The residue remaining at the end of the Bayer process includes the compounds not dissolved by the sodium hydroxide and the compounds formed during the autoclaving, desilication and other stages of the process. The main constitutes are iron, aluminium, silicon, calcium, sodium and titanium compounds but display an extremely wide variation in composition as shown in Table 6.3; this data represents values for commonly used bauxites and the range can be even broader for some bauxites [4, 6].

Table 6.3 Chemical composition, expressed as oxides, commonly found in bauxite residue

Component	Typical range (wt%)
Fe_2O_3	5–60
Al_2O_3	5–30
TiO_2	0–15
CaO	2–14
SiO_2	3–50
Na_2O	1–10
LOM	5–20

In addition to the iron, aluminium, silicon, calcium, sodium and titanium compounds a wide range of other components may also be present at low levels; these will frequently be in the form of metallic oxides e.g. arsenic, beryllium, cadmium, chromium, copper, gallium, lead, manganese, mercury, nickel, potassium, scandium, thorium, uranium, vanadium, zinc, zirconium and rare earth elements (REEs). Non-metallic elements that may occur in the bauxite residue are phosphorus, carbon and sulfur.

The minerals present are complex and comprise some which are present in the bauxite and others that are produced during the autoclaving and the desilication processes. The range of minerals typically found for bauxite residues is shown in Table 6.4 [4, 6, 27].

In addition there are various other minerals sometimes found at low levels including Brookite (an orthorhombic variant of TiO_2), ilmenite ($FeTiO_3$), Carnegieite ($Si_4Al_4Na_4O_{16}$), Dolomite ($CaMg(CO_3)_2$), Hydrogarnet ($Ca_3Al_2(O_4H_4)_3$), various Hydroxycancrinite/Cancrinite ($Na_6Ca_2(Al_6Si_6O_{24})(CO_3)_2{\cdot}2H_2O$), $(Na, Ca)_8(Al, Si)_{12}O_{24}(CO_3){\cdot}4H_2O$), Cancrinite-$NO_3$ ($Na_{7.92}Si_6Al_6O_{31.56}N_{1.74}$), Cancrisilite-$CO_3$ ($Na_{7.86}(AlSiO_4)_6(CO_3)(H_2O)_{3.3}$), Katoite-Si ($Ca_3Al_2(SiO_4)(OH)_8$), Lawsonite ($CaAl_2Si_2O_7(OH)_2{\cdot}H_2O$), Nepheline ($Na_2KAl_4Si_4O_{16}$), Nosean ($Na_8Al_6Si_6O_{24}(SO_4)$), Chamosite ($(Fe, Al, Mg)_3[(Si, Al)_4O_{10}](OH)_2(Fe, Mg)_3(O, OH)_6$), Portlandite ($Ca(OH)_2$), Schaeferite ($(Na_{0.7}Ca_{.2.3})(Mg_{1.85}Mn_{0.15})(VO_4)_{2.88}(PO_4)_{0.12}$)), sodium titanate ($Na_2TiO_3$) and zircon ($ZrSiO_4$).

Table 6.4 Typical range of components found in bauxite residues

Component	Typical range (wt%)
Sodalite ($3Na_2O{\cdot}Al_2O_3{\cdot}2SiO_2{\cdot}0\text{–}2H_2O{\cdot}2NaX$ where X could be CO_3^{2-}, Cl^-, OH^-, SO_4^{2-}, or $Al(OH)_4^-$)	4–40
Al-Goethite ($(Fe, Al)_2O_3{\cdot}nH_2O$)	1–55
Haematite (Fe_2O_3)	10–30
Magnetite (Fe_3O_4)	0–8
Silica (SiO_2) crystalline and amorphous	3–20
Calcium aluminate ($3CaO{\cdot}Al_2O_3{\cdot}6H_2O$)	2–20
Boehmite ($AlOOH$)	0–20
Titanium dioxide (TiO_2) anatase and rutile	2–15
Muscovite ($K_2O{\cdot}3Al_2O_3{\cdot}6SiO_2{\cdot}2H_2O$)	0–15
Calcite ($CaCO_3$)	2–20
Kaolinite ($Al_2O_3{\cdot}2SiO_2.2H_2O$)	0–5
Gibbsite ($Al(OH)_3$)	0–5
Perovskite ($CaTiO_3$)	0–12
Cancrinite ($Na_6[Al_6Si_6O_{24}{\cdot}2CaCO_3$)	0–50
Diaspore ($AlOOH$)	0–5

Sodium is the only element not found in the bauxite itself. Depending on the solubility of the species and the temperature used in the extraction process, some elements will increase in concentration in the bauxite residue relative to the bauxite whilst others will be lower in the bauxite residue. Other than sodium hydroxide, lime is normally the only other inorganic compound introduced during the Bayer process.

A wide variety of organic compounds can also be present, these are derived from vegetable and organic matter in the bauxite/overburden or the use of crystal growth modifiers or flocculants and includes carbohydrates, alcohols, phenols, and the sodium salts of polybasic and hydroxyacids such as humic, fulvic, succinic, acetic or oxalic acids.

The sodium in bauxite residue may be present in a sparingly soluble form, called the desilication product (DSP) or a very soluble form. Small quantities of some of the soluble sodium compounds resulting from the sodium hydroxide used in the extraction process will remain depending on the dewatering and washing systems used. All Bayer alumina refineries aim to maximise the recovery of the valuable caustic soda from the residues in order to reuse it during the extraction process. The residual soluble sodium species, predominantly a mixture of sodium aluminate and sodium carbonate, give rise to an elevated pH for bauxite residue slurries. Over time the residual sodium species are partially neutralised by carbon dioxide from the air to form sodium carbonate and other metal carbonate species; these species will result in a lower pH for the bauxite residue which renders them less hazardous. The desilication product ($3Na_2O \cdot Al_2O_2 \cdot 2SiO_2 \cdot 0\text{–}2H_2O \cdot 2NaX$ where X could be CO_3^{2-}, Cl^-, OH^-, SO_4^{2-}, or $Al(OH)_4^-$) arises from the reaction between sodium aluminate and soluble sodium silicates.

The other factors that are in important when considering uses are the physical characteristics such as particle size distribution and some variable parameters such as moisture content. The median particle size is normally in the range of 2–10 μm, and most commonly in the range 2–4 μm. However, the breadth of particles is both very broad ranging from coarse sandy grains about 1 mm in size down to sub-micron particles for bauxite residues produced from different alumina plants and from different bauxites. The coarse component is predominantly crystalline silica (quartz), whilst the fine components is largely iron oxide. Some alumina refineries separate the different size fractions during processing whilst others do not; the coarse sandy fraction has been given various names, for example "Red Sand™" or "Red Oxide Sand".

6.8.2 *Review of Uses*

The list of applications where bauxite residue has been evaluated and trialled covers almost all of inorganic material science and seeking effective solutions has attracted many researchers from industry, universities, institutes and entrepreneurs [6, 28–79]. Many potential applications have been considered with particular focus on routes focusing on the recovery of elements present in the bauxite residue. Even Bayer himself in his 1892 patent describing the Bayer Process proposed the potential for iron recovery: stating, "The red, iron containing residue after digestion settles well and, with sufficient practice, can be filtered and washed. Due to its high iron and aluminium oxide content, it can be, in an appropriate manner, treated or with other iron ores, be smelted to iron".

Possible applications for bauxite residue can broadly be divided into five categories:

- Recovery of specific components present in the bauxite residue, e.g. iron, titanium, aluminium, lanthanides, scandium, yttrium, and gallium [6, 28–30, 34, 39, 42];
- Use as a major component in the manufacture of another product/material, e.g. cement [31–33];
- Use of the bauxite residue as a component in a building or construction material, e.g. concrete, tiles, insulation, bricks [34, 38–40];
- Use for specific properties of the residue, e.g. soil amelioration or landfill capping [6, 34];
- Conversion of the bauxite residue to a useful material by modifying the compounds present, e.g. Virotec process [35–37].

A list of areas and topics that have been investigated include: Portland cement manufacture [31–33], special cements [41], use in concrete, iron recovery [30, 34–36], titanium recovery [30, 34, 38], use in building panels, bricks, foamed insulating bricks, loft insulation (ceramic wool), tiles, soil amelioration, production of a soil like medium suitable for plant growth, refuse tip capping/site restoration, treatment of acid mine drainage [35–37], road construction [43], dam/levee construction, pigments, glass ceramics. Meanwhile some other uses which have been shown to be technically feasible but not yet exploited to any significant degree are: lanthanides (REEs-rare earth elements) recovery, scandium recovery, gallium recovery, vanadium recovery, cobalt recovery, yttrium recovery, adsorbent of heavy metals, dyes, phosphates, fluoride, water treatment chemical, ceramics, foamed glass, oil drilling or gas extraction proppants, gravel/railway ballast, riprap, calcium and silicon fertiliser, filler for PVC, wood substitute, geopolymers [45], catalysts [46, 47], plasma spray coating of aluminium and copper [48], manufacture of aluminium titanate-Mullite composites for high temperature resistant coatings, composites with epoxides, composites with poly aniline, manufacture of radio-opaque materials for the construction of X-Ray diagnostic and CT scanner rooms, catalyst for hydrocarbon cracking, desulfurisation of flue gas, arsenic removal from water, chromium removal from water. Some applications, such as use in pigments have been successful on a commercial basis but use very small tonnages.

In recent years, the global desire for the more efficient use of resources has led to a greatly increased effort looking at uses of bauxite residue. Increased enthusiasm from industry, more university activity and the input of funds from organisations and bodies such as the EU has led to greater co-operation between industries and academia. Whilst previously there was a focus by producers to use the residue in a single application, there is growing focus on 'total' solutions in order to maximise resource recovery by extracting say the iron fraction, then the scandium or other valuable minor metals giving a residue that can then be used in insulation wool or construction materials.

In 2015, under the Horizon 2020 initiative, the EU agreed to fund 15 PhD students to work on the recovery of materials and the utilisation of bauxite residue. Some Euro 3.7 million was set aside for this European Training Network—MSCA Zero-Waste Valorisation of Bauxite Residue (REDMUD) project. The work focussed on the extraction of iron, aluminium, titanium and rare earth elements (including scandium) as well as the production of new building materials [49].

A European Innovation Partnership has been formed to explore options for using by-products from the aluminium industry, BRAVO (Bauxite Residue and Aluminium Valorisation Operations). This sought to bring together industry with researchers and stakeholders to explore the best available technologies to recover critical raw materials but has not proceeded. Additionally, EU funding of approximately Euro 11.5 million has been allocated to a four year programme starting in May 2018 looking at uses of bauxite residue with other wastes, RemovAL. A particular focus of this project is the installation of pilot plants to evaluate some of the interesting technologies from previous laboratory studies. As part of the H2020 project RemovAl, it is planned to erect a house in the Aspra Spitia area of Greece that will be made entirely out of materials from bauxite residue.

Other European projects that have involved bauxite residue and waste recovery have been ENEXAL (energy-exergy of Aluminium industry) [2010–2014], EURARE (European Rare earth resources) [2013–2017] and three more recent projects are ENSUREAL (Ensuring sustainable alumina production) [2017–2021], SIDEREWIN (Sustainable Electro-wining of Iron) [2017–2022] and SCALE (Scandium–Aluminium in Europe) [2016–2020] a Euro 7 million project to look at the recovery of scandium from bauxite residue.

With the support of EU funds, research groups at KU Leuven in Belgium have been collaborating at the KU Leuven Institute on Sustainable Metals and Minerals (SIM2), on the extraction of metals and rare earths from bauxite residue. The group is part of the SREMat (Sustainable REsources for Engineered Materials) research group that belongs to the SeMPeR (Sustainable Metals Processing and Recycling) group at the department of Materials Engineering. Mixtures of bauxite residue with additives such as carbon, sand and limestone are fired at 1100 to 1200 °C, which causes it to melt, and then poured into water to quench the viscous mass. That results in a glass-like material, which is milled down to a fine powder; an alkaline solution is then added to form a paste which can be moulded to produce bricks, shingles, tiles and building panels. Either dense or porous materials may be manufactured. In the EIT-KIC RawMaterials project RECOVER, a mobile production unit has been constructed in containers, these can be used to manufacture up to 1000 kg a day of final products showing to the public what you can do with bauxite residue and other residues.

In November 2020 another significant EU funded project was launched, ReActiv: Industrial Residue Activation for Sustainable Cement Production research project. One of the world's leading cement companies, LafargeHolcim, in cooperation with seven major alumina producers and 13 other partners across 12 European countries, launched the ambitious 4 year ReActiv project (reactivproject.eu). The ReActiv project will create a novel sustainable symbiotic value chain, linking the by-product of the alumina production industry and the cement production industry. In ReActiv modification will be made to both the alumina production and the cement production side of the chain, in order to link them through the new ReActiv technologies. The latter will modify the properties of the industrial residue, transforming it into a reactive material (with pozzolanic or hydraulic activity) suitable for new, low carbon dioxide footprint, cement products. In this manner, ReActiv proposes a win–win scenario for both industrial sectors ensuring better use of the earth's resources, reducing wastes and carbon dioxide emissions.

At various times targets have been proposed for reuse of bauxite residue, these include: the IAI Alumina Technology Roadmap strategic goal of 20% by 2025; the Chinese Government target of 25% reuse by 2015; the Shandong Province target of 40% reuse by 2015. This compares to an estimated reuse level in China of 10% and globally 2–4%.

6.8.3 *Industrial Scale Up of Uses*

It is estimated that approximately 3 to 4% of the bauxite residue produced annually is used in some way although reliable data is difficult to obtain as it does fluctuate markedly from year to year as the economic situation changes and some uses are internal such as road and dyke construction within a particular alumina refinery so will fluctuate depending on particular expansion programmes.

Current annual estimates collected from various sources are [6]:

- Cement—2,200,000 to 2,700,000 tonnes;
- Raw material/additive in iron and steel production—800,000 to 1,500,000 tonnes;
- Roads (see Fig. 6.38)/landfill capping (see Fig. 6.39)/soil amelioration—250,000 to 500,000 tonnes;
- Construction materials (bricks (see Fig. 6.40), tiles, ceramics etc.)—1,000,000 to 1,400,000 tonnes;
- Glass ceramics, glass fibre—100,000 to 120,000 tonnes
- Other (refractory, adsorbent, acid mine drainage (Virotec), catalyst etc.)—300,000 tonnes.

Based on the above data, this gives a total of approximately 3.4–5.3 million tonnes per year.

Fig. 6.38 Road widening by the Western Australian main roads using red sand™

Fig. 6.39 Use of Bauxaline® in landfill capping in provence

Fig. 6.40 Use of bauxite residue in bricks, JBI, Jamaica

6.8.4 Bauxite Residue Production and Uses in China

China is worth discussing separately as many of the bauxites residues produced are chemically very different from the rest of the world and the producers are devoting considerable effort into searching for and implementing reuse of bauxite residue, much of it being driven by Chinese Government initiatives [50–59]. Alumina production in China has shown a dramatic change in the last 15 years with production increasing from about 2.5 million tonnes of alumina in 2000 to over 70 million tonnes in 2020 [3]; the generation of bauxite residue has grown to an estimated 100 million tonnes a year. There is considerable variation in the amount of bauxite residue generated per tonne of alumina and the compositions vary widely.

The alumina manufacturing routes have historically been very different because of the nature of the indigenous bauxite. Sinter routes or combined Bayer-sinter routes were widespread using indigenous bauxites but are now declining, except for the production of chemical grade aluminas, and the industry has become more dependent on imported bauxites. This change in processing route has significantly changed the characteristics and composition of the bauxite residues being produced. These compositions of the residues continue to change as much of the imported bauxite was from Indonesia and Australia; curtailment of bauxite exports from Indonesia since 2015 is changing the nature of the bauxite residue yet again. Traditionally the alumina plants in the Northern part of China produced a residue very high in calcium and silicon oxides but low in iron oxide making them suitable for cement production whilst those in the south of China have a residue high in iron which makes the recovery of iron the most likely option for them to pursue. The contents of CaO and SiO_2 in bauxite residue from the sintering process and combined process are much higher than that from the Bayer process but the contents of Fe_2O_3 in bauxite residue from the sintering process and combined process are much lower than that from the conventional Bayer process. The main component of the bauxite residue from the sintering process is β-$2CaO{\cdot}SiO_2$, with the mass ratio being close to 50%: the average composition for the sintering plants at Guizhou, Shandong, Shanxi and Zhongzhou is 41% CaO, 22% SiO_2, 8% Al_2O_3, 7.7% Fe_2O_3, 3% TiO, 2% MgO and 2.9% Na_2O. Meanwhile the composition from the process is similar at 45% CaO, 21% SiO_2, 8.1% Al_2O_3, 8.1% Fe_2O_3, 5.1% TiO, 2.2% MgO and 2.8% Na_2O. Meanwhile the main minerals are sodium aluminosilicate, aragonite, calcite, boehmite and perovskite. However, major mineral compositions in Bayer process bauxite residues include haematite (Fe_2O_3), nepernepheline (including natrodavyne, katoite etc.), gibbsite, quartz and other phases.

Bauxite residues from Guangxi and Shandong refineries are high in iron oxide (>35%) and magnetic separation techniques have been used to increase the metal iron concentration to more than 55%. Bauxite residue from combined sintering/Bayer routes are high in larnite (a calcium silicate mineral). Approximately half the bauxite residue produced in China is low iron oxide containing residue from diasporic bauxite, (<8% Fe_2O_3); 25% is high iron oxide containing (>37% Fe_2O_3) from gibbsitic bauxite; 18% is high iron oxide containing (>37% Fe_2O_3) from diasporic bauxite; and 7% is residue from sintering combined routes.

A very strong driving force in China has been Government-imposed legislation requiring that bauxite residue is reused. Since 2005 very large-scale effort to utilise more bauxite residue in many areas with the main effort being devoted to: high intensity magnetic separation to concentrate iron; glass ceramics $CaO–SiO_2–Al_2O_3$; bricks (with lime fly ash); and polymer fillers.

Principle uses in China are [50–52]:

- Sintering/combined route—200,000 to 250,000 t/y in building and construction materials;
- Coarse separated sand fraction from high iron containing residue from gibbsitic or diasporic bauxites—1,000,000 to 1,300,000 t/y building materials, especially bricks, dam construction, insulating materials for furnaces;
- Fine residue fraction from high iron containing residue from gibbsitic or diasporic bauxites—cement 800,000 to 1,000,000 t/y in cement and 800,000 to 1,000,000 t/y in iron and steel production;
- Low iron residue from local diasporic bauxite—100,000 t/y in wall and floor tiles, glass ceramics, glass fibre, wear resistant linings—100,000 t/y with increasing interest in REEs extraction;
- Other bauxite residues—10,000 t/y in mineral fillers for plastics (PVC electrical conduit trunking, PVC drainage pipes), wood substitutes.

Other areas of research in China include geopolymers, rare earth element extraction, wastewater absorption, foamed ceramic bricks, soil additives to reduce acidity and absorb heavy metals, and gas desulfurisation. In 1965 a cement plant was built to consume residue from Shandong Alumina and by the late 1990s over 6 million tonnes of residue were used to make OPC and oil well cements. However, new standards for cement were introduced which restricted sodium level content and the progressive move to imported bauxites led to a reduction in the amount going into cement.

6.8.5 Applications in Cement

Bauxite residue can provide valuable iron and alumina values in the production of Portland cement clinker and it has the added benefit of reducing the carbon dioxide emissions during production. In addition, it can behave as a supplementary cementitious material which is used in mortar or concrete mixes, this can provide: better mechanical properties, more efficient use of resources and lower carbon dioxide emissions [31–33].

The best estimates are that between 2,200,000 and 2,700,000 tonnes of bauxite residue are currently used annually in the production of clinker for Portland cement. The cement plants currently utilising the bauxite residue in clinker production are based in Belarus, China, Cyprus, Georgia, Greece, India, Moldova, Russia and Ukraine.

In mid-2020, Hindalco in India announced that it had achieved 100% utilisation of the bauxite residue from three of its alumina refineries and sales of 2 million tonnes was expected to be sold to some 40 cement plants in India. The bauxite residue is used an alternative to laterite, lithomarge and low-grade bauxite. Bauxite residue from alumina plants in Lanjigarh (state of Orissa) and GCACPL in Mundra (State of Gujarat) are also used in cement clinker production.

The use of bauxite residue in the cement industry in the manufacture of clinker in Europe (Western/Eastern) is estimated at close to 350,000 t/y. The major proportion of this is from the Mykolayiv alumina plant in the Ukraine; the bauxite residue from Mykolayiv is used in cement plants in the Ukraine, Russia, Georgia, Moldova and Belarus. The Mykolayiv refinery blends the residue produced to give the cement plant a consistent feed and the climate allows a reasonably low moisture product to be produced. Additionally, there is growing usage of bauxite residue from Mytilineos's plant in Distomon in cement clinker production at cement plants in Greece and Cyprus; this was estimated to be over 100,000 tonnes in 2020. Significant usage, up to 180,000 t/y was anticipated at a cement plant in Milaki but changing economic circumstances have put increased usage on hold. There has also been some historical usage of bauxite residue from the refinery in Tulcea in a local Romanian cement plant.

There is reportedly small-scale usage of bauxite residue from the Gandja alumina refinery in Azerbaijan in cement.

Modest scale successful utilisation of bauxite residue in Portland cement clinker production was carried out in the mid-1980s in Jamaica at the Caribbean Cement Limited plant in Kingston in which several tens of thousands of bauxite residue from the Aljam alumina plant at Ewarton, some 55 km away. Approximately 4% of residue was added during clinker production with no technical problems noted. The residue was obtained from the recently introduced dry mud stacking approach adopted at

Ewarton and the material stockpiled for a short period to allow it to further dry to a solids content of about 70%. The operation use was discontinued because of transport issues and a drop in the solids content to 65% but is now being reconsidered.

The use of bauxite residue in cement in China was formerly several million tonnes a year [50–52] but this has fallen because of the changes in construction industry standards and a reduction in the number of plants operating a sinter or Bayer-sinter extraction route. It should be noted that the bauxite residue produced from the sinter or Bayer-sinter extraction route is very different chemically from that produced in a conventional Bayer alumina plant.

From the evidence from a number of cement plants that are already using bauxite residue on an industrial scale, bauxite residue can satisfactorily be used in cement clinker manufacture. With the appropriate bauxite residue, typically a usage rate of 3–5% can be accommodated. From the industrial experience, key aspects are:

- A relatively low moisture content—approximately 30% moisture has been used in several plants and this can readily be achieved, some alumina refineries have used a plate and frame filter press to produce a satisfactory product whilst others have achieved a satisfactory level by air drying the bauxite residue;
- A moderately low sodium content, a value of <2.5% Na_2O has been indicated as satisfactory but it will depend on the composition of the other raw materials;
- The appropriate aluminium oxide to iron oxide ratio (an iron oxide to alumina ratio of 0.8:1.2 in the raw mix was found to give the best results in one study although some plants will use the bauxite residue to supplement the iron level whilst others use it for the alumina content); and
- A reasonable proximity of a cement plant (a distance of up to 1200 km has been found to be acceptable in one case but this is exceptional).

Meanwhile, and likely to have an even greater environmental impact in the long term, is the potential to use bauxite residue in supplementary cementious materials. Considerable work has been undertaken on a laboratory/pilot scale exploring opportunities for bauxite residue as a filler in blended cements, sometimes for their pozzolanic activity but also to improve the mechanical properties in blended cement by other mechanisms such as optimising packing density or rheological characteristics. It has been observed at laboratory scale by several groups of researchers that bauxite residue as produced or after calcination, pure or mixed with other additives, may successfully replace clinker in blended cements at dosages between 10 and 30 wt%. If used as produced, this would have a significant impact on the cement industry CO_2 emissions as the clinker production is responsible for 85% of the total emissions in an integrated cement plant. A focus of the ReActiv project discussed above in Sect. 6.8.2, is the calcination of bauxite residue and kaolin to create a supplementary cementious material with good pozzolanic properties.

Certain types of 'special' cement, for example calcium sulfo-aluminate (belite type cement), can also be made utilising bauxite residue, these generate very substantially lower carbon dioxide emission levels during production [41]. Calcium sulfo-aluminate cements require both a higher Fe_2O_3 and Al_2O_3 content in the raw meal which could be partially satisfied by using bauxite residue. There have been some promising studies on using bauxite residue in special cements and iron rich, special setting belite type cements with improved strength when compared to Portland Cement have been made with levels of up to 40 wt% bauxite residue, however, there does not appear to be any industrial usage to date. One of the projects in the MSCA ETN REDMUD programme looked at the use of bauxite residue in calcium sulfo-ferroaluminate cements; the results showed that the production of calcium sulfo-ferroaluminate based clinkers with 50 wt% and 60 wt% of bauxite residue as raw material was feasible.

6.8.6 *Iron and Steel*

The usage of bauxite residue in steel manufacture is of the order of 70,000–100,000 t/y [60, 61, 63, 64], excluding China. As discussed previously, usage in China is estimated at 800,000–1,000,000 t/y. The iron ores that are normally used in iron and steel manufacture have an iron content of typically 55–70% with 66% being available from many good quality sources. Meanwhile for comparison, bauxite residues have a typical range of iron of 3–42%. It is important to consider these contents and realise the difficulty in even closely matching the economics against using virgin iron ore, especially at the current price of iron ore. Some success has been achieved in China, particularly in plants in Southern China where the bauxite residue can have an iron content of approximately 40%. Notably success has also been achieved using magnetic separation techniques as a first stage of processing to concentrate the iron fraction. The bauxite residue material is also wet and has a high sodium content which is a disadvantage in steel production. Recent research in Russia using low alkaline bauxite residue and bentonite has been very effective in raising the iron content, improving agglomeration quality and thereby give greater yields in the production of iron ore pellets [64].

The simultaneous recovery of other metals, for example titanium and aluminium, would improve the economics of using bauxite residue for iron recovery in steel production. The only non-Chinese plant using bauxite residue for making steel is based in the Urals. Several Indian sources of bauxite residue are relatively high in iron, between 30 and 39%, and a considerable amount of work has been done to recover the iron values in the bauxite residue from the NALCO plant using the Romelt process but whilst technically feasible, it was uneconomic because of the high energy costs involved in the process.

6.8.7 *Roads*

Usage of bauxite residue for road building and dyke/levee construction is estimated at 20,000 tonne/year, however, it is difficult to gain an accurate figure as a large percentage is used internally within each alumina site complex, often for roads within the bauxite disposal area.

Alcoa has also supported extensive research into potential beneficial uses for the residue; the most promising option for bauxite residue by-products found in their investigation are Red Sand™ and Alkaloam®. Alcoa has developed a carbonation and wash system to process residue sand, producing a by-product now known as Red Sand™. This is literally a crushed rock that is red in colour. Testing shows Red Sand™ can be used as a general fill material, construction backfill or as a material suitable for road base construction. Figure 6.38 shows the example of using Red Sand™ in the construction of roads in Western Australia; this study was done in conjunction with Curtin University [43]. The target is to replace virgin sand and crushed limestone for sub-grade and top dressing: for each cubic metre of virgin sand replaced there are savings of 4.4 and 2.7 MJ respectively in energy conservation. The benefits include: lower carbon footprint, reduced loss of vegetation as it avoids use of quarries, cost savings, less water usage and reduced eutrophication. Commercialisation of sand from residue would have a range of potential benefits including as a viable substitute for increasingly scarce supplies of quarry sand in the local region, reducing the clearing of natural bushland for sand quarries and reducing the requirement for residue storage facilities. Not only does Red Sand™ have the potential to be a cost-effective alternative to general purpose sand, it has also shown to be high quality construction sand thanks to its excellent drainage and strength characteristics.

6.8.8 *Landfill Capping*

Use of bauxite residue for capping municipal landfills is carried out in France and has been trialled in the USA. Landfill capping trials have been successfully carried out in Louisiana with Cajunite™ where the bauxite residue product was found to have advantages over natural clay. The amounts used in France has varied considerably from year to year but is estimated to be between 40,000 and 100,000 t/y. It can only be undertaken within a relatively small radius of each refinery depending on local transport costs and the availability of other covering/capping materials; the maximum distance that the residue can effectively be transported for this application is estimated to be about 75 km. Figure 6.30 shows the use Bauxaline® from Gardanne in landfill capping in Provence. Municipalities will normally wait until an entire site is full before remediating/capping it, hence the wide variation in usage between years. The bauxite residue must be in a form that can safely and readily be carried in trucks on public roads. Possible concerns are dust from the bauxite residue when dry.

Somewhat related, has been the use as a soil amendment/conditioner for acidic/sandy soils; on large scale trials this has been shown to be safe and beneficial, especially in controlling high levels of phosphorous. However, controversy over two decades has prevented its implementation until recently.

Alkaloam® is the registered trademark of an Alcoa produced soil amendment. In 1993, a proposal for the use of Alkaloam® in the Peel-Harvey coastal plain catchment for trial projects was submitted to the Environmental Protection Authority, by the then Department of Agriculture, and approved (this followed a Public Environmental Review). Since then, trials have consistently demonstrated there are real benefits of adding Alkaloam® to sandy soils which are common in coastal regions of Western Australia. Alkaloam® has proved successful as a soil amendment for nutrient deficient, acid soils and can increase the productivity of farmland.

Alkaloam® is alkaline which increases soil pH in the same way as agricultural lime. While traditional lime can take a number of years to effectively reduce the pH of soil, Alkaloam® can achieve this same result almost straight away. Therefore, Alkaloam® allows farmers to gain maximum benefits from the adjustment in soil pH far sooner than waiting for a soil response over one or more rainy seasons.

6.9 Materials Recovery

The strongly growing demand for rare earth elements (REEs), and the concentration of production in China has led to a renewed interest in the extraction of REEs from bauxite residue [6, 29, 34, 65–77]. The REEs, normally taken to mean the lanthanides plus scandium and yttrium, are essential constituents of modern permanent magnets, nickel metal hydride batteries and lamp phosphors with dramatic growth driven by increasing demand for electric and hybrid cars, wind turbines and compact fluorescent lights. The very strong demand for scandium is driven by its emerging use in aluminium alloys where its ability to control grain size leads to exceptionally high strength but light components. These REEs are normally divided into the light rare earth elements (LREE) covering the elements from lanthanum to samarium and the heavy rare earth elements (HREE) covering europium to lutetium. Studies on the REEs in bauxite has been pursued for many years and was extensive investigated during the 1970s in Italy: REEs are also found in bauxites from Olmedo in Sardinia, Vlasenica in Bosnia Herzegovina, Grebnik in Kosovo and Panassos-Ghiona, Marmara, Evia Island in Greece.

During the Bayer process, the REEs remain undissolved so are concentrated up in the resulting residue. Much work was done during the 1980s on Jamaican bauxite residues using dilute acid as the first stage leaving behind the iron and titanium oxides, the REEs were then recovered by selective precipitation [29]. This development was not pursued but many new projects have started in recent years. These include a joint venture between the Jamaican Bauxite Institute and Nippon Light Metals to construct a pilot plant for extraction from locally arising bauxite residues [67]. The REE content of bauxite residues in Jamaica were found to be up to 2500 mg/kg [68].

A particular focus of the ETN-MSCA REDMUD programme is the recovery of REEs and scandium; most of the work has been carried out on bauxite residue from the Aluminium of Greece (now Mytilineos) refinery and studies have shown that the REE content is in the range 800–1100 mg/kg and it is observed that the highest REEs contents in bauxite residue are most commonly found when karstic bauxite is used as the feed material [69]. Scandium was found to be present at a level of approximately 100 mg/kg in the Aluminium of Greece bauxite residue. Examples of some of the extraction studies undertaken under the ETN-MSCA REDMUD programme include: selective leaching after sulfation roasting at KU Leuven [70]; the use of cation exchange chromatographic techniques at the National Technical University of Athen [71]; the selective uptake of scandium by organic–inorganic hybrid titanium(IV) phosphate materials at the University of Helsinki [72]; improvements in scandium and titanium recovery by minimising silica gel formation by using sulfuric acid in conjunction with hydrogen peroxide at RWTH Aachen University [73]; selective REE digestion using a 'dry' digestion with concentrated hydrochloric acid at KU Leuven [74]; and the recovery of REEs from leachates using supported ionic liquid phases (SILPs) [Hbet-STFSI-PS-DVB] at KU Leuven [75]. In 2016, the British Geological Study and Camborne School of Mines reviewed the potential for REEs in European bauxite residues [65]. Assuming a total REE content of 1000 mg/kg and an extraction efficiency of 50%, then the authors estimated that over 200,000 t of REEs could be available from bauxite residue produced in Europe since 1972. In global terms, over 172,000 t of REEs are sent to bauxite residue disposal ponds which compares to some 100,000 t of REEs produced annually.

Orbite Technologies Inc., a Quebec based materials/mineral processing company has developed an acid extraction route to recover alumina from aluminous clay and other alumina containing materials such as fly ash. Using a modification of this route, they have developed and patented the recovery of alumina, titania, iron oxide, magnesia rare earth elements and other rare metals from bauxite residue. The initial extractions are carried using 6 M hydrochloric acid at temperatures between 140 and 165 °C and a pressure of 60–90 psi; this is used to extract the aluminium fraction, followed by successive recovery of the other required compounds by modified acid strength and temperature steps [77].

6.10 Heavy Metal and Phosphate Removal

The ability of bauxite residues to react with heavy metals, especially from mine and mineral processing sites, has been examined by several groups around the world [34–37] so far with only modest tonnages being sold but several new approaches are emerging.

Virotec, a company that originated in Australia based on work done by David McConchie at Southern Cross University, has widely promoted a product called Bauxsol® since about 2002 for the remediation of contaminated sites and the adsorption of heavy metals particularly from acid mine drainage. Bauxsol® is manufactured by reacting brines with bauxite residue to produce a partially neutralised material; the magnesium to calcium ratio is critical to obtain good heavy metal adsorption; the neutralisation process is called Basecon®. Reaction with brines leads to the formation of minerals such as calcite, brucite, aragonite and hydrotalcites which provide a buffering affect. A plant to produce Bauxsol® is operating at the QAL refinery in Queensland.

In Italy, using bauxite residues from the Eurallumina plant, good heavy metal absorption results were obtained by neutralising the material with seawater. In some formulations, the bauxite residue was mixed with fly ash which improved the absorption of arsenic.

In Korea, work showed how pellets made by heat treating at 600 °C, mixtures of bauxite residue, polypropylene, sodium metasilicate, magnesium chloride and fly ash had good heavy metal absorption, especially lead, copper and cadmium.

Work on bauxite residue from San Ciprian mixed with gypsum was found to have a good capability to remove copper, zinc, nickel and cadmium from waste streams.

Work at the Chinese Academy of Science has shown good benefits in the use of bauxite residue for the adsorption of lead, cadmium and copper.

Recent developments in the UK by REDmedia Technologies with their Redmedite, a pelleted product based on brine treated bauxite residue, has shown success in treating contaminated leachate from Wheal Augusta, a former barytes mine in UK. This work was undertaken by Plymouth University who investigated contaminant removal from leachate and remediation of contaminated soil. The percentage of contaminants removed from the leachate was reported to be: Al—91%, As—99%, Cd—73%, Co—81%, Cr—95%, Cu—99%, Fe—98%, Mn—26%, Ni—28%, Pb—99% and Zn—74%. Improvements in the process based on using gypsum and acid neutralised bauxite residue has resulted in contaminant removal of As— >99%, Ba— >99%, Cd— >99%, Cu—99%, Fe— >99.9%, Ni— >99%, Pb— >99.9% and Zn— >99%.

Pellets of Redmedite and Bauxsol® have proved to be very effective at phosphate removal in sewage treatment and recent improvements in pellet formation are forecast to open up significant tonnages as European regulations on phosphate removal become more stringent.

Recent work on bauxite residues from Aughinish and Gardanne has examined seawater and gypsum treatments of coarse and fine bauxite residues showed the potential benefits of these blends as low-cost adsorbent for phosphorus [78].

Fluorchemie Gmbh have developed and patented (PCT/EP2016/025025-WO2017/157406A) a product called ALFERROCK$^{(R)}$ from bauxite residue which is effective at removing contaminants and undesirable cations and anions from aqueous systems (waste water and drinking water).

6.11 Flame Retardants and Heat Storage

Fluorchemie Gmbh have developed a new flame-retardant additive from bauxite residue, the product is termed MKRS (modified re-carbonised red mud) with the trademark ALFERROCK$^{(R)}$ and has potential applicability in a wide range of polymers (PCT WO2014/000014). One of its particular benefits is the ability to operate over a much broader temperature range, 220–350 °C, than alternative zero halogen inorganic flame retardants such as aluminium hydroxide, boehmite or magnesium hydroxide. In addition to polymer systems where aluminium hydroxide or magnesium hydroxide cannot be used, it has also found to be effective in foamed polymers such as EPS and PUR foams at loadings up to 60%.

In a suitable solid form, calcined ALFERROCK(R) produced from bauxite residue, has a density of approximately 3.93 g/cm^3 and has been found to be very effective as a heat storage medium (WO2017/157664). The material can repeatedly be heated and cooled without deterioration and has a specific thermal capacity in the range of 0.6–0.8 kJ/(kg*K) at 20 °C and 0.9–1.3 kJ/(kg*K) at 726 °C; this enables the material to work effectively in energy storage device to maximise the benefits of solar polar, wind turbines and hydro-electric systems.

6.12 Combination of Wastes to Produce Construction Materials and Minimise Metal Leaching

A British group of entrepreneurs, Purgo Limited, has developed and filed patents [79] covering an innovative approach using bauxite residue in combination with different waste materials in an unconventional way to manufacture construction materials such as building panels, bricks, tiles, aggregates etc. Bauxite residue at a high level, >90%, is mixed with glass and other wastes, compressed and fired to form a variety of useful articles. A merit of the approach is the ability to very substantially minimise the possible leaching of the metal contaminants such as arsenic, cadmium, chromium, lead, mercury, molybdenum and selenium from other wastes which are incorporated with the bauxite residue. Bauxite residue when mixed with potentially toxic/hazardous waste materials treated via the process can meet the 'inert' classification as defined in the Waste Acceptance Criteria Testing standard BS EN 12,457.

6.13 Other Uses

The manufacture of bricks, tiles and other building materials has been shown to be technically possible by many groups of workers from a wide variety of sources of bauxite residue using both fired and chemically bonded methods. Outside China, however, whilst manufacturing plants have started up, production has not continued. Figure 6.40 shows an example of a room constructed in the 1990s at the Jamaica Bauxite Institute using bricks fabricated from bauxite residue.

6.14 Future Trends

Changes within the alumina industry over the last 20 years has meant than many of the relatively small Bayer plants in the world have ceased operation and there has been an inexorable shift to much larger alumina refineries in countries such as Australia, Brazil and India. These large, more modern plants have been able to adopt best practice in managing the bauxite residue they generate. In addition, China has witnessed enormous growth in alumina production during the twenty-first century from just a few million tonnes of alumina per year at the start of the century to over 70 million tonnes in 2020; this is over 50% of world alumina production so their role in the management and the use of bauxite residue has become very important. Recognising this, the Chinese Government has set bold targets for bauxite residue utilisation and the usage rate of bauxite residue in China is already reflecting this.

Indonesia is also expected to show rapid growth in alumina production; for a number of years it was a major exporter of bauxite, but the Indonesian Government imposed a severe restriction on bauxite exports in 2015 with an invitation for alumina producers to start up alumina refineries within Indonesia. These new producers have an opportunity to adopt best practices from the start in managing, storing and using bauxite residue.

Another factor that has influenced changes in the management and use of bauxite residue, was the incident at the bauxite residue ponds adjacent to the Ajka alumina refinery in Hungary in October 2010 when a residue disposal area dam wall was breached releasing a large volume of caustic red mud slurry [6, 80]. The alumina producers, via organisations such as European Aluminium and the International Aluminium Institute (IAI), have since worked collaboratively to look for improved solutions and review best practice guidelines; these have been published in a guideline document [4]. The IAI continues to encourage and fund collaborative effort on improving storage, monitoring regimes, safety standards, looking at improved remediation techniques and opportunities for uses for bauxite residue. Key messages coming out of the best practice reviews have been the drive to dispose of and store bauxite residue in a safer way with lower caustic and higher solids content. These moves will encourage the utilisation of residue as the material produced will be in a more acceptable form for transport, handling and reuse. All the major alumina producers have increased development effort to look at improved and safer management systems, explore innovative remediation techniques and fund additional work both in house and externally looking at new uses.

In addition to the increased development effort by the alumina producers, there has been a resurgence of academic interest in looking for new uses and innovative approaches to long established uses. The EU focus on waste management and the circular economy has led to substantial EU funding under the HORIZON 2020 programme seeking to ensure maximum utilisation of global resources has been a key driver for much of this recent research on bauxite residue. The next phase of funding will be HORIZON EUROPE with a budget of Euro 95.5 billion aimed at research and innovations to help achieve the UN's Sustainable Development Goals;

initiative for better utilisation and management of resources will be a key element. With improvements in technology, and this additional research effort, the next ten years are expected to lead to much greater use of bauxite residue and less bauxite residue going to landfill.

References

1. G. Banvolgyi, The failure of the embankment of red mud reservoir at Ajka (Hungary), in *Proceedings of 36 International ICSOBA Conference, Belem, Brazil, 29 Oct 2018, pp. 387–399*
2. L. England, et al., Recent developments in dewatering technologies for tailings disposals to maximise water recovery, in *SYMPHOS, 5th International Symposium on Innovation and Technology in the Phosphate Industry* (2019)
3. Data published annually by World Aluminium, www.world-aluminium.org
4. World Aluminium and the European Aluminium Association "Bauxite Residue Management: Best Practice", in *International Aluminium Institute*, 10 King Charles II Street, London, SW1Y 4AA, UK. http://bauxite.world-aluminium.org/refining/bauxite-residue-management.html
5. European Aluminium Association, Environmental Profile Report for the European Aluminium Industry (2013)
6. K. Evans, The history, challenges, and new developments in the management and use of bauxite residue. J. Sustain. Metall., 1–16 (2016)
7. K. Evans, E. Nordheim, K. Tsesmelis, Bauxite residue management, in *Light Metals 2012*, ed. by C.E. Suarez (The Minerals, Metals & Materials Society, Warrendale, Pennsylvania, USA, 2012), pp. 63–66
8. International Maritime Organization, Monitoring of the Marine Environment Reports and Dumping Permits Issued, EIA of bauxite residual disposal in the sea (2002)
9. B.G. Purnell, Mud disposal at the Burntisland alumina plant. Light Metals, 157–159 (1986)
10. H.H. Pohland, A.J. Tielens, Design and operation on non-decanted red mud ponds in Ludwigshafen, in *Proceedings of International Conference Bauxite Tailings*, Kingston, Jamaica (1986)
11. E.I. Robinsky, Current status of the sloped thickened tailings disposal system, in *Proceedings of International Conference Bauxite Tailings, Kingston, Jamaica* (1985)
12. J.L. Chandler, The stacking and solar drying process for disposal of bauxite tailings in Jamaica, in *Proceedings of International Conference Bauxite Tailings, Kingston, Jamaica* (1986)
13. R. Bott, T. Langeloh, Process options for the filtration and washing of bauxite residue, in *Bauxite Residue Valorization Conference, Leuven* (2015), pp. 62–70
14. D.J. Cooling, P.S. Hay, L. Guifoyle, Carbonation of bauxite residue, in *Proceedings of the 6th International Alumina Quality Workshop, Brisbane* (2002)
15. L. Guilfoyle, P. Hay, D. Cooling, Use of flue gas for carbonation of bauxite residue, in *Proceedings of the 7th International Alumina Quality Workshop*, ed. by A. McKinnon (AQW Inc., Perth, 2005), pp. 218–220
16. C. Hanahan, D. McConchie, H. Pohl, R. Creelman, M. Clark, C. Stocksiek, Chemistry of seawater neutralization of bauxite refinery residues (red mud). Environ. Eng. Sci. **21**, 125–138 (2004)
17. S.J. Palmer, R.L. Frost, Characterisation of bauxite and seawater neutralised bauxite residue using XRD and vibrational spectroscopic techniques. J. Mater. Sci. **44**, 55–63 (2009)
18. D. Higgins, T. Curtin, R. Courtney, Effectiveness of a constructed wetland for treating alkaline bauxite residue leachate: a 1-year field study. Environ. Sci. Pollut. Res. (2017). https://doi.org/10.1007/s11356-017-8544-1
19. P.A. Lyew-Ayee, S.D. Persaud, K.A. Evans, R.G. Tapp, From red to green—a Regulators and Company partnership in bauxite residue remediation. ICSOBA Sem. Bauxite Residue, Travaux ICSOBA **36**(40), 137–145 (2011)

20. W.B. O'Callaghan, S.C. McDaniel, D.M. Richards, R.E. Reid, Development of a topsoil-free vegetative cover on a former red mud disposal site (Alcan Jamaica Company, 1999)
21. A. Schmalenberger, O. O'Sullivan, J. Gahan, P.D. Cotter, R. Courtney, Bacterial communities established in bauxite residues with different restoration histories. Environ. Sci. Tecnol. **47**(13), 7110–7119 (2013)
22. R. Courtney, T. Harrington, K.A. Byrne, Indicators of soil formation in restored bauxite residues. Ecol. Eng. **58**, 63–68 (2013)
23. R.G. Courtney, G. Mullen, T. Harrington, An evaluation of revegetating success on bauxite residue. Resor. Ecol. **17**(3), 350–358 (2009)
24. R.G. Courtney, T. Harrington, Revegetation strategies for bauxite residue: a case study of Aughinish, Alumina, Ireland. ICSOBA Sem. Bauxite Residue, Travaux ICSOBA **36**(40), 146–153 (2011)
25. T.C. Santini, C. Hinz, A.W. Rate, C.M. Carter, R.J. Gilkes, In situ neutralisation of uncarbonated bauxite residue mud by cross layer leaching with carbonated bauxite residue mud. J. Hazard. Mater. **194**, 119–127 (2011)
26. T.C. Santini, J.L. Kerr, L.A. Warren, Microbially-driven strategies for bioremediation of bauxite residue. J. Hazard. Mater. **293**, 131–157 (2015)
27. F. Feret, Selected applications of Rietveld-XRD analysis for raw materials of the aluminium industry, in *JCPDS-International Centre for Diffraction Data* (2013). ISSN 1097-0002
28. E. Erca, R. Apak, Furnace smelting and extractive metallurgy of red mud: recovery of TiO_2, Al_2O_3 and pig iron. J. Chem. Technol. Biotechnol. **70**(3), 241–246 (1999)
29. Alcan International Limited, United States Patent, USP 5,030,424 (1991)
30. C.R. Borra, B. Blanpain, Y. Pontikes, K. Binnemans, T. Van Gerven, Comparative analysis of processes for recovery of rare earth elements from bauxite residue. JoM. **68**, 2958–2962 (2016)
31. P.E. Tsakiridis, S. Agatzini, L.P. Oustadakis, Red mud addition in the raw meal for the production of Portland cement clinker. J. Hazard. Mater. **B116**, 103–110 (2004)
32. I. Vangelatos, G.N. Angelopoulos, D. Boufounos, Utilization of ferroalumina as raw material in the production of ordinary portland cement. J. Hazard. Mater. **168**, 473–478 (2009)
33. Y. Pontikes, G. Angelopoulos, Bauxite residue in cement and cementitious applications: current status and a possible way forward. Resour. Conserv. Recycl. **73**, 53–63 (2013)
34. C. Klauber, M. Gräfe, G. Power, Bauxite residue issues: II options for residue utilization. Hydrometallurgy **108**(1), 11–32 (2011)
35. D. McConchie, M. Clark, C. Hanahan, F. Davies-McConchie, The use of seawater-neutralised bauxite refinery residues in the management of acid sulphate soils, sulphidic mine tailings and acid mine drainage, in *Proceedings of the 3rd Queensland Environment Conference, Environmental Engineering Society*, ed. by K. Gaul (2000), pp. 201–208
36. D. McConchie, F. Davies-McConchie, M. Clark, L. Fergusson, The use of industrial mineral additives to enhance the performance of bauxsol-based environmental remediation reagents. Paper presented to Australian Industrial Minerals Conference, Brisbane, Australia (2003)
37. L. Fergusson, Commercialisation of environmental technologies derived from alumina refinery residues: a ten-year case history of virotec, Commonwealth Scientific and Industrial Research Organisation (CSIRO), Project ATF-06-3 "Management of Bauxite Residues", Department of Resources, Energy and Tourism (DRET), Commonwealth Government of Australia, Asia-Pacific Partnership on Clean Development
38. B.K. Parekh, W.M. Goldberger, An assessment of technology for the possible utilization of Bayer process muds. Published by the U.S. Environmental Protection Agency, EPA 600/2-76-301
39. R.K. Paramguru, P.C. Rath, V.N. Misra, Trends in red mud utilization—a review. Miner. Process. Extr. Metall. Rev. **26**(1), 1–29 (2005)
40. U.V. Parlikar, P.K. Saka, S.A. Khadilkar, Technological options for effective utilization of bauxite residue (Red mud)—a review, in *International Seminar on Bauxite Residue (RED MUD)*, Goa, India (2011)
41. M. Singh, S.N. Upadhayay, P.M. Prasad, Preparation of special cements from red mud. Waste Manage. **8**, 665–670 (1996)

42. S. Agatzini-Leonardou, P. Oustadakis, P.E. Tsakiridis, C. Markopoulos, Titanium leaching from red mud by diluted sulfuric acid at atmospheric pressure. J. Hazard. Mater. **157**(2–3), 579–586 (2008)
43. W.K. Biswas, D.J. Cooling, Sustainability assessment of red sand^{TM} as a substitute for virgin sand and crushed limestone. J. Ind. Ecol. **17**(5), 756–762 (2013)
44. A. Agrawal, K.K. Sahu, B.D. Pandy, E. Kalkan, Utilization of red mud as a stabilization material for the preparation of clay liners. Eng. Geol. **87**, 220–229 (2006)
45. G. Zhang, J. He, R.P. Gambrell, Synthesis, characterization, and mechanical properties of red mud-based geopolymers. Transp. Res. Record: J. Transp. Res. Board. **2167**, 1–9 (2010)
46. W. Shaobin, H.M. Aug, M.O. Tade, Novel applications of red mud as coagulant, adsorbent and catalyst for environmentally benign processes. Chemosphere **72**(11), 1621–1635 (2008)
47. A.I. Cakici, J. Yanik, S.U. Çar, T. Karayildirim, H. Anil, Utilization of red mud as catalyst in conversion of waste oil and waste plastics to fuel. J. Mater. Cycles Waste Manage. **6**(1), 20–26 (2004)
48. A. Satapathy, H. Sutar, S.C. Mishra, S.K. Sahoo, Characterization of plasma sprayed pure red mud coatings: an analysis. Am. Chem. Sci. J. **3**(2), 151–163 (2013)
49. EU MSCA-ETN—European Training Network for zero-waste valorization of Bauxite Residue (Red Mud) project on; website, etn.redmud.org
50. W. Liu, The developing of red mud utilization in China, in *Bauxite Residue Valorization Conference, Leuven* (2015), pp. 147–156
51. G. Songqing, Y. Zhonglin, Q. Lijuan, Security disposal and comprehensive utilization of bauxite residues. Light Metals, 47–51 (2017)
52. D.Y. Liu, C.S. Wu, Stockpiling and comprehensive utilization of red mud. Res. Prog. Mater. **5**, 1232–1246 (2012).https://doi.org/10.3390/ma5071232
53. Q. Guanzhou, Y. Liu, T. Jiang, Y. Hu, Influence of coal sort on the direct reduction of high-iron-content red mud. J. Cent. South Univ. Technol. **2**(2), 27–31 (1995)
54. M. Fofana, S. Kmet, T. Jakabsk, Treatment of red mud from alumina production by high-intensity magnetic separation. Magn. Electr. Sep. **6**, 243–251 (1995)
55. X. Qinfang, L. Xiaohong, M.E. Schlesinger, J.L. Watson, Low temperature reduction of ferric iron in red mud. Light Metals, 157–162 (2001)
56. L. Zhong, Y.F. Zhang, Sub molten salt method recycling red mud. Chin. J. Nonferrous Met. **18**, 70–73 (2008)
57. X.F. Zheng, Recycling technology of aluminum and sodium from low temperature Bayer progress red mud. Shandong Metall. **32**, 16–17 (2010)
58. W. Liu, J. Yang, B. Xiao, Review on treatment and utilization of bauxite residues in China. Int. J. Min. Process. **93**, 220–231 (2009)
59. X.R. Qiu, Y.Y. Qi, Reasonable utilization of red mud in the cement industry. Cem. Technol. **6**, 103–105 (2011)
60. B. Mishra, A. Staley, D. Kirkpatrick, Recovery and utilization of iron from red mud. Light Metals, 149–156 (2001)
61. B. Mishra, A. Staley, D. Kirkpatrick, Recovery of value-added products from red mud. Miner. Metall. Process. **19**(2), 87–94 (2002)
62. K. Fotini, An innovative geotechnical application of bauxite residue. Electron. J. Geotech. Eng. **13**(G), 1–9 (2008)
63. D.D. Dimas, P. Ioanna, D. Panias, Utilization of alumina red mud for synthesis of inorganic polymeric materials. Miner. Process. Extr. Metall. Rev. **30**(3), 211–239 (2009)
64. G. Podgorodetskiy, V. Gorbunov, A. Panov, S. Gorbachev, Complex Additives on the basis of bauxite residue for intensification of iron-ore sintering and pelletising, in *Bauxite Residue Valorization Conference, Leuven* (2015), pp. 147–156
65. E.A. Deady, E. Mouchos, K. Goodenough, B.J. Williams, F. Wall, A review of the potential for rare-earth element resources from European red muds: examples from Seydisehir, Turkey and Parnassus-Giona, Greece. Min. Mag. **80**(1), 43–61 (2016)
66. C.R. Borra, B. Blanpain, Y. Pontikes, K. Binnemans, T. Van Gerven, Comparative analysis of processes for recovery of rare earths from bauxite residue. JoM **68**, 2958–2962 (2016)

67. Jamaican Information Service (10th October 2014) Jamaica has full ownership of Rare Earths pilot plant. Government of Jamaica
68. A.S. Wagh, W.R. Pinnock, Occurrence of scandium and rare earth elements in Jamaican bauxite wastes. Econ. Geog **82**, 757–761 (1987)
69. J. Vind, A. Malfliet, B. Bart Blanpain, E. Petros, P.E. Tsakiridis, A.H. Tkaczyk, V. Vassiliadou, D. Panias, Rare earth element phases in bauxite residue. Minerals, in press (2018)
70. C.R. Borra, B. Blanpain, Y. Pontikes, K. Binnemans, T. Van Gerven, Selective leaching of rare earths from bauxite residue after sulphation roasting, in *Bauxite Residue Valorization Conference, Leuven* (2015), pp. 301–307
71. L.-A. Tsakanika, M. Ochsenkuhn-Petropoulou, Separation and recovery of rare earths after red mud leaching by cation-exchange chromatography, in *Bauxite Residue Valorization Conference, Leuven* (2015), pp. 309–315
72. W. Zhang, R. Koivula, Highly selective uptake of scandium over iron by organic-inorganic hybrid titanium (IV) phosphate materials relevant to bauxite residue leachate conditions, in *2nd Conference Bauxite Residue Valorisation and Best Practices, Athens* (2018)
73. A. Godze et al., A novel approach for enhanced Sc and Ti leaching efficiency from bauxite residue with suppressed silica gel formation, in *2nd Conference Bauxite Residue Valorisation and Best Practices, Athens* (2018)
74. R.M. Rodolfo et al., Recovery of rare earths from bauxite residue slag by high-pressure acid leaching, in *2nd Conference Bauxite Residue Valorisation and Best Practices, Athens* (2018)
75. D. Avidibegovic et al., Purification of low concentration of rare earths from high concentration impurities in leach liquor of bauxite residue slag by a supported ionic liquid phase, in *2nd Conference Bauxite Residue Valorisation and Best Practices, Athens* (2018)
76. M.O. Petropulu, T. Lyberopulu, G. Parissakis, Selective separation and determination of scandium, yttrium and lanthanides in red mud by a combined ion exchange/solvent extraction method. Anal. Chim. Acta **315**, 231–237 (1995)
77. Orbite Technologies Inc., PCT patent WO 2013/104059
78. P.C. Cusack, M.G. Healy, P.C. Ryan, I.T. Burke, M.T. Lisa, L.M.T. O'Donoghue, E. Ujaczki, R. Courtney, Enhancement of bauxite residue as a low-cost adsorbent for phosphorus in aqueous solution, using seawater and gypsum treatments. J. Clean. Prod. **179**, 217–224 (2018)
79. Purgo Limited, PCT patent WO2017/103594A1—J.P. Sherry, J. Bowles, D. Shanahan, S. Jennings, G. Salt. Method of treatment of bauxite residue
80. W. Maye, I. Burke, Risks, remediation and recovery: lessons for bauxite residue management from Ajka, in *Bauxite Residue Valorization Conference, Leuven* (2015), pp. 35–45

Quentin D. Avery Regional Product Line Manager, Filter Press Technologies.—FLSmidth, Inc. Email: quentin.avery@flsmidth.com

has been actively working with filtration for over 35 years specializing in pressure filtration. He's currently teamed with FLSmidth having joined the company in 2006 as Global Sr. Product Manager for Filter Press Technologies. Over the course of more than three decades in the process equipment industry, Quentin has worked on many difficult pressure filter applications in a very wide range of industries that include filtrations for food, pharmaceutical, chemical, mineral and hydrometallurgical materials. His experience with pressure filters began in his family owned business where as president of Avery Filter Company, Inc., he headed the pressure filtration business and developed his own unique line of automated pressure filters which he successfully marketed throughout North America. Quentin has done extensive pressure filtration lab and pilot testing on red mud in alumina refineries around the world and has presented papers on residue dewatering at ICSOBA, PASTE and other global conferences. He attended Farleigh Dickinson University in New Jersey and the State University of New York and has been an active member of the American Filtration Society.

Manfred Bach Sales Manager Alumina and Pyromet Technology, ENAR. Email: Manfred_bach1@gmx.net

After passing the bachelor's degree in environmental protection, Manfred Bach has started his professional career in the field of solid liquid separation as application engineer with company Pannevis B.V., now part of Outotec. After a short employment at Westinghouse Environmental Services, Germany, he joined the solid liquid separation division of Orenstein and Koppel, Germany, now part of Thyssen-Krupp. In 1998 Manfred joined Dorr-Oliver GmbH in Germany, now part of FLSmidth A.S., Denmark. Based on his previous experience Manfred soon started working on alumina refinery related projects, including filtration, sedimentation and classification applications. Manfred developed the ClariTube polishing filter technology as an alternative to current technologies for security filtration. Manfred today is in the position of a sales manager and responsible for FLSmidth's alumina refinery related activities covering Europe, Russia and Northern Africa.

Stephan Beaulieu Process and Environment Consultant. Email: stephan.beaulieu@augh.com

Stephan Beaulieu is a chemical engineer working in the Alumina world since 1990. He has worked in Canada in the Vaudreuil alumina refinery and currently working at Aughinish refinery in Ireland. He is still passionate about the production of alumina for the wonderful sector of aluminium which is sustainable product that can be infinitely recycled. He has worked in different production and technical roles and more recently, he is working on helping to develop road map for bauxite residue reuse, reducing greenhouse gas emissions and optimisation of refinery.

David Cooling Consultant, Civil Engineering, Retired Alcoa World Alumina. Email: davidJ\jcooling@outlook.com

David joined Alcoa in 1981 as a graduated engineer after completing a Bachelor degree in Civil Engineering at the University of WA. He has since completed a Masters Degree in Applied Science with Curtin University, researching the consolidation behavior of the fine grained bauxite tailings, and a Ph.D. with Melbourne University on improving the sustainability of bauxite residue management.

David is currently the Global Impoundments Governance and Technical Manager with Alcoa Corporation, responsible for coordinating residue development activities across the organization, looking to continuously improve residue storage practices. He has been heavily involved in the preliminary investigations into residue filtration, particularly the initial test work and development of financial and business case assessments.

Ken Evans Consultant—Alumina and Bauxite Residue. Email: ken.evans111@btinternet.com

Since graduating in chemistry in 1968, Ken Evans has worked in material science, mainly in in the bauxite and alumina industry. He joined the Rio Tinto/Alcan/British Aluminium group in 1971 and held a variety of senior technical roles until his 'retirement' from full time employment. Since 2013 he has continued to actively be involved in many aspects of the alumina and bauxite industry as a consultant. The work has encompassed a very wide range of areas including: the production and use of specialty aluminium hydroxides and aluminium oxides encompassing all aspects from basic research through to commercialisation of new products; a major involvement in the acquisition and divestment of businesses in the inorganic's chemicals area; management and use of bauxite residue; the closure of alumina operations and bauxite residue operations. Author of ten international patents on alumina, fire retardants, intumescent materials, water treatment, electronic materials, toothpaste, paint, ceramics. In addition, he has authored numerous papers and presentations on areas relating to the alumina industry.

Fred Schoenbrunn Director for Thickeners, FL Smidth. Fred.Schoenbrunn@flsmidth.com

Fred has been with FL Smidth for 30 years, starting with Eimco Process Equipment in the Technology department and working the last 25 years in sizing and design of thickeners and clarifiers as a product manager and group manager, and currently as the company's Director for Thickeners. Prior to coming to FL Smidth, he spent 5 years as the supervising engineer in a chemical plant. He has been instrumental in the development of the deep cone paste thickener market over the past 20 years and helped develop the standards for paste thickeners. Over the same span he has worked in a broad variety of minerals and industrial applications and had a hand in the design of many of the world's largest and highest capacity thickeners. Among other achievements, he has been involved in the design of tailings thickeners accounting for about 25% of global copper concentrate production.

Fred has authored or co-authored 24 books and papers including articles on thickener design and control for the SME Plant Design Handbook. He is also a contributing author in the latest editions of Perry's Chemical Engineer's Handbook, and has several patents related to thickener design and operation. He is a certified Professional Engineer with a bachelor's degree in Chemical Engineering from Lehigh University.

Chapter 7
Hydrate Precipitation, Classification and Filtration

Dennis R. Audet, Manfred Bach, and Benny E. Raahauge

Abstract The precipitation area is an area of a Bayer plant that is mostly responsible for the production and product quality. The control of the circuit is seen by many as art rather than a science because of its complexity and the many aspect of the process that are difficult to understand easily at first. Another factor that is particular to this area is the long inertia and slow response of the circuit, partly due to the large inventory of seed, equivalent to more than a week of production for most circuit. This chapter covers the science to help understand this area in order to control and optimise it better. Classification of the precipitated hydrate is an integral part of the precipitation circuit operation and the presentation is based on hydrocyclones as primary and secondary classification devices which have replaced conventional thickeners in modern refineries. Fine seed thickening is performed in a conventional thickener. 3-stage classification flowsheets are discussed versus 2- and 1-stage and the influence on final strength of alumina is discussed as a consequence of the number of classification stages. Disc filters with large diameter discs are the standard equipment for coarse and fine seed hydrate filtration. These filters have replaced other vacuum filters, mainly disc filters with smaller disc diameter, due to their significantly higher filtration capacity which is a result of a significantly increased hydraulic capacity of the filter components in combination with a short cycle time. For product hydrate filtration horizontal pan filters are the state of the art technology, allowing for effective cake washing with up to three counter current wash stages, in combination with the application of steam for reducing the cake moisture.

Benny E. Raahauge is deceased.

D. R. Audet (✉)
Director Audet Process Audit, Brisbane, Australia

M. Bach
ENAR, TAYNusstein, Germany

B. E. Raahauge
FLSmidth - Retired, Deceased, Rungsted Kyst, Denmark

B. E. Raahauge and F. S. Williams (eds.), *Smelter Grade Alumina from Bauxite*, Springer Series in Materials Science 320,
https://doi.org/10.1007/978-3-030-88586-1_7

7.1 Hydrate Crystallization Fundamentals

The precipitation area is a critical area of an alumina plant because this is where the production can fluctuate the most, is also responsible for the product quality and it is also poorly understood.

In the early days of the Bayer process, the alumina produced was like flour and done batch wise but as technology evolved it was found out that this type of product was harder to handle than other mineral products that are free flowing and behaves more like sand. This new kind of product was called "sandy" alumina by opposition to the "floury" alumina [1]. This new type of product could be handled more easily during ships loading and unloading and allowed pneumatic transport to be used in the smelters pot rooms.

In the last decades the product specifications have been highly dictated by Health and Safety issues, and the product is asked to be low in fines, less dusty, to improve the workplace environment in the smelters. As such another important quality requirement is to be strong in order not to generate fines during handling.

To produce this kind of product, the processing technology has evolved considerably by the addition of agglomeration stage, cooling sections and conversion to continuous operation for the older plants. But in order to introduce these changes a more in-depth understanding of alumina precipitation was required.

7.1.1 Nucleation, Agglomeration, Breakage and Growth

There are many theories on how gibbsite (alumina hydroxide) really crystallizes, some more proven than others. Unlike other crystallization processes, gibbsite crytallisation is slow and only reach equilibrium solubility after a very long time or under extremely uncontrolled situation. The reaction rate is vastly different to digestion reaction, where solubility is reached in minutes, in precipitation 24–48 h are needed to achieve good productivity. Which explains that much of the liquor residence time in a Bayer plant is in the precipitation tanks.

What is certain is that the reaction is complex and involve several sub-processes: nucleation or birth of new very small crystals, agglomeration and growth [2, 3]. At the same time, the agitation conditions are so strong that breakage of formed crystals happens continuously. Figure 7.1 shows graphically how each subprocess contribute to form the final particles of product size.

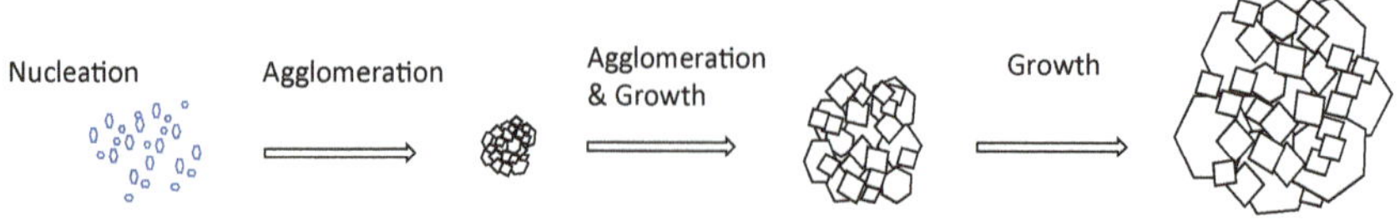

Fig. 7.1 Illustration of the various stages of a particle from nuclei to full size product

Nucleation is by far the fastest reaction in the precipitation circuit and it is needed to maintain a suitable balance of particle. Balance between what is agglomerated (inserted in the structure of other bigger particles), what grows, what is recirculated and what leaves with the product. Steps similar to polymerization to form tiny crystals called clusters which in turn needs to be inserted into a bigger crystal to be able to survive their metastable condition. This means that the presence (and quantity) of seed is crucial and important and particularly crystal surface. Nuclei are freshly generated particles that need to be in synchronization with agglomeration of fines to provide a "sink" for these nuclei [4–6]. Otherwise in the absence of particle of intermediate size catastrophic nucleation is experienced as those particles will grow individually instead of being part of a bigger structure.

In the opposite situation, in shortage of fines, the agglomeration is reduced to less agglomerative conditions: Higher seed charge, and lower temperature. Finally, proper analysis of particle size distribution shows that agglomeration is also happening in the first growth tanks and to some extent contribute to the fines balance [7]. This agglomeration though is harder to control in one direction or another but is believed to contribute to particle strength by making particles more mosaic.

Finally, growth is the sub-processes which is the most prevailing in normal conditions that most particles go through along the precipitation circuit residence time [6, 8]. Alumina hydrate growth rate is slow compared to other crystallisation process with rate of growth of the order of only a few microns per hour in the fastest conditions. The growth conditions and growth rate affect how well hydrate is crystallised, with crystals shape morphology being better formed at slower growth rate.

The precipitation rate is highly dominated by the surface area of the seed which means that smaller particles are more productive although less desired in the product. The opposite is also true as the coarser the particle the less productive they become. Figure 7.2 shows for a typical seed in a refinery the proportion of the productivity that is done by the various size ranges. It is clear that the fines are much more productive than coarser particles as the production from the bin 0–32 μm is nearly as much as from the bin 96–128 μm despite the fact that there are about 6 times less in the seed. This also stresses why a good classification is important to remove the particles that are product size and not producing much.

7.1.2 Kinetics Aspect of These Mechanism

Nucleation, particle growth rate, agglomeration, occluded soda are all different phenomena happening in the precipitation process. All of these sub processes are driven strongly by supersaturation and nearly all to a power of two. Supersaturation in its simplest expression is the difference between the current alumina concentration and the equilibrium solubility at the given conditions. The equilibrium solubility in turn is strongly related to both the liquor composition and temperature [5, 6, 9]. Many expressions have been proposed over the years to represent the supersaturation and

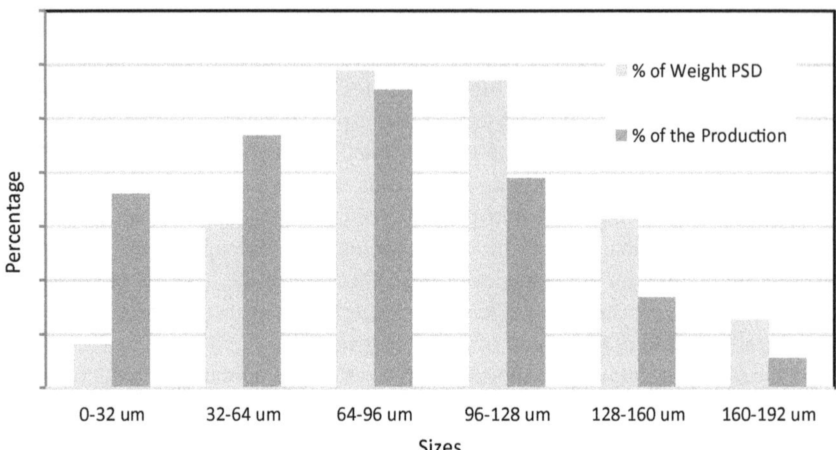

Fig. 7.2 Graph showing the contribution of several size ranges to production and their relative concentration in the seed

general agreement has not been reached so far. Most used are far from theoretical ones due to volumetric/traditional units most probably.

Here are the main expressions proposed and used in various studies:

$$\text{Relative Supersaturation with Free Caustic} = \frac{\left(\frac{A}{FC} - \frac{A^*}{FC^*}\right)}{\frac{A^*}{FC^*}}$$

$$\text{Relative Supersaturation} = \frac{A - A^*}{A^*}$$

$$\text{Supersaturation A to C ratio} = \left(\frac{A}{C} - \frac{A^*}{C^*}\right)$$

$$\text{Supersaturation in g per litre} = A - A^*$$

where

A is the Alumina concentration,
A^* is the Alumina concentration at equilibrium solubility,
C is the caustic concentration,
FC is the Free Caustic concentration calculated by subtracting the stochiometric concentration of caustic that is linked to the dissolved alumina from the caustic concentration.

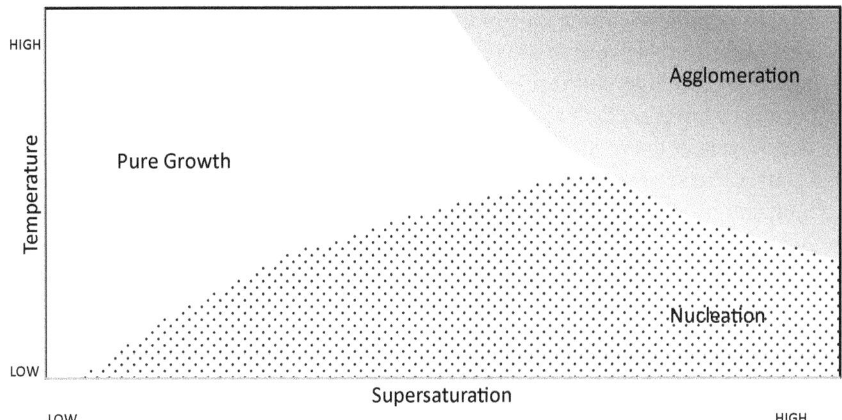

Fig. 7.3 Graphical representation of conditions favouring each of the three sub-process

However, the temperature on its own has also an effect on each these sub-processes but the difference this effect has does not go in the same direction for each of them. Temperature increase will be beneficial for some of the reactions: Agglomeration, Growth but it will be detrimental to Nucleation, by reducing it, and strength (making the product weaker). The graph below is a way to illustrate which combination of temperature and supersaturation favour each of the sub-processes (Fig. 7.3).

At a glance, one can see that growth is the process that is always present and that the two others are more restricted to smaller zone. It has to be said though, that this is meant to express the real processes and that often these are hard to differentiate, particularly when happening at conditions where two processes are happening concurrently, like the high supersaturation and intermediate temperature zone for instance.

Temperature also has a marked impact on the heat of reaction. At digestion conditions the dissolution reaction the energy required is much lower than the opposite reaction, crystallisation, that is happening at much lower temperature. It is not obvious to most that crystallisation is exothermic owing to the open environment and that natural cooling is actually desired. But in warmer locations or in summer for more Nordic places, it is often observed that the middle of the precipitation circuit can become significantly hotter than the first growth tank in circuit with low or little external cooling.

Many models have been published over the year for each of the sub-process: nucleation, growth, agglomeration and occluded soda incorporation. A general trend is that the newer equation is more accurate and cover a wider range, thanks to advance in computer power which enable to fit experimental data to more complex equation with a better theoretical background. However, these models all agree on several points that also correspond with the plant personnel experience:

- Higher alumina to caustic ratio of the liquor feeding agglomeration will give more agglomeration; because of the faster growth and cementing rate.
- Lower seed charge will result in higher degree of agglomeration. This certainly because a lower seed charge will mean a higher in-tank ratio and a faster growth rate to cement the particles together.
- Coarser fine seed is easier to agglomerate, again this will result in higher in-tank ratio and better cementing.
- A minimum residence time is needed to have a strong agglomerate: this is certainly to have a good cementing of the agglomerate that will give these a better chance of surviving the strong hydrodynamic forces.

There are some issues where even the models do not reconcile what is observed in the refinery, the size dependency of the agglomeration process. This is no doubt due to misinterpretation of agglomeration versus net agglomeration. The latter is what one can measure after some of the agglomerates have broken down either because of poor agglomerate strength or collisions with another particle. There is also the fact that particle made of bigger particle will be harder to cement, hence more likely to break later. How much of this breakage is happening depends on a lot of what is happening after agglomeration, namely if there is some strong agitation due to filtration, pumping or re-slurring for instance.

7.1.3 Impact of Organics and Other Impurities on Solubility

Over the years many authors have published equations to predict alumina equilibrium solubility all using different methodology and giving slightly different results. Because the equilibrium solubility plays such an important role for each process involved in the precipitation of alumina, it is important to understand well what influences it. Clearly alumina in pure water is completely insoluble, hence the solubility in the Bayer process is entirely attributable to the use of sodium hydroxide, called caustic for simplification. Less desirable though is the fact that all sodium impurities also contribute to increase in equilibrium solubility [10]. This effect on solubility is happening at both end of the process (digestion and precipitation) but effect is more considerable in precipitation, and consequently causes some productivity losses.

Among these the equation published by Rosenberg and Healy [11] is certainly the most widely used because it is the one with the better theoretical background, using ionic strength, energy of activation, ions activity for instance. This equation expresses all the main component and impurity in the Bayer liquor to their Ionic strength effect and contributions, including the organics impurities.

There are quite a lot of organic impurities that are present in Bayer liquor, all coming mostly from the organics in Bauxite and extracted and decomposed during digestion. Analytically though, they are all considered as one impurity [12]. The most common way to represent them is expressed as Organic carbon concentration due to the fact the method to monitor their concentration cannot differentiate the large range of compounds. Their molecular weight goes from low to high molecular and they are present at different concentrations. Their actions are also quite different too, from no effect (except colouring the liquor) to absorption and slowing down of particle growth rate.

The following Table 7.1 summarises the various effects of the organics present in liquor that are generally classified as Low, Intermediate and High molecular weight.

Hence, one can see that the low molecular weight are the ones that are the most important for the production and likely to cause some losses. The graph below shows how important these compounds can be in the liquor in these analyses carried on nine different liquors [12] (Fig. 7.4).

Table 7.1 Effects of the organics of the three main ranges of molecular weight on the process

Molecular weight	Effect on solubility	Absorbance on particle surface	Effect on liquor colour	Example of compound
Low	High	No	No	Accetate, succinate, formate
Intermediate	Average	Yes, and some compound have poisoning effect	No	Hydro and di-hydroxy benzoic acids
High	Negligible	Mostly absorption	Yes	Fulvate, humates

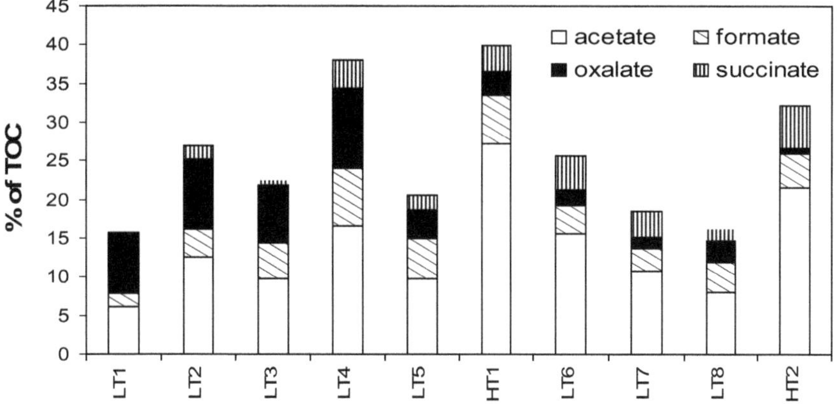

Fig. 7.4 Concentration of the four main low molecular weight organics in nine plant liquor

Finally, the organic impurity that interfere the most with the process is undoubtedly the sodium oxalate because often its concentration goes above its equilibrium solubility and crystallise in very fine needles which render the product both weaker and less pure. Also, fine seed contaminated with oxalate crystals is much harder to agglomerate and it is sometimes required to install a washing system for the Fine seed to remove these crystals by dissolving them in hot water or condensate.

7.2 Product Quality Control

There are many aspects of alumina quality that are important for the smelters to be able to process it efficiently. Some of these properties are results of the operation of the precipitation like the strength and chemical purity and can be somewhat controlled. But there are also properties that are happening during calcination like the SSA, angle of repose and densities (bulk and loose).

7.2.1 Product Quality Control: Alumina Strength

The strength of the product is not a property of the alumina that can be easily targeted and is more a natural consequence of the control parameters used at a given time. By visual observation under microscope, the shape and morphology of the particles can give clues about the expected strength of the alumina hydroxide during calcination [13]. Below are morphologies that represent typical products well, see Fig. 7.5. The mosaic structure as the name suggest is a particle made of largely small particles and a few big ones. This is seen as the ideal and strongest product possible that can withstand strong stress in calcination (see Chap. 10 Calcination, Sect. 10.5.3–10.5.6), or in pneumatic transport once at the smelter without breaking down (see Chap. 11 Alumina Quality, Sect. 11.2.1). The radial type is quite representative of what is seen in most large refinery is made of a well agglomerate core with larger crystallite on the outside. This type is also quite strong although a bit weaker than a mosaic structure. These two are generally quite round in shape which is also an advantage

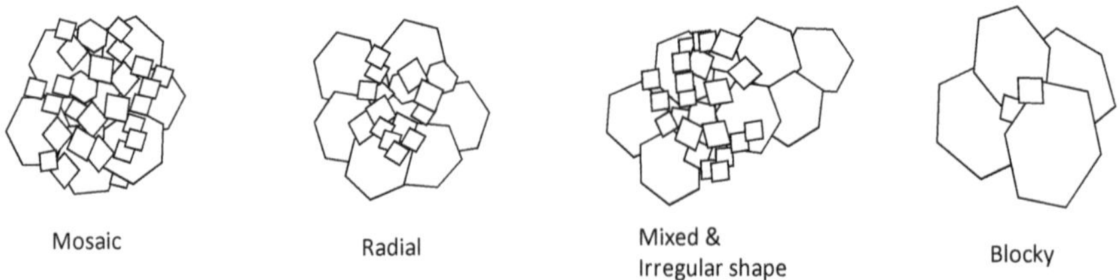

Fig. 7.5 Illustration of four typical types of morphology

to give the particle more strength. The third type is clearly not desirable as it is weaker and is symptomatic of unstable operation where the agglomeration would have been increased, following excess of fines, and have created deform particle. The other morphology type can be somewhat strong depending on the calcination conditions and is generally produced by refineries that have little classification and/or no agglomeration section. This is typical for Chinese refineries [14].

However, it is important to know the consequence that normal control can have on particle morphology and strength. One of the factors whose variations are the most likely to affect the morphology is agglomeration. Well-rounded shape particle is achieved when degree of agglomeration is neither too high nor too low. A need towards higher degree of agglomeration can cause irregular shape particle that would break easily during calcination, but equally excessively low degree of agglomeration can make particle too radial. The second most important factor to influence strength is temperature of the circuit. A cooler circuit with a steeper temperature profile normally gives a stronger product and a more mosaic product.

Finally, apart from the morphology of the product, a third important factor for strength is the presence of sodium oxalate especially after a large oxalate nucleation event. The oxalate crystals have a high affinity to alumina hydroxide and agglomerate easily [15]. The weakness is caused when the alumina particle continues to grow with the foreign crystal inside its structure. During calcination the sodium oxalate, an organic compound, will burn and is likely to break the particle at that moment.

7.2.2 Product Quality: Chemical Purity

Apart from the alumina strength, the chemical purity is important for smelters and can become out of specification once the circuit becomes out of control. The transient conditions in these situation (either Fines generation mode or Oxalate crisis situation) where temperature will significantly be outside the normal range can push the chemical purity out of specs. Higher temperature will bring the silica in product to higher level while lowering the occluded soda: which are both non desirable outcomes. Occluded soda is required at a certain level for good smelter operation to maintain an appropriate level of bath.

On the opposite, cooler temperature will bring up the occluded soda which is not desirable either as it create and excess of electrolytic bath for the smelter. However, colder temperatures have no effect on silica which is an undesirable impurity in any way.

Finally, Calcium and Iron are two impurity that can become out of specification but in most case it is due to problems in digestion (excessive lime dosage) or polishing filtration.

7.3 Precipitation Flow Sheets and Tank Designs

To maximise the production of alumina, the design of the flowsheet is utmost important as it not only dictate the production rate but also the operational stability and the product quality. Apart from the flowsheet, the design of the tanks allows to reduce power requirement and ease of operation. Over the last decades, the computers have become increasingly powerful and enabled to develop mathematical model simulating the process with high level of details and accuracy, thus leading to great improvement in the design and operation.

7.3.1 *Precipitation Flow Sheets Designs*

Many decades ago, the precipitation circuit were all operating in a batch mode. As operation modernised the conversion to continuous operation became the only option to increase production and reduce costs. During that period, there were also two types of product produced, depending on the refinery design, floury and sandy alumina, which were as the name suggest either very fine or coarse (and free flowing). The attraction of the first type of product was its simplicity and lower capital costs but the need for a free-flowing product have directed the development towards the sandy type with many variations in the design. One main difference between the two is the importance that has been given to the agglomeration section to enable to use this section to help agglomerate more fines when needed, and the classification section to separate product and seeds [1, 16].

Figure 7.6 shows the three main types of design existing nowadays, but many variations of these exists. There are several schools of thoughts that favour one approach for the other, each having advantages and disadvantages. In short, one can postulate that the presence of more classification and agglomeration allow for a more flexible control but require more capital.

In some design the alumina rich liquor is split between two tanks at the beginning of the circuit either to improve strength and or reduce the occluded soda impurity of the product [17]. This design was used for instance in the Aughinish refinery with the particularity of having smaller tanks as agglomerators.

The Classification section is where many variations exist in order to reduce the number of equipment while allowing flexibility to control the circuit and bring it back in control when required.

The 3-stage classification serves also to remove what would be product from the fine seed in the refinery that have an agglomeration section. An example of that design is the Gove refinery that was operating in Northern Territory of Australia [1]. The desire to reduce the capital for new refineries is the reasons for going towards the other two designs with two or just one classification stage for the new sites. The Yarwun refinery in Australia is an example with 2-stage classification [18], while most Chinese refineries uses only 1-stage classification [14].

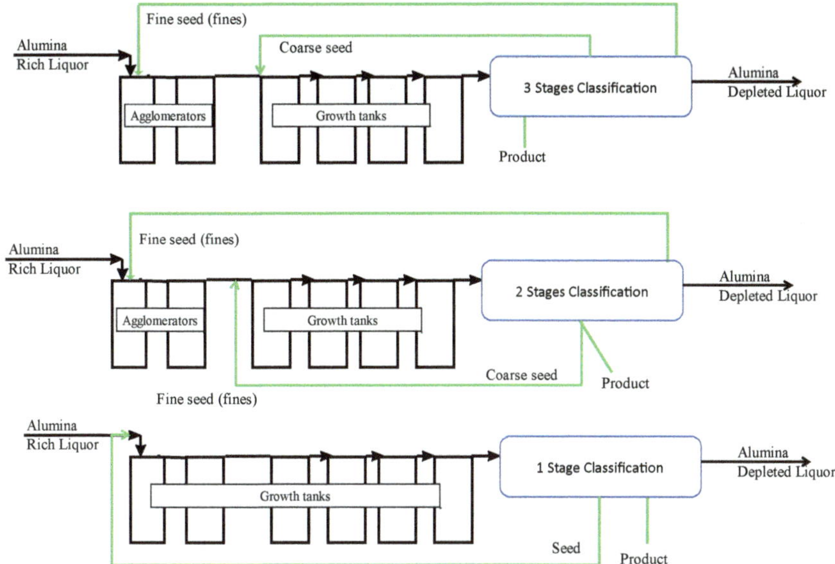

Fig. 7.6 Comparison of the three main types of precipitation circuit

The less classification stages the weaker agglomerated mosaic particle are produced resulting in weaker alumina particles when calcined, (see Chap. 10 Calcination, Sect. 10.5.3–10.5.6).

Finally, in refinery where sodium oxalate is co precipitating [16], the classification section is often used to also wash the fine seed or the product to remove the crystallised oxalate from the surface of the seeds. The oxalate rich stream is then treated to recrystallise and remove this impurity from the circuit.

7.3.2 Retention Time and Temperature Profile Impact on Yield

As mentioned earlier, the precipitation reaction is quite exothermic and in part of the world that have hot weather or seasons it is important to have means of cooling the reacting slurry during the course of the circuit to optimise its productivity. Because the precipitation rate is dependent on both the temperature and the supersaturation, which itself also depends on temperature, there are an optimum temperature for each stages of the precipitation.

So, for a given alumina concentration, this optimum temperature is a relatively soft optimum for the lower temperatures but at the beginning of the circuit the optimum is much pronounced and approaching it can yield to significant production gain (Fig. 7.7).

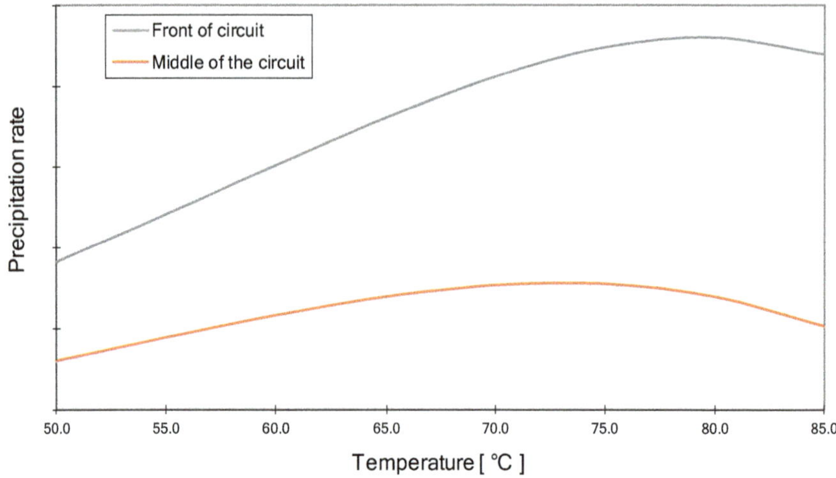

Fig. 7.7 Example of effect of the temperature at the beginning and middle of the circuit

As such the use of a given temperature profile is not seen as a "control tool" but more as a mean of optimization [9, 19, 20]. Generally, the refinery cannot match well the optimum as it would require an extensive cooling and control system but rather operate in the vicinity of it, keeping in mind that operating below allow to sustain the number of fines required and being above slows down that generation rate. Figure 7.8 illustrate this well showing a typical optimum profile. It is important to state that there is not a single profile that would work for every refinery as it is dependent on too many factors like liquor purity, caustic concentration, seed charge,

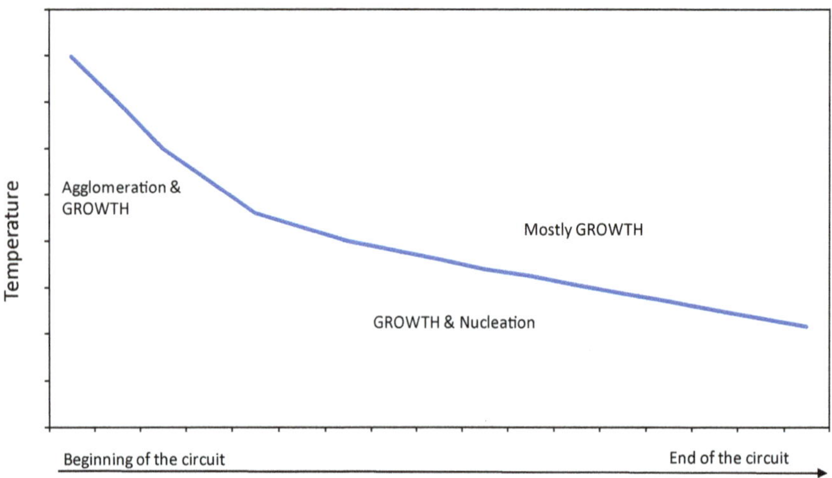

Fig. 7.8 Typical temperature profile and type of crystallisation on both sides of the optimum

etc. However, in period of start-up or in a recovery from unplanned shutdown the temperature profile can be used to drive the circuit back into a desirable condition.

In adjusting the temperature profile though, it should be kept in mind that the total amount of cooling is a finite quantity and is dependent on the operation of the cooling tower and ambient conditions prevailing at a given time.

7.3.3 Mathematical Modelling of Precipitation Flow Sheet

Because of the complexity of the precipitation process, it has long been agreed that the best way to optimise both production and product quality was to use mathematical modelling. There is considerable amount of literature that has been published on how to determine the equations required to describe the process accurately and to solve the equations in an efficient way [4, 5, 7, 9]. The determination of the equations requires considerable amount of experimental work and proper experimental design. However, recently a novel approach where no experimental tests are involved but rather a statistical approach is used to cast an old equation into a new equation form that is better. This method works only for the basic equations like solubility and growth rate for instance.

Depending on which aspect of the product quality is planned to be optimised, the model can be limited to simulate production and productivity. However, when size distribution prediction is part of the optimisation, a more complex type of model is required that is called Population Balance Modelling. In this approach, the solids in each streams and tanks are simulated using a range of particle numbers of various sizes. Population balance modelling has allowed some great changes to be done in some refinery, [6, 18] but they do require important effort to develop and appropriate resources to maintain and use.

While in the beginning of the modelling of the precipitation process most models needed to also build the program to solve the equations, nowadays there are a few options to solve the equations without having to construct a programming platform, like SysCAD or Aspen for instance.

With computer getting faster and faster, it is also possible to use simpler tools like spreadsheets to do optimisation [21]. This approach can be quite effective to optimise the temperature profile and the cooling required at the various heat exchanger along the circuit. Another use for simpler tool like this is to quantify the optimum caustic concentration to use in the refinery. This could be done with more complicated tools but because the optimum is quite soft, the optimum does not require to be determined with great precision, as seen in the example below (Fig. 7.9). Another reason is that it is likely to change with liquor and seed variations.

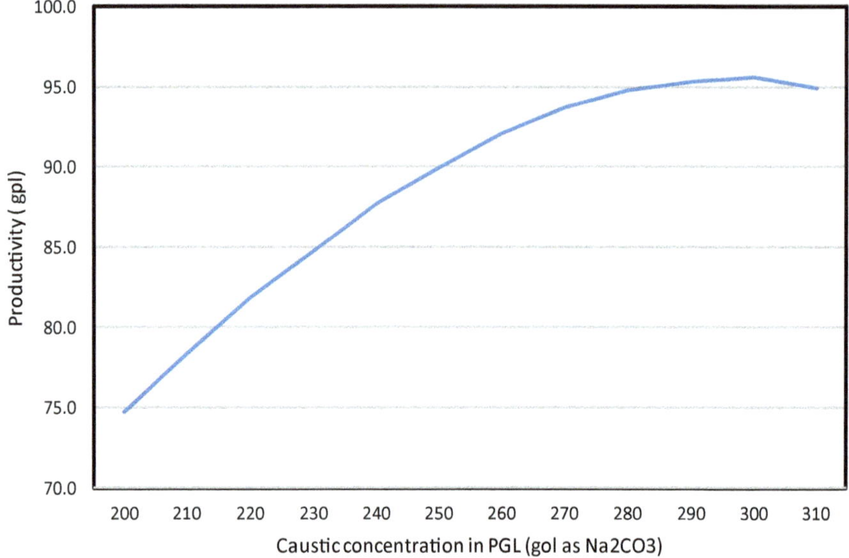

Fig. 7.9 Example of optimum caustic curve determined by mathematical modelling

7.3.4 Precipitator Tank Design Options: Agitation and Mixing Hydrodynamic

An important challenge in the design of the precipitators is due to the fact that alumina hydrate particles are much heavier than typical mineral oxide, and consequently will settle more rapidly to the bottom of the tank; the density of alumina hydroxide is 2.42 compared to many crystals being closer to unity.

Seed suspension is also important to maximize the precipitation and minimize scaling. Scaling rate in the first tanks of the circuit is high and is worse with agitation system that are too gentle [22]. The quality of the suspension also plays an important role, on the crystal morphology and precipitation rate.

For the design of the tanks in modern refineries, there are basically two schools of thoughts: use of a series of agitator on a long shaft from bottom to top of the tank or the use of an internal draft tube with a single agitator [23].

In most implementations of the first type, the use of special design agitator like Ekato allow substantial reduction in the energy required to maintain a good suspension. The exact amount of agitator and spacing along the main shaft is also part of this optimisation to reduce the energy requirements.

The second type has been championed by agitator manufacturer Lightning [24] which has optimised the design for tanks of larger size. In this design, a large tubular insert (called draft -tube) allow the agitator to be placed just a few meters below the surface of the tank just above the entrance of the draft—tube. The massive downward flow coming out of the bottom of the tube at the lower section of the tank is what allows the solids to re-suspend by moving upward in the annular section of the tank—basically outside the draft-tube. One of the claims apart from the reduction in energy is the better ability to re-suspend the suspension in situation of power outage. The use of slots of various size and position at the lower part of the draft-tube is what makes this possible.

There is another option that has been studied as an alternative, the swirl agitator which have the benefit of the draft tube without requiring the installation of a draft tube [25]. The special design of this agitator is creating an intense agitation on the upper part of wall of the tank, which is where the scale normally is forming the faster and manage to keep the adequate suspension. So far, this technology has been trialled and installed in one location but is yet to become proven and more installed.

Finally, it is important to highlight that the argument for homogenous suspension could be challenge for two reasons: (1) As discussed earlier, the alumina precipitation rate is particularly slow which would indicate that the diffusivity of the ion to the crystal would not be the driving factor. (2) This point has been demonstrated in an experiment carried out in laboratory where a very poorly agitated suspension was precipitating as fast as a well agitated one [14].

7.3.5 Productivity Calculations

At small scale it is relatively easy to calculate or measure the productivity of the precipitation reactions, either by chemical concentrations measurements or directly by balance of the solids involved in the reaction before and after. However, at plant scale it is significantly harder because of the number of entrants to the reaction and the difficulty to simply do a mass balance.

While the production rate is by opposition rather easy to know from the weighing the physical amount of product coming out of calcination at any given time. As the productivity is the production per volume of liquor circulating, expressed in gram per litter or kg per cubic meter, a reference flow of liquor is needed whether it is the rich liquor (often called Green or Pregnant liquor) depends on the refinery.

Chemical Productivity

The productivity is normally calculated using the difference in composition of the combined streams at the beginning of the circuit and of liquor leaving the circuit. One complication arises by the fact that liquor reduces in volume (increases in density) as the alumina hydrate is precipitated out. This is due to some of the water getting incorporated in the alumina hydroxide structure. For this reason, it is better to use a representation of the alumina concentration that is ratioed to the caustic, which does not vary during the reaction. The amount of solids in the liquor leaving the precipitation circuit are normally subtracted as this is irrecuperable production loss, but these solids will resolubilise during digestion. Hence the simplest equation that can be used for the productivity is:

$$\text{Productivity} = (C_{\text{Caustic}} * \left(\text{Ratio Ratio}_{A/C}^{\text{Init}} - \text{Ratio}_{A/C}^{\text{Final}}\right) - \text{Solids in Liquor to Digestion}$$

Physical Productivity

In older refineries, the rich liquor flow is rather well known but in newer refineries it is composed of a concentrated flow and some dilutions streams. It would normally only need to have well maintained flowmeters to get a good accuracy of the combined flow but because of the nature of these highly supersaturated flows, flowmeters for this duty tend to scale rapidly from the moment they have just been cleaned. Therefore, the reference flow is often calculated using flows of liquor leaving the precipitation circuit going to digestion.

As mentioned earlier the liquor density changes as it precipitates which means that rich liquor streams will always be larger flows and that an equation using these, and the production rate will inevitably be smaller. The general equation using physical production is:

$$\text{Productivity} = \frac{\text{Production rate}}{\text{Liquor Flow}} - \text{Solids to Digestion}$$

7.4 Hydrate Classification

Classification of precipitated hydrate is an indispensable part of any precipitation flowsheet as previously described. However, the classification part can be designed with 1–3 classification stages with significant impact on stability of the precipitation process, hydrate product quality with respect to particle size distribution and strength as calcined alumina, and capital investment. A precipitation circuit with 3-stage classification is shown in Fig. 7.10. This type of flowsheet is preferable with respect to control and stabilization of the precipitation circuit.

The typical precipitation flowsheet in Fig. 7.10 do not include important aspects such as interstage cooling to control the temperature profile across the precipitation tanks. Neither is washing of product hydrate to recover soda, or treatment of fine seed hydrates to remove sodium oxalate shown prior to feeding the fine seed into the agglomeration part.

7.4.1 Hydrocyclones

Classification of precipitated hydrate into product, coarse and fine seed has been done by primary, secondary and tertiary thickeners [26] until the advent of hydro cyclones. Figure 7.11 shows the main dimensions of a hydro cyclone.

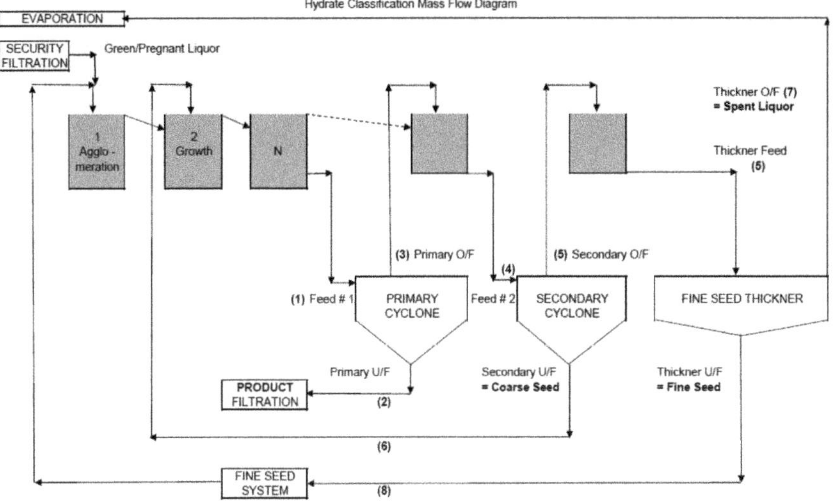

Fig. 7.10 Typical hydrate precipitation flowsheet with 3-stage classification

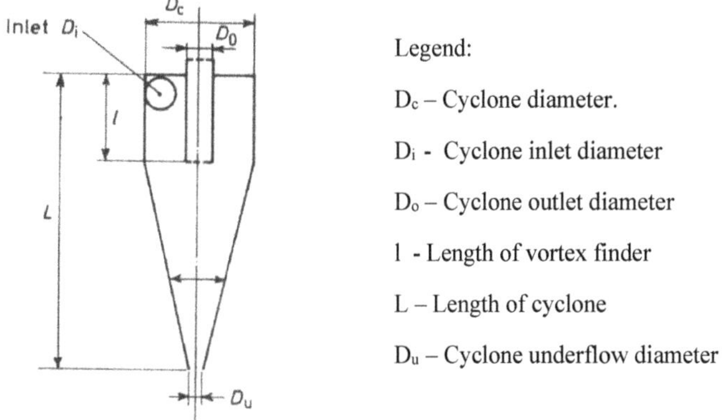

Fig. 7.11 Typical hydrocyclone with main dimensions [27]

Cyclones should not be designed for the size of the flow only. Achieving the desired cut size (d_{50}) use of the right size cyclone with a small diameter D_c, and with the length L, long as suitable is of advantage.

In general, it can be shown by dimensional analysis [27] that the cut size d_{50} depends on several variables, such as liquor and solids properties as well as the hydro cyclone design itself:

$$d_{50}/D_c = f(Re_i, \Delta\rho/\rho)$$

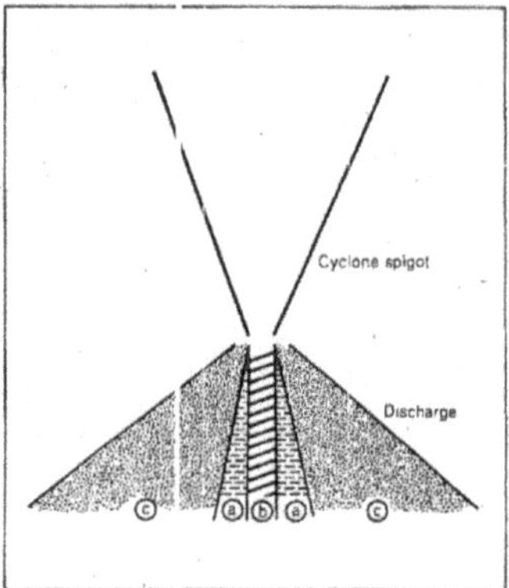

Legend:

Effect of Spigot size on cyclone underflow:

Zone (a): Correct operation

Zone (b): Roping – spigot is too small.

Zone (c): Excessively dilute – Spigot is too large.

Fig. 7.12 Cyclone underflow discharge patterns [28]

where,

$Re_i = (Q\rho/D_c\ \mu)$, Reynolds number for the cyclone inlet.
$Q =$ Volumetric flow of the inlet solid–liquid suspension/slurry.
$\rho =$ Density of the liquor phase.
$\Delta\rho = \rho_s - \rho$, Density difference between solid and liquid phase.

I addition to the cyclone itself the "spigot" or "Apex" must also be correctly designed, see Fig. 7.12.

Larger diameter cyclones are used for obtaining a coarser cut size. The efficiency of the solid liquid separation depends also very much on the Particle Size Distribution in the feed slurry.

The vendors of hydro cyclones have all developed their own sizing procedures [29, 30].

As mentioned in this book hydro cyclones have the following applications in the Bayer process:

i. Beneficiation of bauxite prior to feeding the refinery—see Chap. 2.
ii. Classification of bauxite slurry from grinding mill in closed-loop grinding circuit—see Chap. 3.
iii. De-sanding of bauxite residue/red mud in digestion slurry—see Chap. 5.
iv. Classification of precipitated hydrate into product and seed hydrate—this chapter.

As shown in the flowsheet in Fig. 7.10, 3-stage classification circuit hydro cyclones are used for primary and secondary classification. Hydro cyclones have replaced large thickeners because these are designed for a specific flowrate and are not easily adjustable to fit variations in plant flow rate. Fluctuations result in a varying product size, and important quality parameter for smelters. Thickeners have a limited capacity and are a bottleneck to any significant expansion of the plant. Furthermore, using hydro cyclones reduces the footprint of the precipitation circuit and is the more cost effective than thickeners.

7.4.1.1 Primary Classification of Product Hydrate

Large capacities can be achieved by using many small cyclones in parallel. Figure 7.13 shows a cluster of hydro cyclones for primary classification.

Feed to the primary classification cyclone from the last precipitation tank can have a solids concentration up to 590 gpl (up to 37 wt. % solids). The underflow may reach more than 1000 gpl (>57 wt.% solids) with a mass recovery to the underflow of 40–60% subject to plant location. Overflow concentration is not so importance as it is mixed with other streams, but it is important that there is a minimum of fine particles, say <2 wt.% below 45 microns in the underflow which is product hydrate for calcination.

Fig. 7.13 FLSmidth Krebs gMAX® 10—10.5 cyclone cluster at Queensland Alumina

The performance of the hydro cyclone depends on many parameters. According to Buntenbach [30], the feed parameters for hydro cyclone clusters is:

- Solids content in the feed: 400–500 gpL
- Amount of particles <45 μm: 20–6 wt.%

The efficiency of the classification and thus the amount of particles <45 μm in the underflow product hydrate is influenced by:

- The geometry of the hydrocyclone.
- Particle Size Distribution (PSD) of the solids feed.
- Pressure drop.
- Achievable solids content of the underflow.

The Mass Recovery to the Underflow, Rw, can be calculated from the below formula [30]:

$$Rw = [(1 - Wo/Wf)/(1 - Wo/Wu)]$$

where: Wo = solids content of overflow, Wf = solids content of feed, and Wu = solids content of underflow.

7.4.1.2 Secondary Classification of Hydrate Seed

Feed to the secondary cyclone comprises by-pass from the precipitation tank and the overflow from the primary cyclone. The feed solids concentration can be up to 400 gpl (up to 27 wt.% solids). The underflow may reach more than 900 gpl (>53 wt.% solids) and the overflow with a solid's concentration of <50 gpl (4 wt.% solids) and maximum recovery to underflow. The underflow is used as coarse seed and the overflow with the fine seed is concentrated in a fine seed thickener after removal of sodium oxalate that has a negative effect on the precipitation process as per Sect. 7.1.3.

7.4.2 Fine Seed Thickener

Fine Seed Thickeners are often used in alumina plants to separate spent liquor from fine hydrate seed which is returned to the precipitation circuit for controlling hydrate crystal growth. The feed to the fine seed thickeners is the fine seed slurry received from the hydro cyclone overflow from the secondary hydrate classification step. Often also filtrate from the product hydrate filtration is added to the fine seed thickeners. In many cases flocculant is added to the thickener feed stream at low dosages for accelerating the settling of the fine seed particles and achieving an overflow within the desired solids concentration, typically less than 1 g/l. Flocculants also can have a lubricating effect on the thickened hydrate slurry resulting in a lower torque compared to non-flocculated hydrate slurries at similar densities The thickened underflow in most cases is pumped to fine seed filters before the solids are returned to the precipitation tanks. The overflow (spent liquor) is pumped to the evaporation area. The underflow solids concentration can easily reach 50 w/w%.

Fine Seed thickener tanks typically are providing a bottom slope between 10 and 20 degrees and a rather shallow side wall depth of a few meters only. The diameter of fine seed thickener tanks, commonly in the range of 30–50 m, involves a center column tank design. The center column is supporting the rake drive, the rake mechanism with centre cage including scale load, the lifting device, the feedwell and a support frame for carrying the roof load. Rake mechanisms typically are of pipe design with sufficiently large, posted rake blades.

Alternate rake designs for fine seed thickeners are CableTorq rakes. These are using a double-hinge pin attachment of each of the two rake arms in combination with a cable suspension system, allowing for automatic and self-regulating arm lifting. Both arms are lifting independently from each other, with the centre scraper remaining in normal raking position, even when the arms are lifting (Figs. 7.14, 7.15 and 7.16).

The built up of hydrate scales is a major issue to be considered in the design of the fine seed thickener tank and rake. Therefore, it is important to provide a rake design that offers minimum surfaces for built up of scales and that is cleaning friendly (Fig. 7.17).

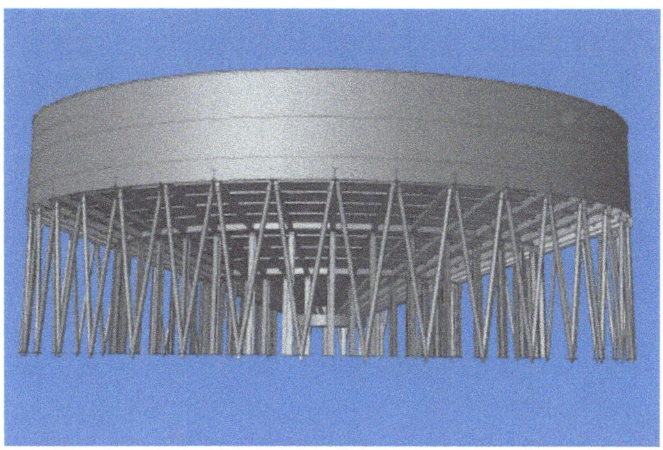

Fig. 7.14 Typical design of a fine seed thickener with center discharge

Fig. 7.15 Typical underflow discharge of a fine seed thickener with CableTorq rake mechanism

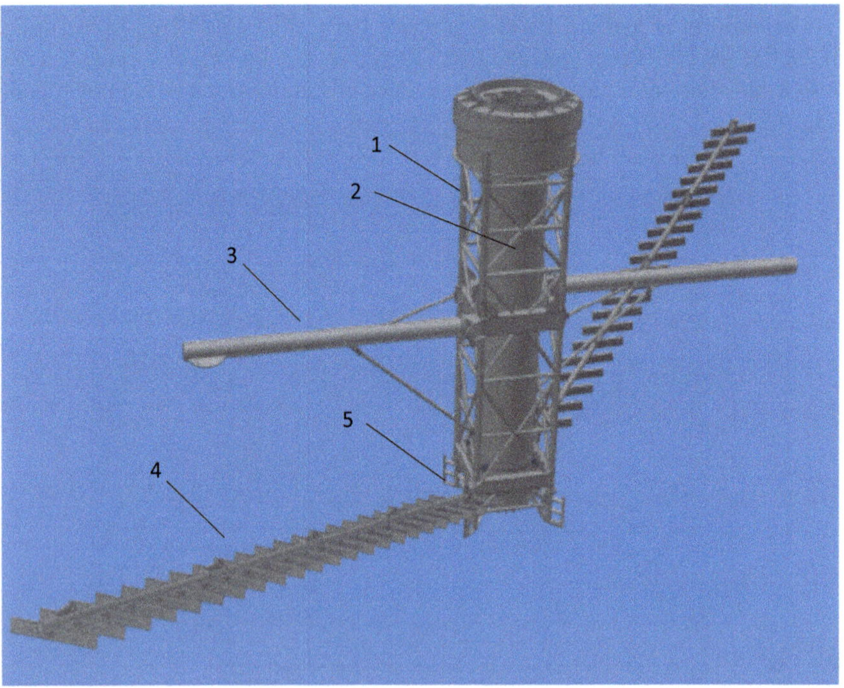

Fig. 7.16 Design of CableTorq rake mechanism comprising: center cage (1), center column (2), torque arm (3), rake arm (4), center scrapers (5). The rake arms are connected by cables to the torque arms allowing the rake arms to lift and in case of an obstacle and still push mud to the center discharge

Fig. 7.17 Scale built-up on a conventional rake arm in truss design

Fine seed thickeners often require storage capacity for overflow liquor. This is accomplished by adding the required volume to the cylindrical section of the tank, together with an appropriate circumferential overflow launder.

Modern rake drives are providing 4-point contact bearings, designed to handle tilting forces and driven by high efficiency planetary drives driven by electrical or hydraulic motors. The rake drives can be rotated in reverse if required and are provided with a permanent rake torque measurement.

Also, peripheral rake drives have been installed for larger fine seed thickeners with diameter of up to 50 m (Fig. 7.18).

7.5 Seed and Product Hydrate Filtration Options

7.5.1 Seed Preparation

Vacuum filtration mainly on disc and pan filters is an essential unit operation in an alumina refinery. Vacuum disc filters are widely used for filtration of coarse seed hydrate, vacuum pan filters for product hydrate. Mainly in the last two decades the development of these filters has been driven by the growing plant capacities and the

Fig. 7.18 Rake drive in center column thickener design type MX-2000 A2 manufactured by FLSmidth. The rake drive is comprising two sets of planetary gear boxes with electrical motors (1) and an electrically operated lifting device (2)

requirement to the filter manufacturers to process higher flowrates on one filter. As a result, the various disc filter manufacturers developed filters mainly used in the alumina industry with up to 172 m^2 filtration area, comprising discs with a diameter of up to 6 m and with up to six discs in one unit. Similarly, the size of vacuum pan filters increased from 51 to 71 m^2 in one unit, combined with improved cake washing and cake dewatering performance.

Within the hydrate classification flowsheets fine and coarse seed hydrate is returned from the classification area back to the agglomeration and growth tanks after removing most of the liquor by vacuum filtration.

As far as vacuum filtration is concerned coarse and fine seed hydrate slurries have quite different filtration behaviour due to their difference in particle size distribution and feed solids concentration. Coarse seed is an extremely fast filtering media comprising a D50 particle size of approx. 80–100 μm together with a high solids concentration in the range of >700 g/l. Fine seed hydrate on the other hand is filtering at much lower specific filtration rate, due to higher content of fines in combination with a much lower feed solids concentration.

In most alumina refineries coarse and fine seed slurries both are filtered on vacuum disc filters. The advantage of a vacuum disc filter is that it provides the largest filtration area per square meter footprint compared to any other vacuum filter. This is an important parameter since the flow rates to the filtration unit are high and the required filtration area for seed filtration is usually quite large.

For coarse seed filtration a specific filtration rate of 5–6 t/h dry solids per square meter filtration area is a widely used standard, resulting in a filtrate flow rate of approx. 10 m^3/h per square meter filtration area. The requirement to operate efficiently at such high specific flowrates has resulted in unique high-capacity filter concepts offered to the market by various vendors. The High-Capacity Disc Filter, also known as Large Diameter Disc Filter, has been designed along this requirement and provides very large hydraulic channels for filtrate and air at minimum pressure losses. To keep the pressure loss low, the number of discs has been reduced significantly and the diameter of the discs has been increased compared to disc filters known at the time when this generation of disc filters has been developed. As Large Diameter Disc Filters are operating at high peripheral speeds of up to 100 m/min, the slurry in the filter trough can be kept in suspension by the mixing energy introduced by the rotating discs which are provided with few stationary mixing elements. An agitator, as it is required for other designs with smaller disc diameter, is not required in this case. The disc diameter of large diameter disc filters for coarse seed can reach up to 6.0 m. Up to 4 discs are provided on the filter shaft, resulting in a total filtration area of up to 190 m^2. To reach the maximum filtration capacity, these disc filters are providing up to 40 sectors on one disc. The sectors are connected to sloped channels designed for high volumetric flows for efficient filtrate removal and cake dewatering. For cake discharge an efficient cake blowback pulse is provided. The blowback pulse provides enough energy to release the filter cake from the filter media and to force the cake to drop into the cake discharge chute. The filter media, which in the standard design is a textile bag, is then sucked back to the filter sector by a short vacuum pulse. The cake moisture achieved for coarse seed is in the range of 15%–17%, depending on the particle size distribution and type and condition of the filter media. Wire mesh designs are available as an alternative to textile filter bags. Obviously wire mesh designs achieve significantly higher media life at significantly reduced filter maintenance.[1]

Fine seed hydrate slurries are filtered at a specific filtration rate in the range of approx. 1.0–1.5 t/h dry solids per square meter filtration area. For fine seed filtration standard disc filters such as the FLSmidth Eimco AgiDisc Filter are in operation. Depending on the number of discs the shaft will be provided with two suction heads to keep pressure losses at a minimum and allow for efficient filtrate discharge. AgiDisc filters are manufactured with a disc diameter of up to 3.7 m and up to 300 m^2 filtration area. They can be operated with up to 15 discs. In many plants fine seed is also filtered on Large Diameter Disc Filters thereby reducing the number of different filter designs and minimizing spare parts to take on stock since many parts are identical for both filter designs.

Characteristics of Large Diameter Disc Filters

Large Diameter Disc Filters are made of several parallel discs installed on a common main shaft with up to 40 individual vacuum channels, depending on the filter supplier. The channels are either of pipe or trapezoidal shape and are connected to the main

[1] FLSmidth Meliplate.

valve. The diameter of the filter discs can be up to 6.0 m with a filtration area of up to 47.5 m^2 in a single disc. Compared to all other vacuum filter designs, Large Diameter Disc Filters are achieving the largest filtration area and the lowest invest cost per square meter footprint.

On a Large Diameter Disc Filter each disc is rotating in its own narrow feed trough. When the slurry if fed to the individual troughs, it is kept in suspension just by the agitation effect of the disc paddles itself which are installed to the disc rim. The effective mixing of the feed slurry ins important to maintain a homogenous slurry which is necessary to achieve a uniform cake thickness, which is a major parameter to achieve an overall good performance of the filter, including a low cake moisture and an effective cake discharge by a short blow-back pulse.

As the filter shaft rotates, the filter discs are submerged into the feed slurry. Vacuum is applied to the sectors as soon as they enter the slurry and are fully submerged, starting the cake to form on the surface of the filter media. After the sector leaves the slurry, the vacuum applied pulls air and filtrate through the inside of the sector into and along the channel of the main shaft and out through the large diameter valve to the vacuum receivers (Fig. 7.19).

Vacuum cut-off occurs just before the dewatered cake-laden sector approaches the discharge point. Pressurized air then loosens the cake from the filter media and guided deflector blades direct the cake as it falls through the wide discharge chutes for removal. The slurry level in each trough can be adjusted to allow for higher filtration capacity or lower cake moisture (Fig. 7.20).

The Large Diameter Disc Filter is operated with a variable speed disc drive and level control measurement allowing to adjust the filter at the optimum point of operation.

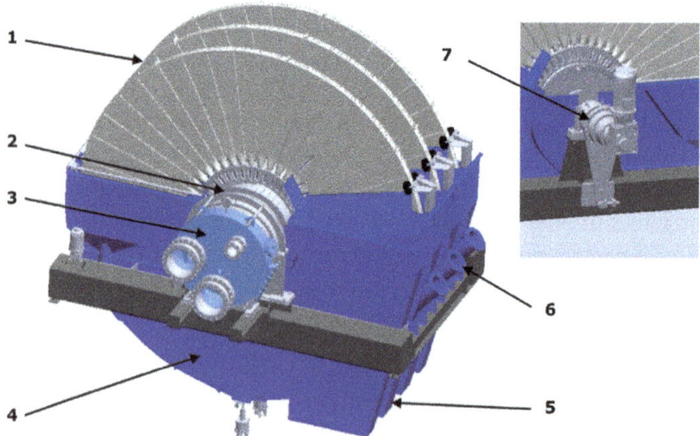

Fig. 7.19 Large diameter disc filter, valve side view: discs (1), shaft (2), control valve (3), trough with bottom drain valves (4), cake discharge chutes (5), slurry feed nozzles (6); drive side view: filter drive motor with gear box (7)—courtesy of FLSmidth

Fig. 7.20 Cake discharge from large diameter disc filter sectors by blowback

A circumferential rim is provided on each disc for parallel adjustment of the single sectors. The sectors are connected to the main shaft typically by a bayonet device, sealed with a rubber sleeve. The thickness of each sector is increasing progressively from the outer peripheral to the outlet nozzle (trapezoidal section). For drainage of the filtrate the filter cloth is supported by a waved wire net. The filter discs are designed to allow easy sector handling for filter bag replacement and re-installation (Fig. 7.21).

Fig. 7.21 Installation of large diameter disc filter on top of precipitation tanks

7.5.2 *Washing Efficiency and Flow Sheets*

Product hydrate filtration is a major process step in the Bayer process to provide product hydrate to the alumina calciner at the required quality and quantity. Main quality parameters that can be influenced by the product hydrate filtration process are cake moisture and leachable caustic soda levels. A low cake moisture is required to achieve a low fuel consumption in the alumina calciner, low caustic soda levels are required to minimize soda losses and protect the refractory of the calciner. Plants which are also producing chemical grade hydrate are even more sensitive to low caustic soda levels since this is a main quality parameter of their final product.

Vacuum pan filter versus vacuum drum filter

For product hydrate filtration the filter designs used in the industry are vacuum drum filters and vacuum pan filters. Both are suitable for filtration of a relatively coarse media in combination with intensive cake wash and can achieve low cake moistures. However, in the last 20–30 years mainly vacuum pan filters have been considered for greenfield alumina refinery projects, only few plants were still using vacuum drum filter technology. This trend is following the recognition of the industry that vacuum pan filters are offering many operational and process advantages for product hydrate filtration flowsheets which are discussed in the following below and which are also described in a joint paper from Aluminium Oxid Stade and FLSmidth, that has been presented during TMS 2009 in San Francisco [31].

Due to the filtration area of a vacuum pan filter having a flat horizontal orientation, the filtration processes taking place on the filter pan are supported by the gravitational force all the time. Looking at the geometry and operating mode of a vacuum drum filter comprising a cylindrical filtration surface it is found that cake formation starts at a point when the direction of cake formation and filtrate flow is diametrically to the gravitational force. While in a vacuum pan filter design [32] care is taken that the filtrate flow to the control valve is supported by the sloped filter pan bottom, the filtrate in a vacuum drum filter first needs to be lifted to the control valve level by the vacuum applied. The control valve of a vacuum pan filter on the other hand is arranged at the lowest point of the filter (Figs. 7.22 and 7.23).

Another distinct difference between both filter designs is that on a vacuum pan filter the cake is discharged by a screw, while on a vacuum drum filter the filter cake leaves the drum in the direction of gravity supported by a stationary scraper and after applying a short blow-back.

The cake discharge from the filter media of the vacuum drum filter is nearly complete, whereby on a vacuum pan filter the screw discharge method is leaving a

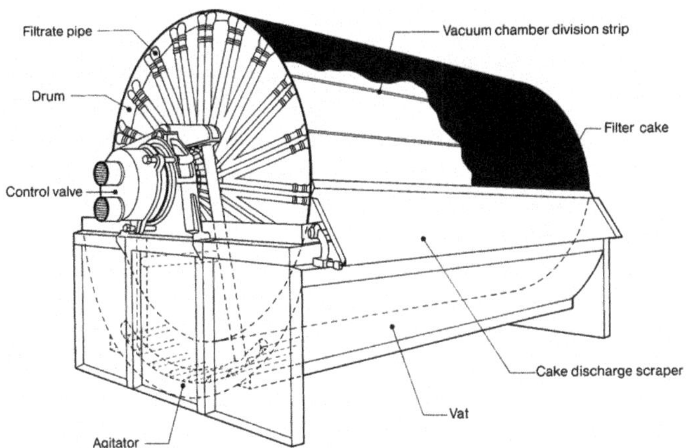

Fig. 7.22 Vacuum drum filter with external filtrate piping (view from valve side)

Fig. 7.23 Vacuum drum filter with external filtrate pipes and cake wash pipes with spray nozzles. (view from valve side)—courtesy of FLSmidth

approx. 10 mm thick heel cake layer on the filter media of the pan. The heel cake layer left on the filter media is then mixed with the newly incoming feed slurry in the slurry feed zone. Since the heel cake layer is higher in caustic and moisture than the filter cake on top which is removed by the scroll, this process is positive for the quality of the product hydrate and a distinct advantage compared to the operation of a vacuum drum filter.

The horizontal design of a vacuum pan filter is a significant advantage for efficient cake washing. The quantity of wash water that can be applied to the filter cake on a vacuum drum filter is limited and needs to be controlled well, since access water will just run down into the filter vat and will not be utilized for cake washing. On a vacuum pan filter the horizontal geometry is supporting the efficient use of the wash water applied onto the filter cake as much better control of the distribution of the wash water is possible. Another unique differentiation of both filter systems is that the geometry of the vacuum pan filter allows for counter current cake washing with two or three wash stages. Since the geometry of a vacuum drum filter is not favourable for counter current washing these filters are typically installed with a single cake wash. On a vacuum pan filter the wash water flow rate can be changed over a wide range and still completely be utilized for the cake wash process. Also wash water distribution is less critical on a vacuum pan filter since—for the same filtration area—the wash water needs to be distributed over a approx. 40% shorter

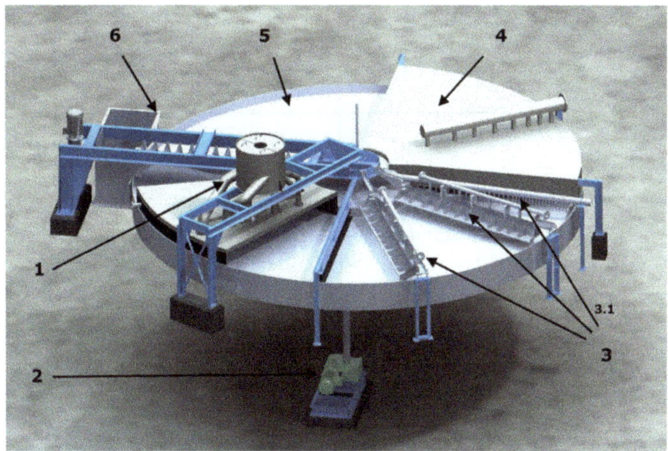

Fig. 7.24 Pan filter, comprising: slurry feed device (1), pan drive (2), cake wash devices including one spray pipe and two overflow boxes (3), steam hood (4), filter pan (5), cake discharge scroll with cake discharge chute (6)

distance compared to the vacuum drum filter wash.[2] As a result, the risk of non-washed spots is much higher on a vacuum drum filter than on a vacuum pan filter for the same wash ratio (Fig. 7.24).

The wash water flow rate, wash water temperature and method of application, all are having an impact to performance of the vacuum pan filter. Cake washing efficiency on product hydrate is inherently very high on a vacuum pan filter: 99% washing efficiency can be achieved with a 2.5 displacement wash. Thereby thicker cakes, i.e., 60–90 mm wash more efficiently and reliably. The desired residual 0.05% Na_2O concentration in the filter cake, based on dry hydrate, can theoretically be obtained by a single stage wash of 0.2 m^3 of condensate per t of dry hydrate for a 50–75 mm thick discharged filter cake. Thinner cakes wash less reliably (Fig. 7.25).

Tests indicate that, with feed liquor concentration below about 120 g/l Na_2CO_3, the pan filter can achieve cake at 0.01–0.05% Na_2O using a wash ratio of 0.2–0.4 l/kg dry hydrate. This is one of the reasons why often a pre-filtration stage was introduced ahead of the final product hydrate filter. The pre-filtration typically is carried out on large diameter disc filters followed by a mixing tank to re-slurry the filter cake with filtrate from the product hydrate filter.

[2] Example based on a 3.8 m cake width of a 71 m^2 vacuum pan filter, compared to 5.3 m cake width of a drum filter with same filtration area and 4.2 m drum diameter.

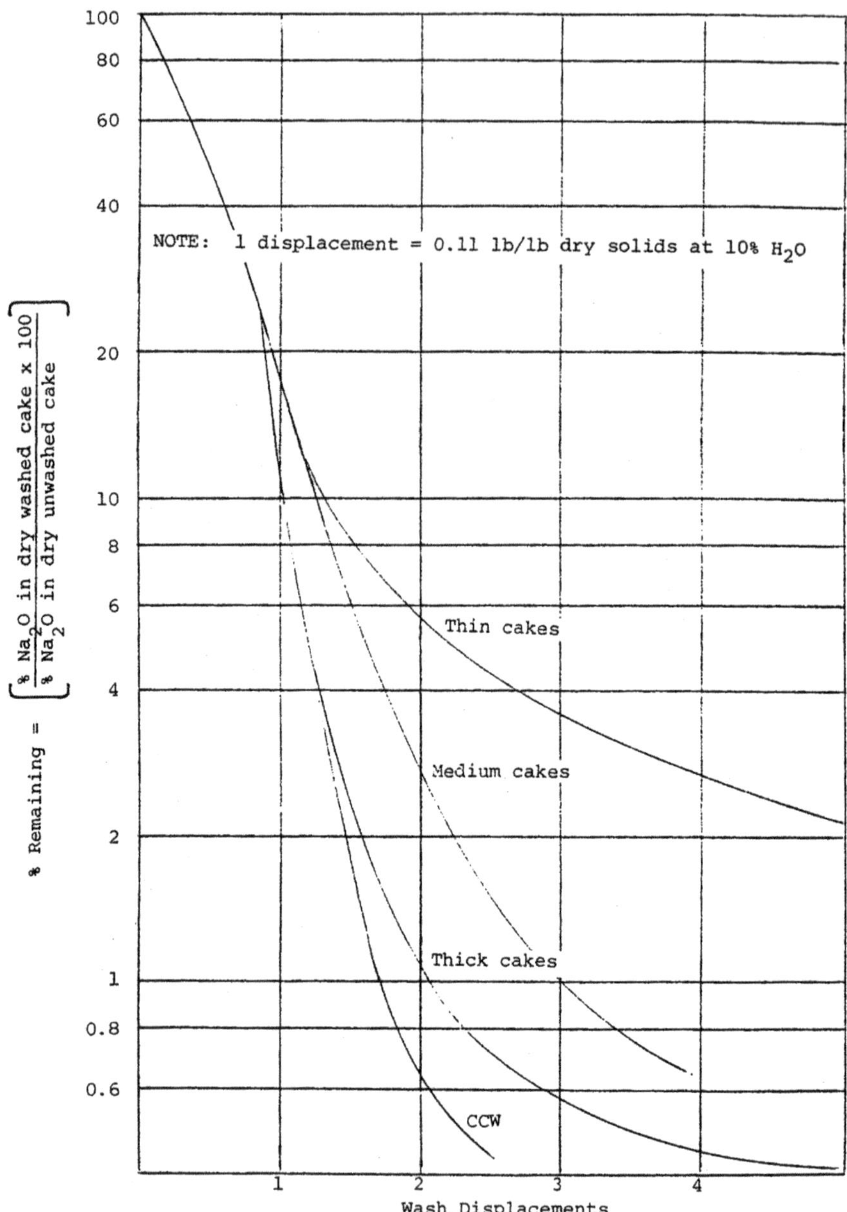

Fig. 7.25 Washing efficiency of discharged filter cakes, excluding heel cake layer—courtesy FLSmidth

References

1. O. Tschamper, *Improvements by the New Alusuisse Process for Producing Coarse Aluminium Hydrate in the Bayer Process* (AIME Annual Meeting, Chicago, 1981)
2. C. Misra, E.T. White, Kinetics of crystallization of aluminium trihydroxide from seeded caustic aluminate solutions. Chem. Eng. Progr. Sympos. Ser. **110**, 53 (1970)
3. T.J. Soar, A.R. Gerson, Nucleation mechanisms of gibbsite as suggested by computational molecular modelling, in *Fifth International Alumina Quality Workshop, Bunbury* (1999)
4. J. Mordini, B. Cristol, Mathematical model of alumina Trihydrate precipitation from Bayer aluminate liquors. Paper presented at the 4th Yugoslav international symposium of aluminium, Titograd, April 1982
5. W.J. Crama, J. Visser, Modelling and computer simulation of Alumina trihydrate precipitation. *Light Metals* 73–82 (1994)
6. S. Veesler, R. Boistelle, Growth kinetics of hydragillite $Al(OH)_3$ from caustic soda solutions. J. Cryst. Growth **142**, 177–183 (1994)
7. D. Ilievski, E.T. White, Modelling Bayer precipitation with agglomeration, in *Essential Readings in Light Metals*, ed. by D. Donaldson, B.E. Raahauge (Springer International Publishing, Cham, Switzerland, 2017), pp. 509–515
8. E.T. White, S.H. Bateman, Effect of caustic concentration on the growth rate of $Al(0H_3)$ particles. *Light Metals* 157 (1988)
9. D.R. Audet, J.E. Larocque, Hyprod simulation: optimization of productivity and quality of Bayer alumina precipitation system, in *Light Metals. Proceedings of Sessions, 121st TMS* (1992)
10. A. Lectard, F. Nicolas, Influence of mineral and organic impurities on the alumina trihydrate precipitation yield in the Bayer process. *Light Metals* 123–142 (1983)
11. S.P. Rosenberg, S.J. Healy, A thermodynamic model for gibbsite solubility in Bayer liquors, in *Fourth International Alumina Quality Workshop, Darwin* (1996)
12. F. Picard, D. Audet, H. Boily, J. Larocque, *Identification of Hydrate Active Organics (HAO) Present in Spent Bayer Liquors by State-of-the-Art Analytical Methods* (AQW, 2002)
13. M.N. Pons, V. Plagnieux et al., Comparison of methods for the characterisation by image analysis of crystalline agglomerates: the case of gibbsite. Powder Technol. (2005)
14. Q. Wang, Y. Jiang, Optimization of precipitation of Chinese refinery, in *Proceedings of the 10th International Alumina Quality Workshop, Perth* (2015)
15. A. O'Connell, J. Vaughan, S. Beaulieu, Maintaining an oxalate free precipitation circuit at Aughinish alumina, in *Proceedings of the 6th International Alumina Quality Workshop, Brisbane* (2002)
16. S.G. Howard, Operation of the Alusuisse precipitation process at Gove, N.T. Australia. *Light Metals* (1988)
17. B.A. Hiscox, C.D. Ellis, J.E. Larocque, D.R. Audet, Process for precipitating alumina from Bayer process liquor, US Patent 5,158,577 (1992)
18. D.R. Audet, R. Little, *Stabilisation of Product Quality at Yarwun Alumina Refinery* (AQW, Perth, 2012)
19. R. Den Hond, I. Hiralal et al., *Alumina Yield in The Bayer Process Past, Present And Prospects, TMS* (The Minerals, Metals & Materials Society, 2007)
20. D.R. Audet, J.E. Larocque, Plant trial of a new precipitation line up based on prediction from the alcan precipitation simulation, in *3rd International Alumina Quality Workshop (AQW), Hunter Valey* (1993)
21. D.R. Audet, *Optimization of Alumina Refinery Production using Spreadsheets Models Add-in: Development of Models with Low Experimental Data* (ICSOBA, Quebec Canada, 2016)
22. T. Kumaresan, K. Bhor et al., Alumina hydrate suspension in the draft tube precipitator design—batch versus continuous operation, Travaux 46, in *Proceedings of 35th International ICSOBA Conference, Hamburg, Germany*, 2–5 Oct 2017
23. V. Esquerre, H. Barrué, Hydrodynamics of dense hydrate suspensions in precipitators, in *Proceedings of the 6th International Alumina Quality Workshop, Brisbane* (2002)

24. J.A. Shaw, The design of draft tube circulators. Proc. Australasia Inst. Min. Metall. **283**, 47–58 (1982)
25. D. Stegink, T. Davis et al., Operating experience with improved precipitation agitation system, in *Proceedings of the 9th International Alumina Quality Workshop, Perth* (2012)
26. S. Perra, Eurallumina progress in precipitation control of aluminium tri-hydrate. *Light Metals* (1985)
27. L. Svarovsky, *Hydrocyclones, Chapter 6 in Solid-Liquid Separation, Chemical Engineering Series* (Butterworths, 1977)
28. B.A. Wills, Factors affecting hydrocyclone performance. Mining J. (1980)
29. R.A. Arterburn, *The Sizing and Selection of Hydro cyclones.* (FLSmidth Krebs)
30. S. Buntenbach, *Influence of Process Parameters and Hydro cyclones on the Product Quality of Prepitation* (ICSOBA, 2008), pp. 176–188
31. B. Petersen, M. Bach, R. Arpe, *The World's Largest Hydrate Pan Filter: Engineering Improvements and Experience* (TMS, San Francisco, 2009)
32. Horizontal pan filter brochure, FLSmidth

Dennis R. Audet Director Audet Process Audit. Email: denis_audet@hotmail.com

Denis Audet joined Alcan in 1997 years ago straight after finishing his PhD in Energy storage in Compiegne's UTC (France). He first works at the Arvida Research Centre for 4 years before being sent to Aughinish Alumina in Ireland for 3 years. On his return to Canada at the Arvida Research Centre he worked on a variety of project. The project he is most famous for outside the company is the Precipitation model that includes population balance that have been sold and used in many locations around the world.

He then moved to Banbury, England to help start another research group for 3 years. Following the merger with Alusuisse, in 2002 he was then involved in the start-up of the R&D group in Brisbane that became the Queensland R&D Centre (or QRDC).

He was appointed manager of the Centre in 2010 and has continued to help the Centre to continue its growth until his pre-retirement 2015. Since then, he has been working as a consultant for several Chemical companies, in particular for Alumina plant where is renown to survey or improve precipitation circuit.

Manfred Bach Sales Manager Alumina and Pyromet Technology, ENAR. Email: Manfred_bach1@gmx.net

After passing the bachelor's degree in environmental protection, Manfred Bach has started his professional career in the field of solid liquid separation as application engineer with company Pannevis B.V., now part of Outotec. After a short employment at Westinghouse Environmental Services, Germany, he joined the solid liquid separation division of Orenstein and Koppel, Germany, now part of Thyssen-Krupp. In 1998 Manfred joined Dorr-Oliver GmbH in Germany, now part of FLSmidth A.S., Denmark. Based on his previous experience Manfred soon started working on alumina refinery related projects, including filtration, sedimentation and classification applications. Manfred developed the ClariTube polishing filter technology as an alternative to current technologies for security filtration. Manfred today is in the position of a sales manager and responsible for FLSmidth's alumina refinery related activities covering Europe, Russia and Northern Africa.

Benny E. Raahauge Deceased—Owner—Director Raahauge-SGA ApS, Denmark.

Benny received his MSc. Chemical Engineering from the Danish Technical University in November 1972. Start working with programming minicomputers controlling the raw material mixing for production of Cement Clinker, December 1972 in Process Technical Department, FLSmidth, Cement Division, Copenhagen, Denmark.

From 1975-76, Benny worked as process and plant engineer in a Danish Sugar Factory before re-joining FLSmidth as R&D engineer and later R&D Manager in the Mining Division in Copenhagen. Benny was assigned the task to develop the stationary Gas Suspension Technology for Smelter Grade Alumina (SGA) to replace the rotary kilns.

After successful commissioning of the first 1000 tpd Gas Suspension Calciner (GSC) unit for alumina at Hindalco Industries, India, in 1986, Benny worked in the role as General Manager-Pyro & Alumina Technology based in Copenhagen. This assignment included the responsibility for design, marketing, sales and commissioning of GSC units for Alumina, culminating with commissioning of 3 x 4500 tpd GSC units at Queensland Alumina 2004-05, the world's largest stationary calciners.

During +44 years employment with FLSmidth & Co. contributed with many papers on calcination to international ICSOBA, Alumina Quality Workshop and TMS meetings until retiring in 2018.

Together with Don Donaldson, Benny was Lead-Editor to "Essential Readings in Light Metals—Volume 1—Alumina & Bauxite" with Fred Williams as Co-Editor, Published by Wieley & Sons. Inc. Copyright © 2013 by The Minerals, Metals & Materials Society (TMS).

In 2018 Benny was editor and contributor to Chapter 12.1 Alumina, in the "SME Mineral Processing & Extractive Metallurgy Handbook", Published by Society for Mining, Metallurgy & Exploration (SME), Copyright © 2019.

Chapter 8
Bayer Process Impurities and Their Management

Steven J. Healy

Abstract Impurities introduced into or generated by the Bayer alumina production process can have a detrimental effect on the efficiencies or productivity of industrial alumina production, and/or alumina hydrate or alumina product quality. Significant complexity and cost can be added to the Bayer process to manage them. This chapter looks at the main impurities and the common strategies and industrial processes for management of their impact on smelting grade alumina's production cost, quality and environmental footprint.

8.1 Introduction

Because it represents the vast majority of alumina production, this chapter is limited to impurity management in the *Bayer* process for the production of metallurgical (smelting grade) alumina. Stepping outside these boundaries would dramatically increase an already broad scope.

In the Bayer refinery context, the term "impurity" may refer to substances introduced into or generated by the process which have a detrimental effect on the cost, efficiency or productivity of alumina refineries, or the subsequent reduction of alumina to aluminium in the Hall Héroult process. Significant complexity and cost can be added to the Bayer process to manage them.

From the broadest viewpoint, the Bayer process exists to separate the alumina content of aluminous minerals in bauxite such as gibbsite, boehmite and diaspore from anything which is not. Beyond this dissolution of alumina and its separation from 'bauxite' impurities, the dissolved alumina is then transformed by reprecipitation into gibbsite and calcined to an alumina product with chemical and physical properties that facilitate safe and efficient transport to, and reduction in Hall Héroult aluminium production cells.

In considering the mining of bauxite, minerals which are not "available" alumina under Bayer digestion conditions add to the cost of mining, transport and processing.

S. J. Healy (✉)
Alumina Technology Consultant, Brisbane, QLD, Australia

B. E. Raahauge and F. S. Williams (eds.), *Smelter Grade Alumina from Bauxite*, Springer Series in Materials Science 320,
https://doi.org/10.1007/978-3-030-88586-1_8

Consequently, the processes of exploration, mine planning, grade control and some forms of bauxite beneficiation are preliminary impurity management activities. Being outside of the Bayer process, these activities are not discussed here, even if they can be complimentary to, or competitive with refinery-based impurity management.

Once part of a refinery's bauxite feed, non-alumina yielding minerals are entrained in the bauxite digestion/residue stream, and by physical or chemical means prevented from being suspended or dissolved (apart from minor concentrations) in the pregnant liquor crossing over to the 'white' side of the Bayer process. Those that remain solid after digestion are physically separated, by settling of the bulk in primary decantation, and then by polishing filtration of the residual minor mass (typically < 1 g/L) of suspended solids in pregnant liquor unable to be settled. Some minerals transform in digestion into other minerals or compounds with different physical and chemical properties, which can either simplify or complicate their management.

This collection of minerals and other compounds constitute bauxite residue (or 'red mud'), which could be argued to be a refinery's greatest impurity input by mass, and the separation and management of bauxite residue an alumina refinery's largest impurity management process. It is also its greatest contributor to a refinery's environmental footprint. The disposal of impurities separated by processes other than primary decantation, or otherwise mixed into the bauxite residue stream can have adverse environmental consequences and is often subject to environmental review and licensing. The management of these impurity removal products needs to be considered with this in mind. While impurity management processes which contribute to residue are discussed, the separation, washing and management of residue is left to others to examine in detail.

Once dissolved, impurities are accumulated in a refinery's liquor inventory subject to a dynamic balance of inputs (e.g., bauxite, caustic, lime, additives, raw and recycled water) and outputs (e.g., solid and liquid residue streams, specific removal processes, with DSP, and with product). Variations in any of these inputs and outputs can shift this dynamic balance, changing liquor impurity concentrations.

Some impurities impact more than one aspect of a refinery; its productivity, product quality, and/or environmental footprint, and their control method or tolerance level can vary depending on their most significant effect. For many impurities, their solubility in refinery liquor is very small and their only impact is their contamination of product gibbsite and alumina.

Its chemistry being key to an efficient Bayer process, sodium is an interesting case where, when associated with some anions (hydroxide and aluminate for example), it is productive, while when associated with other anions is un- or counter-productive. Its inclusion in gibbsite and alumina products for example, up to certain concentrations, is not problematic in reduction cell chemistry, but in excess of these values is unwelcome. The incorporation of sodium in product by several mechanisms is discussed with the relevant impurity counterions, and by other mechanisms better discussed in the precipitation topic. The subject of soda in product has been studied in numerous works (e.g. Sang [1] Grocott and Rosenberg [2], Armstrong [3] and Vernon et al. [4]).

The references cited here are in the public domain and are familiar to the author, or commonly used, but are only a minor subset of those that could be used to background the various topics covered. The libraries of many alumina producers are full of excellent technical studies, some of which have never found their way into public domain. Studies on some of the subjects covered here have been ongoing for many decades, and while it is convenient to refer to recent publications, it is common to find similar conclusions in many works from different countries and in different languages. The reader is encouraged to explore nuances in the vast literature available.

8.2 Bauxite and Impurities

To extract aluminous minerals, bauxite is milled to an optimum size before digestion in a strong caustic soda (sodium hydroxide or NaOH) solution or "Bayer liquor" at temperatures typically between 105 and 280 °C. Due to the lower solubility of most metals' hydroxides compared to aluminium under these conditions, this results in selective dissolution of alumina minerals, leaving other elements with less soluble hydroxides largely undissolved, either unchanged, or transformed into other insoluble compounds.

Relatively minor amounts of most of these other elements are solubilised, typically (but not always) as oxy-hydroxide or hydroxide anions. In the cases where their solubility does not limit these elements' accumulation, without an outlet of the same magnitude these ions accumulate to inconvenient concentrations in the recirculating liquor inventory. Apart from their various effects on Bayer unit operations, the formation of these anionic species consumes hydroxide ions, requiring replacement by fresh caustic soda addition to maintain the liquor hydroxide (caustic) concentration.

Impurities such as silica, iron, titanium, carbonate, sulfate, oxalate and various organic compounds are common to most refinery liquors due to their common occurrence in bauxite. Others which do not normally accumulate to significant concentrations such as (for example) beryllium and zinc are related to abnormally high occurrence of less common minerals in bauxites from varying local geologies.

Elemental compositions of some economically important bauxites (Table 8.1) were summarised by Rosenberg [5], referencing various sources. Bauxite composition can vary widely within a region and within a deposit. A product bauxite's composition is also determined by bauxite mine quality targets which, to a large extent, drive mine planning, and can evolve over the economic life of a deposit. While this variation is important to remember, looking at a typical regional bauxite's chemical and mineralogical character still provides a useful average of the available alumina and a range of important impurities. These typical values provide an indication of the impurity management requirement for processing of the bauxites from these regions. Examples cited by Rosenberg and many of his comments on the subject are restated here with their references, with some additional qualifications or variations.

Table 8.1 Composition of various bauxite deposits [5]

Species	Guinea	Brazil	Weipa	Greece	Trombetas	W. Australia (1)	W. Australia (2)	India	Jamaica
Al_2O_3 Tot	49.41	53.4	55.7	56.6	53	37.2	38	48.5	46.8
SiO_2 Tot	1.7	4.2	5.8	4.2	5.1	15.5	2.8	2.7	1.98
Fe_2O_3 Tot	19.06	13.9	11.9	21.9	13.9	24.5	35	17	19.2
CaO	0.01		0.01		0.01	0.01	0.01	0.02	1.13
C inorganic	0.06	0.01	0.03	0.41			0.1		0.4
C organic	0.11	0.03	0.23	0.08		0.2	0.2	0.11	0.2
C total	0.17						0.3		0.6
S total	0.032	0.04	0.03		0.06	0.09	0.1	0.05	0.08
P_2O_5	0.14	0.02	0.07		0.01	0.02	0.015	0.13	0.38
TiO_2	2.94	1.1	2.7	2.8	1.2	2	2.7	5.32	2.25
Na_2O	0.01	0.02	0.01		0.02	0.004		0.02	0.03
MnO	0.02	0.01	0.03			0.006	0.003	0.04	0.32
K_2O	0.03	0	0.01		0.01	0.06		0.03	0.01
MgO	0.01	0.01	0		0.005	0.006		0.05	0.08
V_2O_5	0.05	0.04	0.06		0.05	0.06	0.05	0.19	0.11
ZnO	0	0	0			0.002	0.003	0.008	0.03
F						0.08	0.06		
LOI	26.7	27	23.2	12.2	27	19.8	21	25.7	26

Karst bauxites (e.g., China and Greece), are generally high in inorganic carbon due to formation over carbonate minerals. Guinean, Indian and especially Jamaican bauxites are relatively high in phosphorus. Western Australian bauxites are formed from two quite different parent rock types, but laterized by similar processes which contributes to their elevated organic carbon and sulfur.

Titanium and vanadium occur in most bauxites, but some with vanadium concentrations high enough to support dedicated removal processes, and to produce a vanadium concentrate which can be sold for further processing [6].

Jamaican bauxites have a high and very fine goethite content, and are consequently more difficult to settle in clarification [7]. Organic carbon (and associated oxalate), zinc and manganese present difficulties in processing. See and Feret [8] propose that these latter elements are extracted from zinc-substituted lithiophorite ($(Al, Li)MnO_2(OH)_2$), common in Jamaican deposits.

Bauxites from Guinea are comparatively low in most common impurities, but still require some attention to (for example) oxalate, organic and inorganic carbon and iron. Its low reactive silica means impurity removal with DSP is proportionately lower than bauxites with higher reactive silica. Although its organic carbon content is generally lower than (for example) bauxites from Australia, Asia and Surinam, there is generally a requirement to remove oxalate, and organic carbon usually accumulates to levels which have an impact on product quality and liquor productivity. When digested at high temperature (to recover boehmitic and goethitic alumina), its goethite content can result in elevated iron in product. Its consumption refineries also tend to produce alumina with higher gallium content. Compared to lateritic bauxites from other regions, its inorganic carbon content is elevated, increasing the cost of carbonate control and/or inhibiting liquor productivity. The Bel Air deposit in the Boffa region of Guinea is unique in its andalusite ($AlSiO_5$) content, and having little or no boehmite [9], illustrating that even in the same province, generalisations have their limits.

The bauxites from northern Australia (e.g., Gove and Weipa) are generally lower in goethite, so elevated iron in product is generally not a major issue. Some areas of these deposits can be low in total iron, and where process conditions increase its post-digestion liquor concentration, this rule is stretched. Their relatively high organic carbon content demands more rigorous oxalate and organic carbon management than bauxites from most other regions.

In the case of some bauxites with low impurity inputs, the basic refinery processes (residue separation, DSP precipitation, product removal and lime addition at various locations), are enough to keep most liquor impurities below the requirement or economic justification for management. Some Brazilian and many diasporic bauxites for example, where their organic and inorganic impurity inputs are low and reactive silica sufficiently high, may not require specific impurity removal processes, or only operated intermittently. For many other bauxites this is not the case, and refineries have been sufficiently motivated to implement specific removal technologies or strategies.

To the extent that it is possible for a refinery, the selective use of bauxites that have low impurity inputs, and to some extent, higher reactive silica relative to available alumina, remains one of the simplest means of controlling many impurities.

8.3 Liquor "Purge"

Until the "Sandy Alumina" standard was widely introduced, and ultimately demanded of alumina refineries [10] and until residue washing equipment and circuits were efficient enough to reduce residue liquor losses, the quantities of impurities (along with alumina and caustic soda) leaving the Bayer refinery with residue often resulted in stable and manageable impurity balances. In many cases this outlet for impurities was sufficient to limit impurity accumulation to the point where installation of dedicated impurity removal processes was not critical and/or economic.

The 'Sandy' alumina standard brought a requirement for better size (and consequently often oxalate) control in refineries. Around the same time, better, lower cost synthetic flocculants and more advanced decanter and washer technologies allowed washing mud more efficiently. These developments reduced the losses of dissolved salts with residue liquor significantly, resulting in not only greater recovery of soda and alumina, but the increased accumulation of liquor impurities. One way of mitigating this accumulation was not to optimise the performance of the mud washing circuits, leaving them to perform below their potential. This has often been the lowest cost solution (certainly in capital investment terms) for impurity management and the maintenance of liquor productivity.

Driven by the prize of saving soda and alumina, and the challenges of the cost, availability and social acceptability of increasing residue storage footprints, the industry has moved towards higher density and lower liquor/caustic contents in bauxite residue. Following this, the practice of adjusting liquor losses with residue to manage impurities has become less attractive. A greater interest and economic justification for dedicated impurity removal processes has followed.

An alternative to purging of liquor with residue is the transfer of significant volumes of liquor to other refineries or industries less concerned about the impurity levels. This rarely occurs, and tends to be opportunistic, often related to refinery shutdowns and commissioning. Where an opportunity is presented and economics are supportive, this form of purge addresses both impurity management and environmental objectives.

Jamieson [11], suggested reacting Bayer liquor with a reactive silica source such as silica fume or fly ash to produce geopolymer products or to simply entomb impurities. Any such safe and productive use has more to recommend it than simple deposition in a residue storage area, but probably due to economics, is not known to be practiced.

8.4 Liquor Neutralisation and Discharge

Where liquor purge is practiced, and particularly in high rainfall areas, this liquor can be diluted by rainwater and because it retains a relatively high pH, accumulate to problematic volumes. Beyond the constraint of the volume that can be returned to the refinery process while maintaining the refinery water balance, this liquor contains impurities, which are detrimental to refinery liquor productivity. To maintain manageable inventories of this weak liquor, and to avoid the return of impurities to the Bayer process, a number of processes for remediation of alkalinity, dissolved metals and other environmentally harmful attributes prior to environmental discharge have been developed and practiced. Although they function as a water-balance fix and environmental (and containment pond capital cost) solution, in the majority of cases these processes assist with impurity management.

The most common neutralisation processes practiced are variations on acid, seawater and CO_2 treatments. Each of these processes developed have their merits and limitations. They are not examined in detail in this work because they are not impurity management processes as such, but rather facilitators for liquor purge.

8.5 Impurity Removal with Product

All refineries have some level of liquor losses with product alumina [1–4]. As product gibbsite is washed before calcination, the liquor in contact with solids is effectively diluted, and some of this dilute liquor is forwarded as filter cake moisture with the product gibbsite to calcination. The impurities associated with this liquor then leave the process either as precipitated salts in the solid product, or as vapor or dust in the calciner stack gases. This liquor loss is minor, but in some impurity balances, it can be significant.

In addition to the dilute liquor entrained with gibbsite, other impurities are removed with product, precipitated with or adsorbed to product gibbsite. This is generally considered a negative result, since these liquor impurities become alumina product impurities, and are negative in some way to alumina utilisation. Caustic soda is also lost. Oxalate and other organic compounds, iron, silica, calcium, titanium and phosphorus are some examples. In some cases, these impurities have relatively small circulating masses, so the small amount removed with product can be significant in controlling their accumulation in liquor. The trade-off between maintaining a product quality specification and utilising this removal mechanism to avoid the need for a dedicated removal process can sometimes be a delicate balance.

Table 8.2 shows typical product impurity specifications for most common elements of concern. The minor element content of the bauxite(s) being processed can vary widely, so for example, while in one refinery, a specific impurity removal with product may contribute significantly to a balance, in another it may be only a small fraction of the bauxite input and require additional management measures.

Table 8.2 Typical SGA alumina chemical quality specification

Element (as oxide)	Weight % in SGA
Na_2O	<0.40
Fe_2O_3	<0.020
SiO_2	<0.020
V_2O_5	<0.001
CaO	<0.35
TiO_2	<0.002
P_2O_5	<0.002
ZnO	<0.001
Ga_2O_3	<0.001

8.6 Silica and Other Impurity Removal with Desilication Product

Silica enters the refinery as several bauxite minerals, most commonly as kaolinite and quartz. The dissolution of these minerals in Bayer liquor results in supersaturation of the silicate ion and the requirement for its removal by precipitating Desilication Product (DSP). If not precipitated from liquor in predesilication, digestion or pre-settling residence time, DSP and other silica scales precipitate on heat exchange and other equipment surfaces, and with gibbsite in the precipitation area, as silica contamination in product. The detailed discussion of DSP chemistry and optimum desilication conditions will be left for others to explore.

DSP precipitation results in the majority of a refinery's caustic soda consumption, a major operating cost. Kaolinite dissolution and the precipitation of sodalite DSP is summarised in the following equations (8.1 and 8.2):

$$3Al_2Si_2O_6(OH)_4 + 18NaOH \rightarrow 6Na_2SiO_3 + 6NaAl(OH)_4 + 3H_2O \quad (8.1)$$

$$6Na_2SiO_3 + 6NaAl(OH)_4 + Na_2X \rightarrow Na_6[Al_6Si_6O_{24}] \cdot Na_2X + 12NaOH + 6H_2O \quad (8.2)$$

where: Na_2X indicates the sodium salt of sulfate, carbonate, chloride and other anions.

Many studies to reduce DSP caustic consumption by minimizing silica dissolution and/or DSP formation, or by lowering the stoichiometric sodium in the reaction products are in the open literature, though few have been adopted. Lime addition at high temperature can form calcium DSPs and lower the soda cost, and is practiced, where possible. A comprehensive review of these studies and associated chemistry was published by Smith, and is a good source of detail [12].

One approach to managing silica as an impurity is kinetically minimising its dissolution. A process was developed and reported by Harato et al. [13], which relies on the higher dissolution rate of gibbsite compared to kaolin. This differential dissolution principle is illustrated in Fig. 8.1. To exploit this difference, bauxite slurry and liquor which has been heated separately, is mixed at a suitable digestion temperature (typically ~130 °C) to react long enough to achieve nearly complete gibbsite dissolution (2–5 min), while limiting kaolin dissolution to less than 50%. The liquor and bauxite residue must be rapidly separated, and the liquor separately desilicated. The thickened slurry containing the partially reacted kaolin is passed to the mud washing circuit, where further silica dissolution must also be constrained.

Since the time available for DSP precipitation in the heaters and digestor is short, the digestion liquor contains a high silicate concentration, requiring a separate seeded desilication vessel, and subsequent flash system and DSP separator. An alternative is redirecting the pregnant liquor from this 'fast' digestion to a secondary digestion where the alumina concentration is topped up (since it is limited by temperature and time in the fast digestion step), and the liquor is more completely desilicated. A reduction in soda loss of more than 50% is claimed to have been achieved in

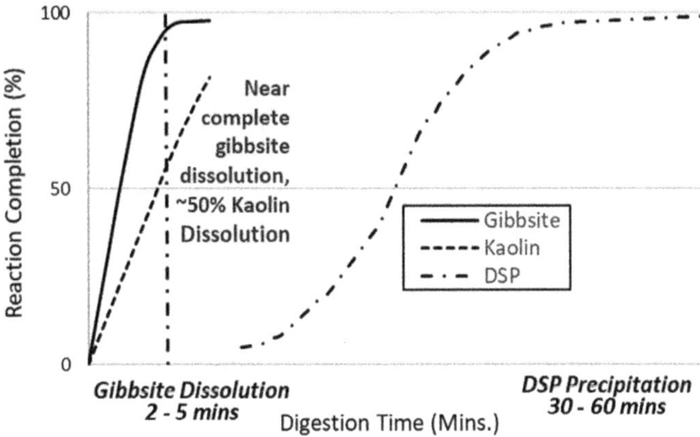

Fig. 8.1 Principle of differential alumina and silica dissolution rates [13]

routine operation with this process. However, it should be noted that the alumina extraction efficiency is slightly lower with this system, clarification of the liquor can be problematic, and due to the temperature required, the process is unsuitable for boehmitic bauxite. A simple arrangement to illustrate the process is depicted in Fig. 8.2.

DSP precipitation is also a significant element of every refinery's sulfate carbonate and chloride, (amongst other minor) impurity balances. These ions are intercalated in the sodalite lattices, or in both the cages and channels of cancrinite. Because

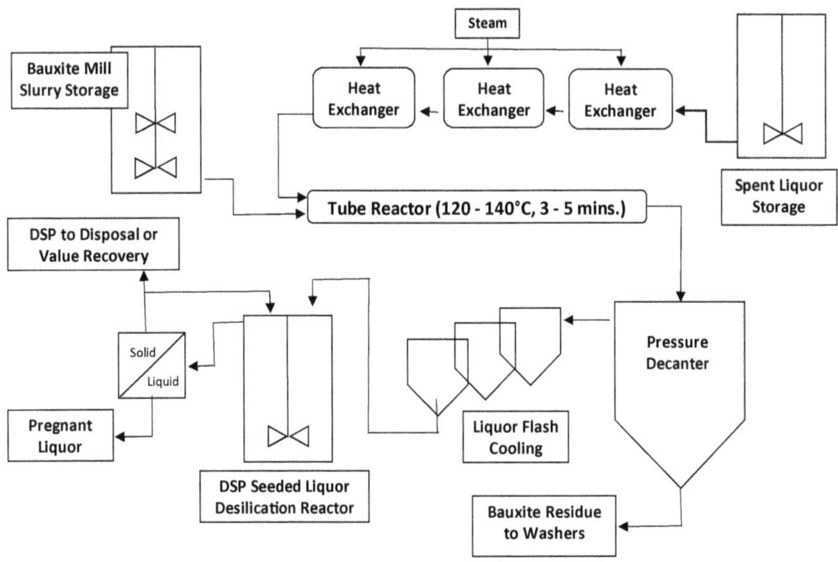

Fig. 8.2 Simple arrangement for differential dissolution

the sodalite structure appears to template around the sulfate anion, sulfate is the favoured anion for inclusion [14]. For this reason, DSP precipitation is a primary impurity removal process for these anions. Where bauxite reactive silica is higher, other removal processes may not be required, or installed at lower capacity because of the use of low reactive silica bauxites.

Some refineries have deliberately increased their bauxite reactive silica as an impurity management strategy, notwithstanding its effect on caustic consumption [15]. While low caustic consumption is an economic advantage, very low reactive silica bauxites can result in higher liquor impurity concentrations, with the attendant loss of liquor productivity.

When silica (typically as precipitated DSP) accumulates in the precipitation area's gibbsite seed inventory above acceptable concentrations, it can lead to uncontrolled silica precipitation and product above the silica specification. The cause is usually some combination of high silica supersaturation and tanks that are poorly suspended and/or left off-line to cool for too long. The remedy can be expensive, sometimes requiring the re-digestion of a large proportion of the gibbsite seed inventory, with the associated production loss. For this reason, monitoring and good management of liquor desilication and cleaning and maintenance of precipitators is a key control measure.

8.7 Scalping Precipitation

Within the first hour(s) of the Bayer precipitation circuit's residence time, many product impurities which cross from the red side to the white side either dissolved, as a colloid, or suspended in pregnant liquor, are co-precipitated or otherwise incorporated in the growing or agglomerating gibbsite seed. In doing so, these impurities are removed from the liquor stream, and no longer contribute to impurities included in subsequent gibbsite precipitation. This mechanism can be employed as a removal process if the solid product from this early precipitation is removed and sacrificially re-digested or becomes a higher impurity product side-stream. If the high impurity gibbsite is re-digested, there is an undesirable loss of liquor productivity, while alternative higher impurity products require profit margins that compensate for the additional cost of the process.

This process has been studied and has found limited practice. There were no easily found and referenced studies in the open literature (although the determined search may find some), but for a given pregnant liquor impurity profile and precipitation flowsheet, the flowsheet requirements, economics and product quality implications can be determined with some relatively simple laboratory work. Because refineries are usually designed to produce acceptable quality product (and despite significant quality excursions), their product will generally find a market, the scalping process is of more interest for low impurity hydrate or alumina products where price premiums apply. Needless to say, for metallurgical alumina production, it is preferred that product impurities can be maintained without this additional cost and complexity.

8.8 Carbon

8.8.1 *Carbonate*

Carbonate is a major and universal impurity in Bayer refinery liquors. It is principally from 3 sources; a product of the decomposition of bauxite or other organics; as mineral carbonates with bauxite; or formed when atmospheric carbon dioxide is dissolved into dilute refinery liquor. The quantitative contribution of each of these mechanisms to the carbonate balance vary with bauxite carbon (organic and inorganic) and digestion conditions, and with the residue washing, thickening and storage technologies and operating practices. Organic and other process additives can also contribute carbonate, but since the adoption of synthetic flocculants these are generally minor.

Without a means of removal, carbonate can accumulate to high concentrations, limiting liquor productivity by increasing total sodium (or ionic strength) and limiting the optimum hydroxide/ aluminate concentration. As with other impurities which grossly limit productivity, carbonate removal and replacement with hydroxide ions (caustic soda) is necessary to maintain Bayer liquor productivity.

Sodium carbonate has quite high solubility even in strong Bayer liquors, and unless at higher concentrations, supersaturation by evaporation is required for its precipitation. To remove carbonate by precipitation without evaporation, a cation (or cations) is required to form a much less soluble salt than Na_2CO_3. For reasons of cost and availability, and because it results in other beneficial reactions, calcium from slaked lime ($Ca(OH)_2$) addition is the industry preferred choice, although not without complications. Because aluminate is available to form less soluble calcium aluminates, the efficient precipitation of calcium carbonate by slaked lime addition requires optimum conditions.

Although many options have been studied to limit organic and inorganic carbon in refinery bauxite feeds, their cost means that only simple ones, such as selective mining practices, wet or dry screening or hydrocyclone fines separation are often applied. Various forms of bauxite roasting, for example, are effective for organic carbon removal, but have high energy and capital costs, and may result in mineral transformations which may reduce alumina recovery in digestion [16, 17].

Some bauxites are relatively high in inorganic carbon (usually as calcium carbonate), which is dissolved under digestion conditions. Calcium associates with a range of available anions (including aluminate) forming compounds with a lower solubility than calcium carbonate such as for example, TCA ($Ca_3Al_2(OH)_{12}$), apatites ($3Ca_3[PO_4]_2 \cdot Ca[F, Cl, OH]_2$), and at high temperature, perovskite ($CaTiO_3$). Equation 8.3 shows the formation of TCA, a common and successful competitor to carbonate for the calcium ion.

$$3CaCO_3 + 2NaAl(OH)_4 + 4NaOH \rightarrow Ca_3Al_2(OH)_{12} + 3Na_2CO_3 \qquad (8.3)$$

Where a refinery has a supernatant liquor decanted from residue and progressively expressed with its consolidation or washed by rain from its surface to a catchment lake, atmospheric CO_2 is dissolved into the resultant weak liquor. Carbonation of the hydroxide ion (and loss of causticity) is the undesirable result (8.4).

$$CO_2 + H_2O \leftrightarrow H_2CO_3 \leftrightarrow HCO_3^- + H^+ \leftrightarrow CO_3^{2-} + 2H^+ \tag{8.4}$$

If this weak liquor is recycled, this carbonate becomes an input into the main refinery liquor inventory. The return of this carbonated weak liquor to the refinery has marginal economics, and treatment and environmental discharge is also commonly used as a carbonate (and other impurity) management practice (see previous comments on liquor purge and neutralisation).

CO_2 has low solubility in the high ionic strength and temperatures of strong Bayer liquor, and coupled with the low liquor surface area exposed to the atmosphere in refinery vessels, atmospheric carbonation of strong liquors in the refinery is very minor.

Managing carbonate to a target concentration is usually achieved through a combination of losses with DSP, liquor losses with bauxite residue, and a causticisation or carbonate precipitation and removal process. Most practiced carbonate removal processes are side-stream, although some refineries with relatively low carbonate inputs and high DSP losses (from high reactive silica bauxites) occasionally practice 'inside' causticisation, which was a more common practice prior to development of side stream processes. Some refineries with high DSP or liquor losses run their side-stream causticisers in campaigns or at lower than design capacities.

The adoption of press filters for residue reduces the volume of weak liquor disposed to less than 30 wt.%, which is close to or above bauxite residue's long-term consolidated density. Where employed, the volume of dilute liquor expressed from residue consolidation is substantially reduced, if not practically eliminated. While primarily installed to facilitate better residue management, the adoption of press filters significantly reduces the opportunity for atmospheric carbonation and can be considered one part of a carbonate and caustic consumption reduction strategy. However, it also significantly reduces liquor purge with associated impurities (including carbonate), increasing accumulation. The net result of these opposite effects needs to be determined for every case but may often result in increased liquor carbonate and other impurities.

8.8.1.1 Causticisation

Causticisation' (or 'Caustification') in a washer overflow liquor is one of the two most commonly practiced process to remove carbonate from Bayer liquor. Its objective is the precipitation of calcium carbonate ($CaCO_3$) by reaction of liquor carbonate with calcium hydroxide ('slaked lime' or $Ca(OH)_2$) under optimum conditions. Equation 8.6 is a simple expression of what would happen in the absence of aluminate and other interfering ions, but this simple reaction is not favoured in Bayer liquor.

$$Ca(OH)_2 + Na_2CO_3 \rightarrow CaCO_3 + 2NaOH \quad (8.5)$$

The chemistry of causticisation determines that at low carbonate concentrations these processes work less efficiently in terms of absolute carbonate removal, lime efficiency and alumina consumption. Conversely, Na_2CO_3 removal rates as high as 20–30 kg of per tonne of alumina are not unusual in side stream causticisation where the bauxite being processed has low extractable alumina and high organic and/or inorganic carbon content [5].

Side-stream causticisation relies on the precipitation of calcium carbonate following the addition of calcium hydroxide, but the aluminate ion complicates this reaction. It has been demonstrated [18], 19 that the reaction proceeds first by the rapid formation of a quaternary calcium aluminate species (hydrocalumite), either as the hemi-carbonate or monocarbonate, as shown below:

$$4Ca(OH)_2 + 2Al(OH)_4^- + CO_3^{2-} \rightarrow Ca_4[Al(OH)_6]_2 \cdot CO_3 + 4OH^- \quad (8.6)$$

The hydrocalumite intermediate then goes on to form primarily calcium carbonate or TCA, with the outcome depending strongly upon the reaction conditions.

$$Ca_4[Al(OH)_6]_2CO_3 + 3CO_3^{2-} \rightarrow 4CaCO_3 + 2Al(OH)_4^- + 4OH^- \quad (8.7)$$

$$3Ca_4[Al(OH)_6]_2 \cdot CO_3 + 2Al(OH)_4^- + 4OH^- \rightarrow 4Ca_3[Al(OH)_6]_2 + 3CO_3^{2-} \quad (8.8)$$

The combination of (8.6) and (8.7) gives the net causticisation reaction, while (8.6) and (8.8) gives the competing TCA formation reaction.

Side-stream causticisation uses the clarified overflow from one of the mud washers as the carbonate source and reacts it with milk of lime in a stirred reactor at typical temperatures of between 90 and 105 °C, for up to several hours. Typically, the efficiency of this process is not high, and lime efficiencies (the lime that actually removes carbonate, expressed as a percentage of the lime added) of 45–55% are common. The remaining lime is either converted to TCA or remains unreacted within the core of some lime particles.

There are 3 equipment elements utilised for common side-stream causticisation. A shell and tube heat exchanger or a contact heater heated by live steam is required to heat the washer overflow liquor to above 95 °C (for highest efficiency). A Continuous Stirred Tank Reactor(s) for residence time of 1 h or more, and a thickener(s) to separate calcium solids from the causticised liquor. Reactors may be in parallel or series, and reactors and settlers arranged to allow taking offline for cleaning and maintenance while allowing the facility to continue operating. A simpler (and lower capital cost) version is a single reactor and settler, and this arrangement is depicted in Fig. 8.3.

Causticisation facilities are common technology in Bayer refineries and their optimum operation widely understood. If designed well and process targets met and

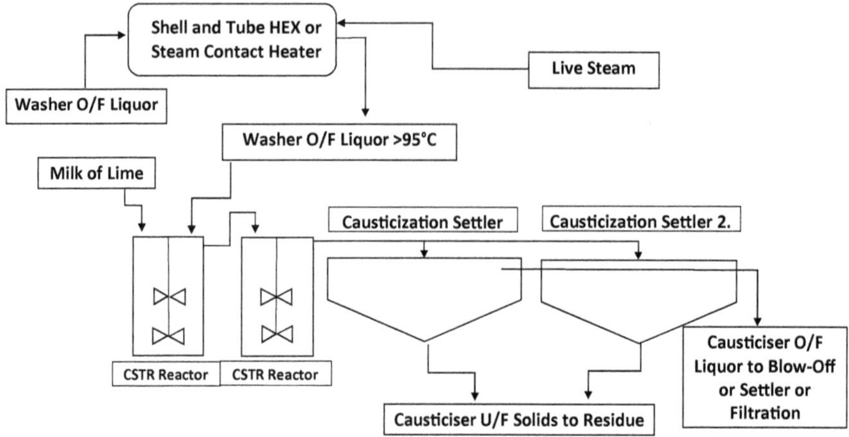

Fig. 8.3 Simple side-stream carbonate process

maintained, if equipment is kept clean, and flows and concentrations kept steady, the results are predictable and economically attractive.

Prior to the widespread adoption of "side-stream" causticisation, "inside" causticisation, or the addition of burnt or slaked lime to the bauxite mills or injected into the digestion feed stream, so it is present during digestion, was a common practice. Although some causticisation may occur as a by-product of lime addition for other purposes, it is no longer widely practiced for the purposes of efficient bulk carbonate removal.

8.8.1.2 Recent Causticisation Improvements

Poor causticisation efficiency affects refinery productivity directly, through alumina losses with TCA, and indirectly via a lower than optimal caustic concentration. Improved causticisation efficiency can be achieved by increasing the temperature of the causticisation reaction, which increases the rate of conversion of hydrocalumite to calcium carbonate, minimizing the opportunity for TCA to form, as well as shifting the calcium aluminate/calcium carbonate equilibrium to higher C/S values.

Improved causticisation processes have been patented and industrialised in Western Australia by Alcoa [20] and Worsley Alumina [21]. Alumina losses due to TCA are small in these facilities, due to the high lime efficiency (>90%).

Roach et al. [18] showed that the achievable carbonate concentration in this system can be predicted from the equilibrium between TCA and $CaCO_3$ in the liquor, as a function of the S (or Total Alkalinity) concentration.

By performing causticisation at high temperature, for a short period and separating the solids before, or immediately after cooling, a high carbonate removal and lime utilisation efficiency can be achieved. A patent based on this concept [20], was industrialised. Figure 8.4 illustrates a simple equipment arrangement to deliver the patent's claims.

Rosenberg et al. [19] demonstrated that the addition of suitable complexing agents such as gluconate or sucrose could substantially reduce the rate of TCA formation and allow a new pseudo-equilibrium to be achieved, allowing both higher C/S and lime efficiency to be achieved. The relationship between C/S and S, both with and without a complexing agent (TCA inhibitor), is shown in Fig. 8.5.

A 2 stage causticisation process was patented by Rosenberg et al. [21]. Very high causticisation efficiency is achieved by preparing hydrocalumite under low temperature conditions where its formation is favored over TCA. By then taking

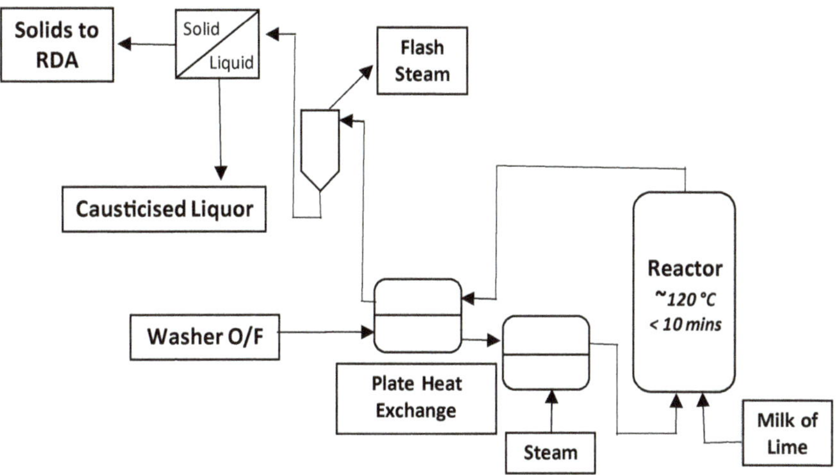

Fig. 8.4 Simple expression of Alcoa 'Equilibrium' Causticisation [20]

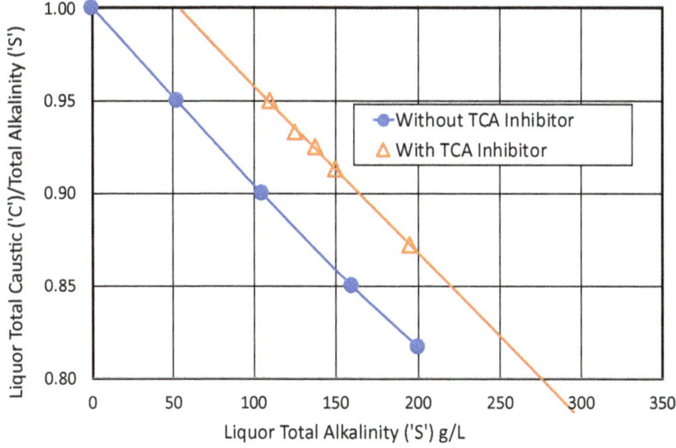

Fig. 8.5 Equilibrium C/S showing the effect of a TCA inhibitor (from [19])

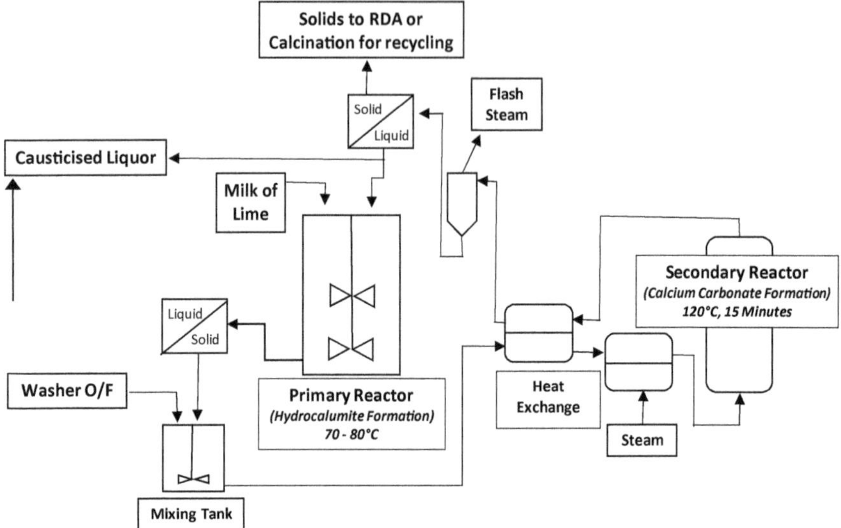

Fig. 8.6 Preferred embodiment of the Worsley 2 stage causticisation process [21]

these solids to high temperature with washer overflow liquor for a short period before separating the solids from the causticized liquor, the $Ca(OH)_2$ is used highly efficiently, and very high causticities are achieved. This process has been successfully industrialised, although no public disclosures of its performance or process details could be found. Figure 8.6 shows the patent's preferred embodiment.

These processes may be more widely adopted after expiry of patents and confidentiality agreements, and a wider understanding of these processes in the global industry established. Given the high specific organic carbon inputs with Western Australian bauxites, as with liquor burning, the economics of these innovative processes may be more difficult to justify where lower impurity bauxites are consumed.

The study and proposal of barium compounds for the removal of carbonate and other impurities has been regularly revisited in the literature and patent history of alumina producing nations and companies and may have found limited practice historically. As a more recent example, a Pechiney patent by Mercier, Magrone and Dealbriges [22], proposes the use of barium hydroxide or aluminate to remove carbonate, sulfate, vanadate and phosphate. This process suggests the recycling of barium after calcination with an alumina source at around 1000 °C. Issues common to barium processes, the cost of suitable barium reagents and/or the high energy cost in recycling by calcination, its toxicity and its mobility in the Bayer process, have inhibited its common practice.

8.8.1.3 Salting Out

Evaporation of spent liquor to high ionic strengths results in sodium carbonate precipitation as anhydrous sodium carbonate, and this has also been quite commonly practiced for carbonate removal. Where sodium oxalate is present and sufficiently above its solubility and metastability limit, it also can be removed. Processes where carbonate, oxalate and sodium fluoro-phosphate and fluoro-vanadate are precipitated and removed have been often reported [23], 24. Because of the relatively low carbonate yield, requiring large spent liquor flows to be evaporated, causticisation is more usually practiced where there are high carbonate process inputs.

These are mostly based on "salting out" in which the liquor to be purified is deep evaporated to an 'S' or 'TA' (Total Alkalinity) concentration of typically 450–550 g/L, in a multi-effect or falling film evaporator, and the precipitate that forms is separated in a thickener or filter, or both. The removal of oxalate is often another target of this process, in which case the cake will also contain significant amounts of sodium oxalate and some organics. The organics are typically humates which results in a viscous sticky cake, and can cause filtration problems, with poor cake separation and blocking of the discharge chutes [25]. Where sulfate is present, a double salt such as Burkeite ($Na_2CO_3 \cdot 2Na_2SO_4$) may precipitate [5].

Cooling liquor to temperatures below those in the Bayer process, to increase supersaturation of a number of sodium salts, including carbonate and sulfate has been studied. Ostap and Bartok patented a process for carbonate and sulfate removal at temperatures down to -15 °C [26]. The carbonate solubility measured in Bayer liquor claimed in the patent is reproduced in Fig. 8.7.

With all salting out processes, the productivity depends on the liquor composition, and the process economics depend on this productivity and efficient separation of salts and liquor. In the case above of cooling, the energy needed to lower and again raise liquor temperatures might normally be a prohibitive cost, and where this is provided at low cost from the environment or as a by-product of nearby industrial processes, the overall economics may be transformed.

Many processes for removing carbonate have been studied over the history of the Bayer Process. A process referred to as "drowning out" (Malito and Rogers [27]), requires the addition of a suitable organic solvent, such as methanol or propanol, which results in precipitation of carbonate, oxalate and other salts. Due to poor solvent recovery and the risks with flammable vapors, this process has never been implemented.

8.8.2 Oxalate

Of all Bayer process impurities, few attract more attention than the simplest dicarboxylate. While some oxalate enters the Bayer process with bauxite as oxalic acid adsorbed on various minerals [28], the majority is created by the oxidation and

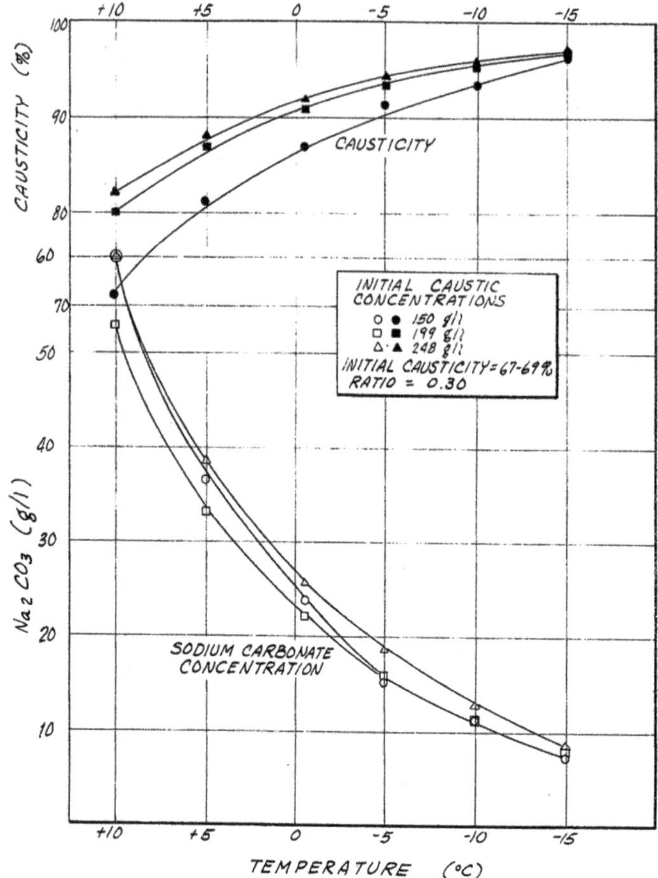

Fig. 8.7 Effect of temperature on Na_2CO_3 concentration and causticity of refinery liquor of between 150 and 248 c/L caustic [26]

pyrolysis of bauxite organics under Bayer digestion conditions. The rate of its formation in digestion is principally related to the bauxite organic carbon concentration and composition, and to the conditions of time, temperature and the presence of oxygen and catalytic compounds under which they are processed. Mechanisms for this and other organic degradation reactions have been proposed and documented (e.g. Costine et al. [29]).

Beyond the extraction and breakdown of organics in bauxite through digestion, the liquor organics inventory continues to produce more oxalate with each digestion cycle, as demonstrated by Power and Tichbon [30]. The contribution of this breakdown to oxalate formation depends on the temperature and concentration of the organics and other factors. The breakdown of other organic additives (flocculants, lubricants, defoamers, dewatering aids) can also produce oxalate and contribute to the circulating concentration.

Table 8.3 Some oxalate generation rates from a range of bauxites [30]

Bauxite source	Oxalate generated			
	$Na_2C_2O_4$/t dry bauxite (kg)		$Na_2C_2O_4$/t Avail. Al_2O_3 (kg)	
	143 °C	230 °C	143 °C	230 °C
Arkansas	0.5	0.5	1.4	1.4
Suriname	0.6	0.9	1.2	1.8
Jamaican	0.4	1.9	0.9	4.1
Gove	1.5	1.6	3.3	3.4
Weipa	1.5	1.6	3.3	3.1
Kimberleys	0.4	0.5	1	1.3
Darling Range	0.9	0.6	2.7	2.1
Wagina Island	1.1	2.0	2.7	4.7
Jamaica	0.6	0.8	1.4	1.7
USSR 1 (diaspore)	0.1	0.2	1.7	0.5

Power and Tichbon reported some oxalate generation rates for a range of bauxites at low (143 °C) and high temperature (230 °C) digestion conditions [30]. The results (Table 8.3) are also expressed as impurity inputs/t of available alumina, a more useful value on which to make comparisons between bauxites. These generation rates are useful in closing refinery mass balances and predicting the required oxalate removal capacity for a given bauxite feed. They can be determined relatively simply by laboratory digestion tests.

Being relatively high in organic carbon, most lateritic bauxites require some level of oxalate management. Karst bauxites, however, often do not accumulate oxalate to a level requiring a dedicated removal process. Some lateritic bauxites, for example Trombetas from Brazil, have very low organic carbon (< 0.05%) and generate commensurately low levels of oxalate. Where refineries process them exclusively, or in considered bauxite blends, they can avoid reaching the oxalate precipitation concentration in the refinery precipitation area, often referred to as the 'critical oxalate concentration', or 'COC'.

The need to rigorously manage oxalate is prompted by its effect when co-precipitated with gibbsite in the precipitation area. Its subsequent impact on gibbsite nucleation, and the implication of this effect on alumina product size and size stability is a major inconvenience in refineries. This precipitation can occur in a controlled way, or despite rigorous measures to prevent it.

8.8.2.1 Oxalate Removal

Beyond the soluble losses with residue and product, the management of sodium oxalate in alumina refineries generally requires its removal from process liquor by precipitation (8.9).

$$2Na^{+}(aq) + C_2O_4^{2-}(aq) \rightarrow Na_2C_2O_4(solid) \quad (8.9)$$

The main strategies for removing sodium oxalate from refinery liquors are (1) crystallisation from a liquor side-stream or (2) co-precipitation with gibbsite in the precipitation circuit, then dissolution in hot condensate from deliquored gibbsite seed. The choice between these approaches is sometimes explained as related to the magnitude of oxalate to be managed or product qualities targeted, but in practice seems more related to precipitation area design heritages. There are examples where different organisations use different approaches for the same bauxite, whether higher or lower in oxalate or organics. Each refinery owner may consider their oxalate management choice is the most cost effective and produces the best alumina product quality. Certainly, both can be effective, and pros and cons can be argued for each.

Oxalate can be converted to carbonate under high temperature Wet Oxidation conditions [31], but because it is applied to reduce overall organics, it will be examined in that section.

Oxalate Stabilisation

Sodium oxalate is stabilised in liquor above its thermodynamic solubility by the adsorption of Bayer organics and/or surfactants onto existing sodium oxalate seed and to forming nuclei. These adsorbents are sometimes called "Oxalate Active Organics" (OAOs). The oxalate precipitation rate under the effect of OAOs can be so slow that it appears the solubility equilibrium has been reached. The concentration at which this occurs has been referred to as the "apparent oxalate solubility", but is more accurately a metastable supersaturated state, where precipitation kinetics are slowed by orders of magnitude due to seed deactivation.

This Apparent Solubility or metastability limit is a function of liquor supersaturation relative to the thermodynamic sodium oxalate solubility, but also the surface area of oxalate seed, and the humate and other seed deactivator concentrations. For typical Bayer residence times, metastability is reduced by removal of the adsorbing species, increased oxalate supersaturation, and/or increased active seed surface area. This effect is seen in both in oxalate co-precipitation and side stream removal. In both cases supersaturated liquor oxalate is inhibited from precipitation by the lack of active seed surface and will only begin to precipitate after reaching a critical oxalate concentration (COC) or supersaturation.

Atkins and Grocott demonstrated this effect in evaporated spent liquor. Evaporation beyond liquor specific concentrations results in increased surfactant concentrations and related oxalate metastability. Figure 8.8 shows that oxalate removal is most effective at the curve's minimum. Further evaporation results in decreased oxalate removal, and the curve minimum was shifted down and to the right with decreasing humate concentration [32].

When sodium oxalate is precipitated from liquor, these OAOs are adsorbed on its surfaces, which is greater at the increased evaporated liquor concentrations

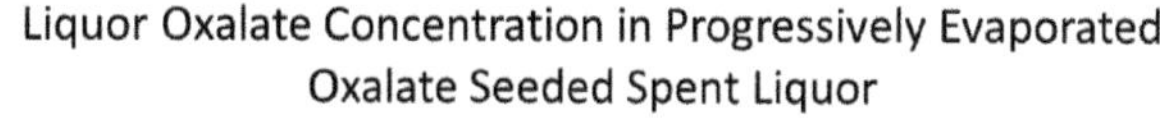

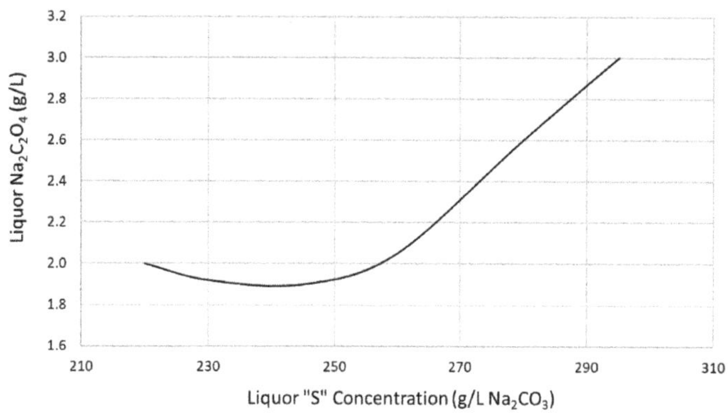

Fig. 8.8 Oxalate stability in evaporated liquor [32]

and increased OAO concentration. Consequently, the precipitation of oxalate also becomes a removal process for oxalate precipitation inhibitors.

To reduce this oxalate stability, Poly-DADMAC (poly diallyl dimethyl ammonium chloride), a water-soluble, cationic, polymeric quaternary ammonium compound, can be used to remove humates by addition to mud settlers or washers. The resin formed by reaction with humates adsorbs on residue and is disposed with it [33]. Poly-DADMAC use is relatively costly, but for refineries with a large oxalate and humate input and/or where product color is important (as with specialty hydrate applications), it is sometimes used. Poly-DADMAC exchanges humates for chloride ions, increasing chloride in liquor.

Ion exchange resins may also be used, but are relatively expensive in capital, consumables and cleaning and maintenance terms. The real cost of resins has been dropping over time and they may at some point be a more economically interesting option.

Williams and Perrotta [34] demonstrated the removal of oxalate poisons with various adsorbents, including layered-double hydroxides, magnesia, activated carbon and ESP dust, with varying success. While technically feasible, for cost and complexity reasons the use of adsorbents for oxalate destabilisation is not widely practiced at an industrial scale.

Side Stream Removal

Spent liquor evaporation before and/or after separating the side-stream to increase oxalate supersaturation is a general requirement of side-stream oxalate removal. A sub-stream of spent liquor large enough to yield the required removal rate is separated from the main flow to a reactor with sodium oxalate seed. The degree of

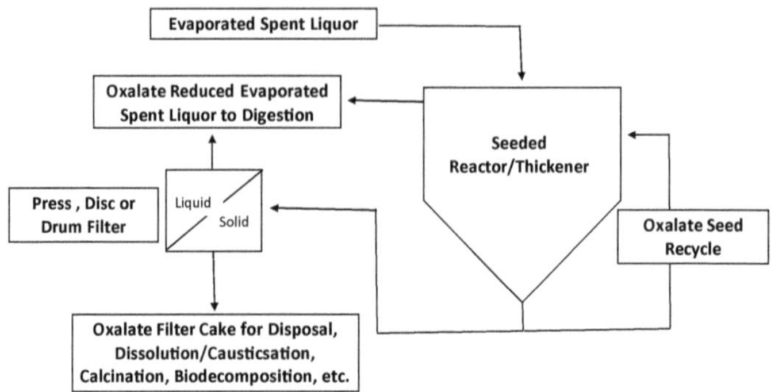

Fig. 8.9 Simple oxalate reactor/thickener arrangement

liquor evaporation varies from process to process, depending on a range of design and operating factors. Apart from the oxalate stabilization discussed above, excessive evaporation results in viscous liquor and often the precipitation of other fine salts, rendering separation by thickening and/or filtration inefficient, limiting the liquor concentration target.

Different types of reactors can be used for the oxalate precipitation. In one approach, the reactor is also a solids retention classifier, where precipitation of oxalate coarse enough not to overflow is key, otherwise a large part of the oxalate precipitated can find its way back to digestion with the spent liquor, resulting in much lower removal rate than that precipitated. A simple reactor/thickener arrangement is pictured in Fig. 8.9.

Delivering process conditions to achieve a sufficiently coarse oxalate seed inventory is an important element of process design and operation in most of the successful oxalate reactor systems industrialised to date. Where seed is filtered from treated liquor (possibly after thickening), high filtrate solids may also result in inefficient removal. The improved performance and reduced cost of press filters may replace other modes of filtration in future due to their relatively low cake moistures and filtrate solids.

Ultrasonic waves in a pipe reactor [35] were reported to overcome oxalate stabilization by Sonocrystallisation through local cavitation. Bubble collapse creates high localized temperature and pressure, causing oxalate nucleation. Pilot trials showed a reduction in the oxalate concentration from 2.5% w/v down to 1.4% w/v, but the filed patent was left to lapse, suggesting that the process was not reliable and/or economic [36].

Salting-out, another common means of removing oxalate from a side-stream, was discussed previously for carbonate. Some aspects of its use for oxalate removal were described by Gnyra and Lever [37], and the complexity and compromises that need to be made in its routine operation were discussed by O'Connell et al. [38]. Approaches for oxalate precipitation based on liquor chilling or "drowning out" have been described [39], but energy costs and solvent recovery make these methods' commercialisation unlikely.

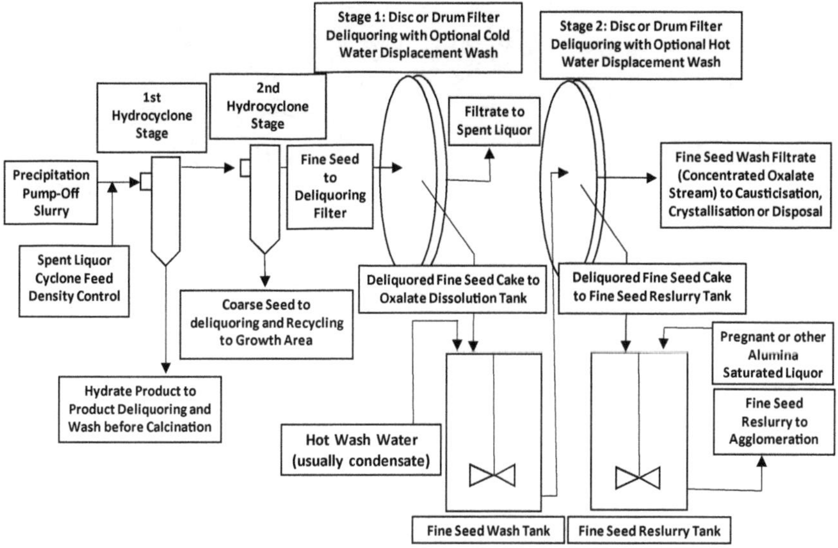

Fig. 8.10 Simple seed classification and fine seed wash arrangement

Co-precipitation and Seed Wash

In co-precipitation, sodium oxalate is precipitated with gibbsite in the growth phase of precipitation. Due to the increased gibbsite nucleation rates caused by solid oxalate, seed classification (into fine, coarse and product size fractions), fine seed deliquoring and washing (where oxalate is dissolved) and an agglomeration phase are generally required to keep the circuit and product suitably coarse or 'sandy'. Solid oxalate is avoided in the agglomeration phase due to its impacts on product quality, including oxalate occlusion [40]. The paper by Power and Tichbon is a good source for the effect of oxalate on the precipitation process [30], while Reyhani et al. demonstrate oxalate's ability to seed gibbsite nucleation [41].

When precipitation gibbsite seed is classified, whether in gravity seed thickeners or hydrocyclones, solid phase sodium oxalate is usually concentrated with fine seed. Following classification, fine seed is deliquored on a filter (usually disc or drum), to remove as much associated spent liquor as possible.

Washing the deliquored cake follows (usually in hot condensate) to dissolve the sodium oxalate. The lower the residual liquor in the cake before washing, the higher the achievable oxalate and lower the caustic concentrations in the seed wash water. This high oxalate/low soda (and alumina) composition facilitates more efficient calcium oxalate precipitation by lime addition, both in terms of oxalate removed and lime consumed. Where the concentrated oxalate stream is recrystallized, sent to residue or biologically treated to convert oxalate to carbonate, processes are more efficient, and soda and alumina losses are minimised by the higher oxalate to caustic ratio. A simplified classification and seed washing scheme is pictured in Fig. 8.10.

Fig. 8.11 **a** Sodium oxalate needles, and **b** precipitated in ball morphology [40]

The oxalate-rich solution resulting from washing the deliquored seed, is either sent directly to an oxalate disposal step, or recrystallized in a side-stream process. Oxalate size and morphology, as well as the gibbsite size distribution can inhibit the deliquoring step, leaving the cake high in strong liquor, and resulting in excessive water being required to dissolve the oxalate, and producing a low oxalate to soda ratio in the wash liquor. This has negative implications for oxalate causticisation efficiency and on the efficiency of any secondary oxalate crystallisation step.

Oxalate Balls

Sodium oxalate's crystal morphology is altered by Bayer liquor organics. Surfactants can also affect oxalate crystal morphology, changing its form when precipitated from Bayer liquor. Depending on oxalate supersaturation, the concentration and morphology of any oxalate seed present, and the concentration and type of surfactants present, a range of morphologies between very fine needles to networks of needles to spheres (balls) can be precipitated. Figure 8.11 illustrates a common needle morphology precipitated from Bayer liquor and spherulitic oxalate.

Roe, Owen and Jankowski of Nalco Chemicals reported the effect of Crystal Growth Modifiers (used primarily to coarsen gibbsite product in Bayer precipitation), on oxalate morphology [42]. Nalco (now Ecolab) have since done extensive development work to optimise products for oxalate ball formation and their products and many publications on the subject are a good starting point for those interested in examining this approach to oxalate removal [43].

Using this effect to produce easily settled or filtered coarse or spherical oxalate crystals, improving removal efficiency, was explored by Keeney and Grocott [44], and also by Morton [45].

In side-stream removal, the objective is to produce oxalate balls and separate them by filtration (possibly after thickening). The advantage is in improving the separation efficiency of the usually fine sodium oxalate needle precipitate, which is difficult to thicken or filter. The addition of polyacrylate to a salting out process where oxalate is precipitated is known to improve filterability of the precipitated salts, and this can often be attributable to better oxalate morphology.

In a co-precipitation environment, these balls need to be coarse enough to be separated from precipitation liquor and gibbsite seed by screening, typically requiring particle diameters of >500 μm. This can reduce the need for seed washing, a relatively costly and equipment intensive process step. Oxalate ball formation and removal was successfully practiced for more than 30 years at the Sherwin refinery in Texas, USA.

A suitable refinery flowsheet and liquor chemistry (particularly organics) is an important prerequisite for success. Ongoing development of the additives and experience may see this approach become more widely practiced due to its simplicity and low capital cost.

Oxalate Disposal

Sodium oxalate is toxic, with a lethal oral dose in human adults of between 15 and 30 g [46]. It is also chemically stable. Its disposal as a solid cake in residue areas is prohibited in a number of jurisdictions. Whether sent with residue to storage as a concentrated liquor stream or as a solid, it can be mobilized and recrystallize with other salts at the surface of stored residue due to capillary action, where it can then contribute to the toxicity of airborne dust raised from residue areas. Rain on residue can dissolve it and this run-off can join residue decant and under-drainage return to the refinery or enter the local environment in managed or accidental discharges. For the above reasons, secure disposal or increasingly, conversion to carbonate by calcination or bacterial digestion, is often practiced. Calcium oxalate is also toxic, but its disposal with the main residue stream seems to be more generally acceptable, and more widely practiced.

Roasting

One practice has been to roast sodium oxalate cake to convert it to sodium carbonate (8.10):

$$Na_2C_2O_4 + \frac{1}{2}O_2 \xrightarrow{>280\,^{\circ}\mathrm{C}} Na_2CO_3 \quad (8.10)$$

Roasting is practiced at a higher temperature than the 280 °C, often quoted as the threshold for decomposition, to improve the speed of conversion. The subsequent decomposition of the sodium carbonate produced to sodium oxide can occur at a higher calcination temperature, but the higher energy requirement prejudices economic soda recovery. The calcined solids can be dissolved, causticised and returned to the process, or disposed with residue. The presence of organics in the oxalate cake calcined may result in odour, and together with alkaline dusts have been known to raise concerns in the refinery and local communities [5].

An alternative form of calcination is to add an oxalate cake to liquor burner or SLC feed, where sodium oxalate is first converted to carbonate and then to sodium aluminate to recover the sodium. The high cost of liquor burning and concerns about emissions has limited this option to the few refineries with the technology installed.

Recovery of the sodium values as hydroxide from a concentrated sodium oxalate stream or precipitated cake is possible, but not always operationally or economically interesting. Causticisation with lime is the most widely practiced process for soda recovery, and often as importantly, oxalate immobilization. Sodium oxalate reacts with calcium hydroxide to form calcium oxalate monohydrate (whewellite) (8.11) which is often disposed with the residue stream.

$$Ca(OH)_2 + Na_2C_2O_4 + H_2O \rightarrow CaC_2O_4 \cdot H_2O + 2NaOH \tag{8.11}$$

However, the sodium oxalate is usually presented to process options in fine seed wash liquors, where the presence of aluminate and carbonate results in calcium aluminate and calcium carbonate precipitates [18, 40], resulting in low lime and oxalate removal efficiency.

The causticisation of sodium oxalate is sometimes referred to in the industry as ‘oxalate destruction’. It is not. If after causticisation, the calcium oxalate product is mixed with more concentrated Bayer liquors, calcium compounds less soluble than oxalate under these conditions precipitate, and the oxalate ion returned to liquor by the same chemistry that makes causticisation inefficient. Directing causticisation solids with their associated liquor too high in the Washer train for example (to recover more of the higher causticity liquor) is avoided due to this effect.

The low efficiency of oxalate causticisation inspired the development of a two-stage process [47] where aluminate is first removed from the solution by reaction with lime to produce hydrocalumite according to (8.3). The hydrocalumite is separated by filtration and sent to the carbonate causticiser, while the clarified oxalate-rich solution is treated with lime to produce calcium oxalate monohydrate. Lime efficiencies of up to 80% were claimed for this process, but the process has yet to be industrialised.

Bio-degradation

Where pH, salinity and nutrient availability are supportive, oxalate is naturally and readily metabolized to carbonate and bicarbonate by commonly occurring bacteria. These bacteria have for decades been known to occur and be active around Bayer refineries in accidentally or deliberately remediating oxalate in dilute Bayer liquor in residue areas or surrounding environments.

These bacteria have been employed to treat oxalate-rich, low causticity streams. Brassinga et al. [48] proposed acidification of an oxalate rich stream to bring the pH to a near-neutral range to encourage the bacterial action. Morton adapted the process and bacteria for alkaline solutions by either continuous or batch operation [49].

Fig. 8.12 Continuous biological oxalate destruction unit at Alcoa Kwinana, Australia (IAI website)

A batch system was operated at the Worsley refinery for many years but required constant intervention to operate reliably and was discontinued [5]. A 1993 Alcan International patent indicated an existing operation in a Jamaican refinery [50]. All these processes were batch, utilised oxalate cake that was precipitated from Bayer liquor, relied on conditioning locally occurring bacteria to an alkaline environment and supplementing with nutrients to keep the bacteria active.

Alcoa have subsequently developed and installed a continuous process using Moving Bed Bio Reactor (MBBR) technology, commonly used in wastewater treatment. This facility processes a large proportion of the sodium oxalate produced at several Alcoa sites [51]. Figure 8.12 pictures the reactor and facility. The sodium carbonate and bicarbonate rich stream produced may be returned to the mud washers, with a proportion leaving with residue liquor and the rest causticised in side-stream carbonate causticisation. This development is clearly the most sophisticated and sustainable of all the processes developed to date for oxalate disposal.

Photocatalytic oxidation of sodium oxalate in air and pure oxygen, with and without TiO_2 catalyst, has been studied [52, 53]. The need to acidify to obtain a suitable oxidation potential, and the expense of the catalyst, has prevented industrialization.

8.8.3 Total Organic Carbon

There have been many publications on the origins and reactions of Bayer organics and the complex chemistry and process implications. Power and Loh [54] comprehensively reviewed the literature, providing a good set of references, and summarised the reported process impacts of organics [55]. Liquor organics have multiple and widespread effects. The high concentrations of organic sodium salts impact aluminate solubility and their adsorption on active surfaces can greatly affect reaction rates (particularly in precipitation). Changing the morphology of the various precipitates, and increased product impurity incorporation are other important impacts [4].

HMW organics have an effect on the colder Bayer unit operations due to their strong adsorption on most surfaces. They are also the major parent species from which simpler organics are produced during cyclic liquor re-digestion. While many solutions have been proposed for removing HMW organics, the practical removal and management of detrimental effects have so far been more agricultural.

Adsorbents such as activated carbon [56], magnesia [57], dolomite, hydrotalcite [58], TCA, alumina and other caustic-stable materials [34] have been studied, and used to some extent. These adsorbents are often fine materials to provide high surface area, and consequently difficult to separate after liquor treatment. The material consumption and associated cost is high, may not remove the most detrimental organics, and have very little impact of overall liquor TOC. Some refineries have found it useful, either continuously or in campaigns, but is not common industry practice.

While the extensive work on organics has provided an understanding of their chemistry and balances, most of the work has not informed major breakthroughs in practical removal processes. Apart from those working on new solutions, the deep and detailed Bayer organics literature has not so far informed practical choices in removal or management, which to date is more a matter of choosing the most suitable of the few available options. Apart from liquor purge, and input minimization, the industrially proven, economically interesting removal choices today are wet oxidation or liquor calcination.

8.8.3.1 Minimisation of Bauxite TOC

The reduction of TOC in feed bauxite is an obvious way to reduce liquor organics, and processes in mine planning, mining and beneficiation can make a significant impact on product bauxite from a given mine and are practiced [59].

Where refineries are not committed to a single bauxite supply, buying a lower TOC bauxite is also a simple management strategy and is practiced.

8.8.3.2 Organics Tolerance

When examining organic management options, it usually goes unstated that tolerance has been (and remains), the default and most practiced organic management strategy in the Bayer alumina industry. It seems obvious from the low adoption rate of available technologies, apart from those refineries processing high organic bauxites (such as those in Western Australia), that these options present less compelling economic cases than accepting and managing the various impacts of organics on a refinery's operation and economics.

Operating with high organics (and other impurities) requires a good understanding of Bayer chemistry and good control over operating, cleaning and maintenance practices. Refineries operating under this constraint learn to optimise liquor productivity with a given liquor organic inventory and manage around the other negative implications such as higher equipment scaling rates, liquor boiling point, viscosity, foam, and the usual spectrum of effects reported. Most refineries are designed and/or have a production plan with an assumption of a steady state organic (and other impurity) concentrations. Even when an economically attractive way of removing organic impurities (and improving productivity) is identified, the capital requirement, the risks in technology implementation and the requirement for debottlenecking to realise the improved productivity, means that the option may not be more attractive than simply expanding production with more of the same impurity constrained production.

8.8.3.3 Wet Oxidation

Wet oxidation (W.O.) relies on the injection of oxygen or air into Bayer liquor at elevated temperatures (typically > 230 °C). Wet Oxidation is practiced in Kraft liquors in the pulp and paper industries where there is a long technical history, which has provided some background for the Bayer process application. The few examples of Bayer Process industrialisation have been practiced on digestion slurries in tube or autoclave digestion, or on a spent liquor side- stream (a similar process to that practiced in Kraft liquors). The longest successful application of the technology is at AOS Stade, Germany [60]. The Stade refinery in Germany with its tube digestion technology (in which W.O. is performed) is pictured in Fig. 8.13.

The presence of oxygen at temperature results in breakdown of the larger organic molecules, ultimately producing a range of smaller "refractory" organic compounds (such as acetate and oxalate) and sodium carbonate. Oxygen injection is an augmentation of oxidative processes which occur to some extent under non-oxygen enriched digestion conditions [30]. While wet oxidation proceeds under low temperature digestion conditions, the slower reaction rates and overall organic conversion rates

Fig. 8.13 AOS Stade Alumina Refinery, Germany (Courtesy of AOS Stade)

do not justify the costs and it is not presently practiced. Although catalysed low temperature W.O. (to increase the rate of organics conversion) has been studied [61], no industrialisation has been reported, although it could be argued that W.O. practiced in the presence of bauxite residue is already catalysed to some extent by some of the minerals present.

Oxalate and carbonate created by wet oxidation are managed by the usual processes, at the usual cost. The smaller organic compounds have higher sodium to carbon stoichiometry, so oxidation of larger molecules results in an increased caustic loss [30], 60. The smaller organic compounds (acetate is a good example), are more difficult to break down into carbonate. Concerns have been raised about W.O. in that these more refractory and higher stoichiometric sodium compounds accumulate disproportionately in the spectrum of Bayer liquor organics. Under Stade conditions, where digestion temperatures are among the highest in the industry, the acetate trends reported suggest these are broken down or mechanistically by-passed, even if at a slower rate than larger and less stable compounds [31].

Many of the organic compounds present in Bayer liquors will undergo either base-catalysed oxidation by water or oxygen at elevated temperatures, evolving hydrogen [29, 62]. The explosive risk of an oxygen and hydrogen mix in a closed system at elevated temperatures is an important consideration. Gas accumulation in digestion equipment led to an explosion at Stade in 1982 [60], and since this was reported, the study of W.O. process installation includes the rigorous management of this risk. The oxygen concentration in the Stade process is managed below the explosive limit for H_2/O_2 mixtures.

At least one case of the W.O of a liquor side-stream being industrialised is reported [63]. Using a process developed by ALUTERV-FKI of Hungary, a wet air oxidation unit processing a side-stream of 40 m^3/h of strong spent liquor was commissioned at Hindalco's Renukoot refinery in 1986, after being developed in bench and pilot scale testing. In 1987, the first full year of operation, the liquor total organic carbon TOC) decreased by 20–25%, and liquor humates by ~50%. Although effective in oxidizing liquor organics, the resulting oxalate, carbonate and fluoride precipitate was difficult to filter, and increased oxalate resulting from the oxidation was difficult to manage. Eventually, alternatives to manage organics and their impacts were preferred. Seeding and controlling salt precipitation conditions could in future improve the solids filterability, and given the advances in filtration technology, the option may be revived at some point, subject to the same general costs and constraints of Bayer liquor W.O. The ALUTERV-FKI pilot arrangement for the process is illustrated in Fig. 8.14.

The costs of oxygen, oxalate and carbonate removal and higher organic soda effects of W.O. are traded off against the improved liquor productivity and other benefits. Although W.O. is one of the few operating organics management solutions, the costs, process implications and risks have limited installation to only a few cases.

Alternative oxidants, such as hydrogen peroxide, ozone [37] and manganese dioxide [64], have been studied, as has a submerged plasma torch [65], but cost efficiency, safety and other disadvantages have discouraged industrialisation.

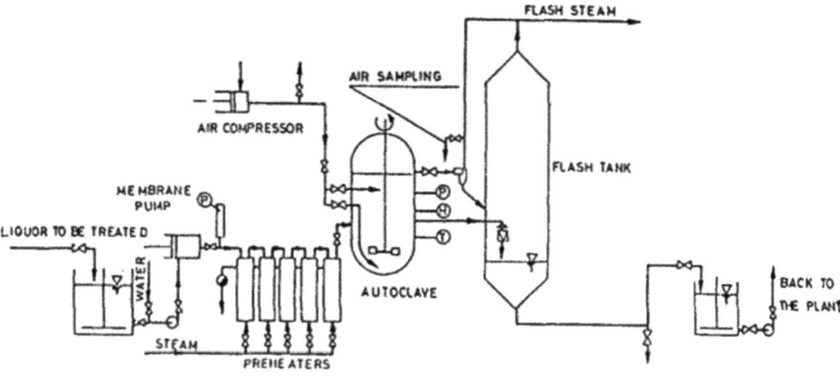

Fig. 8.14 Flowsheet of wet air oxidation pilot designed by Aluterv-FKI [63]

8.8.3.4 Liquor Burning

If a side stream of spent liquor is evaporated to dry solids and calcined to temperatures above 300 °C, the organic salts are largely oxidized into sodium carbonate (for example 8.10). A process commonly referred to as "Liquor Burning" was developed by Yamada and Shibue [66] where the calcination temperature is raised to around 1000 °C and a molar equivalent to the sodium concentration of either alumina or gibbsite is added to the evaporated liquor prior to calcination. Under these conditions, the sodium carbonate that form reacts with the alumina source to produce sodium aluminate. A block flow diagram of the process from the patent is pictured in Fig. 8.15. Aluminate can then be leached from the sinter in liquor, and the alumina rich liquor recirculated, while the leached residue can join the bauxite residue stream. A simplified summary of the chemistry follows in (8.12) and (8.13).

$$Na_2O + Al_2O_3 \xrightarrow{>850\,^\circ C} 2NaAlO_2 \tag{8.12}$$

$$NaAlO_2 + 2H_2O \rightarrow NaAl(OH)_4 \tag{8.13}$$

Liquor burning has been industrialised in Japan, and in Western Australia where the world's largest units operate in Alcoa's Kwinana and Wagerup operations, and the Worsley Alumina refinery.

The Worsley Alumina Liquor Burning process is described by Prinsloo in [67]. The process starts with evaporation of 40 m^3/h of spent liquor from 310 to 560 g/l caustic. The evaporated stream then flows along with a fine hydrate/alumina to a pug mixer with slight alumina excess to the target Al_2O_3:Na_2O ratio. This slurry is fed with dried product from the downstream process to a paddle mixer. This mixture flows into the 800 °C off gases from the downstream rotary kiln. The mixture is entrained with the gas stream through a rotor (cage mill) and dried while going up a vertical drying duct. The temperature in the drier is controlled at 260 °C.

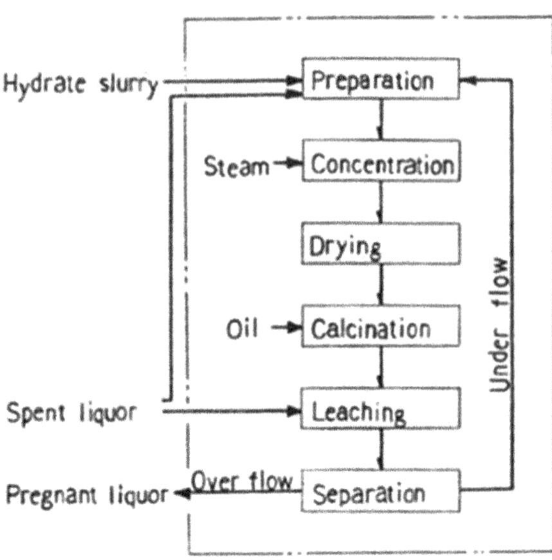

Fig. 8.15 Liquor burning block flow diagram [66]

The cage mill breaks up the bigger lumps that fall back from the vertical drier duct. The dried solids are separated from the gas through a primary cyclone, a multiclone and a 4-cell bag house. The gas stream is released to atmosphere through a fan and a 60-m-high stack. The solids from the underflows of the multiclone, bag house and part of the primary cyclone are directed to the paddle mixer to be mixed with the slurry coming from the pug mixer. The remainder of the primary cyclone underflow is the product stream fed to the rotary kiln. The 58 m long, 3.75 m diameter Krupp-Polysius rotary kiln is fired with natural gas.

The hot solids are air cooled in a fluidized bed, with the cooler air used as secondary air in the kiln firing. After cooling, the (300 °C) solids overflow into a leach tank and are mixed with spent liquor. Worsley liquor contains a high sulfate concentration, which reports as solid sodium sulfate in the Liquor Burner kiln discharge. The conditions in the leach tank prevent sodium sulfate dissolution. The slurry, containing both sodium sulfate solids and the excess alumina/hydrate solids, is fed to four filter presses. The press liquor filtrate is the final product stream returned to the refinery. The filter cake discharges to a dissolver tank where it is slurried with water dissolving sodium sulfate. The resulting slurry goes to filters where the residual alumina/ hydrate is separated. The liquor is pumped to the residue disposal area and the alumina/hydrate solids are mixed with spent liquor and pumped as evaporator feed to the evaporator. The process is summarized in Fig. 8.16.

8.8.3.5 Solid Liquid Calcination

The solid–liquid calcination process [68] adopts the same sodium/alumina chemistry as liquor Burning (8.12), but utilises a fluidized bed calciner, and proposes

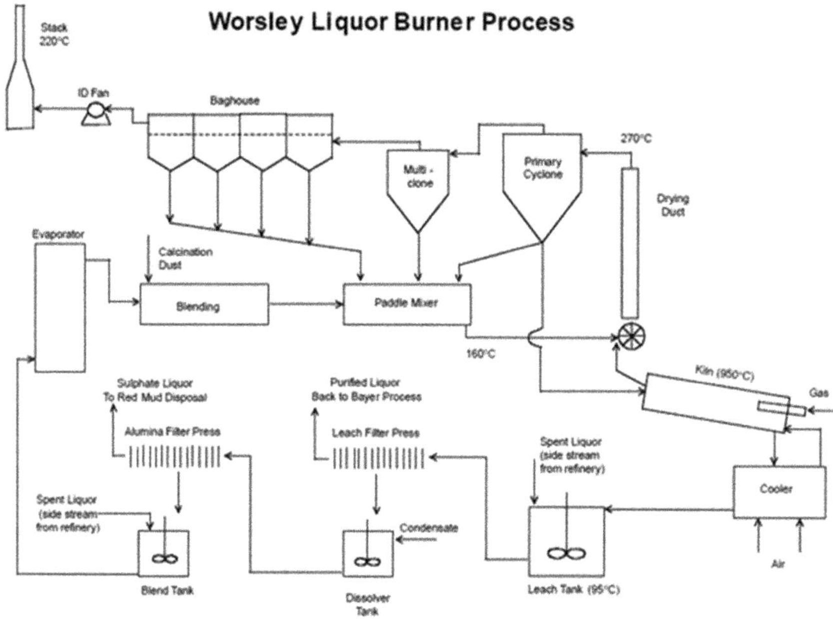

Fig. 8.16 Worsley liquor burning process [67]

the use of bauxite as an alumina source. The iron (as hematite, goethite or other minerals) provides a similar transformation path for soda from organics and carbonate to be causticised (8.14 and 8.15). The economics of the SLC process are potentially enhanced by the production resulting from the sintering of bauxite and soda and subsequent leaching of aluminate. The energy efficiency of the Gas Suspension Drier and Fluid Bed Calciner combination also enhance the economics.

$$Na_2O + Fe_2O_3 \xrightarrow{>850\,^{\circ}C} 2NaFeO_2 \tag{8.14}$$

$$2NaFeO_2 + 4H_2O \rightarrow Fe_2O_3 \cdot 3H_2O + 2NaOH \tag{8.15}$$

The only commissioned industrial SLC unit has operated in the San Ciprian refinery in Spain since the late 1990s. SLC units were constructed in Gove, Australia, but never operated [69]. Figure 8.17 pictures the SLC unit in San Ciprian next to its flowsheet (Table 8.4).

8.8.3.6 Liquor Burning and SLC Emissions

The large Liquor Burning units installed in Western Australia have raised local stakeholder concerns about emissions [5]. Residents close to Alcoa's Wagerup refinery

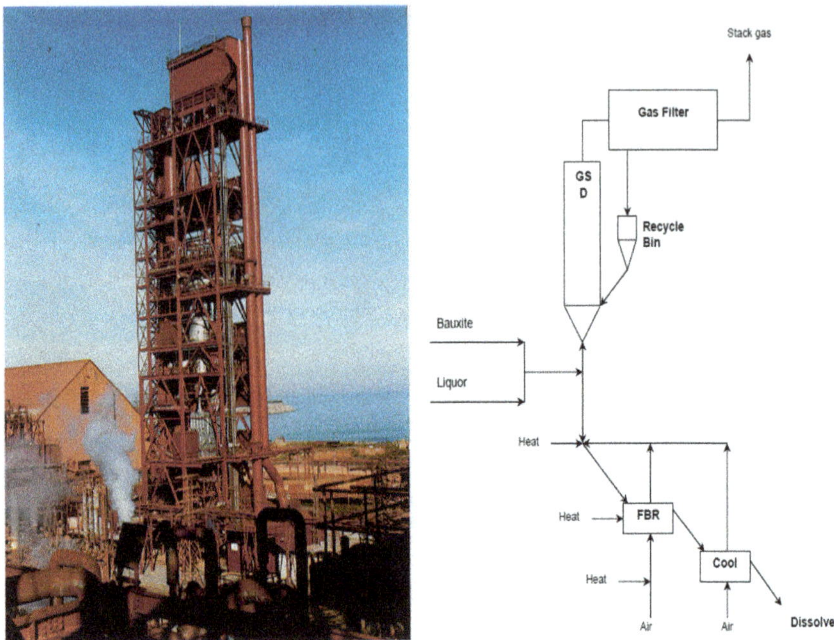

Fig. 8.17 Picture (left) and flowsheet (right) of SLC unit operating in Spain (Courtesy of FLSmidth)

reported unpleasant odors and ill-health soon after the unit's commissioning. The community reaction drew negative national and international attention, and the resulting dispute between the company and community has since been made a case study in sociology and business ethics [70]. Extensive studies of the emissions and possible equipment modifications to minimise them were subsequently undertaken.

While the emission of dusts and odors in Liquor Burners has been controversial, the SLC unit in Spain has apparently been operated without comment or concern. There may be a difference between the two processes in terms of emissions due to the different design of the calcination units and the resulting temperature profiles of gas and solid.

Thermal oxidizers are believed to have been retrofitted on one or more of the Liquor Burning units to minimise any emissions.

Both the Liquor Burning and SLC processes are relatively energy and capital intensive, making these processes sensitive to energy costs, carbon pricing (where it applies) and general emissions regulation, and these factors seem to weigh against their wider implementation.

Table 8.4 Success criteria for the SLC Unit in Spain [68]

Parameter	Success criteria	Achievements	Remarks
Availability	> 85% of available time	Achieved	During 30 days of first 6 month
Organic destruction	> 95%	Achieved	Average TOC < 0.1% in product
Dust emission	< 50 mg/m^3 (dry 11% O_2)	Achieved	After rectifying Bag House
TOC emission	Odor free	Achievable with modified calciner burners	150 ppm at 11% O_2
Recovery of Na & Al	R_{Na} * R_{Al} > 0.85	Almost achieved	Needs optimization of Na/Al in feed
Capacity[a]	(1) Consume all salt cake during commissioning period (2) 80 tpd Clinker	(1) Achieved (2) Achieved	(1) Salt cake storage needed to cover stops of SLC unit (2) 120 tpd design capacity limited by drying capacity
Specific heat consumption	2025 kcal/kg Clinker[b]	Achieved	< 1500 kcal/kg Clinker
Specific power consumption	170 kWh/ton Clinker[b]	Achieved	< 120 kcal/kg Clinker

[a] Capacity during test with ESP dust from the calciners replacing bauxite reached 115 tpd, primarily due to less water required to make the feed slurry pumpable
[b] Worst case scenarios

8.8.3.7 Future Organic Removal Options

Solvent Extraction

A recent patent [71] utilising ionic solvents has raised the possibility of phase separation of a range of organic and inorganic impurities from Bayer liquor. This technology is still in-development and is presently un-industrialised, so its full potential remains relatively unexplored and unproven. With its low energy requirement, relatively small and simple equipment, and selectivity, it has desirable attributes. Further development and mass production of ionic liquids may deliver more selective and lower cost solvents.

Nanofiltration

The development of caustic-stable poly-sulfone membranes, prompted a study of the nanofiltration separation of HMW organics from Bayer liquor. Poor permeate flux rates, rapid fouling and poor flux recovery after cleaning have prevented a practical application to date led most of these efforts to be abandoned [72]. Ceramic membranes allow for better flux recovery and membrane life, and their ongoing development and improvements in polymer membranes may in future yield a viable option.

Biological

Biological processes based on microbial degradation of organics have been suggested [73]. Some recent developments aim to be an alternatives to wet oxidation and liquor burning, but none have been industrialised [74].

8.9 Sulfate

Sulfur is common in most bauxites and can be extracted directly as the sulfate anion, for example from kaolin or iron sulfates like jarosite. Sulfide minerals (pyrite or marcasite) may also be oxidized. The sulfate anion is readily extracted from many of these minerals, typically with precipitation of the corresponding hydroxide, with a significant amount being released upon first contact with spent liquor. Bauxite organics also contain sulfur. When digested, they are responsible for a range of compounds characteristic of alumina refinery odor, and some of this organic sulfur is oxidized to sulfate [5].

Sulfate minerals such as magnesium sulfate (gypsum) are common impurities in lime and are dissolved in Bayer liquor. Sulfuric acid, principally used for cleaning silica scales, is another source of sulfate, and thousands of tonnes can be consumed in a refinery per year. When spent it is collected and stored separately in dedicated areas or neutralized with lime and/or mixed with bauxite residue for disposal. Residual acid in tanks and pipes after cleaning and flushing, errors with acid cleaning line-ups and accidental releases to the refinery's sump systems or process water are common inputs to the refinery liquor inventory.

Where bauxites are high in zinc, the sulfide anion is used to precipitate ZnS to control Zn in metallurgical alumina. A substantial proportion of the sulfide oxidizes to form anions such as sulfate, sulfite and thiosulfate which accumulate in liquor.

For most refineries, the two major removal mechanisms are with entrained liquor in the bauxite residue (with some return to the refinery via the catchment lake where one exists) and by intercalation in DSP, where sulfate is the preferred anion.

If in high concentrations, salting-out sulfate from evaporated spent liquor is possible. As described in the carbonate section, where sulfate is at high concentrations, the primary salt that is precipitated can be Burkeite. Again, salting-out usually produces a slurry of viscous liquor and fine precipitates, often made sticky by humate contamination, that is difficult to separate [75]. Precipitation using Barium has been suggested [22], but most other processes studied are variations to salting-out.

As discussed in the carbonate section, cooling liquor to between −10 and −15 °C was studied by Ostap and Bartok [26], where the decahydrates of sodium sulfate and sodium carbonate crystallise without HMW organics causing too many difficulties. Again, the cost of cooling and reheating liquor is the key economic factor. Figure 8.18 is a reproduction of their sulfate solubility results.

Combining salting-out with an organic removal process, such as liquor-burning, has the advantage of avoiding organic contamination of the precipitate [76]. A process based on this patent was operated industrially for many years. Very fine precipitates made filtration of the cake difficult, and it was subsequently abandoned [5]. Figure 8.19 is the preferred embodiment of the patent.

If liquor fluoride is elevated, $Na_2SO_4 \cdot NaF$ (kogarkoite) can form. Kogarkoite precipitation was studied by Wilson et al. [77], finding its solubility lower than burkeite or anhydrous sodium sulfate, and could precipitate first in

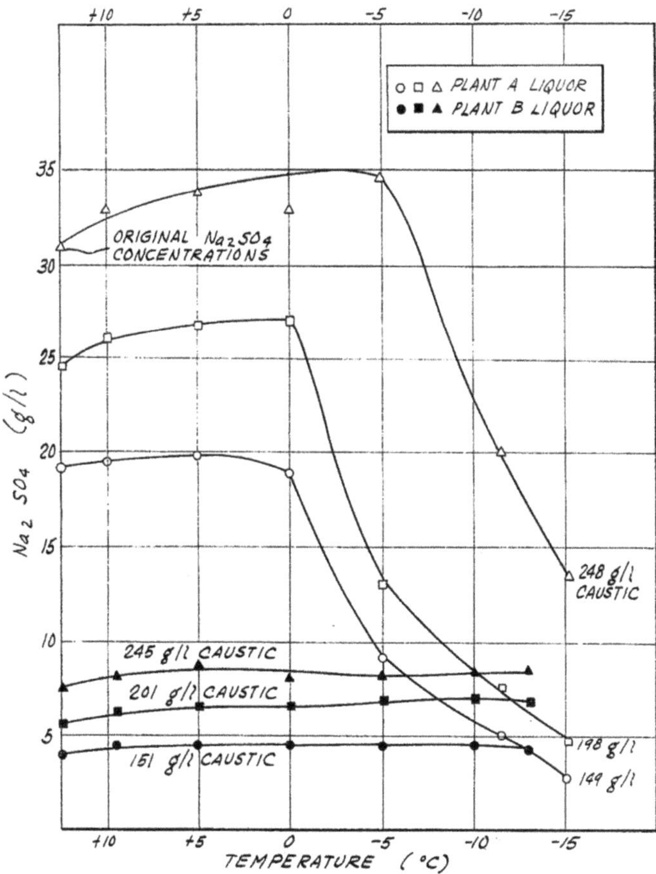

Fig. 8.18 Effect of temperature on Na_2SO_4 concentration in refinery liquor of between 150 and 248 g/L caustic [26]

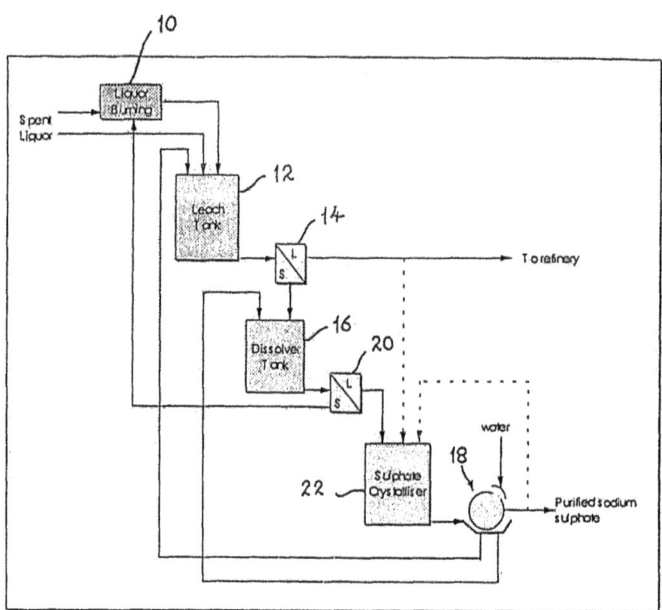

Fig. 8.19 Preferred embodiment of Worsley patent combining liquor burning and salting-out to remove sulfate and other impurities [76]

a salting out process. Kogarkoite scales have been found in several refineries' evaporators [77]. The solubility of kogarkoite suggests removal with less evaporated liquor is possible, potentially avoiding the difficulty of settling or filtration after organic contamination [5].

Sulfate and other anions can be intercalated into layered double hydroxides (LDH's). O'Hare et al. [78] devised a process based on a lithium-aluminium LDH that avoids alumina losses. By exchanging the intercalated anion with carbonate before calcining and re-hydrating the resultant product, the LDH could theoretically be recycled, reducing reagent costs. While never exploited commercially, this remains a possible future option.

Rosenberg et al. developed a simple to implement hydrocalumite LDH based process for removal of oxalate and sulfate from Bayer liquors [40]. The LDH precipitate containing the impurity anions is deliquored and the solids discarded. While demonstrated at a pilot scale, the alumina and lime consumption is expensive and has never been industrialised [5].

8.10 Halides

8.10.1 Chloride

Prior to the move to membrane production technology (substantially reducing chloride in product), the largest feed of chloride to Bayer refineries was as an impurity in caustic soda produced in diaphragm production technology. Caustic soda product today is a minor chloride input to Bayer liquor and is not generally a cause of serious accumulation.

Fresh water inputs to a refinery can obviously be a source, and a small chloride presence is common in bauxites. Hydrochloric acid used for equipment cleaning is mixed with liquor after incomplete post-cleaning rinses, with spills through sumps and with cleaning acid line-up errors. Where poly-DADMAC is used, chloride is also exchanged into liquor for HMW organics.

Due to the high solubility of chloride compounds, it is a difficult impurity anion to remove by precipitation. Between the switch to membrane caustic, removal by intercalation in DSP, the losses with residue and product, and avoidance of other inputs, refineries rarely accumulate high chloride concentrations. In any case, there is presently no other practical cost-effective removal process available, and refineries obliged to live with its accumulated concentration.

Apart from reducing liquor's productive potential by increasing ionic strength (limiting causticity), chloride does not trigger scaling or other negative process effects so is not a major problem impurity.

8.10.2 Fluoride

Fluoride is found in most bauxites, substituted for hydroxide in gibbsite or boehmite, and in the various clay components that constitute the reactive silica, or in various other fluoride-containing minerals such as fluorspar [79]. It is present in most refinery liquors, although rarely analysed. Outlets are with residue (as entrained liquor, some incorporation into DSP and TCA from filter aid or spent causticiser solids), or with gibbsite product (substituted for the hydroxyl ion), from where it may be emitted as HF in calciner stack gases.

The liquor concentration is usually low, and only obvious in the formation of double salts with anions such as vanadate, arsenate and sulfate [76]. These salts have comparatively low solubility and can scale the surfaces of evaporators, reducing evaporator efficiency [80]. In extreme cases, they can also co-crystallise with gibbsite in the precipitation circuit, resulting in serious product quality issues similar to the familiar oxalate "shower", in which gibbsite crystals nucleate on the surfaces of the double salt in the high supersaturation stages of precipitation [76].

At the ETI refinery in Turkey, impure NaF precipitated as scales in the tanks and pipes downstream of the evaporators [80]. A removal system where evaporated liquor was treated with fresh caustic to salt out the NaF was implemented. The precipitate was thickened and then filtered, and a relatively pure NaF cake obtained.

A removal process developed at the Worsley refinery in Western Australia relies on the substitution of fluoride for hydroxide in the aluminate ion. By manipulating the reaction conditions to encourage a high degree of this substitution, fluoride is preferentially removed when slaked lime reacts to form TCA [81].

Some fluoride is removed as fluorhydroxyapatite and fluorapatite when lime is added in digestion.

8.10.3 Other Halides

Small quantities of bromide and iodide may also be present in bauxite in but their concentrations rarely become significant, and have little process or product quality impact, and many refineries will be unaware of their presence in their liquor.

If they do cause issues, their presence will be felt in emissions from other impurity removal processes, such as liquor burning. In the liquor burner, these halides are evaporated to dryness and are present as their sodium salts. In the liquor burner kiln, or in the subsequent baghouses, they can be oxidized to elemental bromine or iodine and be discharged with the off-gases. In extreme cases, this has resulted in visible colored plumes. In at least one location this necessitated the installation of a halide gas scrubber, to reduce the halides back to their respective anions [5].

Boron is used in aluminum and its alloys as a grain refiner and to improve conductivity by precipitating vanadium, titanium, chromium, and molybdenum. Boron can be used alone (at levels of 0.005–0.1%) as a grain refiner during solidification but becomes more effective when used with an excess of titanium. Commercial grain refiners commonly contain titanium and boron in a 5-to-l ratio [82].

8.11 Iron

Iron is introduced with bauxite as hematite (Fe_2O_3) or goethite (FeOOH), sometimes called alumino-goethite due to common substitution of aluminium in the goethite lattice), or less common iron minerals (e.g., jarosite, or iron sulfides). The vast majority of this iron is undissolved and removed as a dominant element of residue to storage, or in some cases for recovery for various applications.

Iron is however solubilised to a degree in digestion, more so under high temperature conditions. Iron in liquor was correlated by Roach and Jameison to a range of parameters including digestion temperature, total iron in bauxite, hematite and goethite concentration and free caustic concentration [83].

Its negative impact comes from its association with gibbsite product in the precipitation area increasing the iron content of the product alumina. This iron ends up in aluminium metal, and above 0.010 or 0.020%, this alumina is un-welcome in smelters, particularly those that produce high quality metal. Iron in alumina is not the only iron input to smelter pots, steel anode stubs are also a major contributor to iron in metal, and strict control of iron in alumina is only part of the challenge of producing low iron aluminium. Reduced iron in liquor (and subsequently in alumina product) may be achieved in high temperature digestion by lime addition. Due to competing calcium reactions, there is usually a threshold below which there is no response to lime addition.

The concentration of iron in liquor after digestion is also related to the mineral phase present. Numerous studies on the subject indicate that a higher proportion of goethite in the bauxite, the higher the liquor iron after digestion. Goethite transforms hydrothermally to hematite, and the reaction rate increases between low digestion (where only a little occurs) and high temperature digestion, at the top end of which transformation is largely complete, up to the constraints of the related chemistry. During this transformation, aluminium substituted in the goethite matrix is released and becomes available.

$$FeOOH \rightarrow Fe_2O_3 + H_2O \tag{8.16}$$

$$2Fe_{1-x}Al_xOOH \rightarrow Fe_{2(1-x)}Al_{2x}O_3 + H_2O \tag{8.17}$$

8.12 Zinc

Zinc is detrimental to aluminium electrolysis [84], reducing the current efficiency and giving a spangled appearance to the metal surface. In aluminium metal, Zn exceeding 0.01% makes fragile metal. Extrusion billet for "bright" applications such as auto-trim, and for foil products are most sensitive.

The specification for ZnO in metallurgical alumina has become increasingly tighter. In 1982, Thé proposed 0.023% ZnO[85], while in 2015 Suss et al. advocated 0.01% ZnO in alumina [86].

Zn-containing minerals identified in bauxites include sphalerite (ZnS), woodruffite ($2(Zn, Mn)0.5MnO_2 \cdot 4H_2O$), gahnite ($ZnAl_2O_4$), zincblende (β-ZnS), smithsonite ($ZnCO_3$) and zincophorite ($Al(Zn_xMn_{1-x})O_2(OH)_2$, x: 0.02–0.24). Zn has also been detected along with other minerals. Sphalerite is considered the most significant Zn-containing mineral [87].

Some typical Zinc content in karst bauxites (as wt. % ZnO), are: Jamaica (0.02–0.05%), Montenegro (0.04%), Hungary (0.01%), and Russia (0.02–0.075%). ZnO content in lateritic bauxites is generally significantly less than in karst bauxites. Some typical figures are: Guinea (0.002%), India (0.007%), and Brazil (0.008%) [87].

ZnS dissolution in Bayer liquor to form the hydroxyzincate ion is described in (8.18).

$$ZnS + 4NaOH \leftrightarrow Na_2Zn(OH)_4 + Na_2S \tag{8.18}$$

Dissolution is strongly temperature dependent. In low temperature digestion, only 10–20% of Zn dissolves. Since these refineries process lateritic ores (with low Zn content, they have low zinc in liquor. At high temperature, Zn dissolution is usually > 50%, in some cases nearly 100% [88]. About 90% of Zn entering the precipitation area reports to the alumina product [89].

Almost all zinc control methods are based on the low solubility of ZnS in alkaline solutions. The method developed by Hrishikesan et al. [90] in 1969 and still in common use involves the addition of sodium sulfide to digestor blow-off slurry, followed by a short period of slurry retention to allow adsorption of the ZnS particles on the mud.

$$Zn(OH)_4^{2-} + S^{2-} \rightarrow ZnS + 4OH^- \quad (8.19)$$

A variation on this theme was patented by Magrone [91], in which sodium sulfide was added into the settler feed, and zinc sulfide was added to the settlers to act as seed, removing the need for a separate holding tank. In both these processes, the zinc sulfide departs with the bauxite residue.

Another process still in use was developed by Bird and Vance [92], in which sodium zincate is first formed and then reacted with sodium sulfide to produce high surface area zinc sulfide crystals, which act as seed. A slight excess of sodium sulfide is maintained in the solution, and the slurry is then injected into the clarifier overflow where it removes the zincate present in the liquor. The solids are removed in the liquor filters and discarded with residue.

Addition of sodium sulfide prior to pregnant liquor security filtration to convert the excess hydroxo-zincate to insoluble zinc sulfide is an obvious way to reduce the amount of ZnO in the alumina product [93]. Other metals, such as Cu and Pb react simultaneously with hydroxo-zincate, and non-soluble metallic sulfides form as well. This is why no clear stoichiometry between sulfide and ZnO-content in liquor has be established [90]. The detrimental effect of Na_2S addition is that the unreacted sulfide gradually oxidises to sulfate which accumulates in liquor and is harmful to precipitation [94]. Another issue is corrosion caused by the high level of Na_2S.

US Patent 4,282,191 [92] suggests using freshly synthesized ZnS seed along with Na_2S to form ZnS from sodium zincate or Zn-hydroxide formed during the digestion of bauxite. The ZnS seed is synthetized from ZnO (from an outside source) and Na_2S. The Na_2S is added to the liquor in excess when ZnS is synthetised, so that after formation of the ZnS seed, in combination with the residual Na_2S already present in the settler overflow, an Na_2S concentration of at least 0.15 g/l, preferably between about 0.15 and about 0.18 g/l be maintained. The resulting ZnS is then removed by filtration of the pregnant liquor. The ZnO content of the pregnant liquor could thereby be reduced from 20–30 to 5–15 mg/l.

The US Patent 6,352,675 by Malito proposes that dithiocarbonate and/or dithiocarbonate compounds have been found effective to remove metals such as zinc from Bayer liquor by the formation of complexes [95].

In a process described by Suss et al. [96], and practiced at the Kamensk Uralsky alumina refinery, elemental sulfur is introduced to bauxite grinding at 2–4 kg/t of alumina. A sulfide ion (S^{2-}) concentration of ~0.25 g/l is achieved, leading to formation of ZnS (sphalerite) during digestion. The advantage of this process is the relative low cost of sulfur, but a variety of oxidized sulfur species (such as sulfite, sulfate and thiosulfate) also form and adds to the impurity level of the refinery. VAMI recently optimised the use of elemental sulfur for the removal of ZnO by addition to first washer liquor, the thiosulfate largely converts to sulfide and sulfite in the course of digestion and ZnS seed is formed.

8.13 Vanadium

Vanadium is a common element in bauxites, with V_2O_5 contents ranging from 0.05 0.25% [97]. It is extracted from source minerals into liquor during digestion and can co-precipitate with gibbsite to problematic levels if accumulated in liquor.

Vanadium can also create a nuisance by precipitating as sodium fluoro-vanadate in the lower temperature / higher soda areas of a refinery, such as the cold end of evaporation. This salt can precipitate in combination with similar sodium fluoro-salts (fluoro-phosphate, fluoro-arsenate), depending on the liquor concentrations of fluoride, phosphorus and arsenic. The solubility study of Sipos et al. summarise relevant publications and patents, discussing speciation, chemistry and some published removal methods [98]. This study precipitated the identified phases NaF (Villiaumite), $Na_7F(VO_4)_2 \cdot xH_2O$, $Na_7F(PO_4)_2 \cdot xH_2O$ and $Na_7F(AsO_4)_2 \cdot xH_2O$ from evaporated spent liquor from the Gove refinery.

Some Indian ores are high in Vanadium content, to the extent that it is necessary to remove it via a salting-out process using spent liquor. This process is aided by the presence of fluoride in the liquor stream, as the fluoride-vanadate double salt is much less soluble than vanadate itself. This process has been practiced at a number of Indian refineries [6].

For most other refineries, vanadium is removed primarily via incorporation into TCA [99], and via soluble losses with residue. A process based on removal with TCA [81] has been implemented at the Worsley refinery.

Beyond its nuisance to the Bayer and Hall-Héroult processes, vanadium has value and is in demand as a steel alloying additive and its oxide (V_2O_5) has many applications as catalyst.

There is usually 10 to 200 ppm V in commercial-grade aluminum, and because it lowers conductivity, it generally needs to be precipitated from electrical conductor alloys with boron [82].

8.14 Phosphorus

All bauxites contain some caustic soluble phosphates. The level can range from 0.03% to up to 3% as P_2O_5. Identified Phosphate minerals include calcium phosphate

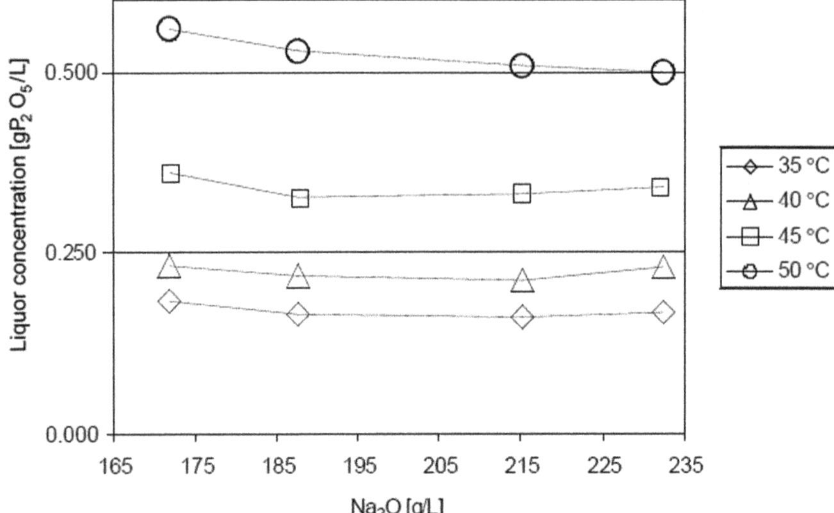

Fig. 8.20 Solubility of $Na_7F(PO_4)_2 \cdot xH_2O$ presented as liquor [P_2O_5] at different Gove spent liquor temperatures and caustic concentrations [98]

($Ca_3[PO_4]_2$), crandallite ($CaAl_3[PO_4]_2[OH]_5 \cdot H_2O$) and apatite ($3Ca_3[PO_4]_2 \cdot Ca[F, Cl, OH]_2$) [100].

Phosphorus has a negative impact on current efficiency in alumina reduction cells and must be controlled in metallurgical alumina to a typical specification of $<0.002\%$.

P in alumina product is controlled through residue losses (solid and liquor), as a product impurity, in residue solid and liquor losses, and the formation of carbonate apatite ($3Ca_3[PO_4]_2 \cdot CaCO_3$) through the addition of lime in a number of locations on the red side of the Bayer process. For high temperature refineries, lime is usually added in digestion; in the mills for double stream processes, and in the residence after heating for single stream processes (due to the formation of calcium titanate scales such as cassite or perovskite). For low temperature refineries, lime is often added to digestion via the or directly into autoclaves, or after digestion blow-off.

Where fluoride and phosphorus concentrations in liquor are high enough, sodium fluoro-phosphate can precipitate ($Na_7F(PO_4)_2 \cdot xH_2O$), along with analogous sodium-fluoro double salts [98]. Figure 8.20 reproduces the results of the Sipos et al. salting out study in Gove spent liquor, and the solubility behavior that underlies the phosphate removal opportunity.

8.15 Mercury

As a terrestrial, aquatic/marine and atmospheric pollutant, mercury emissions have attracted the interest of regulators and the alumina industry. Good introductions to mercury in the Bayer process were published by Dobbs et al. [101], and Smith [102] identifying sources, concentrations in refinery process materials, speciation and emissions from various points.

Mercury concentrations in unidentified bauxites with a range of almost 3 orders of magnitude, from 1200 to 20 ppb were reported by Smith [102].

Condensed mercury is often collected in traps in condensate lines where after reduction and vaporisation in digestion, it is condensed with steam in the condensate system. The balance of mercury then leaves the refinery with residue and as a vent gas from the digestion area.

Mercury has been used at the level of 0.05% in sacrificial anodes used to protect steel structures. Other than for this use, mercury in aluminum or in contact with it as a metal or a salt will cause rapid corrosion of most aluminum alloys [82].

8.16 Beryllium

Beryllium is present in some bauxites at trace levels. Some of the beryllium in bauxite dissolves during digestion, resulting in trace concentrations in liquor streams. A proportion of this liquor beryllium is included in hydrate during gibbsite precipitation and remaining in the alumina product after calcination. Where beryllium is elevated in alumina, it can accumulate in reduction cells to parts-per-million (ppm) levels, where operating practices (breaking the pot crust is suggested as the prime cause) can results in dusts containing beryllium.

The potential risk of over-exposure to beryllium in the potroom is unacceptable and where beryllium is elevated, it is monitored and managed. Management of the dust generated in the potroom and changing to lower beryllium bauxites/aluminas are presently the only proven management processes.

Beryllium is used in aluminum alloys containing magnesium to reduce oxidation at elevated temperatures. Up to 0.1% Be is used in aluminizing baths for steel to improve adhesion of the aluminum film and restrict the formation of the deleterious iron-aluminum complex [82].

8.17 Gallium

Some level of gallium is common in bauxites, with higher levels in some regions than others. It is generally not a problem impurity, although Shaw et al. [103] suggested in 1996 that gallium would increasingly become an issue for smelters. The study found that its inclusion in product was probably due to isomorphic substitution in gibbsite, and linearly related to gallium's concentration in liquor up to the highest liquor concentration measured, 1800 ppm (as Ga_2O_3). A correlation (8.20) was presented.

$$\left(\frac{Ga}{Al}\right)_{ppt} \times 10^6 = K \times Ga_{Liq}\,(\text{as } Ga_2O_3) \qquad (8.20)$$

Apart from the usual losses with product and the residue stream, gallium can be removed by solvent extraction or ion exchange resins, and it can be profitable. Like many of the rare metals, gallium is a small market and bringing on a new production stream can have a major impact on the market price and turn a gallium removal facility from a positive economic proposition to negative one, and the size of this market and related supply/demand equation has limited its wider application.

As an impurity in aluminium, gallium is usually present at levels of 0.001–0.02%. At these levels its effect on mechanical properties is quite small. At the 0.2% level, it has been found to affect the corrosion characteristics and the response to etching and brightening of some alloys [82].

8.18 Titanium

Smith summarises titanium chemistry well in [104], which is a good reference for the subject. Titanium commonly occurs in bauxites usually as anatase and rutile (polymorphs of TiO_2) and usually in the range of 1–3%, although it can be higher in some bauxites. Anatase dissolves more readily than rutile in digestion to varying extents depending on digestion temperature, caustic and other ions present. Dissolution results in the titanate ion, sometimes expressed as TiO_3^{2-}, although it is more likely to exist (like alumina) as a hydroxy complex (e.g., $Ti(OH)_3^-$).

The phases precipitated when slaked lime is added to liquor has been demonstrated and appears to be via a transition compound referred to as calcium hydroxytitanate (CTH). CTH is believed to transform into perovskite ($CaTiO_3$) and titanium substituted TCA ($Ca_3(Al, Fe)_2[Si, TiO_4]_n(OH)_{12-4n}$), although there remains some differences in evidence and opinion on the exact phases and mechanisms. Stoichiometrically, the equation for the precipitation of perovskite could look like the following;

$$Ca(OH)_2 + 2TiO_2 \rightarrow CaTi_2O_4(OH)_2 \qquad (8.21)$$

The precipitation of (or transformation of precipitated CTH to), perovskite interferes with the dissolution of boehmite in high temperature digestion, and other reactions. It also forms hard scales which are difficult to remove.

In the absence of lime, sodium titanate compounds precipitate (usually forming scales) such as the sodium-titanium-iron compound freudenbergite (generic form $Na_2Fe_3 + 2Ti_6O_{16}$ [105]) and are responsible for some of the sodium measured in muds from high temperature processes.

Titanium accumulation in liquor is minimized by the usual solid and liquor losses with residue (including removed scales) and product, and the lime reactions discussed above, and usually keep titanium from exceeding product quality specifications.

8.19 Lithium

The lightest of light (and alkali) metals, lithium is known to associate and substitute for sodium in sodium aluminate ($LiAl(OH)_4$) [106]. In high temperature refineries using diaspore elevated Li in product.

According to Huang [107], increasing the lithium ion concentration in solution improved the precipitation rate and lithium content in precipitated gibbsite. The presence of the lithium ion significantly increased the total particle number due to the preferential nucleation. Elevating temperature or reducing lithium ion concentration decreased lithium content and reduced the fine particle amount in gibbsite. The lithium ion also changed the morphology by promoting the growth of thc (110) and (200) faces, producing bar- or flake-shaped gibbsite attached to coarse particles, and led to the poor alumina strength.

The level of lithium in metal is typically of the order of a few ppm, but at a level of less than 5 ppm, can promote the discoloration (blue corrosion) of aluminum foil under humid conditions. Traces of lithium greatly increase the oxidation rate of molten aluminum and alter the surface characteristics of wrought products [82].

8.20 Impurity Removal and Environmental Footprint

In most refinery impurity removal processes, the removal by-products are either mixed with the bauxite residue stream destined for the Residue Disposal Area (RDA) or stored separately in dedicated areas. With the increasing effort to minimise bauxite residue in storage, new options for recycling, re-processing or utilisation of these by-products are being developed, and known ones are being re-examined.

One of the greatest barriers to utilisation of bauxite residue is its high pH and sodium content. More efficient washing will address this, but alternative impurity removal processes must be found to maintain the impurity balance of many refineries where improved washing is achieved.

The solid alkalinity of bauxite residue derives largely from the presence of DSP and TCA, both of which buffer the residue to high pH values, even after neutralization at discharge. In the case of DSP, this also constitutes a mechanism for the slow release of soluble sodium into the residue as the DSP slowly dissolves. While the minimization of DSP has been a goal of the industry for many decades, for residue use it is necessary to either reduce its input or find new ways to manage its behavior in residue.

Alternatives already exist for TCA. The high efficiency causticisation processes described earlier can greatly reduce TCA content in residue, and other sources, such as filter aid, can be isolated from the residue and even recycled. The practice of sodium oxalate causticisation, due to its inefficiency, also contributes to the presence of unstable quaternary calcium aluminate species in residue, along with unreacted lime.

The recent development of more efficient and operable biological oxalate destruction provides a means of reducing this.

The presence of zinc sulfide can result in odours (due to biological activity) and contribute to the toxic content of the residue (although it could be argued that this is simply returning the zinc to its source) [5]. Similarly, sulfates, when present with an organic source such as oxalate, can be reduced and create odour problems.

Acknowledgements Steve Rosenberg's paper from the 2017 ICSOBA conference in Hamburg [5], provides a good summary of aspects of the origin and management of common Bayer impurities, with a good collection of references. Rather than reinvent this well researched work, this chapter takes this paper as a starting point.

Thanks to György Banvolgyi who provided a very complete and detailed section on zinc, which has been edited to the level of detail of the rest of the sections. György's ICSOBA 2017 paper on this subject is more detailed and recommended reading.

Thanks to Benny Raahauge for his very thorough and detailed contribution on Solid Liquid Calcination. Benny provided a more detailed initial draft, which has again been edited here to match the resolution of the rest of the chapter. Benny's many publications are recommended for greater detail on this topic.

References

1. J.V. Sang, Factors affecting residual Na_2O in precipitation products, in *Light Metals 1984*, pp. 223–236
2. S.C. Grocott, S.P. Rosenberg, Soda in alumina. Possible mechanisms for soda incorporation in Smelter grade alumina for the 1990's and beyond, in *International Alumina Quality Workshop*, pp. 271–287 (1988). http://www.aqw.com.au
3. L. Armstrong, Bound soda incorporation during hydrate precipitation, in *Third International Alumina Quality Workshop*, Hunter Valley, NSW, pp. 282–291. http://www.aqw.com.au
4. C. Vernon, et al., Soda incorporation during hydrate precipitation, in *Essential Readings in Light Metals*, ed. By D. Donaldson, B.E. Raahauge (Wiley, Hoboken, NJ, 2013), pp. 602–607
5. S.P. Rosenberg, Impurity removal in the Bayer process, in *35th International Conference and Exhibition of ICSOBA*, Hamburg, 2–5 October 2017
6. N.K. Kshatriya, P.K.N. Raghavan, Improvement in vanadium salt separation, in *International Conference on Non-Ferrous Metals*, Pune, India, 8–9 July 2005, Tech 15/1–15/9
7. G.I.D. Roach, Effect of particle characteristics on the solids density of Bayer mud slurries, in *Essential Readings in Light Metals*, ed. by D. Donaldson and B.E. Raahauge (Wiley, Hoboken, NJ, 2013), pp. 417–424
8. F.R. Feret, J. See, Occurrence and characterization of Zn and Mn in bauxite, in *Light Metals 2006*, pp. 41–45
9. Z. Yin et al., An Introduction to Bel Air Bauxite, in *ICSOBA 2018*
10. S. Minai et al., Sandy alumina conversion, in *Light Metals, 1978*, p. 95
11. E.J. Jamieson, Method for management of contaminants in alkaline process liquors. International patent WO2008017109 (2008)
12. P. Smith, The processing of high silica bauxites—review of existing and potential processes. Hydrometallurgy **98**(1–2), 162–176 (2009)
13. T. Harato et al., The development of a new Bayer process that reduces the desilication loss of soda by 50% compared to the conventional process, in *Alumina Quality Workshop*, pp. 311–320 (1996). http://www.aqw.com.au

14. P. Smith et al., Understanding growth of DSP in the presence of inorganic ions, in *Alumina Quality Workshop*, pp. 191–194, http://www.aqw.com.au
15. D. Thomas, L. Armstrong, T. Leong, Impact of liquor causticity on plant operation, in *Alumina Quality Workshop* (2002). http://www.aqw.com.au/
16. A. Rijkeboer, Bauxite Processing, US patent 5,279,645 (1992)
17. M. Hollitt, J. Kisler, B. Raahauge, The Comalco bauxite activation process, in *Alumina Quality Workshop*, pp. 115–122 (2002). http://www.aqw.com.au
18. G.I.D. Roach, The equilibrium approach to causticisation for optimising liquor causticity, in *Essential Readings in Light Metals* (Wiley, Hoboken, NJ, 2013), pp. 228–234.
19. S.P. Rosenberg, et al., Some aspects of calcium chemistry in the Bayer process, in *Essential Readings in Light Metals* (Wiley, Hoboken, NJ, 2013), pp. 210–216
20. G.D. Roach et al, Method for the causticisation of Bayer process solutions. Canadian patent CA2596897 (2007)
21. S.P. Rosenberg et al., Improved Bayer causticisation. Australian patent 1999058412
22. H Mercier, R Magrone, J Dealbriges, Purification of solutions circulating in the Bayer cycle, US patent 4,101,629, 1978
23. M. O'Dea, A. O'Connell, Control of product size and strength with challenging impurity balance, in *ICSOBA and XXV Conference Aluminium of Siberia*, Krasnoyarsk (2019)
24. K. Yamada et al., Method for the removal of impurities from sodium aluminate solution. European patent, EP0013407B1 (1980)
25. J.L. Porter, Process of purifying caustic aluminate liquors. US Patent 2889982 (1959)
26. S. Ostap, D. Bartok, Removal of sodium carbonate and sodium sulfate from Bayer solutions. US Patent 3,508,884 (1970)
27. J.T. Malito, G.C. Rogers Jr, Purification of impure Bayer process liquors. US patent 4,430,310 (1984)
28. J.S. Bhatti et al., Influence of soil organic matter removal and pH on oxalate sorption onto a spodic horizon. Soil Sci. Am. **62** (1998)
29. A. Costine et al., Understanding the origin and mechanism of hydrogen production in the Bayer process, in *Alumina Quality Workshop*, pp. 208–213 (2012). http://www.aqw.com.au
30. G.P. Power, W. Tichbon, Sodium oxalate in the Bayer process: its origin and effects, in *Alumina Quality Workshop*, pp. 99–115, http://www.aqw.com.au
31. A. Lalla, R. Arpe, 12 years of experience with wet oxidation, in *Light Metals 2002*
32. P. Atkins, S.C. Grocott, The liquid anion exchange process for organics removal, in *Light Metals 1993*, pp. 151–157
33. W.J. Roe, J.T. Malito, Purification of Bayer process liquors. US Patent, 4,578,255 (1986)
34. F.S. Williams, A.J. Perrotta, Enhanced oxalate removal utilizing the multi-functional purox process, in *Light Metals 1998*, pp. 81–87
35. G. Ruecroft et al., Improving the Bayer process by power ultrasound induced crystallization (sonocrystallization) of key impurities, in *Light Metals 2005*, pp. 163–166
36. D. Hipkiss, Removal of contaminants from a Bayer liquor. International Patent WO2007/066143 B2 (2007)
37. R. Gnyra, G. Lever, Review of Bayer organics-oxalate control processes, in *Essential Readings in Light Metals*, pp. 278–283 (2013)
38. A. O'Connell et al., Maintaining an oxalate free precipitation circuit at Aughinish Alumina, in *Alumina Quality Workshop*, pp. 270–273 (2002). http://www.aqw.com.au
39. S.C. Grocott, I.R. Harrison, Two new oxalate removal processes, in *Alumina Quality Workshop*, pp. 424–435 (1996). http://www.aqw.com.au
40. T.S. Li et al., The influence of solid phase oxalate SPO) on gibbsite secondary nucleation in synthetic caustic-aluminate solution, in *Alumina Quality Workshop*, pp. 165–173 (2015). http://www.aqw.com.au
41. M. Reyhani et al, Gibbsite nucleation at sodium oxalate surfaces, 14th International Symposium on Industrial Crystallisation - Cambridge 12–16 September 1999
42. W.J. Roe, D.O. Owen, J.A. Jankowski, Crystal growth modification: practical and theoretical considerations for the Bayer process, in *Alumina Quality Workshop* (1988). http://www.aqw.com.au

43. R.T. Chester, C.Q. He, J.D. Kildea, *Management and Control of Sodium Oxalate Precipitation in the Bayer Process* (Nalco/Ecolab publication, 2001 reprint)
44. M.E. Keeney, S.C. Grocott Crystallisation of sodium oxalates, Australian patent AU 1988024528, 1988
45. R.A. Morton, Hydrate precipitation and oxalate removal. US Patent 1993050778 (1993)
46. Fischer Scientific Sodium Oxalate Material Safety Data Sheet, CAS# 62-76-0, Revision Revision #15 Date: 15 Feb 2008
47. S.P. Rosenberg et al., Process for the removal of oxalate and or sulfate from Bayer liquors. US Patent 7,244,404B2 (2007)
48. R.D. Brassinga et al., Biodegradation of oxalate ions in aqueous solution. Australian Patent 1989039465 (1989)
49. R.A. Morton et al., Biological disposal of oxalates. Australian Patent, AU 1991073182 (1991)
50. D. Chinloy, J. Doucet, M.A. McKenzie, K. The, Improvements in processes for the Alkaline Biodegradation of Organic Impurities, EP 0659169 B1 (1993)
51. N.J. McSweeney, *The Microbiology of Oxalate Degradation in Bioreactors Treating Bayer Liquor Organics Wastes* (School of Biomedical, Biomolecular Sciences, University of Western Australia, 2011), p. 149
52. H. Wang, A.A. Adesina, Photocatalytic causticization of sodium oxalate using commercial TiO_2 particles. Appl. Catal. B **14**, 241–247 (1997)
53. J. Bangun, A.A. Adesina, The photodegradation kinetics of aqueous sodium oxalate solution using Tio_2 catalyst. Appl. Catal. A **175**, 221–235 (1998)
54. G. Power, J. Loh, Organic compounds in the processing of lateritic bauxites to alumina. Hydrometallurgy **105**(1–2), 1–29 (2010)
55. G. Power et al., Organic compounds in the processing of lateritic bauxites to alumina part 2: effects of organics in the Bayer process. Hydrometallurgy **127–128**, 125–149 (2012)
56. B. Gnyra, Removal of oxalate from Bayer process liquor. US Patent 4,275,043 (1991)
57. B. Schepers et al., Method for removing harmful organic compounds from aluminate liquors of the Bayer process. US Patent 4046855 (1977)
58. W.A. Nigro, G.A. O'Neill, Method for reducing the amount of colorants in a caustic liquor. US Patent 5,068,095 (1991)
59. S. Beaulieu, Impurities control—from a mine to a plant, in *ICSOBA 2017*
60. D.W. Arnswald et al., Removal of organic carbon from Bayer liquor by wet oxidation in tube digesters, in *Essential Readings in Light Metals*, pp. 304–308 (2013)
61. S. Bhargava, D.B. Akolekar, J. Tordio, S. Eyer, Organics removal from alumina industrial liquor: application of the catalytic wet oxidation process to natural and synthetic Bayer liquors, in *Alumina Quality Workshop* (2002). http://www.aqw.com.au
62. J. Dong et al., Wet oxidation of Bayer liquor organics: reaction mechanisms, in *Light Metals 2010*, pp. 209–213
63. J. Zöldi, K. Solymár, I. Feyér, Organic removal and its effect on precipitation and alumina quality, in *Alumina Quality Workshop*, pp. 24–37 (1988). http://www.aqw.com.au
64. D.A. Swinkels, K. Chouzadjian, Removal of organics from Bayer process streams. US Patent 4836990 (1989)
65. L. Armstrong, G. Soucy, Oxidation of Bayer liquor organics with submerged plasma, in *Light Metals 2007*, pp. 145–149
66. K. Yamada, Y. Shibue, Method for treatment of a Bayer liquor. US Patent 4,280,987 (1981)
67. P. Prinsloo, The application of cerafil® ceramic filter elements for particulate removal on a liquor burning application for Worsley Alumina Pty Ltd, in *Alumina Quality Workshop*, pp. 346–349 (2002). http://www.aqw.com.au
68. J.M. Alvarado et al., Solid-liquid calcination (SLC) process for liquor burning, in *Alumina Quality Workshop*, pp. 383–392 (1996). http://www.aqw.com.au
69. R.L. Clegg, L.G. Armstrong, Development of liquor purification at Alcan Gove, in *Alumina Quality Workshop*, pp. 137–142 (2005). http://www.aqw.com.au
70. M. Brueckner, M.A. Mamun, Living downwind from corporate social responsibility: a community perspective on corporate practice. Business Ethics: A European Review **19**(4), 326–348 (2010)

71. J. Lean, S. Griffin, M. Taylor, Methods and compositions for the removal of impurities from an impurity-loaded organic salt. US Patent 8435411 (2013)
72. A.R. Gillespie, Bayer liquor purification by nano-filtration membranes, in *Alumina Quality Workshop*, pp. 149–152 (2005). http://www.aqw.com.au
73. R. Chinloy et al., Processes for the alkaline biodegradation of organic impurities. US Patent 5,271,844 (1993)
74. A.J. McKinnon, C.L. Baker, Process for the destruction of organics in Bayer process streams. US Patent 20140051153 (2014)
75. A.A. Anderson, Process of purifying caustic aluminate liquors. US Patent 2,806,766 (1957)
76. D.M. Southwood-Jones, S.P. Rosenberg, Removal of impurities in Bayer process. Australian Patent 1995020347 (1995)
77. D.J. Wilson et al, Fluoride chemistry in the Bayer process, in *Alumina Quality Workshop*, pp. 281–287 (2002)
78. D.M. O'Hare, et al., Process for removing impurities from bauxite. US Patent 6,479,024 (2002)
79. J.J. Thomas et al., Fluoride content of clay minerals and argillaceous earth minerals. Clays Clay Miner. **25**, 278–284 (1977)
80. E Savkilioglu et al, The control of fluoride concentration in ETI aluminum Bayer refinery liquor, in *Light Metals 2013*, pp. 183–186
81. S.P. Rosenberg, et al., Process for the removal of anionic impurities from caustic aluminate solutions. US Patent 7,691,348 B2 (2010)
82. Total Materia website, http://www.totalmateria.com/Article55.htm. Accessed 14 Jan 2020
83. G. Roach, E. Jamieson, Effect of bauxite and digestion conditions on iron in SGA, in *Alumina Quality Workshop*, pp. 340–345 (2002). http://www.aqw.com.au
84. K.I. Verghese, The impact of impurities on the Bayer process, in *Proceedings of 8th International Light Metal Congress*, Leoben-Vienna, pp. 42–46 (1987)
85. P.J. The, Removal of copper and zinc from Bayer liquor by process filtration, in *Light Metals 1982*, pp. 129–141
86. A. Suss, I. Paromova, N. Kuznetsova, A. Panov, Optimization of elemental sulfur utilization of zinc removal from Bayer cycle, in *10th Alumina Quality Workshop*, pp. 383–388 (2015)
87. G. Bánvölgyi, A review of zinc in bauxites, the Bayer process, alumina and aluminium, in *ICSOBA Conference*, Travaux No 46, pp. 197–204
88. S. Ostap, Effect of bauxite mineralogy on its processing characteristics, Bauxite, Los Angeles, pp. 651–671 (1984)
89. E.B. Teas, J.J. Kotte, The effect of impurities on process efficiency and methods for impurity control and removal, in *Proceedings of JBI—JGS Symposium 1980*, "Bauxite/Alumina Industry in the Americas", pp. 100–129
90. K.G. Hrishikesan et al., Treatment of Bayer process digester slurry. US patent 3469935 (1969)
91. R. Magrone, Process for removing zinc from aluminate liquor. UK patent 1373843A (1974)
92. R.D. Bird, H.R. Vance, Zinc removal from aluminate solutions. US patent 4282191 (1981)
93. K.G. Hrishikesan, Zinc Removal, US Patent 3,445,186, 1969
94. K.I. Verghese, The impact of impurities on the Bayer process, in *Light Metals 1987*, pp. 42–46
95. J. Malito, Process for removing heavy metals from caustic fluid stream. US Patent 6,352,675 (2002)
96. A. Suss, I. Paromova, N. Kuznetsova, A. Panov, Optimization of elemental sulfur utilization of zinc removal from Bayer cycle, in *10th Alumina Quality Workshop*, pp. 383–388 (2015). http://www.aqw.com.au
97. P. Meshram et al., Extraction of V_2O_5 from Bayer sludge: an overview. IBAAS **3**, 170–183 (2014)
98. G. Sipos et al., Solubility concentrations of arsenic, fluorine, phosphorus and vanadium in evaporated Bayer liquor, in *Alumina Quality Workshop*, Brisbane (2002). http://www.aqw.com.au
99. P. Smith, Reactions of lime under high temperature Bayer digestion conditions. Hydrometallurgy **170**, 16–23 (2017)

100. S. Sankaranarayanan, Effect of bauxite characteristics and process conditions on control of silica and phosphate in Bayer liquor, in *ICSOBA 2017*
101. C. Dobbs et al, Mercury emissions in the Bayer process—an overview, in *Alumina Quality Workshop*, pp. 199–204 (2005). http://www.aqw.com.au
102. P. Smith, C. Klauber, C. Vernon, *Mercury in the Bayer Process—DMIR—1852* (A.J. Parker CRC Hydrometallurgy, March 2002)
103. M. Shaw, P. Smith, R. Cornell, Gallium incorporation into precipitated gibbsite, in *Alumina Quality Workshop* (1996). http://www.aqw.com.au
104. P. Smith, Reactions of lime under high temperature Bayer digestion conditions, CSIRO Mineral Resources Flagship (2010)
105. J.D. Cashiona et al., Freudenbergite—a new example of electron hopping, in *Proceedings—38th Annual Condensed Matter and Materials*, Auckland, NZ (2014)
106. M. Wellington, F. Valcin, Impact of Bayer process liquor impurities on causticization. Ind. Eng. Chem. Res. **46**(15), 5094–5099 (2007). https://doi.org/10.1021/ie070012u
107. W. Huang et al., Effect of lithium ion on seed precipitation from sodium aluminate solution. Trans. Nonferrous Metals Soc. China **29**(6), 1323–1331 (2019)

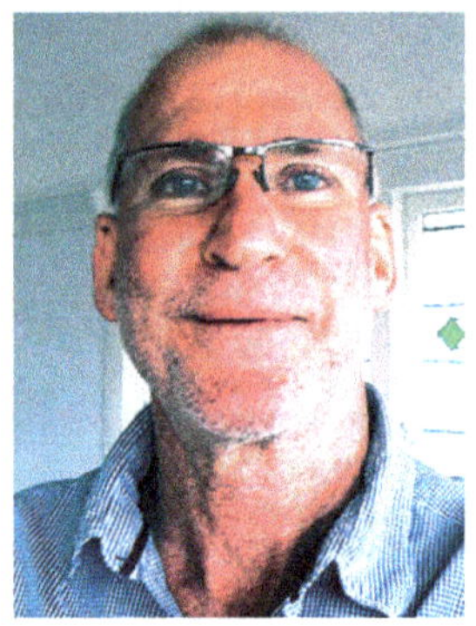

Steven J. Healy Alumina Industry Consultant. Email: sjhealy56@gmail.com

Steve Healy has worked in the Alumina industry for more than 30 years, starting his journey with Aloca in Western Australia, before moving to Worsley Alumina and then Nabalco (the Gove refinery). He was brought into Alcan and its rich technical history as it acquired the Al Group, then being part of the Pechiney technical integration after Alcan's acquisition of the French Aluminium producer, both organizations with their own long histories of innovation. His final full-time employer was Rio Tinto, following the acquisition of Alcan, where he worked until retirement from full time work in 2016. His longest role with Rio was as General Manager of B&A R&D, and his last role was Chief Technologist in the B&A Technology group. He was a member of the IAI Bauxite and Alumina Committee, serving as Chair and Co-Chair and the Australian Alumina Technical Panel. He has been a subject Chair at ICSOBA, a reviewer for TMS, and Organising Committee Chair at the Alumina Quality Workshop. He has authored more than a hundred internal company reports on varying subjects and reviewed and edited an even larger number. Since 2006 he has been working part time as a consulting Bauxite and Alumina technologist and has been particularly active in China. In his various positions he had an opportunity to see many different refineries, technical cultures and technical heritages and has been involved in a wide range of projects from business improvement to blue sky R&D, working with partners from research institutions to operations.

Chapter 9
Bayer Process—Water and Energy Balance

Robert LaMacchia and Raphael Costa

Abstract This chapter aims to illustrate and discuss the inputs and outputs for both water and energy in the context of the Bayer process, highlighting throughout the sections their interconnectedness. It breaks down these inputs and outputs in light of the unit operations of the Bayer process and discusses the reasons as to the variations in consumption figures, based on refinery design and other raw material considerations. The section on water shows how different technological developments, in conjunction with the bauxite to be processed, can drive much of the Bayer loop water balance. Explanations are given as to why boehmitic bauxites have the potential to drive somewhat different water balances compared to gibbsitic bauxites, owing to higher digestion temperatures and the removal of water within their precipitation circuits. Going through the history of the Bayer process and the changes brought out by modern equipment, examples show how the alumina refining process has undergone significant change over time. From a design standpoint, examples discussing caustic washing systems for precipitators highlight the significant impact this critical system can have on both equipment availability as well as the plant water balance. Beyond the Bayer loop, the greater water balance of the refinery is considered, showing how much of this balance is a slave to the climatic conditions of the refinery, but also highlighting how residue disposal technology has significantly changed the net balance through adjusting the areas for water catchment versus evaporation. Within energy, a significant focus is placed on the importance of precipitation yield of alumina in the Bayer liquor, in determining a plant energy efficiency, and how this can perhaps help to explain the variability in different plant energy numbers. The quality of the bauxite is also discussed, including the analysis of energy requirements in high temperature digestion refineries, or boehmitic/diasporic bauxite operators. The subsequent section on clarification reinforces the importance of yield further, emphasizing how residue handling solutions in different parts of clarification have both hindered and helped yield and recovery. The sections on heat

R. LaMacchia (✉)
Alcoa World Alumina, Perth, Australia
e-mail: robert.lamacchia@alcoa.com

R. Costa
Hydro Bauxite and Alumina, Rio de Janeiro, Brazil

B. E. Raahauge and F. S. Williams (eds.), *Smelter Grade Alumina from Bauxite*, Springer Series in Materials Science 320,
https://doi.org/10.1007/978-3-030-88586-1_9

interchange and precipitation show how modern design improvements such as plate heat exchange between green and spent liquor as well as inter stage cooling have unlocked even more potential in energy efficiency.

9.1 Refinery Water Balance

Water is an essential resource to the production of alumina by the Bayer process, used in a variety of different ways that guarantee both economic as well as environmental outcomes are satisfied for alumina producers. Further, as the Bayer process has evolved since its inception, the extent to which water is used and in what ways has changed, always with the intent of improving efficiencies of unit operation.

9.1.1 Bayer Process Water

Within the refinery battery limits, the Bayer process itself contains the greater layer of complexity with regard to understanding and optimizing water usage. This is partly borne out of the cyclical nature of the liquor circuit, as shown in greater detail in Fig. 9.1 [1], which demands that any introduced water is removed. The water usage is intimately tied with the energy demand and soda consumption of an alumina refinery, so many of the water related process decisions must be made in light of the entire process, as opposed to a unit operation in isolation.

When discussing the use of water in an alumina refinery, it is critical to understand where the battery limits of the process are. This point is illustrated in Fig. 9.2 where the Bayer process is highlighted. The design and operation of a refinery is contingent on its access to local water sources and a good understanding of the local rainfall patterns. To better appreciate these complexities, both internal to the Bayer process as well as the external refinery envelope, it is necessary to breakdown each component and understand how they interact with the system as a whole.

An overview of the key inputs and outputs of water into the Bayer process are shown in Table 9.1 [2]. While it is not the intent of this section to completely review the Bayer process at great detail, the best approach to understand the inputs and outputs that drive the water balance is to understand in each unit process, how and why water enters and exits the circuit.

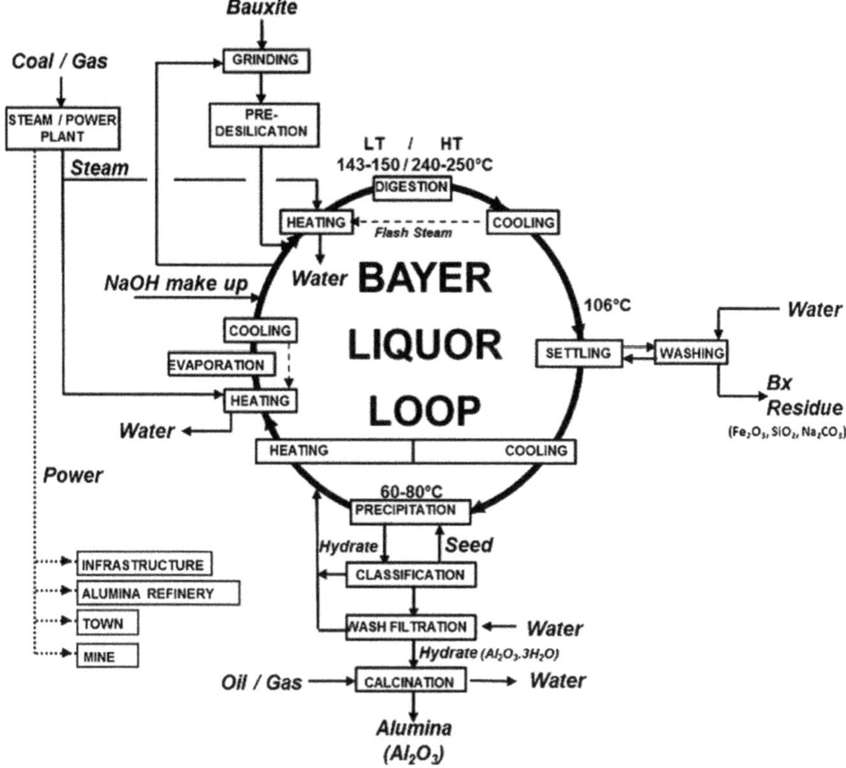

Fig. 9.1 A generic Bayer process liquor loop illustrating unit processes and general water and energy flows [1]

9.1.1.1 Digestion

Raw Materials

Water enters the process with the key raw material, bauxite, by way of free moisture. Associated with any bauxite source is some component of water, whose quantity depends on climate and geography, or the bauxite beneficiation processes. The bauxite moisture is normally a specified parameter, that the refinery utilizes in the design (or forecast) of their associated water balance. A unique practice, at Hydro's Alunorte refinery in Brazil, is to filter a bauxite slurry using hyperbaric filtration. The Paragominas bauxite mine, which is situated more than 240 km away, was evaluated for rail transport, a more conventional approach, but the economics favored pipeline transfer at approximately 50% solids loading [3]. Not only does this put the bauxite moisture specification in the refinery's hands, it also provides a much larger quantity of water than is typically dealt with in bauxite handling. At this location (the only

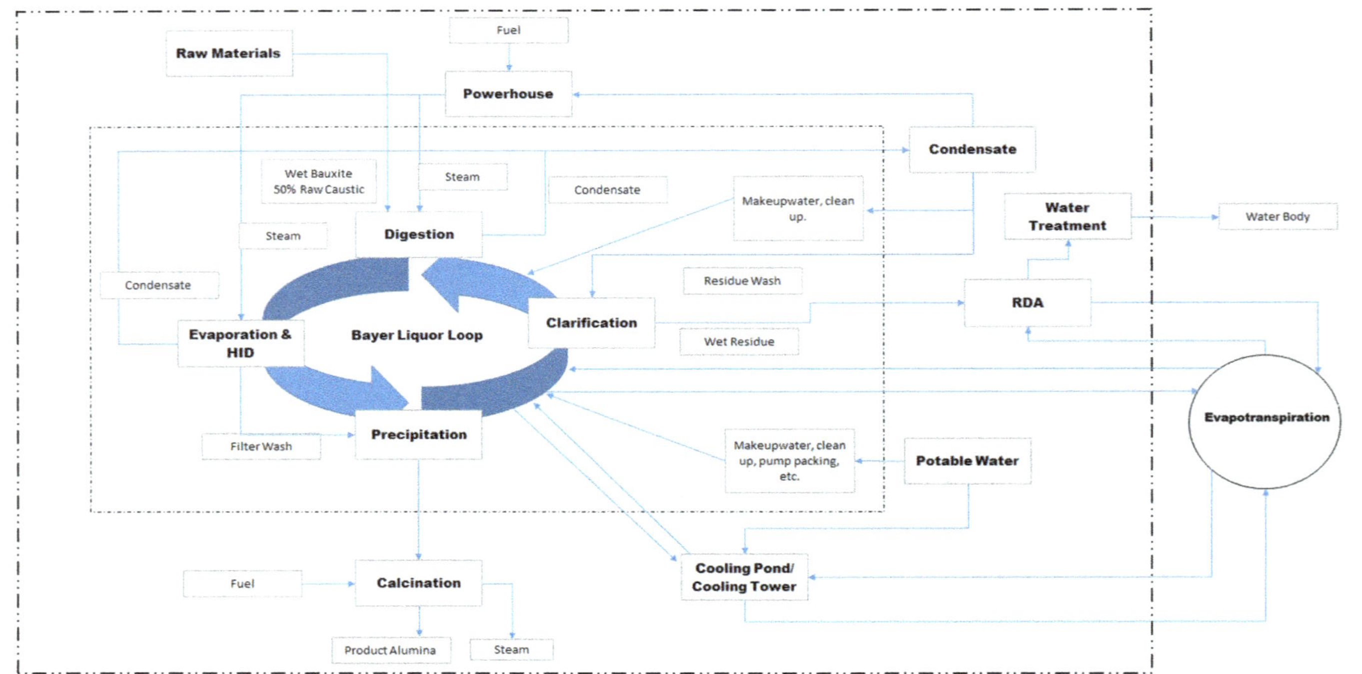

Fig. 9.2 Example alumina refinery water balance exporting excess water via water treatment. The inner envelope represents the main water processes within the Bayer liquor loop, while the outer boundary is for the entire location

Table 9.1 Generic water inputs and outputs to the Bayer process [2]

Inputs	Outputs
Residue wash	Digestion evaporation
$Al(OH)_3$ wash	Evaporator building evaporation
Free moisture in bauxite Injected steam	Heat interchange evaporation Boehmite/diaspore digestion-gibbsite precipitation[a]
Purge water	Moisture in $Al(OH)_3$ to calcination
50% NaOH	Liquor in residue slurry to RDA
Cleanup and makeup water	Evaporation from solution surfaces
Uncontrolled dilution/rainfall	Spills

[a]Only applicable to refineries digesting boehmitic/gibbsitic bauxites where water is consumed in the digestion reaction, but not liberated in the precipitation. See the sub section on digestion in 9.1.1.1

one in the world), the bauxite slurry is filtered in hyperbaric filters achieving a specification of ~ 14% moisture. The filtrate from the filters is thickened (as the filtrate typically contains ~ 4 gpL solids) and the water treated prior to discharge in the local river. This operation of filtering can be seen as external to the Bayer process water balance envelope, as the water that is separated does not enter the process loop; only the product of the filtration does (i.e., the humid bauxite).

The other essential raw material in the Bayer process, sodium hydroxide, also adds water to the process. Modern practice utilizes 50% solution of sodium hydroxide, termed raw caustic, which adds a substantial amount of water. Older practice in fact avoided this water addition by utilizing soda ash in conjunction with lime to manufacture sodium hydroxide in situ according to:

$$Na_2CO_{3(s)} + CaO_{(s)} + H_2O_{(l)} \rightarrow 2NaOH_{(aq)} + CaCO_{3(s)}$$

With the two materials dosed into batch digesters prior to heating. Such practice has largely ceased due to the lime and energy penalty associated with the production of caustic soda this way.

Pre-desilication

To initiate the process, the bauxite is slurried in spent liquor, typically in some form of grinding circuit to reduce the particle size to an acceptable level to guarantee sufficient extraction of the alumina in the subsequent digestion step. Post milling, the slurry is usually (but not in all refineries) sent to predesilication, where the slurry is to be heated, typically from a temperature of ~ 70 to ~ 100 °C to ensure adequate desilication [4]. Historically (and even in modern practice [5]), this heating has been achieved by way of direct steam injection (i.e. dilution), using either live steam from the powerhouse, or from the digestion flash tanks. In more modern practice, this

Fig. 9.3 Indirect bauxite slurry heaters at Alunorte alumina refinery [6]

dilution has been mitigated by way of indirect slurry heating using pipe in pipe heaters (see the example from Alunorte shown in Fig. 9.3 [6]). The original design utilizing steam injection was borne out of a concern that the highly viscous slurry would scale too fast in an indirect heating scheme. As the section on energy will show, this concern has been surmounted by the relatively recent developments of thick slurry heating [7] (originally reported by Kumar [8] as discussed by Thomas [9]). Perhaps the key difficulty for slurry heating via pipe in pipe technology (aside from scaling concerns) is the size discrepancy required (versus live steam injection via contacting heating or conventional shell and tube heating), as illustrated in the size comparison shown in Fig. 9.4 [9]. The decision is not simple and needs to be evaluated case by case, based on capital and operating expenditure considerations for the site and bauxite source in question. The most recently built refinery to date, Ma'aden, has elected to utilize contact steam, while Alunorte, also a relatively new refinery (and the largest in the Western World), has elected to use indirect slurry heaters; demonstrating that the decision is not trivial and is based on local economic and environmental considerations.

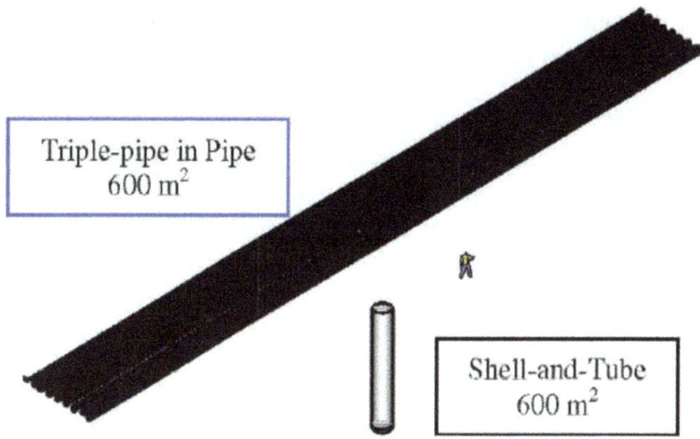

Fig. 9.4 Relative sizes required for equal area heaters using pipe-in-pipe versus conventional shell and tube [9]. Note—for the same heat transfer, a contact heater would require even smaller area

In pre-desilication, the silica chemistry may in fact release a small quantity of water depending on the stoichiometry of desilication product utilized, per the following reaction schemes:

Kaolin Dissolution

$$Al_2O_3 \cdot 2SiO_2 \cdot 2H_2O_{(s)} + 6NaOH_{(aq)} \rightarrow$$
$$2Na\left[Al(OH)_4\right]_{(aq)} + 2Na_2SiO_{3(aq)} + H_2O_{(l)}$$

Desilication Product (DSP) precipitation:

$$6Na\left[Al(OH)_4\right]_{(aq)} + 6Na_2SiO_{3(aq)} + Na_2X_{(aq)} + (y-6)H_2O_{(l)} \rightarrow$$
$$3Na_2O \cdot 3Al_2O_3 \cdot 6SiO_2 \cdot Na_2X \cdot yH_2O_{(s)} + 12NaOH_{(aq)}$$

Net reaction:

$$3[Al_2O_3 \cdot 2SiO_2 \cdot 2H_2O]_{(s)} + 6NaOH_{(aq)} + Na_2X_{(aq)} \rightarrow$$
$$3Na_2O \cdot 3Al_2O_3 \cdot 6SiO_2 \cdot Na_2X \cdot yH_2O_{(s)} + (9-y)H_2O_{(l)}$$

where $X = SO_4^{2-}, CO_3^{2-}, 2Cl^-, 2Al(OH)_4^-$; depending on the concentration profile of the anionic impurities in the liquor stream [10].

The value for y has not been confirmed in the literature, and various sources report values between 0 and 6 (y = 1 [2], y = 0 [11], y = 6 [12, 13]), but consensus seems to prefer y to reflect the lack of certainty [14, 15]. The material impact on the water addition to the process of this inaccuracy is not a great significance owing to the lower mass percentage of kaolinite in the bauxite feed (maximum kaolinite weight percent typically around 12%).

Digestion

In the digestion of the aluminous minerals themselves, the bauxite plus spent liquor slurry must be digested at a temperature dictated by the type of alumina mineral in the bauxite. Of the three main alumina minerals (gibbsite—γ $Al(OH)_3$, boehmite—γ AlOOH and diaspore—α AlOOH), gibbsite requires the least aggressive conditions, ~ 150 °C, followed by boehmite ~ 230–250 °C then Diaspore > 270 °C [16, 17]. In the case of boehmitic and diasporic bauxites, there is a water consumption according to the following reaction scheme:

$$AlOOH_{(s)} + NaOH_{(aq)} + H_2O_{(l)} \rightarrow Na\left[Al(OH)_4\right]_{(aq)}$$

Which does not happen for a gibbsitic bauxite, that reacts according to the following:

$$Al(OH)_{3(s)} + NaOH_{(aq)} \rightarrow Na\left[Al(OH)_4\right]_{(aq)}$$

Note that in some texts, the dissolution product of aluminous minerals has been described as

$$Al(OH)_{3(s)} + NaOH_{(aq)} \rightarrow NaAlO_{2(aq)} + 2H_2O_{(l)}$$

Suggesting a liberation of water for all digestions; however, through various forms of spectroscopy, this reaction product is no longer considered significant and the aluminate ion ($Al(OH)_4$), is considered predominant at Bayer conditions [18]. Note that the material impact of considering gibbsite dissolution via the incorrect reaction is essentially zero, as the production of water is recovered by gibbsite precipitation which proceed via the reverse of the same reaction. Regarding the impact of alumina dissolution on the water balance, it can be seen that in the case of the boehmitic/diasporic bauxites, water is actually removed from the process. Water is used in the digestion to form sodium aluminate and is not returned to the liquor circuit (because boehmite/diaspore is not produced in precipitation; gibbsite is). The reaction scheme for this macro-process is:

$$AlOOH_{(s)} + NaOH_{(aq)} + H_2O_{(l)} \rightarrow Na\left[Al(OH)_4\right]_{(aq)} \rightarrow Al(OH)_{3(s)} + NaOH_{(aq)}$$

If the net of this reaction is taken, i.e.:

$$AlOOH_{(s)} + H_2O_{(l)} \rightarrow Al(OH)_{3(s)}$$

the processing of these boehmitic/diasporic bauxites (also known as monohydrated bauxites) is tantamount to adding water from the Bayer liquor loop to the alumina mineral.

The parameters for temperature of dissolution are not absolute, and in many ways, the temperature setting in digestion is also set by the target yield of a refinery in conjunction with the caustic concentration. To illustrate this point, India's largest refinery operates at atmospheric pressure processing gibbsitic bauxite [19]. This may seem to fly against the comments made regarding digestion temperature, but the concept of digestion temperature cannot be divorced from the target A/TC (or A/C depending on alumina producer nomenclature), commonly known as the blow off ratio. The higher this ratio, the higher the potential yield of the refinery (or productivity per liter of circulating liquor). To illustrate with actual numbers—while a 150 °C digester (gibbsitic) refinery running a 300 TC concentration in spent liquor may be able to obtain a A/TC of 0.750, the Indian refinery counterpart running at 105 °C may only sustain a blow off ratio of 0.550. Without increasing the temperature, or increasing the caustic concentration considerably, the higher ratio will simply not be accessible as it exceeds the thermodynamic limit at atmospheric conditions [20]. Thus, it is still possible to run the refinery at the reduced temperature.

For the purposes of discussing the water balance within a refinery, a digester is an evaporator (if the temperature of digestion is above boiling point) as Fig. 9.5 illustrates [21]. Post digestion, the slurry is discharged into flash tanks which operate at gradually lower pressures allowing the slurry to evaporate an amount of steam dictated by the vapor liquid equilibrium. Ultimately, the slurry finishes this process at atmospheric pressure, in what is commonly known as the blow off tank. For the purposes of the water balance, the number of flash tanks and the efficiency to which steam is removed to heat exchangers heating up the incoming slurry/liquid is not consequential. Obviously, for energy efficiency these details are important, but to understand the amount of water being removed, the key parameter is the digestion temperature. If the heat transfer efficiency is poor, or the heater areas are undersized, this will manifest itself in the final flash tank, the blow off tank, evacuating a large amount of steam to the atmosphere.

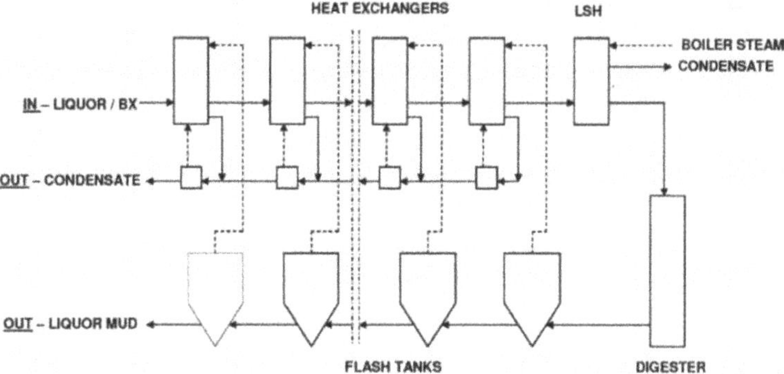

Fig. 9.5 Generic scheme of a (single stream) digestion unit with arbitrary number of flash tanks. Adapted from Donaldson [21]

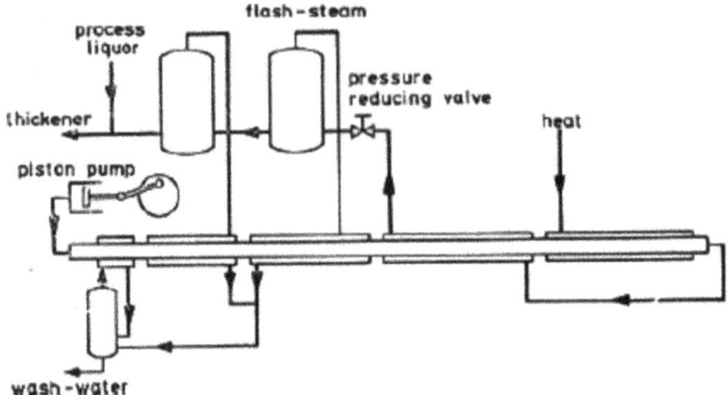

Fig. 9.6 Unique digestion flowsheet [22] highlighting tube high temperature digestion and the sacrifice of one flashing stage to allow the digested slurry to intimately exchange heat with the incoming colder slurry

The other key aspect to the water balance is the mode of heating utilized in the last liquor/slurry heater. Historically, but even in some modern refineries (e.g., Ma'aden), the last heater has been heated via contact steam, i.e., directly injecting the steam; which reduces the evaporating capacity of the digester as the water injected needs to be removed. The more efficient mode (from a water removal perspective) is to utilize indirect heating via conventional shell and tube heating per the previous heating stages. As the approach temperature (and U factor) is lower for the indirect option, the required pressure of the powerhouse steam is higher. The differences of these two modes of operation are further expanded in the energy section. In any case, both modes of operation will remove water from the liquor stream, with a greater impact associated with indirect heating.

In the context of tube digestion, the principles regarding water removal are the same. A unique (older) flowsheet has been reported which would reduce the evaporation extent in digestion as shown in Fig. 9.6 [22].

Here it can be seen that one stage of flashing has been given up allowing the digested slurry to more intimately exchange heat with the incoming slurry. Note that for this particular flow sheet, the process in question is dissolving diasporic bauxite which necessitates digester temperatures in excess of 270 °C. Thus, while some of the evaporating capacity has been lost, the substantial amount of evaporation available in going from ~ 250 °C to atmospheric boiling point (~105 °C) would most likely justify the reduction in evaporation (notwithstanding the benefit in energy performance associated with the change in heat transfer).

9.1.1.2 Clarification

Within the clarification circuit, the largest input to the water balance (and for the entire refinery) is the washing of the residue to minimize soluble soda losses. If water is not added in this part of the process, and the digester discharge slurry were simply separated, a significant amount of caustic soda (and associated soluble alumina) would be rejected with the residue leaving to the disposal area. To combat this, a 'chain' of Counter Current Decantation (CCD) washers is installed, which lowers the concentration of the caustic soda leaving with the residue (see for example [23, 24]). The more wash water utilized in the train, the less soluble soda that is lost. Further to this, the water added will need to be removed. There exists a balance between the soda and energy costs of the process that is largely driven by how much water is added to the process. Evaporation capacity has a key role, as one cannot simply choose to save as much soda as possible by adding an exceptional amount of water to the process, without the corresponding evaporation capacity to remove this added water.

To best understand how the CCD circuit impacts the soluble soda loss in conjunction with the water addition, it is useful to review the hypothetical CCD circuit shown in Fig. 9.7.

Here a five-stage circuit (one thickener with four washers) is separating a digester discharge slurry with ~ 5% solids and a total alkali concentration of ~ 270 gpL. The washing medium here is pure condensate (as opposed to weak caustic solution, i.e. lake water), and the mixing efficiencies are assumed to be 100%, for the sake of illustration.

Figure 9.8 and Table 9.2 show how the concentration profile, as well as the soluble soda loss are impacted down the train, with two key variables changed—the washer underflow solids concentration and the wash ratio, or the tons of washing water added to the last washer, per ton of residue. The quantitative values themselves are not as important as the qualitative trends here. It can be seen that for two conditions with essentially the same TA leaving the washer train, the soda loss can be hugely different, owing to the difference in solids concentration. Thus, while the amount of water added is key to the soluble soda loss reduction; another enabler that has changed throughout the history of the industry is the solids concentration of the material leaving the Bayer process to go to the residue storage area.

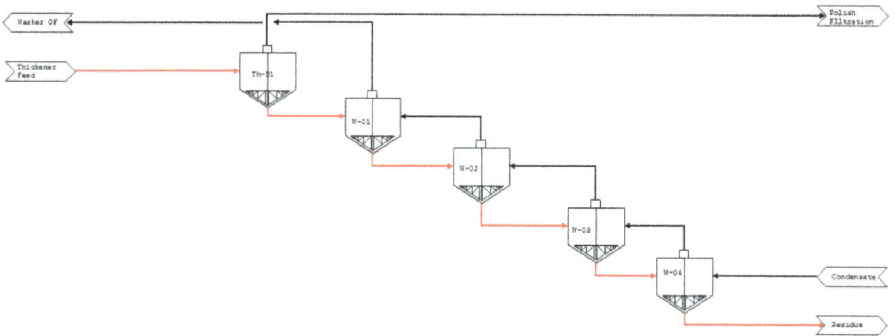

Fig. 9.7 Hypothetical CCD circuit with digester discharge TA 270 gpL. Red indicates slurry flow, while black indicates condensate or liquor

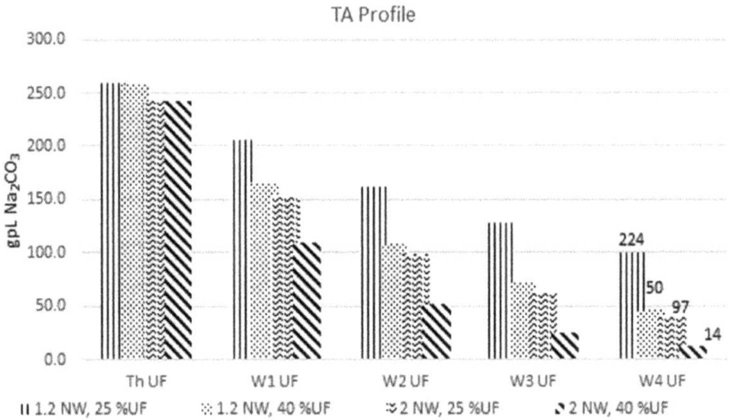

Fig. 9.8 TA Profile for hypothetical mud washing circuit. Numbers above W4 UF indicate the soluble soda loss leaving the refinery, in kg NaOH/t Al_2O_3 production. NW = Net wash

Table 9.2 TA profile and soluble soda loss for hypothetical CCD circuit with differing UF solids concentrations and net washes

Net wash (t H_2O/t res)	1.2	1.2	2	2
UF solids concentration (mass %)	25	40	25	40
Soluble soda loss (kg NaOH/t Al_2O_3)	224	50	97	14
TA profile	gpL Na_2CO_3			
Th UF	259	257	242	242
W1 UF	205	166	154	109
W2 UF	162	108	98	52
W3 UF	127	71	63	25
W4 UF	100	46	41	12
Soluble soda loss profile	kg NaOH/t Al_2O_3			
Th UF	502	225	468	212
W1 UF	417	157	321	109
W2 UF	343	108	218	55
W3 UF	278	74	146	28

In older refineries, where solids concentration leaving the washer was approximately 35% maximum, the clarification circuit could easily be a thickener and washer train spanning 8 vessels or more, to ensure that the soda lost leaving the last washer was as low as feasible. This approach, which required a significant capital investment as well as land area, was essentially constrained by solids concentration; being a function of the geometry of the vessels used, in conjunction with flocculation technology and pumping capacity of the time. This form of residue disposal is typically dubbed 'wet disposal'. In the 1980s [25–27] alumina producers converged

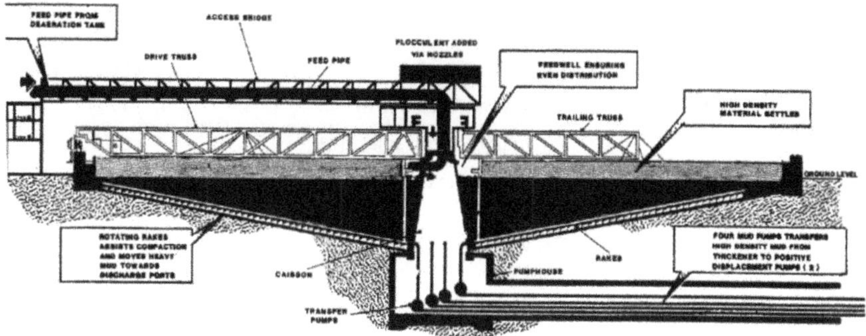

Fig. 9.9 Super thickener for residue mud, installed at Pinjarra alumina refinery in 1986, as reported by Cooling [25]

on different thickening techniques to increase this solids concentration, allowing the residue to be disposed of at a significantly higher solids loading. The Alcoa approach favored 'super-thickeners' [25] which were new designs based on the conventional washer/thickener skeleton but with much larger diameters, that allowed for much higher solids loading; see an example in Fig. 9.9. The Alcan approach favored the use of deep cone thickeners to achieve the same end [26]. Concurrent to these advances, direct filtration of the residue, which had been present in the industry for producing 'wet stacking' ready residue already, simultaneously improved, with modern day drum filters achieving ~ 60% solids concentration [28, 29].

It is important to realize that these technological advancements were not developed for soda/energy purposes in isolation—the main benefit was to convert from a wet stacking residue disposal to a dry stacking disposal, whereby the thixotropic high solids content residue slurry is layered in the residue area and allowed to dry prior to subsequent layer deposition [27]. This, in conjunction with mud farming techniques (see for example [30])—whereby the underlaying moist residue is re-exposed to the surface for further drying, has enabled a much more benign residue area and increased the capacity of the residue area substantially. Further to this, the ability of adding to existing residue capacity was also made easier [25].

In the last decade, the industry has seen a subsequent development in this field, achieving even lower moisture contents, targeting as high as 80% solids concentration through pressure filtration technologies (see for example- [28, 29, 31–33]. This new disposal technique, dubbed 'dry disposal', removes the thixotropic nature of the residue and allows for a sandier, even lower soda residue.

When considering the increase in solids concentration afforded by technological advances in solid liquid separation, in a brownfield operation, there is an added layer of complexity that must be addressed. If the net wash (as defined earlier, the tonnes of H_2O per tonne of residue) is maintained the same after substantially increasing the solids concentration of the residue leaving the process, the soluble soda loss will go down, but the net dilution, that is the volume of water entering the refinery via mud washing (or difference between wash water and liquid leaving with underflow), will

Table 9.3 Net dilution differences associated with maintaining the same net wash as solids content in residue increases

Parameter	Drum filtration	Press filtration
Solids content (%)	64	80
Solids flow (Basis) (t/h)	1	1
Dilute liquor with solids (t/h)	0.56	0.25
Vol flow out (SG solids ~ 3.3) [m^3/h]	0.87	0.55
Net wash (t H_2O/t dry residue)	1	
Vol flow In (m^3/h)	1	
Net dilution (m^3/h)	0.13	0.45

go up. This is illustrated by example for the drum filter versus press filter example, relevant to Alunorte per the work of de Miranda et al. [29] in Table 9.3.

Thus, without changing the net wash, the evaporation rate of the refinery would have to increase. Fortunately, owing to differences in washing efficiency of the filtration technologies at different solids concentration, it is typical that the net wash can in fact reduce for the same (or improved) soluble soda loss. Thus, one may think that the advent of this sort of filtration improvements allows for simultaneous reduction of soluble soda loss without increasing the requirement on evaporation. More important in this analysis though, is the impact on the liquor impurities.

By reducing the amount of liquor leaving with the residue (in Table 9.3, the flow out reduced from 0.56 to 0.25 t/h), the bleed for impurities to leave the process is also reduced, meaning that their concentration in the main liquor can in fact increase, and negatively affect the liquor yield. Consideration of these technologies which reduce the water demand, lower the soluble soda loss and generate a less hazardous residue, must be weighed together with the entire flow sheet to ensure that production rate and product quality are not negatively impacted.

Note that this section has focused only on the CCD component of the clarification process, ignoring sand removal altogether. Sand removal, via sand traps (or cyclones) and classifiers are an integral part to some refineries; namely those with large quantities of coarse solid material in their residue. However, from a water balance perspective, the concepts regarding water balance are essentially the same. The main extra consideration for refineries running sand removal is the split of washing water to the CCD Circuit and the sand classifying equipment itself. The alumina producer defines the split such that the total soluble soda loss considering both the mud and sand streams is minimized. This analysis is completed utilizing process modelling software together with the plant's flowsheet and the particle size distribution of the residue.

9.1.1.3 Heat Interchange and Precipitation

Within the heat interchange part of the process, the hot green liquor transfers heat to the returning spent liquor from precipitation. The main ways that this heat transfer can be affected include: (1) using a shell and tube heater with flash tank arrangement, operating under partial vacuum (see Fig. 9.10), or (2) using plate heat exchangers with green and spent liquor flowing counter current; or (3) some combination of the two. If operated using shell and tube heat exchangers with flash tanks, water is removed from the process. Counterarguments suggesting the use of plate heat exchange center around the improved energy performance. This balance will be discussed further in the subsequent section on the Bayer heat balance.

To achieve the last part of trim control on the green liquor temperature, refineries can also use a final heat exchanger with an external water source—either a cooling pond or cooling tower exchanging heat as a barometric condenser, or as a plate.

The purpose of these cooling steps in green liquor (aside from heating the counter current spent liquor stream) are to destabilize the liquor, increasing the driving force for precipitation to increase yield. In utilizing coolers that remove water from this stream, the concentration of caustic increases and acts to undo some of this destabilizing. The net impact is reduced stability, however cooling via non-evaporative means allow for more potential to be realized. This is the key argument against flash cooling green liquor in heat interchange and in trim control of green liquor temperature feeding temperature. This argument continues into the precipitation building, where the precipitating slurry is cooled between some precipitators to further destabilize the slurry. Again, this can be done by flashing means or by plate exchange—in this case, the predominant mode favors the plate option (see for example—[1, 34]) with a cooling pond/cooling tower, or in some cases, the spent liquor. Historically, there was some hesitation to utilize plate exchangers for such a high solids content slurry,

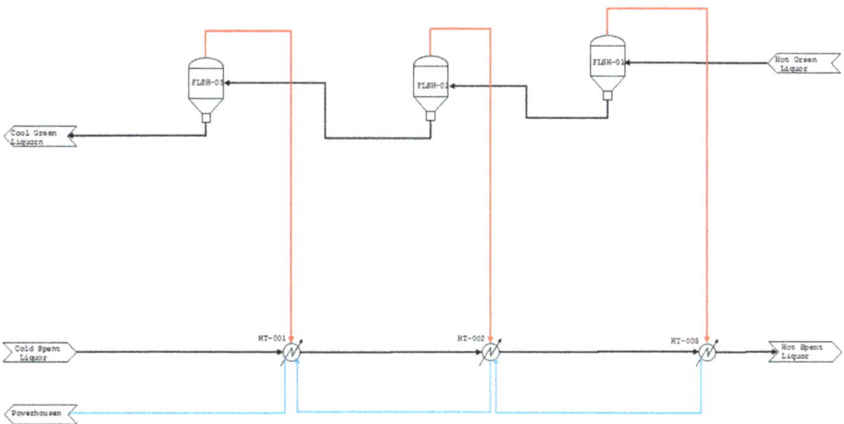

Fig. 9.10 Example Heat interchange building utilizing shell and tube heaters with flash tanks. The black arrows indicate liquor flows, red indicates steam and blue indicates condensate

Table 9.4 Throughput and wash water improvements associated with upgrade from drum filtration to pan filtration of hydrate [35]

	Original drum filters	Designed pan filter
Manufacturer	Dorr Oliver Eimco	FLSmidth
Filter area (m^2)	80	71
Specific capacity ($kg/(m^2.h)$)	1200	2584
Capacity (t/hr dry hydrate)	96	183.5
Density of slurry (g/L)	1320	1350
Water ratio (m^3 water/t hydrate)	0.8	0.4
Steam (5 bar) (t/h)	2	N/A

as seen in precipitation, due to the perceived scaling/plugging risk. However, based on developments in uniquely designed plate exchangers (platulaires, manufactured by Barriquand), first reported in the 1990s [34],[1] this concern was overcome.

The other key addition of water to precipitation (and indeed the process) is to the filtration of the coarse product from the precipitation building that will feed calcination. If the product is not washed sufficiently, the soda content in the product will exceed the specification. Unlike the residue washing in clarification, there is less room for operational flexibility here as the product specification of the refinery is not a free variable in the process. The other key objective is to reduce the moisture as much as possible to ensure for efficient calciner operation (i.e. energy should not be wasted boiling free water out of the hydrate feed). The hydrate must leave with some moisture, and within the liquor loop this is seen as an exit point for water from the process.

The key improvements that have allowed for reduction in required dosage of water typically come from filter improvements themselves. Excellent examples from the industry show replacement of drum filtration technology with pan filters to (among other benefits) reduce the water ratio (i.e. m^3/t dry hydrate) by 50%, for the Aluminum Oxid Stade (AOS) refinery in Germany using a FLSmidth design, as shown in Table 9.4 [35]. Similar improvements have been noted upgrading a three-stage drum filter to pan filters using BOKELA designs [36].

Another potential addition of water addition within Precipitation to the Bayer loop is in the form of caustic make up in precipitation. While a great deal of equipment all throughout the process is caustic washed, the greatest usage, based on the size of the vessels used, is necessarily precipitation. As a precipitator is on-line, alumina scale deposits on the tank's walls (and all exposed surfaces) reducing the effective volume, to the point where the tanks must be taken out of service for caustic cleaning [37]. After cleaning, the "spent" (used) caustic solution, if too high in A/TC for subsequent cleaning, is discharged back into the process; if the A/TC still has some

[1] Thomas [9] suggests ~ 1985 but this is based on an unofficial personal communication.

potential for dissolution, it can be reused for subsequent cleaning. The exact limits for this decision are not necessarily relevant to the water balance but are defined by solubility considerations at the temperature of the caustic wash, which is constrained to atmospheric boiling point, as the precipitators are not pressure vessels. This is not in fact an extra water addition to that already mentioned, as caustic soda water ingress was already discussed in the section on Raw Materials. However, raw caustic cannot be used 'as is' for caustic cleaning as the risk for caustic embrittlement of the tanks is too high [38, 39]. To combat this, the solution must be diluted, to a concentration that is safe from a stress corrosion cracking point of view, normally defined by each refinery, due to a lack of consensus in the open literature about these limits [40]. There are two main ways that this dilution can be achieved:

1. Use fresh water/lake water and add to the water inputs of the process.
2. Use spent liquor.

This, as always, is a design decision that must be made, looking at whether the refinery can handle the added ingress of water comfortably, or requires extra evaporation capacity to safely manage. If option 2 is chosen, i.e. the refinery cannot handle the added water for removal (and does not want to add extra evaporation capacity), then it must be acknowledged that the time to clean tanks will necessarily be longer than option 1—the starting solution in 1 will have an initial A/TC of 0, while that of option 2 will be greater than 0 and less than the spent liquor A/TC (typically ~ 0.4). For a 95 °C cleaning with 400 TC solution, the maximum practical A/TC will be near 0.5 (Rosenberg-Healy [20] solubility 0.597)—the further from this practical limit, the faster the dissolution. Electing option two may then lead to the decision to invest in more precipitators than is strictly required to handle the slower maintenance schedule.

9.1.1.4 Evaporation

The evaporation building in many refineries is typically a series of shell and tube heat exchangers connected to flash tanks, working to essentially close the water balance, if needed. In a shell and tube evaporation building, live steam is added (indirectly, of course) to the final heater/s to guarantee the liquor has sufficient temperature and pressure to commence flashing in the subsequent flash tanks (often in excess of seven effects, or stages). Figure 9.11 shows a possible layout for the evaporation process. While some steam is required, the economy of the evaporator (i.e., the mass of water evaporated per mass of steam added) can be very good, with values of 3 t/t not uncommon. The comment "if needed" regarding evaporation buildings, must be stated as it is true that not all refineries have needed an evaporation to close the water balance. Hudson et al. note that for a large part of its operation, an Alcoa refinery (presumably of mid-twentieth century vintage or earlier) in the US did not require an evaporation building while running a low temperature digestion process on a high-grade bauxite [2]. Wargalla and Brandt [17] demonstrated that the evaporation building in Alumina Oxid Stade (AOS) (running a high temperature digestion for a

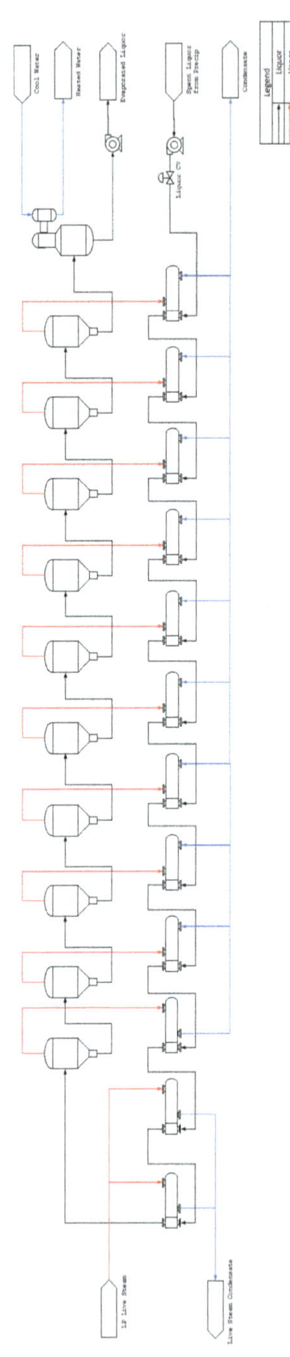

Fig. 9.11 A possible flowsheet for a shell and tube heater-flash tank evaporation process

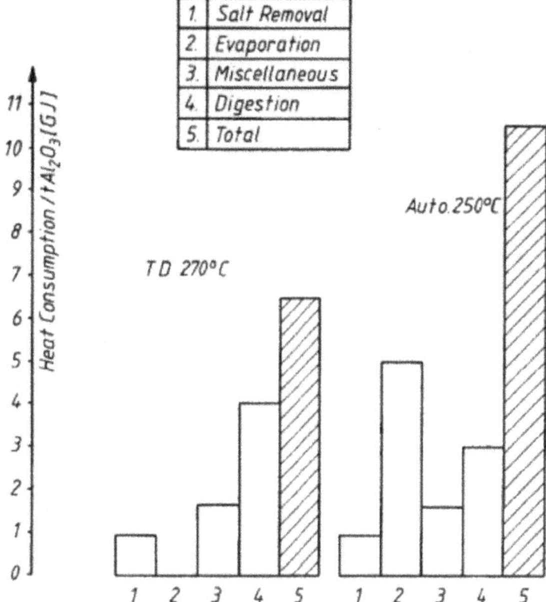

Fig. 9.12 Energy usage for AOS refinery comparing tube digestion with autoclave; showing that the evaporation building could be removed

diasporic bauxite) could be removed through adjustment of the plant concentration and digestion temperature and vessel, as is demonstrated in Fig. 9.12.

This point is not made to suggest that the goal of a refinery should be to remove the evaporation building; the takeaway is that like most operating parameters in a refinery, the need for and extent of evaporation required to close the water balance, depends. It depends on the type of bauxite being processed, the nature of solid–liquid separation and the target soluble soda loss, the way in which heat exchange is conducted between green and spent liquor and the technology utilized for hydrate filtration prior to calcination at the very least.

9.1.1.5 "Other" Water Ingress

In looking at the final generic input to the Bayer loop water balance, the term other is used to capture the ancillary extra uses as well as uses that are less standard and used only in some locations. Firstly, the genuine ancillary uses, which includes:

Hose water

Be it from a condensate source or a lake water source, hose water is needed for the maintenance and hygiene of an area. As lines and tanks are drained for routine maintenance, operational staff must use hose water to maintain area hygiene and get all of the liquor/slurry to the sump areas.

Many alumina producers utilize Kelly filters or vertical pressure filters to 'polish' the green liquor overflow from the thickeners in clarification. These filters which are operated in batches, require cleaning (typically with hose water) after each batch to dislodge the deposited filter cake off the filter cloth. This (admittedly small) contribution to the water balance can be prevented against through use of the much older Alcoa Process Filter or sand filter [41]; or alternatively, the state of the art fully automated vertical leaf filter, which thickens the dislodged mud cake and discharges without hose water (provided by Gaudfrin as a Diastar filter [42], or Bokela as a Backflush Filter [43].

Having noted these legitimate uses of hose water, there remains a host of interesting applications for hose water that have been observed by alumina producers that are not sanctioned. Examples include sealing pumps (beyond the packing water supply) to avoid repacking a pump; maintaining a seal in a vacuum pot; assisting a weak pump by thinning out the feed solids with hose water; replacing injection liquor that is supposed to thin out feed solids due to injection line blockages; to name but a few. In many ways, this is the area that a refinery can truly control in each day of its operation through diligent operational and maintenance practices.

Packing and purge water

Many refiners utilize centrifugal pumps that require packing assemblies to ensure that the Bayer liquor does not backflow out of the pump. This packing material must be cooled to prevent friction, with clean water, which inevitably enters the process liquor. The packing material wears and must be replaced; otherwise, the water demand increases and unnecessarily dilutes the process stream. A similar argument exists for purge water which is necessary to protect instrumentation from the liquor stream. The alternative to the packed pump is to use mechanically sealed pumps, and many refineries may tend this way. The transition can be perceived as expensive (for brownfield operations) and maintenance practices are critical (and expensive if not done correctly) [44].

Rainfall

If rain can fall into open top tanks or areas that direct to sumps which re-direct into the process, then this can add substantial dilution to the process. Regarding bunded areas, in high rainfall climates two options are suggested in the open literature [45]:

(1) Prevent rain from hitting the bunded area through roofing (although horizontal rain provides a challenge).
(2) Prevent process liquor from hitting the bunded area, by rerouting overflows. This would mean that rain ingress into the bunded area could be safely discarded to the environment after checking against environmental contamination (e.g., online conductivity for caustic soda monitoring).

Maintenance Activities

Examples requiring water in maintenance include high pressure water blasting as well as drilling of tubular heaters. The water discharging from these activities may ultimately hit a bunded area and re-route into the process. These activities are necessary but there can be ways to save water ingress. A good example of this is noted by Praveen [46] who found that a US alumina refinery using water blasting and drilling equipment had the supply water tanks overflowing potable water at a rate of ~ 2.3 m^3/h as the tank had to be kept full. The addition of a float valve to regulate this usage was a simple solution estimated to save $13 k per year.

Contaminated Condensate

For condensate to be routed to the powerhouse for boiler make up or precipitation for hydrate washing, it must be of sufficiently high quality, with specifications defined around conductivity monitored online (as a surrogate for caustic soda concentration). If these specifications are exceeded, the condensate needs to be replaced by another water source (commonly demineralized water for boilers) and the plant overall water usage increases. While it is not the intention of the refinery to produce contaminated condensate, this can happen in the digestion and evaporation areas if: (i) a flash tank does not separate the steam from the slurry correctly or (ii) a tube breaks and spills process fluid into the shell side. A flash tank 'carryover' event can be caused by scale within the flash tank shedding due to thermal shock and blocking the outlet, or perhaps during a heater swap if sudden changes are made. A more common and gradual carryover event occurs simply due to higher vapor velocities not allowing caustic to be separated at the vapor outlet of the flash tank. These elevated vapor velocities are the byproduct of scale formation at the vapor outlet, which builds up over time. If some form of carryover does occur, a flash tank may need to be bypassed, or a digester taken down (depending on the operational flexibility of the design) to remove the blockage. If the maintenance schedule for flash tank cleanings is well defined and adhered to, this phenomenon should be avoidable. Recent developments in flash tank design, favoring bottom entry, suggest that this residual life can be substantially increased versus the side entry conventional design [47]. Particular aspects of contaminated condensate generation and prevention are discussed further in the section on Energy.

Lime, Additives, Oxalate Control

As Smith well notes, "lime is often considered as a panacea" in the Bayer process [48]. Its use is ubiquitous and finds application in many different areas. Regarding the water balance, each one of these additions is an addition of water as the lime is added to the process as a dilute caustic or water slurry at ~ 250 gpL solids concentration. Regarding the impact on the water balance, associated with adding a hydrated lime water/weak caustic slurry to the process, the governing reaction is simply:

$$CaO_{(s)} + H_2O_{(l)} \rightarrow Ca(OH)_{2(s)}$$

While this reaction can be seen to be consuming water, the addition of slaked lime to the process as a slurry (with the concentration dictated by the temperature controller of the slaker as the slaking reaction is exothermic and the slaker operates at atmospheric pressure) means that water is in fact added to the process.

Other additives utilized in the process include flocculants to assist in settling of red mud slurries as well as hydrate slurries; crystal growth modifiers in precipitation; rheology modifiers for residue transport or bauxite slurry to digestion throughput to name the most predominant ones. These additives are mentioned here for completeness, but in the context of the water balance, their concentrations are so small (i.e. ppm) that the impact is not substantial.

Oxalate control is also noted in this section as there does exist an application where water is required for oxalate control. Not all refineries are impacted by the 'organics' problems which haunt the Bayer process; and that there are a host of different ways for handling the issue, which are internal to an alumina producer's intellectual property. A strategy that has been employed is coprecipitation of oxalate (see for example [49]). In this process, oxalate is co-precipitated with gibbsite. In the seed filtration step where the fine seed is returned to the initial agglomeration tanks of the precipitation circuit, the seed is washed with water which dissolves the oxalate into the filtrate, prior to sending the solid seed to the precipitation circuit. This oxalate enriched filtrate is then handled separately; but the impact of the water wash adds to the water load that must be removed by evaporation.

9.1.2 Greater Refinery Water Balance

In considering the water balance of the outer envelope or boundary of Fig. 9.2, what happens to the water that is leaving the liquor loop of the Bayer process? To answer this, it is useful to consider a breakdown of the inputs and outputs just discussed as shown in Fig. 9.13. It can be seen in this simple example that except for the bauxite residue moisture and product moisture, almost all the other condensate outputs will return to the refinery for mud wash and hydrate wash activities. Further, it must be realized that in some refineries (typically dual stream digestion), a portion of the

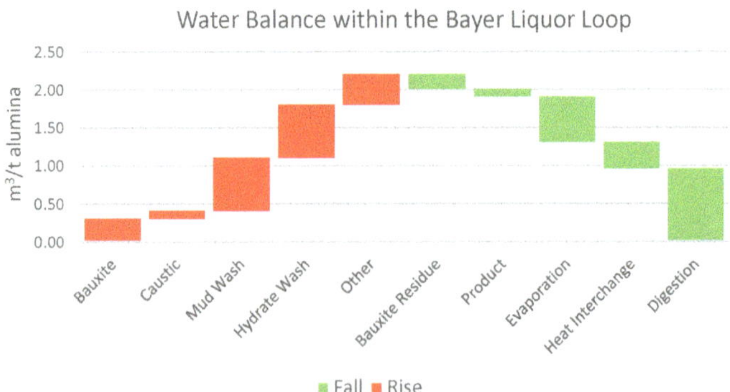

Fig. 9.13 An example Bayer process water balance (derived from data presented by Wischnewski et al. [6]), showing the key inputs and outputs

digestion water 'leaving' is exiting the process as steam out of the blow off tank. In the context of this particular example, any excess condensate will have to leave the refinery and go to water treatment.

The product moisture is easily accounted for—it goes to calcination where the moisture in the hydrate is converted to steam and exits to the atmosphere. The residue moisture deposits in the residue area, where a modern refinery will conduct dry stacking that allows for some water to evaporate and the remainder to be collected by a drainage system (normally an under-drain system). One of the interesting developments of transitioning from wet stacking to dry stacking is the increase in the mismatch in a refinery's evaporative area versus the rainfall catchment area. This is shown schematically in Fig. 9.14 [50].

The rainfall can collect on the entire surface of the dry stacked area and drain to a smaller catchment area. Understanding where exactly water goes and how it gets there (i.e., quantifying—sedimentation, consolidation, desiccation, and rain infiltration versus runoff) has recently started to get attention as described by Li and Harris in work for RTA [51]. This approach can be particularly useful for determining necessary surge volumes for decant ponds. As Cooling describes [50], the runoff water is stored in decant ponds during the winter, higher rainfall months, with the under-drainage supplying the wash water to the clarification circuit. The transition then occurs in the dry summer months where evaporation is high, allowing for the stored run off water to be used, with additional freshwater as needed. The approach

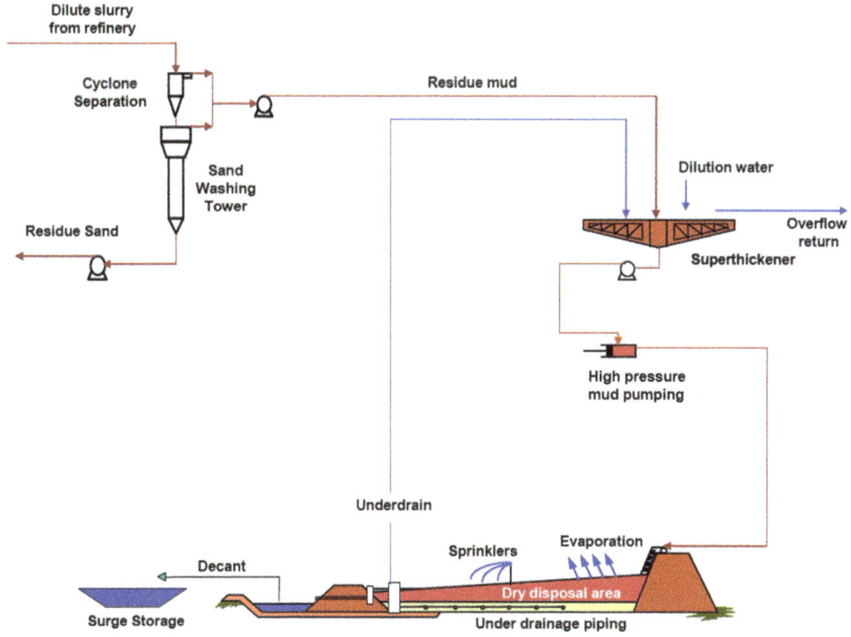

Fig. 9.14 Schematic of dry disposal showing much greater catchment area versus evaporative [50]

Table 9.5 Water permit limitations set by the Federal national pollutant discharge elimination system and texas pollutant discharge elimination system applicable to Pt. Comfort Alumina refinery discharging to Lavaca Bay

Component	Daily Average	Daily Max
Flow (km^3/day)	16.4	24.5
pH	6.0–9.0	6.0–9.0
TOC (mg/L)	55	75
TSS (mg/L)	101.8	209.8
As (mg/L)	0.1	0.3
Cr (mg/L)	0.1	0.36
F (mg/L)	37.2	78.2
Ni (mg/L)	0.14	0.47
Se (mg/L)	0.1	0.3
Zn (mg/L)	0.37	1.2

TOC total organic carbon, *TSS* total suspended solids [52]

will be climate specific with significant analysis of the location's rainfall data to guarantee safe design of surge areas.

An interesting case study is provided by Smith et al. [52] in describing how the transition from wet stacking to dry stacking at Alcoa's Pt. Comfort refinery forced the introduction of water treatment facilities. Given a fixed refinery footprint (being a brown field operation) not allowing for (easy) water surge increase, combined with a weather system that can favor brief and intense storms, the need for water treatment was essentially forced on the refinery. Water treatment for discharge to a local water body is not done in refineries to a common standard, largely due to different legislations and proximity to water bodies. The Texas case study referenced had to consider various heavy metals and Total Organic Carbon against locally defined allowable limits, as shown in Table 9.5.

Beyond the acid treatment methods, or acid plus additive methods, seawater neutralization is another option (both for water discharge as well as solid residue neutralization) with practical refinery application demonstrated by Yarwun, QAL and Eurallumina [53].[2] Simply, the residue slurry is mixed with seawater, which due to its large buffering capacity allows the residue to be deposited as a dilute slurry (with pH ~ 8.6). The supernatant liquid from this operation can then be discharged to the main water body with no impact on local ecology [54]. Simultaneous to this, there is no need for the conventional geosynthetic-clay lining at the bottom of the disposal area; under drainage mixes with the groundwater, ultimately discharging to the same water body.

Another water management that may need to happen outside of the process liquor loop regards cooling water for barometric condensers (or process coolers). If the cooling water comes from a cooling pond, then all that is required should be environmental monitoring of the material returning to the pond, to ensure it is not contaminated from the process stream. For a cooling tower, standard cooling tower practices

[2] Referenced by Cooling [50].

must be followed whereby the cooling water is purged and replenished (with a clean water source, e.g., potable or surface water). Recent reports from additive suppliers have suggested that addition of proprietary chemicals can allow tighter control of the water quality as well as the water requirements for replacing purged material [55].

9.1.3 Some Operational Considerations

Beyond the central ideas regarding the design of how the refinery is supposed to work, it is worth mentioning some important points relating to the actual running of the refinery day to day. In regard to guaranteeing the water balance stays under operational control, adherence to the maintenance strategy and schedule is of great importance.

Precipitators (and other large tanks) coming in and out of service can have significant impacts on the plant volume (as well as the plant's caustic concentration) and must be respected accordingly. Not taking large tanks out of service as planned, in conjunction with heavy rain events can spell trouble for a refinery that can be hard to get out of. To push that particular scenario a little further, it is helpful to understand what *can* happen in a refinery to accommodate such uncomfortable operational positions.

In the scenario where a large rain event coincides with delaying addition of a tank into the live circuit, the plant volume is now compromised and must be shrunk. How can this be done? The discipline with managing the ancillary water usages (e.g., hose water usage, packing water usage, etc.) should be brought to the foreground and minimized as much as possible. If there is no spare evaporation capacity, a refinery may elect to increase digestion temperatures to achieve the same end as a spare evaporator (albeit to a lesser extent). This will be almost immediately limited by over pressure protection limits in digestion, so there will not be too much 'room' for extra evaporation. The digester is not as efficient an evaporator as the evaporation building, so this is a more deleterious move on the site's energy usage. If the site could increase blow off ratio simultaneously, i.e., capitalizing on the thermodynamic ceiling being lifted by the temperature, then this may in fact be a useful strategy. However, it is very unlikely that a site will have available yield waiting to be taken due to a water balance problem!

Returning to controlling the water situation, the other option is to reduce the net dilution (or net wash) in the residue washing CCD train. This will cost a higher soluble soda loss. In the event that the net dilution is reduced substantially, the washer train concentration profile increases, which will ultimately translate to higher concentration material feeding precipitation. This can potentially negatively affect yield as well as product quality. To avoid this, a refinery could simultaneously reduce the spent liquor caustic concentration and blow off ratio in digestion to maintain stability in precipitation; and by now the production has almost certainly been impacted.

9.1.4 Concluding Remarks on Water Balance

It is hoped that this section has served to outline some of both the complexity and history associated with water balancing a Bayer refinery. An appreciation of how the water gets in, where it gets out and, most importantly, where it is desirable for it to get out is necessary to grasp the refinery process dynamic. This dynamic is more complex than the water balance itself, due to the flow on effect of subtle variations in the water balance, as it was illustrated in the section on Operational Considerations. If the water balance as designed is not respected and maintained per its design, it inevitably will affect production and a host of other metrics, owing to the cyclical nature of the Bayer liquor loop.

Beyond the inherent complexities of the Bayer loop, which are governed by different unit operation technology choices as well as bauxite, the greater water balance discussion has aimed to show that the "best" approach or design is a misnomer. A global design cannot account for (not always) subtle local variations in key metrics that dictate water requirements. The local geography, the average evapotranspiration and its variance, the bauxite treated as well as the local environmental legal requirements are but a few choice examples that illustrate the need to customize the overarching philosophy into a unique solution.

What perhaps will become clearer in the next section is the inherent overlap between the water balance and the Bayer processes energy balance, which is often a slave to the water balance in certain areas. It is this at times complex interconnectedness, that can control even more of the key decisions to be made when designing and operating efficient refineries.

9.2 Bayer Process Energy Balance

Understanding and optimizing the energy usage within the Bayer process is one of the most important challenges, given its substantial contribution to the operating cost, combined with the environmental impact associated with wasting fossil fuels. On top of this, energy commodities face the greatest potential for cost volatility when compared to the other major contributors to a refinery's OPEX (namely bauxite and caustic soda). In a similar vein to the treatment applied to the water balance, the energy balance can be broken down to understand the key contributors to the liquor loop's energy usage. What will become clear in each section is the importance of liquor yield in optimizing a refinery's energy usage. Obtaining more alumina from the same volume of liquor inherently reduces the amount of heat that must be applied per tonne of alumina produced. Note that the energy associated with calcination and the powerhouse is not included in this section and is treated elsewhere.

To set the backdrop for the industry, the common metric for energy performance is the GJ/t. The typical range for the industry (using the Bayer process to refine

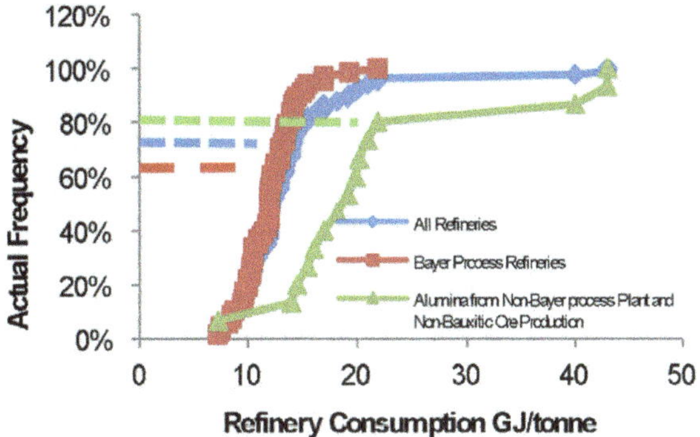

Fig. 9.15 Distribution of alumina refinery energy consumption (2013), showing both Bayer and non-Bayer refineries [56]

alumina) is between 7.2 and 21.9 GJ/t, with the full range of values for 2013 shown in Fig. 9.15 [56].

9.2.1 Digestion and Evaporation

9.2.1.1 Heaters

Digestion and evaporation are the greatest consumers of energy within the liquor circuit as these are the areas where substantial heating of the liquor stream is conducted. Owing to a design skeleton which is over 80 years old, the energy added to the digester is largely recovered and used to heat incoming feed, by virtue of the Alcoa patented Continuously Regenerative Digester (CRD) [57]. Its close relative is of course the Continuously Regenerative Evaporator (CRE) which is used in many Evaporation circuits to control a plant's water balance. The original patent represented the move from batch digestion to a continuous process; but with the flash vapor directly contacting the incoming liquid to transfer heat, and thereby reducing the evaporative capacity of the digester. A subsequent patent soon followed [58] which removed the contact steam component and allowed the heat transfer from flashing to occur through indirect means; the original flow diagram is shown in Fig. 9.16.

Dual Versus Single Stream

In the original CRD, the material being heated indirectly was only the liquor. Since the advent of pre-desilication into the Bayer process, (formally ~ 1968 [59]), the bauxite slurry needed to be heated from its approximate temperature of 75 to > 90 °C to ensure the silica reaction extents are satisfactory. As noted earlier, this was done by contact steam generated from the flash tank. From an energetic standpoint, the critical flaw of the dual stream digester (e.g., liquor is in one stream, bauxite slurry is in a separate stream) is that the heat source in the CRD, namely the digested slurry, is larger than the heat sink, which is ostensibly the liquor stream (as the bauxite slurry

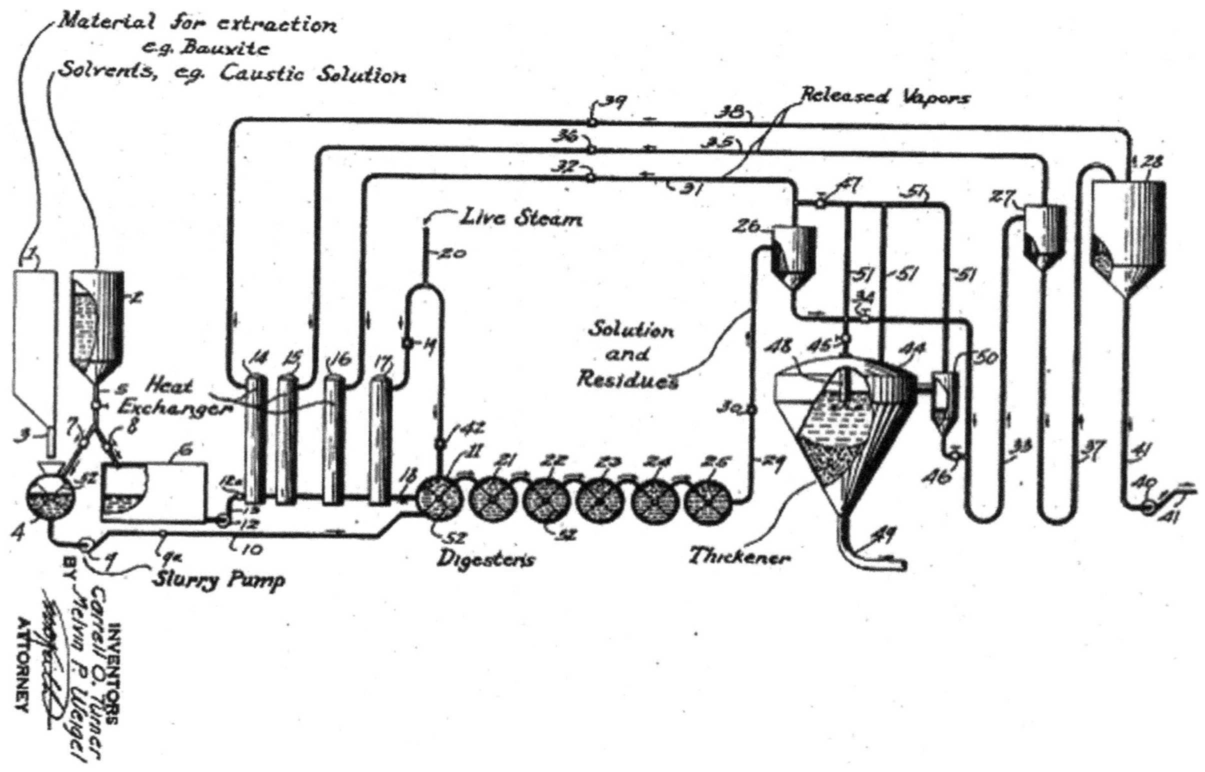

Fig. 9.16 Original CRD for digestion with indirect heating of incoming liquor [58]

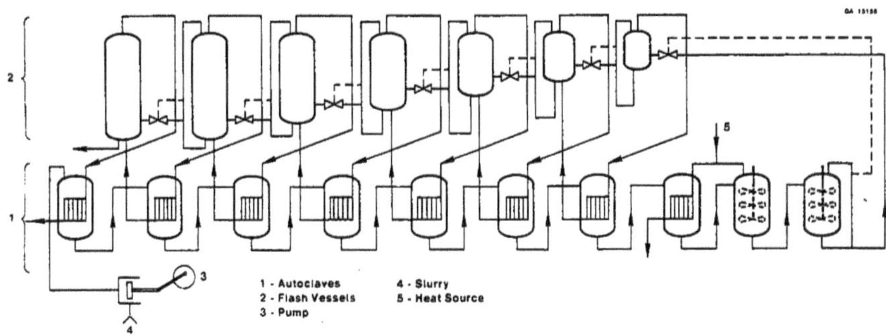

Fig. 9.17 European approach to digester design using autoclaves with steam coils inside

sink is not as great a contribution). This would then lead to a significant quantity of steam being ejected out of the blow off tank, a prime indicator that energy is being wasted.

The solution to this problem is to mix the bauxite slurry with the liquor at the start of digestion to ensure that the source and sink are similarly sized, and less heat is wasted. This solution is dubbed 'single stream' digestion. Single stream digestion can be achieved, maintaining the existing CRD design, by simply adding bauxite slurry to the shell and tube heaters with the spent liquor. This is one approach to single stream, that is quite prevalent in Bayer refineries today, and reportedly was initially utilized circa 1965 [9],[3] heating slurry to 200 °C.

While the CRD heating-digester system is probably the most ubiquitous method today (counting both dual and single stream versions), other European designs are noted in the historical literature for handling harder to dissolve diasporic bauxites. Hungarian refiners (Aluterv) utilized stirred autoclaves with steam going through coils inside of the autoclaves and slurry flowing in the 'shell' of this heat exchanger, as shown in Fig. 9.17 [60, 61].

While this approach does balance the heat source and sink, it seems that the design inherently suffers a lower U factor (< 0.9 kW/(m^2 K)) compared to the shell and tube counterpart (~ 2 kW/(m^2 K) (see data of O'Neill [62] as an example)). Further to this, it is not clear how comparable the scaling of these heaters compared to conventional shell and tube would be.

Another more popular single stream approach is tube digestion single stream digestion, originally developed for the German company VAW at their Aluminum Oxid Stade (AOS) facility, 1967–1968 as was illustrated in Fig. 9.6 earlier [22]. Here the fundamental difference to the heating regime (not the digesters) is the fact that the slurry is flowing in a 'pipe-in-pipe' arrangement, with up to three parallel slurry pipes inside of one larger steam-condensing pipe. This can take up a significant amount of area (as already noted in Fig. 9.4), but this is offset by the reduced scaling risk.

[3] This comment within the publication by Thomas is from a personal communication that cannot be independently verified.

Table 9.6 Heater types with perceived advantages and disadvantages as reported by Thomas [9]

Heater	Advantage	Disadvantage
Dual stream—shell tube	Compact, low cost per area	Heat source does not match sink
Single stream—shell and tube	Off-the shelf. Easier to retrofit brownfield with dual stream	Scaling risk compounded by erosion and plugging risk and higher pressure drop
Single stream—pipe-in-pipe	Long service life	High capital cost
Single stream—autoclave	Pressure rating lower than others	Agitator seals require high pressure High scaling rate

In the excellent historical perspective provided by Thomas [9], a summary of the perceived advantages and disadvantages of the different heater designs is provided, which is recreated in Table 9.6.

9.2.1.2 Energy Requirements

In considering the energy required in a digestion process, there is a natural tendency to suppose that a higher digestion temperature will in fact require more energy. This idea has been debunked many times: see for example the work of Henrickson [16] noting that RTA's Yarwun refinery which runs a high temperature digestion process using ~ 270 °C yet reports less than 10 GJ/t energy usage.

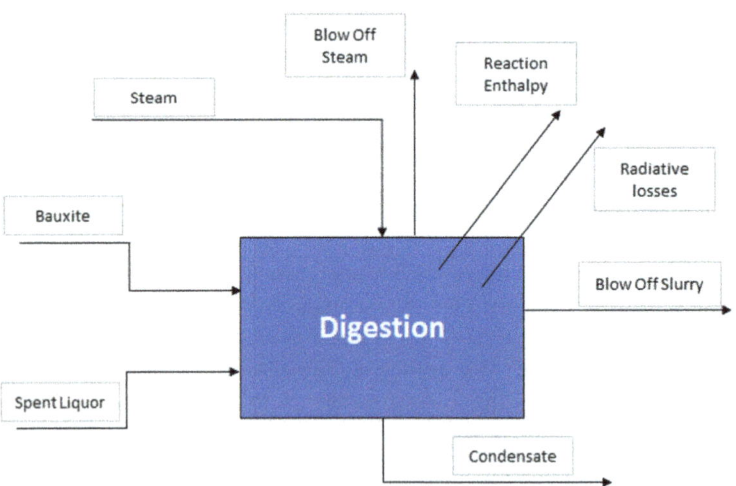

Fig. 9.18 Block flow diagram for a digestion process

When one considers the basic block flow diagram for digestion in Fig. 9.18, it can be seen that what is energetically required is to heat liquor and bauxite up from their mixed temperature, say 70 °C to the temperature of the blow off slurry (i.e. 105 °C), plus the enthalpy of reaction plus radiative losses associated with the equipment's surface area. The remainder of the input energy should at first glance, be recoverable.

Perhaps this representation begs the question—if one ignores the reaction enthalpy and radiative losses (for now), if the bauxite plus spent liquor entered into digestion at, say 105 °C, would all of the input energy be recoverable? It is this line of questioning which adds insight into the mechanics of the design process.

It is the task of the design engineer to specify both the heater areas to use as well as the number of flash tanks. For the sake of demonstration, assume only one flash with slurry entering the first heater at 105 °C and the corresponding blow off tank operating at 105 °C (and atmospheric pressure, obviously). The slurry will not pick up any heat from the flash tank, irrespective of how huge the heater area is. The live steam heater would have to add ~ 40 °C and then post digestion this would get rejected to atmosphere out of the blow off tank.

In fact, even if the slurry entered heater at, say 101 °C, the heater would still not pick up any energy due to the boiling point elevation (BPE) of the liquor stream. For a condensing heater with superheated steam, the heat transfer is based on the saturation temperature [63] as it is the condensing of steam to condensate that provides the thermodynamic driving force for transfer. The saturation temperature of steam is defined by the pressure, but the temperature of the liquor leaving is defined by this saturation temperature plus the BPE that is afforded the liquor by the dissolved salt concentration. The BPE in this example is 5 °C, but it is not a constant as the concentration of dissolved salts vary within a refinery (depending on the strength of the liquor stream) and also from refinery to refinery. For refineries such as that found in the north and northeast of Brazil where the liquor stream is relatively free of impurities and the free caustic concentration quite high, the value will be different to lower caustic concentration refineries with higher impurity profiles (such as that found in the Western Australia refineries). The functional relationship to quantify this phenomena is commonly an empirical relationship developed in house by alumina producers, but some correlations have been published in the open literature (see for example, the Dewey correlation [64] which is the SysCAD process modelling package default [65]).

Acknowledging the existence of the BPE and realizing that the shell's condensing temperature must be the saturation temperature for steam at atmospheric conditions (i.e., 100 °C), slurry would in fact be cooled by the condensing steam! Thus, this simple example should demonstrate that at a minimum, the system must provide enough energy to account for the BPE. To illustrate this point another way—imagine that a digester discharge entered a flash tank at the exact same pressure it had in the digester, the steam would exit the flash tank at the digester's temperature, but it would condense at this temperature less the BPE. Thus again, at a theoretical minimum, the live steam source will have to provide the BPE.

The design engineer cannot solely use heater area without consideration for the number of flash tanks. To elaborate a little further, suppose that the slurry was to come in at 25 °C, allowing for a substantial LMTD. No amount of heater area is going to remove the upper ceiling of the exit temperature, i.e., the flash tank temperature minus the BPE. In the impossible limit of this example, the live steam would still have to heat the exiting slurry by the same ~ 40 °C noted in the example with a 105 °C feed temperature.

Following this line of reasoning, the mandatory BPE amount of heating is a theoretical minimum that can never be practically achieved. To only require this temperature pickup would necessitate an unreasonably large number of flash tanks. Further, as a flash tank is added, the possible minimum will tend to this limit with diminishing returns as extra stages are added.

A point to note is that the 'feed' to a digester could never be the blow off temperature minus the boiling point elevation. While the heat recovered in the spent liquor circuit can be exceptional, the feed can never get the requisite temperature because:

(1) The green liquor which transfers heat to the spent liquor is not the same temperature as the blow off temperature due to the cooling effect of the wash water up the wash train (see latter section on clarification).
(2) The spent liquor is not the only material feeding the digester as fresh bauxite needs to be added which is typically at 25 °C.

Thus, the feed temperature is defined by how well the spent liquor heat is recovered from the green liquor in conjunction with the quantity of bauxite required to hit the blow off ratio target. This should reinforce (one of) the unfair advantages that a refinery with a high-grade bauxite source and/or extraction conditions has over the lower grade counterpart; the digester does not need to use as much energy heating the solid material, as less solid material is required.

As an interesting tangent, the original US alumina plants, originally designed low temperature refineries with three flash tanks [2]; during an era where energy prices were incredibly low and represented ~ 5% of the OPEX (in 1965 [16]). As time has marched forward and energy prices have increased substantially, to the point where the energy consumption can easily reach 40% of the OPEX, this number has climbed in some refineries to five flash tanks. This point is mentioned to show that there are no hard rules (unfortunately) and these sorts of decisions must be made in light of the local and temporal economics, in conjunction with forecasts of commodity price trends.

9.2.1.3 The Enthalpy of Dissolution

Having defined the bare minimum heating requirement, the enthalpy of dissolution can be introduced into the energy balance. Unfortunately, when the literature is consulted, there seems to be a lack of consensus regarding the correct value for the enthalpy of dissolution at 25 °C. Perhaps the most famous work within the industry is that of Langa [66], which shows a strong experimental approach in conjunction with use of reputable sources in the National Bureau of Standards as well as the work of Hemingway and Robie [67]. However, fundamentally, the value arrived at for the

Table 9.7 Standard enthalpies of formation for relevant constituents in the dissolution reactions

Species	$\Delta H_{298.15\ K}$ (kJ/mol)	References
$Al(OH)_{3(s)}$	– 1293.13	Königsberger et al. [69]
$Al(OH)_4{}^-{}_{(aq)}$	– 1500.65	ibid
$AlOOH_{(s)}$	– 995.5	ibid
$OH^-{}_{(aq)}$	– 229.9	[70]
$H_2O_{(l)}$	– 285.5	[70]

heat of dissolution assumes the following dissolution reaction.

$$Al(OH)_{3(s)} + OH^-_{(aq)} \rightarrow NaAlO_{2(aq)} + 2H_2O_{(l)}$$

Which has been shown to be inaccurate [18].

Songqing et al. [68] states the enthalpy of dissolution as 0.66 GJ/t Al_2O_3, without any substantiating data, making the critique of this value difficult. It is believed that the best value to use is the value recommended by Königsberger et al. [69]. This work has consolidated values from various sources across the literature, in conjunction with other thermodynamic measures to yield the most comprehensive and physically consistent properties model for Bayer liquor. According to the following reaction scheme:

$$Al(OH)_{3(s)} + OH^-_{(aq)} \rightarrow Al(OH)^-_{4(aq)}$$

$$AlOOH_{(s)} + OH^-_{(aq)} + H_2O_{(l)} \rightarrow Al(OH\ )^-_{4(aq)}$$

The key enthalpy of formations noted by Königsberger together with a standard value for the hydroxide ion [70] are shown in Table 9.7.

Using a gibbsite enthalpy of formation in agreement with that of Wesolowski [71] and Hemingway et al. [67], an aluminate ion enthalpy of formation in agreement with Wesolowski [71], together with the aforementioned standard value for hydroxide ion, the enthalpy of dissolution of gibbsite at 25 °C is taken as 0.44 GJ/t Al_2O_3. Similarly, the enthalpy of dissolution of boehmite at 25 °C is taken as 0.202 GJ/t Al_2O_3.

However, this does not translate to the heat lost in the digester. The value at temperature, T, is determined according to:

$$\Delta H_{T_i} = \Delta H_{25} + \int_{T_i}^{25} C_{p,r} dT + \int_{25}^{T_f} C_{p,p} dT$$

where

ΔH_{T_i} is the enthalpy of the reaction at the target digestion temperature

ΔH_{25} is the aforementioned heat of reaction at 25 °C

$C_{p,r}$ is the heat capacity of the reactants

$C_{p,p}$ is the heat capacity of the products.

Note that to get at this value precisely necessitates knowledge of the heat capacity functions which are again, determined by alumina refiners in-house (though some do exist in the open literature [66]). The best approach would be to utilize the Pitzer model developed by Königsberger et al. [69], as their properties model would most likely give the best values for the liquor heat capacities before and after. This has not been published in the literature and application of the said properties model requires custom software. Heat capacities for the bauxite will vary depending on the bauxite source, and again, it is assumed that this would be dealt with by alumina producers using in house functions or algorithms. As a very crude estimate, in lieu of anything else, one may choose to simply use the value at 25 °C as a first approximation.

9.2.1.4 Determining the U Factor

Given that the health of the heaters in digestion and evaporation are so important, it is essential that their heat transfer is monitored throughout the heater's life (i.e., both between cycles and over the entire tube life of the heater). Some example applications of heat transfer monitoring include, but are not limited to:

1. Monitoring of the heat transfer after acid cleaning to guarantee that the cleaning procedures have returned the heater to a satisfactory condition.
2. Monitoring of the long-term heat transfer to determine if shell side scaling is occurring and reducing heat transfer independent of acid cleaning/tube maintenance.
3. Daily troubleshooting
4. Evaluation of process changes to improve the energy performance of the building.

While the obvious choice for heat transfer monitoring, is the ubiquitous U-factor or heat transfer coefficient (HTC), the exact nature of the calculation under Bayer conditions is worth review. Perhaps the definitive published work on the topic is that of O'Neill [62], whose work is summarized here. The general strategy advocated is universal for liquor heaters.

For the system illustrated in Fig. 9.19, the U factor is defined per the classical treatment as:

$$U_i = \frac{1}{RES_L + RES_S + RES_T + Res_{ST}}$$

$$U_i = \frac{1}{\frac{r_i}{h_i \cdot r_s(t)} + \frac{r_i \cdot \ln\left(\frac{r_i}{r_S(t)}\right)}{k_s} + \frac{r_i \cdot \ln\left(\frac{r_O}{r_i}\right)}{k_m} + \frac{r_i}{r_o \cdot h_o}}$$

where:

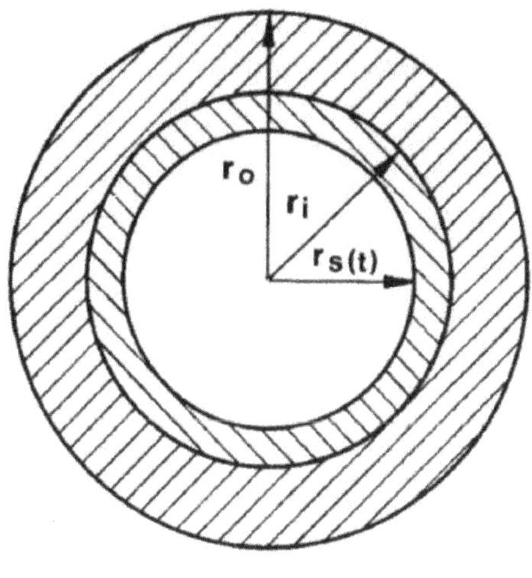

Fig. 9.19 Schematic of a heater tube with DSP scale as utilized by O'Neill in his derivation of a time dependent U-factor model [62]

r_i is the internal radius of the heater tube, m

h_i is the liquor side heat transfer coefficient W/(m^2 K)

$r_{s(t)}$ is the radius of the heater tube with DSP scale, m

k_s is the thermal conductivity of DSP scale, W/(m K)

r_o is the outer radius of the heater tube, m

k_m is the thermal conductivity of the heater tube material (e.g. according to one reference [72], the thermal conductivity of low carbon steel (max 1.5% C), is 36 W/(m K))

h_o is the shell side heat transfer coefficient, W/(m^2 K)

RES_L, RES_S, RES_T, RES_{ST} represent the heat resistance of the liquor, scale, tube and steam respectively.

The total heat transferred on the tube side is:

$$Q_i = U_i \cdot A_i \cdot LMTD$$

where:

A_i is the internal area of the heater tube, m^2.

LMTD is the log mean temperature difference, °C, defined as:

$$LMTD = \frac{T_{out} - T_{in}}{\ln\left(\frac{T_{st} - T_{in}}{T_{st} - T_{out}}\right)}$$

where:

T_{out} is the liquor temperature leaving the heater, °C
T_{in} is the liquor temperature entering the heater, °C
T_{st} is the steam temperature, which is the condensing temperature of the steam defined by the pressure, or, for flash tank generated steam, the measured temperature minus the boiling point elevation of the slurry. This point regarding the saturation temperature versus measured temperature is not explicitly stated by O'Neill.

While r_i, r_o, k_m, are constants, the difficulty (or opportunity for variation) comes from obtaining values for the liquor side heat transfer coefficient, the shell side heat transfer coefficient and the radius of scale.

Liquor Side Resistance

The liquor side heat transfer coefficient used by O'Neill (referencing the work of Kays and Crawford [73]) is:

$$h_i = \left(5 + 0.015 \cdot Re^a \cdot Pr^b\right) \cdot \left(\frac{k_L}{2r_s(t)}\right)$$

where:

$$\mathrm{Pr} = \frac{\mu_L \cdot C_{PL}}{k_L}$$

$$\mathrm{Re} = \frac{2 \cdot r_s(t) \cdot G}{\mu_L}$$

where:

G is the mass flux of liquor, kg/(s m^2)
C_{PL} is the heat capacity of the liquor, J/(kg K)
k_L is the thermal conductivity of the liquor, W/(m K)
μ_L is the dynamic viscosity of the liquor, Pa s.

$$a = 0.88 - \frac{0.24}{4 + Pr}$$

$$b = 0.333 + 0.5 \cdot \mathrm{e}^{(-0.6.Pr)}$$

Use of this equation necessitates a formula for the viscosity, heat capacity and thermal conductivity of the liquor. While correlations are available in the literature for viscosity and heat capacity (see for example the SysCAD recommendations [65]), unfortunately, no empirical data is available openly for the thermal conductivity.

More recent work (see for example Müller-Steinhagen et al. [74], or Duncan et al. [75]) have suggested correlations of the following form (based on the Gnielinski [76] equation):

$$Nu = \frac{\frac{f}{8} \cdot (Re - 1000) \cdot Pr}{1 + 12.7 \cdot \sqrt{\frac{f}{8}}} \cdot \left(\mathrm{Pr}^{\frac{2}{3}} - 1\right) \cdot \left(1 + \left(\frac{2r_i}{L}\right)^{\frac{2}{3}}\right) \cdot \left(\frac{\mathrm{Pr}}{\mathrm{Pr}_{\mathrm{Wall}}}\right)^{0.11}$$

where:

$$f = \left(1.82 \log_{10} Re - 1.64\right)^{-2}$$

L is the length of the heater tube, m

$\mathrm{Pr}_{\mathrm{Wall}}$ is the Prandtl number evaluated at the wall conditions (explained further below).

While considerably more complex, no direct comparison of the two correlations has been made, but it is assumed that the second correlation is more accurate.[4]

Shell Side Resistance

For the shell side heat transfer coefficient, the classical Kern equation is suggested by O'Neill:

$$h_o = C_{Nus} \cdot \left(\frac{k_{cnd}^3 \cdot \rho_{cnd}^2 \cdot \lambda \cdot g}{L \cdot \mu_{cnd} \cdot (T_{Sat} - T_{Wall})}\right)^{\frac{1}{4}}$$

where:

C_{Nus} is a constant, 0.943 for vertical tubes, 0.725 for horizontal

k_{cnd} is the thermal conductivity of water, W/(m K)

ρ_{cnd} is the specific gravity of water, kg/m^3

μ_{cnd} is the dynamic viscosity of water, Pa s

λ is the heat of vaporization (J/kg) in going from steam to condensate at the condensing temperature, T_{sat}

g is gravitational acceleration, m/s^2

L is the tube length for vertical tubes or tube diameter for horizontal tubes, m.

All the condensate properties are evaluated at the film temperature:

$$T_f = 0.5 \cdot (T_{Wall} + T_{Sat})$$

where:

T_{Wall} is the wall temperature, calculated according to:

$$T_{Wall} = T_{sat} - \left(T_{sat} - T_{avg}\right) \cdot \left(\frac{RES_{ST}}{RES_{TOT}}\right)$$

where:

T_{avg} is the average liquor temperature:

---

[4] As the author's published after O'Neill and reference his work, use the scale formation rate equation but not the other heat transfer equations.

$$T_{avg} = 0.5 \cdot (T_{in} + T_{out})$$

The shell side heat transfer calculations are necessarily iterative as the wall temperature depends on the heat transfer coefficient and vice versa, so an iterative loop needs to be implemented. At the time O'Neill's paper was written (1986), this may have been an involved programming exercise, but is easily calculatable using spreadsheet software on modern computers.

Scale Resistance

The radius of scale at time t, is calculated by O'Neill, assuming the following rate equation for DSP precipitation:

$$\frac{dC_S}{dt} = -k \cdot (C_S - C_{S\infty})^2$$

where:

C_s is the liquor concentration of soluble SiO_2

$C_{S\infty}$ is the DSP solubility concentration of SiO_2, based on the Oku-Yamada correlation [77]:

$$C_{S\infty} = 2.7x10^{-5} \cdot N \cdot Al$$

where:

N is the liquor caustic concentration, expressed as g/L Na_2O

Al is liquor alumina concentration, expressed as g/L Al_2O_3.

k is the rate constant, with Arrhenius dependence according to;

$$k = 244341.e^{-\frac{6166.4}{T_K}}$$

where:

T_K is the temperature in Kelvin. While not explicitly stated, based on the logic flow diagram in the original work, it is assumed that the temperature used is the T_{avg}.

When propagated to calculate the radius change at time t, the analytical solution for the radius is:

$$r_{new} = r_{old}.e^{\frac{-k \cdot (C_S - C_{S\infty}) \cdot (t_{new} - t_{old})}{2 \cdot f_{Si} \cdot \rho_S}}$$

where:

t_{new} and t_{old} are the future time being solved for and the start time respectively

f_{Si} is the fraction of silica in DSP, which depends slightly on the type of DSP used (i.e. for the generic structure $3Na_2O \cdot 3Al_2O_3 \cdot 6SiO_2 \cdot Na_2X \cdot 6H_2O$,[5] X can be: SO_4, CO_3, 2Cl, $2Al(OH)_4$. In theory, based on the work of Riley et al. [10], the net

[5] The number of moles of water in the DSP structure has not been well defined either, with values varying between 2 and 6.

weight of DSP could be determined based on the distribution of the different types of DSP formed, but the coefficients for the pertinent equations are not published in the open literature). O'Neill does not make it clear which type of DSP is being utilized.

ρ_s is the density of DSP scale, in g/L.

In the equation for RES_S, the only non-constant is the radius of the scale grown, so equipped with this formula, the resistance due to scale can be solved for.

While a good example of the clever use of a desilication rate equation to obtain the scaling rate inside the tubes, it must be stated that the certainty regarding the accuracy of the rate equation must be called into question. DSP precipitation kinetics is a rich field with an abundance of research, identifying conflicting rate equations (see for example: Oku and Yamada [77], Cousineau and Fulford [78], Duncan et al. [79], Barnes et al. [80], LaMacchia [81]). Further, the rate equation supplied by O'Neill does not come with any experimental data to verify its utility in this application, making it impossible to substantiate. Having noted this, similar rate equations were used by Jamialahmadi and Müller-Steinhagen [82] and Jamialahmadi et al. [74, 83] with good agreement between prediction and measured data.

The main point of contention that would most effect the determined U-factors is the rate constant and more importantly, the activation energy. It is generally accepted (according to aforementioned references) that the order with respect to the concentration term is 2. Since the work of Barnes et al. [80], the direction has shifted to the more appropriate form of the equation using the unitless supersaturation term, i.e.

$$\frac{d\sigma}{dt} = -k \cdot \sigma^2$$

$$\sigma = \frac{C_S}{C_{S\infty}} - 1$$

Further, it would seem, based on the review of the different available DSP SiO_2 solubility correlations by Jamialahmadi and Müller-Steinhagen [84], that the form proposed by Hewett et al. (cited in [84]) fits best:

$$C_{S\infty} = 1.44 \times 10^{-5} \cdot \mathrm{Al}^2 + 1.85 \times 10^{-4} \cdot T + 2.97 \times 10^{-4} \cdot c(\mathrm{NaOH})$$

where:

T is the temperature, in °C

c(NaOH) is the concentration of NaOH, in gpL NaOH.

While Jamialahmadi and Müller-Steinhagen [84] provided a better fitting correlation, when the fits are considered against the number of fitting parameters, the Hewett equation seems the most accurate while simultaneously, the simplest.

As stated from the outset, the general method outline by O'Neill remains the most appropriate, but the specificities with regard to physical property determination, empirical correlations to use and DSP precipitation rate equation have not been openly published and remain areas for further consolidation and verification.

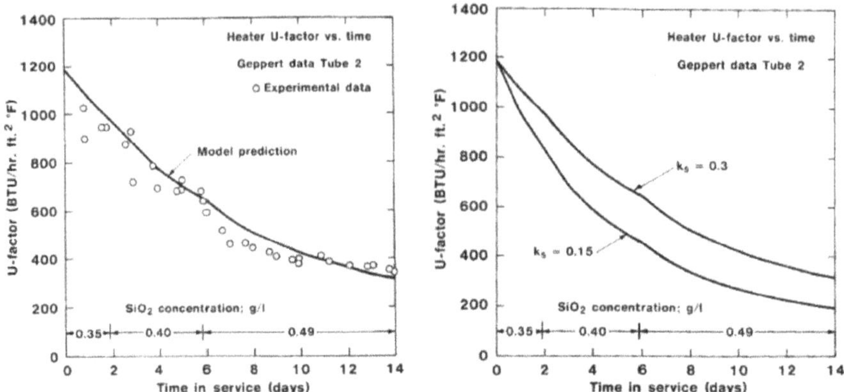

Fig. 9.20 Comparison of U-Factor model performance against laboratory data (left) and model sensitivity to scale thermal conductivity (scale thermal conductivity is in imperial units, i.e., BTU/(h ft.°F)) (right)

While O'Neill's analysis fits the laboratory data of Geppert quite well, as shown in Fig. 9.20 (left), the sensitivity to the 'true' scale thermal conductivity used is also highlighted in the same figure (right). The values tested by O'Neill, translate to 0.259 and 0.519 W/(m K) (0.15 and 0.3 Btu/h ft.°F), yet a variety of values have been suggested in the literature. While Müller-Steinhagen and Jamialahmadi suggest 0.3 W/(m K) [74], Spitzer et al. [85] submit that this would correspond to only ~ 0.2 mm of scale for a 50% reduction in heat transfer, which seems too low a scale thickness for such an impact. While values of 1.2–1.7 W/(m K) have been reported [86] (and dense sodalite can measure as high as 3 W/m K, in line with dense silicates [87]), Spitzer et al. [85] suggest that the value is probably close to ~ 0.5 W/(m K) owing to some combination between the DSP and the liquor (around 0.7–0.8 W/m K). This value of ~ 0.5 gels well with O'Neill's first estimate; but the point regarding sensitivity to the value must be recognized in any of the associated modelling activities.

When discussing thermal conductivities of DSP scale and precipitation kinetics in heater tubes, it is perhaps not so surprising that a variety of values are reported, due to the ambiguous nature of DSP, in a heater. In low temperature digestion ($T < 150$ °C) and pre-desilication, the general consensus is that sodalite is forming (see for example [10]) while high temperature digests ($T > 235$ °C) can start forming cancrinite (see for example [88]); heaters (even of the low temperature variety) can be a mixture of the two phases, depending on time on line. The work of Barnes et al. [89] (see Fig. 9.21) suggests that given enough time, the sodalite that initially deposits on the heater wall will re-dissolve and form cancrinite. It is possible that this interplay of (at least) two phases, with different precipitation kinetics [14, 80, 90] can contribute partially to the difference in desilication rates and possibly the thermal conductivities reported in the literature.

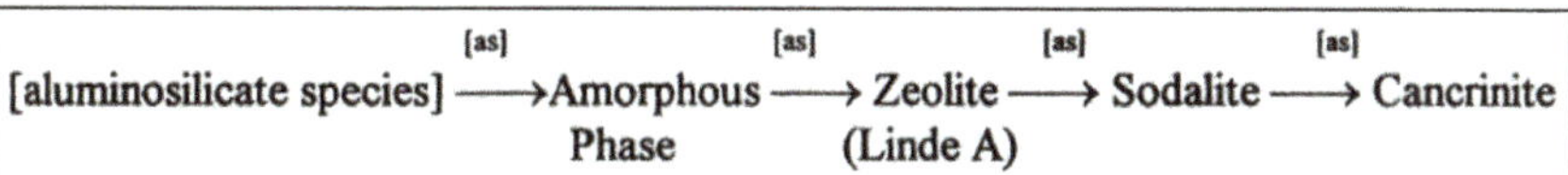

Fig. 9.21 Proposed mechanism of scale formation by Barnes et al. [89], that terminates in cancrinite given sufficient holding time and relative silica supersaturation. [as] refers to the soluble aluminosilicate species

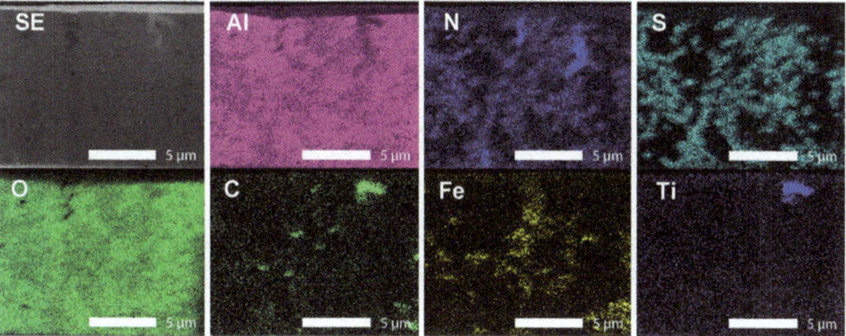

Fig. 9.22 Elemental mapping of a focused ion beam generated section of a single stream plant's scale [92]. Note that based on the text, the N refers to sodium (not nitrogen), S refers to silicon (not sulfur) and C refers to calcium (not carbon)

In the case of refineries running single stream digestion, determining the U-factor of the heaters is made considerably more difficult by the fact that the gibbsite dissolution (and to a lesser extent, kaolin dissolution) is occurring—consuming heat and changing the proportions of liquid and solid in the slurry. In the absence of a complete kinetic model for these reactions, applied to a plug flow reactor model of the heater tubes, predicting the U-Factors accurately is constrained by a refinery's ability to predict the reaction extents in each heater. While sample campaigns of heater-tube discharge can provide a useful snapshot, the reaction extents found at these conditions can hardly be expected to hold all the time, given that the plant concentrations will change and more obviously, the heater temperatures, as they age and come on and off line. Further, it is probably unreasonable to approximate this impact as negligible given the rapid rate of gibbsite dissolution. Based on independent gibbsite extraction investigations, it is perhaps not unreasonable to estimate that some 70–80% of the gibbsite dissolution has occurred prior to slurry arriving at the first digester (see for example the Sumitomo digestion process [91] as a basis for this assertion).[6] A further complication of resolving U-factors in single stream heaters is that the scale is not solely DSP. While DSP seems to be definitively present, recent literature has suggested (perhaps obviously) that other elements can also be found deposited on the tube walls [92] (see Fig. 9.22).

[6] Not valid if the A/C target is too close to, or above the thermodynamic solubility.

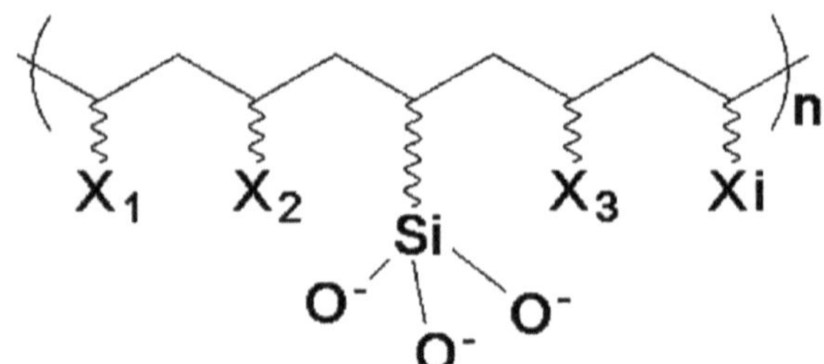

Fig. 9.23 Generic structure of MaxHT [85]

Therefore, predicting the rate of formation of this mixed material and identifying the correct thermal conductivity to use (together with the gibbsite dissolution) remain opportunities for future investigations.

MaxHT

One of the most significant developments in the energy efficiency of digestion and evaporation operations in recent times has been the development of Cytec's (now Solvay's) scale inhibitor, MaxHT (becoming industrially relevant (openly) ~ 2005). As the product is the intellectual property of Solvay, the details are not very specific, but the inventors have made it clear that the reagent does not function by coating the steel surface of the tubes [93]. The patent and other literature suggests a generic polymeric structure as shown in Fig. 9.23, highlighting the importance of the silicon containing functionality, and the fact that a "family" of reagents is included in the MaxHT definition. The mechanism is thought to be via prevention of the DSP nuclei reaching the critical size for scale to form on the heater tubes.

While the exact nature of the mechanism and compound are not openly available, the performance of the reagent has been well established, as highlighted in Fig. 9.24 which shows the difference in heat transfer coefficient decay with and without the reagent added at an unspecified refinery. Some other data from QAL, published by Kiriazis et al. [94], shown in Fig. 9.25, may suggest that the effect is not perhaps as stark as that of Fig. 9.24, but the reduction in scale formation noted with this product is irrefutable based on the current body of data available in the literature, together with common plant experience.

Beyond some commentary regarding the importance of suitable dosing with respect to mixing [85], little work has been published regarding the control of the concentration of MaxHT required. In their analysis of the scale formation rate (via the reciprocal heat transfer coefficient rate), Kiriazis et al. [94] have made some progress in developing the understanding here. In Fig. 9.26, the reciprocal decay rate has been plotted against a measure of the silica supersaturation, in a form basically consistent with the aforementioned DSP precipitation kinetics.

This plot shows that the rate of decay is different with dosage of MaxHT, but the rate is well explained by the difference in plant DSP silica supersaturation, not MaxHT dosage. This data may suggest (preliminarily) that the lower dosages (> 6 ppm) are equivalent to the higher dosages, which could offer a significant savings opportunity for MaxHT users.

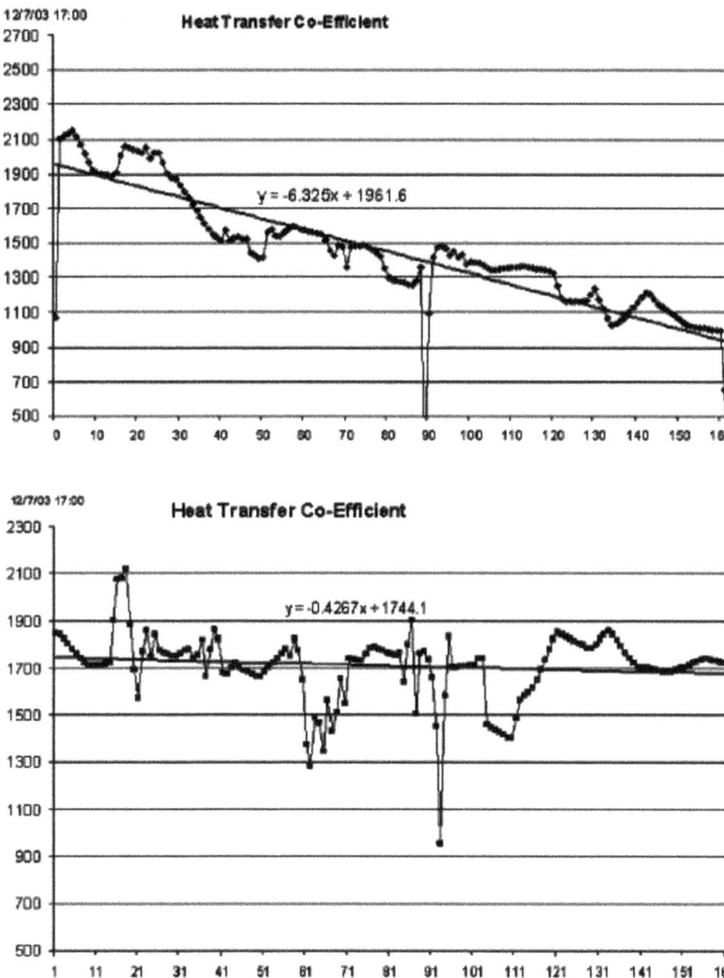

Fig. 9.24 Heat transfer coefficient (in W/(m^2 K)) versus hours on line without (top) and with (bottom) MaxHT added [93]

Another possible positive benefit of MaxHT is the reduced resistance to flow imparted by the different type of DSP formed by adding the antiscalant. Figure 9.27 [94] shows the difference of the scales formed with and without MaxHT, highlighting a much "rougher" surface for standard heater scale. This difference was offered as a potential explanation for the ability to sustain higher flow, despite relatively high scale thicknesses in the heater, not accessible without MaxHT. This point is particular important with regard to the potential production benefit of dosing MaxHT. In running a bank of heaters, while it is not ideal for the heat transfer of the heaters to reduce, the immediately important variable to guarantee the plant's daily production is the liquor exit/digester feed temperature. In the event that heat transfer reduces in the

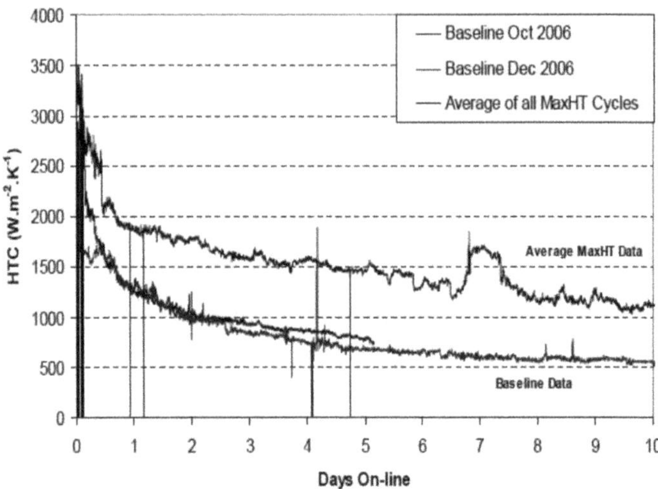

Fig. 9.25 Heat transfer coefficient decay evaluating MaxHT's efficacy at QAL [94]

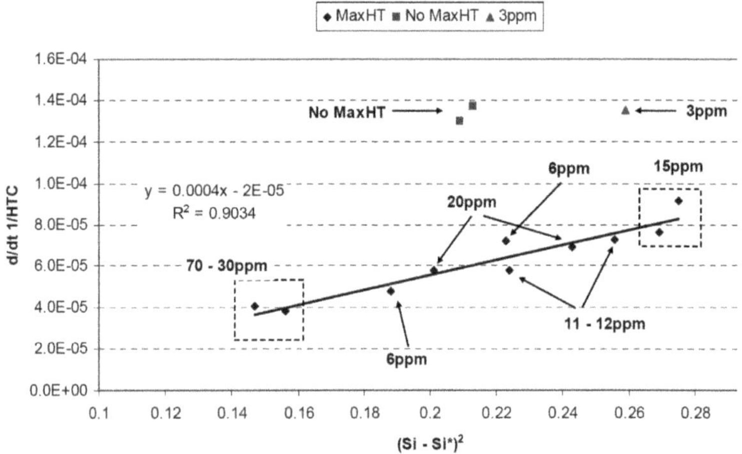

Fig. 9.26 Impact of silica supersaturation and MaxHT dosage on scaling rate in heaters [94]

flash-vapor fed heaters, the temperature of the liquor can be maintained by increasing the live steam flow to the final heater. While not an indefinite solution, this avoids flow restrictions in the process, which is obviously of critical day-to-day production importance. As scale forms, both the heat transfer, and the flow transfer reduces; the latter being explained by the extra pressure drop of the reducing heater tube diameter. In the same way that the live steam flow is increased to offset the poor heat transfer, to combat the extra flow resistance of DSP, the liquor pump can increase its spinning frequency and therefore it's outlet pressure (if a Variable Frequency Drive (VFD) pump; if a fixed frequency pump, a control valve will be opening to achieve the same

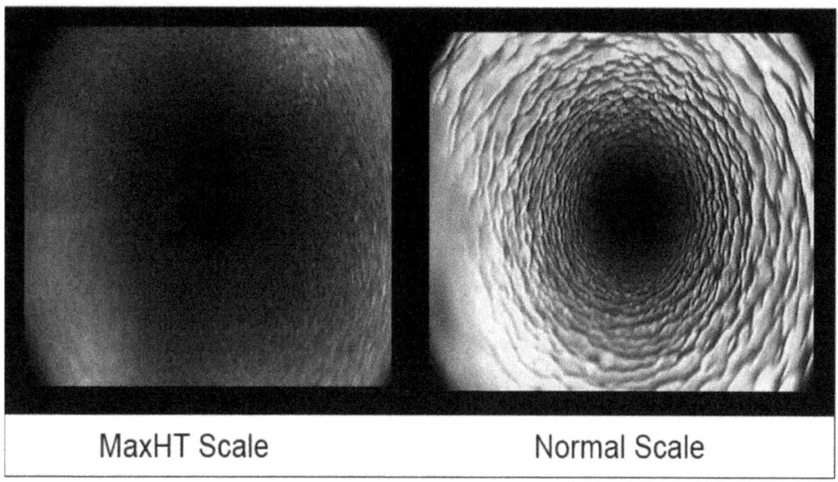

Fig. 9.27 Comparison of different DSP scales formed with and without MaxHT present in a liquor heater [94]

effect). In the same way as the live steam valve solving poor heat transfer, this solution is not long term as the pressure will hit a critical limit (typically determined by the digester or evaporator's Over Pressure Protection (OPP) limit). For safety reasons, to prevent the risk of an exploding vessel, the digester (and evaporators) will have pressure limits that cannot be exceeded. As the pressure output of the pumps pushing liquor through these buildings increases, these values will get closer to these limits, until the point that they cannot be safely increased anymore (perhaps via an electrical fail safe, or operational procedure). If the pressure is allowed to continue to increase, pressure relief valves may start to lift and discharge liquor/slurry into relief tanks to prevent the vessels from building more pressure and exploding. By forming a smoother DSP, the pressure requirement for the same thickness of scale is lower, allowing for more operational longevity prior to a heater swap. A schematic showing how the use of MaxHT can assist with both temperature and flow limitations is illustrated in Fig. 9.28.

Solvay are the sole distributors of the heater antiscalnt in the Bayer process. Other chemical suppliers to the Bayer process have attempted to formulate alternatives [95], but they have not as yet reached the same commercial success, it would seem. Until the patent expires in 2024 [96], or until a competitor develops an economic alternative, it seems the product will remain in circulation based on its proven efficacy.

In the move to single stream digestion, where all of the slurry flows through the heaters, an antiscalant has not been developed (or rather, the results have not been published in the open literature). This is perhaps partially because the scale is not entirely DSP as stated above, and the loss of performance of MaxHT in the presence of solids. Developments have been made to handle a very low solids concentration, commercially identified as MaxHT 500 [85], but this product is suited for evaporator

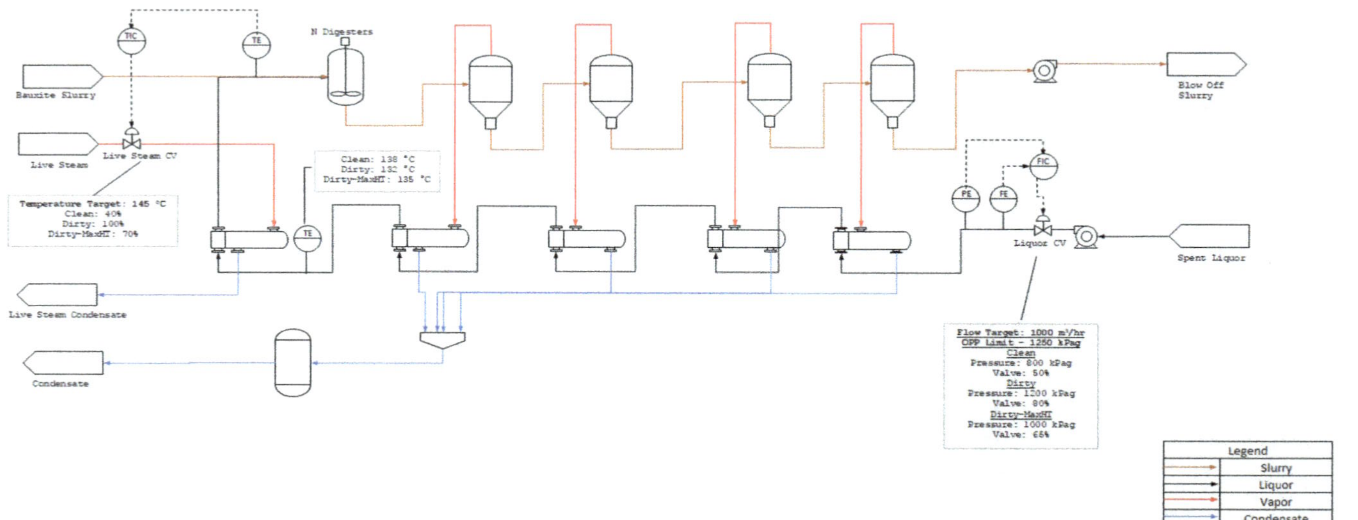

Fig. 9.28 Simplified example low temperature digestion PFD (with some instrumentation/controls and arbitrary flows/temperatures/pressures/valve positions) demonstrating improved flow performance due to MaxHT addition

feed with higher-than-normal classification overflow gibbsite solids. For comparison, a catastrophic evaporator feed solids concentration would be ~ 5 gpL, versus conventional digester feed solids concentration of ~ 200 gpL. The development of scale inhibitors in the single stream applications (and indeed, all aspects of the Bayer process) remains an area for future research with immense economic potential.

9.2.1.5 Guaranteeing the U-Factor

In discussions regarding digestion/evaporation energy efficiency, the focus with respect to flash tanks is centered on how many are required and the associated areas of the connected heat exchangers. However, a point worthy of mention is the need to ensure that the flash tank-heater systems are operating satisfactorily with regard to condensate quality. If the condensate generated by the flashing process contains too much particulate matter or impurities (e.g., aluminum, iron, silica, caustic and organics [97]), the condensate presents a risk for the boiler (and hydrate washing) which it conventionally feeds to continuously generate the steam (and sometimes power) that the refinery requires. Further to this, the impurities and particularly the solid material that may be entrained with the vapor can deposit on the surfaces of the shell and the interconnecting pipework between the flash tank and the heater, ultimately reducing the heat transfer. In the event that a heater is making out of specification condensate (commonly monitored online via conductivity meter (for example: [98, 99])), a refinery will reroute the condensate to a different area (e.g. the last washer) until the problem can be resolved. As the reserves of the water feeding the boilers can reduce substantially, this will need to be supplemented by a separate purchased water source (e.g., a reverse osmosis unit) which obviously presents a direct cost.

Given the potential volume and energy impacts of producing out of specification condensate, the question naturally arises as to how this can happen and how it can be prevented. Failure modes include:

Flash Tank Carry Over

When discussing out of specification condensate due to flash tank carryover, it is important to realize that this can happen through different ways, some controllable through robust operational practice, others preventable through sound maintenance schedules. In all cases, the vapor phase is not satisfactorily separating from the liquid phase; but to differing extents based on the particular failure mode. During any flash tank's operation, the liquor is supersaturated with respect to gibbsite (as the temperature is less than the digestion temperature) and a driving force for "scale" formation exists. Combined with stagnant flow zones [100], conditions favor gibbsite scale, cemented with mud solids, to form all throughout the vessel. This effectively reduces the vessel volume, restricting the slurry flow out as well as increasing the vapor velocity through the dis-entrainment section of the flash tank.

Acknowledging this basic scale formation mechanism, some of the failure modes can be explained:

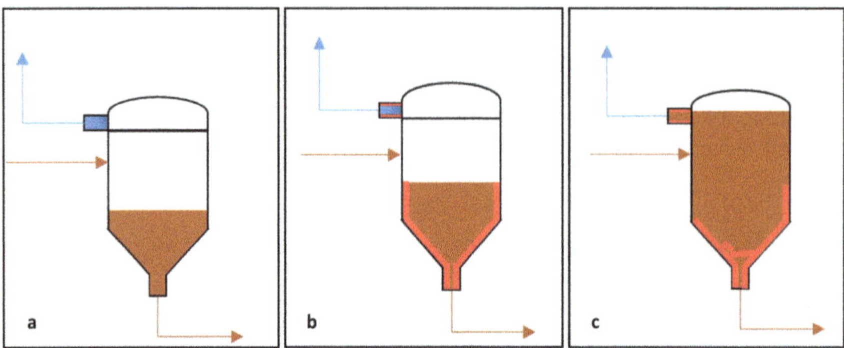

Fig. 9.29 Different phases of a flash tank's operation. **a** Clean usage; **b** usage after some scale formation causing elevated level; **c** flash tank shedding event causing obstruction of outlet and aggressive carry over

- Scale accumulation near tank outlet—shown schematically in Fig. 9.29, part b. As scale deposits on the outlet and walls, there is naturally more resistance to flow and the level of the flash tank correspondingly increases. This reduction in the dis-entrainment area can over time cause the condensate quality to drop. At the bottom of the tank, this scale is thought to be largely "settled" scale (versus "growth" scale) [101]; with the distribution determined by how easy it is for the slurry to flow from the bottom of the vessel to the outlet pipe.
- Scale build up near vapor outlet accelerates vapor velocity—also shown schematically in Fig. 9.29, part b. While the scale growth in this part of the tank should be less than that of the slurry outlet, scale can still deposit here, at a rate dependent on vessel design. The restriction here also increases the vapor velocity which reduces the dis-entrainment efficiency. The presence of scale here forces the pressure of the steam to drop prior to arriving at the shell to condense into water. This reduction in pressure reduces the heat transfer of the steam, which also penalizes the energy efficiency, (on top of the condensate quality issues).
- Scale shedding from flash tank walls and restricting exit—shown schematically in Fig. 9.29, part c. This failure mode requires some sort of process upset to occur, for example—a thermal shock is imparted onto the scale, such that some of the solid expands/contracts at an uneven rate causing breakage, or part of the solid is suddenly swept off the wall by an accelerated slurry flow. The probability of a failure increases as the scale grows, which agrees with plant experience that the risk for an event is greater as the vessel's residual life diminishes. Some possible operational causes include sudden changes in plant flow or perhaps a poorly executed heater cleaning turnaround, which can cause a pressure (and temperature) swing in the flash tank.[7] As with many operational nuances, the

[7] This last point is largely dependent on plant layout: some refineries are designed to handle a single heater swap at a time, others change an entire bank at a time, while others may not be able to handle a swap and manage this through running a spare digester unit.

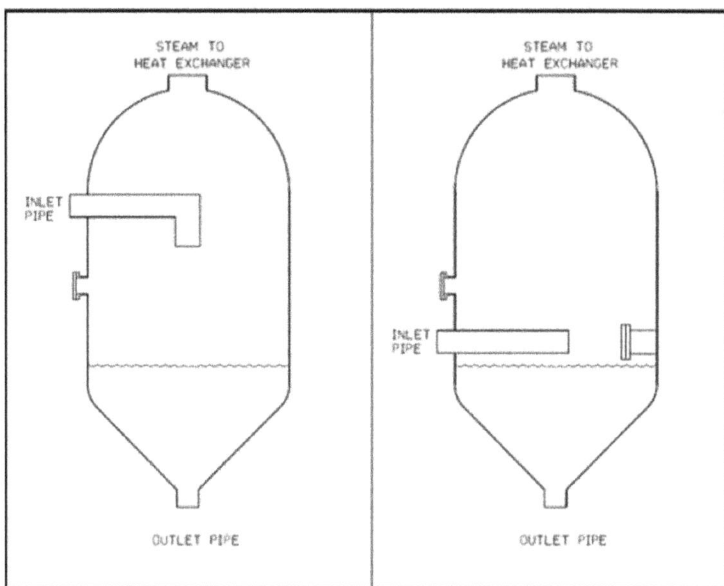

Fig. 9.30 Side entry flash tank designs [47]

cause can be dependent on the specifics of how the plant has been designed. Whatever the exact cause of the shedding event, as stated before, the probability of an occurrence tends to increase with the operational age of the vessel. When this does happen, the vessel (or bank of vessels) may need to be bypassed to mechanically clear scale from the loop, or outlet of the flash tank.

Based on the failure modes for flash tanks to generate out of specification condensate, it would seem that the key driver for safe operation is a robust maintenance program that takes the vessels out of operation at the required interval to clear the settled and growth scale. Beyond this basic approach, developments have been made in flash tank design over the years to extend the campaign life of flash tanks and improve condensate quality, thereby increasing operability without increasing the risk for a condensate contamination and all of the other associated problems (e.g., poor heat transfer, production losses, etc.). While most of the specifics of flash tank design, borne out of operational experience, are not publicly available for IP reasons, a review of some of the developments in flash tank design has been provided by Haneman and Wang [47]. Another important point related to flash tank design is the focus on maintaining the health of the interconnecting pipework between the flash tanks, which can be the rate limiting component in modern flash tank residual life. This is because the flash phenomenon typically starts in the interconnecting pipework, and this is a very erosive phenomenon due to the sudden expansion promoting slurry acceleration.

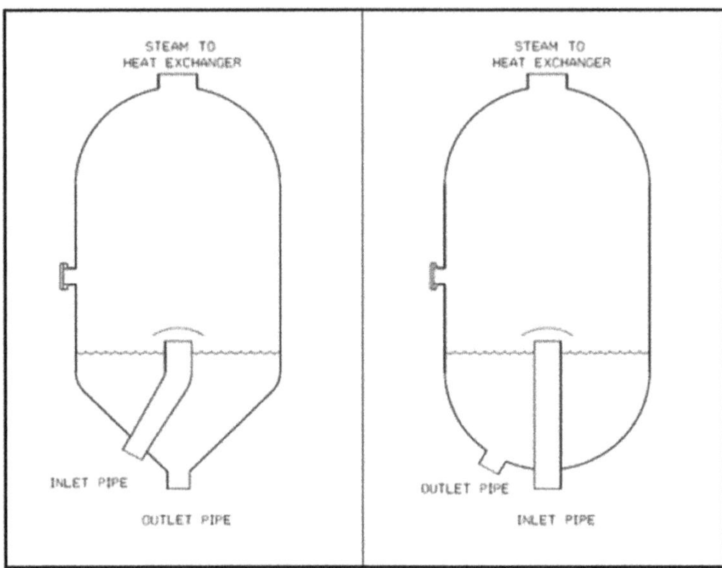

Fig. 9.31 Bottom entry flash tank designs [47]

The earlier designs favored the side entry shown in Fig. 9.30, part a. Unless, the vessels are staged (i.e., the first flash tank is at higher altitude than second and so forth) the slurry will start flashing early on, exposing a lot of the interconnecting pipework to three phase flow. This can be mitigated somewhat by using the part b design of Fig. 9.30, but both designs do not have a good vapor disengagement path and as such, require some extra attributes in the upper part of the vessel to ensure a tortuous pathway for the steam.

The initial bottom entry designs (see Fig. 9.31) reduced the wear issues (or staging requirements) associated with the side entry designs as the up flow does not exceed the slurry level of the preceding flash tank.[8] Further, owing to the up-flow of the slurry allowing for better dis-entrainment, this design was reported to improve condensate quality by 5–10 times and reduced wall scale from 300 mm in 12 months to 25–50 mm in an 18-month campaign. The main limitation noted by Haneman and Wang, is the elbow of Fig. 9.31a; or the uneven distribution of slurry at the bottom of the vessel in Fig. 9.31b requiring replacement or descale respectively at a 12–18-month frequency.

To combat these deficiencies, the authors have proposed the Central Inlet Annular Discharge design of Fig. 9.32 (for both conical and round bottom flash tanks), which was validated by CFD and laboratory simulation as the most robust design (however the authors do not confirm whether the design has been implemented successfully on the plant scale). The campaign life of this design has not been reported to date.

Some more recent work in the patent literature, has suggested that creating a swirling flow (through introduction of a swirling assembly in the feed pipe) as the

[8] The pressure head is all that is preventing the slurry from further flashing.

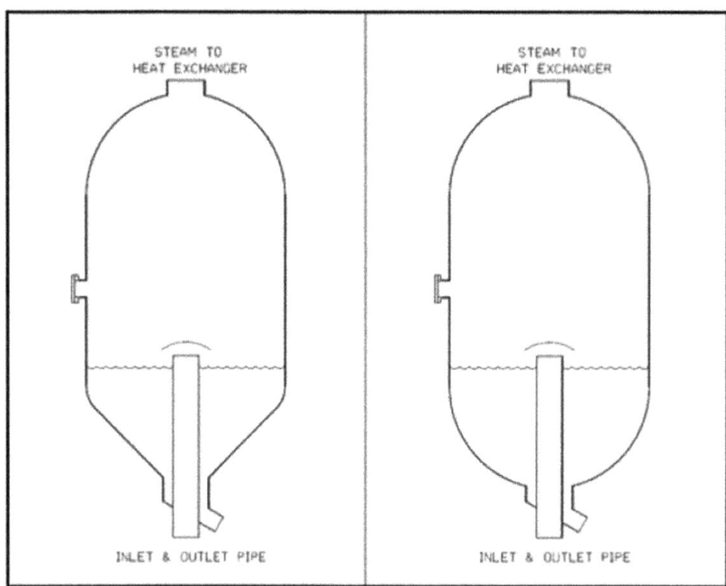

Fig. 9.32 Bottom entry "central inlet annular discharge" flash tank designs [47]

feed to a flash tank can potentially increase vessel life [102]. It is believed that the swirling flow feeding the flash tank will ensure a high velocity at the flash tank walls and bottom to substantially eliminate process scale. It is not clear how this assembly will weather the aggressive flashing conditions, and it is not clear whether the assembly has been tested industrially as yet; but developments in reducing flash tank scale and extending vessel life remain a focus in ongoing research.

Determining the exact onset of flashing in the interconnecting pipework is not trivial. It is possible for pipework with a lot of bends, valves, etc. for flashing to occur earlier than anticipated (perhaps even in a control valve, acting to wear it prematurely, see [103]). To prevent against this and ensure as robust a design as possible, it is necessary to analyze the pressure profile of the system in question. Various methods exist beyond that traditionally advocated based on empirical studies As with the flash tank design, a lot of the specifics as to how this is done are not public in the open literature, but again, Hatch have published some useful information regarding some of the details of their calculation methodology [103–105].

Heater Tube Rupture

Heater tubes are typically made from low carbon steel [106] and accordingly, corrode (and erode, in the case of slurry heaters) over their life in a Bayer refinery, to the point where the tube should be replaced according to a preventative maintenance schedule. In the event that the tube ruptures before it is replaced, the higher-pressure liquor (or slurry) can bleed into the shell and contaminate the condensate.

The operational response must be quick as the water stocks feeding the powerhouse will diminish (depending on which heater has failed and how the plant is set up) and the plant volume may start to increase as this condensate which is supposed to leave the process is forced to (a) go to effluent treatment or (b) stay inside the Bayer loop (typically being rerouted to the washer train, supplanting clean condensate or lake water used for "net dilution"). Upon taking the heater out of service, the culprit heater tube/s after identification[9] can be either replaced or plugged. The plugging obviously reduces the surface area for heat transfer but is (i) faster and (ii) can also be a good indicator of the overall health of a heater; identifying a pass with greater wear problems or being used as a metric to define when the heater must be completely re-tubed.

While less work has been reported regarding the wear mechanisms and preventative measures associated with slurry heaters and their maintenance (e.g., water blasting/drilling as well as acid cleaning), some literature does exist regarding liquor heater tube failure mechanisms and prevention. For liquor heaters, the main wear mechanism is by corrosion; however, this is not normally attributed to the continued exposure to caustic solutions that pass through the tubes during operation, as the caustic solution tends to form a passivation oxide layer, soon after initial flow is established (see [107] and references therein). The corrosion is therefore due to the necessary acid cleaning that is done to the heaters periodically to combat the buildup of DSP scale which grows on the heater tubes at different rates depending on liquor concentrations, and more importantly temperature (see for example: [62, 74, 79]). By washing the heater tubes with a dilute H_2SO_4 solution, the DSP on the tube walls is thought to dissolve according to[10] [108]:

$$\begin{aligned} &3Na_2O \cdot 3Al_2O_3 \cdot 6SiO_2 \cdot Na_2CO_3 \cdot 6H_2O_{(s)} + 13H_2SO_{4(aq)} \\ &\quad \rightarrow 4Na_2SO_{4(aq)} + 3Al_2(SO_4)_{3(aq)} + 6Si(OH)_{4(aq)} + 7H_2O_{(l)} + CO_{2(g)} \end{aligned}$$

According to Kane et al. [107], it is the acid cleaning and initial flow of caustic solution which causes the greatest amount of corrosion. It is thus typical for refiners to utilize some sort of corrosion inhibitor in the acid washing stage to minimize the corrosion in this stage. To minimize corrosion upon first caustic exposure, Kane et al. suggest a pre-passivation treatment whereby the heater is exposed to a low flow and low temperature spent liquor for some ~ 12 h prior to full flow and return to service. A schematic of the approximate different corrosion rates and exposure times is shown in Fig. 9.33.

[9] Possibly identified by removing a heater head, pumping water into the shell and waiting for flow from the holed tube to reveal itself.

[10] Note that in the context of slurry heaters, while mechanical cleaning is required, the maintenance is finalized with an acid wash after mechanical acid cleaning to ensure a bare metal surface. In the case of slurry heaters which have had lime added to the process slurry, H_2SO_4 is not as efficient, presumably due to the unwanted formation of solid calcium sulfate. Thus, for these heaters, HCl is used (see for example: [132]).

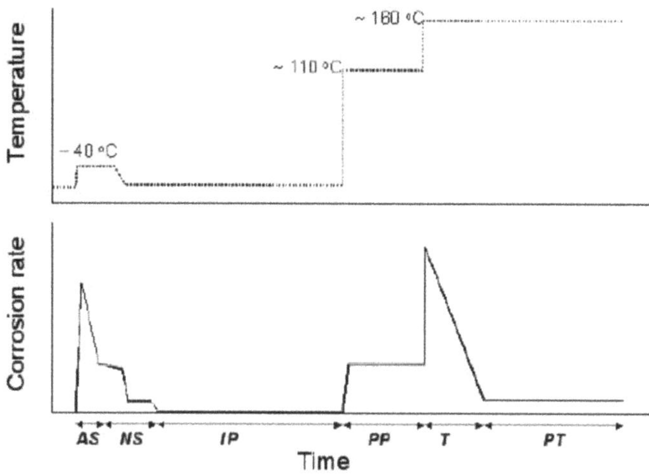

Fig. 9.33 Schematic of corrosion rate and exposure time for different phases of a liquor heater at QAL [107]. Legend: AS—acid shot (acid wash); NS—neutralization shot (wash with caustic post acid cleaning); IP—interim period (heater waiting for return to service); PP—pre-passivation (low flow, low temperature exposure to spent liquor to create passivation film on tube wall); T—transient (full exposure to plant flow/temperature; PT—post transient (stable period after transient period where corrosion rate is minimal)

In the same work, the authors tested an alternative inhibitor to minimize the amount of corrosion in the pre-passivation stage (while simultaneously minimizing acid corrosion during cleaning), using a linear polarization resistance measurement to monitor the corrosion rate online. Unfortunately, due to the high amount of scatter in the data, the benefits of the second additive were not conclusive and minimizing corrosion in this stage remains an area of ongoing research.

Fortin and Breault [108] studied the efficacy and rate of acid dissolution under conditions of differing flow, temperature, acid concentration and inhibitor. While showing the clear benefit of higher flow, temperature and acid concentration, the authors also reported the unusual observation that the efficiency of the acid clean reduces at higher temperatures when inhibitor is present.[11] Further, elemental analyses of DSP growth rings were shown, highlighting thin carbon layers between DSP layers; an unexpected result which the authors hypothesized could have been due to the interaction of the (organic based) inhibitor reacting with the DSP surface. Again, this particular facet of tube cleaning remains unresolved, but essential to understand as optimal heat transfer (as well as flow through the heaters) post cleaning is largely contingent on achieving a bare metal surface, free of DSP scale residue. Their work informs that the balance of needs regarding cleaning efficiency and corrosion resistance need to be considered in choosing the cleaning parameters, as they are, in some instances, clearly in conflict.

[11] The exact ID of the inhibitor was not identified.

Another subtle aspect regarding extending heater tube residual life, is the nature of the material utilized. In the work of del Aguila [106], the impact of different heat treatments on different heater tubes (as manufactured from steel mills) is reported with regard to both microstructure as well as corrosion rate. By varying heating treatments (e.g., temperature and exposure time) as well as cooling rates, it was possible to form the necessary "spheroidised cementite" microstructure which was reported as reducing the corrosion to one third of the untreated tube's rate. Beyond the fundamental data shown, this work also highlighted the importance of refiners working with their steel mill suppliers to produce the tubes in an economic way that obtains the necessary corrosion resistance.

9.2.1.6 The Importance of Indirect Factors

While energy saving measures specific to reducing steam consumption or extending vessel life are critical for a refinery's efficient operation, often when comparison is made between different refineries or step change improvements are noted in a refinery's energy usage, indirect factors, specific to the nature of the flow sheet are often the culprit. To better demonstrate this point more specifically, a dummy digestion flow sheet has been constructed (see Fig. 9.34), using the process simulation software, SysCAD [109]. This is but one example of a possible flowsheet (essentially the same as that of Fig. 9.28), and myriad other options exist, but this specific example can be used to understand the relative importance of different parameters in a general sense. Briefly, the liquor feeds through the tubes of the shell and tube heaters which are being heated by the vapor from the flash tanks. The final heater is a "live steam heater" being heated by an external steam source, which essentially guarantees the digester temperature of the slurry. The bauxite slurry has been set up such that it can feed through the heaters with the spent liquor (i.e. single stream digestion, but without any reaction allowed for in the heaters) or separately (i.e. dual stream digestion). Subsequent to digestion, the slurry passes through four flash tanks, with the final flash tank pressure fixed to atmospheric. The digester feed temperature and blow off ratio are controlled by PID controller, whereby the live steam flow is adjusted to attain the user-set temperature set point and the bauxite flow is used in the same way to maintain the user-set blow off ratio. Aside from the subtle flow sheet difference with regard to bauxite slurry entry, the other variables detailed here relate to process set points and bauxite quality.

In all modeling exercises such as this one, some parameters must be set to fixed values—this is the core reason that this example is considered a dummy case as these values are essentially arbitrary in this discussion. For completeness, a summary of the fixed values associated with the flow sheet of Fig. 9.34 are shown in Table 9.8.

The parameters varied to highlight the impact of these so-called indirect factors include:

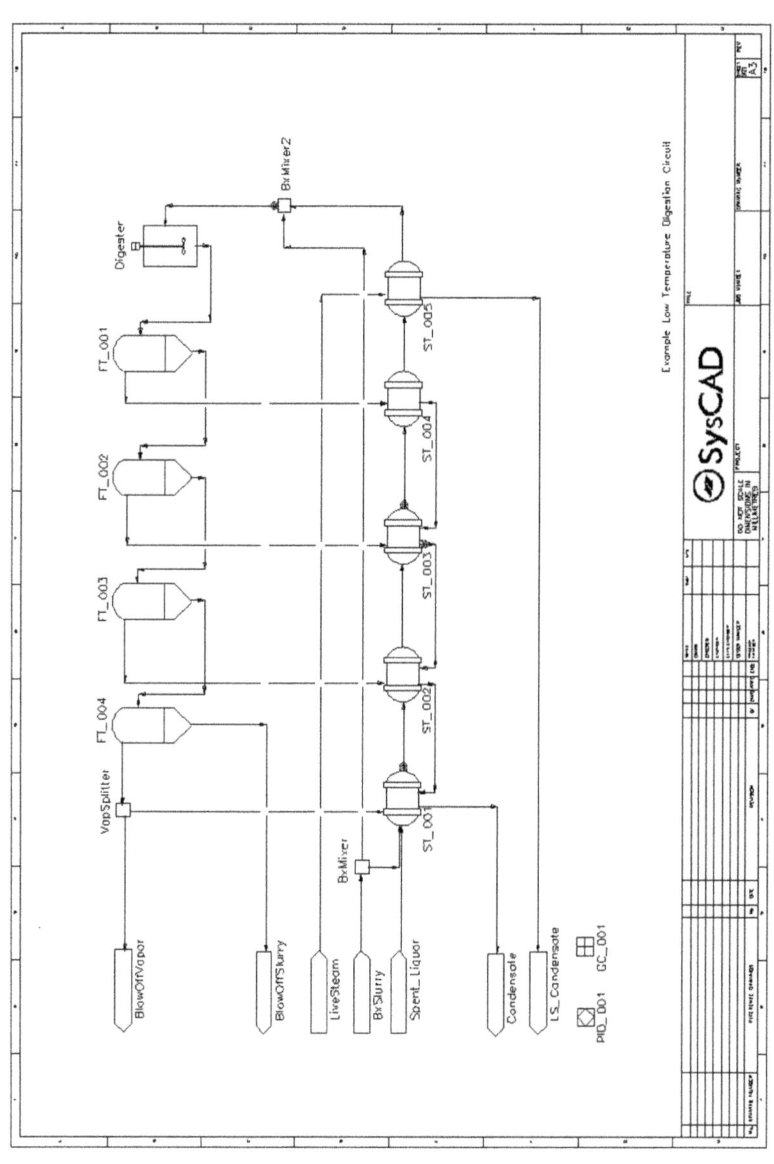

Fig. 9.34 Sample digestion flowsheet, built in SysCAD 9.3 (build 137) [109]

Table 9.8 Summary of fixed values associated with SysCAD simulations

Parameter	Value
Bauxite	
Re. SiO_2	4.5%
Anatase	5.02%
Solids concentration	50%
Temperature	90 °C
Spent liquor	
SiO_2	1.6 gpL
TC/TA	0.9
A/TC	0.4
Na_2SO_4	1 gpL
NaCl	1 gpL
NaF	0 gpL
Oxalate	1 gpL
TOC	5 gpL
Liquor flow	3 tph
Temperature	86 °C
Reaction	
Kaolin dissolution	99%
Digester discharge SiO_2 concentration	1.62 gpL
Heat transfer	
Heater U-factor	2 kW/(m^2 K)
Heater areas	1.5 m^2
Live steam pressure	860 kPa
Live steam superheat	10 °C

Note Spent liquor is the same for liquor in bauxite slurry). All reactive silica is kaolin

- Bauxite quality—30% versus 50% Av. Al_2O_3 (gibbsitic bauxite only). Note that the hematite content of the bauxite was allowed to float to accommodate the changing gibbsite content.
- Blow Off Ratio (A/TC)—0.720 versus 0.750, as well as 0.62.
- Digester Temperature—145 versus 155 °C.
- Gibbsite Extraction—96% versus 99%.
- Liquor Caustic Concentration—250 versus 300 gpL (as Na_2CO_3).

In evaluating the energy performance, the universal metric—the GJ/t has been used, where (following the notation of Fig. 9.34):

$$E = H_{LiveSteam} \cdot \dot{Q}_{m\ LiveSteam} - \left(H_{LS\_Condensate} \cdot \dot{Q}_{m\ LS\_Condensate} + H_{Condensate} \cdot \dot{Q}_{m\ Condensate}\right)$$

where:

E is the energy used, in GJ/h

H is the enthalpy in GJ/t

$\dot{Q}_m$ is the mass flow in t/h.

The tonnes of alumina produced is simply:

$$\dot{Q}_{m\ Digester\ Al_2O_3} = \dot{Q}_{m\ BlowOffSlurry\ Liquor\ Al_2O_3} - \left(\dot{Q}_{m\ BxSlurry\ Liquor\ Al_2O_3} + \dot{Q}_{m\ \ Spent\_Liquor\ Liquor\ Al_2O_3}\right)$$

The ratio of these two values is only the digester GJ/t and ignores the fact that some liquor alumina can be lost in going through clarification, and perhaps not all of this alumina will be recovered as gibbsite in precipitation. While important salient points, it is not the absolute energy that is relevant here, but the relative impact of the aforementioned variables. A summary chart of the relative energy usages for different flow sheets and digester setpoints is presented in Fig. 9.35.

The higher-grade bauxite users have an advantage—by putting different bauxites in the exact same unit operation, with the same targets for temperature and production, the energy usage is worse for the lower Av. Al_2O_3 source. Qualitatively, there is more material to heat up for the lower Av. Al_2O_3 scenario. We see that the difference is less pronounced in the single stream case as the bauxite can actually absorb some of the extra heat it demands by virtue of flowing through the tubes.

We can also see that the higher blow off ratio scenarios seem to require less energy, but the data shown here is a simplification of the reality. As the A/TC increases, it

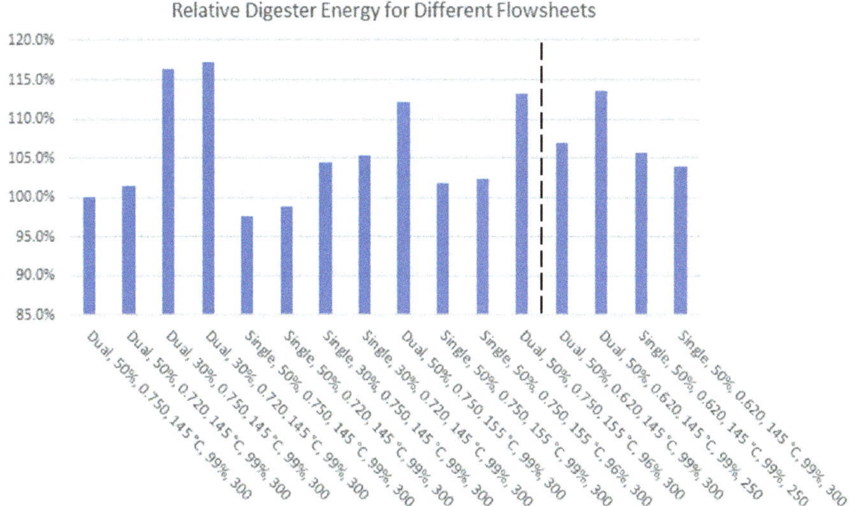

Fig. 9.35 Relative energy performance of different digester flow sheets. The ID of the x-axis is as follows: dual/single stream, Av. Al_2O_3, Blow Off A/TC, digester feed temperature, gibbsite extraction, spent liquor TC (in gpL Na_2CO_3)

gets closer to the thermodynamic limit, which must reduce the extraction extent somewhat; in the scenarios comparing blow off ratios above, the extraction is fixed, which is not strictly correct. A similar observation can be made for the increased temperature scenarios. On the face, they may seem like an unnecessary waste of energy, but that is because *only* the temperature has been raised. The main drive for a temperature increase would either be to increase the extraction for the same production, or to increase the blow off ratio (production) at the same extraction.

A lot of the dual versus single stream differences need not be as stark, or even real at all. The differences exist here as the exact same flowsheet has been applied to both. For the design engineer putting together a dual stream circuit, the solution here would be to increase the heater surface area. On this point, for all of the single stream scenarios, no blow off steam was being lost to atmosphere, versus the dual stream counterpart where the heat sink is not as great.

The comparison between the 99 and 96% shows the utility of better use of the bauxite supply (beyond the more substantial savings in bauxite). Achieving such a significant improvement without any serious change in the other parameters tested (A/TC, temperature) is particularly difficult in practical terms, notwithstanding benefits of extra holding time in digesters and reducing the particle size distribution of the feed bauxite.

For high temperature digestion flowsheets, it has been established that < 10 GJ/t is definitely achievable, taking the Yarwun refinery as a published example [110, 111] as well as Aluminum Oxid Stade GmbH [112].[12] Despite the reality that lower energy usage is possible, it would seem that the path to stepping down to the lower levels lay in achieving greater yield, by both increasing the blow off ratio and the caustic concentration.

The dual stream 250 TC, 0.620 A/TC simulation represents a historical condition for the high temperature flow sheets (remembering again that the point to be illustrated here is the impact of yield, as the amount of energy to heat up one liter of the slurry is not largely different if the flowsheet is properly designed with regard to heater areas and number of flash tanks). The A/TC target is considerably lower, (even at elevated digestion temperatures) owing to the lower solubility of boehmite (and diaspore) compared to gibbsite as Fig. 9.36 shows. The solution would be to increase the caustic concentration, as yield is the product of the ratio difference with the caustic concentration. However, in the dual stream circuits, the maximum caustic concentration is limited by the stress corrosion cracking (SCC) resistance of the mild steel heater tubes, which reduces with temperature (the actual limit is not 250 at these conditions but is used here for demonstrative purposes). By introducing single stream digestion, which dilutes the liquor with lower TC slurry, the liquor caustic concentration can in fact go up without hitting the same SCC limit in the heaters. The bauxite slurry typically has lower caustic concentration liquor as the bauxite is received with some humidity that acts to dilute the liquor which mixes with it at the start of the process. Thus, the step to single stream digestion not only provides an

---

[12] AOS uniquely operates a molten salt heater to achieve their final digestion temperature of 270 °C, which reportedly assists in their excellent energy performance.

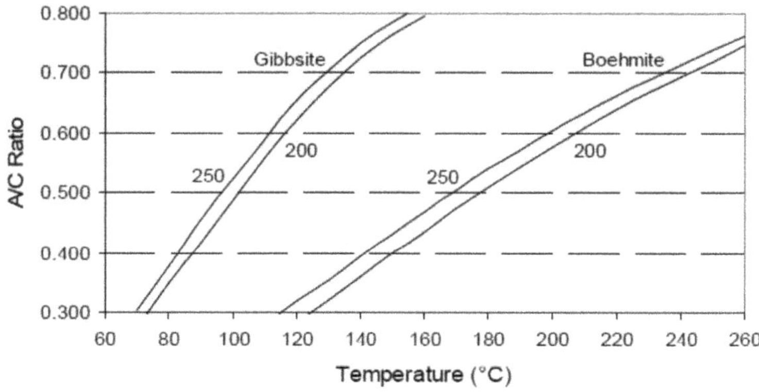

Fig. 9.36 Gibbsite and boehmite solubility curves as a function of temperature at fixed caustic concentrations [113]

extra heat sink to improve energy efficiency, it also reduces the heater tube caustic concentration which allows the caustic concentration of the plant to increase. This breakdown of relative improvements enabled by single streaming and then raising the caustic concentration is illustrated at the end (final three conditions) of Fig. 9.36.

The next enabler is that by having a higher caustic concentration, the thermodynamic solubility limit increases which allows for the A/TC target to be lifted (without extraction penalty). This final improvement is presumably what allows the state-of-the-art high temperature refineries to achieve first quartile energy performance. It must be noted that the way that single streaming is achieved (in high temperature refineries) is typically pipe in pipe, versus shell and tube. The perceived advantages and disadvantages of the pipe-in-pipe versus shell and tube approaches have been discussed and are listed in Table 9.6. From an extraction standpoint, there is also an extraction benefit in that the pipe-in-pipe design allows for plug flow conditions with all of the solid material experiencing the same residence time. This point can be difficult to achieve in CSTR environments particularly with coarse material feed that tends to settle out and leave the vessel with a lower residence time. This guarantee of constant residence time, particularly for the coarse particles, allows for better extraction which further improves energy performance as was highlighted in Fig. 9.35.

After noting the potential energy capability of the higher temperature digestion flow sheets, there are some points to consider which still favor low temperature digestion, if the bauxite is available. Firstly, the capital cost is higher for high temperature flowsheets—not just because more flash tanks are required, but because the pressure rating of the associated higher temperature vessels is also more costly. Further to this, the maintenance cost and experience requirement may be somewhat higher and more complex than the low temperature refiners. As an interesting point—in their discussion of the pipe-in-pipe flowsheet employed at startup of Yarwun, de Boer et al. [110] espoused the advantage of the jacketed pipe heater (JPH) suggesting that they do not experience blockages. In a paper three years later, Fig. 9.37 was presented

Fig. 9.37 Scaled heater pipe (JPH) from Yarwun alumina refinery [111]

[111], showing that they could indeed block if given the correct operational upset. In the same paper, Austin and Perry [111] revealed that other operational challenges such as non-condensable flooding of the heater shells had plagued the operation.

Further to these issues, the work of AOS has highlighted the importance of a robust pigging-then-acid-clean schedule to ensure the tubes are kept as clean as possible, by removing extremely hard titanate scale that has proven a challenge for many high temperature operators [112]. The specifics of the pigging design presented are proprietary and not all refiners have familiarity and experience in optimal descale technique. The recent design of Ma'aden alumina refinery has elected to use contact heaters for the final heating stages to avoid the challenges of this scale [5], which will penalize the digestion energy usage, suggesting that the problems with scale have not been fully resolved by all refiners. These issues regarding hard to remove scale could potentially be reduced by running at lower temperatures, but it is those same higher temperatures which ensure that high A/TC can be achieved, which is instrumental to the lower energy usage.

To accommodate "lower" high temperature digestion, suitable for boehmitic bauxites that avoid scaling issues while still allowing for high liquor productivity, some unique flowsheets have been implemented, that are worth mentioning.

The earliest reference to a "double digestion" process identified is shown in Fig. 9.38 [114] by the British Aluminum Co Ltd in their Burntisland facility in Scotland (now closed). The process contains a low temperature digestion, followed by flashing to atmospheric pressure and then high temperature digestion of the boehmite enriched residue. The pregnant liquor from both digests ultimately report to desilicators to ensure the product silica is in specification, but no doubt penalizing the alumina recovery due to auto-precipitation of the dissolved gibbsite. This flowsheet

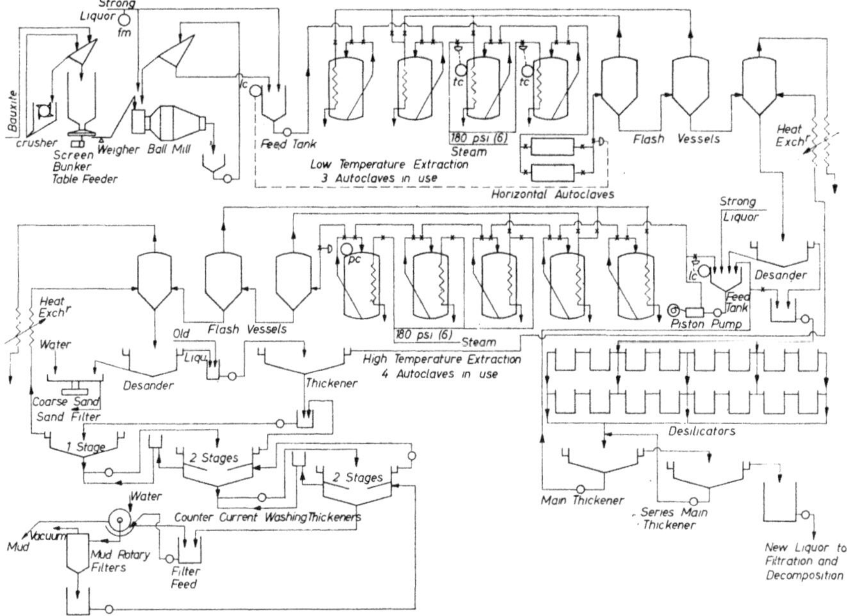

Fig. 9.38 A historical "double digestion" flowsheet proposed by British Aluminum Co Ltd. at their (now closed) Burntisland facility [114]

in and of itself does not demonstrate an energy saving but is included for the historical perspective.

External to this historical reference, the main version of the double digestion flowsheet [52, 115] processed a moderately boehmitic bauxite (~ 5% boehmite alumina) again by first digesting at low temperature, flashing to atmospheric conditions and then separating the boehmite rich residue from the green liquor. This green liquor could flow on to precipitation with a A/TC dictated by gibbsite solubility (discounting for any gibbsite to boehmite reversion which happened in the low temperature digestion (see for example: [116])). The boehmite enriched residue could then be fed to the high temperature circuit, which could not obtain as high an A/TC but guaranteed that most of the boehmite alumina (and all the unextracted/autoprecipitated gibbsite) was extracted. Again, the point is not necessarily the saving in steam associated with lower temperature digestion, but the gibbsite solubility of the low temperature circuit allowing for a higher net yield.

The next significant development in the double digestion technology sphere involves the development of the pressure decanter by Alcan, (now Rio Tinto Alcan), and subsequent digestion development. It should be pointed out, that the pressure decanter is not strictly necessary (depending on the bauxite) but can greatly assist in retrofitting as well as assist in energy improvements. Practically speaking, flashing down the low temperature digestion slurry, then reheating for high temperature dissolution, will incur an energy penalty (beyond the obvious inefficiency associated with

BPE). Use of a pressure decanter to perform the separation and allow the hot low temperature residue slurry to flow to the high temperature circuit has been reported to save some energy. Further to these obvious benefits of the pressure decanter, Kumar et al. [115] provided three unique double digestion flowsheets (see Fig. 9.39) utilizing

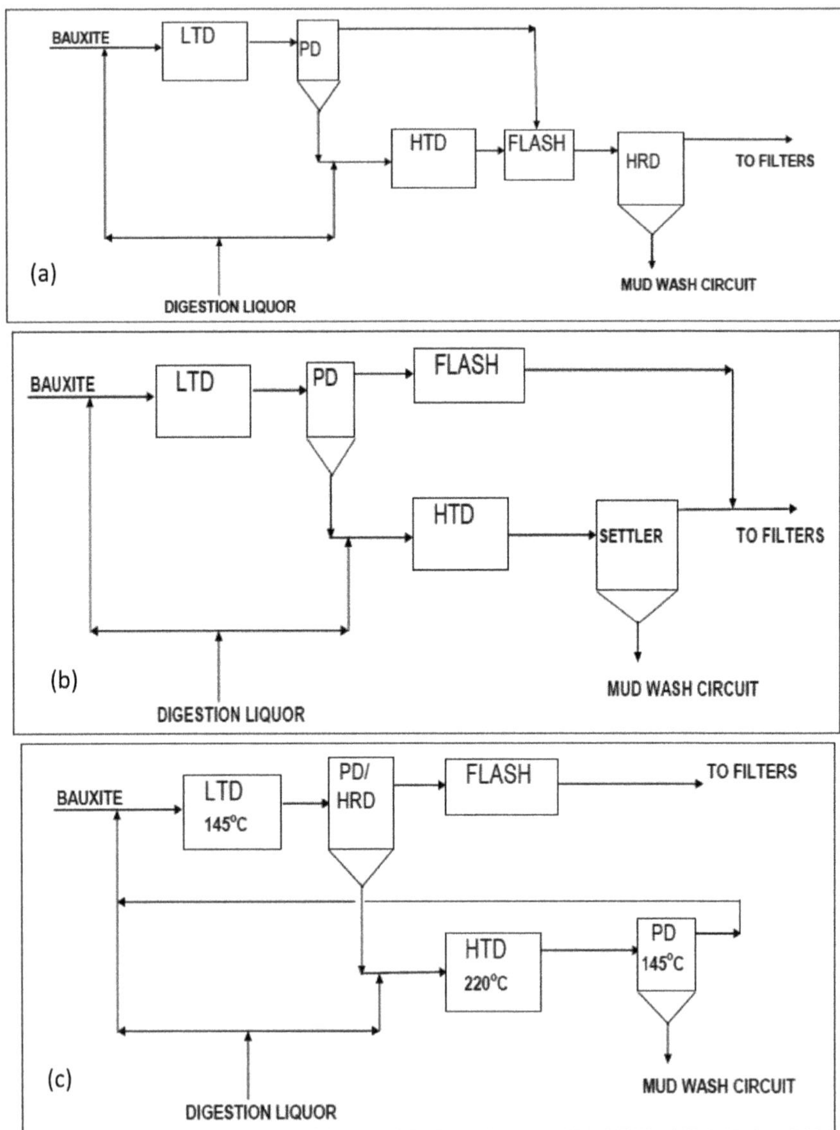

Fig. 9.39 Different "double digestion" flowsheets proposed by Kumar et al. **a** cross flow; **b** parallel and **c** counter-current

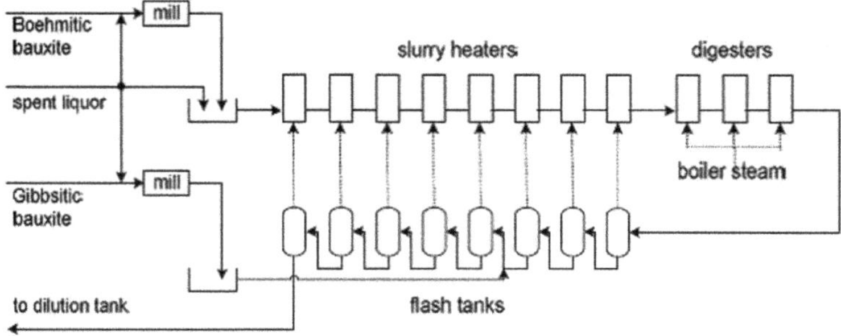

Fig. 9.40 A sweetening flowsheet using gibbsitic bauxite with boehmitic bauxite [113]

the pressure decanter, which have been implemented at full plant scale (brownfields operations).

Cross flow double digestion was implemented in the Gramercy alumina refinery (formerly owned by Kaiser, now Noranda) (see [117]). This double digest replaced a "standard" high temperature digestion flowsheet (which underwent a catastrophic explosion in the digestion area that damaged much of the plant [118]) and improved both energy (via yield, presumably) as well as bauxite extraction. One can speculate that the A/TC in the high temperature circuit had been pushed higher to obtain more yield but suffered an extraction loss by targeting too close to the boehmite solubility limit. The crossflow concept allows the high temperature A/TC to be relaxed (and extraction increased), and the yield made up in the LTD section. The Parallel double digestion flowsheet is very similar to the crossflow sheet, and while the authors have noted it has been implemented at a low temperature plant, forecasting higher boehmite bauxite, the location has not been identified. The counter current double digest flowsheet has presumably been implemented at Gove operations[13] and represents perhaps the most innovative design, requiring a relatively "low" high temperature digestion (~ 220 °C) to secure good extraction, yield and energy.

A simpler brownfield process for improving yield and energy is the sweetening process, which seems to have first mention in a patent attributed to Porter (of Kaiser) [119]. A schematic of the process is shown in Fig. 9.40, highlighting a possible retrofit, whereby a gibbsitic bauxite is added to the high temperature discharging slurry. The hotter flash tanks provide enough temperature and residence time to increase the yield of the liquor without providing an extraction penalty. The lynchpin in this technology is a separate bauxite source, with negligible boehmite. If the same bauxite source were added, it would present as a (probably) prohibitive extraction loss—the boehmite in the bauxite would not be extracted and the boehmite may seed subsequent boehmite reversion of the dissolved gibbsite, as has already been discussed. Ter Weer [1] has estimated an energy savings of 0.7–1.1 GJ/t by way of a 6–10 gpL yield increase, associated with implementation of a sweetening process

[13] No longer running.

to a theoretical high temperature digestion plant running with 65 gpL yield and 9.7 GJ/t energy usage base line.

9.2.2 Clarification

Unlike the previous section where operational and maintenance opportunities exist for refiners to improve their energy performance, clarification energy usage is largely controlled by the specific design employed, with regard to number of vessels as well as how the mud washing is done. As is perhaps obvious, the temperature of the wash water is an opportunity for some energy savings but is set by the design—if using hot condensate versus ambient lake water, more energy stays in the Bayer loop. Similarly, as the solids concentration of the underflow stream (or filtered slurry) increases (e.g., the advent of residue filtration achieving as low as 20–25% moisture), more material stays in the loop and less energy is lost. Further, if the number of vessels increases, the energy of the thickener overflow stream (and the energy) can be higher. The number of vessels is not so straightforward however, as the radiative losses will simultaneously increase with the number. In choosing the number of vessels to use, the soluble soda loss will drive the design process with the energy impact a byproduct, as the material impact on the energy is not as significant as that of the soda.

The principal energy saving opportunity within clarification is to minimize yield and bauxite recovery losses, for the same reasons discussed in 9.2.1.6 The Importance of Indirect Factors. In this vein, there are two main handles within clarification: (i) reduce auto-precipitation of gibbsite in the thickeners and washers and (ii) prevent gibbsite scaling from occurring in the polish filtration. One way to prevent alumina losses through the clarification circuit is to reduce the A/TC target in digestion, as the stability will increase as the supersaturation of gibbsite decreases. This is not an attractive solution for energy saving as the yield, and corresponding plant throughput, will reduce. Summarizing:

1. High A/TC translates to higher yield and better energy performance.
2. Too high A/TC can translate to poor alumina extraction which will negatively impact the alumina recovery from bauxite.
3. High A/TC can translate to high auto-precipitation losses in the CCD circuit and/or scaling in filter building which reduces yield, alumina recovery and possibly plant flow.

Balancing the yield, extraction, recovery and flow is essential to choosing the most appropriate A/TC.[14] Establishing the best A/TC is not trivial, and combinations of extraction models as well as alumina reversion models are necessary for the best analysis (see for example Pei et al. [120] regarding extraction models and Kiriazis [121] for gibbsite reversion model considerations).

[14] The refinery must also consider any possible limitations in precipitation associated with the A/TC which may affect product quality.

Fig. 9.41 Evolution of thickeners—left: flat bottom thickeners/washers (1960s vintage); right: traditional thickeners in background and modern high-rate thickeners (Alcan Deep Cone®) in foreground [126]

Note that in the context of reducing gibbsite auto-precipitation (or reversion) in the CCD circuit, an option may be to reduce the residence time of the mud in the vessel by reducing mud levels or solids concentrations. However, this cannot be considered in a vacuum, as such reductions will negatively impact the soda loss in the vessel/circuit. Thus, alumina losses within the CCD circuit must be considered against soluble soda losses to determine optimal set points.

Two key design improvements which have allowed for reduction in alumina reversion as well as alumina scaling in polish filtration are the High Rate Thickener/Alcan Deep Cone Thickener [122] and the Gaudfrin Diastar Filter [123] (Fig. 9.42). A comparison of the older solid–liquid separation technologies versus the state-of-the-art decanters is shown in Fig. 9.41. Here we can see older flat bottom washers (left) as well as the large diameter cone bottom washers (right, background) together with the state-of-the-art deep cone vessels (right, foreground). Essentially by increasing the pitch of the cone angle as well as increasing the aspect ratio of the vessels, the solids residence time can be minimized, thereby reducing auto-precipitation losses.

The Gaudfrin Diastar filter potentially allows for higher A/TC in digestion by simply reducing the residence time of the filtration. Rather than batch cycle times of 5–10 h (typically associated with Kelly filters or Vertical Pressure Filters), the Diastar can operate for < 1 h filtration times in a continuous fashion. As the liquor being filtered is supersaturated with respect to Gibbsite, by reducing the filtration cycle time, the propensity for gibbsite scale growth, which can blind filters and seriously reduce plant flow, is reduced. As other literature has noted [124], combined with other competitor filter operators manufacturing very similar equipment [125], this technology represents the current state of the art, with the potential to increase plant yield and therefore reduce energy consumption.

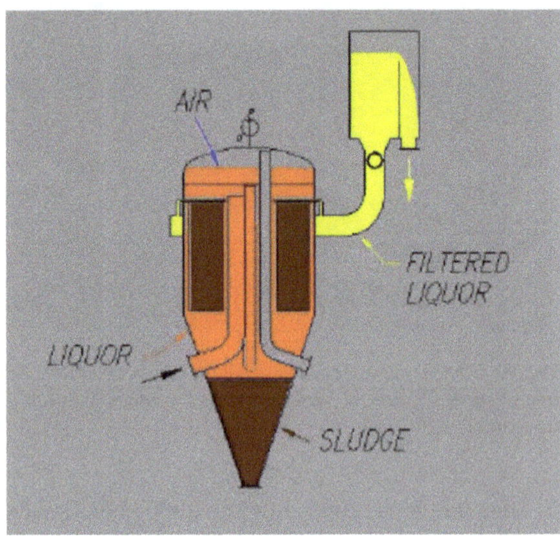

Fig. 9.42 Gaudfrin diastar filter—schematic of operation

9.2.3 *Heat Interchange*

The energy contained in the polished green liquor from clarification is not useful in precipitation as a reduction in temperature will allow for more liquor instability where it is required in precipitation, to increase yield. Thus, for the purpose of actually producing gibbsite, energy from the green liquor must be removed. To ensure that this energy given up goes to a useful sink, as much of the heat as is practical is given to the spent liquor stream that is leaving precipitation and returning to digestion. For further trim control, the green liquor can also be cooled subsequent to this heat transfer with spent liquor via a more controllable source, such as a barometric condenser, using water from a local body or a cooling tower system. It is best if the heat exchanged between spent liquor and green liquor is maximized, versus the trim control, as the energy remains within the Bayer liquor loop this way. To achieve this heat exchange, two principal equipment choices exist in refineries, namely—shell and tube with flash tank heat exchange and plate heat exchange. The modern selection, as noted, for example, by Ter Weer [1], would seem to favor the plate exchange form for two principal reasons:

1. The heat exchange does not suffer the loss associated with boiling point elevation.
2. Direct heat exchange via plates does not evaporate the green liquor side, which causes the caustic concentration to rise and stabilize the liquor.

The principal counter arguments favoring the use of flash with shell and tube systems include:

1. A more robust system with regard to scaling.

2. Condensate generated can assist with the water balance and reduce dependency on evaporation.

In respect to the increased scaling rate of plate exchangers versus flash systems (e.g., ~ 1 week versus ~ >4 weeks for heater tubes and many months for flash tanks), this can be combatted with a suitable sparing and maintenance strategy. Regarding the water balance, presumably this can be combatted by increasing the capacity of the evaporation building at the design stage. For refiners attempting to retrofit plates into an existing shell and tube heat interchange building, this can prove a challenge as one would need to include some extra evaporation capacity associated with the upgrade, making the upgrade somewhat more expensive than perhaps initially forecasted.

After precipitation, the spent liquor can proceed to the evaporation building first, or heat interchange, or perhaps some combination of the two. In determining, which is the best design to pursue, we need to be cognizant of the metrics we want to optimize. In considerations regarding heat Interchange and evaporation together, the key metrics can be summarized as:

- Mass of water removed
- Evaporation economy (t evaporated/t steam added)
- Green liquor temperature feeding precipitation
- Spent liquor temperature feeding digestion.

In determining which flowsheet to pursue, the design engineer associates value to each of these metrics, and these may differ on a case by case basis. For example, if HID is prior to evaporation, perhaps the green liquor feeding precipitation can be 0.5 °C cooler, but the evaporation economy worsens by 5%. In determining, which is correct, we should evaluate whether the yield benefit sufficiently offsets the poorer evaporation economy. As with all such analyses, the evaluation is reliant on the quality of the physical properties employed. A large part of the analysis associated with this design question is dependent on enthalpy/heat capacity and boiling point elevation formulae, as these correlations will largely dictate evaporation amounts, stream temperatures, etc. This sort of evaluation needs to be conducted using a full plant process model as the resultant subtle differences are complex to evaluate. This discussion intends to show the main metrics to consider in the evaluation and highlight some of the challenges of this aspect of the design process.

9.2.4 Precipitation

The precipitation circuit, like the clarification circuit, can guarantee good energy performance by maximizing the liquor yield. While many of the options are strictly about increasing yield, some options do directly involve energy.

The yield-only options include discussion of the number of precipitators to include as well as perhaps seed filtration to minimize the amount of spent liquor recycling to the front of the precipitation circuit, which seems to have entered into the industry's consciousness in the early 1990s (see for example, reference of seed filtration by the Worsley alumina refinery in Western Australia [127]).

The other development in increasing yield, which is intimately involved with energy contained within the Bayer loop, involves inter-stage cooling. During the precipitation circuit, the alumina in liquor diminishes as gibbsite is precipitated, getting closer and closer to the solubility limit. To encourage further precipitation, the temperature of the slurry can be reduced to ensure that the thermodynamic solubility reduces as well. However, the temperature controls not just the thermodynamics, but it also controls the kinetics by virtue of an Arrhenius term. Thus, for an adequate understanding of the correct approach with regard to temperature profiles throughout precipitation circuits to maximize yield, a suitable kinetic/yield model is required which captures the thermodynamic and kinetic effects.[15] As with all such process models within the industry, no doubt many are maintained as trade secret, but various approaches can be found within the literature (see for example Cornell et al. [128] for a summary of some available). In any case, alumina producers seem to have unanimously decided that inter-stage cooling or cooling of the precipitating slurry in certain choice locations in the precipitation circuit, is a yield enhancing option (noted by one author as between 2 and 5 gpL benefit [113]). The exact mechanism by which this is achieved has evolved with different producers favoring different methods. An early example of inter-stage cooling was provided by Alcan in the form of an in-tank plate running cooling water, shown in Fig. 9.43. As the name explains, the cooler is placed within the precipitator itself, which renders some weaknesses—if the abrasive slurry creates a pinhole in the heat exchanger, the slurry will be diluted significantly impacting the yield. Further, the only way to correct the leak is to drain the extremely large tank and repair, which can significantly affect a plant's volume/water management if occurring on an unexpected basis.

Another approach, utilized by Worsley [127], incorporated flash cooling with a barometric condenser/cooling tower design. This method would have guaranteed the cooling, and unlike the in-tank design does not carry the potential maintenance weakness but does reduce the utility of the cooling by flashing. The flash removes water from the slurry, which increases the caustic concentration simultaneous to cooling, which partially offsets the destabilizing effect intended. This choice for a less than optimal cooling was presumably driven by a fear of scaling—the precipitation slurry is unstable and at high solids concentration, encouraging a combination of scale growth and slurry plugging potentially. In avoiding a direct heating method, the flashing method would have perhaps been considered a safer option.

Subsequent to this development, Alumina de Greece, with the help of Barriquand technologies, successfully utilized Platulaire plate heat exchangers (see a schematic in Fig. 9.44) to directly cool the slurry without the evaporating penalty [34]. Based on the success of the plate exchanger application, which secures the cooling without evaporation and is more easily maintainable than the earlier "in-tank" design, these

[15] Notwithstanding possible impacts on oxalate stability and precipitation kinetics if its precipitation is to be avoided.

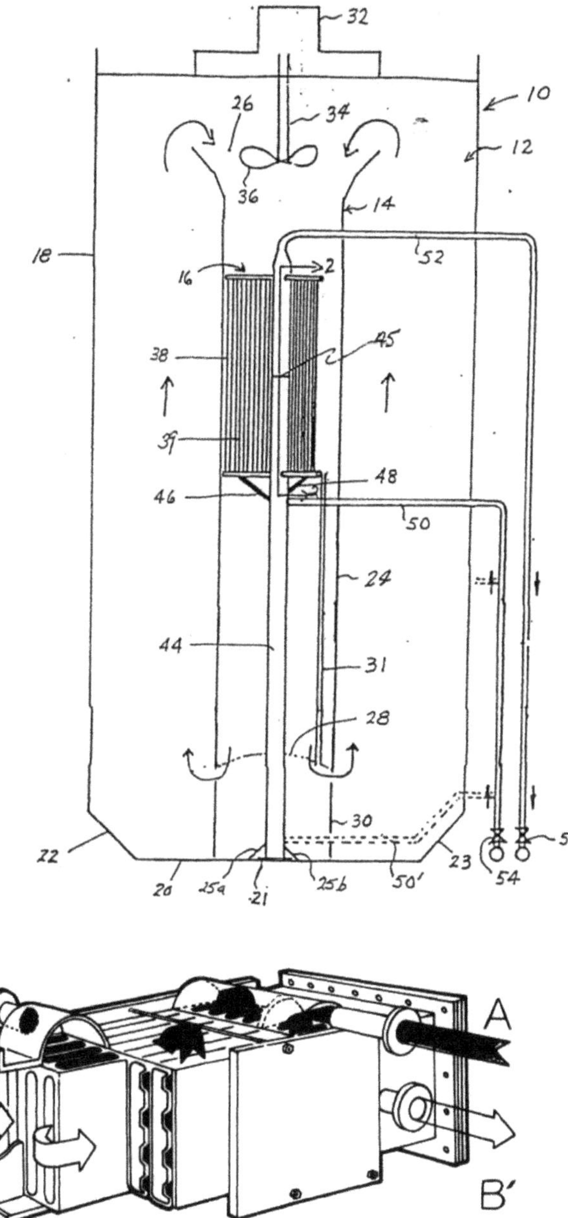

Fig. 9.43 In-tank interstage cooler as patented by Alcan [129]

Fig. 9.44 Barriquand plate heat exchanger incorporated into Alumina de Greece for interstage cooling of precipitation slurry [34]

direct methods for heat exchange appear to be the best application for cooling the precipitation slurry and securing additional yield.

Perhaps the more pertinent question to consider from an energy conservation perspective is what to use for exchanging heat with the green liquor. It seems that the vast majority of plants choose to use cooling water and cooling tower systems. While this may seem somewhat wasteful, given that the spent liquor leaving precipitation needs thermal energy; the impact on yield typically dictates the colder water stream being selected. One tonne of green liquor (~ 95 °C) will correspond to approximately 0.9 tonne of spent liquor (~ 60 °C, for a 300 gpL TC, ~ 100 gpL yield process); this limits the amount of practical cooling that can be done on the green liquor. When compared with a 25 °C cooling water stream that can have the flow adjusted as needed, the choice becomes clearer. Note that the energy usage can in fact go down through the introduction of such a system—if the rate of energy being removed from the Bayer loop is less than the rate of alumina being produced because of the cooling, the impact will be a net improvement on energy consumption. Some unique flowsheets have been proposed, for example that of Simpson et al. [130], demonstrating the ability to use spent liquor as a cooling fluid. While their results clearly show an improvement in steam saving associated with spent liquor duty downstream of precipitation, the impact on yield was not included in the analysis. To summarize, the ultimate goal from a purely energy saving standpoint should be to allow heat exchange between the green and spent liquor within precipitation; however, the cooling capacity of a controllable water stream may in fact offset the energy lost by increasing the plant yield by a greater amount.

9.2.5 Concluding Remarks on Energy Balance

As discussed at the start of this section, the energy usage of a Bayer refinery should be a focal point for alumina producers because of its significant contribution to both a refinery's bottom line as well as the environment. Further to this, its optimization is not a trivial undertaking as it clearly overlaps with the production rate, the soda consumption and the water balance of the refinery.

This section has also highlighted how and why differences in energy usage can occur between refineries, largely due to differences in the raw materials which shape decisions towards different flowsheet designs. Universal to all designs though, it is the critical role of the adequate selection of heat transfer equipment, with a principal focus on its maintenance to guarantee that good efficiencies are being regularly achieved, despite the renowned scaling problems that plague the Bayer process. The other fundamental enabler to lower energy is the Bayer process alumina liquor yield. It has been demonstrated that this key metric can often act as a surrogate for the energy performance of a refinery. All the actions that allow for improved yield, typically enable reductions in a plant's energy usage.

The Bayer process liquor cycle is a loop, and the goal is to ensure that the energy stays within the loop, with heat exchange occurring between different parts of the

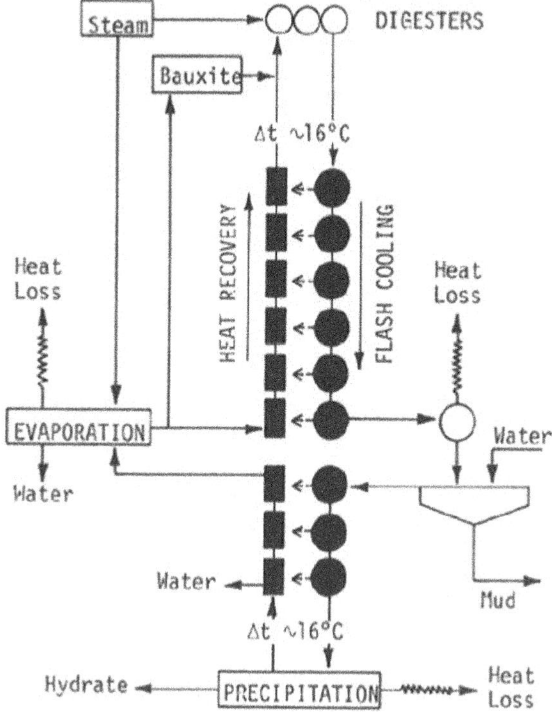

Fig. 9.45 Generic illustration of the Bayer process as a large heat exchanger [131] ideally transferring energy from green/pregnant liquor to spent liquor

cycle. The diagram shown in Fig. 9.45, provided by Donaldson [131], simplifying the Bayer process into a large heat exchanger, highlights this idea well. Wherever possible the heat is exchanged from green liquor to spent liquor, unless it makes sense to do otherwise.

Following the same liquor loop logic, the diagram also emphasizes that the function of evaporation is simply to remove water from the liquor stream. If the liquor leaves hotter than it entered, more heat will be wasted in the digestion (via steam leaving out the final blow off tank to atmosphere) as their correct heat sink will not demand as much energy. Conversely, if the liquor leaving evaporation is cooler, then more steam than necessary will need to be added in digestion—again a waste. Both of these concepts are illustrated, again by Donaldson [131], in the novel diagrams shown in Fig. 9.46, which serve as a good vehicle to emphasize the way energy transfer should be thought of in the Bayer process.

Bayer operators must always be seeking new opportunities to reduce and optimize energy consumption, with particular respect paid to equipment care and maintenance. Fundamentally though, it is the realization of the need to transfer energy within the liquor loop; combined with the earlier considerations regarding maximizing liquor yield that will guarantee the energy usage is optimized for the plant in question, irrespective of the limitations imposed by the bauxite quality.

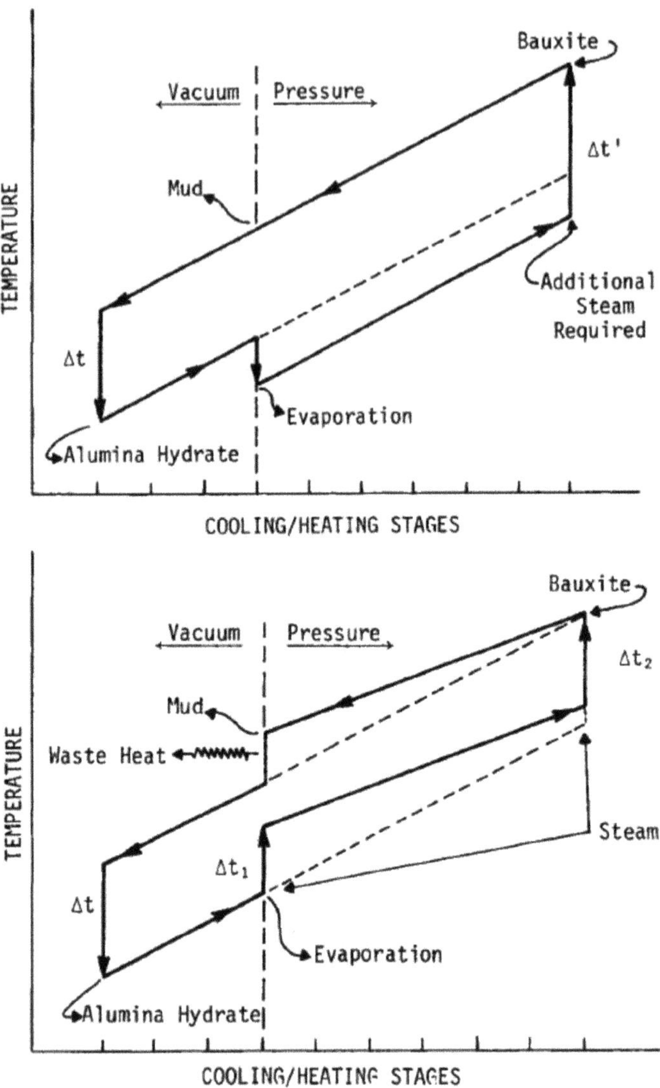

Fig. 9.46 Temperature versus heat transfer stages plots [131] highlighting both a Bayer circuit where evaporation liquor discharge is colder than the inlet (top) and hotter than the inlet (bottom)

References

1. P.-H. Ter Weer, Relationship between liquor yield, plant capacity increases, and energy savings in alumina refining. J. Metals **66**(9), 1939–1943 (2014)
2. L.K. Hudson, C. Misra, A.J. Perrotta, K. Wefers, F.J. Williams, Aluminum oxide, in *Ullmann's Encyclopedia of Industrial Chemistry*, (Weinheim, Wiley-VCH, 2012), pp. 607–644
3. A. Oliveira, J. Dutra, J. Aldi, Alunorte bauxite dewatering station—a unique experience, in *Light Metlas* (2008)

4. J.J. Kotte, Bayer digestion and predigestion desilication reactor design, in *Essential Readings in Light Metals: Alumina and Bauxite*, ed. by D. Donaldson, B.E. Raahauge (Wiley Inc., 2013), pp. 331–349
5. B. Haneman, B.R. Lindsay, Design, startup and operation of the new digestion facility at the Ma'aden alumina refinery, in *Light Metals*, Orlando (2015)
6. R. Wischnewski, C.M. de Azevedo Jr, E.L.S. Moraes, A.B. Monteiro, Alunorte global energy efficiency, in *Light Metals*, San Diego (2011)
7. M. Edwards, R. Kelly, D. deBoer, New technology for indirect thick slurry heating system, in *8th Alumina Quality Workshop*, Darwin (2008)
8. S. Kumar, Indirect bauxite slurry heating system for the Bayer process, in *Light Metals*, Las Vegas (1989)
9. D. Thomas, Heat transfer in the Bayer process, in *Essential Readings in Light Metals: Alumina and Bauxite*, ed. By D. Donaldson, B.E. Raahauge (Wiley Inc., 2013), pp. 705–710
10. G. Riley, P. Smith, D. Binet, R. Pennifold, Plant impurity balances and impurity inclusion in DSP, in *5th Alumina Quality Workshop*, Bunbury (1999)
11. P. Smith, The processing of high silica bauxites—review of existing and potential processes. Hydrometallurgy **98**, 162–176 (2009)
12. G. Bánvölgyi, A.C. Tóth, I. Tassy, In situ formation of sodium aluminum hydrosilicate from kaolinite, in *Light Metals*, New Orleans (1991)
13. P. Smith, C. Wingate, L. De Silva, Mobility of included soda in sodalite, in *8th International Alumina Quality Workshop*, Darwin (2008)
14. M.C. Barnes, J. Addai-Mensha, A.R. Gerson, The kinetics of desilication of synthetic spent Bayer liquor seeded with cancrinite and cancrinite/sodalite mixed-phase Crystals. J. Cryst. Growth **200**, 251–264 (1999)
15. L. Shi, S. Ruan, J. Li, A.R. Gerson, Desilication of low alumina to caustic liquor seeded with sodalite or cancrinite. Hydrometallurgy (2016)
16. L. Henrickson, The need for energy efficiency in Bayer refining, in *Essential Reading in Light Metals: Alumina and Bauxite*, ed. by D. Donaldson, B.E. Raahauge (Wiley Inc., 2013), pp. 691–696
17. G. Wargalla, W. Brandt, Processing of diaspore bauxites, in *Essential Readings in Light Metals: Alumina and Bauxite*, ed. by D. Donaldson, B.E. Raahauge (Wiley Inc., 2016), pp. 393–401
18. P. Sipos, The structure of Al(III) in strongly alkaline aluminate solutions—a review. J. Mol. Liq. **146**, 1–14 (2009)
19. A. Nandi, H. Mahadevan, T.R. Ramachandran, Bauxite-alumina industry of India, in *2nd Asian Bauxite and Alumina Conference*, Marina Mandarin (2012)
20. S.P. Rosenerg, S.J. Healy, A thermodynamic model for gibbsite solubility in Bayer liquors, in *4th International Alumina Quality Workshop*, Darwin (1996)
21. D.J. Donaldosn, Perspective on Bayer process energy, in *Light Metals*, San Diego (2011)
22. K. Bielfeldt, Practical experiences with the tube digester. J Metals 48–54 (1968)
23. D.A. Dahlstrom, R.C. Bennet, R.C. Emmet, P. Harriot, T. Laros, W. Leung, C. McCleary, S.A. Miller, B. Morey, J.Y. Oldshue, G. Priday, C.E. Silverglatt, J.S. Slottee, J.C. Smith, D.B. Todd, Section 18 liquid-solid operations and equipment, in *Perry's Chemical Engineer's Handbook Seventh edition*, ed. by D.W. Green, J.O. Maloney, New York, McGraw Hill (1997)
24. H.F. Scandrett, Equations for calculating recovery of soluble values in a countercurrent decantation washing system, in *Essential Readings in Light Metals: Alumina & Bauxite*, ed. by D. Donaldson, B.E. Raahauge (Wiley Inc., 2016), pp. 879–884
25. D.J. Cooling, Developments in the disposal of residue from the alumina refining industry, in *Essential Reading in Light Metals: Alumina and Bauxite*, ed. by D. Donaldson, B.E. Raahauge (Wiley Inc., 2013), pp. 927–932
26. M.-J. Bélanger, Red mud stacking, in *Essential Readings in Light Metals: Alumina and Bauxite*, ed. by D. Donaldson, B.E. Raahauge (Wiley Sons Inc., 2016), pp. 944–950
27. E.I. Robinsky, Thickened discharge—a new approach to tailings disposal. CIM Bull. **68**(764), 47–53 (1975)

28. J.-M. Rosseaux, B. Langlois, D. Boufounos, A. Cuneo-Raffaelli, Bauxite residue filtraiton experience in Gardanne and aluminium of Greece alumina plant, in *8th International Alumina Quality Workshop*, Darwin (2008)
29. M.M. de Castro, R. Wischnewski, L.G. Corrêa, J.R.A. Filho, A new concept for dry disposal of Alunorte's bauxite residue, in *9th International Alumina Quality Workshop*, Perth (2012)
30. L.D. Munro, D. Smirk, Optimising bauxite residue deliquoring and consolidation, in *9th International Alumina Quality Workshop*, Perth (2012)
31. R. Bott, T. Langeloh, J. Hahn, R. Arpe, G. Vollmers, A. Lalla, A dry bauxite residue by Hi-bar steam pressure filtration, in *6th International Alumina Quality Workshop*, Brisbane (2002)
32. M.M. Castro, C.R.A. Trindade, R.G. Pantoja, E.R.Q. Alves, A.R. Martins, A new technology for dry disposal of Alunorte's bauxite residue, in *Light Metals*, San Antonio (2013)
33. K. Nery, J. Avila, M. Scarminio, L. Bittar, R. Moreno, R. Sono, Study of alternative technologies for residue disposal (red mud), in *Light Metals*, San Diego (2014)
34. N. Denaxas, A. Zervos, Saving energy in alumina-aluminium plants, in *International Conference on energy Efficiency in Process Technology*, Athens (1992)
35. B. Petersen, M. Bach, R. Arpe, The world's largest hydrate pan filter: engineering improvements and experiences, in *Light Metals*, San Francisco (2009)
36. J. Hahn, R. Bott, T. Langeloh, Plant experience with replacement of a three stemp drum filiter plant by a one step pan filter at Al-hydrate product filtration, in *9th International Alumina Quality Workshop*, Perth (2012)
37. G.I.D. Roach, J.B. Cornell, Scaling in Bayer plants, in *13th Australian Chemical Engineering Conference*, Perth (1985)
38. H.W. Schmidt, P.J. Genger, G. Heinemann, C.F. Pgacar, E.H. Wynche, Stress corrosion cracking in alkaline solutions. Corrosion **7**(9), 295–302 (1951)
39. R.N. Parkins, Stress corrosion cracking and hydrogen embrittlement of iron base alloys, in *NACE*, Houston (1977)
40. G. Roach, G. Rubenis, P. Mastin, R. Byrne, Caustic embrittlement in the Bayer industry, in *6th International ALumina Quality Workshop*, Brisbane (2002)
41. L.K. Hudson, Alcoa's process filter, in *Light Metals*, Dallas (1974)
42. Gaudfrin, "Diastar Filter," Gaudfrin, [Online]. Available: http://www.gaudfrin.com/the-diastar-filter/. [Accessed 13 Sept 2017]
43. Bokela, "Backflush Filters," BOKELA GmbH Karlsruhe, [Online]. Available: http://www.bokela.de/en/technologies/pressure-filtration/backflush-filters/general-info.html. [Accessed 13 Sept 2017].
44. AESSEAL, what is a mechanical seal? [Online]. Available: http://www.aesseal.com/en/resources/academy/what-is-a-mechanical-seal. [Accessed 13 Sept 2017]
45. D. Thomas, B. Pei, Alumina refinery design for climate extremes, in *9th International Alumina Quality Workshop*, Perth (2012)
46. P. Duggal, Water conservation study in a Bayer process plant (1994)
47. B. Haneman, A. Wang, Optimising digestion flash tank design for the alumina industry, in *9th International Alumina Quality Workshop*, Perth (2012)
48. P. Smith, Reactions of lime under high temperature Bayer digestion conditions, in *10th International Alumina Quality Workshop*, Perth (2015)
49. D. Audet, R. Little, N.J. Hofstee, Stabilisation of product quality at Yarwun alumina refinery, in *9th International Alumina Quality Workshop*, Perth (2012)
50. D.J. Cooling, Improving the sustainability of residue management practices —Alcoa World Alumina Australia, in *Paste*, Perth (2007)
51. H. Li, S. Harris, A water and mud management model for bauxite residue disposal, in *9th International Alumina Quality Workshop*, Perth (2012)
52. D. Smith, F. Williams, S. Moffatt, Wastewater treatment methods, in *Essential Readings in Light Metals: Alumina & Bauxite*, ed. by D. Donaldson, B.E. Raahauge (Wiley Inc, 2016), pp. 685–690
53. A. Pasupulatey, *Visit Report EurAlumina Refinery* (2002)

54. D. McConchier, P. Saenger, R. Fawkes, An environmental assessment of the use of seawater to neutralise bauxite refinery wastes, in *International Symposium on Extraction and Processing for the Treatment and Minimisation of Wastes* (1996)
55. S. Adano, J. Kildea, S. Millington, P. Ross, Opportunities for water savings and water quality improvements in the Bayer process, in *9th International Alumina Quality Workshop*, Perth (2012)
56. A.A. Scarsella, S. Noack, E. Gasafi, C. Klett, A. Koschnick, Energy in alumina refining: setting new limits, in *Light Metals*, Orlando (2015)
57. M.P. Weigel, *Method and Apparatus for Continuous Digestion*. United States Patent 2056993 (1936)
58. C.O. Turner, M.P. Weigel, *Continuous Digestion Process*. United States Patent 2107919 (1938)
59. C.R. Robert, *Method for Extracting Alumina from its Ores*. United States Patent 3413087 (1968)
60. A. Juhaz, Some apparative improvement solutions at the Bayer process in Hungary, in *Light Metals*, Dallas (1974)
61. G. Winkahus, Measurements by means of radio isotopes in industrial tube digestion and autoclave lines of bauxite plants, in *Second Internationa Symposium of ICSOBA*, Leoben (1971)
62. G. O'Neill, Prediction of heat exchanger-heat transfer coefficient decay due to fouling, in *Essential Readings in Light Metals: Alumina and Bauxite*, ed. by D. Donaldson, B.E. Raahauge (Wiley Inc., 2013), pp. 697–805
63. D. Kern, *Process Heat Transfer* (McGraw-Hill, Sao Paulo, 1965)
64. J.K.L. Dewey, Boiling point rise of Bayer liquors, in *Light Metals*, Chicago (1981)
65. SysCAD, *Alumina 3 Bayer Species Model*, SysCAD, 18 May 2017. [Online]. Available: https://help.syscad.net/index.php?title=Alumina_3_Bayer_Species_Model. [Accessed 24 May 2017].
66. J.M. Langa, The heat of dissolution of gibbsite at Bayer digesiton temperatures, in *Essential Readings in Light Metals: Alumina and Bauxite*, ed. by D. Donaldson, B.E. Raahauge (Wiley Inc, 2013), pp. 170–175
67. B.S. Hemingway, R.A. Robie, Enthalpies of formation of low albite (NaAlSi3O8), Gibbsite (Al(OH3), and NaAlO2; Revised Values for dH, 298°K and dG 298°K of alumina-silicate minerals. J. Res. U.S. Geol. Surv. 413–429 (1977)
68. G. Songqing, Q. Lijuan, Y. Zhonglin, Energy consumption in Bayer process, in *Light Metals* (2007)
69. E. Königsberger, P.M. May, G. Hefter, Comprehensive model of synthetic Bayer liquors. Part 3. Sodium aluminate solutions and the solubility of gibbsite and boehmite. Monatsh. Chem. **137**, 1139–1149 (2006)
70. "Standard Enthalpies of Formation & Standard Entropies of Common Compounds," [Online]. Available: http://www.mrbigler.com/misc/energy-of-formation.PDF. [Accessed 14 Sept 2017]
71. D.J. Wesolowski, Aluminum speciation and equilibria in aqueous solution: I. The solubility of gibbsite in the systen Ma-K-Cl-OH-Al(OH)4 from 0 to 100 °C. Geochim. Cosmochim. Acta **56**, 1065–1091 (1992)
72. The Engineering ToolBox, Thermal Conducitivity of Metals, December 2017. [Online]. Available: https://www.engineeringtoolbox.com/thermal-conductivity-metals-d_858.html
73. W.M. Kays, M.E. Crawford, *Convective Heat and Mass Transfer* (McGraw-Hill Book Company, New York, 1980), p. 245
74. H. Müller-Steinhagen, M. Jamialahmadi, B. Robson, Understanding and mitigating heat exchanger fouling in Bauxite refineries. J Metals 36–41 (1994)
75. A. Duncan, M. Groemping, B. Welch, H. Müller-Steinhagen, The effect of silica, temperature, velocity, and particulates on heat transfer to spent Bayer liquor, in *Essential Readings in Light Metals: Alumina and Bauxite*, ed. by D. Donaldson, B.E. Raahauge (John Inc., 2013), pp. 124–131
76. V. Gnielinski, New equations for heat and mass transfer in turbulent pipe and channel flow. Int. Chem. Eng. **16**, 359–368 (1976)

77. T. Oku, K. Yamada, The dissolution rate of quartz and the rate of desilication in Bayer liquor, in *Light Metals*, New York (1971)
78. P.G. Cousineau, G.D. Fulford, Aspects of the desilication of Bayer liquors, in *Light Metals*, Denver (1987)
79. A. Duncan, H. Müller-Steinhagen, B. Verity, B. Welch, An investigation of desilication kinetics in spent Bayer liquor, in *Light Metals*, Las Vegas (1995)
80. M.C. Barnes, J. Addai-Mensah, A.R. Gerson, The kinetics of desilication of synthetic spent Bayer liquor and sodalite crystal growth. Colloids Surf., A **147**, 283–295 (1999)
81. R. LaMacchia, Toward a better understanding of desilication product (DSP) precipitation kinetics, in *Proceedings of the 9th International Alumina Quality Workshop*, Perth (2012)
82. M. Jamialahmadi, H. Müller-Steinhagen, Numerical model for the prediction of heat transfer and scale profiles in Bayer process live steam heat exchangers, in *Light Metals*, San Diego (1992)
83. H. Müller-Steinhagen, M. Jamialahmadi, B.J. Robson, An investigation into the formation of DSP on heat transfer surfaces, in *Light Metals*, Denver (1993)
84. J. Jamialahmadi, H. Müller-Steinhagen, Determining silica solubility in Bayer process liquor. J. Metals 44–49 (1998)
85. D. Spitzer, O. Chamberlain, C. Franz, M. Lewellyn, Q. Dai, Max HT sodalite scale inhibitor: plant experience and impact on the process, in *Light Metals*, New Orleans (2008)
86. K. Yamada, M. Yoshihara, S. Tasaka, Properties of scale in Bayer process, in *Light Metals*, New York (1985)
87. A.J.H. McGaughey, M. Kaviany, Thermal conductivity decomposition and analysis using molecular dynamics simulations. Part II. Complex silica structures. Int. J. Heat Mass Transf. **47**, 1799–1816 (2004)
88. M. Loan, B. Loughlin, J. Haines, D. Croker, M. Fennell, B.K. Hodnett, In situ time-resolved synchrotron diffraction studies of high temperature Bayer digestion, in *Proceedings of the 7th International Alumina Quality Workshop*, Perth (2005)
89. M.C. Barnes, J. Addai-Mensah, A.R. Gerson, The mechanism of the sodalite-to-cancrinite phase transformation in synthetic spent Bayer liquor. Microporous Mesoporous Mater. **31**, 287–302 (1999)
90. M.C. Barnes, J. Addai-Mensah, A.R. Gerson, The kinetics of desilication of synthetic spent Bayer liquor seeded with pure sodalite, pure cancrinite and dimporphic phase mixture crystals, in *Light Metals* (1999)
91. T. Harato, T. Ishida, H. Kato, M. Inami, K. Ishibashi, Y. Kumagae, M. Murakami, The development of a new Bayer process that reduces the desilication loss of soda by 50% compared to conventional process, in *Proceedings of the 4th International Alumina Quality Workshop*, Darwin (1996)
92. N. Kawashima, L. Shi, N. Xu, J. Li, A.R. Gerson, Characterisation of single-stream bayer plant heat exchanger scale. Hydrometallurgy **159**, 75–86 (2016)
93. D. Spitzer, A. Rothenberg, H. Heitner, F. Kula, M. Lewellyn, O. Chamberlain, Q. Dai, C. Franz, A real solution to sodalite scaling problems, in *7th International Alumina Quality Workshop*, Perth (2005)
94. M. Kiriazis, J. Gill, D. Stegink, Evaluation of Max HT at Queensland Alumina Ltd., in *9th International Alumina Quality Workshop*, Perth (2012)
95. A. Aboagye, J. Kildea, T. La, E. Phillips, Management and control of silica in the Bayer process, in *Proceedings of the 9th International Alumina Quality Workshop*, Perth (2012)
96. D.P. Spitzer, A.S. Rothenberg, H.I. Heitner, F. Kula, Method of preventing or reducing aluminosilicate scale in a Bayer process. Patent US20040011744A1, 22 Jan 2004
97. E.R. Yawn, *Process for Reducin Contaminants in Condensate Resulting from the Conversion of Bauxite to Alumina*. United States Patent US 7,264,729 B2, 4 Sept 2007
98. S. Ainscow, Falling film technology for liquor concentration, experiences from a recent start-up, in *Proceedings of the 6th International Alumina Quality Workshop*, Brisbane (2002)
99. C. Satish Kumar, T. Banerjee, U. Giri, R. Pradha, R. Saha, P. Pattnaik, Advanced process control in the evaporation unit, in *Light Metals*, San Diego (2011)
100. J. Wu, G. Lane, I. Livk, B. Nguyen, L. Graham, D. Stegink, T. Davis, Swirl flow agitation for scale suppression. Int. J. Miner. Process. **112–113**, 19–29 (2012)

101. G.I.D. Roach, J.B. Cornell, Scaling in Bayer plants, in *The AusIMM Conference*, Perth, (1996)
102. C.G. Coleman, J. Wu, Flash Tanks. US Patent 20110199854A1 (2011)
103. Q.K. Tran, Applications of three-phase flashing flow in design and study alumina digestion flash trains, in *Proceedings of the 7th International Alumina Quality Workshop*, Perth (2005)
104. Q.-K. Tran, H. Tran, Momentum change in digestion relief headers, in *Proceedings of the 6th International Alumina Quality Workshop*, Brisbane (2002)
105. Q.-K. Tran, M. Reynolds, Sizing of relief valves for two-phase flow in the Bayer process, in *Proceedings of the 5th International Alumina Quality Workshop*, Bunbury (1999)
106. D. del Aguila, Heat treatment of low carbon steel (LCS) tubes to improve corrosion resistance, in *Proceedings of the 8th International Alumina Quality Workshop*, Perth (2008)
107. A.R. Kane, M.J. Battams, J.N. Connor, A. Xavier, D. del Aguila, Improved pre-passivation of mild steel heat exchanger tubes, in *Proceedings of the 7th International Alumina Quality Workshop*, Perth (2005)
108. S. Fortin, R. Breault, Bayer process heat exchangers cleaning efficiency: optimizing the acid dissolution of sodalite scale, in *Heat Exchanger Fouling and Cleaning: Fundamentals and Applications*, Santa Fe (2003)
109. KWA Kenwalt Australia, *SysCAD*, Perth (2017)
110. D.J. de Boer, M. Edwards, P. McIntosh, Design, start-up and operational aspects of the new digestion process at comalco alumina refinery, gladstone, Queensland, in *7th International Alumina Quality Workshop*, Perth (2005)
111. S. Austin, S. Perry, Digestion efficiency at RTA Yarwun alumina refinery, in *Light Metals*, New Orleans (2008)
112. F. Wolters, M. Schütt, The continuous optimization of the primary energy input at AOS—a story of success, in *Proceedings of the 35th International ICSOBA Conference*, Hamburg (2017)
113. R. den Hond, I. Hiralal, A. Rijkeboer, Alumina yield in the Bayer process past, present and prospects, in *Light Metals*, Orlando (2007)
114. V.E. Davies, J.E. Lloyd, D. Macfie, History of Bauxite processing in British aluminium, in *ICSOBA*, Banska-Bystrica—Žiarnad Hronbom (1972)
115. S. Kumar, B. Hogan, J.-P. Laredde, J. Laurier, G. Forte, Double digestion process: an energy efficient option for treating boehmitic bauxites, in *8th International Alumina Quality Workshop*, Darwin (2008)
116. J.-M. Lamerant, Y. Perret, Boehmitic reversion in a double digestion process on a bauxite containing, in *Light Metals*, Seattle (2002)
117. J. Doucet, C. Hendriks, E. Patz, G. Forté, S. Kumar, R. Paradis, Pressure decantation technology: the Kaiser Gramercy experience, in *Proceedings of the 6th International Alumina Quality Workshop*, Brisbane (2002)
118. W.P. Harrington, G.W. Harrell, B.J. Cohea Jr, Inspection techniques for digestion pressure relief system, in *Light Metals*, Orlando (2007)
119. J.L. Porter, Process for the production of alumina. US Patent 2701752, 8 Feb 1955
120. B. Pei, D. Taylor, D. Thomas, Extraction kinetics of Weipa Monohydrate bauxite, in *Proceedings of the 6th Alumina Quality Workshop*, Brisbane (2002)
121. M. Kiriazis, Settler and washer alumina reversion, in *Proceedings of the 7th Aumina Quality Workshop*, Perth (2005)@@@
122. P.F. Bagatto, D.L. Puxley, *Method of and Apparatus for Thickening Red Muds Derived from Bauxite and Similar Slurries*. US Patent 4830507, 16 May 1989
123. Gaudfrin, *Gaudfrin Diastar Filter for Aluminate Liquor Clarification*, Saint-Germain-en-Laye (2017)@@@
124. R. Shaw, A. Duncan, M. Crosdale, Improvements in smelter grade alumina quality at clarendon alumina works, in *Light Metals*, San Diego (2011)
125. R. Bott, T. Langeloh, J. Hahn, Advanced pregnant liquor purification with a new backflush filter, in *Proceedings of the 8th International Alumina Quality Workshop*, Darwin (2008)
126. T.J. Laros, F.A. Baczek, Selection of sedimentation equipment for the Bayer process an overview of past and present technology, in *Light Metals*, San Francisco (2009)

127. F.N. Newchurch, K.E. Moretto, Modern refinery design for quality alumina a Worsley case study, in *Proceedings of the Second Alumina Quality Workshop*, Perth (1990)
128. R.M. Cornell, D.S. Pannett, N.S. Sullivan, P.C. Clarke, C.M. Bailey, Precipitation of Gibbsite: development of a new rate equation, in *Proceedings of the 5th International Alumina Quality Workshop*, Bunbury (1999)
129. J. Bouchard, B. Hiscox, C. Ellis, *Improved Precipitation Vessel*. Europe Patent 0467617A2, 15 July 1991
130. M. Simpson, S. Hial, R.S. Singh, Optimization of heat recovery from the precipitation circuit, in *Light Metals*, San Diego (2011)
131. D.J. Donaldson, Energy savings in the Bayer process. J. Metals 37–41 (1981)
132. T. Ogawa, A. Nishimura, H. Sasaki, T. Ishida, T. Harato, Effect of acid cleaning using inhibitor in a tubular digestion and in a vapor recompression evaporator at Sumitomo's Alumina refinery, in *Proceedings of the 7th International Alumina Quality Workshop*, Perth (2005)

Robert LaMacchia Senior Research Scientist, Alcoa World Alumina, Perth—Australia. Email: Robert.Lamacchia@alcoa.com

Rob is currently a Senior Research Scientist within Alcoa's Continuous Improvement Centre of Excellence. He is involved in various research projects related to the Bayer process for production of alumina, as well as new technologies and commercialization of new products derived from a Bayer plant. Prior to this he was a Research Consultant in Hydro's Bauxite and Alumina Research and Development group, based in Belem do Para, Brasil and a Senior Staff Red Side Engineer in Alcoa's Point Comfort (TX, USA) alumina refinery. He has an undergraduate degree in chemistry (University of Western Australia) and a master's in chemical engineering (Curtin University).

Raphael Costa Technology Director, Hydro Bauxite & Alumina, Rio de Janeiro—Brazil. Email: Raphael.Costa@hydro.com

Raphael is the technology director for the Bauxite and Alumina business unit of Hydro, responsible for technology selection and development, the R&D group portfolio and the technical support to operations, including the bauxite mining and beneficiation operations at Paragominas, the alumina refining operations at Alunorte and the Mineração Rio do Norte (MRN) J. V. Raphael is also responsible for the areas of digital transformation and business systems. He is a member of the board of directors of the Companhia de Alumina do Para (CAP) alumina refinery project. He is also a member of the Paragominas bauxite mine board of directors, and the Alunorte alumina refinery board of directors. Raphael has more than 30 years of experience in the aluminum industry, working in various technical and managerial positions with Alcoa and Hydro, in Brazil, US, Spain and Australia. He also has 4 years of experience in the steel minimill industry in the US, working for the steelmaking company Gerdau. He has a chemical engineering degree from the University of Campinas (UNICAMP) in Brazil, and a master's in business administration (MBA) degree from the University of Houston-Victoria, in the US.

Chapter 10
Production of Smelter Grade Alumina (SGA) by Calcination

Benny E. Raahauge

Abstract This chapter covers the changes in calcination technology, energy consumption, air pollution control and quality of SGA developed over time since the end of the 2nd World War and up until today. The demand for SGA with high specific surface area ranging from 50 to 80 m^2/g (BET method) originated from the alumina smelter customers driven by the increasing need to capture HF gas emitted from the smelter cell's by so-called Dry-Scrubbing with virgin alumina in Gas Treatment Centers (GTC) located at the alumina smelter. The physio-chemisorbed HF on the virgin alumina collected in the GTC as secondary alumina, was then fed to the alumina smelter cells, or pots, via pot feeding systems. This development starting around 1970 made the Floury type SGA, with very low specific surface area around 5 m^2/g, obsolete, and Floury SGA was gradually replaced with so-called Sandy type SGA. The oil crisis around 1972 accelerated the development, scale-up and commercialization of stationary calciners for production of SGA initiated by Alcoa in the mid 1950 ties. The 25–30% reduction in specific thermal energy consumption in Rotary Kilns, when compared to Stationary Calciners was a very strong driver. The retention time in the Rotary Kiln was reduced from hours to minutes in Fluidized Bed calciners and from minutes to seconds in Gas Suspension or Flash calciners. Simultaneously with the above development several Rotary Kilns were retrofitted with pre-heater cyclones, and in one case also with a calciner furnaces, to reduce the specific thermal energy consumption and increase the production capacity of SGA. The largest Rotary Kilns for production of SGA had a design capacity of 1400 tpd SGA. This capacity was exceeded several times at Queensland Alumina, Australia, where the new Gas Suspension Calciners installed in 2002–2004 was designed for 4500 tpd SGA, driven by economy of scale. Today, the preferred design capacity of stationary calciners is around 3500 tpd SGA to match the single train production capacity of the Bayer Process circuit in modern alumina refineries. However, the above technology shifts have not come without new challenges to the Bayer

Benny E. Raahauge is deceased.

B. E. Raahauge (✉)
FLSmidth A/S - Retired, Deceased, General Manager–Pyro and Alumina Technology, Valby, Staden København, Denmark

B. E. Raahauge and F. S. Williams (eds.), *Smelter Grade Alumina from Bauxite*, Springer Series in Materials Science 320,
https://doi.org/10.1007/978-3-030-88586-1_10

process design. One such major challenge is to produce a hydrate quality, which upon calcination to SGA in stationary calciners, results in a particle size distribution and strength as SGA, that meets the specification of the smelters. Driven by environmental requirements of reduced dust emission Queensland Alumina, Australia, as the first in the alumina industry, decided in 2002, to install Baghouses or Fabric filters on their new Gas Suspension Calciners instead of Electrostatic Precipitators (ESPs), in order to avoid excessive dust emission during a power failure. Up until today heavy fuel oil, natural gas and coal gas ($CO + H_2$) is used as fuel in stationary calciners. But the challenges laying ahead with a predicted increase of the global warming of mother earth, makes hydrogen produced by electrolysis powered with renewable energy the preferred fuel for the not so distant future with potential to make the production of SGA CO_2 free.

10.1 Introduction—From Bauxite to Alumina Produced by Calcination

Alumina is extracted from Bauxite in **Digestion** by concentrated circulating caustic liquor (NaOH).

The cooled liquor with dissolved alumina and caustic is separated from the Bauxite residue (Red mud) by decantation/sedimentation in **Clarification**. After removal of residual particulate matter by polishing filtration, poly-crystalline Aluminium hydroxide ($Al(OH)_3$) is formed in the cooled pregnant liquor on seeded particles in the **Precipitation** circuit (Fig. 10.1).

The diluted spent caustic liquor separated from the precipitated hydrate is heated and sometimes concentrated in **Evaporation** before being reused for extraction of alumina from the incoming Bauxite.

10.1.1 Aluminium Hydroxide or Alumina Tri-hydrate (ATH)

The Poly-crystalline Aluminium hydroxide ($Al(OH)_3$) or Hydrate particles ($Al_2O_3 \cdot 3H_2O$) is produced by precipitation of Aluminium hydroxide on relatively fine-grained solid seed particles of Aluminium hydroxide suspended in a caustic solution:

$$Al(OH)_3(\text{seed particles}) + Al(OH)_4^-(l) \rightarrow Al_2O_3 \cdot 3H_2O(c) + OH^-(l) \quad (10.1)$$

$2Al(OH)_3$ is Equivalent to $Al_2O_3 \cdot 3H_2O$ (C) also Named Alumina Tri-Hydrate (ATH), or just Hydrate.

The external shape of poly-crystalline hydrate particles can be viewed from Scanning Electron Macrograph (SEM) pictures as shown below.

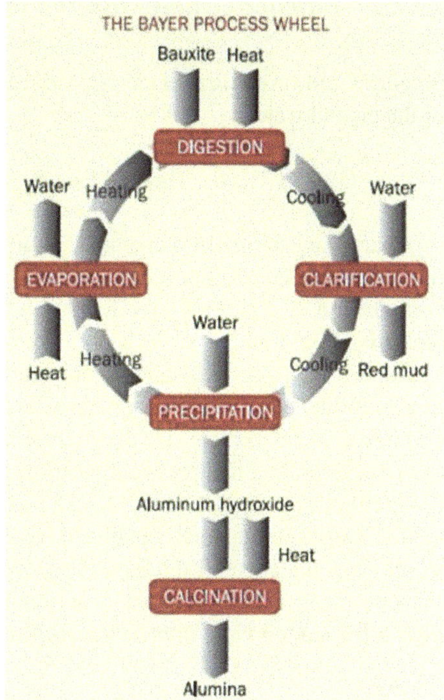

Fig. 10.1 The Bayer process wheel (Courtesy of Gyorgy Banvolgyi)

It is however very difficult to evaluate the internal structure and strength of hydrate particles based on the SEM pictures as shown in Fig. 10.2 and therefore more quantitative measures are used.

Fig. 10.2 Weak hydrate (LHS) and strong hydrate (RHS)

10.1.2 Smelter Grade Alumina (SGA)

Smelter Grade Alumina (SGA) is produced by **Calcination** of Hydrate by removal of the crystal water:

$$Al_2O_3 \cdot 3H_2O(s) + \text{Thermal Energy/Heat} \rightarrow Al_2O_3(s) + 3H_2O(g) \quad (10.2)$$

Apart from the thermo-chemical removal of the crystal water, as measured by the LOI, the calcination process has only minor, if any, influence on the chemical composition of the SGA, save for the increase of the hydrate impurities by about 50% on a %wt basis.

As we shall see below the calcination technology and process conditions is mainly influencing the physical properties of the SGA, like the SSA, including phase composition and particle size distribution to a varying degree.

Around 1970, Smelters were required to lower their emission of HF requiring production of Sandy alumina with SSA of 50–80 m^2/g. The change in alumina quality made Floury alumina unacceptable to most Smelters almost overnight and became a significant challenge for the Bayer Process operator. The Bayer precipitation circuit must be changed to produce a coarser Sandy alumina particle, while maintaining the high precipitation yield of producing Floury alumina [1].

10.1.3 Special or Chemical Grade Alumina (CGA)

Special or Chemical Grade Alumina is much more varied in its chemical (soda), physical and crystallographic properties when compared to SGA. Subject to the applied calcination technology and process conditions, the specific surface area may vary from several hundred square meters per gram to less than 1.0 m^2/g and the content of alpha phase may reach 99.9 for some product applications. Special or Chemical Grade Alumina will not be discussed further in this book, but information can be found elsewhere [2].

10.1.4 Primary Aluminium from SGA by Smelting

Primary aluminium is produced in the Smelters [3] by electro-chemical reduction at about 960 °C. The SGA is dissolved in liquid sodium aluminium fluoride electrolyte, rich in cryolite $3NaF \cdot AlF_3$:

$$Al_2O_3(\text{dissolved}) + 3C(s) + \text{Electrical Energy} \rightarrow 2Al(l) + 3CO(g) \quad (10.3)$$

$$Al_2O_3(\text{dissolved}) + 1.5C(s) + \text{Electrical Energy} \rightarrow 2Al(l) + 1.5CO_2(g) \quad (10.4)$$

The electrical energy consumption "to drive" the above reactions (10.3 and mostly 10.4) is about 11–14 kWh (DC power) per kg liquid primary aluminium metal, at a current efficiency ranging from 97 to 98.5%.

HF is emitted from the electrolytic cells in the pot-room partly due to the hypothetical reaction (10.5) of "residual hydrate" $Al_{10}O_{16}H_2$ or $[Al_2O_3]_5 \cdot H_2O$ [4] in the alumina from the refinery:

$$1\ 1/2[Al_2O_3]_5 \cdot H_2O(s) + Na_3AlF_6(l) + 2C(s) \rightarrow 2Al(l) + 7Al_2O_3](l) + 3HF(g) + CO_2(g) + CO(g) + 3NaF(l) \quad (10.5)$$

At the smelters, virgin SGA with SSA ranging from 60 to 80 m^2/g, is used to capture gaseous HF in the Gas Treatment Centre (GTC), to be further discussed in Chap. 11.

The chemical purity of secondary alumina (after HF capture by SGA) is very important for the electrolytic cell performance and purity of the primary aluminium produced. Alumina with too high content of sodium oxide [5] will require the addition of more aluminium fluoride to maintain the target electrolyte composition with the result that excess electrolyte is produced. The surplus of the cryolite electrolyte produced is made according to reaction (10.6):

$$3Na_2O + 2AlF_3 \rightarrow 2Na_3AlF_6 + Al_2O_3 \quad (10.6)$$

Similarly, calcium oxide produces calcium fluoride by reaction (10.7):

$$3CaO + 2AlF_3 \rightarrow 3CaF_2 + Al_2O_3 \quad (10.7)$$

If the amount of CaO exceeds about 12% of the Na_2O it will result in both an increase in CaF_2 concentration and an increase in the density of the electrolyte [Welch personnel communication 2016].

10.2 Calcination Technology—A Historical Overview

The calcination process of industrial hydrate starts at a temperature exceeding about 245 °C and the formation of Alpha Alumina Phase starts at a temperature exceeding about 900 °C as per reaction (10.8):

$$Al_2O_3 \cdot 3H_2O(s) + Heat \rightarrow'' x - Al_2O_3yOH'' \rightarrow \alpha - Al_2O_3 + Heat \quad (10.8)$$

"x-Al_2O_3yOH" or "γ-Al_2O_3" shall mean various mineral transition phases including X-Ray amorphous alumina phases.

The overall basic (and simplified) thermo-chemical reactions in calcination are almost the same regardless of calcination technology though some differences exist

Table 10.1 Smelter grade alumina (SGA)—properties over time

SGA property	Unit	Floury alumina	Sandy alumina	Typical 2015
Purity	wt%	> 99.3	> 99.3	99.3–99.7
Loss on ignition, 300–1000 °C	wt%	0.1–0.3	0.4–0.8	< 1.0
Gibbsite	wt%	–	–	< 0.5
Bulk density, loose	gpL	800	950	950–1000
Angle of repose	Degree	45	30–32	30–32
Particle size distribution				
> 100 Mesh/150 μm	wt%	–	–	5–10
< 325 Mesh/45 μm		45–60	5–10	Max. 10
Superfines < 20 μm		–	–	1–2
Specific surface area (SSA)	m^2/g	5 (BET)	50–80 (BET)	60–80 (BET)
Alpha phase content	%	85–96	10–30	< 5–10
Attrition index*	wt%	–	–	< 20
Flow funnel test*	S	–	–	85–95
Dust index*	kg/ton	–	–	0.5–1.3

*Not in all Certificate of Analysis—see text for references and further comments

with respect to the physical properties (see Table 10.1) and the mineral phase composition of the resulting SGA particles as we shall discuss below in greater details.

10.2.1 Rotary Kilns for Production of Floury and Sandy Alumina by Calcination.

The first recorded use of rotary kilns for calcining aluminium hydroxide dates to 1935. Rotary kilns have been used for calcination of hydrate to Floury type SGA (mostly in Europe and Asia) or Sandy type SGA (mostly the Americas).

In 1972 the largest 4.3 m dia. × 122 m rotary kiln was supplied with a capacity of 1400 tpd sandy alumina. Today the major use of rotary kilns is to produce highly calcined chemical or specialty grade alumina though a few rotary kilns is still producing Sandy alumina (Figs. 10.3 and 10.4).

The typical rotary kiln is installed with 2.6–4.2% inclination and rotates with about 0.9–1.5 RPM.

The whole process of hydrate drying, pre-heating and calcination takes place in the rotary kiln in counter current flow with the combustion products from combustion of fuel in the burning zone of the kiln.

Final cooling of the SGA takes place in inverted satellite coolers where the alumina is flowing up-hill in counter current flow with the incoming secondary air for combustion of the fuel. The kiln gases are de-dusted in a multi-clone (~ 85% efficiency)

Fig. 10.3 950 tpd Sandy Alumina Rotary Kilns (Ø3.95 × 107 m)—Eurallumina, Sardegna, Italy. Copyright © [1982] by The Minerals, Metals and Materials Society. Used with permission

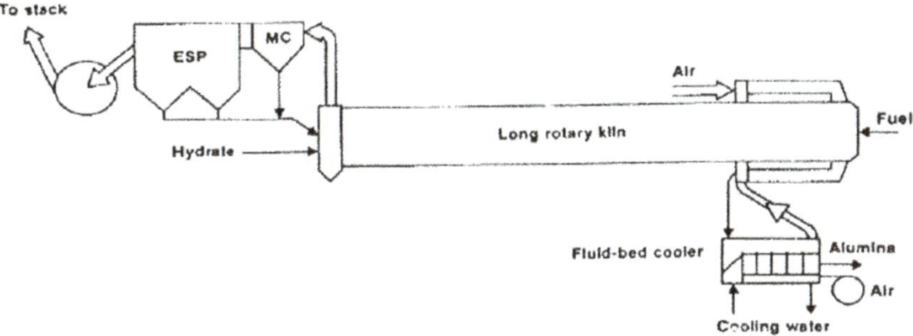

Fig. 10.4 Flowsheet of FLSmidth rotary kiln producing sandy alumina at 4100 kJ (NHV)/kg [53]

followed by an Electrostatic Precipitator. Essentially all the dust and dry hydrate collected is recycled back into the kiln through the internal "Goldberg" duct spiral.

The rotary kiln is a counter current heat exchanger with a very complicated direct heat transfer process taking place between the hot gas, the kiln wall and the moving solid bed (Fig. 10.5). The thermal energy requirement ranges from about 4100 kJ/kg to as much as 6000 kJ/kg or more for small rotary kilns producing floury or highly calcined chemical or specialty grade alumina.

Driven by the oil crisis in the early seventies, several retrofits of rotary kilns to more energy efficient flow sheets were developed, and some with significant increases in calcination capacity (Fig. 10.6).

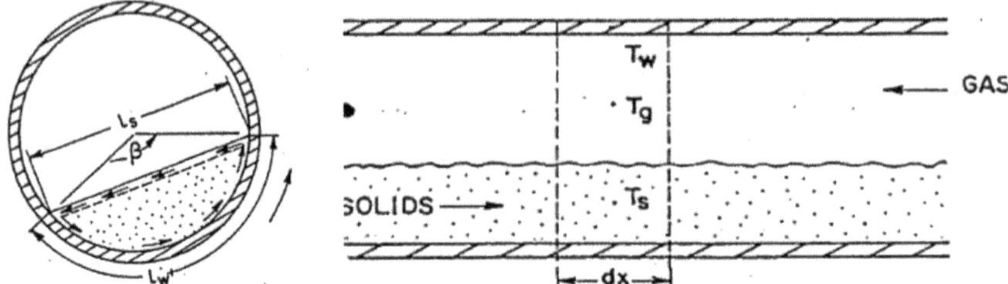

Fig. 10.5 Cross sectional solid movement and counter-current gas-solids flow in rotary kiln

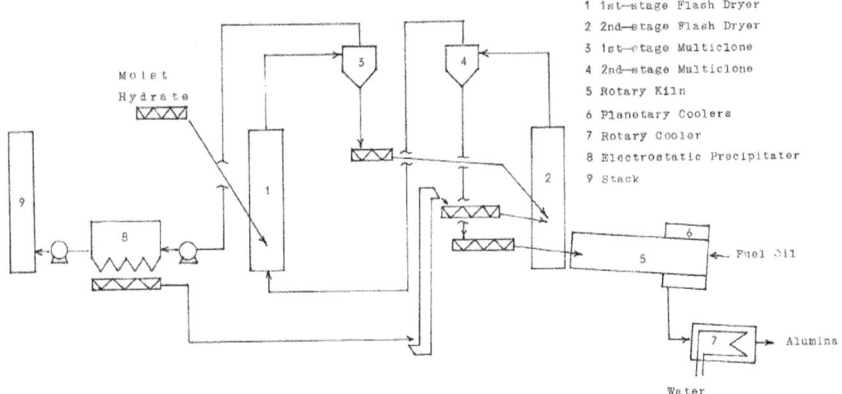

Fig. 10.6 Rotary kiln with two (2) stage of flash dryer/multi-clone stages [6]. Copyright © [1983] by The Minerals, Metals and Materials Society. Used with permission

The specific heat consumption at 10% hydrate moisture was 3280 kJ/Kg SGA with 30% alpha-phase and 65 (m^2/g) SSA. Particle breakdown was 2–3 wt% on 45 μm sieve at an Alumina Attrition Index (AAI) of 10–12%.

By introducing a shell mounted burner, Fig. 10.7, in addition to a flash dryer and pre-calcination stage a 10–33% increase in calcination capacity and 20% reduction in specific fuel consumption was achieved. The alpha-phase/SAA relationship was almost the same as for long rotary kilns and subject only to hydrate quality. The particle breakdown on a 45 μm screen was 2–5 wt% without generation of particles < 15 μm.

The GSC retrofit [8] increased the production capacity by more than 65% and reduced the specific energy consumption to 3215 kJ(LHV)/Kg SGA with 2–5% Alpha-phase, 70 m^2/g SSA and 0.55–0.65% LOI (300–1000 °C). Particle breakdown was the same as for the rotary kiln [8] (Fig. 10.8).

Since the oil crisis in the early 1970s, only stationary calciners have been installed due to the about 25–30% lower fuel consumption per ton alumina compared to Rotary Kilns.

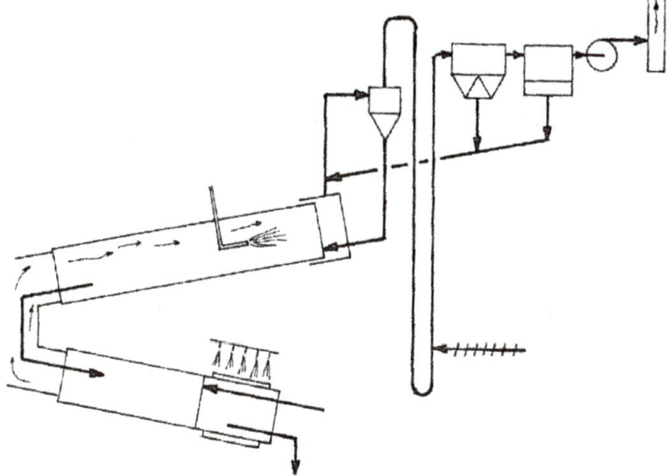

Fig. 10.7 Rotary kiln retrofitted with flash dryer and pre-calcination stage [7]. Copyright © [1983] by The Minerals, Metals and Materials Society. Used with permission

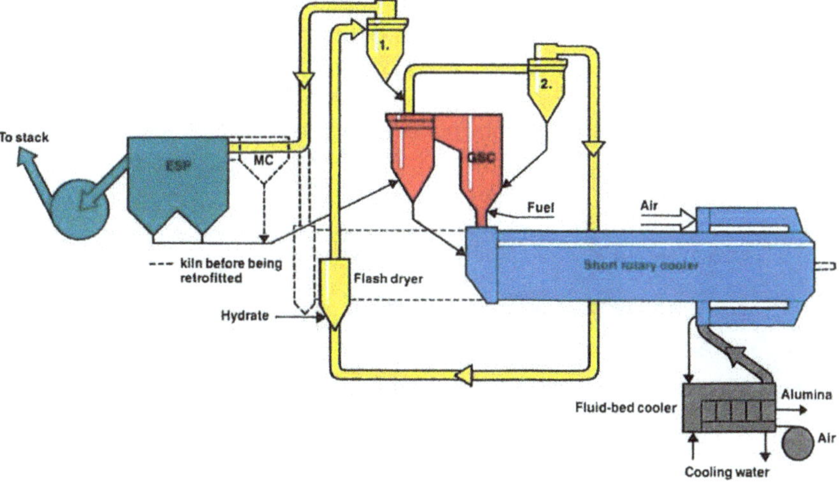

Fig. 10.8 Long rotary kiln retrofitted with gas suspension calciner (GSC) [Curtesy of FLSmidth]

10.2.2 Development of Stationary Calcination Technology

Alcoa, United States, pioneered [9] the development of stationary calciners back in 1945 and the first unit was started up in 1952 (left hand side, Fig. 10.9).

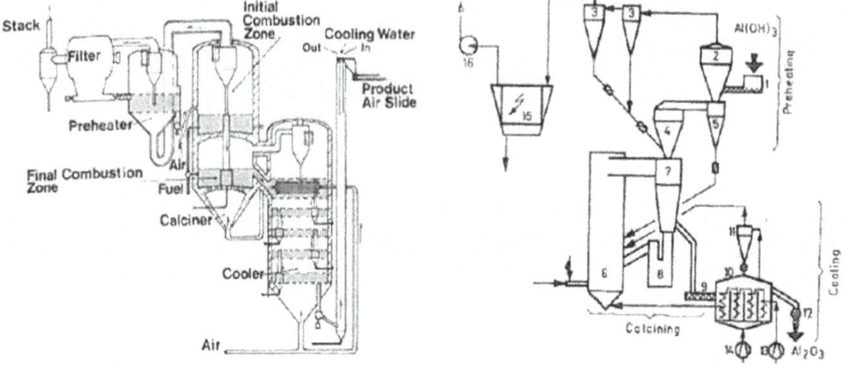

Fig. 10.9 Stationary calciners developed by Alcoa, USA, [9] (lhs) and Lurgi, Germany [10] (rhs). Copyright © [1974, 1973] by The Minerals, Metals and Materials Society. Used with permission

Lurgi, Germany, developed the Expanded or Circulating Fluid-Bed (CFB) calciner (Right Hand Side, Fig. 10.9) in the early 1970s [10]. The first 500 tpd CFB unit was designed for production of Floury alumina [10].

F.L.Smidth, Denmark, invented and introduced the Gas Suspension Pre-Calciner into the cement industry and commissioned the first prototype unit with a rotary kiln in 1976 for production of 4600 tpd cement clinker (Fig. 10.10).

Shortly thereafter the development of the Gas Suspension Calciner (GSC) for alumina was started [11] and the first prototype unit was commissioned at Hindalco, India, in 1986 producing 850 tpd SGA [8, 12].

Fig. 10.10 Cement clinker kiln (lhs) and 3 × 4500 tpd gas suspension calciners for alumina (rhs)

10.2.3 Fundamental Principle of Stationary Calciners/Reactors

When a gas, typically air, is blown vertically upward through a fixed bed of fine particles, contained within vessel, at increasing velocity, the bed of particles will be lifted and lose their contact points and be fluidized or suspended into the gas stream. At gas velocities slightly above the so-called minimum fluidization velocity, a Classical Fluidized Bed is established as seen in Fig. 10.11 "A" with a relatively small quantity of solids being entrained by the fluidizing media and re-cycled back to the fluidized-bed [13] via the underflow of the re-cycling cyclone.

By increasing the mean gas flow, relative to "A", the amount of entrained solids from the fluidized-bed increases and the Circulating Fluid-bed ("B and C") is established with a solid's concentration between the Classical Fluid-bed "A" and the Transport Reactor "D".

As the gas flow and velocity is increased in the vessel the solids concentration becomes more and more diluted until the bed of solids is no longer observable, and the "Transport Reactor" ("D") is formed with/without external solids re-circulation. When the gas velocity increases, the Slip Velocity, defined as the difference between the mean gas and mean solids velocity passes through a maximum.

All the fluidized-bed types "A"–"D", where there is partial or full re-cycle of solids occurring to the fluidization vessel, represents a multi-pass reactor with respect to the solids flow.

The Gas Suspension Calciner, PO4 in Fig. 10.12 (lhs), is a one-pass Transport Reactor with solids flowing downwards at the walls of the reactor vessel. This mode of solids flow is termed the "Shower-Model" and shown in Fig. 10.12 (rhs) with the relatively high solids concentration, Co, in the inlet to the reactor vessel.

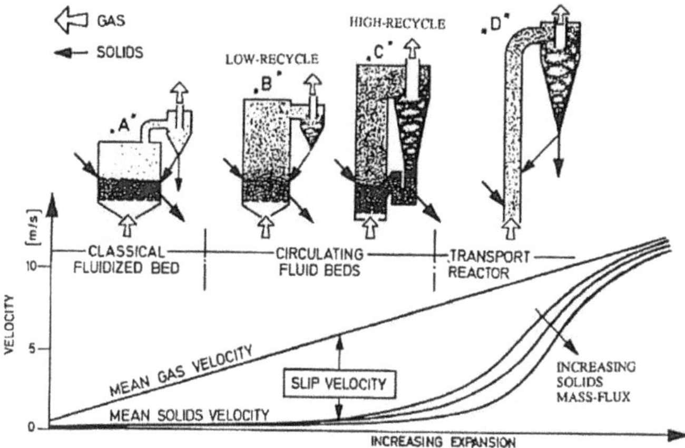

Fig. 10.11 Fluidization or gas suspension of fine solid particles into a vertical up-flowing gas stream

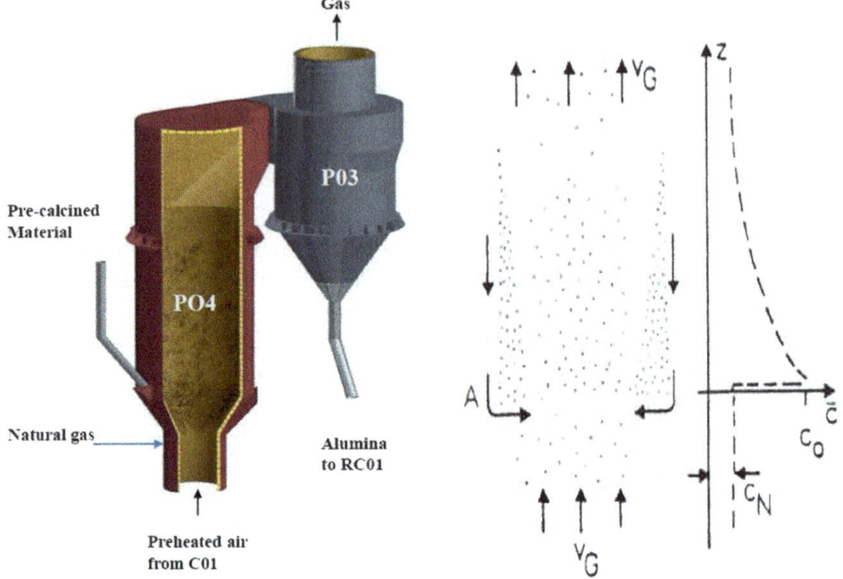

Fig. 10.12 The gas suspension calciner (lhs) and the "shower-model" of solids flow (rhs)

All the reactor designs from the classical fluid-bed to the transport or entrained bed type reactors (Figs. 10.11 and 10.12) can be considered to rely on the same principle of suspension of a continuous mass flow of solid particles dispersed into and surrounded by a vertical upward moving air/gas stream.

The refractory lined Calciner Furnace/Reactor design for simultaneous combustion of fuel and final calcination of partly de-hydrated hydrate to alumina at high temperature (900–1100 °C) is the core piece of high temperature equipment in any stationary calcination plant.

Alcoa, initially in 1945, used the classical fluid-bed technology (top of Fig. 10.13) to provide heat exchange between the solids and gas, as well as enough solids retention time at a calcination temperature of 950 °C to obtain the required degree of calcination with respect to LOI, SSA and content of alpha alumina phase in the SGA. In this design all the air for the combustion of the fuel was passed through the fluid-bed Calcination Furnace, Fig. 10.9 (lhs). However, since the practical fluidization velocity of alumina is less than 0.15 m/s it soon became apparent that the combustion of the fuel had to be separated from the fluid-bed itself and performed in the gas phase of an entrained bed Calciner Furnace/Reactor, (bottom of Fig. 10.13). This would keep the size of the Calciner Furnace within reasonable physical dimensions for larger capacity units. To provide enough solids retention time at 950 °C in the order of minutes, the fluid-bed, or fluidized Holding Vessel, was introduced in the bottom of the Furnace Cyclone separating alumina from the combustion and de-hydration gases, see Fig. 10.13.

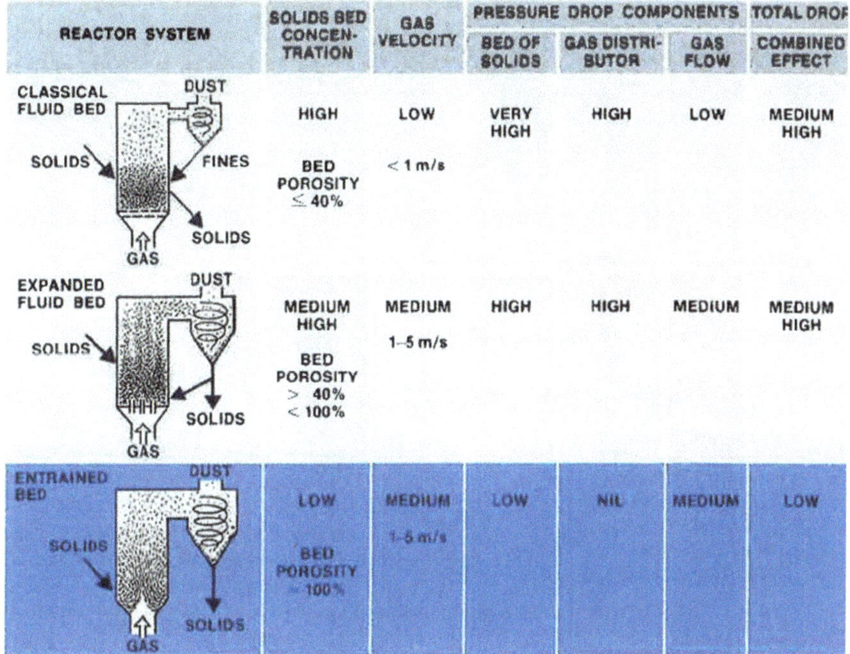

Fig. 10.13 Evolution in calciner furnace/reactor designs [Courtesy of FLSmidth]

Lurgi, developed the multi-pass expanded or Circulating Fluid-Bed (CFB) Calciner Furnace/Reactor (middle of Fig. 10.13) in the early 1970. The 950 °C mixture of partly calcined alumina and SGA is circulated many times over the CFB and Furnace Cyclone to provide enough retention time of the order of minutes. This way the gas handling capacity of the CFB far exceeds the gas handling capacity of classical fluid-bed technology. Dust free primary air is required for fluidization of the re-cycled material into the CFB. The CFB technology is designed with two-stage combustion of fuel in the calcination furnace, see Fig. 10.14.

F.L.Smidth, introduced the GSC Furnace/Reactor which is an entrained bed (bottom of Fig. 10.13) with one (1) single pass only of the GSC Furnace/Reactor and Calciner Furnace cyclone like Alcoa, however without a Holding Vessel in the first place. The GSC calcination temperature is about 1070 °C to obtain the required degree of calcination of the SGA at the short retention time provided of about 12 s. This has been verified experimentally in a 3200 TPD GSC unit for cement operating in India [14].

The CFB and GSC calcinations furnace/reactor models (Fig. 10.14) are designed to operate with a gas velocity in the range 5–10 m/s. This is to ensure economic and safe entrainment of the solids up through the Calcination Furnace into the Furnace/Re-cycle Cyclone at full and partial capacity.

Circulating Fluid-Bed	
Temperature, °C	950
Pressure, kPa (a)	> 101,3
Retention Time, Min.	3 - 5
Nominal Capacity, TPD	≤ 3500

Gas Suspension Calciner		
Holding Vessel	No	Yes
Temperature, °C	1050	< 960
Pressure, kPa (a)	< 101,3	< 101,3
Retention Time, Sec.	10-12	≥ 200
Nominal Capacity, TPD	≤ 4500	≤ 3500

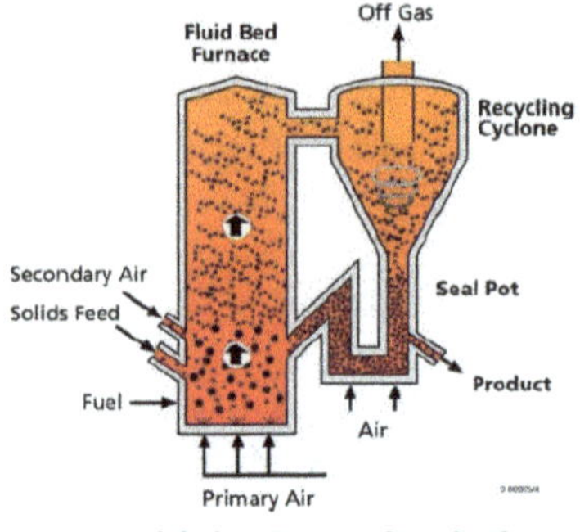

Multiple-Pass Circulating Fluid-Bed Reactor Furnace

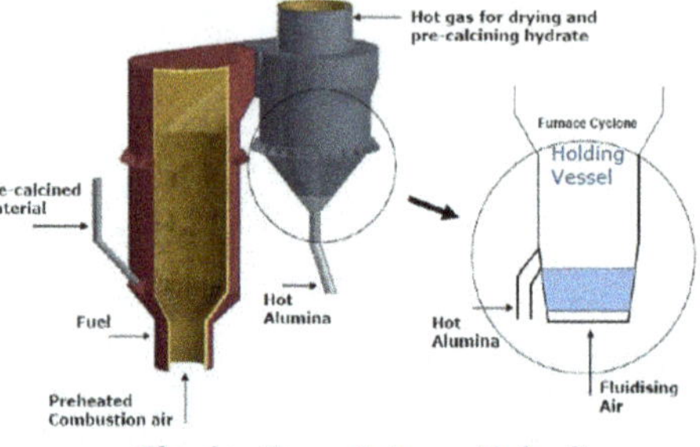

Single-Pass & Low-Velocity Gas Suspension Calciner Furnace wo/w Holding Vessel

Fig. 10.14 Calciner furnace/reactor designs and operating conditions [16]. Copyright © [2011] by The Minerals, Metals and Materials Society. Used with permission

The CFB [15] and GSC Furnace/Reactor designs is the dominating technology in the world today. The major operating differences between these two major reactor technologies are summarized in (Fig. 10.14).

The CFB Calciner Furnace/Reactor is a multi-pass calcination process operated at a pressure exceeding the atmospheric pressure.

The GSC Furnace/Reactor is a single-pass calcination process minimizing the mechanical action on the alumina particles during final calcination. The fluidized Holding Vessel was introduced into the new generation GSC flow sheets following the Alcoa-FLS Calcination Alliance formed in 2001. However, unlike the Alcoa Fluid Flash Calciner Furnace/Reactor, the GSC Furnace/Reactor are operated below the atmospheric pressure. This feature makes the operation of the GSC process relatively safe in the event of emission of hot alumina from holes/openings generated by internal wear of the calciner ducts and/or vessels over time.

10.2.4 Modern Stationary Calciner Flowsheets

10.2.4.1 The FLSmidth Gas Suspension Calciner (GSC)

All modern stationary industrial calcination flow sheets producing SGA are comprised of five main sections:

(1) Venturi Dryer for evaporating of the 6–8% moisture in the wet hydrate filter cake feed followed by gas-solids separation in a cyclone (PO1, see Fig. 10.15).
(2) Pre-Heating/Calcination stage followed by gas-solids separation in a cyclone (PO2);
(3) Calcination Furnace (PO_4) for combustion of fuel followed by a Furnace Cyclone (HVO_3) for gas-solids separation;
(4) Heat Recovery from hot alumina by direct cooling with air for combustion in a multiple stage cyclone heat exchanger (CO_1–CO_4);
(5) Final indirect cooling of alumina with water in a Fluidized Bed Cooler (FBC).

Figure 10.15 shows the flow sheet of the latest FLSmidth Gas Suspension Calcination (GSC) plant design including a fluidized holding vessel in the bottom of the Furnace Cyclone (HVO_3). Fuel for combustion in the Calciner Furnace is added through multiple burner nozzles. Fuels can be Heavy Fuel Oil, Natural Gas or Coal Gas. Coal Gas is mainly used for GSC units in China.

In the GSC units, the air flow for combustion and direct heat recovery from the hot alumina enters at the riser duct to the last cooler cyclone (CO_4). The air entering the riser duct of the last cooler cyclone is pulled in by a single Induced Draft Fan on the exit of the ESP dust collector (on the opposite end of the unit). A Forced Draft Fan supplying low pressure air to the riser duct to the last cooler cyclone can be

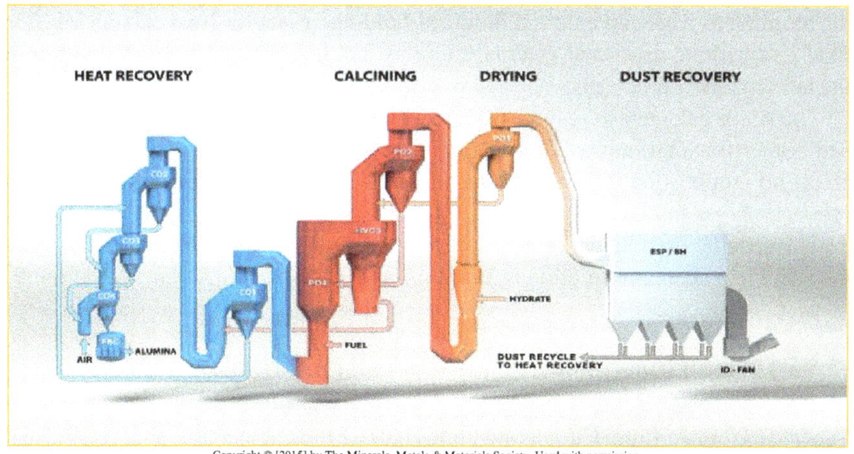

Fig. 10.15 FLSmidth gas suspension calcination flow sheet–capacity 2500–4500 TPD [18]. Copyright © [2015] by The Minerals, Metals and Materials Society. Used with permission

installed from day one, resulting in a reduced specific power consumption. This is due to the relative low inlet air flow compared to the exhaust gas flow from the ESP.

Another option is to install the Forced Draft Fan later with the advantage that the caclnation rate can be increased 10–20%. In the latter case the pressure in the cooler cyclones CO_4 and CO_3 will be above ambient while the remaining part of the calcination circuit will be below ambient pressure.

The dust from the top cyclone (PO_1) overflow is captured in either an Electrostatic Precipitator (ESP) or a Fabric Filter (GSC units only) before being returned to a cyclone cooler stage. Calcination of any Gibbsite originating from the Hydrate feed overflowing the top cyclone (PO_1) is ensured by the temperature at the dust re-cycle point of the multistage cyclone cooler.

Gas Suspension Calciners (GSC) is designed with four (4) stages of direct heat recovery from hot alumina and this is the most energy efficient heat recovery flow sheet compared to other stationary calciners. Average specific heat consumption as low as 2722 kJ(NHV)/kg SGA has been reported for GSC units operating without a fluidized holding vessel at a Calciner Furnace temperature of 1025–1050 °C and a few seconds of solids retention time. GSC units equipped with a fluidized holding vessel adding a few minutes of solids retention time operating at 925–960 °C achieve a specific heat consumption significantly below 2700 kJ (NHV)/kg SGA.

The specific power consumption is about 32 kWh/ton SGA including hydrate filtration and 21 kWh/ton without including hydrate filtration.

10.2.4.2 The Alcoa Fluid Flash Calciner (FFC)

The Alcoa Fluid Flash Calciner Mark VI, shown in Fig. 10.16, uses a fluid-bed hydrate dryer after the top cyclone. The Calciner Furnace is followed by a Furnace Cyclone equipped with a fluidized holding vessel, which in the Mark VI design discharges the alumina to a second external fluidized holding vessel for final calcination. The fluid-bed hydrate dryer and external fluidized holding vessel has been removed in the latest Mark VII design.

The Alcoa calciners have three (3) direct heat recovery stages followed by a Fluid-bed cooler for final indirect alumina cooling with water. A turbo-blower provides all air for combustion and gas moving so that the whole calcination unit operates at pressure above ambient.

The specific thermal energy consumption is about 3025 kJ (NHV)/kg SGA and the specific power consumption is 35 kWh/ton SGA including hydrate filtration [9].

10.2.4.3 The Lurgi/Outotec Circulating Fluid Bed (CFB) Calciner

The latest Outotec/Lurgi Circulating Fluid Bed (CFB) Calciner flow sheet is shown in Fig. 10.17. It has the same principle two-stage front end as the Alcoa and FLSmidth calciner units. The CFB reactor is fed with the major part of the pre-heated/calcined alumina, while a smaller part of the pre-heated/calcined alumina is by-passed to an

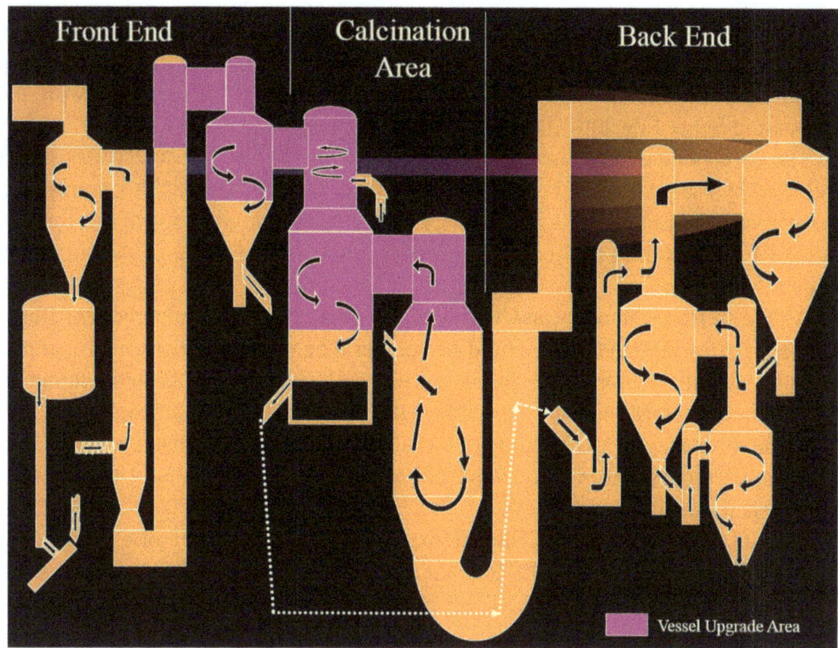

Fig. 10.16 Alcoa Mk III–fluid flash calciner–capacity 1680 TPD SGA [17]

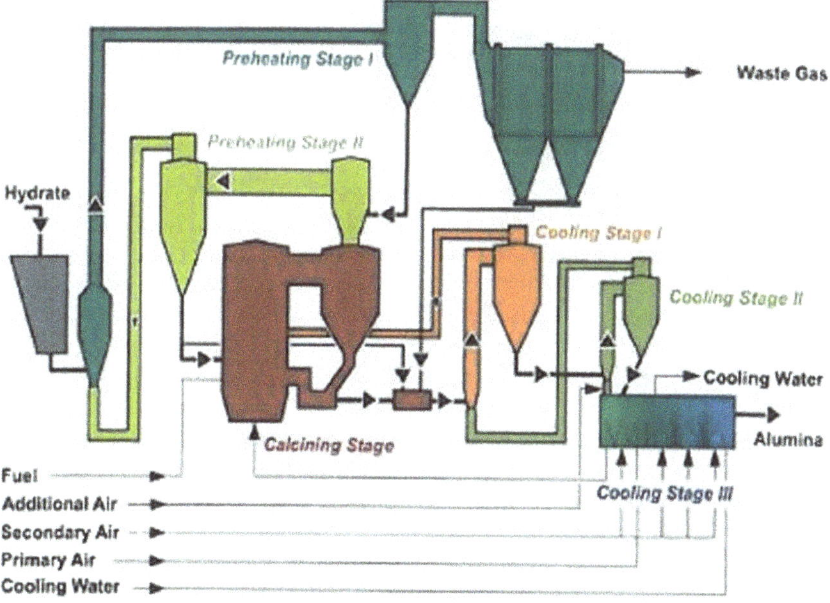

Fig. 10.17 Modern Outotec-circulating fluid bed calciner–capacity 3500 TPD SGA [18]. Copyright © [2015] by The Minerals, Metals and Materials Society. Used with permission

external fluidized holding vessel. In the external fluidized vessel hot alumina from the CFB Reactor is mixed with the low temperature by-passed portion and dust from the Electrostatic Precipitator to ensure calcination of any Gibbsite overflowing the Preheating Stage 1 cyclone into the Electrostatic Precipitator.

The CFB calciner flow sheet is equipped with two (2) direct heat recovery stages followed by a large fluid bed cooler with two sections. The "hot-end" section is used to indirectly preheat the dust-free Primary Air needed for fluidization of the alumina in the CFB Reactor. In the "cold-end" section final cooling of the SGA takes place indirectly with water.

All the air for combustion and fluidization is delivered by several large centrifugal blowers, partly through the large fluid-bed cooler. The whole calciner unit is operated at a pressure above ambient. From the direct cooling stages pre-heated Secondary Air for final combustion is introduced into the CFB reactor down-stream of the fuel injection, primary Air and alumina re-circulation inlet point, so that a two (2) stage combustion process takes place to secure a low formation of NO_x at the elevated CFB Reactor temperature.

The SGA produced in the CFB Calciner flow sheet is a mixture of over-calcined alumina discharged from the CFB Reactor and the under-calcined alumina by-passing it together with the dust from the Electrostatic Precipitator. The so-called "Hydrate by-pass" arrangement is claimed to save fuel for the calcination process and is also practiced in Alcoa calciners.

The specific thermal energy consumption is about 2790 kJ (NHV)/kg SGA [19] in the latest CFB units but have been reduced further to about 2720 kJ (NHV)/kg SGA [20] by retrofitting an indirect fluid-bed dryer into the feed end using steam generated in the "hot-end" of the fluid-bed cooler from surplus sensible heat in the alumina.

10.2.4.4 Sumitomo Suspension Preheater Fluid (SPF) Calciner and FCB Flash Calciner

Sumitomo, Japan, developed their own Suspension Preheater Fluid (SPF) Calciner [21] as seen below in Fig. 10.18 (lhs). The SPF flow sheet has a two (2) stage preheater/calciner system in front of the Calciner Furnace. The Furnace Cyclone is followed by a fluidized holding vessel. There is only one (1) direct alumina cyclone cooling stage in front of the fluid-bed cooler used for indirect heating of primary air.

The air and gas flow through the calciner are provided by a combination of air blowers and Induced Draft Fan located after the Electrostatic Precipitator. The specific thermal energy consumption is about 3060 kJ (NHV)/kg SGA and the specific power consumption is 22 kWh/ton SGA [21].

In 1972, FCB, France, developed their Flash Calciner by adopting the precalcining chamber developed for the cement industry [22]. The first industrial unit was designed after two years of successful testing of a 40 tpd semi-industrial prototype unit feeding 20 electrolysis cells at Aluminium of Greece (ADG).

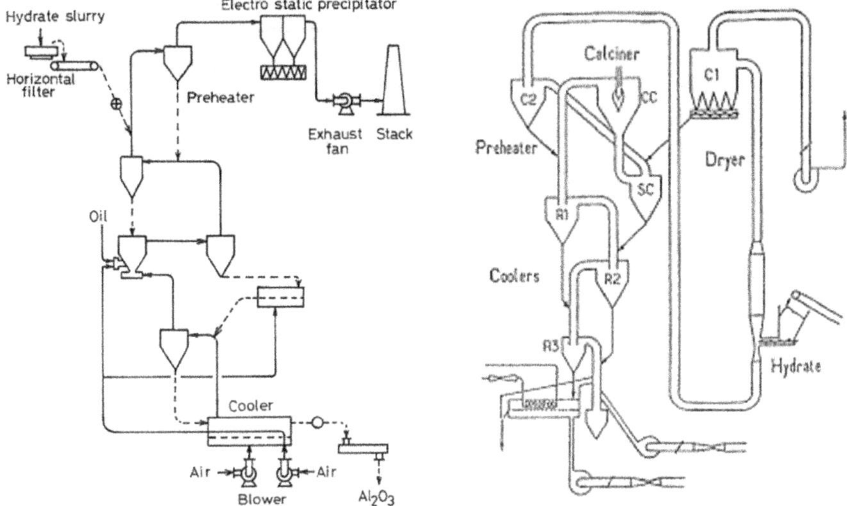

Fig. 10.18 Flow sheets of 600 tpd SPF calciner [21] and 900 tpd FCB flash calciner [22]. Copyright © [1983, 1995] by The Minerals, Metals and Materials Society. Used with permission

The simplified flow sheet of the 900 tpd Flash calciner commissioned at ADG, Greece, in 1984, is shown in Fig. 10.18 (rhs). The FCB Flash Calciner flow sheet has a two (2) stage pre-heater/calciner system in front of the Calciner Furnace (CC). The Furnace Cyclone (SC) is followed by a three (3) stage cyclone cooler (R1–R3) for direct recovery of heat from the hot alumina. Final SGA cooling is done indirectly with water in a fluid-bed cooler.

The air and gas flow through the FCB Flash Calciner is provided by a combination of air blowers and Induced Draft Fan located after the Electrostatic Precipitator. The specific thermal energy consumption is about 2850 kJ (NHV)/kg SGA and the specific power consumption is 17 kWh/ton SGA [22].

10.3 Specific Thermal Energy Consumption in Calciners

Thermal energy consumption in stationary calciners represents a significant part of the overall thermal energy consumption of any alumina refinery (e.g., 37–38% at Alunorte, Brazil [23]).

Consequently, a basic understanding of the elements of the thermal energy balance for stationary calciners and the significance of each element is vital for understanding and minimizing the thermal energy consumption of calcination. Some of the major factors are discussed below.

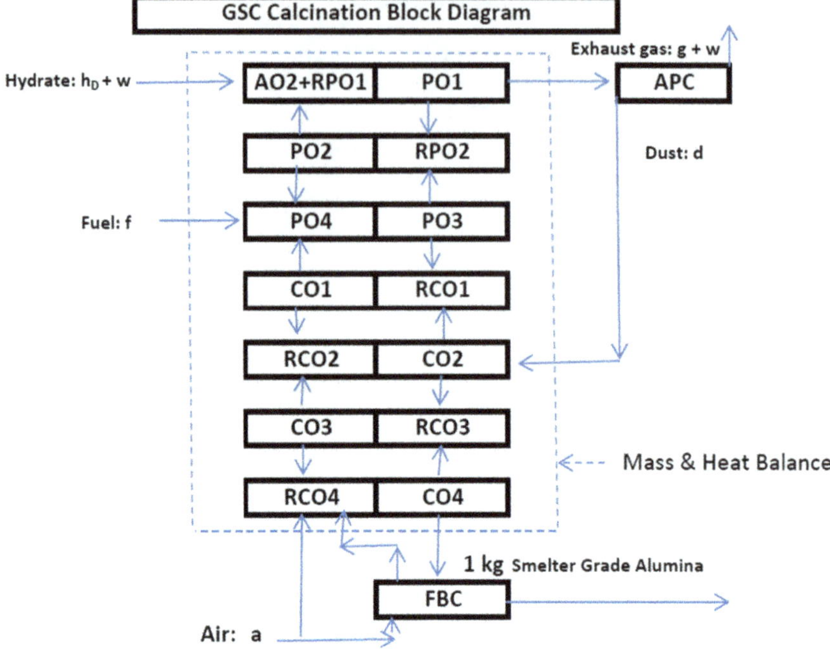

Fig. 10.19 Battery limits for calcination mass and heat balance calculations

With reference to the list of symbols, the steady-state heat balance for the GSC can be formulated (see List of Symbols):

$$\begin{aligned} \mathrm{h_D}\Delta\mathrm{h_{Hyd}} + \mathrm{w}\Delta\mathrm{h_{wl}} + \mathrm{a}\Delta\mathrm{h_{AiR}} + \mathrm{K} &= \mathrm{g}\Delta\mathrm{h_{Gas}} + \mathrm{w}\Delta\mathrm{H_{Cond.}} \\ &+ \mathrm{d}\Delta\mathrm{h_{Dust}} + \Delta\mathrm{h_{SGA}} + \Delta\mathrm{H_{Calcin.}} + \mathrm{q_{Loss}} \end{aligned} \tag{10.9}$$

The heat balance is expressed per ton or kg of SGA produced as per 10.9, within the GSC battery limits shown in Fig. 10.19. It shall be noted from Fig. 10.19 that the heat recovered by the cooling water from indirect cooling of the alumina in the FBC, is *not included* in 10.9 and is shown as outside of the battery limits in Fig. 10.19.

10.3.1 Thermo-Chemistry of Calcination—Review of Standard Heat of Calcination

The Heat of Calcination from the open literature has been studied [24] before any industrial GSC units were in operation and based on pilot plant calcination data. The below calcination reaction model and standard heats of calcination at 25 °C was adopted at the time, assuming the Gibbsite de-hydration path proceeded entirely via Boehmite to the Gamma alumina phase indicated with γ in the reaction schemes below:

$$2Al(OH)_3(c) \rightarrow Al_2O_3 \cdot H_2O(c) + 2H_2O(g) \tag{10.10}$$

$$\begin{aligned}\Delta H_1 &= 23.3 \text{ Kcal/mol} = 228.4 \text{ Kcal/kg}\\ &= 956.2 \text{ KJ/kg Alumina} = 2709 \text{ KJ/kg Water}\end{aligned}$$

$$Al_2O_3 \cdot H_2O(c) \rightarrow \gamma - Al_2O_3(c) + H_2O(g) \tag{10.11}$$

$$\begin{aligned}\Delta H_2 &= 26.1 \text{ Kcal/mol} = 255.9 \text{ Kcal/kg}\\ &= 1071.3 \text{ KJ/kg Alumina} = 6070 \text{ KJ/kg Water}\end{aligned}$$

The heat of calcination for removing the last water molecule increases to ΔHc (25 °C) = 26.1 kcal/mole Alumina = 26.1 kcal/mole H_2O, or 6070 kJ/kg water based on the above assumption and data, which is more than the double of the energy required to remove the first two water molecules.

The total standard heat of de-hydrating Gibbsite to "γ-Al_2O_3" equals 2040 kJ/Kg, or 487 kcal/Kg Alumina and this value was adopted at the time. Alcoa [9] and Lurgi [25] published 472 kcal/Kg, or about 1976 kJ/Kg Alumina for the endothermal de-hydration of Gibbsite according to reaction (10.8). Later a value of about 480 kcal/Kg = 2011 kJ/Kg Alumina was used by FLSmidth as design value moving forward, a number which had already been published by Lurgi (now Outotec) [26].

Alcoa and Lurgi also published the release of 66–67 kcal/Kg, or 278 kJ/Kg Alumina for the exothermal formation of α-Al_2O_3 from γ-Al_2O_3 according to reaction 10.12:

$$\gamma - Al_2O_3(c) \rightarrow \alpha - Al_2O_3(c) \tag{10.12}$$

$$\Delta H_3 = -5.3 \text{ Kcal/mole} = -52.0 \text{ Kcal/kg} = -218 \text{ KJ/kg Alumina}$$

According to investigations if γ-Al_2O_3 is considered either as a super cooled liquid, or disordered crystalline phase [27], the standard exothermic heat of α-Al2O3 formation was determined to be – 14.25 kcal/mole at 298 °C, or – 585 kJ/Kg Alumina. FLSmidth adopted the above conservative ΔH_3 value at 705 °C [28].

10.3.2 *Optimization of Specific Thermal Energy Consumption in Calciners*

There are three major factors in minimizing thermal energy consumption in calciners [29]:

i. Steady-state operation;
i. Number of gas-to-solid/solid-to-gas Direct Heat Transfer (DHT) stages, and;
ii. Design and installation of a cost-efficient refractory lining.

The ***first factor*** of importance is to operate the calciner at steady state rather than in a non-steady state. In Fig. 10.20, a steady state operating calciner shows little potential for reducing the specific energy consumption (SpEC) [2], when compared to un-steady state operation during commissioning of the GSC unit at Plant A.

The ***second factor*** is to design with maximum economical number of gas-to-solid/solid-to-gas Direct Heat Transfer (DHT) stages for pre-calcination and heat recovery from hot alumina discharged from the Holding Vessel in the Furnace Cyclone, HVO_3, see Fig. 10.21.

Of these, the latter design criteria are the most important for the cyclone cooler section owing to the large amount of heat of calcination consumed in the pre-calcination stage, PO_2, as we shall see later.

In Fig. 10.20b, the four (4) stage cyclone cooler, CO_1–CO_4 is illustrated and in Table 10.2, the theoretical impact of number of DHT stages for heat recovery N, is calculated [30] as follows.

The theoretical thermal efficiency, ηHEX, is given by.

$$\eta_{\underset{n=1}{\overset{N}{HEX}}} = \left(Tsi - T_{\underset{n=0}{\overset{N}{N}}}\right)/(Tsi - Tgi) = \Sigma\varphi^{n})/(\Sigma\varphi^{n}), \tag{10.13}$$

where φ = (G/S) (Cpg/Cps), is the heat capacity flow ratio between gas and solids.

G and S is the gas and solids flow rate and Cpg and Cps is their respective specific heat capacities.

The theoretical thermal efficiency, ηHEX (%), increases with number of DHT stages N, in stage wise counter current flow of air and alumina, see (b) in Fig. 10.21 [30].

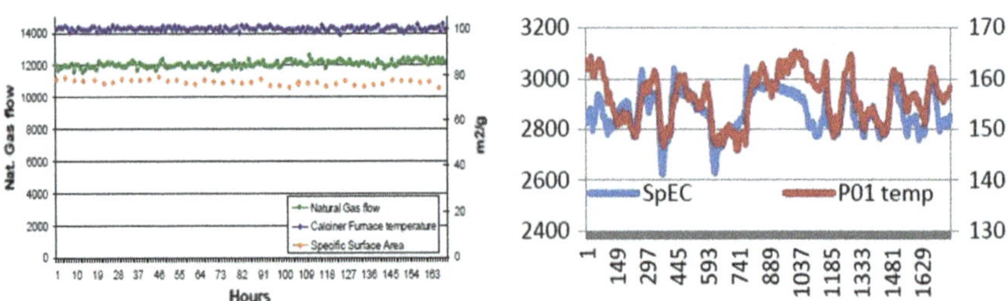

Fig. 10.20 Steady state GSC operation. Un-steady state GSC at commissioning

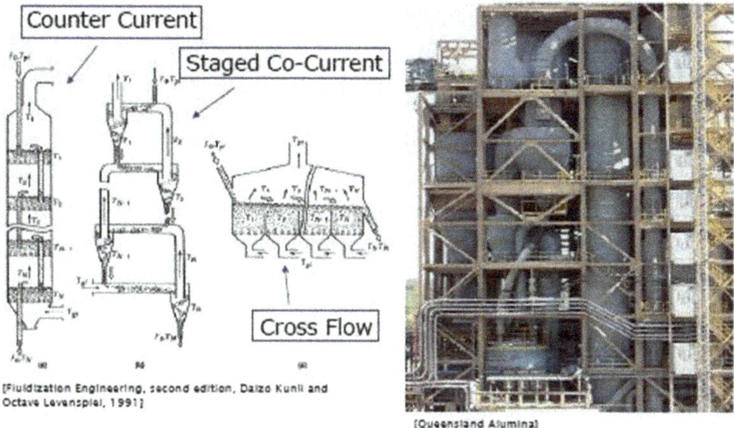

Fig. 10.21 Four (4) direct heat recovery stages [29] in cyclone cooler

Table 10.2 Impact of number of DHR stages and holding vessel

Heat Exchanger Stages	Pre-conditions: $\Phi = 1$, T_{gi}=25°C and T_{FBC}=80°C				
	η_{HEX} (%)	Ts_i (°C)	T_N (°C)	T_1 (°C)	Q_{FBC} (kJ/kg)
N=2	67	950	330	645	220
N=3	75	950	256	719	155
N=4	80	950	210	765	114
N=4	80	1100	240	885	141
N=5	83	1100	208	917	113

$$T_N = Tsi - \eta_{HEX}(Tsi - Tgi) \tag{10.14}$$

$$T_1 = Tgi + \eta_{Gas}(Tsi - Tgi) \text{with } \eta_{Gas} = \eta_{HEX}/\varphi \tag{10.15}$$

$$Q_{FBC} = Cps(T_N - T_{FBC}) \text{in kJ/kg Alumina.} \tag{10.16}$$

Using the formulas 10.14–10.16, the theoretical alumina inlet temperature to the fluid-bed cooler, T_N, pre-heat temperature of the air for combustion, T_1, and heat loss to the fluid bed cooling water, QFBC, can be estimated in response to different:

(a) Cyclone cooler designs with different number of DHT units (N), and;

Operating calcination temperature, T_{si}, in the Holding Vessel in the Furnace Cyclone, HVO_3; In Table 10.2 it is assumed that:

- A gas–solid heat-capacity-flow ratio, $\varphi = 1$,
- An inlet air temperature, $T_{gi} = 25$ °C, and:
- An alumina discharge temperature from the fluid-bed cooler, $T_{FBC} = 80$ °C,
- Adiabatic condition, i.e. no heat loss to the surroundings from equipment considered.

The theoretical heat loss from alumina to the fluid-bed cooling water decreases significantly with the number of direct heat recovery stages (N) applied. This is illustrated by comparing Q_{FBC} for the CFB calciner with N = 2, with stationary calciners having a Holding Vessel ($T_{si} = 950$ °C) and N = 3 or 4 direct gas–solid heat recovery stages. For N = 4, it is seen that increasing the calcination temperature, T_{si}, with approximately 150 °C by removing the Holding Vessel (Fig. 10.14, rhs), increases the heat loss from alumina to the fluid-bed cooling water by nearly (141 – 114) / 141 = 19%. This disadvantage can only be eliminated by introducing a 5th direct heat recovery stage as seen for N = 5.

The ***third factor*** of importance is an economic refractory design to minimize heat loss to the surroundings. From an operational point of view, the heat balance of the refractory lining is always in a dynamic state. This is due to the changing ambient conditions each 24 h day, 365 days a year, regardless of the stationary calcination process being operated at steady state itself. For this, and several additional reasons to be discussed below, it is impossible to always design a refractory lining to the same minimum specific heat loss value.

The surface heat loss from the surface of the calciner plant equipment, ducts and piping, qLoss, comprises three (3) elements; Radiation, Forced and Natural Convection losses expressed in kJ per kg Alumina:

$$q_{Los} = \sum q_{Radiation} + \sum q_{Forced\ Convection} + \sum q_{Natural\ Convection} \tag{10.17}$$

The main complexities involved are:

i. Unsteady-state heat balance of all of the refractory lined vessel and duct elements all the time;
ii. Complicated fluid-dynamics around the calciner vessels, ducts and pipes;
iii. Hot spots formed by refractory anchors in refractory lined vessels and ducts (Fig. 10.22);
iv. Complicated geometry and radiation exchange pattern between the vessels and ducts;
v. Extended heat transfer surface area of vessels and ducts by plate stiffeners and connecting flanges;
vi. Use of anchors and stainless-steel needles in selected refractory areas increasing the effective thermal conductivity of the refractory lining;

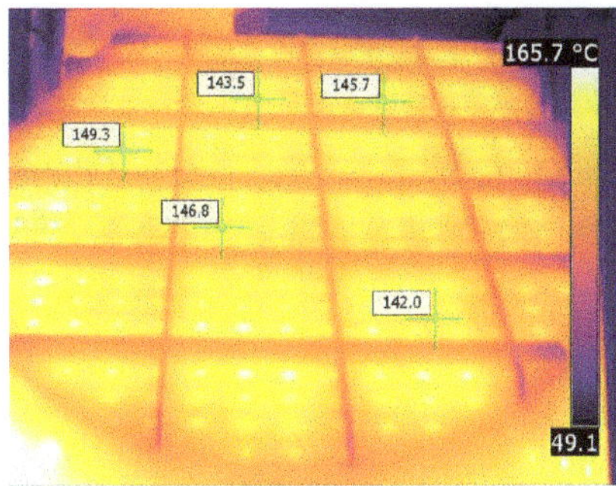

Fig. 10.22 Infrared thermography survey showing a complex heat loss situation

vii. Quality of refractory materials selected, their installation and dehydration before the application;

The calculation of each of the of the above heat loss contributions is well understood but can only be done by using a number of very simplifying assumptions. Consequently, it makes little sense to do very detailed theoretical surface temperature calculations rather than rely on experience including direct surface temperature measurements for identification of high heat loss areas, see Fig. 10.22.

10.3.2.1 Operational Experience and Thermal Energy Consumption in GSC Units

The methodology for simulation/calculation of GSC Mass and Heat Balances was as follows:

(1) ***Plant Data*** was maintained as reported by the GSC plant. This included; LOI of Alumina and free moisture of hydrate feed, fuel rate and specific heat consumption (if calculated), air intake rate (or measured Oxygen level) and measured temperature profile.
(2) The calculated/simulated temperature profile from stage wise ***Mass and Heat Balances***, were matched with measured plant temperatures by adjusting the specific surface heat lost from each stage of the GSC unit and the calcination capacity.

The production of alumina is seldom weighed directly but calculated from a standard specific heat consumption applied to all calciners at the refinery, or calculated from the measured wet hydrate feed rate, moisture content and LOI of the smelter grade alumina. Very often the hydrate feed weigh belt has not been calibrated recently, so both of the above methods are rather uncertain. Neither may the specific energy consumption be determined for each individual calciner unit. However, the measured temperatures are relatively accurate, subject to correct thermocouple installation in the first place and regular maintenance over time.

Because of the simulations the following can be observed:

***Plant A and B Design versus Plant Data*:**

Very conservative (high) exit gas and alumina temperatures were assumed during design resulting in estimated excessive sensible heat losses with exhaust gas and alumina;

M&H Balance versus Design:

Higher surface heat losses than used during design were partly compensated by lower sensible heat losses and heat of calcination from the revised calcination model to be discussed below. The *Design* specific thermal energy consumption exceeds the realized specific thermal energy consumption values by 4–5%.

***Plant A Data versus M&H Balance*:**

Higher calcination rate of 3192 tpd reported versus 3157 tpd calculated. Reported Specific Thermal Heat Consumption of 2663 kJ (NHV)/kg matches simulated value of 2662 (NHV) kJ/kg;

***Plant B Data versus M&H Balance*:**

A higher calcination rate than designed for of 3564 tpd SGA was calculated based on applying an "average specific heat consumption" of 2718 kJ/kg SGA for all operating calciners in the refinery. A higher calcination rate of 3695 tpd SGA was calculated with the specific heat consumption of 2671 kJ (NHV)/kg SGA estimated for the simulated GSC unit.

In view of the above, it can be concluded that further reduction in specific thermal energy consumption of GSC units can be achieved by improved refractory design and thus lower surface heat losses.

10.3.2.2 Optimization of Specific Thermal Energy Consumption in CFB Units

Outotec [31] followed a more radical approach compared to FLSmidth as modern CFB Calciners has a relatively poor heat recovery from hot alumina discharged from the CFB reactor, Figs. 10.14 and 10.17. The reasons being:

(i) The combustion air for cooling the alumina and recover heat is split in primary and secondary air, allowing relatively less air available for cooling the hot alumina discharged from the CFB reactor, Fig. 10.17.
(ii) Equipped with only two (2) direct alumina-to-air Direct Heat Transfer (DHT) stages for recovery of heat from the hot alumina.

In view of the above, Outotec introduced the below design changes to the new generation CFB units named the Circo Cal process by adding:

- One (1) additional direct alumina-to-air Direct Heat Transfer (DHT) stage;

Table 10.3 Comparative mass and heat balance data for gas suspension calciner units

Comparative Temperature Profiles and Heat Balances							
Parameter	Unit	Value	Value	Value	Value	Value	Value
Source	MIPD	Design	Plant Data	M&H Bal.	Design	Plant Data	M&H Bal.
Data	-	A 10-12-2007	A 03-07-2015	A 03-03-2017	B 26-11-2007	B 01/03-04-2014	B 03-03-2014
Alumina Capacity	TPD	3200	3192	3157	3500	3564	3695
LOIA (300-1000C)	%	0,9	0,8	0,8	0,9	0,81	0,81
Hydrate Feed Moisture	%	6,0	6,1	6,1	7,0	4,2	4,2
PO1 O/F (Gas)	Celsius	161	150	150	155	145	145
PO2 Oxygen, Wet	%	2,5	3,2	3,2	2,6	2,5	2,5
PO4 Outlet	Celsius	960	910	910	960	969	969
CO4 U/F (Alumina)	Celsius	166	126	126	175	135	135
Specific Heat Cons. NG(LHV)	kJ/kg Ala	2783	2663	2662	2805	2718[+]	2671
Surface Loss + Heat of Evap & Calcination		2379	NM	2319	2410	NM	2353
Total Sensible Heat Input	kJ/kg Ala	90	NC	89	96	NC	87
Total Sensible Heat Output	kJ/kg Ala	495	NC	433	490	NC	407
Total Sensible Heat Lost	kJ/kg Ala	405	NC	344	394	NC	320

(+) From average of several Stationary Calciners;

- One (1) additional direct pre-heating gas-to-hydrate Direct Heat Transfer (DHT) stage;
- Fluid-bed hydrate dryer stage;
- Simplified cooler design.

Outotec claims a potential specific thermal energy consumption below 2600 kJ (NHV)/kg alumina.

However, Petersen reported about 2720 kJ (NHV)/kg SGA [20] obtained by retrofitting of the AOS Stade Calciner # 2 and Perander reports 2680 kJ (NHV)/kg SGA [31]. In any case the partial retrofitting of the AOS Stade Calciner # 2 is nothing but on level with the specific thermal energy consumption reported form existing operating GSC units, see Table 10.3 and Fig. 10.23.

The 500 tpd CFB Calciner # 2 at AO Stade, Germany, has been partially retrofitted with the "Circo Cal" technology as seen in Fig. 10.24.

The maximum potential improvement by adding the missing 3rd direct alumina-to-air Direct Heat Transfer (DHT) stage at the AOS Stade Calciner # 2, can be done for new units, but will reduce the heat recovery from the FBC and reduce the heat available for drying hydrate. It is therefore difficult to see how a value below 2600 kJ/kg SGA can be obtained.

Furthermore, realizing the full potential of the "Circo Cal" technology may not be possible for CFB Calciners using HFO as fuel, with 2.5–3.5% Sulfur and acid dew point in the range 140–150 °C, raising serious corrosion issues in ESP's and the stack to be mitigated.

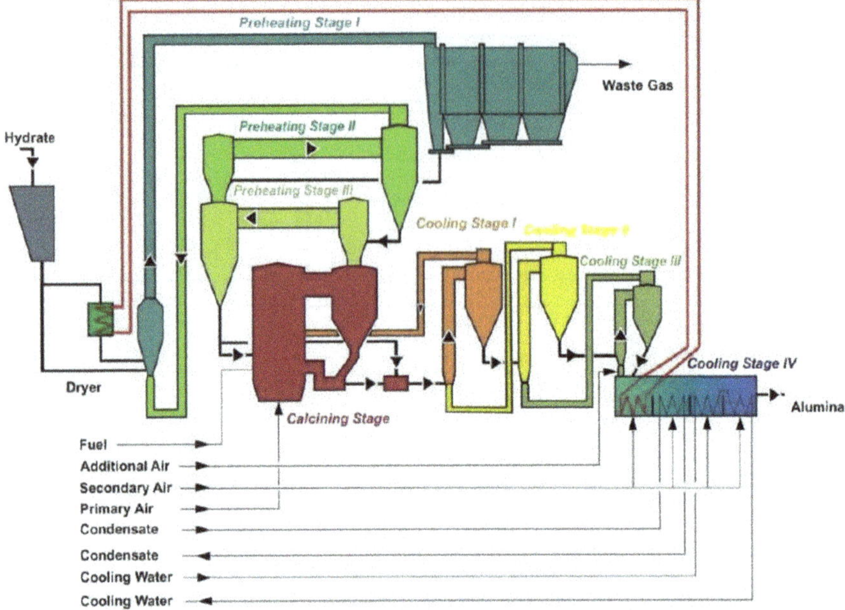

Fig. 10.23 Flow sheet of Outotec new generation CFB calciner "Circo-Cal" [31]

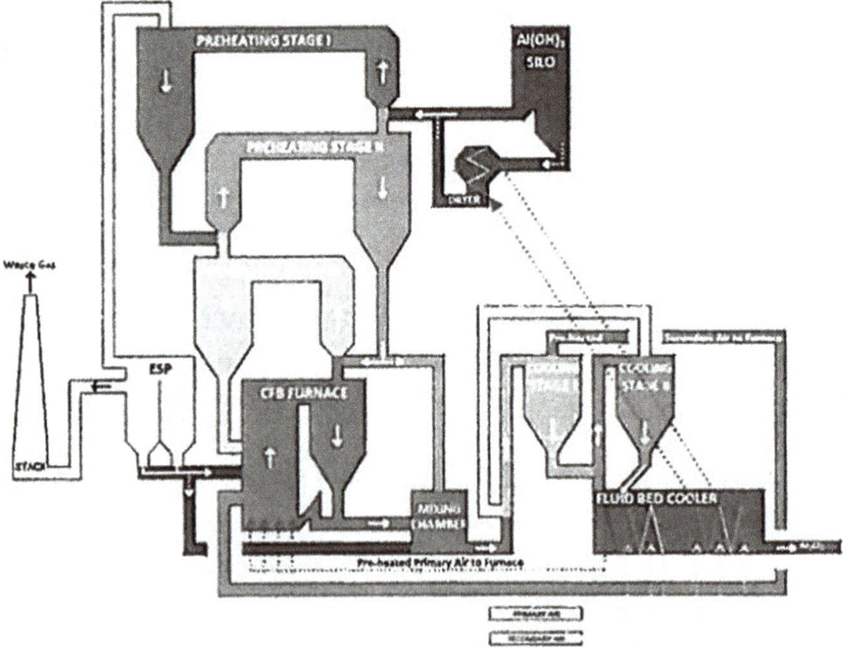

Fig. 10.24 Flow sheet of Outotec retrofit of stade, Germany CFB calciner "Circo-Cal" [20]

10.3.3 Combustion Fundamentals—Net Heating Value and Air to Fuel Ratio [29]

The heat for calcination is provided by the combustion of fuel in the Calciner Furnace (PO_4 in Fig. 10.12). The fuel is introduced in the inlet duct to the calciner furnace through burner nozzles located on the diameter of the inlet duct, Fig. 10.25.

Production of alumina by calcination requires a relatively pure fuel with respect to solids ruling out almost all solid fuels. Combustion of Methane, CH_4, is discussed in some detail below to illustrate the general concept of the Low or Net Heating Value (NHV), ΔHNHV, of the fuel being used:

$$CH_4(g) + 2O_2(g) = CO_2(g) + 2H_2O(g) + \Delta H_{NHV} \qquad (10.18)$$

Since all calciners operate with a stack temperature in excess of 100 °C, the water formed during combustion (and calcination) leaves the stack as water vapor. This means that only the Low or Net Heating Value, ΔH_{NHV}, of the fuel is available for the calcination process.

$$2H_2O(g) = 2H_2O(l) + \Delta H_{Cond.} \qquad (10.19)$$

If the water vapor is condensed, then the heat of condensation, $\Delta H_{Cond.}$ is released in addition to Net Heat Value, ΔH_{NHV}:

$$CH_4(g) + 2O_2(g) = CO_2(g) + 2H_2O(l) + \Delta H_{NHV} + \Delta H_{Cond.} \qquad (10.20)$$

Adding reaction (10.18) and reaction (10.19), and thus adding the Low or Net Heat Value, ΔH_{NHV}, of the fuel and the heat of water vapor condensation, ΔH_{Cond}. The High or Gross Heating Value of the fuel is obtained:

$$\Delta H_{HHV} = \Delta H_{GHV} = \Delta H_{NHV} + \Delta H_{Cond.} \qquad (10.21)$$

Fig. 10.25 On-line inspection of burner nozzles (lhs). Coal gas burner nozzles (rhs)

Table 10.4 Stoichiometric air demand, combustion products and heat of combustion

Fuel	v_{ol} (Air)	CO_2	H_2O	N_2	v_{oz} (Gas)	ΔH_{NHV}	ΔH_{GHV}	$\Delta H_G / \Delta H_N$
Unit	Nm^3/Nm^3	Vol%	Vol%	Vol%	Nm^3/Nm^3	kcal/Nm^3	kcal/Nm^3	–
CH_4	9.57	9.5	18.9	71.6	10.57	8560	9499	1110
Nat. Gas	9.71	10.1	18.3	71.6	10.76	8750	9676	1106
Coal Gas	1.21	17.4	8.5	74.1	2.02	1375	1456	1059
HFO	10.68 (*)	14.4	10.8	74.8	11.29 (*)	9680(**)	10,253(**)	1059

(*)Nm^3/kg, (**)Kcal/kg

The heat recovered by condensation of the water vapor (Vol.% H_2O) in the combustion products at 1 bar pressure and 25 °C is:

$$\Delta H_{Cond.} = 585^{*}\mathrm{Vol} - \%H_2O^{*}v^{*}_{oz}18/2240\left[\mathrm{kcal/Nm^3 Fuel\ Gas}\right] \tag{10.22}$$

Table 10.4 shows some typical fuel data for selected fuel gasses and HFO [29], where the difference between Net and Gross Heating Values between HFO and Natural Gas are clearly demonstrated.

Any modern and safe calcination system is equipped with a Burner Management System (BMS) to supervise safe combustion conditions in the plant by calculating and monitoring the minimum safe operating Air-to-Fuel Ratio (AFR) from measured flow rates of air for combustion and fuel flow [29].

The minimum safe AFR is defined as follows:

$$\begin{aligned} &\mathrm{Min.\ Safe\ AFR} > \mathrm{Stoichiometric\ Air\ Requirement} \\ &+5\% = 1.05^{*}v_{ol} = \mathrm{AFR\ Trip\ Condition} \end{aligned} \tag{10.23}$$

$$\mathrm{Minimum\ Operational\ AFR} > \mathrm{AFR\ Trip\ Condition} + 5\% = \mathrm{AFR\ Alarm\ Point} \tag{10.24}$$

As an example, using the value for natural gas in Table 10.2 we find:

$$\mathrm{Operational\ AFR} > 9.71 \times (1.05) \times (1.05) = 10.7\left(\mathrm{Nm^3 Air/Nm^3 Natural\ Gas}\right) \tag{10.25}$$

Due to the important safety function of the BMS, air and fuel flow into the calciner furnace, PO_4, is always measured continuously with the required accuracy and redundancy to safely guard against explosion risk.

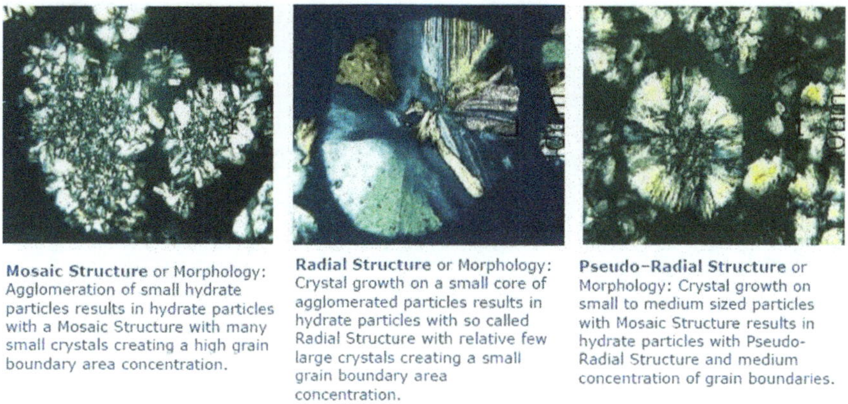

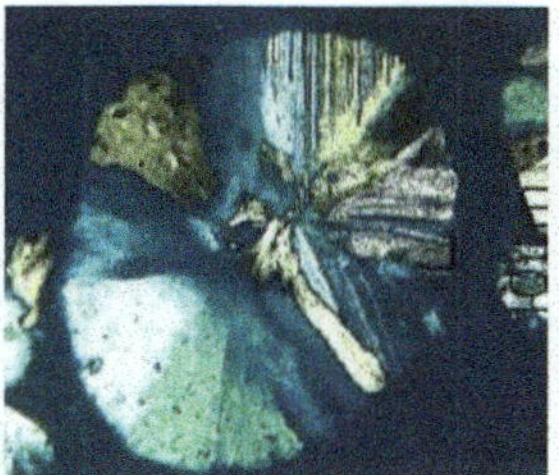

Mosaic Structure or Morphology: Agglomeration of small hydrate particles results in hydrate particles with a Mosaic Structure with many small crystals creating a high grain boundary area concentration.

Radial Structure or Morphology: Crystal growth on a small core of agglomerated particles results in hydrate particles with so called Radial Structure with relative few large crystals creating a small grain boundary area concentration.

Pseudo-Radial Structure or Morphology: Crystal growth on small to medium sized particles with Mosaic Structure results in hydrate particles with Pseudo-Radial Structure and medium concentration of grain boundaries.

Fig. 10.26 Morphology of hydrate particles produced in the Bayer precipitation circuit

10.4 Calcination Chemistry—Observations, Theory and Practical Application

10.4.1 Properties of Alumina Tri-hydrate ($Al_2O_3 \cdot 3H_2O$)

Each calcination unit in the Alumina Refinery receives Gibbsite or Aluminium hydroxide, $Al(OH)_3$ or Alumina Tri-Hydrate, $Al_2O_3 \cdot 3H_2O$, or just Hydrate. The particles are a polycrystalline solid produced in the Precipitation circuit of the Bayer Process.

Subject to hydrate precipitation conditions various hydrate particle morphologies are obtained as shown and explained in Fig. 10.26.

10.4.2 Observations of De-hydration Pathways

The surface changes of de-hydration of hydrate feed particles to SGA can be seen in Fig. 10.27, where a lot of fissures can be observed between layers of alumina perpendicular to the c-axis of the Gibbsite crystals making up each hydrate particle.

The overall basic thermos-chemical reactions are almost the same in rotary kilns and stationary calciners though some differences exist with respect to the composition of mineral phases in the resulting smelter grade alumina particles.

$$Al_2O_3 \cdot 3H_2O(s) + \text{Heat} \rightarrow'' x - Al_2O_3yOH'' \rightarrow \alpha - Al_2O_3 + \text{Heat} \quad (10.26)$$

As seen from the above reaction (10.26), calcination removes water and changes the chemical and physical properties of the hydrate into those of alumina as seen in Table 10.1. The final reaction is formation of α–alumina from "xy–Alumina"

Fig. 10.27 SEM picture of hydrate feed particles (left) and SGA particles (right) × 1000

takes place through several intermediate phase changes. This is the subject of the discussion below.

10.4.3 Changes in Crystal Structure

The ideal crystal structure of the various intermediate mineral phases in "xy–Alumina" in reaction (10.26) is well known as shown in Fig. 10.28 [32].

However, the ideal crystal structure is seldom found in industrial practice, where a significant amount of low crystalline or X-Ray amorphous phases are observed [4]. The phase composition found in practice is to some extent subject to applied calcination technology, Table 10.5, where γ-Al_2O_3 in the SGA is modeled as a defect spinel structure: $H_2(Al_{3.0})(Al_{4.5})(Al_{1.0},Al_{1.5})O_{16}$, or just: $H_2Al_{10}O_{16}$ including some –OH groups.

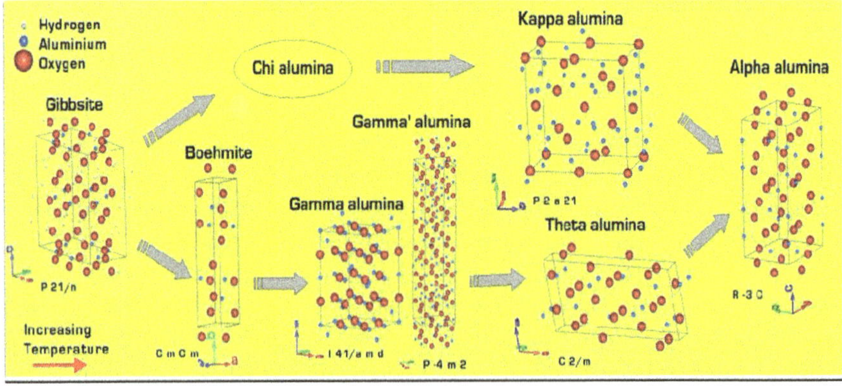

Fig. 10.28 Theoretical change in crystal structure during de-hydration of Gibbsite [32]

Table 10.5 Wt% of alpha, gamma and low crystalline phases in SGA samples [4]

Phase	Rotary Kiln				Fluid Bed Calciner		
	#1	#2	#3	#4	#1	#2	#3
Alpha	19.6	4.8	8.2	7.6	0.7	1.8	1.5
Gamma	45.9	56.3	54.7	59.1	62.3	56.8	59.1
Low cry	34.5	38.9	37.1	33.3	37.0	41.4	39.4

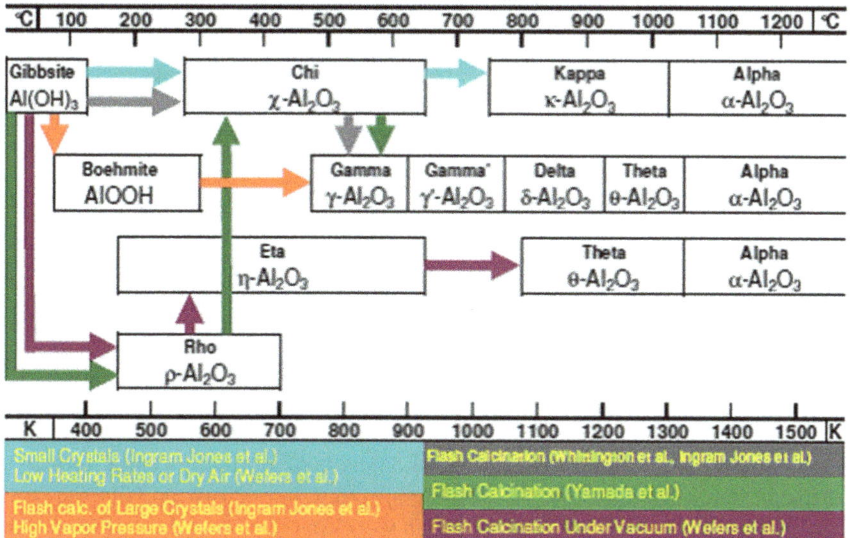

Fig. 10.29 Phase changes during calcination of Gibbsite to alpha phase alumina [15]

Over time many phase transition studies have been made and they are summarized and updated in Fig. 10.29 by Perander [15]. With exception of the Flash Calcination work performed by Yamada [33], all other data are based on laboratory calcination work.

10.4.4 Laboratory/Bench-Scale De-hydration Studies

The de-hydration pathways and phase changes taken place during calcination of Gibbsite/Hydrate to Alumina that were summarized above have been studied in many calcination laboratories tests. Some of these studies will be discussed in more detail below.

Table 10.6 Review of open literature sources on Gibbsite dehydration to boehmite

Study Reference	Equipment	De-hydration of Gibbsite/Bayer hydrate			Particle size	Boehmite phase
		Temp. (°C)	Time (s)	PH_2O (kPa)	Micron	wt%
Rouquerol [34]	DTA	180–205	41.000	0.07	0.2	5
Yamada [33]	Furnace Flash	300–500 385	0.2	NM	Mean 55–90	5
Perlmutter [35]	TG	175	1000	1.07–3.2	106–125	1
Rozic [36]	Flash	610–670	0.4–0.8	NM	< 10	NM
Alcoa [37]	Pressure	500	3600	800	10–200	Max 50/25
Das et al. [38]	DTA	250	750	NM	1.5 and 100	Yes
Ingram-Jones [39]	Furnace flash	300–1200 900–1200	7200 0.6–2.5	NM	Med = 0.5 Med = 14	Yes
Illevski [40]	Furnace	970	8–18,000	NM	Med = 91	0–20
Wang [41]	Furnace	550–650	60	NM	0.5–200	2.6–6.1
FLS R&D	GSC	375	0.5	NM	10–200	4.8–5.7

Rouquerol [34] showed that significant amounts of Boehmite can be formed at relatively low water vapor pressure and temperature in micro-sized Gibbsite particles. Yamada [33] reported a very thorough work covering several calcination technologies, including a 600 tpd Stationary Calciner as shown in Fig. 10.18 and Table 10.6.

Therefore, this work is still the most important, and it stands out even today, since most other work in the open literature has been conducted in laboratory and/or bench-scale units confirming the work of Yamada [33] within the uncertainties of scale-up.

Rozic [36] performed flash calcination of fine gibbsite particles in a pneumatic transport reactor, Fig. 10.30.

The dehydration was shown to be first order with respect to disappearance of the hydrate. The energy of activation was estimated to be 66.5 kJ/mole from classical Arrhenius theory according to reaction 10.27

$$\begin{aligned} & Al_2O_3 \cdot 3H_2O(s) + \text{Heat} \rightarrow x - Al_2O_3(s) + (1 - x) \\ & (Al_2O_3 \cdot 3H_2O)(s) + (3x - (1 - x)) \cdot H_2O(g) \end{aligned} \tag{10.27}$$

No measure of formation of Boehmite or other alumina phases were reported. At 640 °C and 0.75 s retention time, Rozic [36] reports about a pseudo amorphous alumina (X-Ray) with SSA = 250 m^2/g, with 0.46 cm^3/g pore volume and a pore width of 2–50 nm (Fig. 10.31).

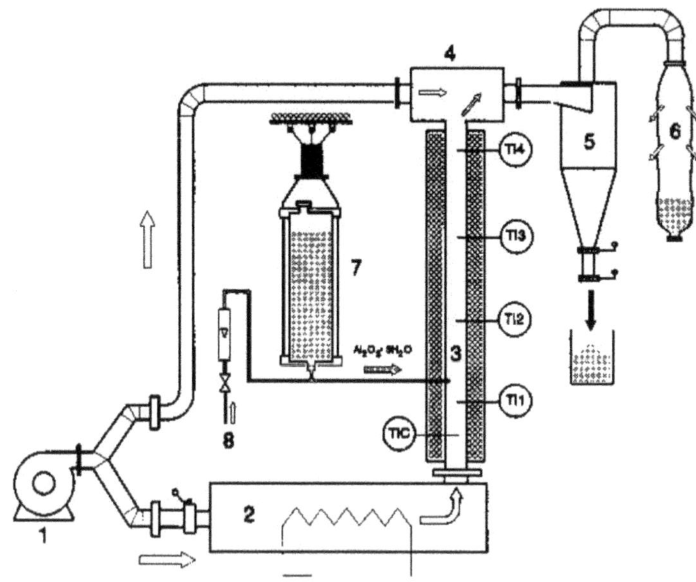

Fig. 1. Schematic diagram of the pilot-scale reactor: 1) blower for air supply, 2) electrical preheaters for air, 3) pneumatic reactor, 4) mixing chamber for warm and cold flow, 5) cyclone, 6) bag filter, 7) vibrating feeding device for gibbsite, 8) compressed air flow for gibbsite transport into the reactor, TIC – temperature inlet controller, TI – thermocouples.

Fig. 10.30 Bench scale pneumatic transport reactor [36]

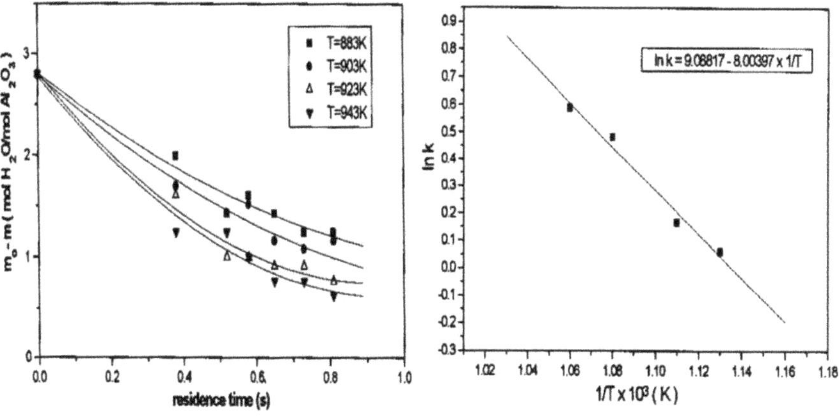

Fig. 10.31 De-hydration rate data and determination of energy of activation [36]

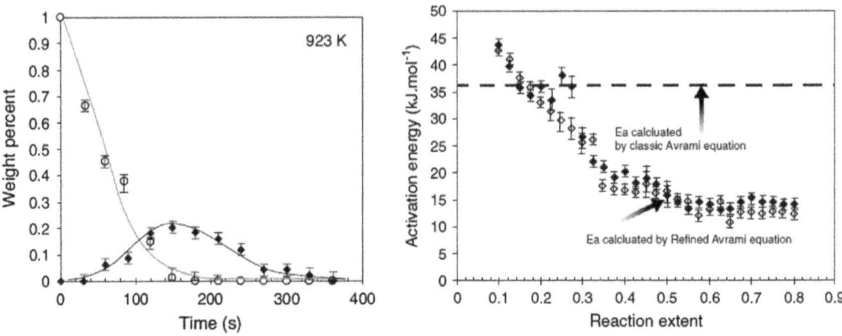

Fig. 10.32 Gibbsite conversion directly to X-ray amorphous alumina-partly via Boehmite [41]

Wang [41] studied the kinetics of Gibbsite de-hydration to X-Ray amorphous alumina in a crucible placed in a furnace, showing that some Boehmite is formed before it subsequently de-hydrates to X-Ray amorphous alumina. The energy of activation was evaluated using the Refined Avrami Method (RAM) (Fig. 10.32).

According to Wang [41], the X-Ray amorphous phases are formed either via the Boehmite phase with high activation energy (~44 kJ/mole) at low Gibbsite conversion, 10.28 or primarily directly from Gibbsite at higher conversion of Gibbsite due to the low energy of activation (~ 14 kJ/mole), 10.29 [11]:

$$\begin{aligned} 2\mathrm{Al(OH)_3}(\mathbf{c}) + \mathrm{Heat} &\rightarrow \mathrm{Al_2O_3 \cdot H_2O}(\mathbf{c}) + 2\mathrm{H_2O}(\mathbf{g}) \\ &+ \mathrm{Heat} \rightarrow \mathrm{Al_2O_3}(\mathbf{a}) + \mathrm{H_2O}(\mathbf{g}) \end{aligned} \tag{10.28}$$

$$2\mathrm{Al(OH)_3}(\mathbf{c}) + \mathrm{Heat} \rightarrow \mathrm{Al_2O_3}(\mathbf{a}) + 3\mathrm{H_2O}(\mathbf{g}) \tag{10.29}$$

where Al_2O_3 (**a**) is the X-Ray amorphous phase formed. The higher energy of activation via Boehmite has implications for the heat of calcination as we shall see later.

When comparing the kinetics investigated by Rozic [36] and Wang [41] it becomes clear that:

(1) The calcination method and thus heating rates are very different and influenced by particle size;
(2) The estimated Energy of Activation and Frequency Factor are very much different too;
(3) The 1st Order reaction constant is much larger, by a factor ~ 10^2, in the work of Rozic [36].

In summary, the De-hydration pathway at high heating rates documented by Yamada [33] is probably the most relevant although some improvements in analytical procedures have been made over time that could alter the conclusions.

Fig. 10.33 FLS minerals bench-scale/pilot calcinations unit [54]

FLS Minerals have calcined several different hydrates in our bench-scale calcinations unit over time and conducted an extensive R&D program investigating particle breakdown during calcination. Neither of the kinetics reported are as fast as observed by FLSmidth's own investigations in either bench-scale or full-scale calcination units. A drawing of the bench-scale GSC unit is shown in Fig. 10.33.

However, as indicated in Fig. 10.40 and further discussed below, the "true" calcination mechanism depends on several other interacting factors than just relatively simple calcination kinetics.

The reaction model by Rozic above [36] suggest that all three water molecules are removed in one single reaction as illustrated by the LOI versus calcination temperature below, Fig. 10.34.

Experimental data have been obtained of LOI versus calcining temperature by FLS in a GSC pilot plant as shown in Fig. 10.34, verifying that removal of the last water molecule (LOI $< 15\%$) requires much higher temperature than the two first water molecules. The experimental data confirms removal of the first two water molecules starts at 245 °C and ends with some "theoretical fraction", e_{TH} (1-y), of Mono-Hydrate or Boehmite at approximately 329 °C, with a corresponding LOI $= 15\%$.

In the cause of the de-hydration reactions, the concentration of neighboring $-OH^-$ groups decreases and the rate of removing the crystal water decreases to a situation where "loose or migrating" $-OH^-$ group have to find a more stable "bound" or other

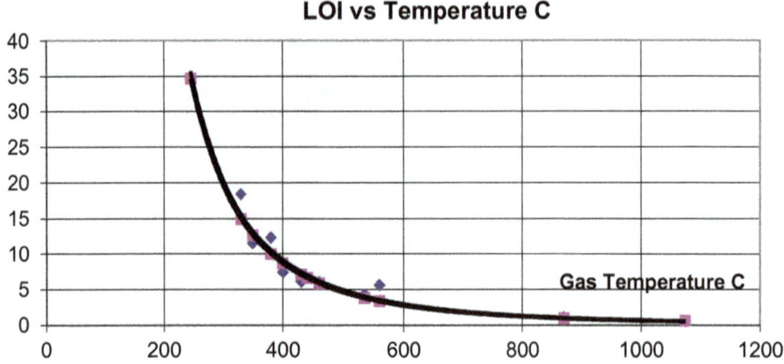

Fig. 10.34 Experimental pilot plant data of LOI versus calcining temperature

"loose or migrating" -OH^- group or partner in order to form a new water molecule leaving behind an atom of oxygen -O^{-2} "bound" to the solid matrix.

It is thus reasonable from a practical and theoretical point to assume that the rate of heat transfer to the particles are the rate controlling process in the initial or pre-calcination phase for removal of the water molecules, while solid state diffusion becomes rate determining for partial to complete removal of the remaining water molecules.

Alcoa Laboratories [37] presented in 1986 their newly invented pressure calcination process and the results from their bench and pilot scale test work. The calcination of hydrate takes place in two stages. In the first stage hydrate is indirectly calcined at 500 °C in a "shell and tube" type reactor by heat exchange with hot combustion products on the shell side supplied from the second calcination stage. At about 96% conversion of Gibbsite, up to 50% Boehmite phase is formed in the first pressure calcination stage resulting in a "relatively dense" and strong alumina particle compared to atmospherically calcined hydrate, see Fig. 10.35.

The SSA of pressure calcined hydrate was equal to about 70 m^2/g and did not change significantly during the final second stage atmospheric calcination at about 750–850 °C.

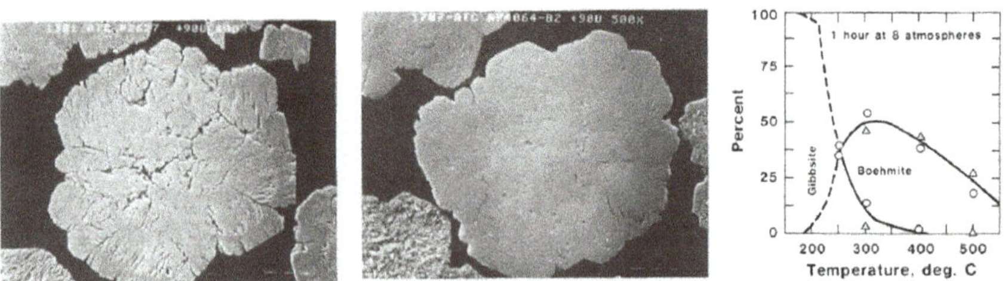

Fig. 10.35 Atmospherically (lhs)—pressure calcined alumina (m)—Boehmite formation (rhs) [37]. Copyright © [1986] by The Minerals, Metals and Materials Society. Used with permission

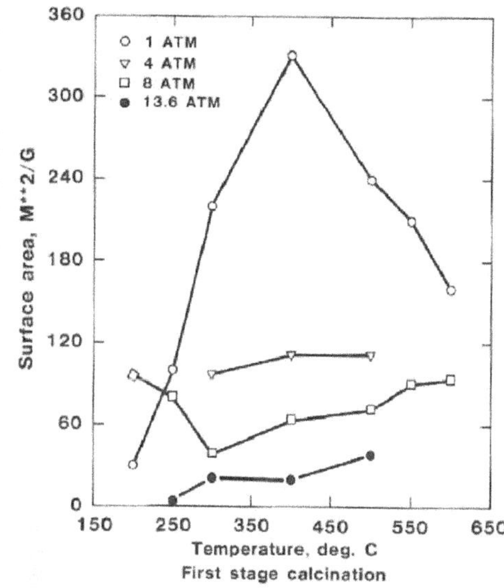

Fig. 10.36 Pilot pressure calciner (lhs) and SSA versus temperature and pressure (rhs) [37]. Copyright © [1986] by The Minerals, Metals and Materials Society. Used with permission

The final alumina product consists mainly of γ-Al_2O_3 with an LOI < 1% and a Forsythe-Hertwig Attrition Index ranging from 2 to 4% on 325 Mesh (45 μm).

The packed hydrate bed in the bottom of the reactor tube, Fig. 10.36 (lhs), is self-fluidizing into a bubbling fluid-bed due to the release of steam. About 80% of the reactor height is fluidized by the released steam.

About 85% of the crystalline water is recovered as process steam at 8 atm (120 psig).

Unexpectedly, a 11% reduction in calcination heat requirement to reach the final LOI of about 1% was realized. The total heat requirement for the pressure calcination process was calculated to 2.8 GJ/ton Al_2O_3 of which about 1.3 GJ/ton can be recovered as process steam resulting in a net calcination heat requirement of only 1.5 GJ/ton Al_2O_3.

Despite the favorable test result with respect to a strong alumina particle with low Attrition Index, SSA and LOI as a commercial SGA and a low net calcination heat requirement, the projected market for new pressure calciner technology has not been able to justify the commercialization of the pressure calcination process [42]. This will also require proper integration into the Bayer process digestion circuit to fully utilize the energy saving potential from the generated process steam.

10.4.5 Industrial Scale De-hydration and Calcination

The only industrial scale exception to the laboratory/bench-scale studies reported above is the 600 tpd Suspension Preheater Flash (SPF) calcination work reported by Yamada et al. [33] The calcination work reported took place in a full scale 600 tpd industrial calcination facility equipped with a gas suspension type pre-heater/calciner system (lhs Fig. 10.18), almost like the FLS GSC flow sheet. The industrial scale results of Yamada are compared with FLSmidth results from 2500 to 4500 tpd GSC units with and without Holding Vessel.

The data in Table 10.7 refers to samples from the underflow of cyclone PO_2 in the pre-calcination stage in which dry hydrate is discharged into riser duct, RPO_2 to cyclone PO_2, and mixed with hot gases from the high temperature GSC Furnace Cyclone PO3 or HVO3 in GSC units equipped with Holding Vessel (see Fig. 10.15 and 10.37.

Table 10.7 Boehmite from Gibbsite de-hydration in industrial calciners

Study Reference	Equipment	Pre-calcination of Bayer hydrate			Particle size	Boehmite
		Temp. (°C)	Time (s)	PH_2O (kPa)	micron	wt%
Yamada [33]	SPF	380–1100	120–180	–	55–90	3
FLSmidth	GSC	339–48/299*	2–3	30–45	10–200	2–4/8*

*GSC with holding vessel

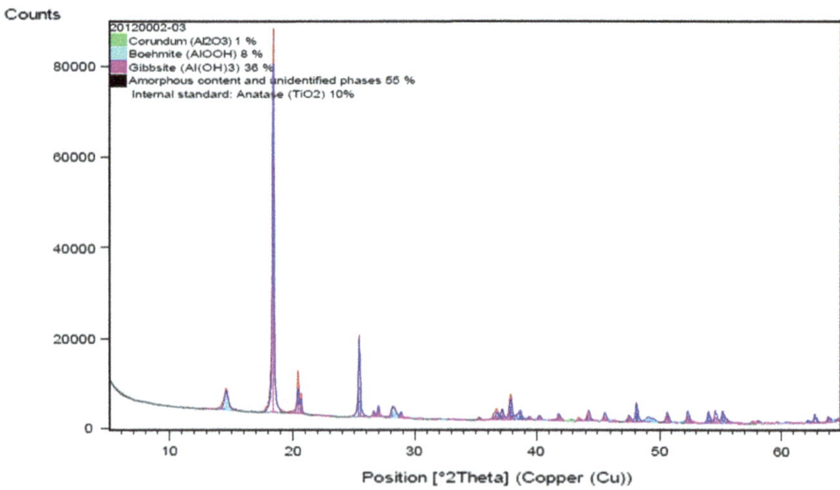

Fig. 10.37 Partly calcined gibbsite from cyclone PO_2 underflow at plant C [Courtesy of FLSmidth]

Table 10.8 Typical GSC operating data in pre-calcination

GSC plant unit	Capacity	Holding vessel	Temperatures °C		Gibbsite	Boehmite	X-Ray Amorph
	TPD		PO_2U/F	PO_3O/F	%	%	%*
A (HFO)	2500	No	316	~ 1100	9	2	89
B (NG)	4530	No	306	992	29	4	67
C (NG)	3200	Yes	299	845	36	8	56

*Including 1% alpha-phase from dust

As shown in Table 10.8, mainly X-ray amorphous alumina phase are observed in the underflow of cyclone PO_2 (pre-calcined feed material to the GSC Furnace) and 2–8% Boehmite phase, only.

The content of Boehmite increases with decreasing PO_2 U/F temperature driven by a lower calcination temperature in the GSC Furnace (PO_4) caused by the introduction of the Holding Vessel. This shows that the peak value of Boehmite formed in RPO_2/PO_2 may be 8%, or more, at a temperature below 300 °C. The formation of X-Ray amorphous alumina continues all the way through the calcination process with formation of some γ- and α-alumina phases in the final SGA product as seen from below Figs. 10.38 and 10.39 (path b).

In view of the above observations, the calcination of Gibbsite in the GSC flow sheet follows mainly the "**Red Calcination**" route in Fig. 10.38 with respect to the identifiable crystalline phases.

Perander et al. [32] used Environmental Scanning Electron Microscopy (ESEM) and Transmission Electron Microscopy (TEM) to show that the initial formation of

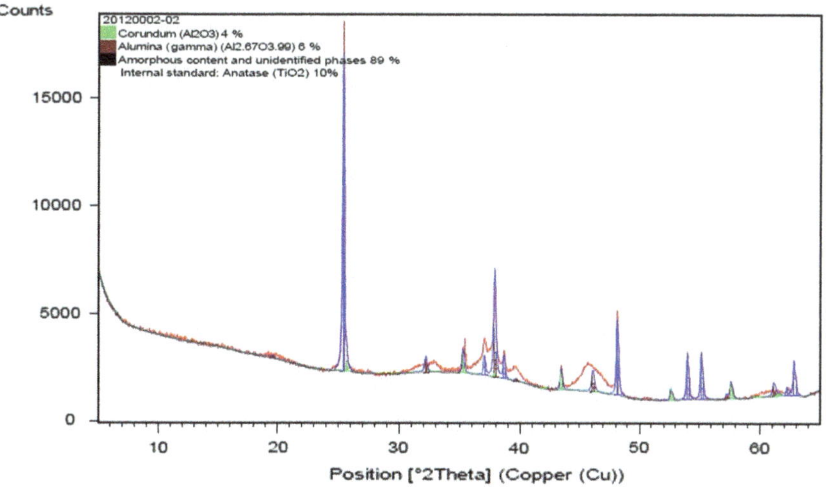

Fig. 10.38 Alumina phase compositions of SGA from gas suspension calciner [43]. Copyright © [2013] by The Minerals, Metals and Materials Society. Used with permission

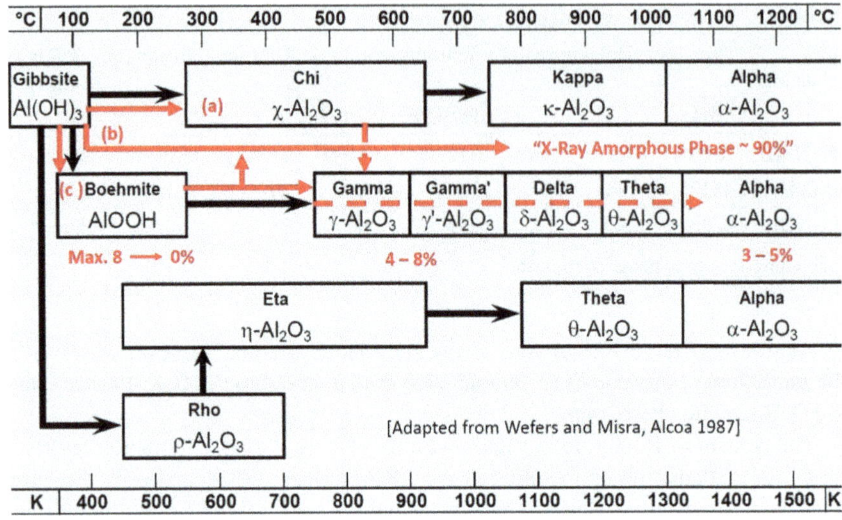

Fig. 10.39 Alumina phase changes during calcination of hydrate in GSC

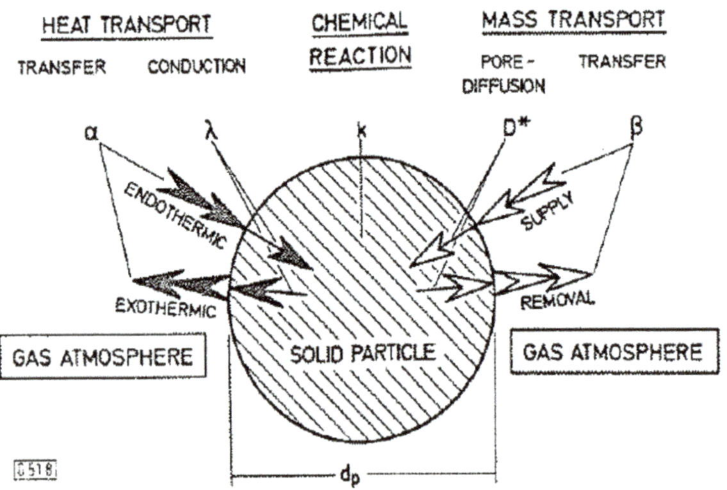

Fig. 10.40 Physio–chemical rate processes in gas suspension calcination of tri-hydrate [13]

alpha phase takes place on the surface of the transition alumina particles in Fluid Bed and Flash Calcination pilot units. This is no surprise as the temperature at the particle surface is higher than at the calcination front/sites in the interior of the particles.

This observation leads to a hypothesis of three different reaction path to alpha alumina. One for the outer shell, or crystal layer, of the particles via Chi-alumina (see Fig. 10.39 (path a)). Below the outer shell most of the Gibbsite is de-hydrated directly to X-Ray amorphous phases (see Fig. 10.39 (path b)), while in the core or interior part of the Gibbsite particles de-hydration takes place via some Boehmite phase formation (see Fig. 10.39 (path c)). However, as shall be shown later in Sect. 10.5.3.3, the formation of alpha phase on the shell do not prevent migration of water vapor out of the particles to the extent that a water vapor pressure is **not** built up high enough to cause an explosive shattering of the particles, contrary to what is postulated by Perander [32].

In summary, the detailed phase changes during calcination as known today are rather complex. During initial heating, a minor amount of Boehmite is formed at high energy of activation, say above 175 °C according to Perlmutter [35], if hydrothermal conditions (P_{H2O} > 1.07 kPa) can prevail in some crystals making up the poly-crystal hydrate or Gibbsite particle being heated. However, the formation of new Boehmite phase's stops as soon as pores are formed allowing the water vapor pressure to be reduced with trapped water vapor escaping and the hydrothermal conditions disappearing. Since the De-hydration of Gibbsite and final calcination to SGA take place through X-Ray amorphous phases, the use of the thermo-chemical calcination model assuming a crystalline calcination pathway must be revised.

10.4.6 Theoretical Rate Processes in Gas Suspension Calcination

Based on the above observed chemical and physical changes the following theoretical rate processes in Fig. 10.40 [13] are considered:

(i) Heat transfer rate from the gas phase to the external surface of the solid particle dependents on the heat transfer coefficient (α in Fig. 10.40). This rate process is important for initial removal of the first moles of water in reaction (10.29) and heating up the solid particles in the calcination furnace PO_4.

(ii) Heat conduction rate from the surface of the solid particle to the calcination front/sites dependents on the heat conductivity of the solid (λ in Fig. 10.40). Heat conduction resistance cannot be entirely neglected owing to the porosity of the calcined part of the relatively small particle sizes < 250 μm of hydrate and alumina.

(iii) The chemical reaction depends on the reaction rate constant (k in Fig. 10.40) for the following simplified calcinations reactions:

$$Al_2O_3 \cdot 3H_2O(s) + \text{Heat} \rightarrow 3H_2O(g) +'' \gamma - Al_2O_3'' \tag{10.29}$$

$$\gamma\text{-}Al_2O_3 \rightarrow \rho\text{-},\ \chi\text{-},\ \gamma\text{-},\ \delta\text{-},\ \theta\text{-},\ \alpha\text{-}Al_2O_3 + \text{Heat} \tag{10.30}$$

However, while the pre-calcination reaction in the RPO_2 riser to cyclone PO_2 is assumed to be initially controlled by the external heat transfer rate (i) to the particle surface, the chemical reactions, 10.29 partly takes place in cyclone PO_2 and the final chemical reactions (10.30) take place in the Calcination Furnace, PO_4.

The removal of residual crystal water in hydrate, 10.29, forms some non-stoichiometric alumina compound.

The final formation of α-alumina phases from γ-alumina phase are exothermic and assumed to be controlled by the reaction kinetics only.

(iv) Diffusion rate of water vapor through the internal pores to the external surface of the solid particles depends on the pore diffusion coefficient D*in Fig. 10.40. The pore diffusion resistance is assumed to be negligible in the calcined solid particle owing to the large specific surface area generated during pre-calcination and the relatively small particle sizes $< 250\ \mu m$ of hydrate and alumina.

(v) The mass transfer rate of water vapor from the external surface area of the solid particles depends on the mass transfer coefficient β in Fig. 10.40. During the calcination reaction (10.29), the partial pressure of water vapor at the particle surface builds up to a level exceeding the partial water vapor pressure in the gas phase, such that the mass transfer rate is equivalent with the rate of heat transfer from the gas to the particle.

The qualitative process analysis and assumptions made above leads to the formulation of the quantitative calcination model for simulation.

10.4.6.1 Practical Applications: Assumptions and Modeling

From the above observed de-hydration/calcination process pathways, the below chemical reaction model is assumed/formulated.

Assumption # 1: When solid Gibbsite, $Al_2O_3 \cdot 3H_2O$ (s), or Alumina Tri-Hydrate (ATH) is heated it is mainly converted directly to X-Ray amorphous alumina phase (XRA), some χ-alumina and some into solid Boehmite, $Al_2O_3 \cdot H_2O$, or Alumina Mono-Hydrate (AMH).

Assumption # 2: The fraction of crystalline solid Boehmite, $Al_2O_3 \cdot H_2O$, or AMH formed inside the Gibbsite crystals, is subsequently converted to a mixture of XRA, non-stoichiometric alumina compound and maybe crystalline γ-alumina, in the GSC Reactor and/or Holding Vessel, if present.

Assumption # 3: Subsequently, some of the XRA may be converted to a mixture of poorly crystalline transition phase alumina's (γ-, ρ-, χ-, γ-, δ-, ϴ- Al_2O_3) in the GSC Reactor and/or Holding Vessel, if present.

Assumption #4: The residual LOI in the below reactions may be thought of as sitting as relatively isolated "-OH" groups in the non-stoichiometric alumina compound, i.e. as a defect spinel structure: $H_2(Al_{3.0})\ (Al_{4.5})\ (Al_{1.0}, Al_{1.5})\ O_{16}$, or just: $H_2Al_{10}O_{16}$.

Assumption #5: A fraction of the γ-alumina phase coming from the χ-alumina formed in small particles (< 10 μm), or surface crystals, is converted to α-alumina phase in the GSC Reactor and/or Holding Vessel, if present.

No ATH or AMH can be detected in the SGA unless "alumina" dust has been mixed directly into the alumina discharged into the Fluid-bed cooler.

With the above assumptions based on observed data the below calcination reaction scheme can be proposed:

$$Al_2O_3 \cdot 3H_2O(c) \overset{k_{BX}}{\rightarrow} Al_2O_3H_2O(c) + 2H_2O(g) \overset{k_{GX}}{\rightarrow} Al_2O_3(a) + 3H_2O(g) \quad (10.31)$$

$$Al_2O_3 \cdot 3H_2O(\mathbf{c}) \overset{k_{GX}}{\rightarrow} Al_2O_3(\mathbf{a}) + 3H_2O(\mathbf{g}) \quad (10.32)$$

In Fig. 10.40 the above reaction model has been simulated using energies of activation as determined by Wang [41] and Plant C data (Table 10.8).

$$Al_2O_3 \cdot 3H_2O(\mathbf{c}) \rightarrow \chi - Al_2O_3 \rightarrow \gamma - Al_2O_3 \quad (10.33A)$$

$$\begin{aligned} &Al_2O_3 \cdot H_2O(\mathbf{c}) \rightarrow x\gamma - Al_2O_3(\mathbf{c}) + (1-x)'' \\ &Al_2O_3 \cdot yH_2O''(\mathbf{c}) + (1-x)(1-y)H_2O(\mathbf{g}) \end{aligned} \quad (10.33B)$$

Reactions in 10.31 and 10.33B illustrates the irreversible de-hydration of Boehmite into a mixture of X-Ray amorphous alumina phases, crystalline γ-alumina and non-stoichiometric alumina compounds. (Fig. 10.41)

$$Al_2O_3(\mathbf{a}) \rightarrow \gamma\text{-}, \gamma'\text{-}, \text{-}\delta\text{-}, \theta\text{-}Al_2O_3 \quad (10.34)$$

Reaction (10.34) illustrates the transition of some of the X-Ray amorphous alumina into different, but poorly crystalline solid transition phases.

$$\theta\text{-}Al_2O_3 \rightarrow \alpha\text{-}Al_2O_3 \quad (10.35)$$

Reaction (10.35) shows the final formation of some α-phase alumina, the only thermodynamic stable alumina phase.

It is perceivable that most of the γ-alumina converted into the stable α-alumina phase is, most likely, coming from the χ-alumina phase formed in small particles (< 10 μm) or surface crystals converted into γ-alumina as these are directly exposed to the hotter gas atmosphere, see Fig. 10.42.

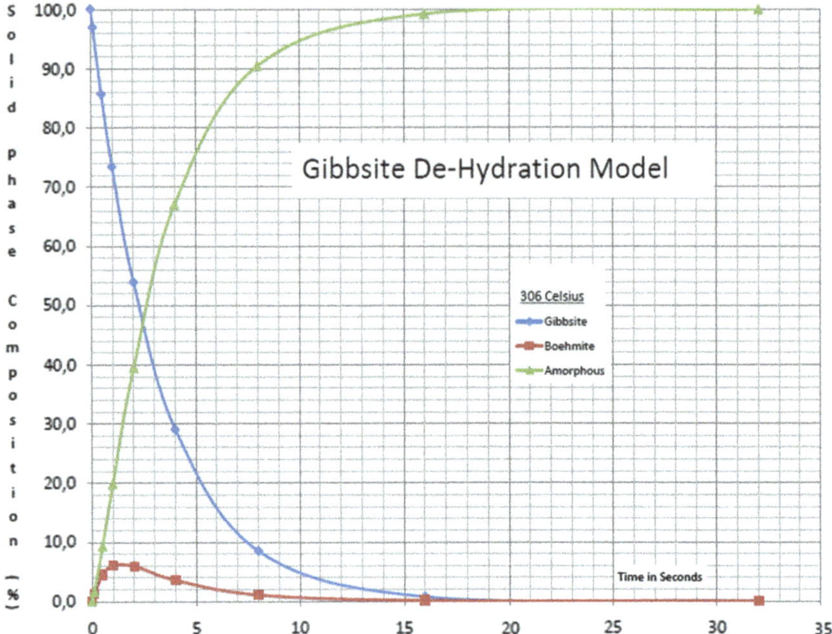

Fig. 10.41 Kinetic Gibbsite De-hydration model from reaction equations 10.31 and 10.32 based on plant B data

Fig. 10.42 α–Al_2O_3 phase formation in small particles (< 10 μm) or surface crystals [32]

10.4.6.2 De-hydration/Pre-calcination

The Degree of Calcination, $e_{TH}(T)$, versus calcining temperature is shown in Fig. 10.43 for the simple reaction (10.27). The reaction starts at temperatures above 245 °C. De-hydration during pre-calcination is a very fast process, initially governed

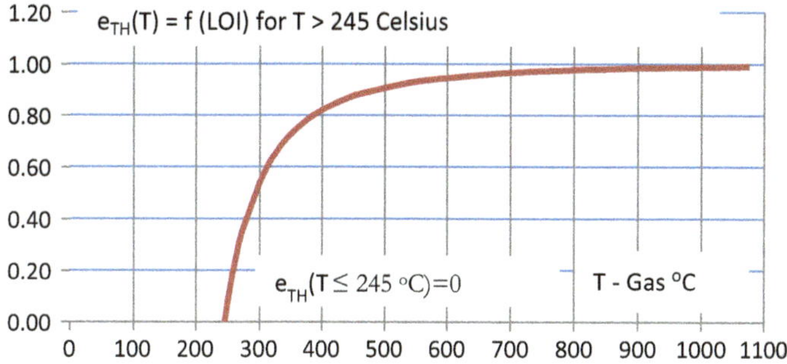

Fig. 10.43 Degree of calcination from experimental LOI versus temperature data

by the rate of heat transfer from the hot gas to the relatively cold, but dry hydrate particles. The de-hydration process slows down as the chemical reaction becomes rate limiting and temperature equilibration between the surface of the pre-calcined particles and the gas phase is approached. In process modeling the de-hydration process is treated as a quasi-stationary process since retention time and temperature differentials changes very little from case to case.

10.4.6.3 Alpha-Phase Formation

Rapidly heated Gibbsite is seen to require the least time for forming 20% alpha-phase, Fig. 10.44. The Alpha-phase formation in a particle, can be described by the

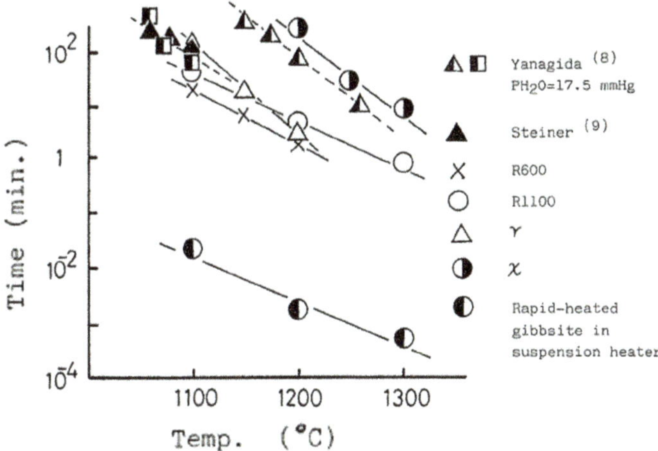

Fig. 10.44 Formation of 20% alpha-phase from gibbsite in suspension preheater [33]. Copyright © [1984] by The Minerals, Metals and Materials Society. Used with permission

1st-Order Uniform Reaction Model for a Porous Solid of Unchanging Particle Size" independent of particle size [11, 30]. Weight fraction α-alumina phase formed per second is:

$$dX\alpha/dt = k\alpha \cdot (1 - X\alpha) \quad (10.36)$$

where: $X\alpha$ = weight fraction α-alumina phase formed at time t.

$$k\alpha = A \cdot \exp(B \cdot (1 - To/Tp)) \quad (10.37)$$

where To and Tp is a reference and actual process temperatures, respectively in Kelvin. The above kinetic parameters A and B have been obtained from operating full scale GSC units at To. Yamada has reported the time required for forming 20% alpha-phase as a function of temperature and subject to different heating methods and rates as shown in Fig. 10.44.

For modeling purposes, consider that each particle is a small batch reactor travelling through the GSC Reactor Furnace and/or Holding Vessel, where the Alpha-phase formation depends on the temperature and Retention Time Distribution of the particles only. Assuming each particle has the same degree of pre-calcination when it enters the GSC and that solids in the reactor vessel are perfectly mixed, the below theoretical Furnace Reactor model applies for N equal sized reactors operating at the same temperature:

$$1 - X\alpha(\theta, T) = (1 + k\alpha \cdot \theta/N)^{-N} = 1/(1 + k\alpha \cdot \theta), \text{ for } N = 1 \quad (10.38)$$

The mean solids retention time, θ, is estimated from plant data [14] from their" Shower Model" of the Calciner Furnace operating at 1379 °K.

The same alpha-phase formation model applies to the Fluidized Holding Vessel operating at a different solid retention time and temperature.

10.4.6.4 Predicting Degree of Calcination of SGA

The FL Smidth overall proprietary calcination model can predict the degree of calcination of SGA as shown in Table 10.9.

The proprietary calcination model is used to decide and specify the operating conditions of the GSC Furnace Reactor and Holding Vessel prior to running the Mass- and Energy Balance calculation as described in Sect. 3.2.1 Operational Experience and Thermal Energy Consumption Comparison.

Table 10.9 Predicting degree of calcination of SGA

Case	1	2
Parameter	GSC-PO_4	HVO_3-Bed
GSC temperature (°C)	950	920
Solids retention time (s)	10	126
%–325 mesh in alumina (6% as base)	6	6
Alumina quality prediction		
Alpha phase (%)	0.0	4.2
SSA (m^2/g)	> 120	75.5
LOI (wt%)	> 1.0	0.87

10.5 Smelter Grade Alumina Quality from Stationary Calciners

Smelter Grade Alumina (SGA) is produced by calcination of Aluminium hydroxide as described above, however only a limited number of alumina quality parameters can be influenced by the calcination process as outlined in Table 10.10.

10.5.1 SGA Chemical Composition as Influenced by the Calcination Process

From a chemical impurity point of view, the Calcination process has a minor, or no impact, on the purity of SGA with a few exceptions.

The overall major impact on impurity level is from the concentration factor of about 156/102 equal to the mole weight ratio between aluminium hydroxide and alumina as a result of removing the crystal water by calcination. Another major impact may come from contamination of the SGA with sulfur (Na2SO4) and vanadium from stationary calciners using heavy fuel oil as their fuel.

A minor impact of increased SiO_2 and Fe_2O_3 content in the SGA may come from wear or spalling-off, of the refractory lining installed. With design gas velocities $\leq$ 20 m/s and good refractory work, this is not a problem in stationary calciners [44] when compared to rotary kilns.

A more detailed discussion of the impact that SGA chemical composition has on the Smelting Process will be made in Chap. 11: Alumina Quality, HF Removal, Dissolution and Aluminium Purity.

The calcination process has a major impact on Degree of Calcination (eTH) as seen in the previous section predicting it by modeling the Alpha-phase formation in the GSC Furnace Reactor and Holding Vessel.

However, LOI is the most direct measure of the Degree of Calcination (see Figs. 10.34 and 10.45) and it is controlled by calcination. The uniformity of

Table 10.10 Calcination process influence on SGA Quality

SGA parameter	Major impact	Minor impact	Typical range	Calcination issue
Na_2O	Precipitation	Filtration	0.3–0.5 wt%	–
Fe_2O_3	Digestion	Calcination	0.01–0.04 wt%	Refractory lining
SiO_2	Pre-Desilication	Calcination	0.01–0.03 wt%	Refractory lining
CaO and others	Digestion/Clarification	–	–	–
V_2O_5	Calcination	–	–	Heavy fuel oil
Sulfur	Calcination	–	–	Heavy fuel oil
Alpha Alumina	Calcination	–	2–10 wt%	Time–temperature
MOI (0–300 °C)	Calcination	–	1.4–4.0 wt%	Time–temperature
LOI(300–1000 °C)	Calcination	–	< 1.0 wt%	Time–temperature
Physical properties of SGA				
Spec. surface area	Calcination	–	60–85 m^2/g	Time–temperature
Particle size	Precipitation	Calcination	10–200 μm	Particle breakdown
Crystal size	Precipitation	–	10–200 μm	–
Attrition index	Precipitation	Calcination	5–30%	Heating rate
Bulk density (loose)	Precipitation	Calcination	900–1050 kg/m^3	Particle breakdown
Angle of repose	Precipitation	Calcination	30–32°	Mineralizer, ~ 45°
Flow time	Precipitation	Calcination	54–71 s/kg	Particle breakdown

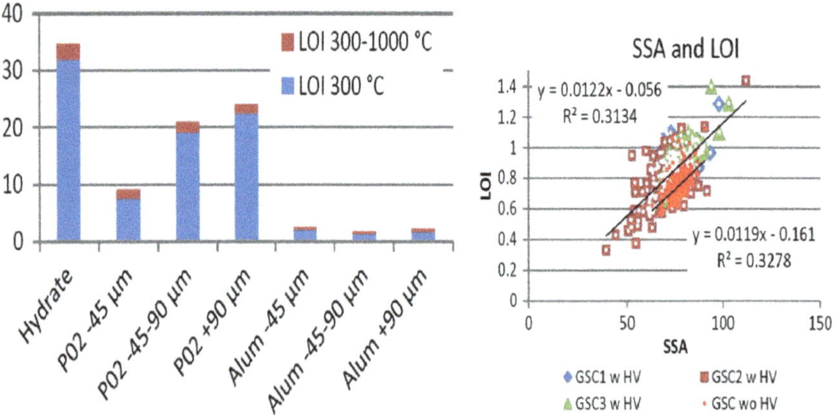

Fig. 10.45 Uniformity of calcination (LOI) and LOI versus SSA correlation for GSC units [43]. Copyright © [2013] by The Minerals, Metals and Materials Society. Used with permission

calcination of LOI (lhs) and LOI versus SSA (rhs) correlation can be seen from Fig. 10.45.

The LOI 300 °C is seen in Fig. 10.45 (lhs) to be very much particle size dependent as expected at the inlet to the Calcination Furnace Reactor (cyclone PO_2 discharge), while it is much more uniform in the final alumina product.

The SSA/LOI-ratio can be estimated from the LOI versus SSA correlation in Fig. 10.45 (rhs) and seen to be slightly dependent on the calcination technology, i.e. GSC with or without a Holding Vessel (HV).

The GSC without a Holding vessel has an SSA/LOI-ratio of about 104 versus an SSA/LOI-ratio of 88 for SGA from a GSC with a Holding Vessel. However, both SGA's have excellent dry scrubbing potential with an SSA/LOI-ratio exceeding 70 ($m^2/g\%$) as further discussed in Chap. 11.

10.5.2 Physical Properties as Influenced by the Calcination Process

During the above-described phase changes, the polycrystalline hydrate particles undergo several other physical changes that takes place as the Gibbsite particle is de-hydrated/calcined into the final SGA product.

As the Loss on Ignition (LOI) decreases the degree of calcination increases. The specific surface area (SSA) increases sharply during pre-calcination and reaches its peak value around 400 °C, Fig. 10.46.

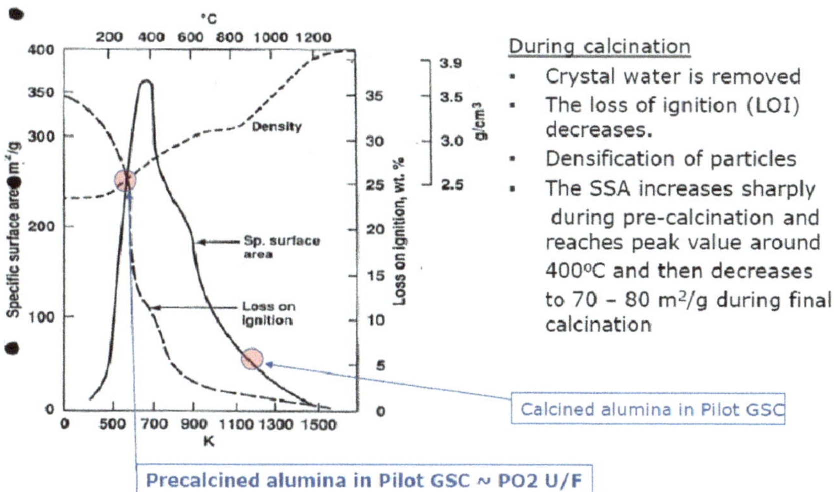

Fig. 10.46 Chemical and physical changes during calcination to "x-Alumina"

10.5.2.1 SSA with Mono-modal and Bi-modal Pore Size Distribution

In practical GSC operation the underflow material from cyclone PO_2 has been measured to have a specific surface area (SSA) of about 110 m^2/g.

As the SSA decreases the pore size increases as well as seen from Fig. 10.47.

The SSA decreases to 75–80 m^2/g with increasing calcination temperature in the Calcination Furnace PO4 from about 950 °C in the GSC with fluidized holding vessel to about 1050 °C in the GSC without fluidized holding vessel. The porous volume developed during calcination can be observed from SEM pictures when comparing hydrate with SGA, Fig. 10.48.

The importance of SSA and pore size distribution will be discussed in much more detail in the next chapter.

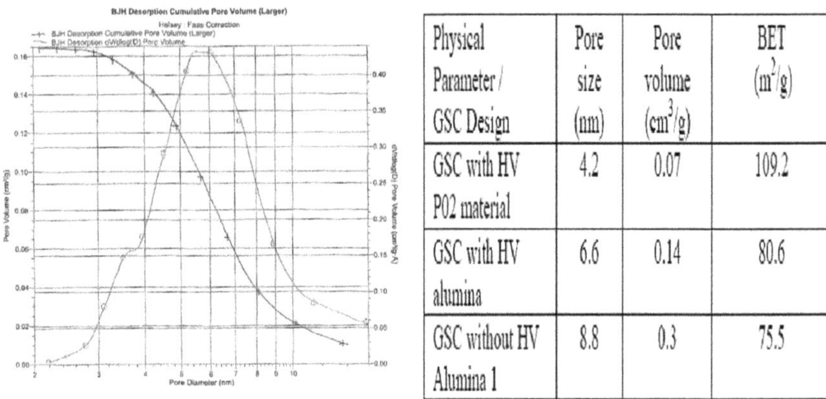

Physical Parameter / GSC Design	Pore size (nm)	Pore volume (cm^3/g)	BET (m^2/g)
GSC with HV P02 material	4.2	0.07	109.2
GSC with HV alumina	6.6	0.14	80.6
GSC without HV Alumina 1	8.8	0.3	75.5

Fig. 10.47 Mono-modal pore size distribution in SGA [43] from gas suspension calciner. Copyright © [2013] by The Minerals, Metals and Materials Society. Used with permission

Fig. 10.48 Polycrystalline hydrate particles: before (lhs) and after calcination (rhs), × 1000

Table 10.11 Gibbsite and degree of calcination in bag house dust and SGA

GSC alumina	Gibbsite wt%	LOI (300–1000 °C) wt%	Alpha %	SSA m^2/g
Bag house dust	13.0	5.87	56	34.7
SGA	0	0.62	3	80.6

However, the calcination process has a major impact on the SSA of the alumina as shown above in Figs. 10.46 and 10.47. In optimizing the SSA it must be realized that it is a compromise between desired alpha-phase content and LOI, as these three parameters are tightly correlated. Since each hydrate/alumina particle can be considered a small batch reactor travelling through the calcination plant while being exposed to almost the same temperature profile and retention time distribution, the resulting SGA can be expected to have a mono-modal Pore Size Distribution as seen in Fig. 10.47 (rhs).

This is also the case, even if the ESP or Bag House dust is mixed into the hot alumina in the cyclone cooler for final calcination as demonstrated from GSC plant data in Table 10.11.

A mono-modal Pore Size Distribution is not obtained in stationary calciners using a Hydrate By-Pass of the Calciner Furnace Reactor to control SSA and/or lower the back-end temperature, and thus the heat lost with the SGA discharged to the Fluid Bed Cooler. The reason being that to meet the SGA bulk specifications with respect to Alpha-phase, LOI (300–1000 °C) and SSA, the By-Passed fraction including ESP or Bag House dust is "under-calcined" as it never reaches the calcination temperature in the Calciner Furnace Reactor when mixed with "over-calcined" alumina from the Calciner Furnace Reactor. Consequently, SGA with a Bi-modal Pore Size Distribution is produced. Such an alumina will not have the maximum SSA available for Dry-scrubbing as discussed further in Chap. 11.

10.5.3 Particle Breakdown During Calcination

Since Calcination in Rotary Kilns was replaced with calcination in Stationary Calciners, following the energy crisis in the 1970'ties, Particle Strength, Size and Breakdown of the Alumina produced has been of major concern in design and operation of the Hydrate Precipitation circuit followed by calcination in Stationary Calciners. Since the development of the GSC technology started by FLSmidth in 1976, the issue of particle breakdown has been a focus of R&D efforts and technical reported by other calciner vendors [21, 45–48]. The results of FLSmidth R&D studies on PB will be described below.

From researching the literature [49] it became clear that material properties, fluid dynamic conditions, solids retention time and degree of calcination (LOI, SSA, Alpha) were the parameters that could affect particle breakdown in stationary calciners. Here, fluid-dynamic conditions are meant to include the combination of local gas velocity, geometry of ducts and vessels as well as design of fluidization nozzle in fluidized beds.

However, there can also be thermal induced stresses in particles during heating that result in particle breakage in the pre-calcination stage due to the anisotropic linear thermal expansion coefficient of gibbsite [50]. Furthermore, as gibbsite converts to boehmite, a change from monoclinic or triclinic crystal system to the rhombic crystal system takes place. The slight change of angles between crystal planes in different crystallites may set up stresses between crystallites making up the particle so that they may break in the pre-calcination stage.

10.5.3.1 Definition of Particle Breakdown (PB) During Calcination

When, hydrate is calcined in industrial or pilot size calciner units at different temperature profiles and gas velocities the Particle Size Distribution (PSD) of the hydrate changes because of PB taken place during calcination. PB during calcinations is calculated from the amounts of fines ($< 45\ \mu m$) and super fines ($< 20\ \mu m$) generated during calcination. PB from samples of hydrate and alumina produced are analyzed by Rotap Sieving to determine the wt% $< 45\ \mu m$ fines and by Malvern laser diffractometric to determine % $< 20\ \mu m$ in hydrate feed and alumina product respectively. The PB is calculated as follows:

$$PB(45\ \mu m) = X_{Alumina} - X_{hydrate} \tag{10.39}$$

$$PB(20\ \mu m) = Y_{Alumina} - Y_{hydrate} \tag{10.40}$$

where: $X_i := $ wt% $< 45\ \mu m$ of hydrate or alumina measured by Rotap Sieving.

$Y_i := \% < 20\ \mu m$ of hydrate or alumina measured by Malvern Laser diffractometry.

In the following, the term PB is used to describe the above explained change in PSD on 45 μm screen to distinguish it from the term attrition as defined below.

10.5.3.2 Early Studies of Particle Breakdown—And the Impact of Solid Sodium Oxalate

In 1980 Brown and Cole [51] presented their paper on "The behavior of sodium oxalate in a Bayer Plant", showing that increasing amounts of occluded sodium oxalate in the precipitated hydrate, from 0.01 to 0.05 wt%, decreased the average particle size of the non-smelter grade alumina produced in a rotary kiln.

Brown and Cole showed how gibbsite can precipitate on needle shaped solid sodium oxalate crystals, Fig. 10.49, and thereby be agglomerated into gibbsite particles containing occluded solid sodium oxalate. Raahauge et al. [52] reported the occlusion of solid sodium oxalate into hydrate particles, Table 10.11, with radial and

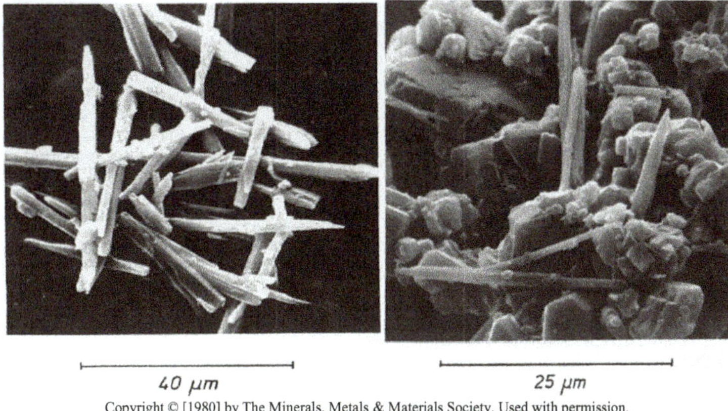

Fig. 10.49 Sodium oxalate particles (lhs) agglomerated into gibbsite particles (rhs) [51]. Copyright © [1980] by The Minerals, Metals and Materials Society. Used with permission

mosaic particle morphology respectively. Radial particles obtain their final size from the precipitation circuit mainly by growth, while mosaic particles obtain their final size mainly by agglomeration, Fig. 10.50 [53].

From Table 10.12, hydrate made up of mainly mosaic particles (0–20% radial) may occlude almost double the amount of solid sodium oxalate as hydrate made up of radial particles (80–100% radial). The occluded oxalate in Gibbsite particles with 40–60% radial structure is in between, making the amount of occluded oxalate appear to be almost proportional to the percentage of mosaic Gibbsite structure.

There seems to be an increase of occluded sodium oxalate up to a maximum of up till 40–60% radial particle structure, where after it decreases.

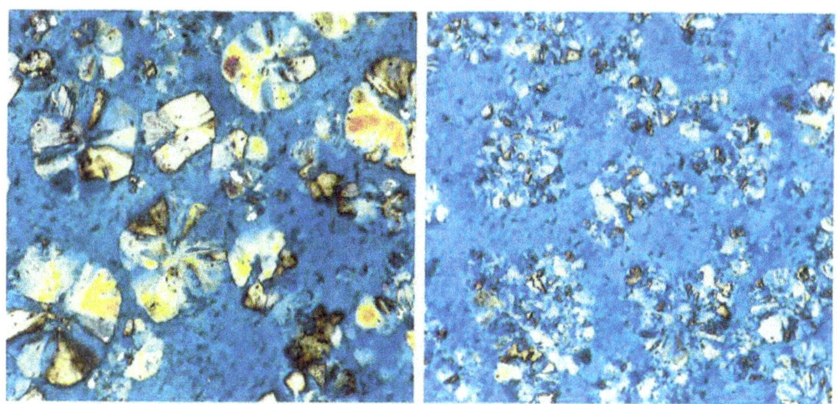

Fig. 10.50 Radial (lhs) particle made by growth and mosaic (rhs) particle made by agglomeration [53]. Copyright © [1982] by The Minerals, Metals and Materials Society. Used with permission

Table 10.12 Occluded Sodium Oxalate versus Hydrate Particle Morphology [53]

Radial structure	%	0–20	20–40
Samples	-	4	5
$(COONa)_2$	min.	0.028	0.028
wt%	mean	0.038	0.037
	max.	0.047	0.057

The observed pilot plant particle breakdown [53] at relatively low gas velocities was found to correlate, Eq. 10.44, with content of solid sodium oxalate and content of radial particle structure of the hydrate with $R^2 = 0.98$.

$$\begin{aligned} &\mathrm{PB(45{-}micron)} = 1,234 \cdot \mathrm{X} - 0.11, \\ &\text{with } \mathrm{X} = \left(10^3 \cdot \mathrm{NaOx}(wt - \%)\right)^{\alpha} / (\mathrm{Rad}(\%))^{\beta} \end{aligned} \tag{10.44}$$

The content of solid sodium oxalate needles was found to increase particle breakdown. The likely reason being that solid sodium oxalate decomposes at about 250 °C and thereby weakens the particle strength to the point where breakdown would occur much more easily (Fig. 10.52).

Around 1980–81, relatively few Bayer plants applied technology for removal of solid oxalate from the fine seed used in agglomeration, when the refinery used rotary kilns for calcination because particle breakdown was not an issue. But with the gradual replacements of rotary kilns with stationary calciners [8] it became clear that hydrate seed washing, Fig. 10.51, and removal of oxalate from the Bayer process was

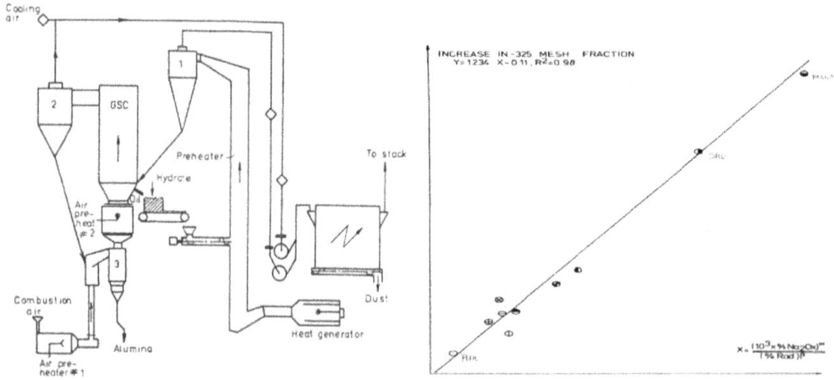

Fig. 10.51 250 kg/h GSC pilot plant (lhs) and particle breakdown correlation (rhs) [53]. Copyright © [1982] by The Minerals, Metals and Materials Society. Used with permission

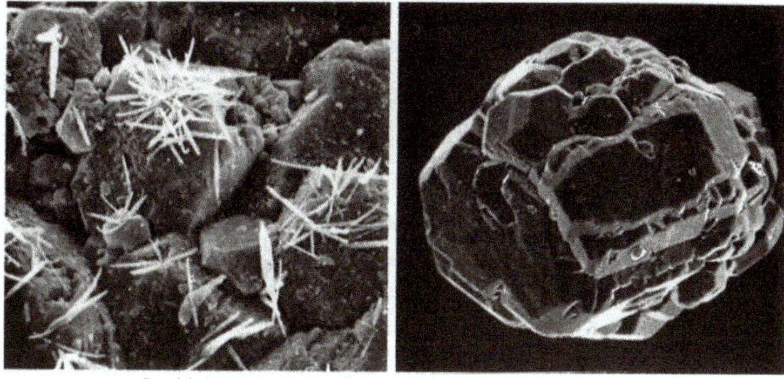

Fig. 10.52 Unwashed seed hydrate (lhs) and washed hydrate (rhs) [53]. Copyright © [1982] by The Minerals, Metals and Materials Society. Used with permission

a must in old and new alumina refineries to mitigate the impact of sodium oxalate on steady state hydrate precipitation and particle breakdown.

10.5.3.3 Recent Particle Breakdown Observation from Pilot Gas Suspension Calcination Test

Raahauge et al. studied and reported particle breakdown in pilot [54] and industrial size GSC plants [16, 55].

The bench scale/pilot GSC unit used for testing is shown in Fig. 10.30, where pre-calcination and complete calcination of hydrate was tested in one single calcination stage. The particle breakdown at complete calcination in the bench scale /pilot GSC unit was compared to the particle breakdown in plant Calciners, Fig. 10.53 [54].

In Fig. 10.53a if the pilot GSC is operated at a gas velocity between 15 and 18 m/s, then the same amount of alumina below the 45 μm particle size will be generated as from a plant Calciner, shown in the column to the very right in Fig. 10.53a. It is therefore reasonable to conclude that the Pilot GSC is representative with respect to

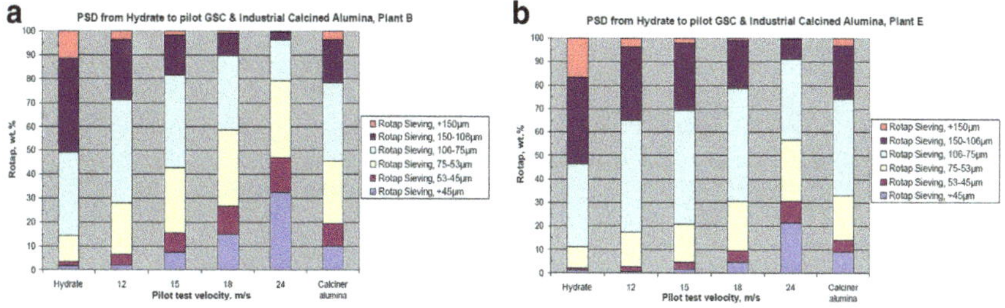

Fig. 10.53 **a** Hydrate and alumina from plant B [54]. **b** Hydrate and alumina from plant E [54]

simulating the fluid dynamic condition in plant Calciners. Comparing the change of particle size distribution (PSD) of plant B (Fig. 10.53a) during calcination with the PSD of plant E hydrate (Fig. 10.53b), it can be deducted that plant E hydrate is more resistant to particle breakdown than Plant B hydrate, because Plant E hydrate can be calcined at 18–24 m/s to breakdown down the same amount as plant B hydrate calcined at 15–18 m/s in the bench scale /pilot GSC unit.

The + 150 μm particle size fractions of both plant B hydrate and plant E hydrate are almost equally sensitive to particle breakdown in the 12–15 m/s test velocity range. In general, considering the changes in PSD from hydrate to alumina, is it possible to identify whether particle size has an influence on the particle breakdown?

It can be seen and concluded that a main contributor of particle breakdown is related to the fraction of coarse hydrate particles (+ 106 μm). The coarse particles are exposed to relative higher centrifugal forces on the wall (and have higher kinetic energy at impact with the walls) compared to smaller particles of the same velocity. This leads to higher collision energy and particle breakdown since the forces binding the different crystals together in the alumina particle in general are of the same order of magnitude.

The calcination test results of different hydrates at 360 °C (pre-calcination) and 1075 °C (final calcination) is shown in Fig. 10.54. In both cases, hydrate is heated from ambient to the calcination temperature in one step in the Bench -scale/Pilot GSC unit, Fig. 10.33.

The results show that the particle breakdown for each individual hydrate tested is a linear function with respect to the cyclone inlet gas velocity raised to the 2nd power (U^2) for the gas velocity range studied. Significantly, all particle breakdown values are ≤ 0, i.e. zero (0) wt-% particle breakdown, at about $U^2 = 140\text{–}150\ m^2/s^2$, i.e., at a test cyclone inlet gas velocity of about 12 m/s.

The experimental fact that no particle breakdown can be measured at a test cyclone inlet gas velocity of about 12 m/s and lower, suggests the following conclusions:

(1) Particle shrinkage reported by Klett et al. [46] has no practical impact on particle breakdown.

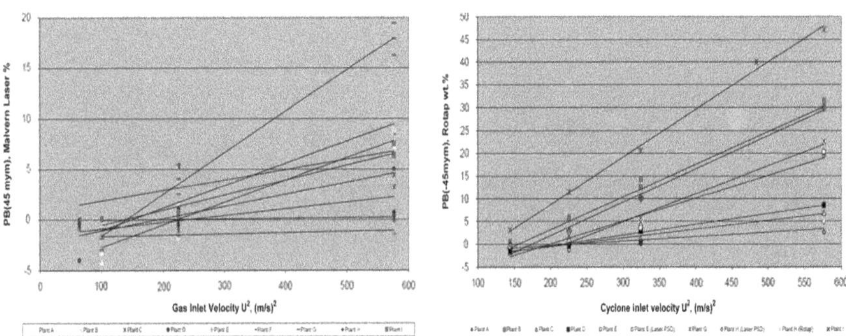

Fig. 10.54 Calcination test of particle breakdown – 360 and 1075 °C [54]

(2) The theory of steam explosion caused by Boehmite phase calcination in the interior of the hydrate particles proposed by Perander et al. [32] has no practical impact on particle breakdown either.
(3) Particle breakdown of hydrate at pre-calcination temperatures at 360 °C is significantly lower than particle breakdown of hydrate at final calcination temperature at 1075 °C.

Apparently, particle breakdown depends on a specific hydrate related factor in addition to the squared-velocity dependence shown in Fig. 10.53 as shall be seen below.

10.5.4 Alumina Particle Strength-Observation from Pilot and Plant Gas Suspension Calcination

There is no generally accepted measurement method of Alumina Particle Strength or "toughness" [56], but the alumina industry has used the Alumina Attrition Index determined by the Forsyth-Hertwig test [57] modified and introduced by Alcoa many years back.

The Alumina Attrition Index (AAI) is calculated as follows:

$$\text{AAI}\ (\%) = 100[1 - (\%\ \text{Alumina} > 45\mu\text{m})_{\text{AFTER}}/(\%\ \text{Alumina} > 45\mu\text{m})_{\text{BEFORE}}] \tag{10.45}$$

In 1985, Zwicker [57] showed how the Fragility of the Alumina, or Alumina Attrition Index (AAI), changed with the calcination temperature and heating rate, Fig. 10.55 (lhs). Sang [58] reported in 1987 the interaction between Hydrate and Alumina Attrition Index versus Hydrate Precipitation parameters such as Initial Ratio and Median Size of Product, Fig. 10.55, (rhs).

The effect of heating rate indicates that the Fragility, or Alumina Attrition Index, will be lower for alumina produced in Stationary Calciners than in Rotary Kilns where the heating rate is relatively low, Table 10.13.

The increase/decrease in fragility or Alumina Attrition Index with calcination temperature and heating rate as influenced by calcination technology has been known for a long time [52] since pilot test work in 1978–79. This has been confirmed by recent R&D work [16, 54, 55] covering both GSC pilot scale and full-scale calcination units.

The Fragility or Alumina Attrition Index has been shown, Fig. 10.56 (lhs), to increase with increasing content of non-mosaic particle structure [54], and decreasing wt% content of soda in the alumina [12], Fig. 10.56 (rhs).

Since particle morphology or structure (see below) and content of soda in the alumina are determined in the hydrate precipitation circuit, it is concluded that the precipitation circuit has the major impact on the resulting Alumina Attrition Index after calcination in rotary kilns and/or stationary calciners.

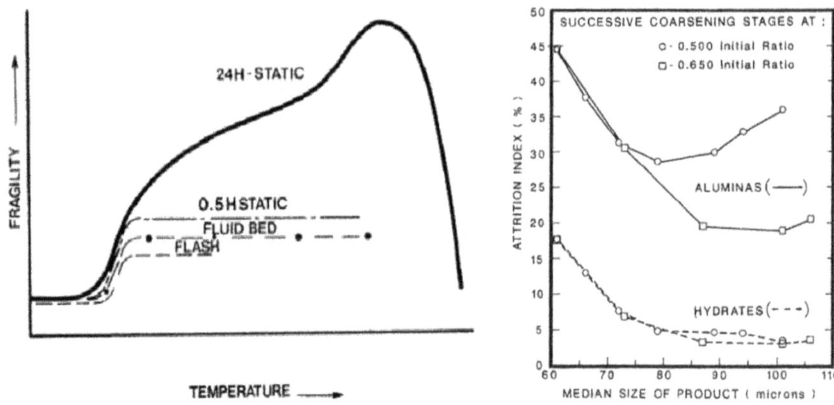

Fig. 10.55 Fragility or alumina attrition index versus calcination temperature [58] and precipitation conditions [59]. Copyright © [1985,1987] by The Minerals, Metals and Materials Society. Used with permission

Table 10.13 Alumina attrition index versus calcination technology [53]

Hydrate source	Rotary Kiln	GSC
A	–	30
B	17	9
C	50	26
D	41	12
E	10	5
F	9	5
G	19	5
H	–	20
I	20	20
K	14	8

Pilot plant studies in 1978–79 of hydrate from 10 different Bayer plants using batch or continuous precipitation technology, or a combination thereof, showed that the attrition index as function of calcination temperature peaked at the pre-calcination stage where about 2 out of 3 water molecules are removed at a temperature of about 270–370 °C [49]. This observation is confirmed as seen from Fig. 10.54, where attrition index of pre-calcined alumina is larger than the attrition index of the alumina calcined at 1070 °C. The difference in heating rates between gas suspension/flash calciners and fluid-bed calciners has a minor impact [58].

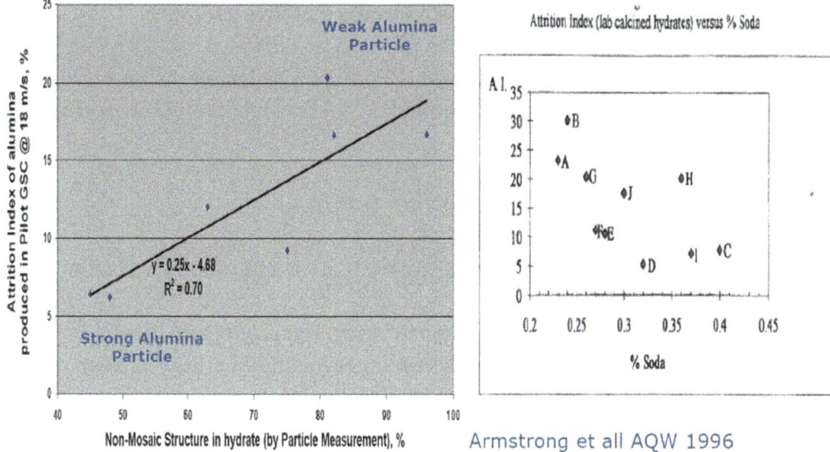

Fig. 10.56 Alumina attrition index (AAI) versus hydrate particle structure and % soda (Na_2O) [54]

There is, however, an impact from heating rate between pilot and industrial Calcination. This can be seen in Fig. 10.57, when comparing AAI of Pilot Calcination at 1075 ° C in one (1) step only (yellow) with Industrial Calcination (light blue). The calcination in industrial plants uses three (3) stages of heating to the final calcination temperature resulting in a higher AAI when compared to Pilot Calcination of hydrate.

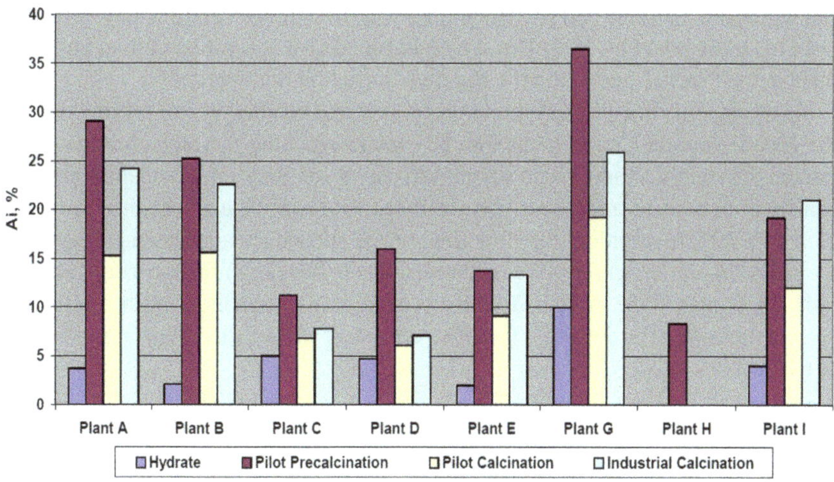

Fig. 10.57 Attrition index for hydrate and alumina from different plants versus calcination conditions [54]

10.5.4.1 Particle Breakdown Versus Alumina Particle Strength.

Sang [59] correlated Particle Breakdown (PB) over a full-scale Alcoa Fluid Flash Calciner plant with the Alumina Attrition Index (AAI) of laboratory calcined hydrate to alumina, with $R^2 = 0.98$:

$$PB(\%) = 0.74 + 0.46^{*}AAI \tag{10.46}$$

In 1992, Lopez [60] reported a similar correlation for a full-scale Circulating Fluid Bed Calciner, $R^2 = 0.74$.

It can therefore be concluded that there is a specific calcination plant correlation between particle breakdown and AAI imbedding the relative constant content of soda in the SGA in any Bayer plant.

Clerin [61] has disclosed a linear correlation between particle breakdown during calcination versus Alumina Attrition Index with $R^2 = 0.715$ from several calcination plants, but without specifying the calcination technology. It is thus concluded that AAI can be the missing specific hydrate related factor in addition to the squared-velocity dependence on particle breakdown.

10.5.4.2 Smelter Impact from SGA with High Alumina Attrition Index (AAI)

There is no general requirement by Smelters with respect to the Alumina Attrition Index in SGA. However, Smelters prefer a strong or "tough" alumina particle that can withstand handling and passage of their Gas Treatment Centre or Dry Scrubbers prior to feeding the Pots [56]. This is to avoid excessive generation of fines and super fines in the Smelter plant with the negative impact on pot operation.

At the Smelter, the primary alumina is stored and handled before passing through the Gas Treatment Centre to capture HF before reaching the individual pots. The impact of primary alumina with a high AAI, say > 18–20%, on the smelter operation can be a significant additional particle breakdown on 45 μm, ranging from 5 to 15% [44, 62]. In addition, the flow time of the secondary alumina increases from the range 54–71 (s/kg) for primary alumina to 71–170 (kg/s) for secondary alumina [44]. In summary the AAI is a measure of alumina particle strength albeit not a perfect measure. The impact on smelter operation will be discussed in more detail in Chap. 11.

10.5.5 Generic Particle Breakdown Model

Based on the alumina hydrate strength, a new ***Generic Particle Breakdown*** model for the Pilot GSC calcination test unit has been discovered and reported [54]. The new independent parameter is defined as; $X = AAI \times U^2$. The empirical curve fit in

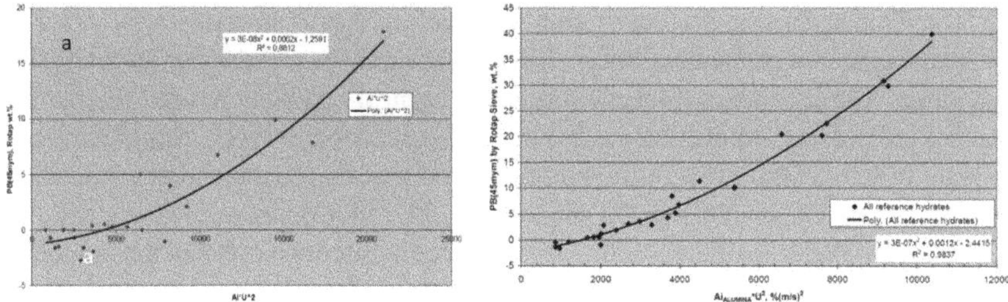

Fig. 10.58 Generic particle breakdown model at 360 °C (lhs) and 1075 °C (rhs) pilot calcination [54]

Fig. 10.58a of hydrate calcined to 360 °C and in Fig. 10.58b of hydrate calcined to 1075 °C covering particle breakdown of all hydrates tested at 12–24 m/s gas velocity in to the cyclone.

The Generic Particle Breakdown model has been established on ***eight (8) reference hydrates*** from different alumina refineries with different hydrate precipitation technologies and it is based on a material property parameter (alumina attrition index) and a GSC design parameter (cyclone inlet gas velocity): The high temperature pilot model correlation found is:

$$\mathrm{PB}(45\mu\mathrm{m}) = 2.7 \times 10^{-7} \times \mathrm{X}^2 + 1.2 \times 10^{-3} \times \mathrm{X} - 2.4\mathrm{wt} - \% \tag{10.47}$$

where: $\mathrm{X} = \mathrm{AAI} \times \mathrm{U}^2$.

AAI = Attrition index of calcined alumina from Pilot GSC test, wt%; U = Pilot test cyclone inlet gas velocity, 12–24 m/s.

Particle breakdown in the pilot plant becomes zero (0) wt-% at $\mathrm{X} = \mathrm{AAI} \cdot \mathrm{U}^2 = 1600$, for say U = 12 m/s at an AAI = 11 wt%, confirming the above conclusions based on the hydrate particle breakdown model as function of the gas velocity squared (U^2) in Fig. 10.54.

However, the Generic Particle Breakdown model also applies at low calcination temperature as well, and it is clearly seen that particle breakdown in the pilot plant is much lower at low temperature than at high calcination temperature. From Fig. 10.55: Particle breakdown at low temperature is about zero (0) wt% for $\mathrm{AAI} \cdot \mathrm{U}^2 = 5000$, say U ~ 16.7 m/s at an AAI = 18 wt%, whereas particle breakdown is equal to about 10% at high calcination temperature in the pilot plant.

The previous correlation (Fig. 10.51 and 10.44) from early 1980s R&D work could not be verified by the R&D work undertaken in 2005, most likely because most Bayer plants were now removing solid sodium oxalate from fine seed before being applied to agglomeration.

The above observation of lower particle breakdown at low pre-calcination temperature versus high final calcination temperature leads to the below ***hypothesis of the particle breakdown mechanism during calcination***.

When a particle is ***heated-up*** by heat transfer from a surrounding hot gas, thermal expansion of the crystals making up the particle starts from the outer shell/particle surface. This expansion provides more "volume" so that more room is provided for thermal expansion of the next lower solids layer in the particle as it continues to be heated up. The heating-up process continues, by heat conduction, until the particle reaches temperature equilibrium as it is leaving the calcination zone. Simultaneously, the rate of temperature rise of the particle is reduced due to the endothermic calcination reactions taken place in the particle [11, 24, 29]. Consequently, very low (see above) or no thermal stresses set-up in the particle interior, in either the tangential or radial direction during heat-up.

When the same very hot alumina particle is subsequently discharged into a much cooler air stream for heat recovery, the particle is ***cooled down*** from the outside. Heat transfer to the cooler air causes the outer particle shell, or surface, to cool and shrink or thermally contract. Simultaneously thermally induced tangential and compressive radial shrinkage stresses are set-up at the particle surface layer and directed towards the particle center.

Because the particle interior cannot shrink fast enough, by conducting heat towards the colder particle surface to relieve the stresses, the compressive thermal stresses in combination with sufficiently large external mechanical impact forces break the particle. Consequently, the larger the temperature drop a hot alumina particle is exposed to, the larger will be the thermal induced stresses and thus the greater the opportunity for particle breakdown.

Since it has been demonstrated that the Pilot GSC is representative for the fluid dynamic conditions in a GSC Plant calciner, there are strong indications that the Generic Particle Breakdown model also applies there, though it is not so easy to verify.

10.5.6 Particle Breakdown in Industrial Calciners

The particle breakdown observed in the industrial calciners at the Alumina refineries supplying the hydrate for pilot testing is shown below in Fig. 10.59 [43].

The particle breakdown seems to increase significantly when the parameter $AI*U^2$ exceeds the range 7500–8000, regardless of the plant considered has a Holding Vessel or not. The observed particle breakdown also corresponds to hydrate with more than 96 wt% > 45 μm.

The parameter $AI * U^2$ is a measure of the friability (AI) and kinetic energy (U^2) of the particles, when passing through the cyclones in the calcinations plant, it seems the particles will break when they exceed a certain threshold value of impact energy. It is the coarse particles that break the most [54], because they are the first to exceed their impact strength threshold, as can be seen from the increase in particle breakdown with increasing size fraction exceeding 45 μm in the hydrate to the calcinations plant, Fig. 10.59 (rhs).

This is not new or surprising information, see Fig. 10.59 (rhs), as it has been shown previously by Sang [59], that over coarsening the hydrate particles to meet the requirement of maximum 10% < 45 μm in the alumina shipped to Smelting [6],

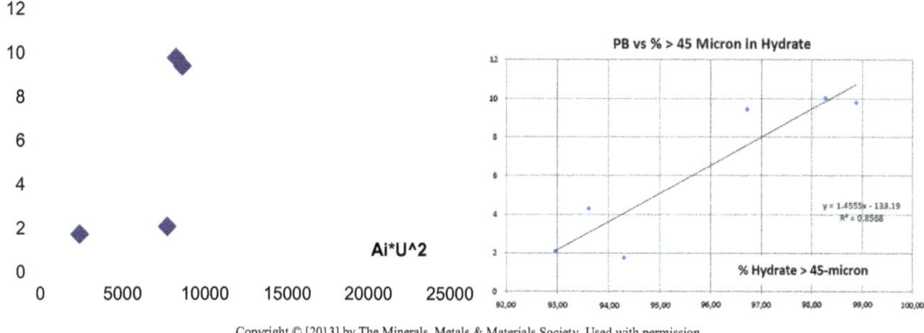

Copyright © [2013] by The Minerals, Metals & Materials Society. Used with permission.

Fig. 10.59 Particle breakdown 45 μm (Y-axis) for industrial calciners [2100–4500 tpd] versus the parameter AI * U^2 (lhs X-axis), and wt% hydrate > 45 μm (rhs X-axis) [43]. Note! AI = AAI. Copyright © [2013] by The Minerals, Metals and Materials Society. Used with permission

results in the precipitation of less strong hydrate particles. Upon calcination these particles, with relative high AAI values, do experience increased particle breakdown, and do not produce a coarser alumina as confirmed during commissioning of GSC units.

10.5.6.1 Relative Particle Breakdown and Alumina Attrition Index of SGA-Rotary Kilns

For confidentially reasons, the % relative Particle Breakdown (PB) will be reported onwards in this chapter. The % relative Particle Breakdown is calculated as a % of the absolute wt% Particle Breakdown on 45 μm sieve (Rotap) observed in a rotary kiln producing alumina with AAI = 17–19%.

As seen from Table 10.14, there seems to be a relationship between Particle Breakdown in Rotary Kilns, Alumina Attrition Index and Hydrate Precipitation Technology [59] (Fig. 10.60).

Table 10.14 Relative particle breakdown (PB) versus alumina attrition index (AAI)

Rel. particle breakdown (PB) versus alumina attrition index (AAI)			
Alumina refinery	Rel. PB (%)	AAI (%)	Precipitation technology
QAL, Australia	~ 150%	22–24	Kaiser aluminium
Eurallumina, Italy	~ 100%	17–19	Impr. Kaiser aluminium
Nabalco, Australia	< 50%	13–15	Alusuisse

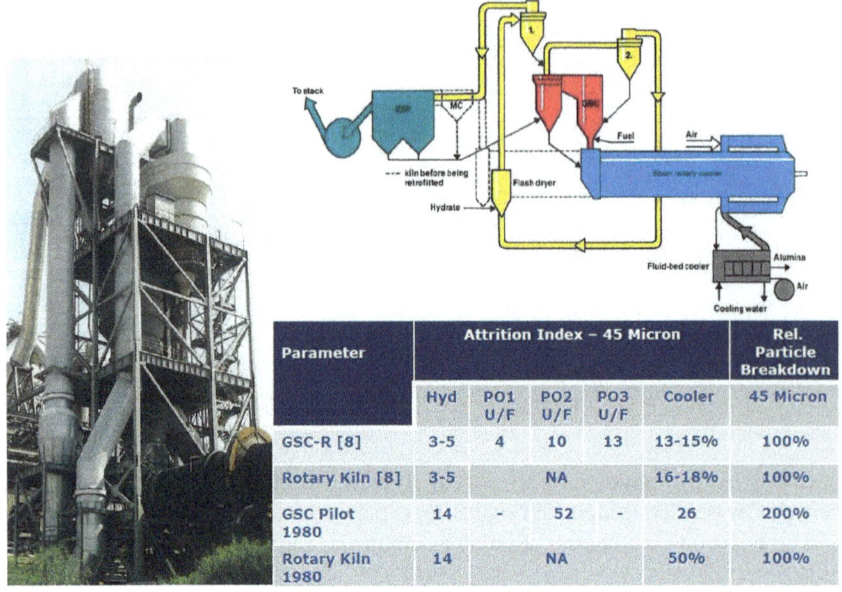

Parameter	Attrition Index - 45 Micron					Rel. Particle Breakdown
	Hyd	PO1 U/F	PO2 U/F	PO3 U/F	Cooler	45 Micron
GSC-R [8]	3-5	4	10	13	13-15%	100%
Rotary Kiln [8]	3-5		NA		16-18%	100%
GSC Pilot 1980	14	-	52	-	26	200%
Rotary Kiln 1980	14		NA		50%	100%

Fig. 10.60 GSC retrofit: hydrate to alumina attrition index and particle breakdown [59, 60]

10.5.6.2 Relative Particle Breakdown and Alumina Attrition Index of SGA–Gas Suspension Calciners

The same hydrate cannot be calcined in two different calciners. It is therefore difficult to compare particle breakdown even in the same refinery owing to variation in hydrate quality [16]. In Fig. 10.59, the relative particle breakdown in the rotary kiln and GSC retrofit was 100%, the same even after improvements in the Hydrate Precipitation circuit at Eurallumina, Italy [8]. The Alumina Attrition Index in the rotary kiln alumina was larger than in the alumina from the GSC retrofitted kiln owing to a higher heating rate in the latter as reported above by the work of Zwicker [58].

The experience with particle breakdown in various calciner designs gathered by FLSmidth over the years is presented in Table 10.15 [55].

It is well known and recognized that minimum particle breakdown is observed in a rotary kiln equipped with a Unax cooler rotating with the kiln. This is due mainly because the hot alumina particles are cooled slowly under gentle mechanical impact forces in that cooler design. Because of that observation, in Table 10.15, the minimum particle breakdown benchmark value for rotary kilns is used to normalize particle breakdown observed in other calciners.

Hydrate N: The GSC Retrofit, with a rotary cooler, Fig. 10.59, and not a cyclone cooler, is seen to have the same relative, and thus low absolute, particle breakdown as the rotary kiln. This observation supports the above conclusion:

Table 10.15 Particle breakdown and strength–industrial GSC units [60]

Calcination technology	Hydrate source	Capacity utilization	Calcination temperature celsius	AAI	Rel. particle breakdown to rotary kiln (%)	Cooler
Rotary Kiln	Hydrate N	100.0	900–1100	16–18	100	Unax(*)
GSC Retrofit	Hydrate N	100.0	1000	13–15	100	Rotary + Unax
GSC without HV	Hydrate P	100.0	1110	6–16	125	4 Stg cyclone
GSC without HV	Hydrate L	101.2	1090	16 ± 3	180	4 Stg cyclone
GSC without HV	Hydrate L	118.6	1045	14	160	4 Stg cyclone
GSC without HV	Hydrate O	102.8	1033	17.6	155	4 Stg cyclone
GSC without HV	Hydrate G	100.0	1050	24–26	300–450	4 Stg cyclone
GSC with HV	Hydrate M	98.7	910	30	250–300	4 Stg cyclone

- Low particle breakdown in the GSC reactor furnace, PO_4, and furnace cyclone, PO_3, itself.
- Minimum mechanical handling of the hot alumina during cool down minimizes particle breakdown;

Hydrate P: Resulting in alumina with AAI ~ 6–16% from GSC without HV, low particle breakdown from low air velocities and low AAI;

Hydrate L: Resulting in alumina with AAI ~ 14–16% from GSC without HV, is seen to have almost no change in particle breakdown despite an increase in capacity utilization following a + 20% calciner capacity upgrade.

Hydrate O: resulting in alumina with AAI ~ 18% from GSC without HV, has particle breakdown comparable with Hydrate L owing to slightly lower air/gas velocities at lower nominal calcination capacity.

Hydrate G: Resulting in alumina with AAI ~ 24–26% from GSC without HV, shows significant higher particle breakdown when compared to other GSC without HV, partly due to higher air velocities.

Hydrate M: Resulting in alumina with AAI ~ 30% from GSC with HV, shows significant higher particle breakdown when compared to other GSC without HV, but not as high as expected versus Hydrate G. The reason being, that the impact from the higher AAI is probably compensated with the lower calcination temperature in a

GSC unit with a Holding Vessel, a re-design of the GSC furnace inlet and lower air velocities.

The conclusion from the data in Table 10.15 clearly suggests that particle breakdown increase with AAI of the alumina which is a result of the applied hydrate precipitation technology [16]. The particle breakdown R&D work by FLSmidth suggests that the majority of particle breakdown (PBD) in full size calciners takes place when cooling the alumina. Particle breakdown is a function of $AAI \cdot U^2$ in the individual cooler cyclones and other critical locations where the alumina is exposed to high air velocities. Future R&D work needs to be focused on reducing the impact from fluid dynamic conditions on particle breakdown. Here "fluid-dynamic conditions" are meant to include the combination of local gas velocity, geometry of ducts and vessels as well as design of fluidization nozzle in fluidized beds on particle breakdown.

10.6 Outlook for a Future CO_2 Free Production of SGA—And Primary Aluminium?

At the entrance to 2019, modern stationary calciners accounts for about 33% of the specific thermal energy consumption for production of SGA, or about 50% of the specific thermal energy consumption of the Bayer process.

The development of alternative energy sources has been moving forward with the growing electrical energy supply from renewable energy sources like solar and wind, as well as hydro power coupled with the more and more cost-efficient production of hydrogen by electrolysis.

In the future it may be possible to replace HFO and natural gas with hydrogen as fuel for the calciners, as well as for the boilers producing steam for the Bayer process. In the foreseeable future, there is no cost competitive alternative to the Bayer Process invented by Carl Josef Bayer while in Russia 1882–1892.

There may, however, become an alternative process to the current carbon electrode smelting process as we know it, invented simultaneously in 1886 by Charles Hall, in USA, and Paul Héroult, in France. On 10 May 2018, Rio Tinto and Alcoa announced [63] a joint R&D project with the ambition to develop the world's first carbon-free aluminium smelting process. Alcoa and RioTinto have formed Elysis, a joint venture company to advance large-scale development and commercialization of the new process with a technology package planned for sale in 2024.

If cost-competitive large-scale production of green hydrogen by electrolysis and carbon-free aluminium production become a reality, the alumina industry and new primary aluminium smelters may be close to CO_2 free by around 2025–2030.

10.7 List of Symbols

a = Air intake rate (Nm^3/ Kg Al_2O_3); d = Dust recycle rate (kg/kg Al_2O_3);

f = Fuel intake rate (Nm^3 Fuel Gas/ kg Al_2O_3) or (kg HFO/ kg Al_2O_3); g = Exhaust gas flow from cyclone PO_1 (Nm^3/ kg Al_2O_3);

h_D = Dry hydrate feed fate (kg/kg Al_2O_3);

K = f ($\Delta h_{Fuel} + \Delta H_{NHV}$), is the specific thermal energy consumption in kJ/kg or MJ/Ton SGA; q_{Forced} Convection = Heat loss by forced convection (kJ/kg Al_2O_3);

q_{Loss} = Sum of heat losses by radiation, forced and natural convection (kJ/kg Al_2O_3); $q_{Natural}$ Convection = Heat loss by natural convection (kJ/kg Al_2O_3);

$q_{Radiation}$ = Heat loss by radiation (kJ/kg Al_2O_3); w = moisture in hydrate (kg/kg Al_2O_3);

Δh_{AiR} = Specific Enthalpy (sensible heat) of ambient air plus air from Fluid Bed Cooler (kJ/Nm^3); ΔH_{Calcin} = Enthalpy of Calcination kJ/kg Al_2O_3 at 25 °C and 1 bar absolute pressure;

$\Delta H_{Cond.}$ = Heat of water vapor condensation/evaporation at 25 °C and 1 bar absolute pressure; Δh_{Dust} = Specific enthalpy (sensible heat) of dust recycled to cyclone CO_2 (kJ/kg Al_2O_3); Δh_{Fuel} = Specific enthalpy (sensible heat) of moist fuel in (kJ/Nm^3 Fuel Gas) or (kJ/kg HFO); Δh_{Gas} = Specific enthalpy (sensible heat) of moist exhaust gas (kJ/Nm^3 Gas);

Δh_{Hyd} = Specific enthalpy (sensible heat) of dry hydrate (kJ/kg Al_2O_3);

ΔH_{NHV} = Net Heating Value of fuel in (kJ/Nm^3 Fuel Gas) or (kJ/kg HFO) at 25 °C and 1 bar absolute pressure;

Δh_{SGA} = Specific enthalpy (sensible heat) of SGA discharged from cyclone CO_4 underflow (kJ/kg Al_2O_3);

Δh_{wl} = Specific enthalpy (sensible heat) of water in dry hydrate (kJ/kg H_2O);

$\Delta h_x = \int C_{px}(T)dT$ from $T_o = 298°$ Kelvin to T° Kelvin, is the specific single-phase enthalpy of a single component i.e. Water (Δh_{wl}) or gas mixture, i.e. like air (Δh_{Air}) or gas products from combustion, evaporation and calcination, (Δh_{Gas}).

Acknowledgements The author is thankful to all those former colleagues at FLSmidth, who over more than 40 years have contributed with their dedicated and hard work has made the development, commercialization, and commissioning of the Gas Suspension Calcination technology for alumina possible.

References

1. O. Tschamper,Improvements by the new Alusuisse process for producing coarse aluminium hydrate in the bayer process, in *Light Metals*, pp. 103–115 (1981)
2. W.C. Sleppy, et al., Non-metallurgical use of alumina and bauxite, in *Light Metals*, pp. 117–124 (1991)
3. B. Welch, Chapter 9.12b "aluminum", in SME mineral processing and extractive metallurgy handbook (2019)

4. T. Ashida, et al., New approaches to phase analysis of smelter grade alumina, in *Light Metals*, pp. 93–96 (2004)
5. S.J. Lindsay, Customer impacts of Na_2O and CaO in smelter grade alumina, in *Light Metals*, pp. 163–167 (2012)
6. M. Ishihare,et al., Conversion of conventional rotary kiln into effective sandy alumina calciner, in *Light Metals*, pp. 143–152 (1983)
7. G. Labriot, Modification of rotary kiln for calcination of alumina and metal hydrates, in *Light Metals*, pp. 153–158 (1983)
8. B.E. Raahauge, et al., Experience with gas suspension calciner for alumina, in *Light Metals* (1991)
9. W.M. Fish, Alumina calcination in the fluid-flash calciner, in *Light Metals*, pp. 673–682 (1974)
10. L. Reh, H. W. Schmidt, Application of circulating fluid bed calciners in large scale aluminaplants, in *Light Metals*, pp. 519–532 (1973)
11. B.E. Raahauge, FLS stationary calciner for alumina—a new simple approach in alumina calcining, in *ICSOBA 1979*, Calgari, Sardegna, Italy (1979)
12. T.A. Venugopalan, Experience with gas suspension calciner alumina, in *Alumina Quality Workshop*, pp. 53–66 (1988)
13. L. Reh, New and efficient high-temperature processes with circulating fluidized bed reactors. Chem. Eng. Technol. **18**, 75–89 (1995)
14. S. Hundebøl, S. Kumar, Retention time of particles in calciners of the cement industry. Zement-Kalk-Gips, Nr. **8**, 422–425 (1987)
15. L.M. Perander, et al., Two perspectives on the evolution and future of alumina, in *Light Metals*, pp. 151–155 (2011)
16. B.E. Raahauge, N. Devarajan, Experience with particle breakdown in gas suspension calciners, in *ICSOBA*, Dubai, UAE (2015)
17. Aughinish Alumina—Up-grade project 2001 introducing FLS Cyclone technology
18. C. Klett, Alumina calcination: a mature technology under review fromsupplier perspective, in *Light Metals*, pp. 79–84 (2015)
19. M. Missalla, et al., Significant improvement of energy efficiency in alunorte's calcination facility, in *Light Metals*, pp. 157–162 (2011)
20. B. Petersen, et al., Application of optimized energy efficient calcination—configuration to AOS stade CFB calciners, in *9th International Alumina Quality Workshop*, pp. 371–374 (2012)
21. K. Yamada, et al., Development of fluid calciner with suspension preheaters, in *Light Metals*, pp. 159–172 (1983)
22. A. Pinoncely, K. Tsouria, FCB flash calciner–10 years of experience, in *Light Metals*, pp. 113–120 (1995)
23. R. Wischnewski, et al., Alunorte global energy efficiency, in *Light Metals*, pp. 179–184 (2011)
24. B.E. Raahauge, et al., Energy saving production of alumina with gas suspension calciner, in *Light Metals*, pp. 173 (2011)
25. H.W. Schmidt, et al., Practical experience with operation of Lurgi/VAW—fluid-bed calciners, in *Light Metals* (1976)
26. H.W. Schmidt, et al., Alumina calcination with the advanced circulating fluid bed technology, in *Light Metals*, pp. 129–135 (1996)
27. W.J. Borer, H. Günthard, Lattice energy, lattice constant, and thermodynamic properties of γ–Al_2O_3. Halvetica Chim. Acta **53**(Face.5), 119–120, 1043–1050 (1970)
28. T. Yokokawa, O.J. Kleppa, A calorimetric study of the transformation of some meta stable modifications of alumina to get to α–Al_2O_3. J. Phys. Chem. **68**(11), 3246 (1964)
29. B.E. Raahauge, Thermal energy consumption in gas suspension calciners, in *ICSOBA* (2017)
30. D. Kunii, O. Levenspiel, *Fluidization Engineering*, 2nd edn (Butterworth-Heinemann Series in Chemical Engineering, 1991)
31. L. Perander, et al., Circo Cal™—Pushing energy efficiency to its limit in circulating fluid bed calcination, in *ICSOBA* (2012)
32. L. Perander, et al. Impact of calciner technologies on smelter grade alumina—micro structure and properties, in *8th International Alumina Quality Workshop*, pp. 103–107 (2008)

33. K. Yamada, et al., Dehydration products of gibbsite by rotary kiln and stationary calciner, in *Light Metals*, pp. 157–171 (1984)
34. J. Rouquerol et al., Thermal decomposition of gibbsite under low pressure—I. Formation of boehmitic phase. J. Catal. **36**, 90–110 (1975)
35. D.D. Perlmutter, L. Canela, Pore structures and kinetics of the thermal decomposition of $Al(OH)_3$. AIChE J. **32**(9) (1986)
36. L. Rozic et al., The kinetics of the partial dehydration of gibbsite to activated alumina in a reactor for pneumatic transport. J. Serb. Chem. Soc. **66**(4), 240–273 (2001)
37. S.W. Sucech, C. Misra, Alcoa pressure calcination process for alumina, in *TMS Light Metals*, pp. 119–124 (1986)
38. Das et al., Thermal decomposition of precipitated fine aluminium trihydroxide. Scand. J. Metall. **33**, 211–219 (2006)
39. V.J. Ingram-Jones, et al., Dehydroxylation sequences of gibbsite and boehmite: study of differences between soak and flash calcination and particle size effects. J. Mater. Chem. **6**(1), 73–79 (1996)
40. B. Wittington, D. Illevski, Determination of the gibbsite dehydration reaction pathway at conditions relevant to Bayer refineries. Chem. Eng. J. **98**, 89–97 (2004)
41. H. Wang, et al., Kinetic modelling of gibbsite dehydration/amorphization in the temperature range 823–923 K. J. Phys. Chem. Solids **67**, 2567–2582 (2006)
42. D. Illevski, New two-stage calcination technology, in *2012, 9th International Alumina Quality Workshop*, pp. 364–370 (2012)
43. S. Wind, B.E. Raahauge, Experience with commissioning new generation gas suspension calciner, in *Light Metals*, pp. 155–162 (2013)
44. A. Taylor, Impacts of the refinery process on the quality of smelter grade alumina, in *2005, 7th International Alumina Quality Workshop*, pp. 103–107 (2005)
45. A. Saatci, et al., Attrition behavior of laboratory calcined alumina from various hydrates and its influence on SG alumina quality and calcination design, in *TMS Light Metals*, pp. 81–86 (2004)
46. C. Klett, et al., Improvement of product quality in circulating fluidized bed calcination, in *Light Metals*, pp. 33–38 (2010)
47. L. Perander, et al., Reducing particle breakage in alumina calcination, in *IBAAS-CHALIECO International Symposium 2013 Proceedings*, pp. 65–70 (2013)
48. L. Perander, et al., Coal gasification and impacts on alumina CFB design, in *Proceedings of the 10th International Alumina Quality Workshop* pp. 243–250 (2015)
49. B.E. Raahauge, J. Nickelsen, Industrial prospects and operational experience with 32 mtpd stationary alumina calciner, in *Light Metals*, pp. 81–104 (1980)
50. S.C. Libby, Decrepitation of calcined abrasive grade bauxite, Moengo, Surinam, in *Light Metals* (1980)
51. N. Brown, T. J. Cole, The behavior of sodium oxalate in Bayer alumina plants, in *Light Metals*, pp. 105–118 (1980)
52. B.E. Raahauge, et al., Application of gas suspension calciner in relation to Bayer hydrate properties, in *The Australasian Institute of Mining and Metallurgy, 1981 Annual Conference*, Sydney NSW (1981)
53. B.E. Raahauge, et al., Energy saving production of alumina with gas suspension calciner, in *111th AIME Annual Meeting*, Dallas, US (1982)
54. S. Wind, C. Jensen-Holm, B.E. Raahauge, Development of particle breakdown and alumna strength during calcination, in *Proceedings of the 9th International Alumina Quality Workshop*, 2012.
55. B.E. Raahauge, N. Devarajan, Experience driven design improvements of gas suspension calciners, in *11th Alumina Quality Workshop International Conference* (2018)
56. S. Lindsay, Attrition of alumina in smelter handling and scrubbing systems, in *Light Metals*, pp. 163–168 (2011)
57. W.L. Forsyth, W.R. Hertwig, Attrition characteristics of fluid cracking catalyst. Ind. Eng. Chem. 1200–1206 (1949)

58. J.D. Zwicker, The generation of fines due to heating of aluminium trihydrate, in *Light Metals*, pp. 373–395 (1985)
59. J.V. Sang, Factors affecting the attrition strength of alumina products, in *Light Metals*, pp. 121–127 (1987)
60. J.E. Lopez, I. Quintero, Evaluation of agglomeration stage conditions to control alumina and hydrate particle breakage, in *Light Metals*, pp. 199–202 (1992)
61. P. Clerin, V. Laurent, Alumina particle breakage in attrition test, in *Light Metals*, pp. 41–47 (2001)
62. S. Chandraskar, et al., Alumina fines' journey from cradle to grave, in *Proceedings of the 7th International Alumina Quality Workshop* (2005)
63. Rio Tinto—Alcoa Media Release 10 May 2018

Benny E. Raahauge Deceased—Owner—Director Raahauge-SGA ApS, Denmark

Benny received his M.Sc. Chemical Engineering from the Danish Technical University in November 1972. Start working with programming minicomputers controlling the raw material mixing for production of Cement Clinker, December 1972 in Process Technical Department, FLSmidth, Cement Division, Copenhagen, Denmark.

From 1975–76, Benny worked as process and plant engineer in a Danish Sugar Factory before re-joining FLSmidth as R&D engineer and later R&D Manager in the Mining Division in Copenhagen. Benny was assigned the task to develop the stationary Gas Suspension Technology for Smelter Grade Alumina (SGA) to replace the rotary kilns.

After successful commissioning of the first 1000 tpd Gas Suspension Calciner (GSC) unit for alumina at Hindalco Industries, India, in 1986, Benny worked in the role as General Manager-Pyro & Alumina Technology based in Copenhagen. This assignment included the responsibility for design, marketing, sales and commissioning of GSC units for Alumina, culminating with commissioning of 3 x 4500 tpd GSC units at Queensland Alumina 2004–05, the world's largest stationary calciners.

During +44 years employment with FLSmidth & Co. contributed with many papers on calcination to international ICSOBA, Alumina Quality Workshop and TMS meetings until retiring in 2018.

Together with Don Donaldson, Benny was Lead-Editor to "Essential Readings in Light Metals—Volume 1—Alumina & Bauxite" with Fred Williams as Co-Editor, Published by Wieley & Sons. Inc. Copyright © 2013 by The Minerals, Metals & Materials Society (TMS).

In 2018 Benny was editor and contributor to Chapter 12.1 Alumina, in the "SME Mineral Processing & Extractive Metallurgy Handbook", Published by Society for Mining, Metallurgy & Exploration (SME), Copyright © 2019.

Chapter 11
Alumina Quality, HF Removal, Dissolution and Aluminum Purity

Stephen J. Lindsay and Pascal Lavoie

Abstract This chapter is intended to serve as a useful reference for those at alumina refineries and aluminum smelters who have interest in technical matters involving usage of Smelter Grade Alumina, or SGA. Accordingly, information on many physical and chemical properties has been categorized into various sections of this chapter that make it straightforward for readers to reference. Clearly the primary purpose of SGA is to serve as feedstock for the production of aluminum metal. However there are a number of additional technical requirements that may vary in degrees of importance for each end user. This chapter will discuss these specific requirements in detail. Much may be expected of this granular, synthetic material that is never fully communicated back to the process engineers on the "red" sides and "white" sides of alumina refineries. In the past it was often possible for suppliers of alumina to meet with their smelting counter-parts at co-located facilities. Factors of importance to the end users were then a part of everyday conversations about process control. With some exceptions those days are now long-gone for most of the western world. Smelters and refineries are now quite often separated by oceans. Tight connections in communications between refineries and smelters have for a large part been lost. Accordingly, in a world in which persons in smelting locations may not know, and may have never visited, a single person at an alumina refinery and vice-versa it becomes useful to know what is important and what is needed by smelting customers, not only what might be "nice to have" in the way of alumina properties. Providing this information to the readers is the primary purpose of this chapter. *Let's begin.*

S. J. Lindsay (✉)
Smelting Specialist Consultant Centre of Excellence, Hatch, Knoxville, USA
e-mail: Stephen.Lindsay@hatch.com

P. Lavoie
Smelting Manufacturing Excellence, Alcoa Aluminum Center of Excellence, Quebec, Canada

B. E. Raahauge and F. S. Williams (eds.), *Smelter Grade Alumina from Bauxite*, Springer Series in Materials Science 320,
https://doi.org/10.1007/978-3-030-88586-1_11

11.1 Introduction

In the strictest sense SGA is not simply Al_2O_3. According to Ashida, Metson, and Hyland [1] $Al_{10}O_{16}H_2$ may be the most accurate approximation for typical SGA. Laymen might regard this as the stoichiometric equivalent of having one residual unit of water spread across five units of aluminum oxide, or as $[Al_2O_3]_5 \cdot H_2O$. SGA also contains impurities that are mostly benign in the reduction process, albeit undesirable to some extent. This includes residual hydroxyl content that remains after $Al(OH)_3$, commonly referred to as alumina tri-hydrate, or $Al_2O_3 \cdot 3H_2O$, has been converted into SGA during calcination. Residual hydroxyls generate hydrogen fluoride, HF, in reduction cells. They also play an important role in dry scrubbing and the recovery of HF from reduction cell off-gas. Residual hydroxyl content is also often credited with enhancing the dissolution rate of alumina in reduction cells.

Fully calcined, alpha-phase alumina has no residual hydroxyl content. It is just as soluble as SGA in the bath of fluoride salts that are used in industrial cells. However, its dissolution rate is much lower than that of SGA. Thus alpha-alumina is more prone to settle to the bottom of industrial cells as a sludge that is consumed slowly and somewhat erratically rather than to go rapidly into solution.

In the practical sense SGA is close enough to Al_2O_3 in composition and properties that it may be considered as such for the design of handling equipment and scheduling of raw material shipments. The latter often relies upon an Alumina Consumption Factor of approximately 1.92 kg Al_2O_3 per kg aluminum produced rather than the theoretical consumption factor of 1.89 kg Al_2O_3/kg Al. The 1.6 wt% difference of these two values is primarily due to the presence of impurities, mostly that of attached water, residual hydroxyls and Na_2O content.

The chemical and physical properties of SGA such as Fe_2O_3 content and particle size distribution are discussed in greater detail later in this chapter. Such parameters are used to define alumina sources of acceptable quality for various aluminum smelters including consideration of metal products.

Definition of product quality is ultimately built around performance requirements. In modern smelters SGA feedstocks are generally expected to;

- Have an acceptable dissolution rate.
- Provide for highly efficient removal of HF from process exhaust gas while limiting HF generation.
- Provide for efficient removal of polycyclic aromatic hydrocarbons, PAHs, from the exhaust gas of anode baking furnaces.
- Contribute manageable levels of impurities into the process or the metal products.
- Avoid excessive production of Hall-Héroult bath through manageable content of Na_2O.
- Avoid any requirements to regularly dilute CaF_2 or contaminants of concern from the electrolyte.
- Pass through handling and pollution control equipment without flowability issues or excessive particle breakage.
- Dispense uniformly and predictably from point feeders.

- Have alpha content that enables acceptable dissolution, proper anode covering and crust formation.
- Have a total cost of use that is acceptable and competitive.

Over the many years since the Bayer process was invented by Carl Josef Bayer, there have been both evolutionary and revolutionary changes in SGA requirements for aluminum smelters.

During the 1970's a proliferation of new regulations required capture and control of fluoride emissions using dry scrubbing technology. The timing of this was coincident with the advent of point-feeding technology that improved the production efficiency of aluminum reduction cells. Both changes then began to shift the industry away from what was known as floury alumina to coarser, sandy alumina.

This new product had a higher specific surface area that facilitated dry scrubbing. It also had a lower alpha-alumina content to facilitate the more rapid dissolution that was required as the use of point feeders displaced side-worked and center bar-breaker feeding technologies. The desire for a coarser granulometry was driven by a need for better flowing, more accurate, predictable dispensing of fluorinated alumina that is higher in fines and superfines content than pure, sandy alumina. This rather rapid shift coupled with greater emphasis on lower energy consumption at refineries has since contributed to an evolutionary change in calcination technology away from rotary calcination units to flash or gas suspension calcination and fluid bed calcination technologies.

Another evolutionary change has been the migration away from co-located alumina refineries and aluminum smelters to large production facilities for each of these products. The largest and most modern alumina refineries in the world today produce more SGA than any individual smelting location might consume. The smelting customers are often located on other continents and producers cannot customize their SGA properties to best suit the needs of each of its multiple clients.

However, this trend has not completely re-shaped the industry. Many recent greenfield developments in China and other nations have also included investment in refineries that predominantly or exclusively serve one captive customer. Alumina refineries with multiple customers tend to focus on a singular product that is designed around the optimization of alumina production and the best utilization of bauxite reserves. Refineries that have a single customer often tune operations around the optimum position for the production of both SGA and aluminum metal.

Evolutionary changes will continue in this industry. These will be primarily focused on obtaining the best value from bauxite reserves, maximizing production, reducing energy consumption and reducing overall costs. Product quality demands, such as for lower Na_2O content, will continue to drive innovation in calcination, precipitation, clarification and alumina handling technologies. Higher quality demands of existing refineries may find some beneficial synergies. However, lower impurities, changes in particle sizing, and factors that affect dry scrubbing such as pore sizing may add expenses that run counter to the optimization strategies of many alumina refineries today. As has been the case in many other industries, necessity will continue to be the driver of lasting change.

11.2 Structure and Composition of Smelter Grade Alumina

11.2.1 Loss on Ignition and Degree of Calcination

Loss on Ignition, or LOI, is one of a few terms used to measure the degree of calcination of SGA. The internationally recognized method for its determination is by ISO 806:2004 [2]. Other properties that are linked to LOI content are; alpha phase content and BET Surface Area. However, the latter should not be used for making a direct inference on the degree of calcination.

After aluminum tri-hydrate, $Al_2O_3 \cdot 3\ H_2O$, has been precipitated by the Bayer process it is approximately 34% by weight as water. As Metson [3] has pointed out it is important to remember that $Al_2O_3 \cdot 3\ H_2O$ does not actually exist. The molecular structure is in the form of aluminum hydroxide $Al(OH)_3$. This frame of reference is truly important as the residual hydroxyl, –OH, units that remain after calcination, coupled with the high surface areas of particles are critical in generating the reactive sites that are desired for capture of HF. Residual hydroxyls also appear to aide more rapid dissolution of alumina in the molten salt electrolyte used in the reduction cells.

The calcination process removes most of these hydroxyls. The primary driving force behind this is the reduction of weight of the material to be shipped. This is a clear economic factor of concern. But, even in cases where refineries and smelters are co-located the aluminum tri-hydrate, or "hydrate" as it is often commonly referred to, is calcined to drive off the same amount of hydroxyl content. Why?

Aluminum tri-hydrate will dissolve in the electrolyte and it can then be reduced to metal. However, the energy required during the calcination step to heat and convert it into Al_2O_3 would remain unchanged. Removal of hydroxyls in reduction cells would require the use of electrical energy, whereas natural gas or oil may be used as the energy source in calcination. In the earliest days of the Bayer process producer gas from coal was used for calcination of alumina [4]. In early days, as it is now, the availability of electrical energy for aluminum smelters often has its limits. It is desirable to use as much of the electrical energy available to a smelter as possible to produce aluminum metal rather than to use it to convert aluminum hydroxide to aluminum oxide.

There is another important factor to consider. If aluminum tri-hydrate were to be fed to a reduction cell the generation of HF gas would increase by orders of magnitude. Typically, SGA has approximately 0.85 wt% LOI, representing its residual hydroxyl content. With approximately 34 wt% LOI aluminum tri-hydrate would increase the generation of HF by roughly forty times the normal contribution of SGA per ton of metal produced. This would neither be practical nor cost effective. It would also strip away hydrogen fluoride, HF, from the electrolyte quite rapidly changing its properties including liquidus temperature. Thus, modern aluminum smelters prefer SGA that has been calcined to the point at which a balance has been achieved between the contribution of LOI to the generation of HF gas and the BET Surface Area that is available to scrub and recover HF from process off-gas.

During the early days of the Hall-Héroult process, long before the development of point feeders, it was preferable to have a low LOI content, typically below 0.5 wt%. The reason for this was also related to fluoride emissions. Prior to the existence of dry scrubbers there was no need for SGA with high surface area and thus no need for residual hydroxyl content to capture HF. The lower the LOI content the lower the HF generation rate. During that era the general view was that LOI represented "paying alumina prices for water". With manual breaking and feeding of pots being predominant only small fractions of rather large batches of alumina would go into solution prior to arriving to the solubility limit of the electrolyte. The major fraction of alumina fed to the pots went to the bottom in the alpha phase. A common specification limit for floury alumina during the 1950's was to have "less than 50% alpha phase content" [5].

It is still desirable to have the lowest LOI content possible, but only so long as the requirement for an adequate amount of surface area has been met to support the needs of the dry scrubbers. Thus, some LOI content is a necessity. However, since high LOI directly translates into high HF generation smelter operators prefer to have high BET Surface Area, of more than 70 m^2/gm, in combination with low LOI content.

This is the facet of greatest importance of LOI in SGA to smelting operations. The higher the LOI content the higher the generation of HF in the reduction cells will be. With higher fluoride evolution from the process more BET Surface Area may be needed to scrub the additional HF. However, these two calcination parameters are positively correlated. There are more active sites for capture of HF on alumina with a higher LOI content.

There are other factors of importance in LOI content. Older calcination units capture superfines in the exhaust from the calcination circuit most often with electro-static-precipitation, ESP, units. The ESP dust that is returned to the process may have 12–15 wt% LOI content due to capture of superfine aluminum tri-hydrate particles that have not been fully calcined. Even though a small fraction of SGA may be comprised of ESP dust it can have a measurable impact on LOI content, at the second decimal place.

One other factor that comes into play is an energy saving method called hydrate by-pass that is employed on most calcination circuits. The general idea behind hydrate by-pass is to add a fraction of aluminum tri-hydrate to calcined alumina as it exits the kiln. This material is quite hot, requiring cooling prior to subsequent handling and storage. Direct additions of hydrate to the hot, kiln discharge material helps to cool it. It also provides a means to partially calcine aluminum tri-hydrate into SGA while using less energy, positively affecting the cost of calcination.

The factors of importance with hydrate by-pass are;

(1) The holding or residence time of the hydrate with the kiln discharge material,
(2) The point at which hydrate enters the process,
(3) The fraction of product that is converted in this way.

The greater the retention time at high temperature the lower the LOI content of the by-pass fraction and the more developed the pore structure for dry scrubbing of HF will be. However, in some older calcination circuits with short retention times increases in hydrate by-pass have been known to drive LOI content above 1.00 wt%. The motivation at the refinery may be; to reduce costs by saving energy, to de-bottleneck the throughput of in the calcination area of the refinery, or to accomplish both. The product quality impacts of hydrate by-pass are further discussed in the section of this chapter that deals with dry scrubbing of HF.

Note that SGA customer vigilance is often lax regarding maximum acceptable LOI content. Greater focus is placed on the other metrics that are related to the degree of calcination, BET Surface Area, alpha-alumina content and perhaps even pore size distribution. When a customer may ask about having lower LOI content the conversation can quickly come to an end when it is suggested that this may also mean lower surface area values. The conversation seldom carries over to ESP dust or the hydrate by-pass material that were included in the production of the product.

It is also worth noting that the ISO definition of LOI between 300 °C and 1000 °C has not always been universally accepted. At one time, various SGA producers reported LOI measured between 300 °C on the low end of the scale and 1000 °C, 1100 °C or 1200 °C on the high end of the scale. While the upper limit of 1000 °C has more or less been universally accepted there is debate about the appropriate point for the lower end of the measurement range.

SGA is usually low in residual Gibbsite content. But the hydroxyl content of $Al(OH)_3$ is quite high. Thus, even small amounts contribute significantly to HF generation from reduction cells. To capture the contribution of Gibbsite to HF generation, it has been proposed that LOI should be measured between 250 °C and 1000 °C [6].

The current standard for LOI measures the loss of hydroxyl units only during heating of transition phases of alumina. From the perspective of those who consume SGA the propensity of the product to generate HF in reduction cells is the parameter of greatest interest. From this perspective LOI from 250 °C to 1000 °C would be a more meaningful metric for SGA customers and it is already being used internally as a metric for at least one major producer of SGA.

11.2.2 Phase Composition and Alpha Alumina Content

SGA that has been calcined to the range of 0.8% ± 0.2 wt% LOI is also commonly known as transition phase alumina. This embraces a range of crystalline and sub-crystalline phases or forms of alumina in the range of transition between synthetic Gibbsite and aluminum oxide in the alpha-phase. Refer to Fig. 11.1. Once heating of aluminum hydroxide commences a series of crystalline and amorphous transition phases form and progressively trend towards a crystalline structure dominated by alpha-phase alumina at high calcination temperatures. To better understand the phase composition and structure of SGA particles we shall follow the thermal progression from aluminum tri-hydrate on through to alpha-phase.

Fig. 11.1 Images of crystalline structure of synthetic aluminum tri-hydrate $Al(OH)_3$, left and crystalline platelets of alpha phase alumina in covering material crust, right [7, 8]

In nature, bauxite has an available-alumina content, the fraction that is readily available for extraction by digestion. Bauxite normally contains Gibbsite, or aluminum tri-hydrate, as the predominant aluminum containing species. It may also contain a significant fraction of Boehmite, or aluminum mono-hydrate, and diaspore, a dense, difficult to digest aluminum oxyhydroxide, α-AlO(OH).

Classical forms of mineral processing would try to recover all three of these molecular forms by removing the impurities and then forming $Al_2O_3 \cdot 3H_2O$, prior to calcination into Al_2O_3. However, this straightforward approach is not practical when it comes to bauxite processing. The pragmatic approach of the Bayer process for making usable product recovers all of the elemental aluminum that is present in soluble phases and then discards the aluminosilicates, desilication product and undigested fractions of monohydrate and diaspore.

The digestion step of the Bayer process uses sodium hydroxide, caustic liquor, to dissolve most of these components into sodium aluminate. Following the clarification step sodium aluminate is seeded to induce precipitation of aluminum tri-hydrate. So, regardless of the digestible form that was found in nature all refinery output enters the calcination step of the Bayer process as synthetic aluminum tri-hydrate, or more accurately $Al(OH)_3$.

The example shown in Fig. 11.2 is of thermo-gravimetric analysis, TGA, for a sample of SGA. Using TGA the sample is heated from room temperature to 1000 °C. The loss in weight vs. temperature and the rate of change per minute are typically both plotted on the same graph to show the overall change as well as to show information on the rate of change in weight.

In Fig. 11.3 this same TGA example has been illustrated to show the temperature ranges where the various alumina phases exist.

MOI, or Moisture on Ignition, appears on the left of the TGA plot. While it is not related to the structure of SGA it does illustrate how much physically combined moisture may be found on alumina that has been exposed to H_2O in air. MOI is a measure of weight loss upon heating from room temperature to 110 °C. This simply

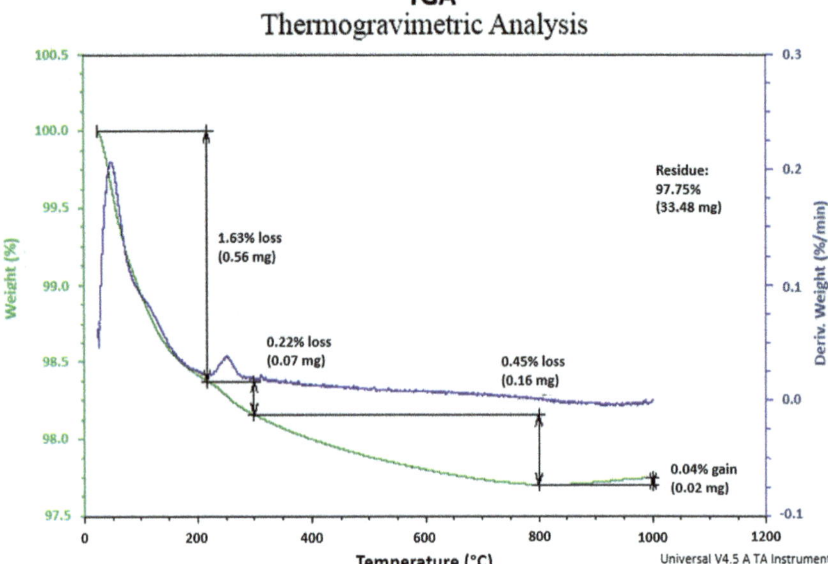

Fig. 11.2 Example of thermogravimetric analysis of smelter grade alumina

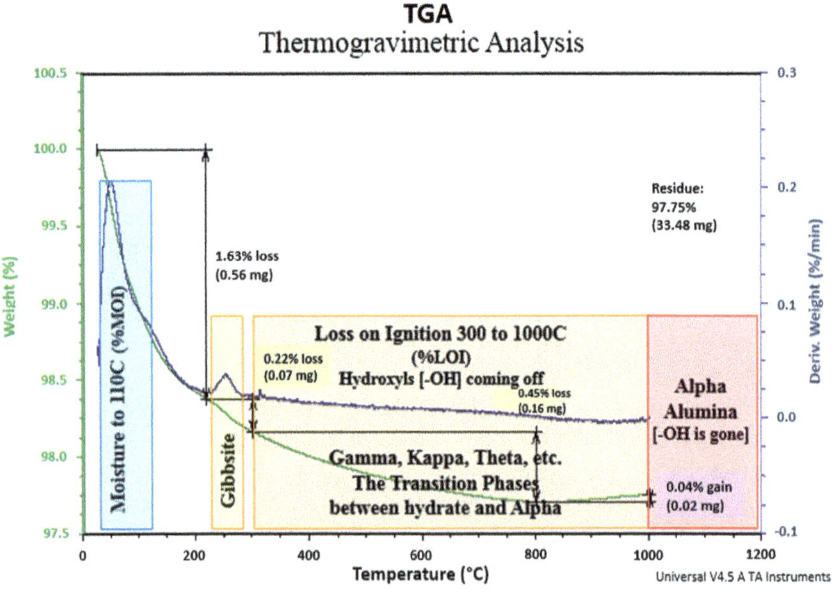

Fig. 11.3 Illustration of various phases of alumina structure versus temperature

defined temperature range captures the greatest rate of change, over 0.1% weight per minute, in the example that is shown.

Note that MOI content is not related to alumina refining. It is also not related to the structure of SGA that is discussed in this section. The ability of SGA to take on and then give up moisture depending upon the humidity of the surrounding air has often been behind the observation that: "The MOI of SGA is not much more than a reflection of the level of humidity in the air of the laboratory when the sample was analyzed." There is much truth to this statement. Accurate sampling and measurements of MOI in product can be quite difficult. Read more about MOI later in this chapter in the section on Efficient HF Removal and Dry-Scrubber Efficiency.

The next shift in the rate of change in weight happens at approximately 250 °C. This is the temperature at which alumina tri-hydrate, or Gibbsite, is converted to alumina monohydrate, or Boehmite, and soon after to the primary transition phase of alumina, called gamma-alumina. This TGA plot shows a small, residual amount of $Al(OH)_3$ being present it the product. This carry-over of Gibbsite usually happens with fine dust that is collected by electro-static precipitator designs that are generally found on older calcination units. However, the point of this graph is to illustrate that the temperature range between 300°–1000 °C is where the various transition phases exist. Above 1000 °C the structure is fully crystalline in the alpha-phase, the most thermodynamically stable form of Al_2O_3.

Keep in mind that the $Al(OH)_3$, or $Al_2O_3 \cdot 3H_2O$, is approximately 34.6% by weight as water. The calcination step and the kiln are designed specifically to remove most of this excess mass. The final 1 wt% of residual hydroxyl content is driven off the molecular structure of transition phase alumina in the range of 300°–1000 °C. Note that with the example of sandy alumina shown in Fig. 11.3 the LOI is low, at just 0.45 wt%.

Many structural phases including gamma and kappa are found in the continuum of transition aluminas [3]. These include other reported forms including theta, delta, epsilon and psi. The subtle distinctions between each structural form are not of great significance to the end user. The main concept to grasp is that as alumina is heated to higher temperatures that residual hydroxyls will continue to be driven off, altering the molecular structure. The irregular order of the structure morphs until it ultimately reaches a well ordered, crystalline structure in the alpha phase. At this point all residual hydroxyls are gone.

The transition to alpha phase is also progressive. If a Certificate of Analysis for SGA were to state that a shipment of alumina has 6% alpha content this does not mean that 94% of the alumina particles are of the various transition phases and that 6% of the particles are in the alpha-phase. It means that the typical alumina particle has approximately 6% by weight of its structure in the alpha phase. Refer to Fig. 11.4.

SGA with high alpha phase content has been observed in both laboratory studies and in reduction cells to have a lower rate of dissolution than SGA with low alpha phase content [10]. In the field it has the reputation of promoting the formation of sludge, or undissolved alumina that settles underneath the liquid metal layer. The common understanding of this is that as transition phase alumina goes into solution it rapidly sheds its residual hydroxyl content forming H_2O and HF as a result of the

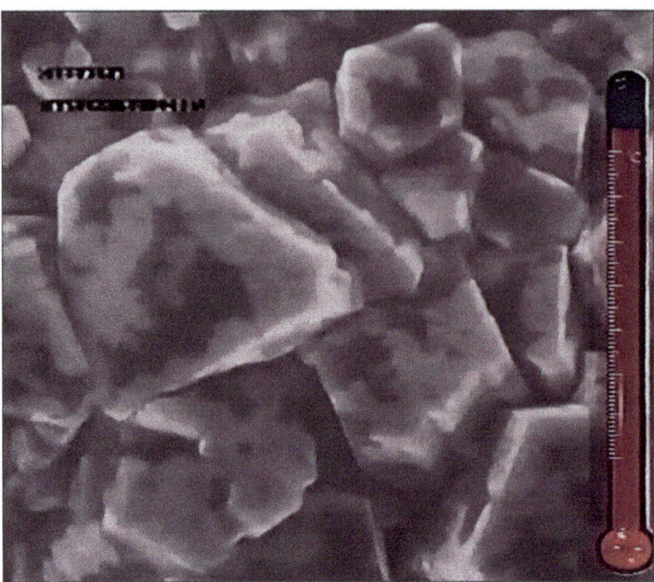

Fig. 11.4 $Al(OH)_3$ particle edges converting to alpha phase (in white) during heating [9]

thermal gradient to which it is subjected. This causes some localized gas formation and agitation that promotes a more rapid dissolution rate. It has also been observed that alumina with both high and low alpha-phase content rapidly goes into solution in laboratory experiments when mechanical stirring is provided. Agitation of alumina particles in the bath promotes more rapid dissolution. However, too much agitation caused by things other than hydroxyls on SGA will negatively affect results in a reduction cell.

At some point between 6 and 10 wt% alpha alumina content it is often observable that sludge formation rates increase in point fed reduction cells. It may have to do with how alpha phase is formed during calcination. It is purely speculative, but at levels above 10 wt% most alumina particles may have something of an exterior "armor" of alpha phase structure surrounding an inner core of transition phase alumina. With more sludge, cells are more prone to instability, increased specific energy consumption, and loss of metal production.

A blend of recycled anode crust material and pure alumina is commonly used as covering material to prevent new carbon anodes from air-burning. High levels of alpha alumina in SGA can detract from the formation of a hard, uniform crust of anode covering material in reduction cells [11]. When the crust over the anodes is weak it is more likely to fracture and fall into the melt during normal pot tending activities. This uncontrolled delivery of solid material into the liquid bath also increases the likelihood of bottom sludge formation and losses in energy efficiency.

Fig. 11.5 Example of platelets of alpha-alumina in side-channel anode covering crust [8]

In the presence of temperatures above 600 °C with fluoride as a catalyst, transition phase alumina converts to interlocking platelets of alpha alumina. This binds the components of the crust together [12] in an effective manner as is illustrated in Fig. 11.5 [8]. A large fraction of the total alumina in anode covering material is converted in this way into alpha-phase alumina [13, 14].

Pure SGA comprises only a fraction of the anode covering material. It typically accounts for 15–25 wt% of the mix. The fraction of SGA that is already in the alpha phase is not available to be converted into interlocking platelets. When the alpha content of SGA is at 10–20 wt% the implication is that the fraction of SGA in transition phases must then be spread very uniformly across the anode covering material in order to avoid zones of weakly bound anode crust.

SGA approaching 20 wt% alpha content can still be found at the remaining refineries that use only rotary kilns for calcination. While this SGA has the same solubility in bath as other alumina sources it does produce noticeably more bottom sludge due to a lower rate of dissolution. A few primary aluminum locations have periodically switched back-and-forth from SGA with high alpha content to SGA with low alpha content. In each case these locations have independently asserted that SGA with a high alpha content, above 10%, leads to losses in current efficiency and production of between 0.25 and 1.0% (most typically 0.5%) versus SGA that has an alpha alumina content of 6 wt% or less. Smelters that have regular experience in adapting during transitions to higher or lower alpha content also report success with being able to regularly use SGA with higher alpha content.

With regard to alpha content there are other items to be considered. Some alumina refineries produce chemical grades of alumina as well as Smelter Grade Alumina. Chemical grade alumina, CGA, products include forms of aluminum hydrate that smelters need not be concerned with unless a rare mistake is made in handling, loading or transport of material. Quite a few chemical grade products are hard burned, very high in alpha phase Al_2O_3. The ESP dust that is collected during calcination of CGA often cannot be added back into the product itself. It is therefore not unusual for the dust from hard burned products to be included with ESP additions made to SGA. Since this is just a small fraction of ESP dust additions there is generally little cause for concern at the smelters so long as it is blended into product. However, smelting clients may wish to know if the refineries that supply them with SGA also produce CGA.

Small amounts of transition material may also be added to SGA. Not to be confused with the transition phases of SGA, transition material is essentially off-grade alumina product. It has been produced while the calcination kiln is in transition from production of SGA to hard burned CGA grades or vice-versa. Transition material may be consumed by adding it to less demanding chemical products or by controlled additions to SGA.

Just as a small fraction of the metal that most smelters produce does not meet quality specifications the same is true for production at alumina refineries. In a smelter the metal from a pot that is very high in iron will be blended in with the metal from other pots in order to keep cast house products within product specifications. Small amounts of transition material and off-grade SGA are metered into SGA product streams by alumina refineries while keeping the product within specifications.

This increases the chance that a small fraction of particles in SGA will be nearly 100% in the alpha-phase. So long as the average alpha-phase content remains below 6% this information may be of interest, but not of importance for aluminum smelters when transition material and hard-burned ESP dust are metered in at controlled rates.

11.2.3 Gibbsite and Chemical Composition

The chemical composition of SGA is typically defined by its impurities. These shall be discussed in greater detail in other parts of this chapter. This section addresses some generalities of how impurities are included in the formation of synthetic Gibbsite that then are carried over into calcined alumina products. The answers have mostly to do with what nature has dealt in the composition of various deposits of bauxite rock. This also has to do with the design of the alumina refinery itself particularly in the areas of clarification, filtration, and precipitation. Then, there is the matter of residual, synthetic Gibbsite that may be present in the final product.

The digestion of naturally formed Gibbsite and the formation of man-made, synthetic, Gibbsite have been the focus of much of this book. It is worth recalling that in layman's terms synthetic Gibbsite may be thought of as approximately 34 wt% water (hydroxyls), 65 wt% alumina and approximately 1wt% impurities.

Any small amount of synthetic Gibbsite that is carried over into SGA will have an impact upon LOI content and also upon the rate of generation of HF in the reduction cells. As illustrated in the section on phase composition we can see that during heating that Gibbsite is first converted to synthetic Boehmite and then to the transition phases of alumina at temperatures above 270 °C. This is generally well below temperatures that are achieved by material that moves via hydrate by-pass around the calcination kilns.

The most common carry-over of Gibbsite into SGA comes with the collection of fines and superfines by electro-static precipitators. Some, or all, of the collected dust may be returned directly into cooler discharge product. This type of technology is more commonly found in older designs of flash or gas suspension calcination units. Older, rotary calcination units have more to be concerned about with alpha phase alumina content than with residual Gibbsite content. The newest designs of calcination units rely primarily upon cyclones to catch this very fine material and then return it to the kiln, rather than to add it directly to cooler discharge product that precedes product storage silos.

The fraction of Gibbsite that may be found in SGA can range from non-detectable amounts to as much as 1.8 wt%. However, less than ten percent of alumina refineries in the western world have more than 0.8 wt% Gibbsite in product. Most have less than 0.5% by weight.

This will have a very small impact upon the value of the SGA itself, although one might argue that alumina prices are being paid for "water". The greatest area of concern for client smelters may center on HF generation rates particularly if conditions are such that the total fluoride content on alumina that is fed to the reduction cells exceeds 1.7% to 2.0 wt% [15]. Above this point highly efficient dry scrubbing begins to noticeably diminish [16] on all but fluid-bed type scrubbers. Thus, additional hydroxyl content in alumina may serve as the tipping point for environmental problems in smelters. This is especially true in the case of some older technology aluminum smelters that have pushed the thermal balance factors of cell operations to the limit in order to boost production or to remain sustainable and competitive.

This might lead the reader to believe that all residual Gibbsite content is undesirable. This is a commonly accepted position. But it must be noted that some of the best results in modern pre-bake cells have come with alumina that has had over 1.0 wt% of Gibbsite content. The metrics of success include world-class current efficiency and anode effect rates. In the case of side-worked cells world-class results in current efficiency were associated with residual Gibbsite content as high as 1.8 wt% and problematic performance has been observed when it was less than 0.5 wt%. How could this be?

Recalling that gas evolution from alumina particles during dissolution is often thought to aid dissolution rates by localized agitation we have one possible answer. While it is far from being conclusive, there is some likelihood that residual Gibbsite content can make up for other deficiencies in overall reduction cell and anode design. This is not to say that residual Gibbsite is essential for peak performance of reduction cells. Innovations with point feeders, alumina feeding logic and slots in anodes have gone a long way towards improving overall alumina dissolution in cells and delivery of world class performance even when the residual Gibbsite content is at non-detectible levels. It is a protracted way of saying that residual Gibbsite may serve as something of an enabler for reduced formation of bottom sludge.

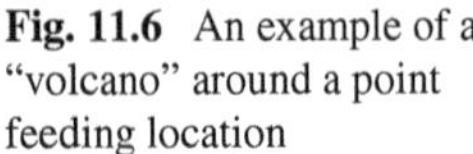
Fig. 11.6 An example of a "volcano" around a point feeding location

There is one other comment on Gibbsite. In some circles an opinion has been formed that residual Gibbsite content causes cones of fine dust to form around point feeder positions. These cones are often referred to as "volcanoes". An example is shown in Fig. 11.6.

Analyses of the dust found in these cones indicate that superfines, minus 20 μm content, is more of a signature property than Gibbsite for point feeding volcano formation. Fine dust can be blown up and out of feeding holes when any point feeder or crust breaker is activated. The pressure in the cavity under the crust then shifts positive relative to atmospheric pressure. This is particularly so on pots with well-sealed crusts. Fine and superfine dust is easily conveyed by gas currents prior to settling out on the crust when the gas velocity suddenly drops after leaving a point feeding hole. Confusion as to Gibbsite being associated with volcano cone formation may be related to superfine ESP dust also often being high in residual Gibbsite content.

Other impurities in SGA are often simply just a reflection of the various trace elements that may be found in various sources of bauxite. The clarification and filtration steps of the Bayer process are not 100% efficient at the removal of impurities, especially those that dissolve to any extent in Bayer liquor. The filtration technology ranging from sand filters on up to latest technologies including Diastar filters can make a large difference in levels of Si, Ti, and trace impurities that are carried through to the precipitation stage. Trace impurities become trapped in the lattice structure as crystals of synthetic Gibbsite nucleate or agglomerate into larger particles.

The overall flow rate of liquor through the process is also commonly found to be correlated to the content of impurities. One might think that low flow rates are beneficial by causing longer residence times in clarification units and the downstream filters that follow. This misses the bigger picture. Refineries that sustain maximum liquor flow rates must do many things right with regard to process control and maintenance. When all is going well the amount of impurities that arrive to each ton of product is often measurably lower than when the refinery is not operating well and liquor flows are at reduced levels.

Thus, prospective customers of SGA are wise to look beyond the specification sheet and the list of typical properties. For many metal product specification limits the range of variation of specific impurities in a major raw material may be much more important than the typical value of any single chemical impurity.

Beyond water and hydroxyl content the major chemical impurities found in SGA are sodium oxide, Na_2O, and calcium oxide, CaO. These are both covered in more detail in the section on impurities of this chapter. Neither is driven by impurities that are found in bauxite.

Sodium oxide content has everything to do with caustic soda from digestion that carries through to the precipitated product. It is either trapped inside the crystalline structure as occluded soda, or as a surface residue that remains on hydrate particles as washable soda, sodium oxide that may be recovered by rinsing with fresh water.

Calcium enters the Bayer circuit in the form of crushed limestone. It is used to causticize the sodium carbonate that forms in water recirculation circuits back into useful sodium hydroxide. It is also used for phosphate control which is very important to most SGA customers [17]. Most of the calcium bearing material settles out and is removed in the clarification process. But a small amount will be carried over into product, especially if the filtration systems are comprised of older technology.

The impacts that each major impurity of concern have on smelting processes and/or metal products may also be found in the section of this chapter that is dedicated specifically to impurities.

11.3 Physical Properties of Smelter Grade Alumina

11.3.1 Particle Size Distribution, Dustiness and Attrition Index

Of all the properties of SGA the one that usually gets the most attention is particle sizing. The focus usually is placed particularly on the content of fines or superfines. Fines are commonly defined as minus 45micron particles. Although the openings of 325 mesh [18] woven wire sieves are slightly larger than the reference standard for wet sieved material. Electroformed screens have openings of 44 μm [19]. Superfines are defined as particles of minus 20 microns in size. These are generally measured using instruments that scatter laser light.

There is nothing that is truly critical about the truncation of the overall particle size distribution, or PSD, at 45 or at 20 μm. At one time, both measures were just the limits at which the smallest particle sizes could be measured with accuracy. In the case of minus 45 μm the original limit of measurement technology was with woven wire screens. The capability of early light scattering devices was at 20 μm particle sizing.

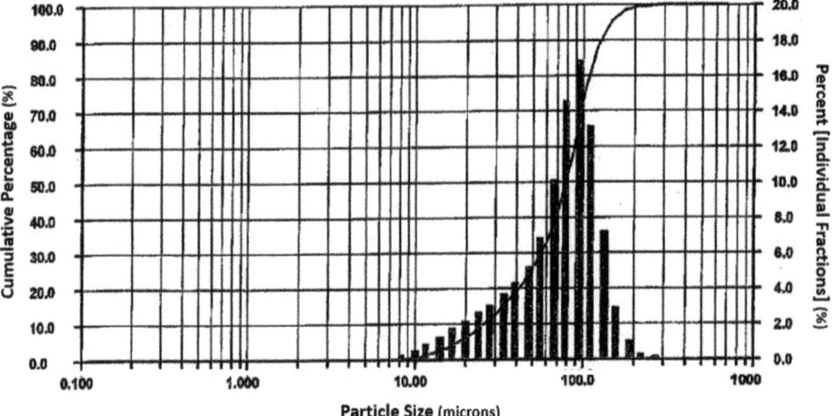

Fig. 11.7 Example of a particle size distribution for one source of SGA

When speaking of the PSD of SGA exactly what size particles are we referring to? A typical particle of SGA is about 100 μm in size, or 1/10th of a millimeter. A superfine particle of 10 μm in diameter is 1/100th of a millimeter. Large particles of SGA, over 150 μm, will not pass the openings of a 100-mesh screen. Also please note that screen sizing is a measure of the second largest dimension of a particle. However, this is a minor point when it comes to the accuracy of measurements using screens.

Figure 11.7 shows an example of a particle size distribution for SGA. The smallest particles are less than 10 μm in size and the largest are greater than 250 μm. The x-scale is logarithmic making the PSD appear to be skewed to the left, towards fines. If this same data were to be plotted on a linear scale the 50th percentile of the distribution, d50, would be just below 80 μm. The 10th percentile, or d10, would be approximately 25 μm and d90 would be approximately 140 μm. This is not very far removed from a Gaussian distribution.

The importance of fine particles has mostly to do with how they affect the packing and flowability of alumina [20] that in turn affects how it is handled, metered, and fed into solution in reduction cells. To look at this properly one must consider the entire PSD rather than just a sub-fraction of it. This point of discussion leads to measurement of flowability which is covered in detail in the next section of this chapter.

Particle sizing and the variations in it determine the packing density of a dry powder. Consider the following example while imagining a large cubic container with no significant edge effects. With a well-organized packing of spheres of uniform diameter, the packing density and void space in the cubic container will not change with the particle size. It would not matter if the container were filled with cannon balls or ball bearings the fraction of void space would be the same. Refer to Fig. 11.8. If we were to mix spheres of various diameters in the same volumetric space the

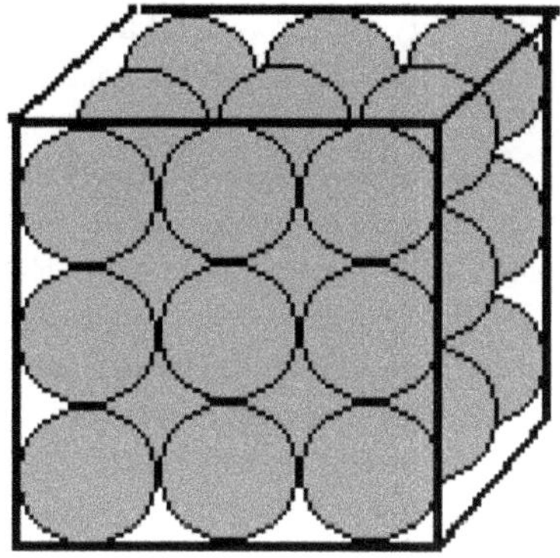

Fig. 11.8 Example of packing density and void space

amount of void space could be greatly reduced, and the packing density would be increased.

Applying this principle, similar results can be observed with commercially available SGA. An alumina source with an overall PSD that is comprised of smaller than average particle diameters [less than 60% of plus 200 mesh fraction, or plus 75 μm] will approach maximum packing density with the addition of a relatively small amount of fine and superfine particles. An alumina source with a PSD that is generally of larger diameter particles [greater than 70% of plus 200 mesh fraction] can accept a larger amount of fine and superfine dust before approaching maximum packing density.

This key point is easily underappreciated at alumina refineries. It extends beyond the range of experience for handling of pure SGA. It truly requires the point of view of a smelting customer. SGA must be able to accept the fine particulates that have been collected by pot line fume control systems in order to satisfy customers that feed dust-laden, fluorinated alumina to pots. Looking only to the fractions of minus 45 μm or minus 20 μm content of pure SGA creates an incomplete assessment of material quality.

Fluorinated alumina is used for more than 95% of the primary aluminum production in the world. Acceptable flowability is needed to have uniform, predictable feed shot sizes. Exhaust gas from the pots carries with it a substantial amount of bath and alumina fines plus superfines. At a minimum 0.8 wt% of the fluorinated alumina delivered to the pots is comprised of this fine material [21]. Approximately half of these particles are alumina fines that were drawn into the fume control system while being delivered to the pot either as feedstock or as a fraction of the anode covering material. The total fraction of fines that end up in fluorinated alumina may be more than a half an order of magnitude higher when attrition impact is also high.

For dry powders including alumina there are metrics that can be used to gauge the proximity of the limit of particle packing before becoming problematic. Hausner ratio, Hr [22], is the result of the packed bulk density of a dry powder divided by its loose bulk density. This metric is used for various types of dry powders including those used in the pharmaceutical industry. The higher the value of Hr, the lower the flowability of the powder. Values over 1.25 Hr are considered to be poor in flowability as measured in grams per second of material passing through a standard funnel. Refer to Fig. 11.9.

Data from various refineries has been compared in Fig. 11.10. It is possible to see that most sources of SGA do not have excessive packing of fine and superfine particles in the bulk aggregate. But some do and those that are over 1.24 Hr may become problematic with segregation of fine particles in storage silos or via attrition of alumina in handling equipment. These borderline sources are also likely to be problematic to specific customers depending upon the amounts of fines and superfines that are present in their fluorinated alumina.

The consideration of segregation in alumina silos introduces another factor that is often overlooked, variation in particle size distribution. One source of SGA with a superfines content that may be considered to be high, greater than 2.0% by weight, is

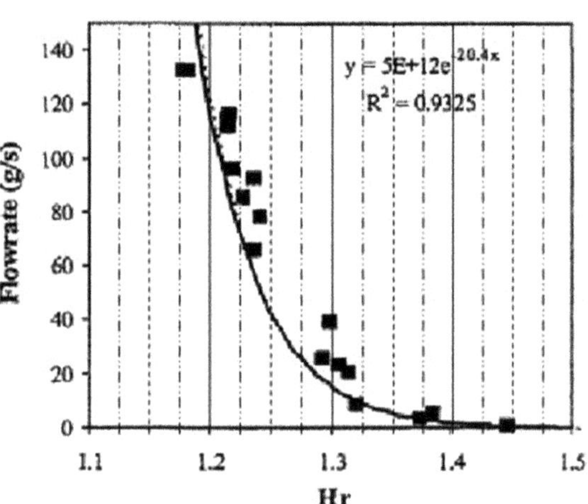

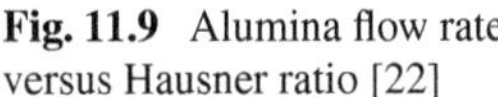
Fig. 11.9 Alumina flow rate versus Hausner ratio [22]

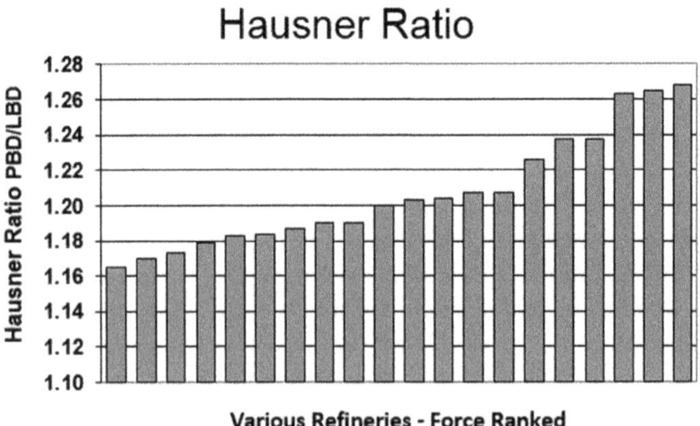

Fig. 11.10 Comparison of Hausner ratios of various alumina refineries

also a preferred source of many aluminum smelters. This same alumina has an overall PSD that is coarse, accommodating more spaces for superfines while maintaining good flowability and Hr values. This source is also a meticulously blended product. Blending avoids large variations in PSD from shipment to shipment and also within individual shipments.

The consistency of PSD at smelting locations is challenging to track, mostly because proper sampling is difficult and is also critical for accurate results. Also, the analytical equipment that is required to make precise determinations can be quite costly. Thus, variation of PSD is often overlooked or ignored by placing faith only in the average value that is reported on a Certificate of Analysis for a shipment of tens of thousands of tons of SGA.

It would be informative if a range of PSD data were provided on the sub-samples that are gathered for each shipment. It would be for the buyers to decide if this information is important enough to pursue on SGA as it is loaded into a ship. However, this information may be of no real value if the storage silos of the smelting customer and the management practices for them tend to cause segregation of fines. Segregation only increases variation of the PSD in alumina feedstock. The methods used by the receiver to unload the vessel may also impact PSD consistency in the storage silos of the smelter [23].

One additional comment on PSD involves a small portion of the plus 100 mesh fraction. As much as 0.04% by weight of an SGA shipment may arrive as grit. See the example shown in Fig. 11.11. The majority of grit is SGA that at some point has become cemented into small particles. The cause of grit appears to be related to efforts to recover settled solids, particularly from the final tanks in precipitation circuits. Particles of grit are usually quite high in residual sodium and silica content indicating that small amounts of settled solids pass through the calcination kiln and into the final product. Note that neither refineries nor smelters generally pass

Fig. 11.11 Example of a handful of alumina grit and debris removed from an alumina distribution box

SGA over screens with openings as fine as 200 or 300 μm to capture and remove this material. Very large vibratory screens and/or low flow rates of SGA would be required to accomplish such separation. Screening devices at smelters are often sized at 800 to 1000 μm for SGA and up to 2000 μm for fluorinated alumina.

Grit can cause routine, nuisance issues with alumina handling and metering systems that have horizontal fluidization elements. It sinks to the bottom of fluidized chambers and can reduce and even choke off flow in extreme cases. Most commonly, grit accumulates in the bottom of level distribution boxes that are designed to uniformly and simultaneously meter SGA to multiple compartments of dry scrubbers. Accumulations must be cleaned out periodically, sometimes more than once per day, to maintain uniform distribution of alumina flow to each compartment of a large dry scrubbing unit. It is not unusual for the amount of grit to vary widely by source of SGA.

The PSD of SGA is also considered when the topic of dustiness is raised. The finest fractions are most likely to become and remain airborne, especially when exposed to windy conditions such as with ship unloading by use of clam shells or other mechanical devices such as articulating bucket elevators. Dustiness is less of a problem when vacuum unloading systems are used.

Dustiness may also be a problem at belt transfer points of refineries, port facilities, or smelting locations. This is becoming less of a problem in modern smelting facilities that often employ fluidized conveyance or dense phase transfer systems for SGA. These systems are enclosed and do not encounter the same issues with air displacement that belt-to-belt transfer points have.

There have been some efforts within the industry to develop Dust Indexes with results measured in kilograms of dust per metric ton of alumina. At this time none of these have become standardized measures in the industry.

The Attrition Index, or A.I., of SGA is a unitless value that is calculated in the following manner.

$$\text{Attrition Index} = ((\% + 44\,\mu\text{m Before} - \% + 44\,\mu\text{m After})/(\% + 44\,\mu\text{m Before})) \times 100$$

"Before" and "After" refer to the percentage of plus 325 mesh fraction in the sample of SGA before and after being placed in an attrition column which uses a jet of pressurized air to impact particles of alumina against a metal plate at the end of the vertical column. Refer to Fig. 11.12. Such testing requires precise control of air pressure and velocity of the gas stream in the column as well as the time of exposure to these conditions. Note that there is a common misperception that Attrition Index testing measures the minus 45 micron content of SGA before and after the test. These tests focus on determinations of the weight percent of larger particles, not the fines.

Consistent particle velocity when impacting a metal plate is the main factor in being able to make consistent determinations of A.I. according to studies performed by Audet and Clegg [24].

Note that there is no single standard method in the industry for determination of A.I. Multiple methods have been in use by various refineries. However, results

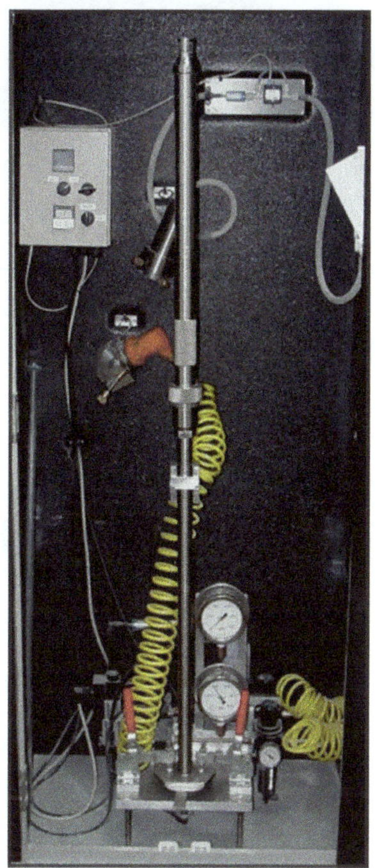

Fig. 11.12 Attrition column testing device

of unpublished comparative analyses indicate that most commonly used methods yield somewhat similar results, generally of ± 3 of the reported value of other methods although some methods can vary by more than 7 points of A.I.

There is one other common misperception about Attrition Index values. This even carries over to the Certificates of Analysis of some alumina refineries. Testing yields an A.I. value, a unitless number. The calculation applies the same unit of measure [plus 44 μm] in both the numerator and the denominator thus cancelling out the common unit of measure. Why is it important? The misperception is that an A.I. value of 15 implies that the SGA was 15% higher in fines content after the attrition test. This is not so. Attrition Index is a ratio of the amount of difference in the plus 44 μm fraction caused by the test to the amount of plus 44 μm material that was in the SGA prior to testing. As an example if a sample of SGA had 92% plus 44 μm [8% minus 44 μm] before the test and 15% less plus 44 μm after the test the A.I. value would be 16.3.

Regardless, the benefits of A.I. data are greatest when comparing an individual alumina refinery to its own historical performance. Comparison of reported values

Fig. 11.13 Mother and daughter particles of alumina shown at 500X magnification

from many refineries begins to lose significance due to the different methods that are in use to measure A.I.

Attrition Index is one way to measure the toughness of alumina particles. Full size grains are referred to as mother particles that are often somewhat spherical or blocky in shape. Broken edges that have been separated from mother particles are called daughter particles. Note that the longest dimension of these shards may be larger than the second largest dimension. Refer to Fig. 11.13. Accordingly, daughter particles may more easily pass through a woven wire or electroformed screen, slightly biasing the intent of measurement of plus 44 μm "diameter" particles before and after an attrition index test.

As with fines and superfines broken fractions of alumina particles can alter the packing density, Hr, and flowability of pure alumina. Unfortunately, there is no good, general guideline for the importance of A.I. that works for all smelters. Studies must first be made to better understand the Attrition Footprint of a smelter and perhaps of its associated port facility as well [21]. Refer to Fig. 11.14.

Some smelter designs have very low Attrition Footprints. These locations may use conveyor belts, air slides, and fluid bed type dry scrubbing equipment. Other locations may use certain types of high energy pneumatic transport systems. Systems with many 90 degree elbows, and/or poorly maintained dry scrubbing equipment can break down the particles of even the toughest SGA. The mid-point for smelting locations appears to be A.I./15.6 = 1.0 wt% of additional minus 20 μm content of alumina particles at the pots. This equates to 2.75 ± 0.25 wt% increase in minus 45 μm content. However, very few smelting locations, perhaps only 10–15%, are near, or at, this mid-point. Thus, conducting an Attrition Mapping exercise is important.

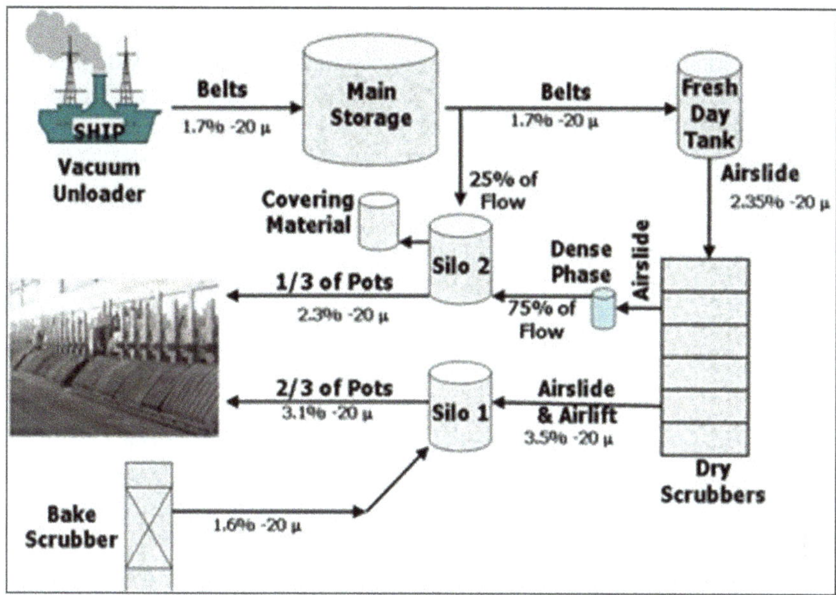

Fig. 11.14 Example of attrition mapping exercise in a smelter

This again raises the point that SGA must be able to accommodate the fines and superfines that are drawn into the fume control systems that then become a portion of fluorinated alumina feedstock. More than 85% of pot exhaust particulate matter is minus 45 μm in diameter and approximately 65% is minus 20 μm material [21, 25]. Alumina handling and dry scrubbing systems that have a high level of attrition can increase the level of minus 20 μm content in fluorinated alumina by more than 2.0 wt% and the minus 45 μm content by more than 10 wt% above fresh SGA baseline levels. Minimum contributions of 0.25 wt% minus 20 μm, 0.40 wt% minus 45 μm content plus 0.4 wt% bath dust still exist for a few older technology smelters. However, achieving such a desirable result generally requires the use of older alumina handling systems and fluid bed type dry scrubbing technology.

It is important to note that while alumina refineries cannot be held accountable for the attrition impact of the equipment of their customers it is wise to make some level of accommodation for fines and superfines that will be in fluorinated alumina. Shipping of product that is consistently above 1.20 Hr, or over 85 s in flow funnel time by the Alcoa method, may also be a good way to invite customer complaints and associated problems.

There are two main factors that appear to drive A.I. at refineries. Precipitation and resulting particle morphology is the primary driver by far [26]. Particles that have some degree of agglomeration tend to be tougher than crystals of aluminum tri-hydrate that grow into a more distinct crystalline shape, or blocky structure.

Another consideration is related to calcination technology and the heating rate of alumina particles. Older technology rotary calcination units generally have the largest negative impact on the A.I. of product due to a slow heating rate. The gentle tumbling of the hydrate and transition phase alumina inside a rotary kiln does little to break particles down into smaller sizes. But, the slow heating rate and longer time at peak temperature increases particle fragility [27].

Even when A.I. is not reported the user can be moderately certain that if the refinery uses rotary kilns for calcination, often evidenced by an alpha phase content of more than 10 wt%, that the attrition index is likely to be high, over 20, and occasionally higher when the precipitation technology does not produce tough alumina particles. Modern calcination technology doesn't have the same thermal treatment of particles. They also often place some focus on particle velocities within calcination circuits to help keep particle breakdown to a minimum.

Over time the importance of A.I. to smelting locations appears to have diminished. Feeding systems of newer smelters are less sensitive to fines content at the pot. In addition, slotted anode technology has helped to reduce sensitivity to fines. The most modern alumina handling equipment is also more likely to have a lower attrition footprint than some earlier designs. At older locations when emphasis is placed on proper cleaning and de-scaling of fluorinated alumina transfer equipment and dry scrubbers the amount of attrition by the smelter can be substantially reduced. Added focus on proper alumina silo management to prevent segregation of fines also reduces the relative importance of A.I. In the 1990's a number of smelters preferred to have alumina with an A.I. of 15 or less. Those same smelters with the improvements referred to above now perform well with A.I. of 20 or less.

11.3.2 Angle of Repose and Flowability

For decades it was common for alumina refineries to report the static angle of repose on Certificates of Analysis. This is no longer a common practice and it is now somewhat rare for angle of repose to be reported at all. However, the practice is still common in Eastern Europe and in Russia.

The angle of repose test is actually a method for the measurement of friction between alumina particles. The larger the angle of repose, the greater the amount of friction. It is a simple, albeit less precise, surrogate measure of alumina packing density, Hausner ratio, or flowability. Specified ranges for SGA are typically between 30° and 36°. The simplicity of the method is its strength. Virtually any lab can perform an angle of repose test using simple and inexpensive equipment.

"It can be determined by pouring the powder into a funnel which is held at a fixed height above a flat base and measuring the angle formed by the powder's conical heap over a horizontal surface" [28]. Alumina is often allowed to fall on a "bulls' eye" target of concentric circles in order to more simply determine the angle of repose using a standard amount of material through a standard funnel at a standard height above the target. Refer to Fig. 11.15.

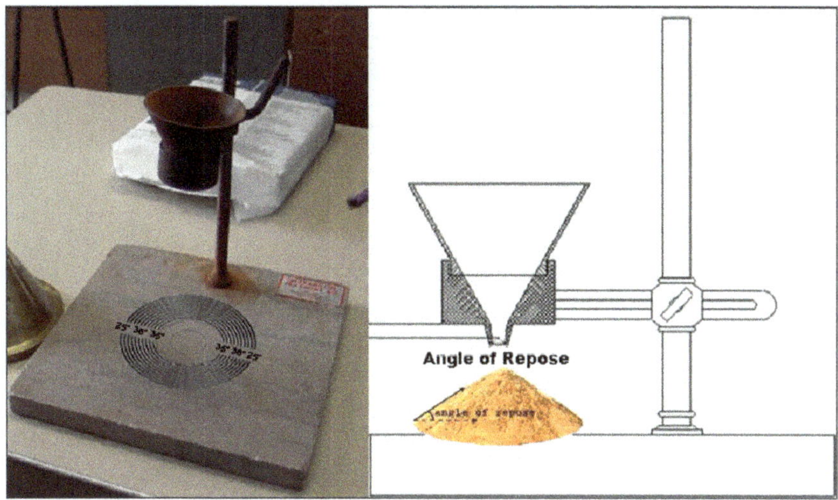

Fig. 11.15 Equipment for determination of angle of repose

"Angles of repose under 30° point to powders with good flowability, 30°–45° show some cohesiveness, 45°–55° show true cohesiveness, and anything above 55° is considered as having very high cohesiveness and of limited flowability" [29].

Most sources of SGA in the Western World have angles of repose that vary from 30 to 36 degrees. See Fig. 11.16.

Related to flowability, the pharmaceutical industry commonly uses the Carr index [30] to measure the compressibility and flowability of dry powder. Pharmaceutical and other dry powder processing industries also use the Hausner ratio [22]. No standard method exists among producers of SGA for the measurement of flowability of alumina.

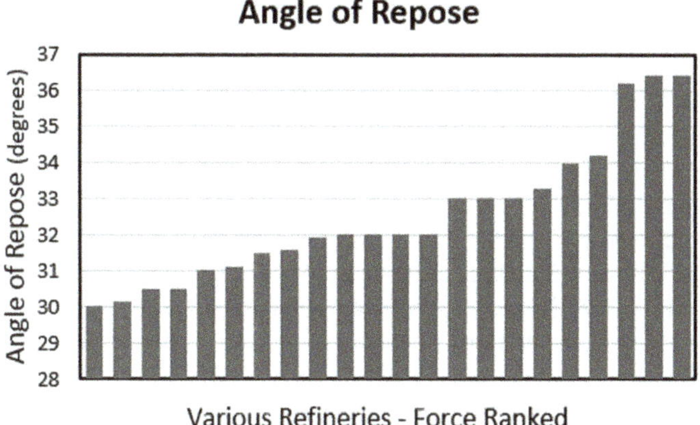

Fig. 11.16 Comparison of the angle of repose of various alumina refineries

Fig. 11.17 Alcoa flow funnel in measurement stand with a scale to the left [33]

However, one large aluminum company as well as other refineries and smelters regularly use a standardized flow funnel [31, 32]. A 4 mm orifice Alcoa Flow Funnel test is common amongst the alumina refineries and a few smelting locations. A larger 6 mm diameter orifice is used at most smelters for the sake of being able to make measurements of both pure and fluorinated alumina using the same instrument.

In the paper Key Physical Properties of Smelter Grade Alumina [33] claims have been made about the importance of flowability as a key physical property of SGA for aluminum smelters. Flow funnel tests are simple and require relatively inexpensive equipment. What is measured is the amount of time that it takes for a sample of a standard amount of SGA to pass through the orifice of the funnel. The flowability of the material is expressed in terms of grams per second. Refer to Fig. 11.17.

The flow funnel test is also simple enough that measurements can be made in the field by operators with a minimum of training although care must be taken with sampling protocols. Measurements that must be lab certified for shipments of SGA have more stringent procedures with strict focus on sample preparation. Flow funnel measurements are also common in the aluminum fluoride industry. Since samples of standard mass are used the results are almost always expressed in the seconds taken for the material to pass through the funnel.

Statistical analysis has determined that the minus 20 μm content of SGA is the property that explains the greatest amount of variation in flowability. It is also the property that explains much of the variation in the maximum dissolution rate of alumina [10]. However, making accurate analyses of multiple alumina samples on a daily basis for minus 20 μm content would require expensive equipment and specially trained lab technicians. At most smelters the ability to analyze samples around the clock and on any day of the week is not practical. Fortunately flow funnel data is a close surrogate to the minus 20 μm content for explaining variations in maximum dissolution rate. It can also be measured easily under a variety of industrial settings at any time of day.

The second factor of importance explaining variation in flow funnel data is the plus 45 μm content or plus 325 mesh fraction. The third factor is the plus 75 μm content, or plus 200 mesh fraction. These three fractions; minus 20 μm, plus 325 mesh and plus 200 mesh collectively explain more than 98% of variation in average results of flow funnel time. This observation is directly connected to the theories behind packing density, Hausner ratio and Carr's index. It clearly demonstrates that placing focus solely upon any one or any two of these size fractions will paint an incomplete picture of the flow characteristics of the overall particle size distribution.

Practical use of flowability data may manifest itself in many forms in smelting operations including the ability to estimate the consistency of alumina received in shipments.

Alumina with lower-than-average flowability or that is highly variable in flowability can create bottlenecks with some alumina handling system components. These include vibratory screens with less than a 10 degree pitch and silos with steep conical bottoms. It may also lead to accumulations of packed fines on the perimeters of large storage silos that later may slough off as slugs of feedstock to downstream equipment or processes.

Slugs of low flowability material can also be observed when they pass through the filtration systems of dry scrubbers. They cause temporal increases in filter pressure drop. With flowability data the proper assignment of the root causes of such problems becomes easier.

Fresh alumina metering systems to dry scrubbers can also be affected by flowability of the alumina or by significant variations in it. For example, deflector plate systems are not highly accurate for metering of alumina flow. Rotary vane or volumetric feeding systems are generally preferred and quite accurate. However, deflector plate metering may be preferred with alumina of low flowability and high fines or superfines content. Alumina that has a high Hausner ratio tends to pack and may not completely fill or discharge from the volumetric segments of a rotary vane feeder. With volumetric feeders superfine material may also flush through the equipment at high flow rates should it become aerated.

Greater risk to operational stability of reduction cells occurs when alumina with a high Hausner ratio does not consistently fill or discharge from the volumetric chambers of point feeders.

The primary benefit of being able to monitor the flowability of SGA over time is to be able to know if the alumina source, or if a certain, specific shipment is compatible with pot feeding equipment and control logic or other equipment and systems. Refer to Fig. 11.18 [33].

Most point-fed reduction cell technologies work well with an Alcoa Flow Funnel Time, AFFT, of 85 s or less using a flow funnel with a 4 mm discharge orifice and a 100 g sample. Many cell technologies begin to experience problems when AFFT values exceed 95 s.

The problems normally manifest themselves as undissolved alumina under the liquid metal layer accumulates. Bottom sludge is much less conductive than pure metal. Accordingly, when it accumulates it changes the current flow vectors inside

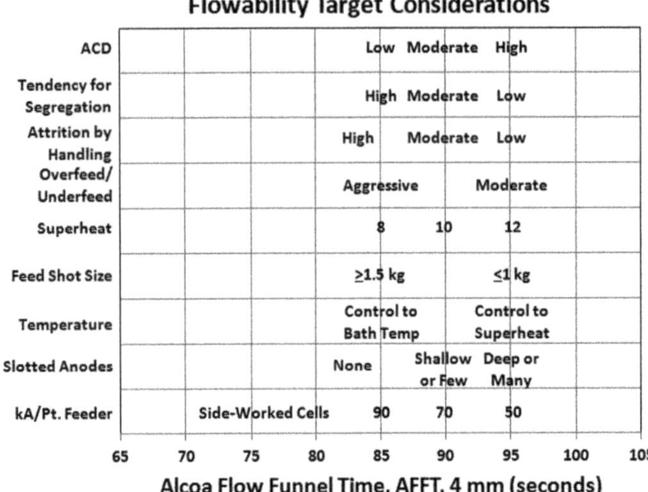

Fig. 11.18 Examples of considerations for flowability targets using a funnel with a 4 mm orifice [33]

the reduction cell. Aluminum is a paramagnetic metal meaning that the electromagnetic fields that follow Fleming's Right-Hand Rule causes the molten metal in a cell to move. Accordingly, the preferred direction of current flow is always vertical and downwards. Deviations from vertical current flow caused by deposits of bottom sludge generate waves in the liquid metal layer destabilizing the cell and causing some of the liquid aluminum to back-react with CO_2 into CO plus Al_2O_3 in solution. To recover stability under such circumstances the anode to cathode distance in the cell is generally increased. This consumes more energy, generates additional heat losses, increases the generation and emissions of fluorides, increases the dissolution of metal in bath, and ultimately leads to additional losses in aluminum production.

The bottom sludge can still go into solution and it eventually will. But, there are no straight-forward metrics to know at any given moment how much alumina is going to the formation of bottom sludge or how much of it is being put into solution in the bath of molten fluoride salts. These uncertainties then lead to a higher likelihood of having an anode effect. When the amount of alumina in solution has nearly been depleted the carbon anodes then begin to react with the molten fluoride salts to form CF_4 and C_2F_6. These are very stable and powerful greenhouse gases that can stay in the atmosphere for thousands of years. Accordingly, substantial efforts have been taken at smelters to greatly reduce the typical rate of anode effect occurrences. Many of these have been accomplished via changes to alumina feed control.

As the flow funnel time for alumina improves, to 85 s or less, flowability problems begin to diminish. This is especially true for smaller, older technology cells that do not have advanced feed delivery systems or alumina handling systems that avoid segregation of fine particles or attrition of SGA. Suffice it to say that the current state of the art for alumina feeding involves the delivery of a cold slug of alumina onto the surface of liquid bath under a variety of conditions that are conducive or not so conducive to mixing and rapid dissolution. Thus, alumina with a high Hausner ratio, or low flowability, may not be easily metered by point feeders. Nor is alumina with poor flowability likely to spread out in a thin layer over the surface of the liquid bath for effective heat transfer and subsequent dissolution.

Flow funnel data also serves other purposes, including characterization of fluorinated alumina that follows dry scrubbing equipment. It is most commonly used for comparative analyses between individual dry scrubbing systems. A fume treatment system with a high flow funnel time likely needs to be de-scaled or otherwise serviced. Fume treatment systems with the lowest and/or least variable flow funnel time are usually in good condition regarding the attrition of alumina particles. In a few instances flow funnel time has also been used to track the amount of scale formation in airlift tubes for fluorinated alumina. However better metrics for this same function also exist.

11.3.3 Bulk Density and Thermal Conductivity

As noted above in the discussion on Particle Size Distribution and Hausner ratio there are two measures of bulk density for Smelter Grade Alumina.

The most common is Loose Bulk Density, or L.B.D. There is also a measure of Packed Bulk Density, or P.B.D. Again, Hausner ratio is simply P.B.D. divided by L.B.D. These are also referred to as tapped (or vibrated) bulk density and un-tapped bulk density, or just as bulk density [34] in ISO Standard 18842:2015.

Not surprisingly, many sources of SGA with higher-than-average fines or superfines content also tend to have above average P.B.D. values and below average L.B.D. values.

Point feeders used in reduction cells rely upon volumetric displacement to deliver shots of alumina to cells. Many alumina feeding systems for dry scrubbers also use volumetric displacement type feeders. When the bulk density of SGA shifts the conveyance rate of alumina per unit time to volumetric feeding systems also changes.

In reduction cells these shifts are usually not of great significance. The alumina feed control systems can adapt relatively quickly to most changes in bulk density. Slight changes in cell resistance are used to approximate the alumina that is in solution in stable reduction cells.

Changes in the rates of rotary vane or volumetric feeders to dry scrubbing systems will also occur without significant interventions. The pots that are serviced by each system will ultimately consume alumina based on mass flow rates rather than volumetric consumption rates. Changes in density may cause deviations in intermediate silo levels. But, these are quickly corrected by adjusting the feeder flow rates.

The largest direct impact of bulk density is with inventory management. Silo dimensions are fixed. When bulk density changes the mass that is kept in storage will also change. The best method for keeping up with these changes is to periodically gather samples for the measurement of tapped and untapped density. While it is

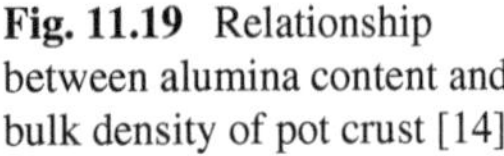

Fig. 11.19 Relationship between alumina content and bulk density of pot crust [14]

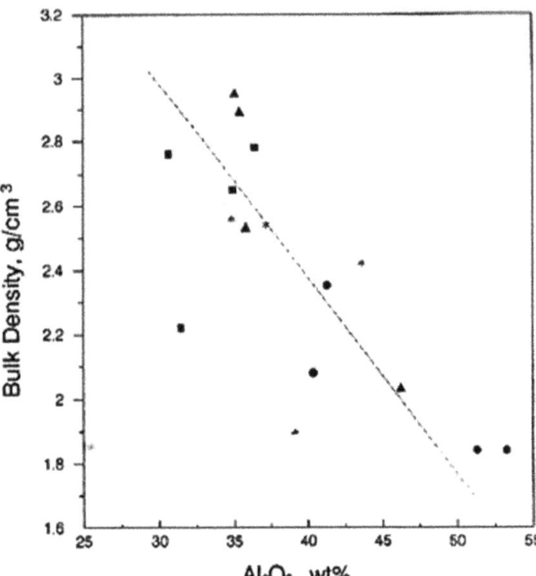

not proven to be completely accurate a good Rule of Thumb for silo inventory management is to apply the average of tapped and untapped density to the volumetric displacement of alumina in storage silos.

The bulk density of crust in reduction cells is related to its alumina content as is shown in Fig. 11.19. This example is for actual pot crust comprised of coarse crust bath of up to 15 mm in diameter mixed with pure alumina at various percentages and measured by mercury intrusion porosimetry.

The thermal conductivity of alumina is low compared to other materials that are used in smelting processes. Anode and cathode materials and certainly aluminum metal all conduct heat more readily than alumina. Accordingly, alumina is sometimes used as a thermal regulator on the frozen crust of reduction cells. It may be used to put more of a blanket on top of cells that strive for low specific energy consumption per ton of metal produced. Refer to the example shown in Fig. 11.20.

On the other hand, alumina may also be kept at low levels on the top crusts of reduction cells that have boosted the current intensity and the conventional options for spillage of excess heat to new limits in order to increase production. This type of cell operations typically opt to control the percent of alumina that is in the mix of anode covering material rather than to use alumina as a top layer blanket. Such operation often also explores ways to reduce the thickness of anode covering material to be able to spill more heat through the top crust. The challenge is to be able to properly protect the carbon anodes from oxidation that is commonly referred to as air-burning using a minimum amount of cover.

The reader should note that the percentage of alumina in crust generally is not a control variable for smelting operations [36]. It is an output of the overall thermal

	Alumina	20%CCB	40%CCB	60%CCB	80%CCB	CCB
Void Fraction	0.685	0.666	0.626	0.495	0.378	0.348

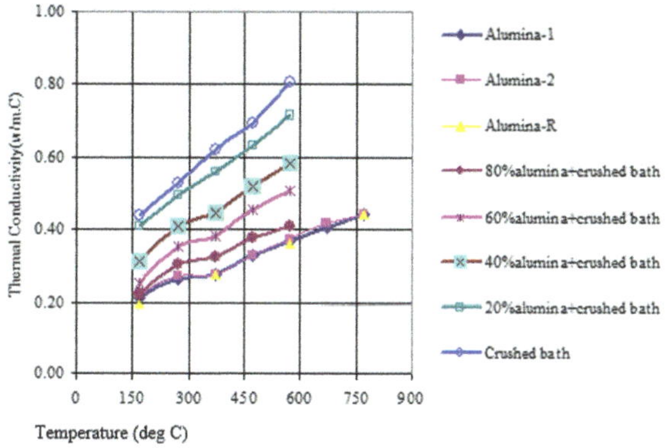

Fig. 11.20 Thermal conductivity and void fraction of pot crust versus concentrations of alumina [35]

balance of the pot lines. Smelters that spill more heat through a unit area of pot crust are more likely to have a higher specific melt rate of crust. In such a circumstance more alumina is also delivered to the liquid bath from pot crust requiring additions of alumina to the cover material mix to be higher.

11.3.4 Specific Surface Area and Pore Size Distribution

One of the more common questions about BET Surface Area is: "What does BET stand for?" The answer is Brunauer, Emmett and Teller [37], the scientists who developed the method for determination of surface area. There is another method for determination of total surface area, the BJH, Barrett-Joyner-Halenda, method that uses a different set of assumptions about the geometry of pores, as being more cylindrical than spherical. However, in the aluminum industry BET Surface Area is used following ISO 8008:2005(en) [38].

As crystals of aluminum tri-hydrate are converted into transition phase SGA the total surface area of particles changes radically, by over two orders of magnitude.

As can been seen in Fig. 11.21 [25] the specific surface area of aluminum tri-hydrate is low, on the order of 10 m^2/gm. The same is true for alpha phase alumina that has been heated to over 1000 °C. With heating that is just above 400 °C in the initial formation of transition phase alumina, the BET Surface Area may rise above 350 m^2/gm. Notice that this occurs as combined hydroxyls are rapidly being driven off. Trace c in Fig. 11.21 illustrates how LOI content is reduced from greater than 30 wt% to less than 5 wt% quite rapidly over a small range of temperature change. To the layman the common explanation is that "water" is rapidly being driven off during

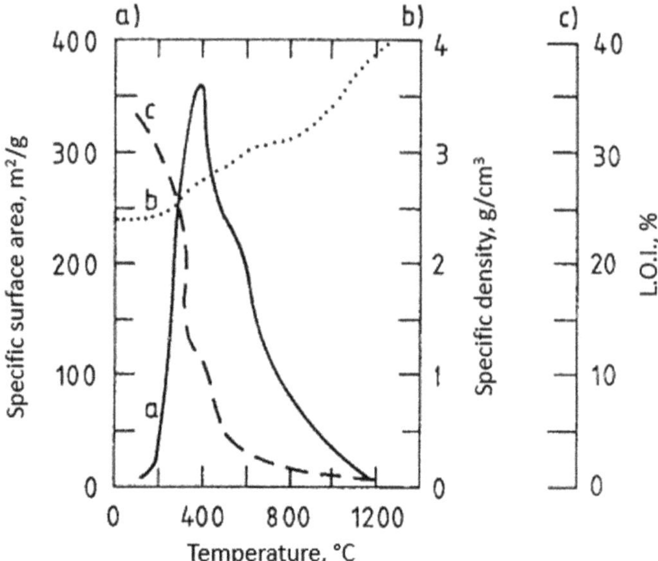

Changes in selected properties during calcination of α-alumina trihydride. a) B.E.T. surface area. (B.E.T. = area determined by N_2 adsorption) b) Density. c) Residual water (or L.O.I.). (L.O.I. = based on weight loss on heating to 1000°C).

Fig. 11.21 From aluminium smelter technology by Welch and Grjotheim [25]

the conversion of aluminum tri-hydrate into Smelter Grade Alumina. The structure of these particles then changes radically as this "water" vapor escapes. Refer to Fig. 11.22. To some extent this explanation is accurate as the time that aluminum

Fig. 11.22 Images of crystalline structure of synthetic aluminum tri-hydrate $Al(OH)_3$, left and SGA, right [7, 16]

Fig. 11.23 Porosity of SGA at various magnifications [39]

hydrate spends in the kiln, often aimed as if to quench the flames, is quite short. Modern calcination is not a slow roasting process. The conversion from $Al(OH)_3$ to $Al_{10}O_{16}H_2$ is a transitioning process during which greater than 97% of the hydroxyls are driven off by heating.

When one examines the plates and fissures shown in the image of the SGA particle in Fig. 11.22 it appears to be evident how the BET Surface Area of alumina can be so high. However, looking under a magnifying glass can be entirely misleading! The closer that you look, the more surface area that you will see, right down to the level of nanometers that can only be observed by scanning electron microscopes, SEMs. Refer to Fig. 11.23.

With highest magnification we begin to see the smallest pores and only then can begin to have some grasp of total surface area. The smallest pores begin to form as Boehmite converts to transition phase alumina at temperatures of approximately 400 °C. With temperatures between 400 and 800 °C the pores continue to increase in size and the net surface area is reduced. The target for BET Surface Area is commonly between 70 and 80 m^2/gm. This target range is achieved near 800 °C for the calcination temperature. Above this temperature the particles then begin to approach the crystalline structure of alpha phase alumina with BET Surface Area that is quite low.

As was illustrated in the section on alpha phase alumina, each particle of SGA is not uniformly heated or held at temperature for the same amount of time in an industrial calcination kiln. Accordingly, there is a distribution of pore sizes and that is to be expected. Sometimes these distributions are Gaussian. At other times they

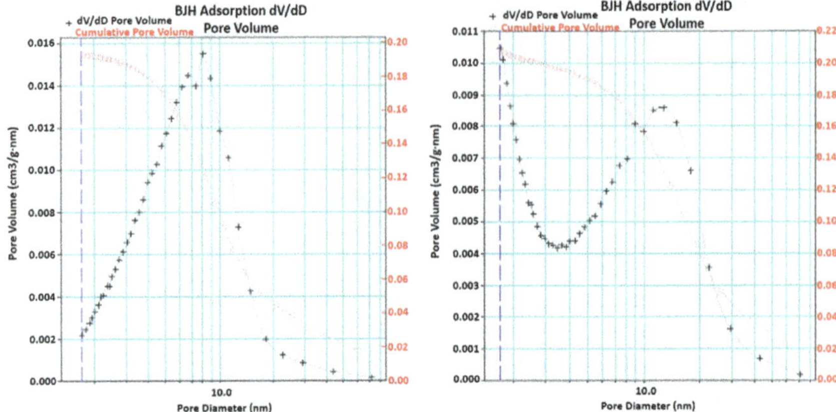

Fig. 11.24 Examples of mono-modal and bi-modal pore size distribution [40]

are bi-modal indicating that more than just one driver of random behavior in creation of nano-pores do exist in some industrial calcination units. Refer to Fig. 11.24.

As early as 2004 Dr. James Metson and his associates were making inquiries as to what might be behind such bi-modal pore size distributions for SGA [40]. The Light Metals 2011 paper titled, "Towards Redefining the Alumina Specifications Sheet—The Case of HF Emissions" [6] was an in-depth review of this topic. Counter intuitive results were reported in which an alumina source with higher BET Surface Area yielded much higher HF concentrations in the chimney of a modern dry scrubbing system. The literature was further advanced with the publication of "Assessing the Role of Smelter Grade Alumina Porosity in the HF Scrubbing Mechanism" in Light Metals 2016 [41]. This contribution made it clear that pores of less than 3 nm in size (BJH method) would become blocked, making internal pore volume of alumina particles inaccessible and ineffective for dry scrubbing of HF from process off-gas. The work of Dando and Lindsay [39] in that same year confirmed that pores of less than 4 nm (BET method) would become closed and inaccessible when exposed to HF gas.

From this collective body of work it clearly appears that pores of less than 3–4 nm in size do not fully contribute to the ability of SGA to scrub HF out of a process gas stream. As can be seen in Fig. 11.24, pore size distributions have some fraction of pores that are below 3–4 nm. The bi-modal distributions generally have a significantly higher fraction of these small nano-pores. Thus, they are less effective for dry scrubbing. This is particularly noticeable when the loading of fluoride on alumina is above a saturation of approximately 0.23 mg F/m^2 of BET Surface Area.

Suggestions have been made that alumina specification sheets need to be redefined to quantify the fraction of BJH method pore sizing that exists below a certain size, such as 3 nm. This is one approach to what may be needed to obtain SGA that is excellent for dry scrubbing. There are other methods that may be considered as well.

One alternative and insightful approach attributed to the Laterrière smelter and the Alcan Research Center in Saguenay, Quebec is simply to take the ratio of BET Surface Area to the LOI content of SGA that has a BET Surface Area of 70 m^2/gm or higher. Numeric ratios of 85 or higher indicate SGA that is acceptable for most dry scrubbers. Numeric ratios of 100 or more indicate SGA that is excellent for dry scrubbing. For SGA with BET Surface Area/(LOI wt%) ratios of less than 85 there have been multiple industrial examples in which the concentration of HF in the chimney of the dry scrubber have increased to two times or four times the normal results. In some rare cases over short durations increases of more than twenty-five fold increases vs. normal HF concentrations have been observed.

What is so insightful about this method is that it incorporates one of the most significant factors contributing to the generation of HF in the reduction cells, the LOI wt% of SGA. It also includes the BET Surface Area that provides the reactive sites for removal of HF by dry scrubbing. Unlike direct measurement of Pore Size Distribution which can be time consuming and costly this method uses two very commonly reported alumina quality parameters. This would then simplify the process of establishing a new, minimum specification. It also acts well to catch SGA with bimodal pore size distributions that are typically higher in LOI content than alumina sources of the same BET Surface Area that have unimodal pore size distributions.

The cause(s) behind formation of bi-modal pore size distributions have not yet been proven using properly designed experimentation. However, there is a substantial amount of unpublished data that indicates that the fraction of hydrate by-pass in SGA and the designs of the hydrate by-bass circuits themselves can be of great importance. Some older calcination unit designs have low retention times for aluminum hydrate that is directed into a holding vessel for alumina as it exits the kiln. Some specific refineries have had production bottlenecks with calcination capacity that have been alleviated via increases in the amount of hydrate by-pass.

Ultimately the reason for concern with bi-modal pore size distributions comes down to a fundamental. The primary aluminum industry has heretofore neither developed nor insisted upon specifications that deliver full satisfaction with the potential for HF generation and dry scrubbing for the recovery of HF. BET Surface Area and LOI have been treated as separate parameters leaving refineries with an additional degree of freedom in finding more cost-effective approaches to energy consumption. Smelters that are willing to accept SGA that has 70 m^2/gm BET Surface Area and 1.0 wt% LOI in the same shipment have no one to blame but themselves if the dry scrubbing results are then found to be unacceptable.

11.4 Efficient HF Removal and Dry-Scrubber Efficiency

11.4.1 Sources of HF from Smelting

One cannot produce HF without H, hydrogen. The reaction to form HF occurs with AlF_3 that is in solution in bath or with $NaAlF_4$ as a vapor at the liquid bath interface [42]. There are three primary sources of hydrogen that react with molten bath, a mix of fluoride salts, in aluminum reduction cells;

(1) Residual hydroxyls plus additional chemisorbed and physiosorbed moisture on alumina particles.
(2) Moisture that is present in air which contacts and reacts with liquid bath.
(3) Residual hydrocarbons from carbon anode blocks.

The least of these sources is residual hydrocarbons from anodes. This is true even in the case of Søderberg anodes that are baked in situ in reduction cells.

When total fluoride generation in primary aluminum smelters has been studied ranges of 24–40 kg F/mt of aluminum produced are most common from point fed prebake cells. Side-worked prebake cells and Søderberg cells generally fall into the range of 18–24 kg F/mt Al. The contribution from anodes is generally less than 10 wt% of the total, or less than 3 kg HF/mt Al produced [42]. The evidence of this, even with horizontal stud and vertical stud, side worked Søderberg cells, is that total %F on fluorinated alumina is quite consistent. Side-worked Søderberg and prebake cells and point-fed Søderberg cells typically have 1.1 ± 0.2 wt% F on fluorinated alumina. The major fraction of this low fluoride content is explained by sources of H in alumina and by contact of liquid bath with humidity in air. Changes in anode paste properties including wet versus dry paste have caused no observable impact upon total fluoride generation from these types of cells. The same is true for unpublished comparisons made between Søderberg locations and side-worked pre-baked locations using the same alumina source.

Total moisture in air will vary with relative humidity at any temperature. The capacity to hold moisture in air, absolute humidity, also increases with increasing temperature. Thus, hot and steamy air in summer months can contain much more total moisture than cold, dry air in winter months. The effect can often be seen by studying trends in the generation rate of HF from pot lines. A good surrogate for this is to follow the weight percentage of F on fluorinated alumina after the pot line dry scrubbers. Refer to Fig. 11.25.

Seasonal variation will in general terms be more evident at smelting locations that are furthest away from the tropics, especially if ambient air can become quite dry in winter months. However, some locations in equatorial or tropical zones may also see differences in fluoride generation between the rainy season and the dry season of the year.

Moisture in the air is the most variable component of total fluoride generation. Part of this variation will depend upon the ability of ambient air to contact the surface of the liquid bath in pots or $NaAlF_4$ vapor immediately above the surface of the liquid

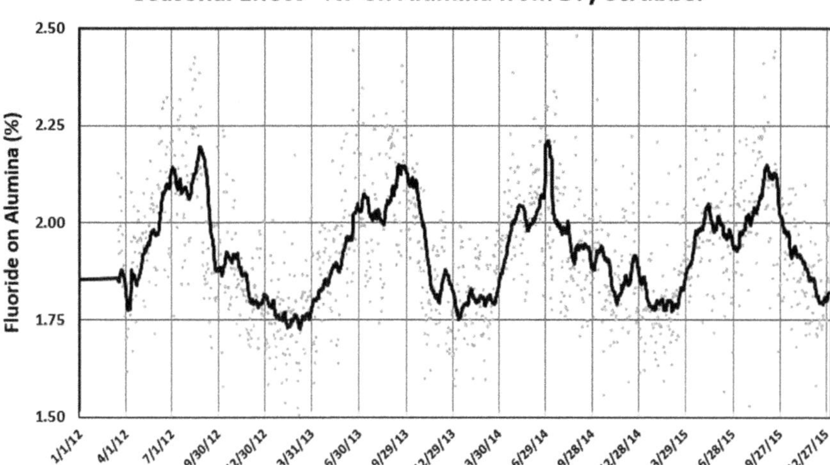

Fig. 11.25 Example of seasonal change in generation of HF in reduction cells

bath. Well crusted reduction cells will have less opportunity for contact with air than cells with substantial amounts of cracks or open holes in the crust. An example of this was shown in Light Metals 2003 in a paper by Slaugenhaupt, Bruggeman, Tarcy and Dando titled the "Effect of Open Holes on Vapor-Phase Fluoride Evolution from Pots" [43]. Refer to Fig. 11.26.

Transition phase alumina, $Al_{10}O_{16}H_2$, with plenty of reactive hydroxyl sites that capture HF, also delivers hydrogen units into the bath. A fraction of these hydroxyl units reacts with HF to eventually form AlF_3 on the surfaces of alumina particles. However, the by-product of the reaction between –OH units and HF is H_2O. Since the alumina surface is hydrophilic this H_2O is then available to be physiosorbed or chemisorbed back on to the surface.

The other stark reality is that dry scrubbers only access approximately 25% of the total hydroxyl sites that are available for reaction. Another way of looking at this would be to say that only 25% of the available BET Surface Area is used to scrub out HF before the reaction kinetics slow down for access to pore volume that is well removed from the surface of alumina particles. SGA does not become fully saturated with fluoride until it reaches approximately 7 wt% F. Refer to Fig. 11.27. However, dry scrubbing efficiency drops off substantially when the fluoride content of alumina increases above approximately 1.8 wt% [42].

Thus, perhaps only one-quarter of the total LOI fraction on alumina is consumed by reaction with HF entering the dry scrubbers. The remainder is available to generate HF in the bath with some efficiency as the particles of alumina go into solution under electrolysis. Comparisons of multiple sources of SGA indicate that levels of LOI over 0.85 wt% are probably counter-productive in as much as dry scrubbing is concerned. Refer to Fig. 11.28. Any overage goes to produce more HF from the process and

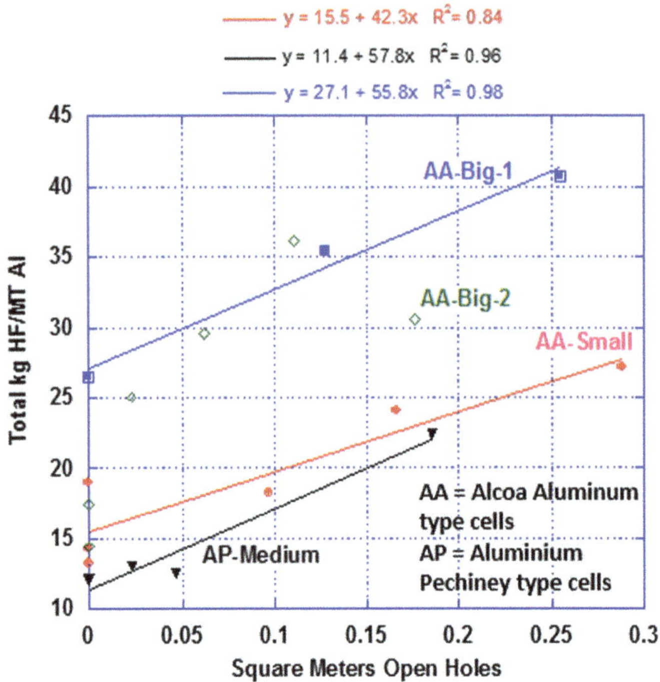

Fig. 11.26 Fluoride evolution versus open hole area in pot crust [43]

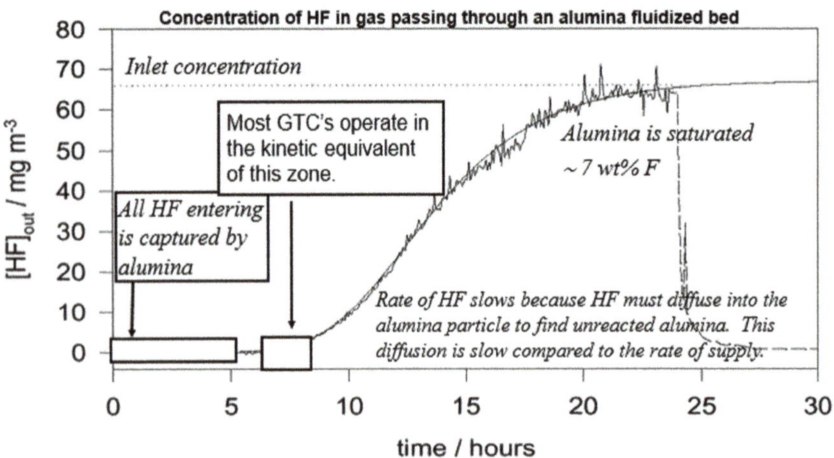

Fig. 11.27 HF breakthrough and saturation curve [42]

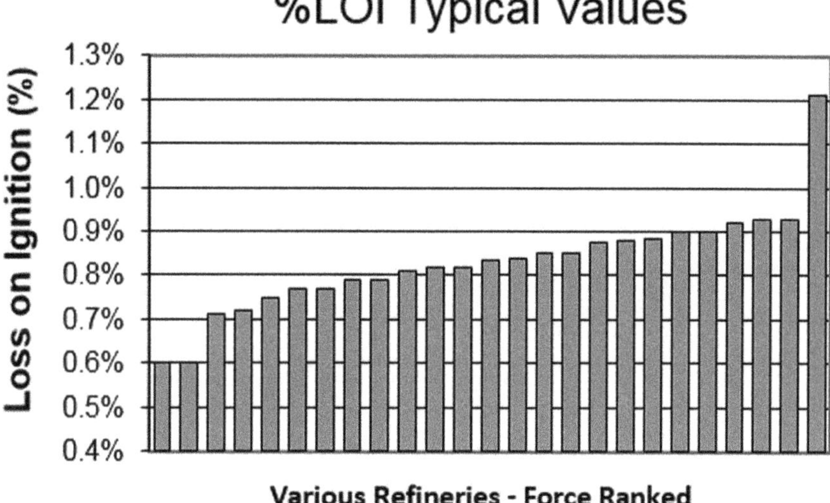

Fig. 11.28 Typical values of LOI content at various alumina refineries

increases the risk that customers will complain about the ability to sustain low levels of HF in the chimneys of the dry scrubbers.

The LOI fraction is not the entire story of hydrogen on alumina. As noted, SGA is hydrophilic. Thus, it attracts moisture, and it comes to equilibrium with the moisture content in air. This ability to come to equilibrium extends to SGA that is loaded into a ship in the tropics and that is consumed on a very cold and dry winter day in climates such as Québec. The humidity of SGA will adjust to its surroundings [40].

Dr. Margaret Hyland's work indicates that each mass unit of structural –OH [LOI content] generates an order of magnitude more HF than each mass unit of physisorbed surface water [44]. Note that this experimentation compared prepared samples of alumina with 0.14 wt% LOI and 13 wt% of added surface water to alumina with 1.2 wt% LOI and 0.3 wt% of physiosorbed moisture. Although enlightening, neither condition is representative of typical SGA that is used in industrial conditions.

When added to molten bath for dissolution, the H_2O content as surface water flashed off rapidly, as soon as the alumina hit the hot surface of the bath. The structural hydroxyl content took time, up to two minutes in experimentations by others [10], generating HF as the alumina went into solution.

Then there is chemisorbed water and the residual hydrate [Gibbsite] content of SGA that comes off at temperatures between 110 and 300 °C. By referring back to the example shown in Fig. 11.3 it can be seen that the amount of dewatering over this temperature range is equivalent to approximately 60 wt% of the MOI losses between room temperature and 110 °C. This chemisorbed water plus Gibbsite can account for more than 25% of the total evolution of fluoride from reduction cells. Refer again to Fig. 11.25.

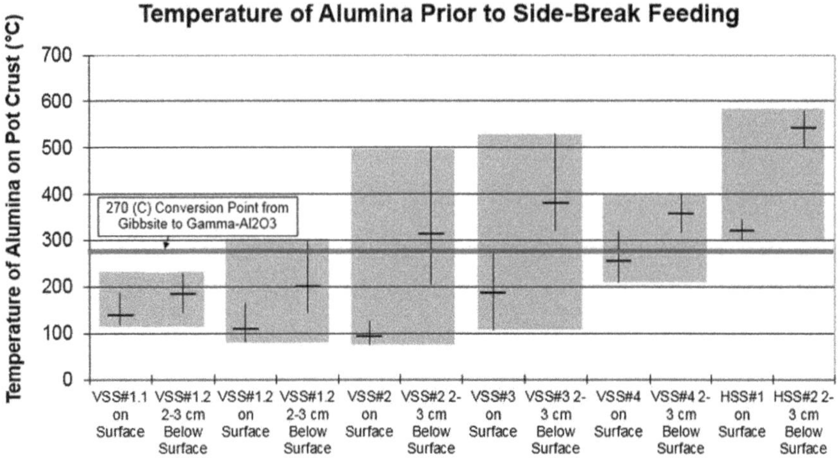

Fig. 11.29 Alumina temperatures at surface and sub-surface on side-worked reduction cells

As additional evidence for this claim, consider situations in which side-worked prebake pots have been converted to point-fed pots. The total fluoride evolution from reduction cells generally increased by approximately 36%, from approximately 22 kg F/ton Al to approximately 30 kg F/ton Al, while using the same sources of SGA. The difference appears to be due primarily to the difference in pre-heating. Refer to Fig. 11.29. Alumina used by side-working of pots can arrive to temperatures that are between 200 and 600 °C prior to feeding whereas alumina to point-fed pots is generally delivered to the bath below 125 °C.

This roasting of fluorinated alumina (or pure alumina in some cases) appears to be sufficient to drive off: all physiosorbed H_2O, the chemisorbed alumina that is loosely bound to reactive sites, and often residual Gibbsite as well.

Side-worked and Søderberg type reduction cells are now largely things of the past. Few remain in operation with the notable exceptions of Russia and Brazil. This being so, considerations are generally made for the most common case, point-fed prebake reduction cells when it comes to generation of HF.

On its way to reduction cells alumina is pre-heated when it passes through gas treatment centers. These commonly operate between 85 and 125 °C depending upon ambient temperature and other factors. Most commonly gas treatment centers operate above the boiling point of water, driving off much of the physiosorbed H_2O. However, most alumina conveyance systems between the gas treatment centers and the reduction cells use various forms of low pressure air to fluidize or otherwise transport the alumina to individual pots. The air that is used in these systems is neither preheated, nor passed through air driers. Any desiccation and much of the increase in temperature that may have occurred in the gas treatment centers may then be undone by conveyance that uses moist, ambient air prior to the arrival of alumina to any point feeding locations.

Reduction cells do have alumina bins that hold between 12 and 24 h of supply. Since operating pots are hot the alumina in the bins are also warm. Alumina is typically fed to pots between 100 and 125 °C. It is unfortunate that sufficient amounts of waste heat from reduction cells have not yet been harnessed to pre-heat alumina to approximately 300 °C prior to being fed to the cell. It would not only provide some small improvement to energy efficiency. It would greatly reduce the evolution of HF from the cells. Few primary metal producers are currently considering such enhancements at this time due to the development costs that would be involved.

To summarize the contributions of SGA to the generation of HF any residual Gibbsite and LOI over 0.85 wt% should be of concern to smelters. The combination of total LOI content plus Gibbsite typically account for more than 40% of the total HF generation of reduction cells.

Concern with excessive HF evolution applies particularly to smelters that have pushed the limits of cell thermal balance, with openings and cracks in pot crust, and at those locations with older, less efficient systems for dry scrubbing of HF gas.

11.4.2 Accessible Specific Surface Area on Primary and Secondary Alumina

A common misconception about SGA is that the finer fractions have more surface area than the coarser particles of alumina. This is incorrect, and for a few reasons.

Refer again to Fig. 11.13. It clearly describes how BET Surface Area is not able to be observed using common microscopes. The foundation of total surface area is built upon pores with diameters of approximately 10 nm, or 100,000 times smaller than a millimeter. A superfine particle of alumina, of 20 microns in size, is 20,000 times larger than the diameter of a typical pore. Thus, even the smallest alumina particles are massive when compared to the scale at which most of the surface area is present.

Electrostatic precipitator, ESP, dust can arrive to 15 wt% LOI due to its residual Gibbsite content, which itself is very low in surface area. ESP dust is also often very high in alpha alumina content, as high as 30 wt%, which also has low BET Surface Area. Accordingly, the BET Surface Area of superfine particles in SGA is often lower than the surface area of larger particles of alumina.

Figure 11.30 by Dr. James Metson illustrates well that large particles of SGA have more specific surface area [40].

Section 11.3.4 of this chapter expounded upon the impact that bi-modal pore size distributions can have upon accessible surface area for dry scrubbing. The relevance of this to effective dry scrubbing of HF is that pores that are smaller than 3 or 4 nm in size can easily become closed off during reaction with HF gas. This limits access to the interior structure and surface area thereby reducing the capacity of SGA to be fully effective at dry scrubbing of HF.

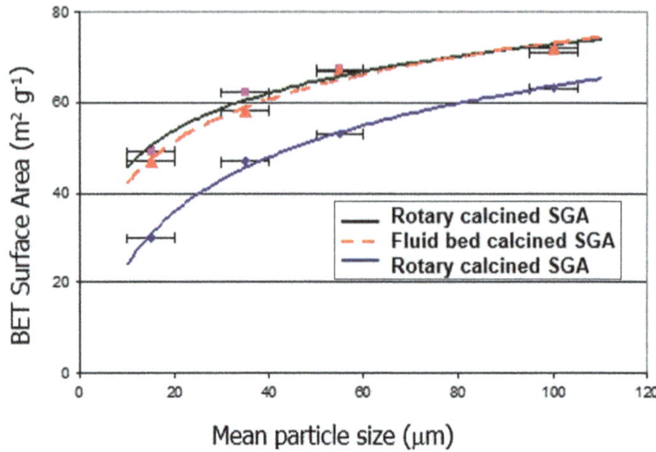

Fig. 11.30 BET surface area versus mean particle diameter of SGA [40]

Likewise, it has been noted that that fluorinated alumina loses its effectiveness for dry scrubbing once approximately 25% of its capacity to react with HF has been exceeded. Pore size distributions of fluorinated alumina have mean pore diameters that are larger than the mean pore diameters of unreacted, or fresh SGA. This is entirely due to the throat sizes of smaller pores becoming restricted or completely blocked. As one would expect, when pores become constricted the measured values of BET Surface Area also diminish [39, 41].

The reaction of HF with easily accessible hydroxyl sites is essentially instantaneous. Once the reaction with HF must extend much beyond surface area that is located near the surfaces of particles the diffusion of HF to internal pores begins to control the rate of reaction. Thus, the majority of BET Surface Area that is inside the internal structure of particles is not easily accessible for rapid and highly efficient scrubbing.

Once fluorinated alumina exceeds 0.23 mg F/m^2 of BET Surface Area in the most common type of dry scrubbers the alumina is considered to be spent and no longer highly effective for HF scrubbing. Fluid bed type dry scrubbers remain effective at loadings of up to 0.30 mg F/m^2 of BET Surface Area [45]. However, only a few of these type dry scrubbers remain in operation due to their higher costs of construction and operation.

11.4.3 The Role of Sulfur Dioxide

Sulfur dioxide, SO_2, also attaches to reactive sites of SGA. However, this is weakly bonded and is easily driven off reactive sites when HF is present. In a paper titled the "The Competitive Adsorption of HF and SO_2 on Smelter Grade Alumina" [39] this phenomenon is fully described.

The key conclusions from this work include:

(1) SO_2 is physically and reversibly adsorbed on SGA, while HF irreversibly, chemically reacts with SGA to form aluminum oxyfluoride and aluminum fluoride hydrates.
(2) SO_2 has no impact on HF reaction with SGA and is quantitatively desorbed from SGA at high HF loadings (greater than 2 wt% F, when solely from HF).
(3) The major fraction of sulfur species present on reacted SGA from commercial alumina scrubbers are sulfur compounds, Na_2SO_3, S_9, and $Na_2(SO_4)$.

As such it might appear that SO_2 has no clear role in dry scrubbing and that it is far from being connected to the physical or chemical parameters of SGA. Such a conclusion would not be correct.

By being able to attach to reactive sites of SGA it then becomes possible to move units of SO_2 exhaust from one location to another within a smelter. This is accomplished either by concentrating SO_2 to remove it more effectively from an exhaust gas stream [46] or to help a location stay within its ground level concentration-based regulatory limits. This is of growing importance in the primary aluminum industry as SO_2 emission regulations tighten.

The parameters of concern for SO_2 include BET Surface Area, Pore Size Distribution and the S content of SGA.

With abundant surface area on SGA sufficient reactive sites may remain that have not reacted with HF. These can then be available for management of SO_2.

SO_2 molecules are much larger in size than HF molecules. Thus, the small pore diameters that inhibit HF scrubbing effectiveness also inhibit the ability of SGA to carry as much SO_2 in two ways. HF that cannot enter constricted pores will migrate to larger pores that could have carried SO_2 molecules and fewer pores of sufficient diameter to carry SO_2 are present.

The sulfur content of SGA is rarely measured or reported by alumina refineries. Its contribution to the SO_2 emissions from a smelter is generally less than 5 wt% of total inputs and is often on the order of just 2 wt% of total inputs. It is not an entirely insignificant fraction of SO_2 emissions, but it has often been overlooked.

Tightening regulations on acceptable mass emissions rates of SO_2 may begin to raise more questions on how the SGA has been calcined. Calcination with heavy fuel oil generally contributes concentrations of sulfur on alumina that can be an order of magnitude higher than SGA that has been calcined with natural gas. It is recognized that the fuel source for calcination is not the only contributor to the sulfur content of alumina. But, it often is the most important factor. It is quite possible that in the not-too-distant future some smelters may require reporting of sulfur content on Certificates of Analysis for SGA. Few refineries currently provide such information.

11.5 Alumina Dissolution and Current Efficiency (by Pascal Lavoie)

11.5.1 Theoretical Alumina Dissolution Process

Primary aluminum is produced through the Hall-Héroult Process by electrochemical reduction of SGA dissolved in a cryolite-based electrolyte (bath).

As the SGA is continuously consumed, modern cells use point feeders to replenish the alumina concentration in the bath. Point feeders periodically dose small amounts of alumina, between 1 and 2 kg depending on the technology, in several locations of the cells. The number of point feeders is dependent on cell size, generally being between two and six.

Failure to replenish the electrolyte with sufficient alumina will result in anode effects, where fluoride species are decomposed instead of alumina, resulting in the formation of perfluoro-carbon gases (PFC) [47]. Anode effects not only result in a waste of energy, but also damage the environment as these stable compounds are potent greenhouse gases. On the other hand, alumina supplied in excessive quantities or going undissolved, leads to sludge formation at the bottom of the cell.

Sludge is usually made of undissolved alpha alumina (either fed as or converted from transition phase within the cell) and frozen cryolite in roughly equal proportions with limited AlF_3 and CaF_2 [48]. Cell automatic controllers attempt to regulate alumina feed to periodically allow for back-feeding of undissolved alumina. However, this method is often not effective for a non-negligible proportion of the hundreds of cells within the smelter. Sludge can therefore accumulate at the bottom to form a resistive layer on the cathode. This leads to instability in the cells resulting in loss of current efficiency (loss of production) and requires higher operating voltage to recover (higher energy consumption).

From a smelter point of view, even though the theoretical solubility of alumina is the same between SGA sources, the practical solubility rate is paramount. That is the ability to dissolve each dose of alumina quickly before it reaches the metal pad and turns into sludge.

Dissolving alumina is relatively challenging because of the conditions under which it is fed. Dissolving sugar in coffee for example is relatively easy. The liquid is hot and unless one adds nine sugars to their coffee, the task is not likely to be insurmountable. Now imagine attempting to dissolve that amount of sugar in a cup of water at 10 °C, with the sugar being chilled 850 °C below the temperature of the water (*with apologies to Lord Kelvin*). You are likely to solidify the contents of your cup before the endeavor is complete. In reduction cells 1 kg of alumina at about 100 °C is added to a liquid at 960 °C which begins to freeze at 950 °C.

Let us look at the fundamental dissolution process. In order to replenish the electrolyte properly, the alumina needs to be:

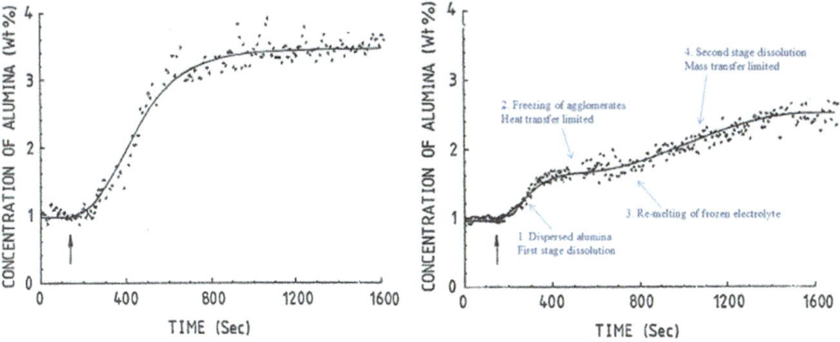

Fig. 11.31 Dissolution curve of a lab cell adapted from [49]

1. Reliably delivered to the liquid bath.
2. Dispersed in the feeding zone and *dissolved rapidly* before it can fall through to the metal pad.
3. Distributed under all the bottom surface of the anodes in the cell maintaining a minimum concentration to avoid anode effects.

When fed to the bath surface, the alumina undergoes a 4-steps process [49]:

1. Upon addition, a fraction of the alumina dose is dispersed and dissolves rather rapidly.
2. Due to limited available sensible heat (given by the bath temperature above its liquidus, referred to as superheat) a frozen layer of electrolyte is formed on the surface of particles. If multiple particles are present, the transition from γ-alumina to α-alumina will form platelets with the frozen electrolyte and result in agglomerations, or rafts.
3. The frozen electrolyte will have to melt before further dissolution can occur.
4. After liquid bath contact is re-established, the dissolution can take place.

Figure 11.31 shows dissolution curves measured by Jain [49] showing clear stages of the dissolution process. The left curve shows a rapidly dispersed and dissolved case, the right a case of poorly dissolving alumina. Given the dissolution conditions were the same but for the alumina that was used, it is clear that the alumina can play a large role in the ability to dissolve or not.

Although there are a multitude of smelter-controlled factors affecting the ability to dissolve alumina quickly, several SGA properties heavily impact the dissolution process. The following section will therefore explore these.

11.5.2 Impact of Phase Composition—LOI

It is widely accepted that α-alumina is intrinsically slower to dissolve. The mechanism proposed is that since α-alumina does not go through phase transitions releasing

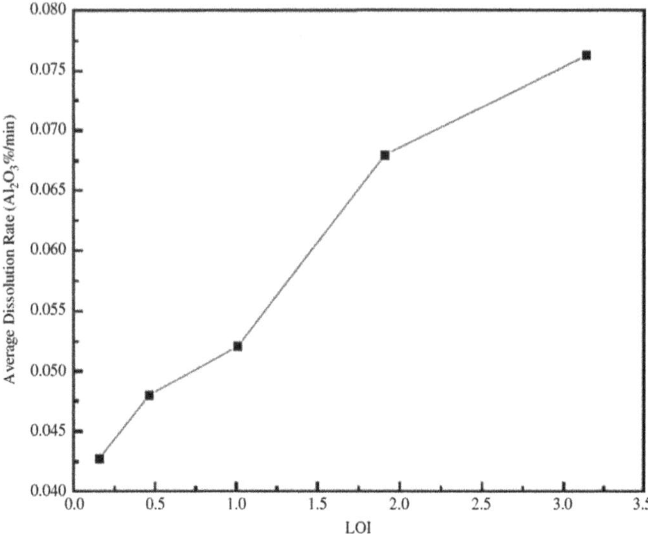

Fig. 11.32 Alumina dissolution rate increases with LOI [53]

hydroxyls, the lack of gas generation associated with hydroxyls release and explosion of particles limit the effective contact area with bath [49]. This trend was confirmed by experiments with pre-treatment of SGA conducted by Dando et al. They show that as the higher hydroxyl phases are gradually eliminated (Gibbsite, Boehmite, gamma), the dissolution rate is increasingly reduced. SGA fully converted to alpha, would also tend to sink immediately upon addition [50]. Since LOI is a good measure of the residual hydroxyls contained in alumina, it is a good indicator of the dispersion potential and the ability for that alumina to be dissolved rapidly. Multiple laboratory experiments have demonstrated increasing of dissolution rate with increasing LOI [51–53]. Yang et al. [53] illustrated this most recently as shown in Fig. 11.32.

Jain also found that attempts at dissolving alpha alumina required a higher amount of superheat to prevent agglomeration and crusting [49]. This also agrees with the theory of less effective contact area restricting both the heat transfer and mass transfer. While approaching the liquidus temperature, dissolution time of a given dose of alpha alumina is increased by a factor of three compared to dissolution in a higher superheat electrolyte.

Dissolving SGA with a significant proportion of alpha alumina is therefore practically feasible but requires careful feed control and higher superheat. This translates to a higher cost to the smelter but also incurs risk of losing feeding control, with the production losses associated with it.

Given the amount of heat necessary to raise alumina to operating temperature and dissolve it, it would seem a good idea to pre-heat the reactant before injecting it in the melt. Indeed, Kobbelvedt found that pre-heating SGA to 600 °C before feeding led to an increased proportion of alumina being dispersed and dissolved rapidly [54].

This may indicate that with pre-heating, the reduction of gas formation promoting dissolution is compensated by the heat contained in the alumina to limit freezing and agglomeration of particles. Fast dissolution with pre-heated alumina was also measured by Jain [55]. However, pre-heating alumina to temperatures that would result in full conversion to alpha is likely to be catastrophic for dissolution in practice as dispersion would not be aided by gas release and lead to rapid agglomeration.

While alpha alumina can be challenging to work with from a smelter point of view, it is not impossible. There are multiple cost implications however, and therefore a decision to use high alpha SGA should be thought through in that regard. Certain reduction cell technologies are actually more "capable" when it comes to handling high alpha SGA. Magnetically uncompensated cells have a higher metal pad velocity which enables higher back-feeding of the sludge. Normally operated at higher anode–cathode distance, they are often operated at higher superheat, which also helps with the dissolution of alpha content. Finally, cells that do not yet operate point feeders may be a good client for high alpha SGA, as the lack of dose weight control and its larger amount is more prone to sludging, the operations are usually designed to handle this better. However, regularly switching between high and low alpha and alumina sources will inevitably induce variation in the process and incur sludging, affecting current efficiency.

11.5.3 Impact of Particle Size Distribution

Fine alumina, especially − 20 μm has been found to have a major impact on dissolution, both in laboratory and industrial cells [52, 56]. Having finer alumina affects flowability and dispersion in bath.

Lindsay [33] re-iterated the correlation between measured alumina flowability and smelter performance. Looking at different plants over time, this work found a strong correlation between the minus 20 μm content of the SGA and the smelter current efficiency. The correlation was also repeatable between plants. The impact of fines is found to be greater when the funnel flow time is high (low alumina flowability, as shown on Fig. 11.33. As mentioned earlier in this chapter, the flowability of alumina affects the time scale of the feeder dose and the fanning on the surface, closely related to the dispersion of the doses.

Bagshaw [51] found that fines dissolved approximately three times slower than SGA but that as the alpha content of the fines increased from 50 to 100%, the dissolution time penalty was only an additional 25%. Industrial SGA being a distribution of phases and particle sizes, the nature of the fines therefore plays a role in the dissolution in smelters. With fine particles containing larger amounts of alpha-phase and/or residual gibbsite compared to the bulk [57], the interaction of fine particles affecting flowability coupled with the presence of high alpha content can make dissolution of fine SGA very problematic.

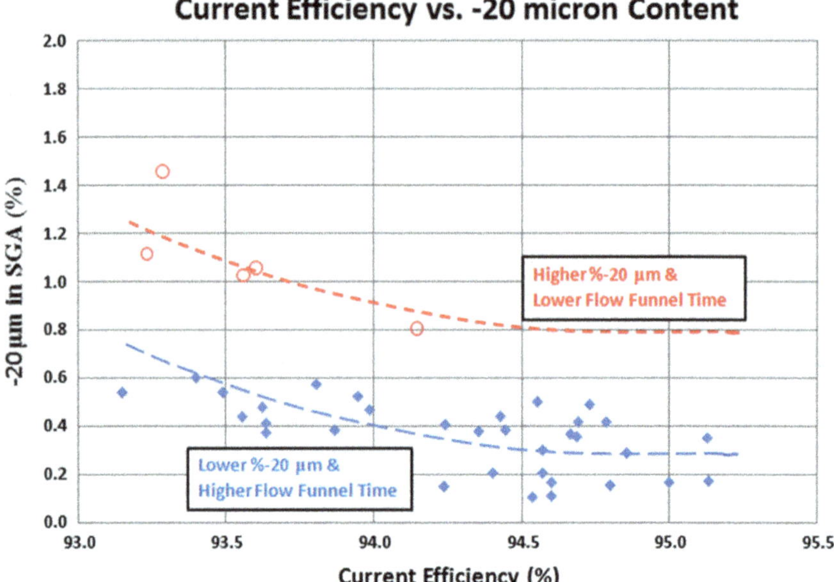

Fig. 11.33 Impact of fines and flowability on current efficiency in one smelter [33]

11.5.4 Effect of Calcination Technology

As explained in Chap. 10, the calcination technology and how it is operated impacts both the particle size and phase composition. As such, it can affect the dissolution rate of the SGA produced.

During calcination, the heating rate and residence time affects the calcination pathway of gibbsite. Rotary kilns tend to produce a larger percentage of alpha alumina compared to gas suspension or circulating fluidized bed calciners. This is notably detrimental to dissolution. Depending upon other properties, smelters using SGA containing < 6% alpha may be in the best position in terms of dissolution. Dissolvability issues can become substantial around 10% alpha in the bulk and large-scale sludging in reduction cells should be expected above 15% alpha in the bulk. As mentioned previously, it is possible to adjust operations to cope with high levels of alpha, and some smelters are better equipped to deal with the type of material produced in rotary kilns. However, switching between radically different materials is bound to upset smelting operations to some extent.

The transitional phases formed being different between calcination technologies (refer to Chap. 10, Fig. 10.39), one might expect some impact on dissolution. Unfortunately, no specific studies have yet been performed to assess the dissolution of these different phases (theta vs. kappa alumina for example). Therefore, everything else being equal, one could refer to the LOI of SGA as a proxy to dissolvability [53] as it is possible that different phases with similar hydroxyl content would dissolve similarly. The desirable LOI range of 0.70–0.85% remains.

Rotary kilns also tend to produce SGA with higher attrition, which also hinders dissolution as particles breakdown in transport and processes prior to being fed in cells. This would be especially problematic in smelters with handling and gas treatment technologies that are hard on particles. e.g., conveying equipment with high velocities, sharp angles, high recirculation in scrubbers, etc.

Through time, calciners have evolved considerably in designs. Modern stationary calciners are more energy efficient and can reliably produce SGA with less than 2% alpha content. As long as residual gibbsite is also not present, dissolvability is maximized while controlling HF generation. Attention should also be put on porosity distribution to ensure HF capture capability is maintained as surface provided by pores < 3–4 nm do not contribute significantly to HF capture [39, 41].

As operation of a hydrate by-pass, which helps reduce the calciner energy consumption, also tends to increase the LOI and SSA of the product, calciners should refrain from compensating with over calcination of the bulk as this would also result in higher concentration of alpha and induce a bi-modal pore size distribution.

The nature of the fines is also problematic. In significant proportions, fines and superfines would certainly be detrimental to dissolution when physical properties of the bulk such as flowability are negatively impacted. But the fine particles are also prone to either over calcination AND under calcination. ESP and bag-house dust are a source of this material. Care should be taken to properly calcine this material and dose it consistently to cells.

Experiences in smelters tend to show that variability in the content (and its impact on flowability) has more detrimental effects than the absolute percentage of fines. Finally, as fines are prone to segregation during transport and storage, slugs of this material periodically going to portions of the reduction line would be most damaging to performance.

11.5.5 Impact of Alumina Feeder Design and Reduction Cell Operations

With everything else being equal in terms of reduction cell operations, the nature of alumina can have a significant impact on the performance of a smelter. However, there are a multitude of factors regarding reduction cell technology, design and operation that also impact the dissolution of alumina [58]. The subject of this book focusing on alumina, only brief summaries of cell aspects and their effects is given here.

Bath Chemistry

Bath chemistry affects the solubility of alumina in the electrolyte. Dissolution rate decreases linearly when alumina concentration goes above 4% in AlF_3 modified bath, or even as low as 3% in high LiF modified bath [49, 59, 60]. Smelters therefore aim to maintain alumina concentration between 2 and 3.5%.

The amount of excess AlF_3 in the bath also affects dissolution as it reduces alumina solubility. Kushel and Welch [52] found a doubling of dissolution time when the excess AlF_3 was increased from 8 to 12%, which is the common range of operation in smelters.

Bath Superheat

Superheat is arguably the most influential smelter factor impacting alumina dissolution in modern reduction cells. Within the aluminum industry, superheat is defined as the bath operating temperature above its liquidus. While not appropriate in thermodynamics terms, it essentially reflects the amount of sensible heat available within the electrolysis zone. In order to properly dissolve an alumina feed shot a minimum superheat of 5–6 °C, if not more, is required locally in the feeding zone. Otherwise, the dissolution process becomes heat-transfer limited with freezing occurring on particles that cannot be re-melted. Agglomeration occurs and leads to sludge formation [55]. Low electrolyte superheat therefore has a large impact on reducing first stage dissolution and promotes second stage, increasing dissolution time by 35% to more than 50% [52, 59]. Even with the best alumina available, lacking the heat energy to dissolve it will result in sludge formation.

Turbulence

Turbulence plays an important part in dissolving alumina as it impacts firstly the local dispersion of the added alumina, then the heat transfer to the remaining agglomerates of grains. It therefore contributes to reducing the re-melting time of agglomerates. Having lower turbulence increases the dissolution time by up to twenty-five-fold under laboratory conditions [52].

Turbulence in the feeding zone comes from the anodic gases released by the reduction process, primarily as CO_2. Cells with anodes that are slotted with the gas directed towards the center channel therefore have higher dissolution rates [33]. It is however important to differentiate gas-induced turbulence in the feeding zones that will help dissolution from large liquid displacement turbulence driven by out-of-control instability that is driven by magneto-hydro-dynamics (MHD). These would rather cause disruption to the bath-metal interface that will result in increased re-oxidation of metal and lower current efficiency in the cell.

Feeder Design

Several designs of point feeders are employed, and they can impact dissolution in multiple ways. The operation of the crust breaker, the role of which is to open the top of the cell so that alumina can be fed onto open bath, has a large impact. The rate of feed of the dose is also important as slowing down the rate leads to higher alumina/bath effective contact area. The height of delivery, that is the velocity of the particles entering the bath, also affects the effective contact area.

The size of the dose certainly has an influence. It might be assumed that smaller doses would be easier to dissolve as the strain on local sensible heat available for dissolution is smaller. However, a smaller dose means to feed at a higher frequency. When the breaker is operated more often, its temperature can then become too high, leading to an accumulation of frozen bath on it and loss of control of the feeder hole condition.

A high number of point feeders relative to the cell amperage can help control this. But a smelter which embarks on an amperage increase program will end up feeding at high frequency, and often into a smaller mass of electrolytes as the anodes are also enlarged. Thus, dissolution will eventually become a bottleneck to amperage increase.

The locations of the feeders can also have an effect. Feeders placed between two anodes in the center channel feed into a smaller bath volume compared to when placed at the corners of four anodes [61]. Also, metal pad movement driven by MHD is the main driver for alumina distribution globally within the cell. Feeders therefore supply alumina for certain anodes following the metal pad flow. This can lead to zones with lower alumina concentration and can result in higher anode effect frequency [62, 63].

Feed Strategy

As briefly mentioned earlier, feed rates are changed to induce resistance changes in the cell. No direct and continuous measure of alumina concentration and bath resistance changes with concentration is available. Lowering the feed rate to "starve" the cell allows tracking the resistance change down to a minimum and then increasing to a desired value corresponding to a given concentration.

The aggressiveness of the feeding strategy affects the local sensible heat available for dissolution, and even the global sensible heat. This can therefore lead to decrease of stage 1 dissolution and result in sludging. However, the overfeed/underfeed around the mean consumption of the cell must be large enough to overcome other variables affecting resistance and sources of unmetered alumina additions such as back-feeding and anode cover contributions.

Cell Operating Parameters and Effect on Feeder Hole Condition

Finally, other cell parameters and operations can also affect dissolution. The complex interactions between design factors, cell parameters, cell operations and alumina properties play together and affect the feeder hole condition and impacts dissolution.

Liquid levels, especially bath height impact the bath volume and sensible heat energy that is available for dissolution. Operations like anode changing drains the sensible heat energy significantly and can have a large impact on the feeder hole condition and the ability to maintain dissolution within a nearby feeder hole.

For an existing smelter, design often comes from the original construction. Amperage increases programs, when faced with dissolution issues, may warrant some changes. This in the past has been done mostly by converting older feeding mechanisms (side works, bar breakers) to point feeders. The capital cost involved by the number of replications and the logistics of the changes, over five to seven years as the cells are replaced, makes design changes challenging. However, understanding how operations and cell parameters interact and affect dissolution for a given smelter or technology is the key to controlling it. Then, changes to dissolution coming from a change in alumina source for example can be anticipated and more or less controlled.

11.6 Impurities Impact upon Aluminum Production, Purity and Properties

11.6.1 Sodium Oxide and Calcium Oxide

The sodium oxide content of SGA is of interest to both refineries and smelters. At refineries control of the level of sodium oxide in hydrate and calcined alumina is tied to yield and thus to overall production. Interest in sodium oxide at aluminum smelters has mostly to do with bath balance, bath chemistry control, and cost [64].

The mass balance around sodium in smelters is dynamic. The bath contained in reduction cells is primarily NaF and AlF_3. Trace levels of Na metal leave with the aluminum produced. A large portion of Na inputs are intercalated into cathodes. The amount depends upon the types of cathode blocks and the age of the reduction cell. Young pots soak up sodium rapidly and old pots have fewer sites left for intercalation. Losses of sodium into cathode blocks also depend upon their composition. Graphite cathodes intercalate less sodium than anthracitic blends with 30 wt% or 50 wt% graphite grain additions. Many modern, high amperage cells have migrated to 100% graphite cathode blocks. Accordingly, such cells produce about 50% more excess bath than cells that used graphite-added, anthracitic blocks [65]. There are other sodium losses in smelters and other net inputs such as with sodium carbonate, or soda ash, additions.

Note that some smelters have an average concentration of Na in aluminum of more than 100 ppm. Other smelters may have less than 50 ppm Na in metal. Those with higher sodium content in metal generally have high levels of current efficiency. However, good current efficiency may also be obtained with moderate levels of Na in aluminum. Note that the sodium content of metal is not linked in any direct way to the Na_2O content of alumina.

High levels of Na_2O in SGA require additional AlF_3 to neutralize it in order to keep the bath chemistry ratio of NaF to AlF_3 within its desired range.

All forms of Na that enter or leave reduction cells help to determine the bath balance of a smelter. Some smelters consume bath. Most generate excess bath. Generally, two main factors determine the overall mass balance. Sodium absorption into the cathode is the large sink for Na and sodium oxide in the alumina is the predominant input of Na.

Smelters with low pot life tend to have losses of Na exceed inputs. Accordingly, they consume bath. These smelters generally prefer Na_2O content of greater than 0.40 wt% in SGA to reduce the amount of bath that must be purchased and distributed. Some smelters are bath-neutral, neither consuming nor producing significant quantities of bath. These generally have long pot life and low Na_2O content (e.g., less than

0.30 wt%). Most smelters produce more bath than they need. This creates market surpluses of bath except when many new smelters are under construction. These plants generally have long pot life, graphite cathode blocks and moderate or high levels of sodium oxide. The balance point for these customers may be less than 0.28% Na_2O in SGA [65].

Excess bath costs much more to produce than it can be sold for. The amount that is recovered by sales of excess bath may only cover 10–15% of its net production costs [65]. This causes smelters that produce much excess bath to call for reductions in sodium oxide content in SGA. As the primary aluminum industry migrates more and more towards graphite cathodes and long pot life as new technology replaces old the call for lower Na_2O content in alumina will only tend to increase.

The concern will not be entirely due to costs. Disposal and storage of excess bath as an industrial waste has been an emerging and growing concern since 2016. Regions such as Québec and some nations including the People's Republic of China have instituted or are considering regulations on how long excess bath materials can be stored on site and/or transported across international borders.

There are few other uses for excess bath outside the primary aluminum industry. Those that exist do not have the potential to consume much volume. Accordingly it is anticipated that pressure upon alumina refineries to find innovative ways to reduce the Na_2O content of SGA is going to increase. It is likely that this issue will soon become another example of evolutionary change for what is expected of SGA by consumers.

Variation of Sodium Oxide Content

Control of bath ratio, or excess fluoride content, is another matter related to the sodium oxide content of alumina. Smelters generally drive to operate with as high an excess fluoride or AlF_3 content as practicable to keep pot operating temperatures low. The solubility of aluminum in bath is temperature dependent and is a major factor in current efficiency optimization, perhaps second only to cell stability. The lower the temperature is, the lower the likelihood for re-oxidation of metal and current efficiency loss [66].

However, if the excess fluoride content becomes too high the cell can become cold enough to suffer loss of stability and operational problems. It's a tightrope that requires consistent Na inputs to balance the NaF content and AlF_3 input that help to keep bath chemistry on target.

When the Na input in SGA varies it affects bath chemistry and the ability of cells to operate at peak efficiency. Thus, smelters prefer low variation of Na_2O in SGA, generally no more than 0.02 wt% variation from one shipment to the next. When shipment to shipment variation exceeds 0.03 wt% it begins to invite short-term operational issues for reduction cells. High variation between shipments at a smelter may also be caused by the use of multiple sources of SGA.

Unfortunately, there is no uniform method or industry standard for measuring and reporting the variation of Na_2O content in alumina. A standard lot size, such as 5000 mt of SGA production, for a sample and the method for gathering samples from each lot would be necessary to establish such a standard. The best available information is range data of shipments which can be very misleading when making comparisons between refineries. Alumina lot sizes can vary from a few hundred tons via conveyor belt to more than 50,000 mt in a ship. Not surprisingly the largest shipments tend to have the smallest ranges of Na_2O content between shipments.

Calcium Oxide

Calcium Oxide content also affects bath chemistry in smelters. Calcium accumulates in bath as CaF_2, or fluorspar, a.k.a. "spar". CaF_2 content of 4.5–6.5 wt% is generally desirable. It reduces the liquidus temperature of the bath and modifies the crust. Too much fluorspar in the bath reduces alumina solubility, making the formation of bottom sludge more likely. Concentrations of 7 wt% or higher on average is considered to be too high. CaF_2 also increases the density of the liquid bath reducing its ability to remain cleanly separated from liquid aluminum that is not greatly different in its specific gravity. As such, high CaO in SGA and high CaF_2 in bath can pose a threat to current efficiency [64].

Although CaO in alumina is a major source it is not the only source of calcium that enters smelters. Anode coke also contains calcium that may account for as much as 30 wt% of the total Ca input from raw materials. Certain coke sources from China can have higher than average calcium levels in them. It is generally a reflection of the hardness of the water that is used to cut green coke into manageable sizing after tar residues have been solidified into coke by thermal treatment in shafts or towers. Calcium residue from the water dries onto the porous, sponge-like, surface area of the green coke.

A fraction of the crude oil sources for petroleum coke have high Total Acid Number (TAN) values. These contribute substantially to the Ca content of coke. This has been particularly true of some sources in China. Poor operation of de-salter units at petrochemical refineries may also contribute to high Ca levels in CPC. However such conditions are not the norm for the industry.

Calcium may also enter from the cast house, downstream of the smelter. Cast houses that use materials such as bone ash to seal the lids of crucibles can contribute significant amounts of Ca to the bath since crucibles are regularly cleaned in the services area of the pot lines with material recovered from the crucibles then flowing back to the pots.

The maximum calcium content of aluminum metal is important for a few product specifications. However, CaO readily accumulates in the bath as CaF_2. Accordingly, there is very little transference of Ca into the metal, typically a few parts per million. It is worth noting that investigations into problems with Ca in metal have generally uncovered issues with control of materials and work practices in the cast house, not the CaF_2 content of bath.

While the CaO level of the SGA is important, the ratio of CaO to Na_2O is the primary metric of interest for the fluorspar content in bath. Excess sodium oxide contributes to new bath generation that dilutes fluorspar. Note that the average CaO/Na_2O ratio involving all alumina shipments over periods of six months or more is the measure of importance. The large amounts of bath in a smelter buffer the impact that most individual shipments of alumina may have on the CaF_2 level in bath.

A CaO to Na_2O ratio of 8% or less in SGA over time will typically cause the CaF_2 content of bath to decrease. A CaO/Na_2O ratio of 10% is generally acceptable for most smelters. With ratios greater than 12% over time many smelters will tend to see increases in CaF_2 levels. Ratios of 14% or higher may then force a smelter to use another source of SGA or to take expensive measures to periodically dilute fluorspar levels in their bath.

If fluorspar levels go too low, below 4.5%, smelters can add relatively inexpensive acid-grade fluorspar to the pots. However, if fluorspar levels go too high the only remedy is generally to dilute the bath at a cost that can easily exceed hundreds of thousands of dollars per year. Acquiring shipments of SGA from a source that has a very low CaO/Na_2O ratio is another way to dilute the CaF_2 content of electrolyte in pots. Sources of SGA in the equatorial region of Brazil are particularly low in CaO/Na_2O ratio.

The ratio CaO to Na_2O is a rather new technical parameter for SGA quality [64]. It was first defined by Gene S. Miller and first reported on by Stephen Lindsay of Alcoa. However, some aluminum producers have taken to reporting the ratio of Na_2O to CaO. As many alumina refineries have a CaO/Na_2O ratio close to 10%, the corresponding Na_2O to CaO ratio is often close to 10. This difference in approach and similarity of the measures can create some confusion. The CaO/Na_2O ratio was created by Miller to define a metric for control of CaF_2 content. As CaF_2 is a minor constituent of bath when compared to the NaF content the appropriate metric is CaO/Na_2O. When this ratio is low, so is the CaF_2 content in bath and vice-versa.

11.6.2 Iron Oxide and Silica

Iron

Iron oxide, Fe_2O_3, content in raw materials is a primary impurity of concern for most metal products. Using a consumption factor of 1.920 t Al_2O_3/t Al, a concentration of 100 ppm of Fe_2O_3 in alumina contributes 134 ppm of Fe to the metal produced. This may not appear to be much when one considers that the iron from alumina contributes only 10–20 wt% of the total iron in the metal at many (not all) smelters. But, unlike pot room process contamination factors that can vary greatly from pot-to-pot on iron impact, the iron oxide content of alumina and other raw materials affects the metal produced from every reduction cell [67].

Consider the example shown in Fig. 11.34. This "visual mass balance" draws attention to the fraction of pots that can produce metal of high purity, less than 650 ppm Fe, or Grade P0406A [99.90% Al purity]. By changing from an SGA source with 0.014 wt% Fe_2O_3 to another source with 0.010 wt% Fe_2O_3 much more of the desired premium product can be produced.

Figure 11.34 also illustrates that the contribution of carbon anodes to the iron content of metal can be similar to the contribution of SGA. Other raw materials contribute very little Fe to the metal.

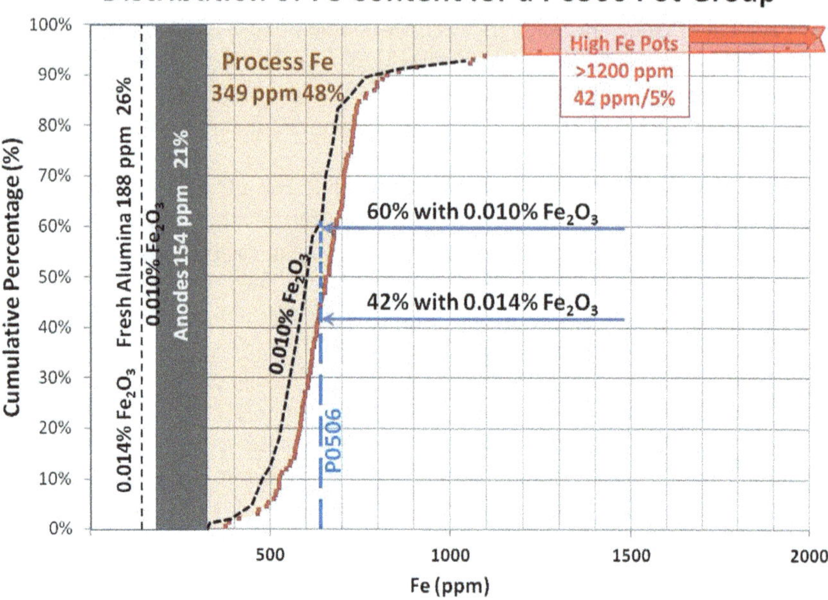

Fig. 11.34 Visual mass balance of iron contamination sources

Minimization of the process iron component is also a key opportunity for improvement. Process iron is driven by factors that have greater variation than the Fe_2O_3 content of alumina. These factors include variations in total liquid levels in reduction cells, the thicknesses of individual anode butts and old or damaged cathodes.

Variation of the Fe_2O_3 content is important as well as it also affects all cells. Building upon the example given in Fig. 11.34 an alumina source with 0.012 wt% Fe_2O_3 and low shipment-to-shipment variation may be preferred to an alumina source with 0.010 wt% Fe_2O_3 and high variation.

In Fig. 11.35 we see such an example of a refinery that is normally quite capable of providing alumina to a producer of P0406A metal. Shipments typically average 0.009–0.012 wt% Fe_2O_3. However, some individual shipments are well above this typical range.

As with other impurities, acceptable levels for metal products will be determined by more than just the average alumina properties. They will more often depend upon the maximum level of an impurity.

With this in mind the guidelines for Fe_2O_3 in SGA is shown in Fig. 11.36. These are not product specification limits. They are general summaries of what aluminum producers will typically seek out for Fe_2O_3 content in SGA in order to be able to produce substantial amounts of the metal products listed.

There is one aspect of the Fe_2O_3 content of SGA that many smelting customers may not understand. There is an inverse relationship that the iron content of bauxite has with the iron content of alumina. Typically, the Fe_2O_3 content of SGA will be

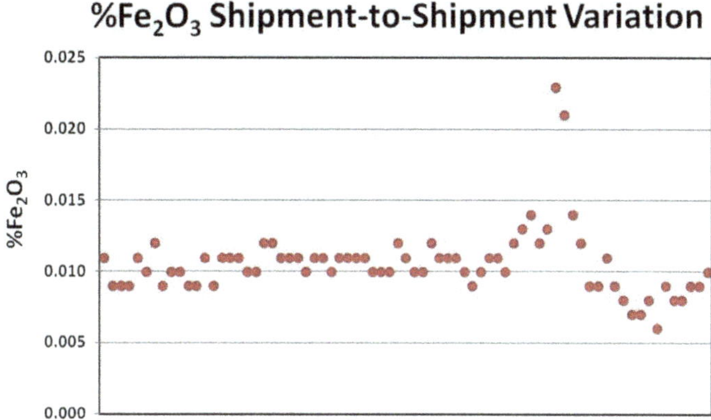

Fig. 11.35 Example of variation from an alumina refinery

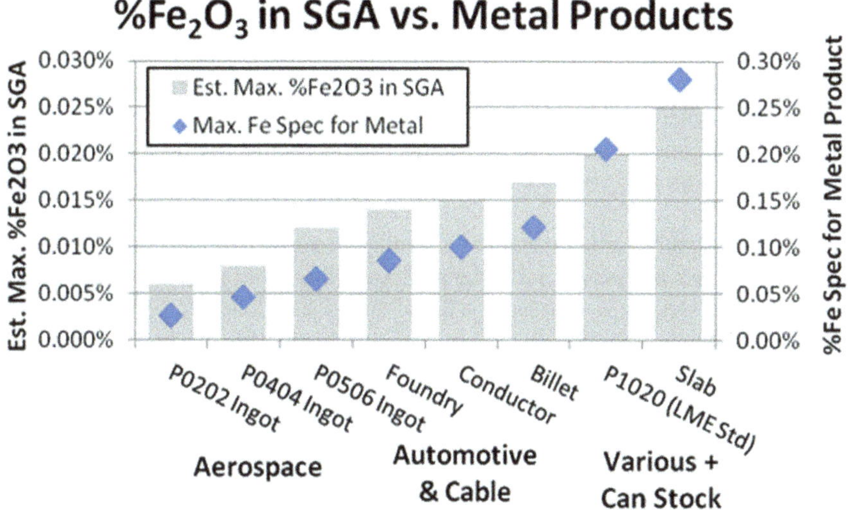

Fig. 11.36 General guidelines for maximum Fe_2O_3 content in SGA [67]

low when for Fe_2O_3 in bauxite is high. This is especially true when the hematite content of bauxite is high, and the Goethite content is low.

Silicon

The SiO_2 content of alumina is also important for metal products. Silica enters with alumina, carbon anodes, and AlF_3. It often also enters from silicon carbide materials that are used in cell lining designs.

Fewer metal products rely upon low silica levels than low iron levels. These are primarily limited to electrical conductor products and to pure metal grades that are suitable for aerospace and electronic products. Refer to Fig. 11.37. The majority

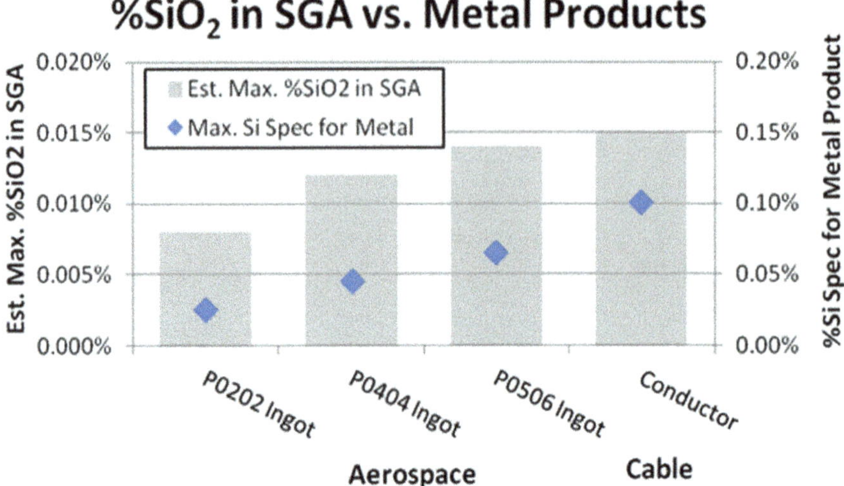

Fig. 11.37 General guidelines for maximum SiO_2 content in SGA [67]

of aluminum smelters in the world, perhaps more than 75%, will have no product specific limitations for which typical contents of SiO_2 in SGA pose a threat. Note that a concentration of 100 ppm of SiO_2 in alumina contributes 90 ppm Si in metal produced [67].

Silicon contamination in metal has a significant negative impact on its electrical conductivity. It is not as harmful at the part per million level as many other metallic impurities. However, Si is typically found in greater concentrations in primary aluminum than all other metallic impurities except iron. It is this higher concentration that is behind its overall negative impact on conductivity. Silicon also detracts from strength and corrosion resistance that are required for aerospace products.

At some level as with other multi-valent elements, Si in metal also detracts from current efficiency [67]. At typical levels of approximately 0.015 wt% SiO_2 in alumina the impact from SGA is below 0.05% in production losses. Typical levels of total Si in metal of 300–400 ppm may account for 0.10–0.14% loss in current efficiency, which is not insignificant to most smelting operations.

As with iron, both the typical and maximum content of SiO_2 in SGA will be of concern. Production of conductor rod and pure metal grades for aerospace and electronic products cannot tolerate spikes in impurities. When such events occur, it can be quite disruptive to the ability to produce these metal products at all.

11.6.3 Phosphorus and Beryllium

Phosphorus

There have been multiple studies in laboratories and in smelters that demonstrate the powerful negative impact that a small amount of phosphorus in bath can have on current efficiency. A concentration of 100 ppm in bath can cause a loss of 0.85 ± 0.20% current efficiency and a similar loss in net aluminum production [68]. See Fig. 11.38.

Phosphorus is multivalent, "hijacking" electrons via redox reactions in the liquid bath. Accordingly, trace levels of phosphorus in major raw materials is a matter of concern to modern smelters that have point feeding and highly efficient dry scrubbing systems. With a partition coefficient between bath and metal of approximately 10:1, a total of 10 ppm P in metal is roughly equivalent to 100 ppm P in bath.

Not all phosphorus enters smelters with SGA. Anode coke, aluminum fluoride, calcium fluoride and other sources such as bone ash and the cast iron that the smelters use to connect anode and cathode blocks to conductor rods must all be considered in an overall mass balance. However, even at levels below 5 ppm of P_2O_5 alumina may account for as much as 50% of the net input of phosphorus to a smelter. See the illustration in Fig. 11.39 for one modern smelter with 5.0 ppm of P_2O_5 in its SGA supply. While being an over-simplification a Rule of Thumb for modern point-fed prebake smelters is that each 1.0 ppm of P_2O_5 in SGA will contribute approximately 10 ppm P to the steady state concentration of the bath and approximately 1 ppm P to the aluminum metal.

From the perspective of smelting operations little can be done with phosphorus except to shed it in a controlled fashion. Phosphorus is quite volatile and passes readily to the dry scrubbing system and to the anode crust circuit where it is captured

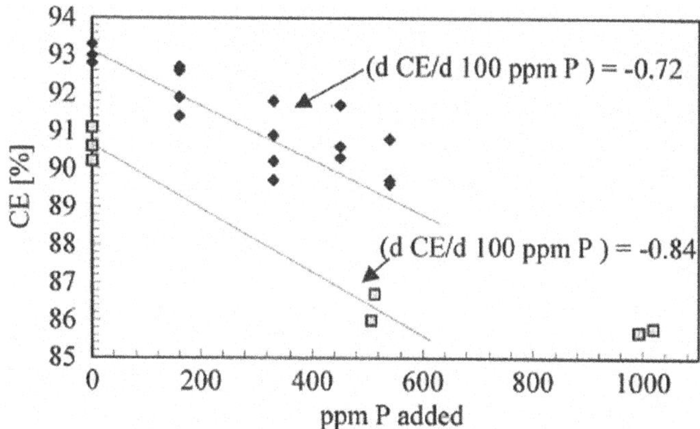

Fig. 11.38 Measured impact of phosphorus in bath upon current efficiency [68]

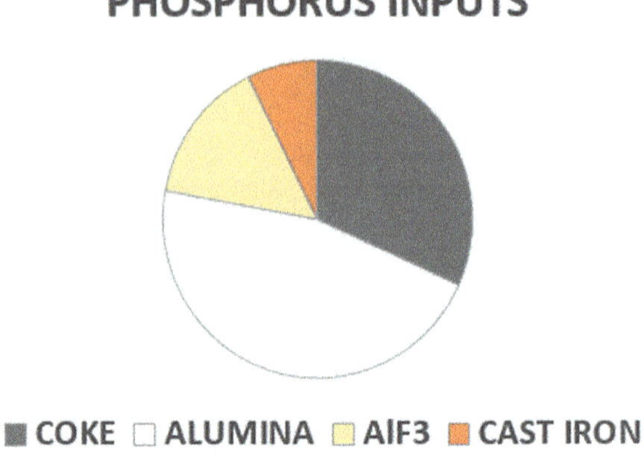

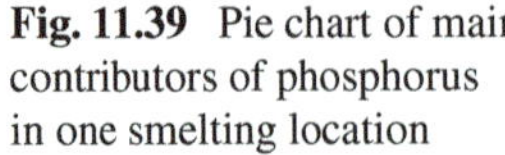
Fig. 11.39 Pie chart of main contributors of phosphorus in one smelting location

and returned to the reduction cells. A large fraction of phosphorus simply exits the cells at trace levels with the metal as it is transferred to the cast house.

It has been observed that the distribution of phosphorus in bath and metal of reduction cells is commonly bi-modal. Pots are often either high or low in phosphorus content with few pots having intermediate levels unless phosphorus contamination is present in abundance. Phosphorus attaches to the surface of anode dust giving it mobility through the anode crust and dry scrubber recirculation circuits [69]. Thus, skimming of carbon dust from pots and proper attention to hole cleaning during anode change both provide opportunities to rapidly purge phosphorus from the process.

Note that metal and bath in Søderberg reduction cells are all but devoid of phosphorus. Carbon dust skimming, fugitive losses, and losses to collection systems such as cyclonic separators account for this difference. Accordingly, older technology smelters with Søderberg cells or side-worked pots are not very sensitive to the level of P_2O_5 in alumina and other raw materials.

Phosphate in SGA is usually tied to two main factors. There are contributions from bauxite that can be observed in hydrate and contributions from calcination if heavy fuel oil is used [70]. Understanding the contributions from each source is important for the control of P_2O_5 in SGA. With heavy fuel oil purchases the phosphate content is normally not included in specification sheets. However, refineries can qualify sources of calcination fuel based on analyses of phosphate levels in product. If low phosphate oil happens to carry a premium in pricing, then separate grades of fuel oil may be able to be used at the boiler house of the refinery with the lower phosphorus oil used in calcination.

Phosphorus in aluminum metal affects both porosity and brittleness. A small fraction of metal products, less than 10% by volume, have tight constraints on phosphorus content. Such constraints are most common with foundry products that are used for wheels on cars and trucks. Constraints also exist for some forms of rolling slab.

Due to its volatility some phosphorus may be lost to the ambient just by holding crucibles for minimum time periods before pouring the metal to ingot furnaces. Phosphorus that remains in the metal can be treated by strontium modification prior to casting to reduce it to acceptable levels. However, there are limits on the amount of strontium that can be added.

Beryllium

Trace levels of beryllium are co-deposited with bauxite in every known source in the western world. Bauxite with the highest concentrations of beryllium can be found in certain regions of China, Jamaica, Greece, and the Balkan Peninsula. High trace level concentrations can also be found in some countries on the Mediterranean rim, Saudi Arabia, the Minas Gerais region of southern Brazil, and islands of the Caribbean. Modest trace level concentrations are found in countries on the Caribbean rim, and the Gove peninsula of Australia. The lowest concentrations of Be are found in bauxite deposits of Western Australia, northern Brazil, Vietnam and the west coast of Africa. Regardless of the source a fraction of the beryllium found in bauxite, often in the range of 25–33 wt%, then arrives to SGA.

Many nations have identified beryllium to be a health risk that is most notably associated with specific allergic reactions, respiratory conditions and chronic beryllium disease. The concentrations in alumina are low enough that over-exposure potential at alumina refineries is highly unlikely. However, this is not the case for all aluminum smelters.

Beryllium tends to be uniformly distributed in bath from pot-to-pot. As alumina containing elemental Be is fed to each pot it rapidly reacts with fluoride and sodium in the bath to form BeF_2 and Na_2BeF_4 [71]. Although it is a generalization, the rule of thumb is that at steady state the elemental concentration of Be in bath will be approximately sixty (60) times greater than the average elemental concentration of Be in SGA that has been used over the prior thirty-six months.

Fortunately for worker health, tests have shown that beryllium in bath is at least partially water soluble, most likely the fraction that is BeF_2. While more study is needed this may help to explain why the incidence rate of beryllium related ailment is so low in the primary aluminum industry vs. industries with worker exposure to metallic beryllium or beryllium oxide. However regulatory agencies have focused primarily on the presence of elemental Be with regard to protective exposure limits.

The primary source of beryllium to aluminum smelters is through the alumina supply. Most often alumina is the sole source of beryllium input. Long-term average levels as low as 0.2 ppm Be in SGA can be of concern to some smelters that are located in regions with the strictest protective standards. Other regions of the world may not be as strict, but precautionary measures are often advisable in circumstances in which the average Be content of all sources of SGA used over a twelve-month period has exceeded or will exceed 0.5 ppm.

Some beryllium may also enter a smelter through imports of bath produced at other smelters that have modest to high trace levels of beryllium in their alumina supply. However, most modern smelters are net producers of excess bath, rather than importers of it.

Beryllium exits smelters in material streams where sodium is also lost. Only about 10 wt% of Be exits with aluminum metal. Much of it eventually exits with spent pot lining material and forms of bath that are disposed of.

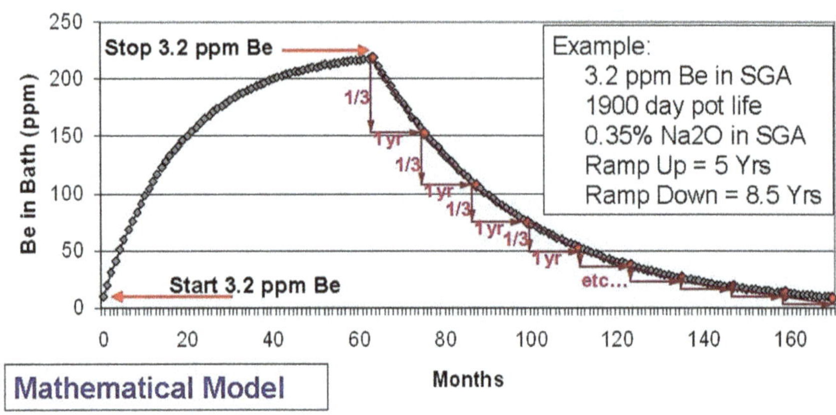

Fig. 11.40 Beryllium concentration and decay model [71]

Levels over 5 ppm Be in alumina are uncommon but do exist in the alumina from a few refineries. Smelters that use these sources over multiple years may also encounter other issues with beryllium in metal. Many products have specification limits of 1.0 ppm maximum for Be.

Once there are relatively high levels of beryllium in bath if the source of alumina then changes, the rate of decay by dilution is slow, taking five to eight years to return to original concentrations. Generally, the dilution rate is such that each year one-third of the gap will be closed between the current concentration of Be in bath and the baseline level of beryllium in bath. See Fig. 11.40 for details.

Once a smelter begins to use an alumina source that has more than a fraction of a part per million of elemental beryllium in it a long-term relationship with concentrated levels of beryllium in pot room bath also begins. The only option to assure low concentrations of beryllium, or other alkaline earth metals, in bath is through rigorous procurement and management of raw materials that are low in these impurities.

At this time there are no processes for removal of beryllium content in pot room bath, alumina or bauxite. Research in this area has identified some technical solutions for bauxite [72]. However, these are currently not commercially viable solutions.

11.6.4 *Vanadium, Sulfur and Other Impurities*

Vanadium

Vanadium pentoxide, V_2O_5, is of modest concern to aluminum smelters, primarily because levels of vanadium in anode coke have been increasing over time. This has a negative impact upon some metal products. As with SiO_2 there is also a slight negative impact of vanadium upon current efficiency. But this is generally well below the small negative impact of Si that is present in higher concentrations in bath and metal.

As with phosphorus, high levels of vanadium in SGA are most commonly associated with refineries that use heavy fuel oil for calcination. However, bauxite can be a significant contributor as well, especially if organic content is high [64].

In the smelter vanadium pentoxide is reduced to vanadium metal in the reduction cells. V_2O_5 at 10 ppm in SGA contributes 11 ppm of V to metal.

Since vanadium is an element that significantly reduces the electrical conductivity of aluminum, some smelting clients may not consider using SGA with more than 0.0015 wt% V_2O_5 on some pot lines. Such locations may also place limits on MnO, Cr_2O_3 and occasionally the ZrO_2 content of SGA.

Vanadium in aluminum is also known to promote coarse titanium particle formation during casting that cannot be removed by thermal homogenization of metal.

In the cast house vanadium and titanium can be precipitated out of metal as borides via additions of boron metal. Doing this also slows the metal preparation processes and reduces cast house output capacity. Precautions must also be taken to prevent the borides from entering metal products as inclusions that can be detrimental to a variety of downstream products.

Sulfur

There are a few points on sulfur that have not already been covered in the section on dry scrubbers and SO_2. It has been noted that very few alumina refineries routinely measure or report the percentage of sulfur in SGA. The reason is that it is difficult to measure and has generally been off the radar screens of most aluminum smelters and alumina refineries with the majority of smelting focus having been placed on the sulfur content of anode coke.

While some sulfur can enter with organics in the bauxite the primary source of sulfur on SGA is considered to come from combustion products of natural gas or heavy oil during calcination. As noted earlier in this chapter the reactive sites on SGA will form weak bonds with SO_2 [39] during calcination. The SO_2 on SGA then remains in place until the alumina is exposed to HF from the exhaust of reduction cells. In effect, some SO_2 that is generated during calcination of SGA at the refinery is then re-located to dry scrubber chimneys of gas treatment centers in aluminum smelters for release or for treatment. Until recently, only about 5% of the primary aluminum industry treats dry scrubber off-gas for SO_2. These locations are primarily in Norway, the Middle East and at some locations in Russia. Most recently there have been many smelters in China that have been adding SO_2 scrubbing systems.

There has also been an emerging regulatory trend on ground level SO_2 concentrations with regulations already in place in a number of countries. Elevated concentrations at ground level present a health hazard. The primary source of SO_2 emissions from aluminum smelters is related to the anode coke. But 2–5% does enter with the SGA and this may add to desire in the primary aluminum industry to see sulfur content reported on Certificates of Analysis.

The data that is available has made it clear that calcination with heavy fuel oil can contribute concentrations of sulfur on alumina that are as much as an order of magnitude higher than SGA that has been calcined with natural gas. See Fig. 11.41.

Zinc

Zinc content of SGA is of concern to many smelters that serve the extrusion billet market. ZnO at 10 ppm in SGA contributes 15.4 ppm of Zn to metal.

This impurity causes spangling, a surface finish defect. It is similar to the surface appearance of galvanized steel but on a much smaller scale. Refer to Fig. 11.42. This is of greatest concern on products such as clear coated automotive trim and Venetian blind stock. For this reason, some smelting clients may not consider using SGA with more than 0.010 wt% ZnO. High levels of ZnO in bauxite and SGA is most commonly associated with Jamaican sources.

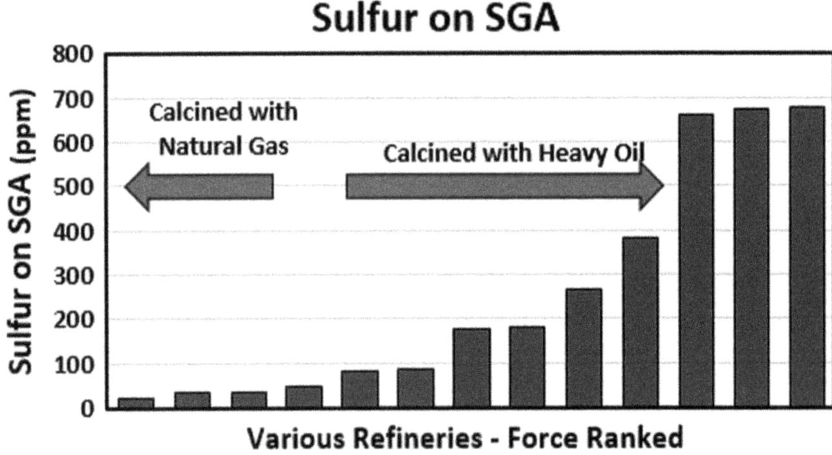

Fig. 11.41 Comparisons of sulfur content of various SGA sources

Fig. 11.42 Sample of alloy metal with a heavily spangled surface

Gallium

Gallium oxide is commonly found in modest levels in all sources of SGA. It is also commonly reported on Certificates of Analysis and in maximum specification limits. With the exception of very high purity ingot for aerospace and electronic applications it is all but inconsequential to aluminum smelters and to downstream metal products.

Titanium

The same is generally true of TiO_2. While there are metal products that have tight Ti limits, sometimes in combination with other trace metals, the TiO_2 content of alumina receives very little attention. This is because most cast houses add small amounts of titanium diboride to wrought alloy products as a grain refiner of the metal.

Variation of TiO_2 in SGA appears to be related to flow rates and residence times in clarification. Significant curtailments in SGA production rates may cause the concentration of TiO_2 in SGA to increase.

Potassium

The content of K_2O in SGA is generally low enough that it seldom gets much attention. Like beryllium, potassium concentrates in the bath at smelters. It is known to have severe negative effects on intercalation swelling of cathodes and upon the life of pot linings. The area of greatest concern for K_2O content is currently in China and to some extent in Russia. Some regional deposits of ore in China are quite high in potassium.

Lithium

Like other alkali metals lithium accumulates in the bath of reduction cells. Unlike Be and K, the effects of LiF in bath are generally beneficial. Lithium has a strong, positive impact on the conductivity of the bath. It also reduces the liquidus temperature of bath enabling lower operating temperatures. This reduces energy consumption and allows the cells to be operated with greater stability. At one time lithium carbonate was commonly added to the bath at some smelters to acquire these benefits. However, the market price of Li_2CO_3 has long-since eclipsed the economic benefits of using this electrolyte additive.

The negative aspect of Li is that levels over a few parts per million are unacceptable for many metal products. Levels greater than 1 ppm also exclude it from some metal products. In general, concentrations below 200 ppm Li_2O in SGA tend to avoid running into issues with most metal products.

Very high levels of Li_2O in alumina, over 500 ppm, have been reported in the Henan province of China [73]. High levels of lithium are also known to exist in bauxite and alumina from Greece and the Balkan Peninsula. Modest levels of lithium are also found in bauxite deposits in Saudi Arabia and in the Minas Gerais region of southern Brazil.

Other Impurities

There are many elements that can be found in bauxite and alumina at trace level concentrations. Invariably these may prompt a few questions of concern by more curious minds at refineries or smelters.

Traces of metals such as Cr, Mn, Co, Ni, Cu, Zr, B, Sn, Sb, Nb, Cd and elements from the Lanthanide series will report to aluminum metal at trace, or non-detectable levels.

Elements including Mg, Sr, Th and U readily form fluorides and will concentrate in the bath, albeit at low levels. Note that the presence of the fluorides of K, Sr, Th and U can make Hall-Héroult Cell bath slightly more radioactive than background radiation levels found in soil. However, granite counter-tops and bananas [potassium] are often more radioactive than Hall-Héroult bath.

11.7 Chapter Summary and Closing Comments

This chapter has attempted to discuss Smelter Grade Alumina from the point of view of users at primary aluminum smelting locations. To achieve this end the many physical and chemical properties discussed herein have been categorized into specific reference units. Some key points have also appeared in more than one of these units to enable more thorough understanding.

The chapter was written to be of use to persons who are involved with alumina refinery operations and quality control as well as those who are involved with smelting operations and raw materials selection.

A lot will happen to alumina after it leaves calcination and before it is fed to a reduction cell. The product quality of SGA as it is loaded to a ship or other container is as far as the refinery can go in exerting oversight of quality. It is just as important for smelting operations and process control to understand their own sub-systems when it comes to alumina handling, dry scrubbing and contributions of process impurities to fluorinated alumina. The Certificate of Analysis for SGA is only a starting point.

This chapter has been focused on that starting point and many of the various physical and chemical properties of SGA have been covered in detail. There has also been focus on definition of what is needed in the way of product quality for smelting operations today. The wide range of what can happen to SGA once it becomes fluorinated alumina feedstock could serve as the subject of a book itself. It is an area that begs more study in the primary aluminum industry.

Nonetheless many readers may want to know what product quality specifications might look like from the point of view of the majority of smelting customers. Needs-based product specifications should not try to lock down every single product quality parameter regardless of its importance. The following reference Table 11.1, tries to capture these needs for the most common types of point-fed, pre-bake reduction cells of > 200 kA on up through the most modern technology of 600 kA + cells.

Table 11.1 Example template for customer specifications for SGA

Specification for smelter grade alumina	
Flowability (s)	≤85 s, Alcoa flow funnel, 100 g, 4 mm orifice
Flowability variation	Standard deviation of flow funnel time per 10,000 metric
Within shipment (s)	Tons of material shipped over the past 12 months
Attrition index (no unit)	≤20 for most customers
BET surface area (m^2/gm)	≥70 and ≤80 m^2/gm
BET surface area/%LOI* (no unit)	≥85* with LOI method changed to (250–1000 °C)
Alpha phase (%)	≤10% for most, ≤6% for some customers
Na_2O (%)	≤0.30%
Na_2O variation (%)	Standard deviation of %Na_2O per 35,000 metric tons of materials shipped over the past 12 months
CaO/Na_2O (%)	≤12
Fe_2O_3 (%)	≤0.015% for most, ≤0.010% for some customers
Peak FeO_3 (%) reference only	Highest value of material shipped over the past 12 months
SiO_2 (%)	≤0.015% for most, ≤0.010% for some customers
Peak SiO_2 (%) reference only	Highest value of material shipped over the past 12 months
P_2O_5 (%)	≤0.010% for most, ≤0.005% for some customers
ZnO (%)	≤0.010% for most customers
Be (ppm)	0.5 ppm for some, ≤0.2 ppm for other customers
S (%)	Some customers may require reporting

The "customer specification" template offered here is sure to be anything but universally accepted. It includes many commonly reported properties. But it also includes measurements made by methods that are not generally recognized in the industry. It even calls for the modification of ISO Method 806:2004 for LOI content determination. However, each line item is tied to needs of smelting customers. Note that many commonly reported impurities of lesser importance have been omitted.

Measurements of particle size fractions, or truncations of the overall particle size distribution, convey little meaning to customers that are concerned with the flowability of alumina and how well it can integrate the fine particles that are added to convert it to fluorinated alumina. Accordingly reporting these size fractions is not necessary so long as a standard measure of flowability is reported.

Combination of Gibbsite with LOI provides a more meaningful metric for HF generation. The ratio of BET Surface Area and LOI content closes the loop on parameters that are important to dry scrubbing.

Few refineries may be able to easily attain less than 0.30 wt% Na_2O in product. However, their customers have no need to produce excess bath at great expense for a material that may soon become widely regulated as an industrial waste product. The specification is only a reflection of the actual need of most SGA customers.

The metrics on product variability are currently not common in the industry. Data on peak concentrations help to define product utility for the production of certain metal products. Data on standard deviations of certain parameters are tied to product stability that customers desire from shipment to shipment or within shipment.

In an industry that is dominated by producer specifications, guaranteed limits and Certificates of Analysis table 11.1 certainly may give a number of producers a more clear focus for the years to come.

We hope that you have found this chapter to be both useful and informative. Our best efforts have been applied to provide you with current, fact-based information that have originated in both published and unpublished works as well as our own extensive experiences as end users of SGA.

References

1. T. Ashida, J. Metson, M. Hyland, New approaches to phase analysis of smelter grade aluminas. Light Metals **2004**, 93–96 (2004)
2. ISO Method 806:2004 Determination of loss of mass at 300 °C and 1000 °C, International Organization for Standardization. https://www.iso.org/standard/36360.html
3. J. Metson, Introduction to smelter grade alumina—Production, distribution and trends, *TMS Short Course on Smelter Grade Alumina*, Charlotte, N.C. (2004), p. 16
4. F. Habashi, Karl Josef Bayer and his time—Part 2. CIM Bull. **97** (1083) (October 2004)
5. P. Homsi, Aluminas for modern smelting, in *Proceedings of the 3rd Australasian Smelting Technology Conference*, Sydney, Australia (November, 1989), pp. 474–506
6. L. Perander, M. Stam, M. Hyland, J. Metson, Towards redefining the alumina specifications sheet—The case of HF emissions. Light Metals **2011**, 285–290 (2011)
7. Photo reprinted from B. Whittington, D. Ilievski, Determination of the gibbsite dehydration reaction pathway at conditions relevant to Bayer refineries. Chemical Eng. J. **98** (1–2), 89–97 with permission from Elsevier to reprint (15 Mar 2004)
8. F. Allard, M. Désilets, M. LesBreux, A. Blais, Chemical characterization and thermodynamic investigation of anode crust used in aluminum electrolysis cells. Light Metals **2015**, 565–570 (2015)
9. Screen capture courtesy of Alcoa Corp. from "Unmasking the Mysteries of Hydrate and Alumina" video produced by Alcoa of Australia Ltd. and The Western Australian Centre for Microscopy at The University of Western Australia, in *Presented at the 4th International Alumina Quality Workshop*, Darwin, Australia, (June, 1996)
10. X. Wang, Alumina dissolution in aluminum smelting electrolyte. Light Metals **2009**, 383–388 (2009)
11. B.E. Raahauge, "Experience with Gas Suspension Calciner for Alumina", TMS Annual Meeting – Alumina & Bauxite Symposium, New Orleans, February, 1991, 1991, this was a late-entry paper that was presented but not included in *Light Metals 1991*, available upon request from the author (February 1991)
12. T. Johnston, N. Richards, Correlation between alumina properties and crusts. Light Metals **1983**, 623–639 (1983)
13. T. Groutso, M. Taylor, A. Hudson, Aspects of crust formation from today's anode cover material. Light Metals **2009**, 405–410 (2009)
14. X. Liu, M. Taylor, S. George, Crust formation and deterioration in industrial cells. Light Metals **1992**, 489–494 (1992)

15. S. Lindsay, Effective techniques to control fluoride emissions. Light Metals **2007**, 199–204 (2007)
16. S. Lindsay, N. Dando, Dry scrubbing for modern pre-bake cells. Light Metals **2009**, 275–280 (2009)
17. N.T. Chaplin, Reaction of lime in sodium aluminate liquors. Light Metals **1971**, 47–61 (1971)
18. W.S. Tyler, *Standard test sieves*. http://wstyler.com/product/test-sieves/#tyler-test-sieves
19. ISO Method 2926:2013 Aluminum oxide used for the production of primary aluminium—Particle size analysis for the range 45–150 μm—Method using electroformed sieves, International Organization for Standardization. https://www/iso.org/standard/52316.html
20. R. Zou, X. Lu, A. Yu, G. Roach, Packing and flowability of alumina powders, in *Proceedings of the 8th International Conference on Bulk Materials Storage, Handling and Transportation*. University of Wollongong, July, 2004, with permission to reprint image 11.ii.3 from Engineers Australia
21. S. Lindsay, Attrition of alumina in smelter handling and scrubbing systems. Light Metals **2011**, 163–168 (2011)
22. H. Hausner, Powder characteristics and their effect on powder processing. Powder Technol. **30**, 3–8 (1981)
23. C. Behrens, The impact of alumina quality variations by loading and un-loading of marine vessels. Light Metals **2005**, 123–125 (2005)
24. D. Audet, R. Clegg, Development of a new attrition index using single impact, in *Proceedings of the 8th International Alumina Quality Workshop* (September, 2008), pp. 117–120
25. K. Grjotheim, B. Welch, *Aluminium smelter technology 2nd edition*, 1988, pp. 32 & 202 Additional acknowledgement to the work of Karl Wefers and Chanakya Misra—Alcoa Laboratories (1987)
26. J.V. Sang, Factors affecting the attrition strength of alumina products. Light Metals **1987**, 121–127 (1987)
27. A.D. Zwicker, The generation of fines due to heating of alumina trihydrate. Light Metals **1985**, 377–396 (1985)
28. M. Minniti de Campos, M. Carmo Ferreira, A comparative analysis of the flow properties between two alumina-based dry powders. Adv. Mater. Sci. Eng. **2013** (519846), 7 p (2013). https://doi.org/10.1155/2013/519846
29. D. Geldart, E.C. Abdullah, A. Hassanpour, L.C. Nwoke, I. Wouters, Characterization of powder flowability using measurement of angle of repose. China Particuology **4**, 104–107 (2006)
30. R. Carr, Evaluating flow properties of solids. Chem Eng. **72**, 163–168 (18 January 1965)
31. H.P. Hsieh, Measurement of flowability and dustiness of alumina. Light Metals **1987**, 139–149 (1987)
32. C.K. Matocha Sr., J.H. Crooks, C.S. Zediak, T. Tejchman, Alumina properties characterization—The flow funnel test incorporating angle of repose and loose bulk density determinations. Light Metals **1991**, 179–185 (1991)
33. S. Lindsay, Key physical properties of smelter grade alumina. Light Metals **2014**, 597–601 (2014)
34. ISO Method 18842:2015 Aluminium oxide primary used for the production of aluminium—Method for the determination of tapped and untapped density. International Organization for Standardization. https://www.iso.org/standard/63550.html
35. M. Taylor, Anode cover material—Science, practice and future needs, in *Proceedings of the 9th Australasian Smelting Technology Conference* (November, 2007), pp. 1–20
36. S. Lindsay, "Anode Covering Material", training module of the TMS Industrial Aluminum Electrolysis course, Longkou, China (September, 2016)
37. S. Brunauer, P.H. Emmett, E. Teller, Adsorption of gases in multimolecular layers. Journal of American Chemistry Society **60**(2), 309–319 (1938)
38. ISO Method 8008:2005(en) Aluminium oxide primary used for the production of aluminium—Determination of specific surface area by nitrogen adsorption. International Organization for Standardization. https://www.iso.org/obp/ui/#iso:std:iso:8008:ed-2:v1:en

39. N. Dando, S. Lindsay, The competitive adsorption of HF and SO_2 on smelter grade alumina. Light Metals **2016**, 527–531 (2016)
40. J. Metson, Making and analysing smelter grade alumina, *TMS Short Course on Smelter Grade Alumina*, Charlotte, N.C. (2004), pp. 12, 15, 31–35
41. G. Agbenyegah, G. McIntosh, M. Hyland, J. Metson, Assessing the role of smelter grade alumina porosity in the HF scrubbing mechanism. Light Metals **2016**, 521–525 (2016)
42. M. Hyland, D. Wong, Module 3A: Fluoride emissions and dry scrubbing basics. Control of Potline Scrubber and Fugitive Emissions for Aluminum Smelters Course, Reykjavik (May, 2017), pp. 8–18, 28
43. M. Slaugenhaupt, J. Bruggeman, G. Tarcy, N. Dando, Effect of open holes on vapor-phase fluoride evolution from pots. Light Metals **2003**, 199–204 (2003)
44. M. Hyland, Alumina properties and dry scrubber performance, *TMS Short Course on Smelter Grade Alumina*, Charlotte, N.C. (2004), pp. 18–19
45. C. Cochran, W. Sleppy, W. Frank, Fumes in aluminum smelting: chemistry of evolution and recovery. JOM (9), 54–57 (1970)
46. S. Lindsay, N. Dando, S. Broek, Predictive formula for the competitive adsorption of HF and SO_2 on smelter grade alumina used in dry scrubbing applications. Light Metals **2017**, 487–493 (2017)
47. J. Thonstad, Critical current densities in cryolite-alumina melts. Electrochim. Acta **12**(9), 1219–1226 (1967)
48. P.-Y. Geay, B.J. Welch, P. Homsi, Sludge in operating aluminum smelting cells. Light Metals **2001**, 541–547 (2001)
49. R.K. Jain, S.B. Tricklebank, B.J. Welch, Interaction of aluminas with aluminum smelting electrolytes. Light Metals **1983**, 609–622 (1983)
50. N. Dando, X. Wang, J. Sorensen, W. Xu, Impact of thermal pretreatment on alumina dissolution rate and HF evolution. Light Metals **2010**, 541–546 (2010)
51. A.N. Bagshaw, G. Kuschel, M.P. Taylor, S.B. Tricklebank, B.J. Welch, Effect of operating conditions on the dissolution of primary and secondary (reacted) alumina powders in electrolytes. Light Metals **1985**(1985), 649–659 (1985)
52. G.I. Kuschel, B.J. Welch, Further studies of alumina dissolution under conditions similar to cell operation. Light Metals **1991**, 299–305 (1991)
53. Y. Yang, B. Gao, X. Hu, Z. Wang, Z. Shi, Influence of LOI on alumina dissolution in molten aluminum electrolyte. Molten Salts Chem. Technol. 77–83 (2011)
54. O. Kobbeltvedt, S. Rolseth, J. Thonstad, *The Dissolution Behaviour of Alumina in Cryolite Bath on a Laboratory Scale and in Point Fed Industrial Cells.* Department of Electrochemistry, Norwegian Institute of Technology, N-7034 Trondheim, Norway SINTEF Materials Technology N 7034 (1995)
55. R.K. Jain, M.P. Taylor, S.B. Tricklebank, B.J. Welch, A study of the relationship between the properties of alumina and its interaction with aluminium smelting electrolytes. Proc. Int. Symp. Molten Salt Chem. Technol., 1st, 59–64 (1983)
56. P. Homsi, Alumina requirements for modern smelters, in *6th Australasian Smelter Technology Conference and Workshop*, Melbourne, Australia (1998), pp. 73–90
57. J.B. Metson, M.M. Hyland, T. Groutso, Alumina phase distribution, structural hydroxyl and performance of smelter grade aluminas in the reduction cell. Light Metals **2005**, 127–131 (2005)
58. P. Lavoie, M.P. Taylor, J.B. Metson, A review of alumina feeding and dissolution factors in aluminum reduction cells. Metall. Mater. Trans. B. **47**(4), 2690–2696 (2016)
59. X. Liu, J.M. Purdie, M.P. Taylor, B.J. Welch, Measurement and modeling of alumina mixing and dissolution for varying electrolyte heat and mass transfer conditions. Light Metals **1991**, 289–298 (1991)
60. Y. Yang, B. Gao, Z. Wang, Z. Shi, X. Hu, The formation and dissolution of crust upon alumina addition into cryolite electrolyte. JOM **67**(9), 2170–2180 (2015)
61. S. Zhan, M. Li, J. Zhou, J. Yang, Y. Zhou, CFD simulation of dissolution process of alumina in an aluminum reduction cell with two-particle phase population balance model. Appl. Therm. Eng. **73**(1), 805–818 (2014)

62. B. Bardet, T. Foetisch, S. Renaudier, J. Rappaz, M. Flueck, M. Picasso, Alumina dissolution modeling in aluminium electrolysis cell considering MHD driven convection and thermal impact. Light Metals **2016**, 315–319 (2016)
63. J. Tessier, K. Cantin, D.T. Magnusson, Investigation of alumina concentration gradients within Hall-Heroult electrolytic bath. Light Metals **2018**, 515–522 (2018)
64. S. Lindsay, SGA requirements in coming years. Light Metals **2005**, 117–122 (2005)
65. S. Lindsay, Crushed tapped bath quality, in *Proceedings of the 5th International Conference on Electrodes and Support Services for Primary Aluminium Smelters* (2011)
66. G. Tarcy, Current efficiency improvements and optimization, TMS Industrial Electrolysis Course, Reykjavik, Module 16 (2013)
67. S. Lindsay, Metallic impurities from the mine to metal products. Light Metals **2013**, 177–181 (2013)
68. P. Solli, *Current efficiency in Aluminium Electrolysis Cells*, Dr. ing. Thesis, NTNU Trondheim, Norway (1993)
69. E. Haugland, G. Haarberg, E. Thisted, J. Thonstad, The behaviour of phosphorus impurities in aluminium electrolysis cells. Light Metals **2001**, 549–553 (2001)
70. S. Lindsay, Facing alumina, bath and metal purity issues, in *Proceedings of the 9th Australasian Smelting Technology Conference* (November, 2007), pp. 5–22
71. S. Lindsay, C. Dobbs, Beryllium in pot room bath. Light Metals **2006**, 95–100 (2006)
72. S. Eyer, M. Nunes, C. Dobbs, A. Russo, K. Burke, The analysis of beryllium in Bayer solids and liquids, in *Proceedings of the 7th International Alumina Quality Workshop*, Perth, Australia (2005), pp. 254–257
73. J. Metson, D. Wong, J. Hung, M. Taylor, Impacts of impurities introduced into the aluminium reduction cell. Light Metals **2013**, 9–13 (2013)

Stephen J. Lindsay Smelting Specialist Consultant Centre of Excellence—Hatch. Email: Stephen.Lindsay@Hatch.com

Stephen Lindsay served in numerous technical and managerial capacities with Alcoa for nearly 40 years prior to his retirement at the end of 2018. His technical responsibilities included alumina-based dry scrubbing for control of emissions across Alcoa's smelting operations worldwide and selection criteria management for Smelter Grade Alumina usage at Alcoa smelters. This extended to the role of leading multiple Customer–Supplier technical teams and alumina quality index scorecard management between each Alcoa refinery and its internal customers. In addition Steve served as a capital projects subject matter expert for alumina handling systems and alumina-based dry scrubbing systems.

He continues to serve as a subject matter expert and consultant to the primary aluminum industry through his affiliation with Hatch. Activities include project support for alumina-based dry scrubbing systems and training workshops on the properties of alumina and its proper handling and storage.

His published articles on the qualities of Smelter Grade Alumina have been included in the proceedings of International Alumina Quality Workshops, Light Metals, the Journal of Metals, and the proceedings of Australasian Smelting Technology Conferences. His patents are related to variable alumina feeding enhancement to emissions control systems and the delivery of alumina to reduction cell feed points.

Steve also served as the Editor of Light Metals in 2011 and was awarded the Distinguished Service Award by the Light Metals Division of TMS in 2016. He has been awarded two Best Papers in Aluminum Reduction Technology by TMS, one of which defined the role that beryllium impurities found in bauxite and alumina can play in downstream metal products and workplace safety precautions.

He also contributes as an instructor on properties of Smelter Grade Alumina, alumina-based dry scrubbing, alumina handling and other topics related to Aluminum Smelting in training courses sponsored by TMS as well as the University of Auckland's Light Metals Research Center.

Steve lives near Knoxville, Tennessee in the USA.

Pascal Lavoie's Director—Smelting Manufacturing Excellence, Alcoa Aluminum Center of Excellence. Email: Pascal.Lavoie @alcoa.com

career is connected with light metals production from the very beginning. He obtained his bachelor's degree in Materials and Metallurgical Engineering from Université Laval, Québec. He then joined Noranda's Magnola magnesium smelter as Process Engineer. When Magnola was curtailed, he moved to Noranda New Madrid smelter as Metallurgical Process Engineer and obtained a Six Sigma black belt certification.

In 2006, Pascal joined the Light Metals Research Centre of the University of Auckland (LMRC) as Manager—International projects and then Chief Engineer where he led teams conducting more than 40 industrially-based R&D projects.

LMRC has a long history of research into smelter grade alumina characteristics and its related performance in the reduction process. Pascal's primary research interest being reduction process control, he focused on alumina feed control and dissolution in industrial cells of all sizes. He led designs and adaptation of alumina feed control strategies in multiple aluminium smelters, improving dissolution and dispersion control of alumina. Pascal has recently completing his doctoral thesis on alumina dissolution. He also taught aluminium reduction process control and alumina feed strategy design at post-graduate level.

Returning to Canada as consultant in 2012, he provided support to smelting operations and engineering for reduction technologies, control and production systems.

Pascal joined Alcoa in 2018, where he now leads technical support for all aspects of aluminum smelting across Alcoa's global manufacturing footprint .

Chapter 12
Alumina Storage and Handling

Carsten Duwe

Abstract The pneumatic handling of Primary Alumina, Secondary Alumina as well as Crushed Bath, Anode cover material and Fluoride (AlF3) is very challenging, caused by the very different and extreme bulk characteristics. Large capacity storage facilities as well as huge transport capacities over long conveying distances are required at the harbor areas and the smelter plants all over the world. The most important State-of-the-Art technologies and components are shown and described in this chapter. The right selection of a competitive system for a smooth and highest reliable pneumatic handling of the different bulk materials is our daily challenging business. Long term experiences and a strong performance are the keys for satisfied clients worldwide, our best partners since decades. The selected references are taken from the list of hundreds of plants worldwide.

12.1 Introduction to Alumina Storage and Handling

The pneumatic handling of mineral bulk materials is worldwide a common and State-of-the-Art technology.

In view to the bulk material transport, pneumatic conveying systems are most of time, more compact, very flexible in view to constructional restrictions, and need much less expensive supporting structure and foundation.

Otherwise, the compressed air generation and supply of an equivalent air volume flow at a certain pressure level is obligatory, as well as the vent air separation and cleaning from dust before entering to ambient.

Very often, dried conveying and fluidizing air is mandatory for the pneumatic handling of hygroscopic or sticky bulk materials.

In general, the compressed air generation is much more energy consuming compared to all kind of drives for mechanical conveyor systems like belt or pipe conveyor, bucket elevators, drag chain or screw conveyors.

C. Duwe (✉)
REEL Mollet GmbH, Pinneberg, Germany
e-mail: Carsten.duwe@reel-moeller.com

B. E. Raahauge and F. S. Williams (eds.), *Smelter Grade Alumina from Bauxite*, Springer Series in Materials Science 320,
https://doi.org/10.1007/978-3-030-88586-1_12

Otherwise, mechanical transport systems are much more wear and maintenance intensive.

The best possible technical and commercial solution for a pneumatic handling system depends strongly on the application, constructional restrictions, and limitations in view to invest and/or operational costs.

Furthermore, it must be considered, that the pneumatic handling is limited in view to the bulk material characteristics regarding e.g. grain size, water content, flow ability and fluid ability.

Therefore, a detail investigation of the bulk material characteristics and the ambient conditions in advance, is obligatory for the best possible selection of a pneumatic or mechanical handling system with highest possible reliability and lowest possible operational costs during lifetime.

12.2 Alumina

Aluminum oxide with the chemical formula Al_2O_3 is commonly called "Alumina" and is a chemical compound of aluminum and oxygen. Aluminum oxide is used to produce aluminum metal in so-called electrolysis cells.

Aluminum oxide, a white fine powder with a density of approx. 3.8–4 g/m^3, is an abrasive bulk material, owing to its hardness.

12.2.1 Bulk Characteristics of Alumina

The different bulk characteristics of Alumina (Al_2O_3), also called as Primary or Fresh Alumina, varies in a certain range.

The following bulk characteristics are for Primary or Fresh Alumina type "sandy".

The typical grain size distribution, defined in client's specifications, is in the range of 5–150 μm.

FLS Hamburg is normally using the following values in weight% for Primary Alumina as delivery condition inside the large capacity port silo after ship unloading:

100% < 1 mm
99.9% < 200 μm
90% < 150 μm
50% < 80 - 90 μm
1–10% < 45 μm ("Fines")
1–3% < 20 μm ("Super fines").

The so-called fines content, commonly defined as the particle fraction < 45 μm could vary up to 20%, depending on bad quality or especially after divers intermediary pneumatic transports e.g. by tanker trucks from the port area to the smelter plant.

Processible alumina, before entering into the electrolyte cell, shall have a fines content of < 30%. Higher fines content in the smelter influences the process of the electrolysis, and consequently, the output of the electrolyte cell and the quality of the aluminum metal.

The apparent bulk density at atmospheric pressure is in the range 0.95–1.05 t/m^3.

The apparent bulk density under pressure, e.g. under bulk material column, is normally up to 1.2 t/m^3.

The angle of repose is in the range of 30–40°.

Specific characteristics:	
Moisture content:	Nil
Fluidization ability:	Very high
Flow ability:	Very high
Abrasiveness:	Very high
Sticky:	No
Explosive:	No

Values to be considered for the design of silos and hoppers:

Bulk density (to determine the required volume): 0.95 t/m^3
Bulk density (for structure design): 1.20 t/m^3
Natural angle of repose on filling: 23–30°.

In principle, the bulk density varies inside a silo, especially inside large capacity silos. The highest bulk density up to 1.2 t/m^3, is expected at the silo bottom area.

12.2.2 Fluidization of Alumina

One of the most important characteristics of Primary Alumina is the flow ability during fluidized conditions (Fig. 12.1).

A nearly linear increase of the pressure p2 as function of the specific fluidizing air amount (blue curve) up to a maximum value at comparably low 0.6 [Nm^3/ (m^2 × min)] is typically for Primary Alumina.

The increase in volume (red measuring points) is strongly distinct up to a specific fluidizing air amount of approx. 1 Nm^3/ (m^2 × min) and becomes stable at specific fluidizing air amounts of approx. 3 Nm^3/ (m^2 × min). This fluidization tests characterizes Primary Alumina as highly fluidizable, consequently, Alumina is well known as free-flowing bulk material.

The typical properties of Smelter Grade or Sandy Alumina [1] is summarized in Table 12.1.

The bulk material characteristics of Secondary Alumina could be very different to Primary Alumina, caused by a much finer grain size distribution as well as by contamination, resulting from the cleaning of the electrolysis cell venting air.

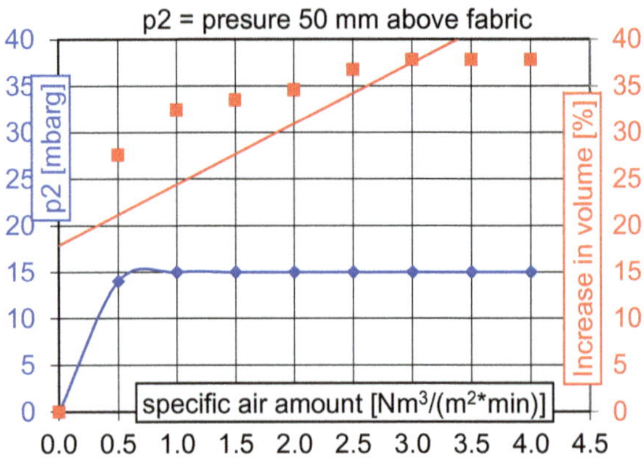

Fig. 12.1 Fluidization test (exemplary)—alumina

Table 12.1 Typical properties of SGA or Sandy alumina [1]

Dry sandy alumina		
Geldart group	[–]	A/B
Bulk density	ρ_{ss} [kg/m^3]	950–1050
Vibrate bulk density	ρ_{SR} [kg/m^3]	1150–1250
Aerated bulk density[a]	$\rho_{S,aer}$ [kg/m^3]	775–875
Solids density	ρ_S [kg/m^3]	4300–4800
Particle density	ρ_p [kg/m^3]	Approx. 2200
Average particle-ϕ at R = 50 wt%	$d_{s,50}$ [μm]	75–95
Particle size distribution: RRSB slope	α_{RRSB} [degree]	65–75
Particle shape acc. FEM 2582	[1]	Mixture of III, IV, V
Angle of slope	α_{SS} [degree]	28–38
Specific surface acc. Blaine	A_{Bleine} [m^2/g]	0.1–0.2
Specific surface acc. BET	A_{BET} [m^2/g]	$\geq$ 50
Minimum fluidization velocity	u_L [m/min]	0.20–0.35
Bed expansion at $V_F = 5\ U_L$	H_{exp}/H_L [1]	1.24–1.28
De-aeration time of 2 kg solid[a,b]	$\Delta\tau_{WS}$ [s]	5–22
Airslide inclination at $\dot{q}_{WS} = 1.0$ m/min	α_{asi} [degree]	$\leq$ 1.5
Internal/effective angle of friction	$\varphi_i \cong \varphi_e$ [degree]	35
Wall friction angle against steel St37	φ_w [degree]	Approx. 21
Jenike flow function	$FF = \sigma_1/f_c$ [1]	$\gg$ 10
Compressibility acc. Carr	$R_C = (\rho_{SR} - \rho_{SS})/\rho_{SR}$ [1]	$\ll$ 0.2

[a]Fluidized with $\dot{q}_{WS} \cong 4.0$ m/min

[b]Diameter of fluidized bed: 100 mm

12.3 Alumina Storage Without Segregation

Many different silo systems are used for the storage of bulk materials since decades. Some are used mainly for low storage capacities and short storage time, such as cone silos with silo diameter of up to approx. 5 m.

Other types are used for so-called large capacity storage facilities at the harbor or in the smelter area.

The large capacity Primary Alumina silo for the Fjardaal Smelter, Island, one of the largest Fresh Alumina silo worldwide, was built for 85,000 t in 2005/2006.

In view to large capacity silos, Anti-segregation equipment shall be foreseen to prevent segregation. Segregation means definitely a change of the bulk material characteristics. The more segregation takes place, the more varies the grain size distribution or the fines content as function of time. The fines content ($x < 45\ \mu m$) as well as the ultra-fines content ($x < 20\ \mu m$) verifiable influences the downstream process of the electrolysis cells. An equivalent Anti-Segregation equipment for the feeding of the silo as well as for the discharge is essential for minimized segregation effects inside a large capacity silo and is State-of-the-Art Technology.

12.3.1 Anti-segregation Equipment

The first step for minimized segregation inside a large capacity silo is an equivalent feeding system.

One of the main characteristics of such a system is a constant bulk material flow via a feeding spider with a small distribution pot in the centre, equipped with one direct outlet to the silo, and 6–8 spider arms, parallel operating air slides to the outer diameter of the silo (outer feeding points). Each spider arm air slide is also equipped with one middle outlet opening (middle feeding points) to the silo.

At the beginning of the feeding process, only the outlets at the end of the 6–8 spider air slides (outer feeding points) are in operation. The centre outlet as well as the 6–8 middle outlets are closed. If the bulk material level at the 6–8 outer discharge points are reached, approx. 85% of the storage capacity, the 6–8 middle discharge points will be opened. After reaching the bulk material level at the middle discharge points, the centre outlet will be opened to fill the silo up to 100% of the maximum filling capacity. The results of such a feeding procedure can be seen in Fig. 12.2.

To prevent segregation during filling of the silos, all 6–8 spider air slides feed, parallel in operation, one so-called Anti- Segregation chutes, each. Please see Fig. 12.3.

The Anti-segregation chutes are equipped with outlet openings to both sides and sloping plates inside.

Therefore, the Fresh Alumina flows down inside the Anti-Segregation chutes from one sloping plate to the next, nearest possible to the silo bottom. The bulk material is not falling down over 30–40 m and will not be distributed over the whole inner cross

Fig. 12.2 Anti-segregation silo feeding "Copyright © Emirates Global Aluminium"

Fig. 12.3 (Left): Silo filling completed and (Right): Anti-segregation chute

section of the silo but flows down via a small and well-defined cross section. The dusting effect during flow down is minimized, consequently, the segregation during filling is minimized.

If the silo gets full, the deepest outlet openings, one after another, get closed by the increasing bulk material level.

If the bulk material level inside the silo reaches the outlet openings at the high end of the Anti-Segregation chutes, the filling step via the Anti-Segregation chutes is finished.

Figures 12.4 show the 85,000 m^3 Fresh Alumina silo with a silo diameter of 55 m for the Fjardaal smelter in Island.

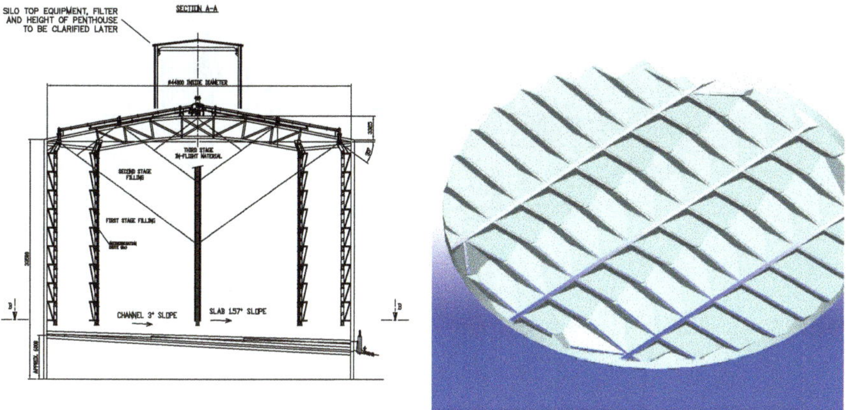

Fig. 12.4 (Left): 85,000 m^3 fresh alumina silo with side discharge (side view) and (Right): multiple point discharge to the side

The bulk material feeding is realized by an eight-arm-spider air slide system size 20 with multiple feeding points (8 + 8 + 1 feeding points) and designed for a feeding capacity of nominal 600 t/h and maximum 660 t/h.

The feeding of the center distribution pot of the spider air slide system is done by a belt conveyor. 8 Anti-Segregation chutes are foreseen for the smooth and controlled flow of the Fresh Alumina into the silo.

3 main 3-level air slides are foreseen for the nine-point-discharge of the silo. The tilted flat silo bottom with saddles from both sides, and divided into 3 fluidization sections per air slide, feed the equivalent discharge air slides.

The 3 main discharge air slides are directly connected with an airlift, designed for nominal 120 t/h discharge capacity.

Controlled sectional fluidization mode and up to 3 main discharge air slides parallel in operation, create a homogenization effect, for balancing segregation of the bulk material, leaving the silo.

The 85,000 m^3 large capacity silo for Fjardaal smelter is successful in operation since 2006.

12.3.2 Alternative Silo Discharge Systems

An alternative discharge system is a large capacity silo with an easier center outlet, combined for instance with an air slide, feeding a screw pump (Fig. 12.5).

Figure 12.6 (left), shows another favored large capacity silo with tunnel discharge, feeding a belt conveyor for downstream transport. Another type of silos is the Multi Discharge Silo (MDS Silo) with silo diameters from 15 to approx. 30 m and net volumes of 4000–40,000 m^3, Fig. 12.6 (right).

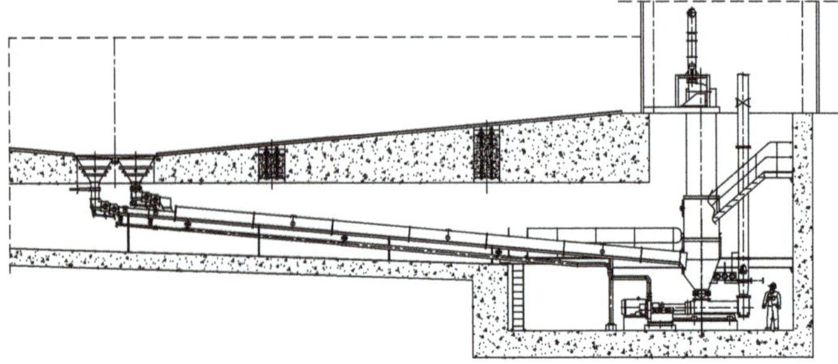

Fig. 12.5 Alumina silo with center discharge

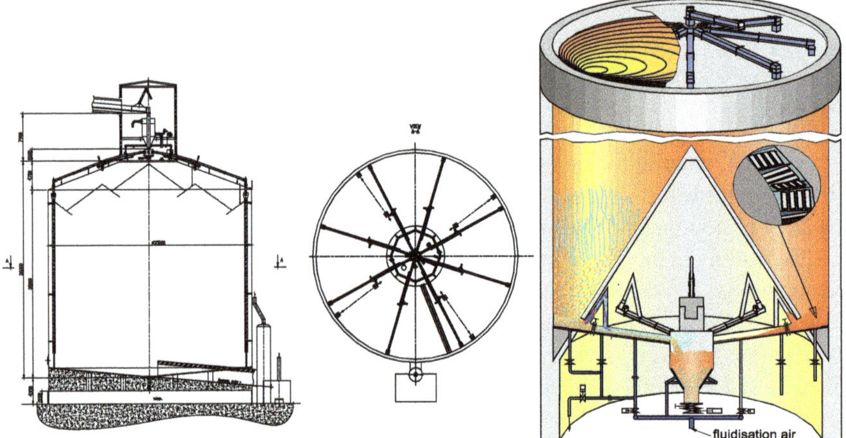

Fig. 12.6 (Left): Alumina silo with "sheep shelter" tunnel discharge and (Right): multi discharge silo (so-called MDS-Silo)

The discharge from the silo will be realized section wise with tilted and oppositely located aeration pads, section wise outlet openings, connected with rotary shut-off valves and downstream located air slides, feeding an intermediate discharge bin in the center under the inner discharge conus of the silo. This intermediate discharge bin is equipped with a fabric filter for de-dusting the vent air, coming from the fluidization of the active silo bottom section(s), the equivalent discharge air slide(s) and the a.m. intermediate discharge bin.

Furthermore, this intermediate discharge hopper bin is equipped with fluidization pads at the conical part for a continuous, smooth and quick flow out of the silo.

The inner silo discharge conus, made of concrete, should prevent bridging of the bulk material at the discharge area. Additionally, the necessary fluidization air amount for the silo bottom section between the silo wall and the lower part of inner discharge conus is minimized.

The low-pressure fluidization air ($p \leq 0.5$ barg) is normally supplied by separate rotary piston blowers and not from a superior high-pressure control air plant net, caused by energetic reasons.

The total height of a Multi Discharge Silo (MDS Silo) is slightly increased, caused by the inner discharge conus. Otherwise, the so-called Dead Stock of the MDS Silo after complete emptying, is comparably low.

12.3.3 Silo Dead Stock

The so-called Dead Stock or Dead Storage of a silo is normally also an important issue for the complete emptying, therefore mostly part of the guarantee values.

Typical values are in the range of ≤ 1–2% of the maximum volumetric value of a large capacity silo.

The smaller the value for the Dead Stock is, the higher the number of necessary fluidization pads at the silo bottom as well as the amount of fluidization air are necessary.

However, a clockwise switch over of one to another fluidization zones, 6–8 zones are normally used, minimizes the necessary amount of fluidization as well as vent air volume. Typically, two opposite fluidization zones are parallel in operation, continuously switched over to the next pair of opposite fluidization zones.

Figure 12.7 shows the Dead stock of not usable Primary Alumina inside a Flat bottom silo. Between the bulk material- hills you can see at Fig. 12.4 (right), the single fluidization pads are installed. The smaller the distance between two adjacent fluidization pads are, the smaller the Dead Stock of non-fluidized and unusable bulk material definitely is.

Figure 12.4 (right) shows an example of a flat bottom of a silo, highly equipped with fluidization pads.

A Multi Discharge Silo (MDS Silo) is a typical medium size long term storage silo, often used also for long term Fly ash storage.

Fig. 12.7 Dead stock of a flat bottom silo "Copyright © Emirates Global Aluminium"

12.4 Alumina Handling Systems

A lot of different pneumatic transport systems are necessary for the Harbour area as well as for the Smelter facilities. Different kind of bulk materials are handled, such as Primary Alumina, Fluorinated Alumina, Crushed Bath, Aluminum fluoride and Anode Cover Material (ACM), a mix of Crushed bath and Alumina (Fig. 12.8).

All these bulk materials have their own bulk characteristics, for instance the grain size distribution, the bulk and solid density, the angle of repose, the fluidization properties and flowability.

The Crushed Bath and the Anode Cover Material are very coarse bulk materials, compared to the Primary Alumina and Aluminium fluoride. Consequently, the design as well as the used equipment of the pneumatic conveying systems are much different.

12.4.1 *Pressure Vessel TURBUFLOW® Dense Phase Conveying*

The pneumatic transport of Primary Alumina is one of the most important conveying systems.

Approx. two tonnes of Primary Alumina are necessary to produce one tonne of Aluminium. Therefore, the necessary unloading, loading, conveying and storage

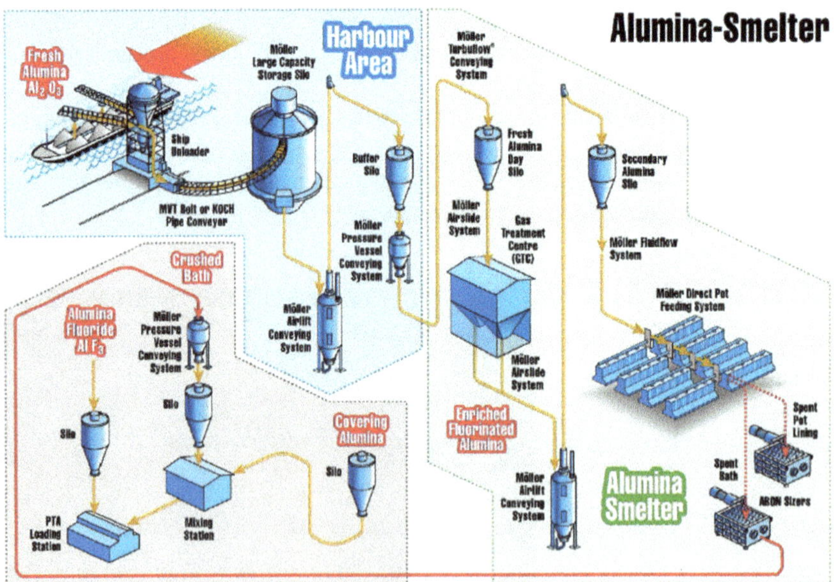

Fig. 12.8 Harbour area and alumina smelter plant (Courtesy of FLSmidth)

capacities for Primary Alumina are comparable high, as well as the conveying distances.

Furthermore, the requirements to these pneumatic conveying systems getting more and more restrictive and specific in the last years. For instance, the prevention of fines generation ($x < 45\ \mu m$), also called as Attrition, is more and more an issue for the pneumatic conveying systems. This is caused by the influence of the fines content of the Fluorinated Alumina in view to the electrolysis process.

Furthermore, the specific energy consumption of the Primary Alumina transport influences the operational costs, and is another challenge for the design of pneumatic conveying systems for Primary Alumina (Fig. 12.9).

The pressure vessel conveying system with the patented TURBUFLOW® conveying pipe is the main transport system for this application since more than 30 years.

Since 2004, the second generation of the TURBUFLOW® pipe-in-pipe technology will be used.

Conveying distances up to 1000 m and more for conveying capacities up to 50 t/h and more are realized and in successful operation since many years.

In principle, the TURBUFLOW® conveying pipe consists of a large outer steel pipe and a small inner steel pipe. The small inner pipe is equipped with frequent flow in and flow out openings, with variable but well defined distance between two openings.

The small inner pipe inside the main outer pipe, is working like a bypass pipe for the conveying air. When the air flow resistance is getting higher than normal inside the outer pipe, caused by larger dunes, the continuously flowing conveying air, bypasses the larger dunes via the small inner pipe.

Downstream the larger dune, the timewise increased conveying air amount inside the inner pipe flows back again into the main pipe via the openings of the inner pipe. The locally high air velocities at the openings of the smaller inner pipe create local turbulences. These locally working turbulences destroy partly too large dunes, to increase again the smooth flow of the bulk material air mix in the outer pipe.

The flow in and flow out openings at the inner bypass pipe are also designed as a flow resistance, to minimize air flow inside the smaller inner pipe during normal operation with normal dunes and normal flow resistance in the outer pipe.

This called "TURBUFLOW" effect works continuously and self-regulating, without any valves or expensive control systems.

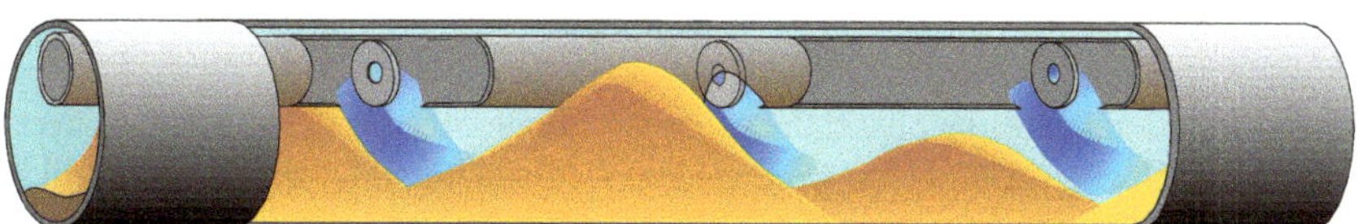

Fig. 12.9 Patented TURBUFLOW® conveying pipe

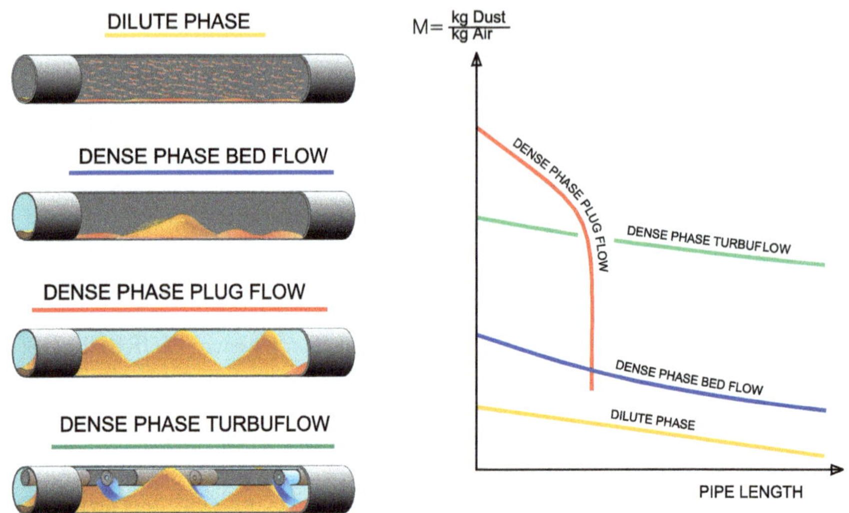

Fig. 12.10 Comparison TURBUFLOW® with normal pipe

This is the reason why the pneumatic transport of the Primary Alumina via the TURBUFLOW® conveying pipe works with very low conveying velocities, minimized energy consumption and generation of fines, and lowest possible maintenance costs.

Compared to a pneumatic transport via normal steel pipes without an inner pipe, the TURBUFLOW® conveying pipe is working at much higher bulk material/air ratios, especially for longer conveying distances (Fig. 12.10).

In principle, this TURBUFLOW effect works continuously as blockage removal, caused by too large dunes.

The TURBUFLOW® conveying pipe is also used for the pneumatic transport of Aluminum fluoride.

The conveying pipe bends for Primary Alumina conveying systems must be medium wear resistant. Most of time, basalt bends with normal radius of curvature reaches lifetimes of much more than 2 years full operation (Fig. 12.11).

One of the main components of a pressure vessel system is the filling valve and the pressure vessel outlet valve. For both applications, so-called Dome valves are State-of-the-Art technology (Fig. 12.12).

A spherical calotte and an inflatable sealing are the main components of a dome valve.

At the end of the filling phase, the spherical calotte cuts through the bulk material column for closing. After closing the spherical calotte, the inflatable sealing will be pressurized for a gas tight shut-off to the filling chute and feeding hopper or silo. The inflatable sealing is a rubber type sealing. The standard type is applicable up to 200 °C, the high temperature types up to 300 °C. The complete control of the sealing is realized by a small control box, directly connected to the pneumatic drive

Fig. 12.11 Dome valve type Möller®

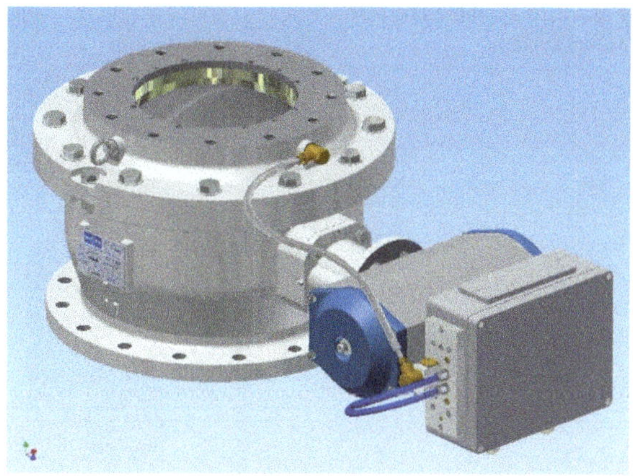

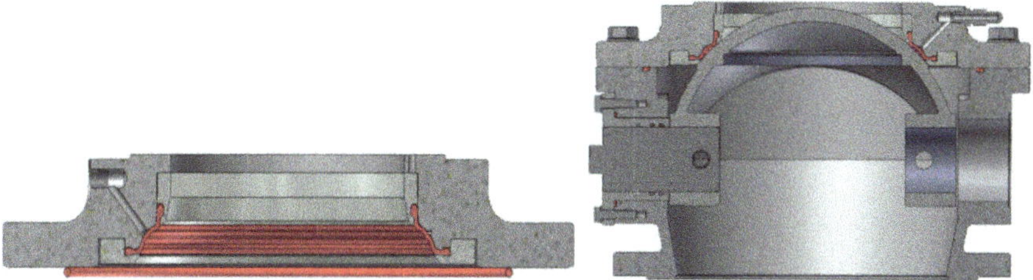

Fig. 12.12 (Left): inflatable rubber sealing and (Right): spherical calotte as closing element

of the Möller® Dome valve. Only the on/off signal as well as the electrical supply is necessary from external.

After reaching the open position, 100% of the cross-sectional area is free for filling the pressure vessel in shortest possible time. Additionally, the risk of bulk material bridges during filling phase of the pressure vessel is minimized. Two important advantages against an ordinary butterfly valve as filling equipment.

However, when using butterfly valves as pressure vessel inlet valves, two butterfly valves in series are necessary. The first one for cutting through the material column, and the second butterfly valve for the gas tight closing.

One favourite type of pressure vessel systems is the so-called Double pressure vessel system. Reference is made to Fig. 12.13.

No pre-hopper is necessary with the necessary more free height, because one pressure vessel is always in the filling phase during operation, and the second pressure vessel is always in conveying phase.

Consequently, up to 55–58 min. per operating hour can be used for the conveying phase. Compared to this, a single pressure vessel system reaches approx. 45 min per operating hour for the conveying phase.

Based on a given necessary average conveying capacity, the gross conveying capacity, the required conveying air amount and the connected cross section of

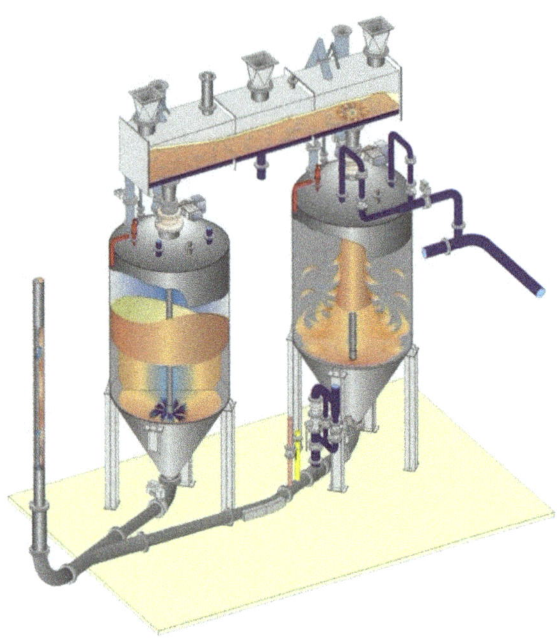

Fig. 12.13 Double pressure vessel system

the conveying pipes of a Double pressure vessel system are more than 20% less, compared to a Single pressure vessel system.

12.4.2 Pressure Vessel Dense Phase Conveying with Standard Pipe

The pneumatic conveying of Crushed Bath and Anode Cover Material works much different to the pneumatic transport of Primary Alumina. Both bulk materials are much coarser than Primary Alumina.

A typical grain size distribution of Crushed Bath is 100% < 20,000 μm, 50% < 3000–5000 μm and 10% < 70 μm. A typical grain size distribution of Primary Alumina is 100% < 200 μm, 50% < 80–90 μm and 2–10% < 45 μm.

Furthermore, Primary Alumina is free flowing and easily fluid able, Crushed Bath is badly flowing and not fluid able.

Because of this, normal conveying pipes without an inner pipe as well as much higher conveying velocities will be used for the pneumatic transport of Crushed Bath and Anode Cover Material (Fig. 12.14).

The pick-up velocities for Crushed Bath and Anode Cover Material are in the range of 20–25 m/s. The pick-up velocities for Primary Alumina with TURBUFLOW® conveying pipe are in the range of 5–8 m/s.

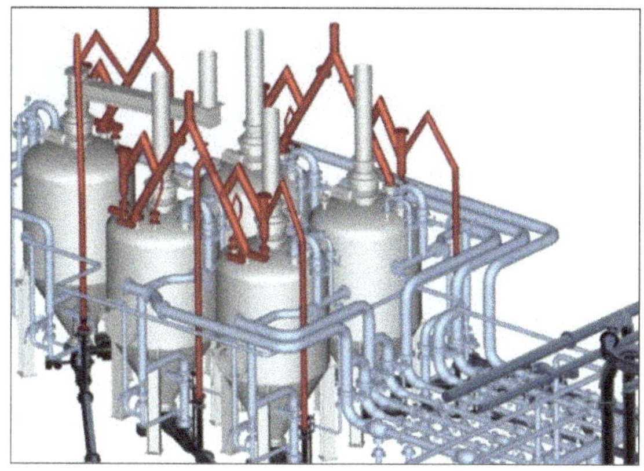

Fig. 12.14 Pressure vessel station with conveying air flow control

The conveying pipe bends for Crushed Bath conveying systems must be highly wear resistant. Most of time, basalt bends are not sufficient, but ceramic lined bends with high radius of curvature.

12.4.3 Airlift Conveying for Vertical Transport

Typical conveying capacities are 500–1200 t/h per ship unloader. Downstream the ship unloader, belt conveyors transporting the Primary Alumina over hundreds of meters to the bottom of the large capacity silos. The necessary pneumatic conveying to the top of the large capacity silos are realized by Airlift systems.

Another application for the pneumatic transport of Primary Alumina is the ultra large capacity feeding of a large capacity silo as part of the ship unloading at the Harbor area (Fig. 12.15).

In principle, the airlift will be feed from the top. The full fluidized bottom of the airlift vessel feed the bulk material to the conveying pipe in the center of the airlift body. A nozzle in the center of the fluidization bottom speeds up the mix of bulk material and conveying air.

A non-return valve upstream the nozzle avoids bulk material flow into the conveying air piping.

During normal operation at design conveying capacity and design pressure, the airlift is approx. 90% filled with bulk material.

If the actual conveying capacity is lower than the design capacity, the bulk material level drops inside the airlift.

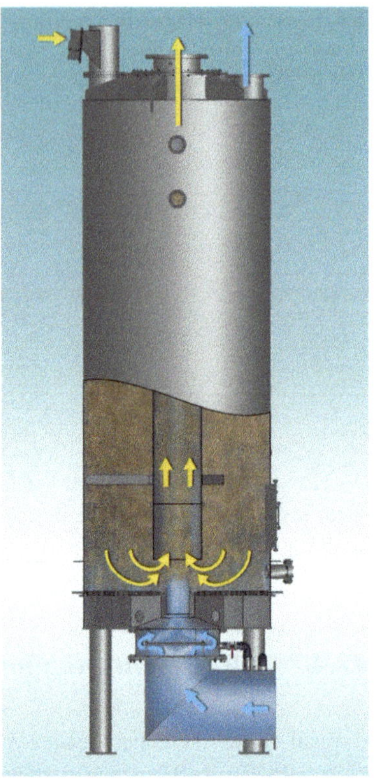

Fig. 12.15 Airlift system for vertical transport

12.4.4 Screw Pump Conveying

The screw pump conveying system is an alternative to a pressure vessel conveying system, but only for the pneumatic transport of fine bulk materials like Primary Alumina or Aluminum fluoride. Definitely not for Crushed bath or Anode Cover Material, caused by the much too coarse grain size distribution.

One of the main advantages is the low necessary free height for a screw pump with feeding hopper. Especially for high conveying capacities > 100 t/h, the necessary free height for a pressure vessel system is much higher.

In addition to this, a screw pump system is cheaper in view to the invest costs, but more expensive in view to the operational and maintenance costs.

The specific energy consumption is much higher, compared with a pressure vessel system with Turbuflow® conveying pipe. Furthermore, the maximum line pressure for a screw pump (≤ 2.0 barg) is much lower than for a pressure vessel system (up to 10 barg).

Consequently, the conveying air amount is much higher, as well as the size of the de-dusting filter on top of the target silo or hopper.

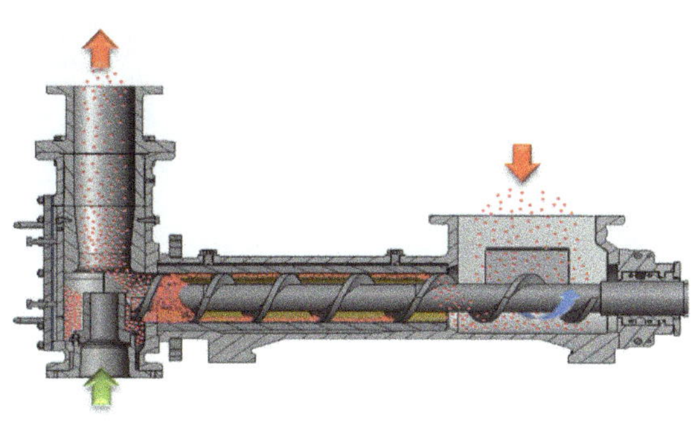

Fig. 12.16 (Left): screw pump type Möller and (Right): FK screw pump type M

In principle, the bulk material will be compacted by the high-speed screw, eg. 985 1/min on a 50 Hz net basis, before entering into the bulk material-air-mixing chamber to the left side of the Fig. 12.16.

This local compacted bulk material must withstand the maximum pressure at the beginning of the conveying pipe, caused by the pressure loss of the pneumatic transport via the complete conveying pipe until the target silo.

A nozzle inside the bulk material-air-mixing chamber speeds up bulk material feed by the screw, to prevent bulk material blockage inside.

The screw pump type Möller is suitable for conveying capacities up to 130 t/h and operating pressure of up to 1.5 barg.

If higher conveying capacities and longer conveying distances are necessary, the so-called FK screw pump with the actual type M are used. Operating pressures in the range of 1.5–2.0 barg are possible to handle, depending on the bulk material characteristics and the application.

The main difference is the flap valve, integrated inside the mixing chamber, to avoid conveying air backflow via the screw, caused by high pulsing overpressure in the conveying pipe.

Additionally, the two bearings are located at both far end sides of the screw pump. This is one of the main design differences to the screw pump type Möller. Otherwise, the screw pump type Möller is more compact and on a lower cost basis.

12.4.5 Air Slide Conveying

The Air slide conveying system is one of the oldest pneumatic conveying system, but still important and state-of-the-art.

In principle, the bulk material will be fluidized by low pressure compressed air (0.3–0.5 barg) from the bottom via a fabric, made of polyester for instance.

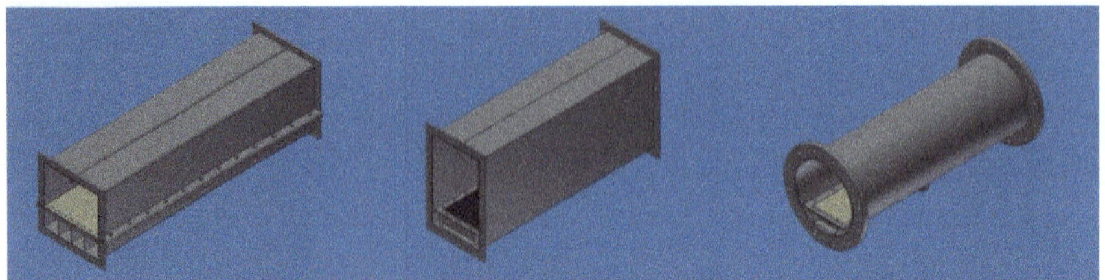

Fig. 12.17 (Left): standard air slide, (Middle): high profile air slide and (Right): pipe air slide

Ordinary air slides are erected with a declination in the range of 5–10° in flow direction, depending on the bulk material characteristics and the required conveying capacity.

Air slide systems are gravity driven transport systems. Caused by very low conveying velocities in the range of 0.5–3.0 m/s, ordinary air slide systems are very smooth operating transport facilities. The particle breakdown (Attrition) is normally very low, as well as the specific energy consumption.

The conveying distance of air slide systems is limited, normally < 100 m, caused by the maximum available free height.

The fluidizing air must be frequently vented to a filter system to limit the air velocities and the equivalent wear inside the air slides. High profile air slide systems are used for longer air slide systems. Reference is made to Fig. 12.17.

A special type of air slides is the so-called pipe air slide. This special kind of air slide will be used for applications with strong over- and under pressure pulsing inside the air slide. A plant combination with an airlift could be such an application.

For all kind of air slides, the range of conveying capacity is remarkable wide. Up to 1200 t/h and more is possible with the equivalent large air slide sizes.

12.4.6 Truck Loading and Unloading

For nearly all used bulk materials, truck loading and unloading systems are used. If the distance from the harbor area to the smelter area is >> 1 km, Fresh Alumina as well as Aluminum fluoride will often be transported by trucks.

But also Crushed Bath and Anode cover material will be transported by trucks from one to another area.

Many different types of trucks for the transport of bulk materials are available. The so-called type tanker truck is shown in Fig. 12.18 and is made for the transport of Fresh Alumina to the Day silo of the Gas treatment centre (GTC).

Some trucks are equipped with on-board compressors for the pneumatic discharge of the bulk material. Other types of trucks need external compressed air supply for discharging.

Fig. 12.18 Truck loading station for primary alumina "Copyright © Emirates Global Aluminium"

The connection between the truck and the downstream conveying pipe will be realized by special wear resistant rubber hoses size DN 80–DN 150, normally 4–6 m long. The connection between the truck and an external compressed air supply will also be realized by rubber hoses DN 65–DN 100.

Pneumatic operating Waggon unloading systems working in a similar way as Truck unloading systems. An external compressed air supply instead of an on-board compressor for waggon unloading is the standard solution. A gravity driven discharge of waggons into an underground hopper is also a common technical solution (Fig. 12.19).

12.5 Alumina Distribution and Direct Pot Feeding

12.5.1 Introduction

After a construction period of two years, in late July 2014, four pneumatic conveying systems were handed over by FL Smidth to Emirates Global Aluminium (EGA) to feed 444 aluminium electrolysis cells (pots) in two pot rooms of the new pot line 3 (Fig. 12.20) at the Emirates Aluminium (EMAL) smelter in Abu Dhabi, UAE.

The conveying systems are four FLUIDFLOW® air slide systems installed without gradient. Each of the four conveying systems is designed to transport 44.4 t/h of

Fig. 12.19 Truck unloading for primary alumina "Copyright © Emirates Global Aluminium"

Fig. 12.20 View along the two pot rooms of EMAL pot line 3. "Copyright © Emirates Global Aluminium"

secondary alumina over a distance of 425 m from the day silo to 111 electrolysis cells each.

The following report will explain the design and function of the world's largest conveyor systems, which feed the aluminium electrolysis cells with the world's largest alumina demand, and illustrate the challenges associated with commissioning.

12.5.2 Pot Feeding Development Steps

Over the past 40 years, the demand for aluminium has risen continuously in many industries. It's scarcely possible to imagine the modern world without aluminium as a working material.

Today, primary aluminium is manufactured in electrolysis cells in huge aluminium smelters. With a maximum production volume of 150 kg/h, the capacity of a single electrolysis cell is relatively low. For economic reasons, numerous electrolysis cells are arranged in a row in pot rooms in order to increase aluminium production.

In the electrolysis cells, secondary alumina is generally converted into primary aluminium. The exhaust gases generated by this process are purified in gas treatment centres through the use of primary alumina as the adsorbent. Enriched with emission particles, the alumina leaves the gas treatment centre as secondary alumina and is temporarily stored in silos, from which it must then be transported to the pots.

In the early days, the electrolysis cells were supplied with alumina using shovels and wheelbarrows. Later, the task was taken over by pot room cranes in the pot rooms. One of the disadvantages of the mechanical conveying systems was the vast amount of dust generated—a burden for operating personnel—the loss of alumina and the discontinuous feeding process.

The first pneumatic pressure vessel pipe conveying systems for direct pot feeding were introduced in the mid-1980s. The disadvantage here is the high conveying speeds required, which lead to severe scaling in the conveying pipeline and particle attrition of the alumina. Today, all modern aluminium smelters are equipped with special air slide systems for direct pot feeding.

12.5.3 Direct Pot Feeding Systems

The first generation of direct pot feeding systems type FLSmidth designed consisted of a pressure vessel conveying system for long-distance transport along the pot room and FLUIDFLOW® pot air slides for the distribution of alumina to the individual electrolysis cells [2]. Starting in 1984, this system type was utilised in numerous locations in Germany, Russia and, most recently, at the Aluminij Mostar aluminium smelter in Bosnia and Herzegovina, for the supply of 256 pots. The disadvantage of this system type is the incidence of scaling in the conveying pipeline as well as particle attrition of the alumina.

For this reason, FLSmidth developed the second-generation direct pot feeding system. Here the pressure vessel conveying system for long-distance transport along the pot room was replaced by a horizontal FLUIDFLOW® distribution air slide system. The distribution of the alumina to the individual pots still takes place with FLUIDFLOW® pot air slides [3]. The first system was put into operation in 2008 for the supply of 44 pots in pot line 8 at the Dubai Aluminium (DUBAL) smelter.

This was followed by installations in Iran (228 pots HORMOZAL pot line 2, IMIDRO), Russia (168 pots, BOGUCHANY Phase 1, RUSAL) and Abu Dhabi (444 pots, EMAL pot line 3, EGA).

Further systems are currently being installed in Venezuela (396 pots, PUERTO ORDAZ, pot lines 3 and 4, CVG ALCASA) and Russia (168 pots BOGUCHANY Phase 2, RUSAL). A further project in Russia is in the early implementation phase (176 pots, TAISHET Phase 1, RUSAL).

12.5.4 System Description and Functionality of Direct Pot Feeding at EMAL Pot Line 3

Pot line 3 at the EMAL aluminium smelter consists of two pot rooms. Each of the pot rooms has 222 electrolysis cells of type DX + installed in a row (Fig. 12.21). The total length of a pot room is roughly 1500 m and is thus the longest in the world.

In total, four identical pot feeding systems were installed by FLS, with each one supplying 111 pots. Each of the four systems is structured as follows (Figs. 12.22 and 12.23).

The secondary alumina from the gas treatment centre is stored in a 900 ton day silo. The bulk material discharge from the silo is done by way of a fluidisation bottom (Fig. 12.24) and flows via a FLUIDFLOW® air slide to a main bin on the pot room wall, which serves as a lock bin and is positioned roughly at the midpoint of the 111 pots.

On the bottom of the main bin, there are two connected FLUIDFLOW® distribution air slides that run parallel to the pot room outer wall in opposing directions and supply all pots with alumina. The longest distribution air slide supplies 60 pots over 390 m and is installed entirely horizontally due to the limited total construction height of less than 8 m. The air supply for fluidisation of the alumina in the distribution air slides is implemented by means of a frequency-controlled rotary blower. The fluidisation air quantity is regulated depending on the fill level at the end of the distribution air slide via the fluidisation pressure. To avoid flow disturbances in the distribution air slides, the fluidisation air is channelled to the gas treatment centre via vent domes at regular intervals. The vent domes are wear- and maintenance-free gravity separators with the task of separating swept-up alumina from the vent air

Fig. 12.21 Electrolysis cells in pot room. “Copyright © Emirates Global Aluminium”

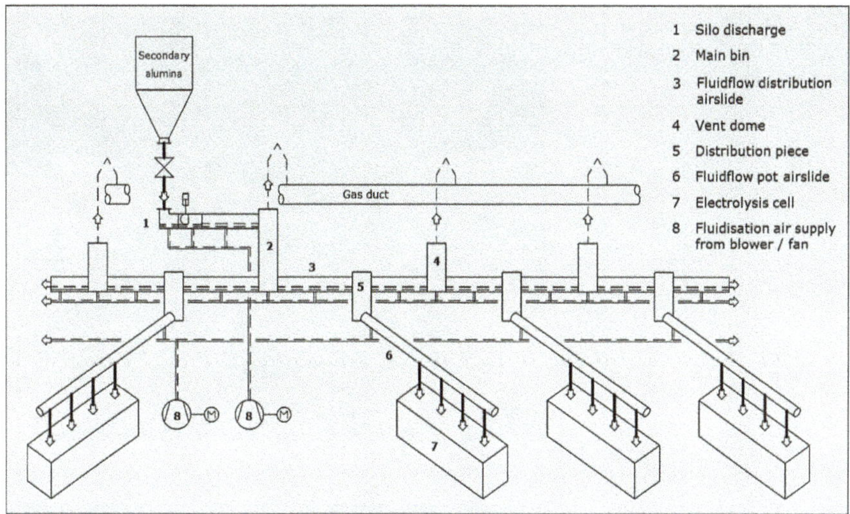

Fig. 12.22 Process flow diagram of direct pot feeding

Fig. 12.23 System configuration of direct pot feeding

in order to minimise dust emissions to the gas duct and maintain a uniform bulk material quality along the conveying distance.

The distribution air slide along the pot room has one separate outlet for each pot. A downpipe is attached to each outlet which runs diagonally into the pot room through the façade and supplies the 111 FLUIDFLOW® pot air slides to the individual pots. Each pot air slide is approx. 16 m long and distributes the alumina through filling spouts into five pot bunkers with a volume of approx. 1 m^3 (Fig. 12.25) each. The pot air slides are also horizontal, installed without gradient. The air quantity required for the fluidisation of the alumina is provided by a frequency-controlled fan. The fluidisation air quantity is controlled by means of a constant fluidisation pressure.

Fig. 12.24 Silo outlet and FLUIDFLOW® air slide to the main bin

Fig. 12.25 FLUIDFLOW® pot air slide

As the air quantity required for fluidisation is extremely low, the vent air can escape into the pot bunker via the filling spouts and flow from there to the gas duct through a connection.

The bulk material supply from the day silo to the pot feeding system as well as the air supply for the FLUIDFLOW® air slides are continuously in operation, ensuring that the entire pot feeding system is always 100% filled with alumina.

The bunkers are always filled to the maximum via the self-regulating filling spouts. If alumina is added from the pot bunkers to the molten bath, the fill level in the bunkers drops and the opening of the filling spout is cleared so that alumina is automatically refilled until the filling spout is automatically closed again by the alumina (Fig. 12.26).

Second-generation direct pot feeding works completely automatically and is absolutely gas tight. During stationary operation, no valves are actuated. Maintenance work and the need for spare parts are minimal due to the design and the system's manner of operation. At less than 1 m/s, the conveying speed of the alumina in the

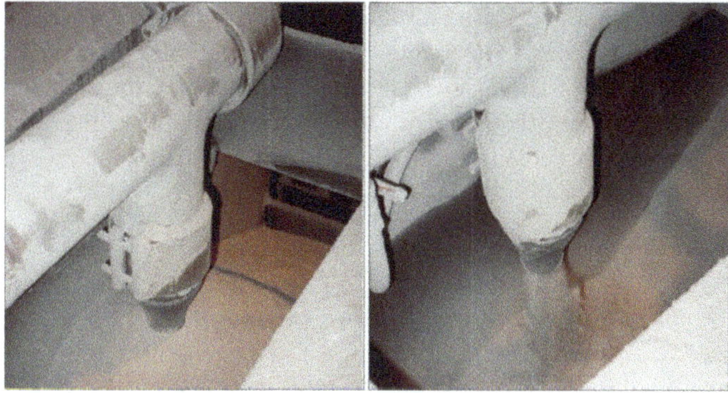

Fig. 12.26 Self-regulating filling spouts in the pot bunker

FLUIDFLOW® air slides is extremely low, which precludes scaling, particle attrition and system wear.

12.5.5 Conveying Principle of the FLUIDFLOW® Distribution Air Slide

In a traditional air slide system, the air slide has a slight gradient to use gravity as the acting force. With a horizontally configured FLUIDFLOW® air slide (Fig. 12.27), this support is naturally missing.

In order to transport the alumina over the conveying distance of 390 m from the main bin along the pot room through the FLUIDFLOW® distribution air slides, a pressure drop generated by particle resistance, among other factors, must be overcome, as in the case in traditional pneumatic pipe transport. To keep the pressure, drop as low as possible, the viscosity of the alumina is significantly reduced by introducing compressed air via a fluidisation fabric. The advantage with alumina is the

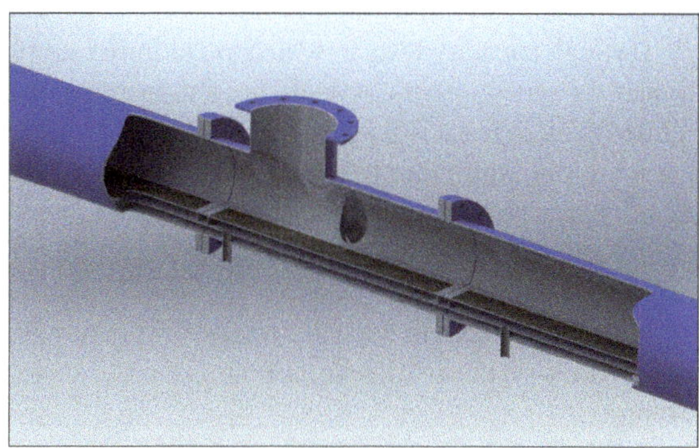

Fig. 12.27 Structure of FLUIDFLOW® air slide

relatively low air quantity required for fluidisation. This makes it possible to convey alumina over large distances with comparably low energy requirements.

The pressure in the FLUIDFLOW® distribution air slide at the main bin is greater than at the end of the distribution air slide. For this reason, the fill level in the vent domes of the main bin decreases towards the end. The main bin serves as an inlet device and seals the system vis a vis the silo.

12.5.6 Experiences During Commissioning

12.5.6.1 Air Distribution Along the Fluidflow® Distribution Air Slide

As mentioned above, it is necessary to reduce the viscosity of the alumina. Therefore, the alumina is fluidised with compressed air. To prevent flow disturbances in the FLUIDFLOW® distribution air slide the fluidisation air must be distributed evenly along the entire 736 m conveying distance. To achieve this, throttle valves are installed at regular intervals in the fluidisation air pipe. Due to the lower pressure level at the end of the FLUIDFLOW® distribution air slide, the throttle valves in this area must be more strongly throttled than at the beginning in the vicinity of the main bin.

Because of the step-by-step commissioning of the electrolysis cells, some difficulties in terms of uniform air distribution along the considerable conveying distance were expected. These difficulties were remedied during the commissioning process using software newly developed for this system type. Using the software, the requisite opening cross-sections for the throttle valves along the conveying distance could be calculated so as to ensure uniform air distribution and adequate supply of alumina to all areas at all times.

12.5.6.2 Conveying Capacity

Contractually, a maximum alumina demand of 400 kg/h per pot was agreed. Thus for 111 electrolysis cells, that amounts to a required total conveying capacity of 44.4 t/h per pot feeding system.

Through capacity tests in which the alumina supply to all pots was initially interrupted for several hours, it could be demonstrated that all pots were simultaneously refilled with at least 740 kg/h. The resupply of alumina to the pots in the area of the main bin was significantly higher yet. Altogether, a total conveying capacity of over 82 t/h was achieved during the capacity test.

If only particular electrolysis cells are refilled, this is done with a conveying capacity of up to 10 t/h.

With these planned capacity reserves, it is possible to interrupt operation of the pot feeding systems for several hours in order to carry out maintenance work, for example.

12.5.6.3 Dust Emissions at the Vent Domes and Segregation Along the FLUIDFLOW® Distribution Air Slide

As described above, it is necessary to ventilate the FLUIDFLOW® distribution air slide at regular intervals in order to prevent flow disturbances in the distribution air slide. For economic reasons (large numbers of vent air points and difficult system accessibility), traditional filter elements with compressed air cleaning were not a viable option. It was therefore decided to use vent domes on the FLUIDFLOW® distribution air slide. The vent domes are wear- and maintenance-free gravity separators with the task of separating swept-up alumina from the vent air in order to minimise dust emissions to the gas duct and maintain a uniform bulk material quality along the conveying distance.

Although no limit values for dust emissions and degree of segregation were defined, upon initial start-up it was discovered that too much fine alumina was emitted with the original vent dome design. This led to an increased dust concentration in the gas duct and high stress on the gas treatment centre. Moreover, the alumina quality along the conveying distance was changed such that the quality at the beginning of the FLUIDFLOW® distribution air slide corresponded to that of the silo, while the fine particle quantity at the end of the conveying distance was significantly reduced.

To remedy this, the design of the vent domes was changed. The height and diameter of the vent domes were increased, significantly increasing the separation rate (Fig. 12.28).

Independent studies by the plant operator of various air slide systems from different providers showed that substantial segregation of the alumina occurs here as well.

Fig. 12.28 (Left): modified vent domes. "Copyright © Emirates Global Aluminium" and (Right): test set-up in the technical centre

Fig. 12.29 Electrolysis cells in pot room of DUBAL pot line 8 "Copyright © Emirates Global Aluminium"

Through the modification of the vent domes, the degree of segregation was significantly reduced and is now substantially lower than those of other system manufacturers despite having the longest conveying distance in the world [4].

Through studies on full-scale test systems at the technical centre of FLSmidth, the degree of separation could be determined for different vent dome geometries and operating modes (Fig. 12.28). For this reason, calculating and guaranteeing the degree of segregation and vent air emissions for future systems is capable (Fig. 12.29).

Based on the test results in the technical centre, in early 2017 FLSmidth was commissioned to reduce the dust emissions of the direct pot feeding system at the aluminium smelter DUBAL pot line 8 by at least 30%. The rebuild of the vent domes was done in summer 2017. After commissioning phase, the resulting dust emission, equivalent to the internal Alumina recirculation rate, was reduced by more than 90%.

12.5.6.4 Regular Partial Draining of the System to Increase Conveying Capacity

After successful system optimization through the modification of the vent domes, it was shown that regular partial draining of the conveying systems increases the flowability of the alumina and thus the conveying capacity.

Multiple times each day, the alumina supply from the silo is interrupted for a certain period of time while the aluminium feed to the pots continues to operate. The result is a targeted partial draining of the system in which fine particles from the vent domes as well as deposited coarse material on the fluidisation fabric are carry out and transported to the electrolysis cells. Thereafter, fresh secondary alumina flows from the silo into the conveying system.

12.5.7 Experience After Two years of System Operation

Prior to expiry of the warranty period, in early 2016 FLSmidth conducted a thorough inspection of the conveying systems for direct pot feeding. At this time the system components had been in operation for between two and two-and-a-half years.

The inspection established the following findings:

- Due to different alumina qualities with different flow qualities, the fluidisation volume flow must be minimally adjusted.
- Regular partial draining of the system as described above is required in order to counteract minimal, natural segregation effects and improve flow behaviour.
- No scaling was found in the FLUIDFLOW® air slides and vent domes.
- The fluidisation fabric showed no signs of wear.

12.6 Summary and Outlook

The supply of ever longer pot rooms with an ever-increasing demand for alumina faces plant manufacturers with ever greater challenges. FLS successfully managed the reliably supply all 444 pots of the longest pot line in the world with alumina in the required quality with four horizontal FLUIDFLOW® air slide systems. At present it appears that the length of future pot rooms will not continue to increase, but larger conveying capacities must be realized and defined, low dust emission limit values as well as degrees of segregation have to be guaranteed contractually.

With the aid of newly developed design software, direct pot feeding systems can be planned, and all relevant system parameters can be defined for various different conveying capacities and system parameters. Emission limit values and degrees of segregation can be calculated on the basis of the technical centre tests and can be guaranteed.

References

1. A. Wolf, P. Hilgraf, *FLUIDCON—A New Pneumatic Conveying System for Alumina* (Light Metals, 2006), pp. 81–87
2. C. Duwe, K. Gelden, The Moeller direct pot feeding system for a smooth and constant pneumatic transport of secondary alumina to electrolytic cells, in *Proceedings of 6th International Alumina Quality Workshop*, pp. 228–232
3. C. Duwe, T. Letz, New generation of FLSmidth Möller direct pot feeding system. Aluminium **84**, 1/2 (2008)
4. Hauschildt, Successful commissioning of world‘s longest pot feeding system. Aluminium 1–2 (2018)

Carsten Duwe Head of Process Department, REEL Möller GmbH, Pinneberg, Germany. Email: Carsten.Duwe@flsmidth.com

Dipl. chem. Ing. Carsten Duwe, graduate engineer (MSc), was born in Bonn, Germany and has been working with FLSmidth Hamburg, Germany since 1997. During his studies in Chemical Engineering at the University of Karlsruhe (now Karlsruhe Institute of Technology) from 1984 to 1990, he has specialized in chemical and mechanical process engineering. From 1991 to 1997 he worked as Sales and Project, R&D and Department Manager at BETH GmbH, Lübeck. During this time, he developed his experiences and know how in the fabric filter business, especially for coal fired boiler and waste incineration applications. He was responsible for the successful development of the so-called Bethpuls Fly-Stream-Adsorption-System. Since 1997 he has been working for Johannes Möller Hamburg, part of the FLSmidth group since 1996. During the following 23 years in the pneumatic handling business, he took over different responsibilities as specialist and manager. From 1999 to 2014 as manager and proxy holder, he was responsible for the technical departments at FLS Miljö Deutschland, later FLSmidth Hamburg. Since 2014, as part of the new organization, he took over responsibility as Head of Process Design and the R&D projects in close co-operation with the Danish mother company in Copenhagen. Different technical publications as well as two mentioning as inventor at the EU patent office in view to an advanced pipe-in-pipe pneumatic transport technology (2003) and fly ash removal systems from ESP (2009).

Chapter 13
Health, Safety and Emissions to the Environment from Alumina Refineries

Brady Haneman and Benny E. Raahauge

Abstract The health and safety of personnel working in industrial facilities must be secured by reasonable protection measures. Likewise, the protection of the environment from the operation of the industrial facilities must meet the emission standards set by the authorities to obtain and maintain the License to Operate these industrial facilities. The prime focus of this chapter considers the Health, Safety and Environmental (HSE) issues from a process point of view. This means that protecting personnel health is limited to a description of the major chemicals that the personnel can be exposed to. Thus, leaving it up to the refinery management to ensure that adequate procedures and protection measures to protect the health and well-being of their personnel during their working hours is enforced. Several cases of hazardous events in alumina refineries, including the bauxite residue storage area, introduces some of the process safety issues that have occurred in the past. From there, an introduction to Functional Safety is made representing a systematic approach to deal with this subject in today's alumina refinery design and capacity expansions. Safety aspects of operating high pressure and temperature equipment in the digestion area are touched upon as well. Sources of the major gas borne emissions to the Environment from the alumina refinery operation within allowable local emission limits is discussed for NO_x, CO, SO_x, VOC, CO_2, and particulate matter. Time has shown that authorities in general tend to tighten the allowable gas borne emissions to the environment and in many places rightfully so. Anticipating that this trend will continue in the future, it is important to understand current days technology with a view to future possibilities for acceptable solutions. In the near-term Natural Gas will be replacing HFO to lower SO_x emissions and the "Carbon Footprint" of Alumina Refineries. In the medium-term, the ESP may be gradually outplaced with Fabric Filters (FF) or ESP/FF hybrids using "Tailor made" catalyst to reduce emissions of NO_x and

Benny E. Raahauge is deceased.

B. Haneman (✉)
Hatch, Brisbane, Australia
e-mail: brady.haneman@hatch.com

B. E. Raahauge
FLSmidth - Retired, Deceased, Haslev, Denmark

B. E. Raahauge and F. S. Williams (eds.), *Smelter Grade Alumina from Bauxite*, Springer Series in Materials Science 320,
https://doi.org/10.1007/978-3-030-88586-1_13

VOC's and eliminate emissions of dust plumes when a power failure happens. In the long-term we may expect to see a gradual replacements of fossil fuels with "Green Hydrogen" to reduce the "Carbon Footprint" of future Alumina Refineries to almost ZERO.

13.1 Introduction

In general, the health and safety of personnel working in industrial facilities must be secured by reasonable protection measures. This is the responsibility of all who are involved in the design, construction, commissioning, operation, maintenance, and management of industrial facilities i.e., like the Bayer Process plant with its associated facilities such as the Boiler/Powerhouse and the Calcination units.

Likewise, the protection of the environment from the operation of the industrial facilities is the responsibility of all who are involved in the design, construction, commissioning, operation, maintenance, and management of industrial facilities to obtain and maintain the License to Operate these industrial facilities.

Having said that, the ultimate responsibility to fulfill the above objective rests with the management of the industrial facilities, which in this case is the management of the Alumina Refinery in question.

The prime focus of this chapter will be on securing Functional Safety and Environmental Impact from the alumina refinery operation within tolerable risk and allowable emission limits.

13.2 Health and Chemicals in the Bayer Process

Building and maintaining a healthy and well-being workforce has many elements and aspects founded on a safe working environment in the alumina refinery.

Whilst elevated temperature fluids will incur burns to contact of the skin, most chemicals employed in the digestion facility and elsewhere in the Bayer circuit are hazardous to health and generally highly corrosive to human contact.

All have potential to cause irreparable health related issues if safety procedures and personal protective equipment (PPE) are not employed. Material Safety Data Sheets (MSDS) provide salient information for any chemical, including hazardous identification information, composition, toxicological effects, first aid treatment and containment advice.

A brief outline of the main chemical agents in the digestion and other alumina refinery facilities is provided below.

13.2.1 Caustic Soda

As the principal lixiviant for alumina dissolution, caustic soda is present throughout the digestion and downstream process facilities. It is miscible in water, colorless and highly corrosive to skin. Swallowing can result in nausea, vomiting, abdominal pain and chemical burns to the gastrointestinal tract. Contact with the eyes can cause corneal burns. Contamination of eyes can result in permanent injury. Contact with skin will result in severe irritation and chemical burns.

Treatment of caustic soda contact has historically utilized immersion and dousing of the affected area with copious quantities of water. Over more recent years, the non-toxic chemical agent Diphoterine, has been employed as a medical treatment to neutralize the caustic in addition to rinsing the affected area with water.

13.2.2 Acids

Acids such as sulfuric acid, are used for chemical cleaning of heat transfer equipment in the digestion and other facilities. Dilute sulfuric acid is prepared from 98% w/w concentrated sulfuric acid.

Sulfuric acid is corrosive to human skin on contact and poisonous on ingestion and inhalation of its vapor. It will cause severe burns leading to necrosis and scarring of tissue. The acid mists are very corrosive and can cause severe irritation and injury if inhaled. The degree and severity of respiratory effects are influenced by the size of the aerosol particulate, deposition site, concentration, and humidity. Inhalation of acid mists may cause severe lung damage and pulmonary oedema (accumulation of fluid in lungs).

13.2.3 Lime

Milk of lime (also known as slaked lime or hydrated lime) may be added in the digestion facility for the control of impurities and to improve extraction efficiency. It is prepared by reacting calcium oxide with water to produce calcium hydroxide. Milk of lime is corrosive to contact of the skin and overexposure may cause alkaline burns and irritant or allergic dermatitis.

In some alumina refineries lime is produced by calcination of limestone. Over exposure to powder from calcium oxide dust (when mixing) may result in severe skin burn, when reacting with moisture in the skin, mucous membrane irritation of the nose and throat, coughing and bronchitis.

PPE should always be used to avoid eye or skin contact and avoiding inhalation from dust generation during preparation.

13.2.4 Volatile Organic Compounds (VOC)/Odors

This subject is discussed in Sect. 13.5.1.4.

13.2.5 Mercury

Whilst present at only trace quantities (ppm) in the bauxite, mercury may be emitted in the digestion area non-condensable gases and other atmospheric release points. Some refineries incorporate sub-cooling of the non-condensable gases emanating from the digestion area to ~50 °C, such that any mercury vapor present is condensed. It may then be collected as liquid mercury for disposal. In elemental form, mercury vapor is highly toxic. Direct contact of liquid mercury with skin will cause irritation and burns. Similarly, direct contact with the eyes will cause irritation and burns.

13.2.6 Hydrogen

Hydrogen gas constitutes approximately 80% of the non-condensable gas emissions from the digestion facility.

It is highly flammable and presents the risk of forming explosive mixtures over a wide range of concentrations from 4 to 74% in air. Areas of confined pipework may accumulate hydrogen gas and high point vents should be considered where appropriate, to enable gas testing prior to hot work.

A detailed discussion of the many elements and aspects involved, such as noise, to ensure a well-being workforce is beyond the scope of this book and will not be further dealt with.

13.3 Alumina Refinery Safety Incidents

The main processing areas in the alumina refinery is the Bayer Process stages as shown in Fig. 13.1 connected by the circulating flow of a caustic solution with a concentration varying from 150 to 350 gpl Na_2CO_3 [2] subject to digestion and precipitation conditions.

Over time, several examples of severe incidents of hazardous events, in or around, Alumina Refineries has been reported, sometime with fatal or less severe consequences for the people involved, but always with significant cost to the owners of the facilities and sometimes the local society.

In the following a few examples are presented covering Digestion, Calcination and the Red Mud Pond.

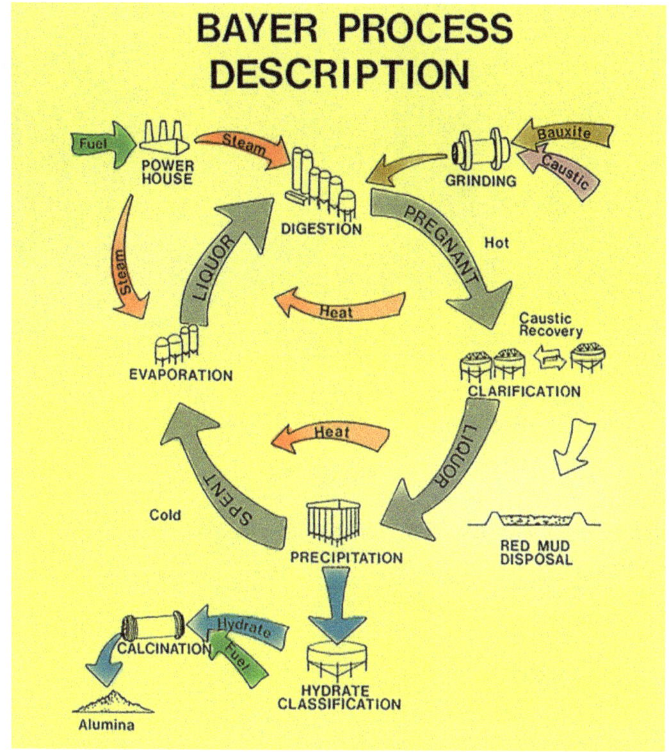

Fig. 13.1 The Bayer process wheel [1]

13.3.1 Digestion

For the digestion facilities in alumina refineries, a heightened awareness of safety considerations is paramount, as the chemicals employed throughout the processing areas are generally under elevated conditions of concentration, pressure, and temperature, see Table 13.1.

On 5. July 1999, an explosion happened in the digestion plant at the Kaiser Alumina Refinery at Gramercy, Louisiana, USA, injuring 29 people (Fig. 13.2).

Table 13.1 Typical digestion conditions subject to bauxite mineralogy

Digestion conditions	Low temperature (typical)	High temperature (typical)
Caustic concentration (as gpl. Na_2CO_3)	150–250	150–350
Temperature, °C	143–145	240–280
Pressure, Ata	<10	<80

Fig. 13.2 Before and after explosion of flash vessels at Kaiser Aluminum Gramercy Refinery 5. July 1999

The investigating body concluded that excessive pressure in several large tanks caused the explosion [3].

13.3.2 Calcination

Outside the Bayer Process Wheel, but still part of the alumina refinery, the Powerhouse, Calcination and Liquor Burning (if applied) process areas comprises high temperature combustion of fuels at 950–1500 °C.

The operating pressure in the Powerhouse, Calcination and Liquor Burning process areas is slightly below or above atmospheric pressure.

Typical fuels such as coal (powerhouse only), heavy fuel, natural gas or coal gas from a coal gasification plant are used.

In October 1991, an explosion happened in an alumina rotary kiln at Queensland Alumina Refinery, Gladstone, Australia [4], Fig. 13.3.

The basic reason for the explosion was a series of events generated by recycling of cold dust from the satellite coolers into the kiln burning zone ultimately resulting in an explosive gas mixture reaching the ESP, where the explosion occurred.

Fortunately, no persons were injured, but extensive damage happened to the ESP, the multi-clone and ducting.

The kiln was re-commissioned in December 1991 after repair and installation of a new ESP.

Fig. 13.3 ESP and ducting after explosion in a rotary kiln at Queensland Alumina, Australia, October 1991

13.3.3 Red Mud Pond

Another area outside to the Bayer Process Wheel, but still part of the alumina refinery, is the red mud disposal area, where history has shown that this area can indeed be subject to both functional safety issues and environmental damage.

On 4 October 2010, the embankment of bauxite residue reservoir 10 at the Ajka Alumina refinery, Hungary, failed, as seen in Fig. 13.4 [5].

The incident caused 8 direct fatalities and 286 people was injured receiving medical care, of which 120 people were hospitalized.

The incident also caused severe damage to the environment. Estimated 1 million m^3 bauxite residue slurry containing 853.000 m^3 supernatant liquor with pH 13 flooded about 1017 ha effecting 367 houses or other buildings (Fig. 13.5).

The Ajka incident increased focus on safe bauxite residue storage technology and its on-going management [6].

13.4 Process Hazard and Risk Management

Many health and safety related work processes are utilized in the evolution of a flowsheet to the as built operational facility. These generally incorporate:

- Design reviews
- Codes and Standards

Fig. 13.4 Ajka Alumina Refinery, Hungary, the embankment of reservoir 10 failed on 10 October 2010

Fig. 13.5 The village of Kolontár, Hungary, flooded with bauxite residue slurry with pH 13 in the liquor

- Functional Safety—Available Best Practice of Process Safety
- Overpressure protection systems.

13.4.1 Design Reviews

Design reviews are used in the plant development to ensure that safe operability and maintainability aspects are incorporated throughout the facilities.

Careful layout of valves and equipment will ensure safe and unencumbered access to working areas for personnel.

Elevated work platforms for maintenance should ensure safe access and egress in the event of a hazardous release of fluids or stored energy. Similarly, the development of tag-out procedures must be re-viewed to incorporate safe isolation of energized equipment for maintenance.

13.4.2 Codes and Standards

There are a multitude of international codes and standards available for reference in the process industry.

These have been developed over many years by industry bodies, practice specialists and governments to reduce both the likelihood of occurrence and severity of health and safety related incidents. In general, these codes and standards set out minimum requirements to be incorporated into the engineering, design, and inspection processes.

13.4.3 Functional Safety—Available Best Practice of Process Safety

Functional safety is based on the analysis of process hazards and management of the risk. This is a basic requirement of today's design of a process plant. A selection of important codes and standards is listed below.

- AS/NZS ISO 31000—Risk Management—Principles and Guidelines
- IEC 61508-0 Functional safety of electrical/electronic/programmable electronic safety-related systems
- IEC 61511-1 Functional safety—Safety instrumented systems for the process industry sector Part 1
- API RP 752—Management of Hazards Associated with Process Plant Permanent Buildings

- API RP 755—Fatigue Risk Management Systems for Personnel in the Refining and Petrochemical Industries
- AS/NZS 2430—Classification of Hazardous Areas
- AS3780 The Storage and Handling of Corrosive Substances.

The standard IEC 61511-1 "Functional Safety—Safety Instrumented Systems (SIS) for the process industry sector Part 1-3", is fundamental to the process industry and represents todays best practice in situations where the process design itself cannot provide sufficient safe operability of a process plant in its original design.

A SIS is typically designed to a tolerable risk level according to the **ALARP** principle, developed in UK [7], and it states that risk shall be reduced to a level "**As Low As Reasonably Possible**".

In quantitative terms, subject to government mandate and company size, typical tolerable risk level may range from 10^{-5} to 10^{-3} fatalities per man-year.

The methodology of implementation of functional safety life cycle activities in a process plant is outlined in the IEC 61511-1 standard. The standard describes how execution of a Safety Instrumented System (SIS) shall be managed (§ 5) and follow the Functional Safety Life Cycle Phases (§ 6) as shown in Fig. 13.6.

As indicated in Fig. 13.6, all the work activities shall be verified in accordance with IEC 61511 §7, with the objective to demonstrate by review, analysis and/or testing that the required outputs satisfy the defined requirements for the appropriate phases.

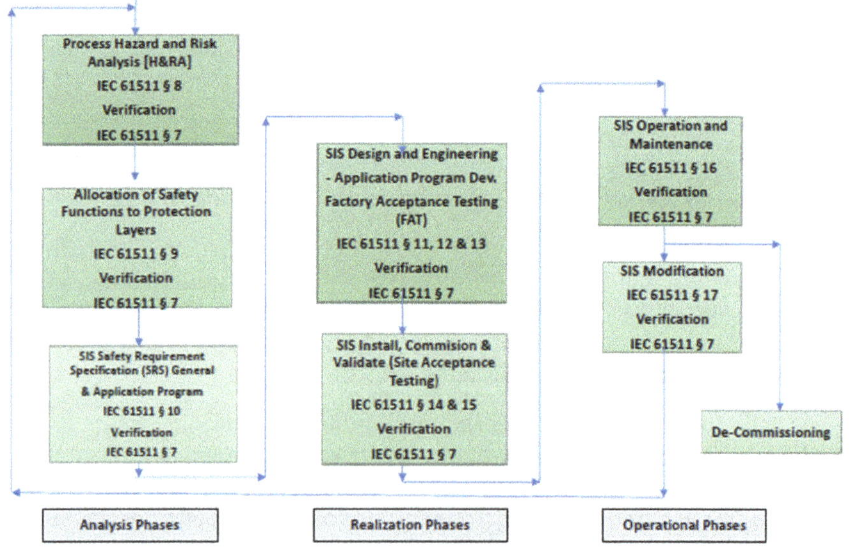

Fig. 13.6 Functional safety life cycle phase diagram (LCPD)

13.4.3.1 Analysis Phase

Process Hazard and Risk Analysis [H&RA]—or HAZOP

The objectives of §8 in IEC 61511-1 are to determine:

- The hazards and hazardous events of the process and associated equipment.
- The sequence of events leading to the hazardous event.
- The process risk associated with the hazardous events.
- Any requirement for risk reduction.
- The safety functions required to achieve the necessary risk reduction.
- If any of the safety functions are Safety Instrumented Functions (SIF).

Allocation of Safety Functions to Protection Layers

The objectives of §9 in IEC 61511-1 are to determine (Fig. 13.7):

- Allocation of safety functions to protection layer.
- Determine the required SIF's.
- Determine for each SIF the associated risk reduction factor or safety integrity requirements.

Safety Instrumented System (SIS)—Safety Requirements Specification (SRS)

The objective of §10 in IEC 61511-1:

- To specify the requirements for the SIS, including any application programs and the architecture of the SIS.

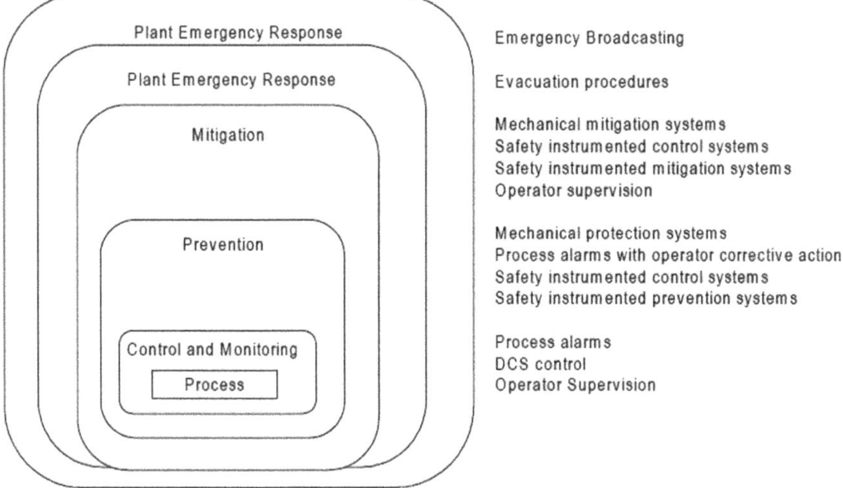

Fig. 13.7 Typical protection layers and risk reduction means [8]

13.4.3.2 Realization Phase

Safety Instrumented System (SIS)—Design and Engineering

The objective of §11 in IEC 61511-1:

- To design one or multiple SIS to provide the SIF and meet the specified integrity requirements (e.g.: associated risk reduction and Safety Integrity Level (SIL)) for the industrial facility under design.

The design of the SIS shall be in accordance with the SIS safety requirement specification (SRS), considering all the additional requirements of §11.

Safety Instrumented System (SIS)—Application program development

The objective of §12 in IEC 61511-1:

- To define the requirements for development of the application program. This includes design, implementation, and requirements for verification by review and testing. Also, the application program methodology and tools must follow the safety manual(s) of the safety PLC used in each SIS.

The application program of the SIS shall be in accordance with the application safety requirements and all requirements of this clause for all SIS up to and including SIL 3.

Factory Acceptance Test (FAT)

The objective of §13 in IEC 61511-1:

- The objective of §13 is to test the devices of the SIS to ensure the requirements defined in the SRS are met.

By testing the logic solver, associated software, and hardware prior to installation, i.e. by simulation input to the interface, errors can be readily identified and corrected before hook-up to the field instrumentation and final control elements at the plant site. The FAT is sometime referred to as an integration test and can be part of the validation in accordance with §15.

SIS Installation and commissioning

The objective of the requirements of §14 are to:

- Install the SIS in according to the specifications and drawings.
- Commission the SIS so that it is ready for final system validation.

The purpose of the commissioning activities is to ensure that each of the SIS devices is individually ready to operate. As specified in the design phase.

SIS Safety validation

The objective of the requirements of §14 are to:

- Validate, through inspection and testing, that the installed and commissioned SIS and its associated SIF(s) achieve the requirements as stated in the Safety Requirement Specification (SRS).

  This activity is sometime referred to the **Site Acceptance Testing (SAT)**.

13.4.3.3 Operational Phase

SIS operation and maintenance

The objective of the requirements of §16 are to ensure that:

- The required SIL and SIF is maintained operation and maintenance.
- The SIS is operated and maintained in a way that sustains the required safety integrity.

Compensating measures that ensure continued safety while the SIS, or part thereof SIF(s), is disabled or degraded due to by-pass (repair and testing) shall be applied with associated operation limits (duration, process parameters etc.). The operator shall be provided with information on the procedures to be applied before and during by-pass and what should be done before the removal of the by-pass and the maximum time allowed to be in the by-pass state. The operating and maintenance procedure can include verification that by-passes are removed after proof testing.

SIS modifications

The objective of the requirements of §17 are to ensure that:

- That modifications to any SIS are carefully planned, reviewed, approved and documented prior to making the change.
- To ensure that the required safety integrity of the SIS is maintained despite of any changes made to the SIS.

Modifications of the Basic Process Control System (BPCS), other equipment, process or operating conditions shall be reviewed to determine whether they are such that the nature or frequency of demands on the SIS will be affected. Those having an adverse effect shall be considered further to determine whether the level of risk reduction will still be sufficient.

SIS de-commission

The objective of the requirements of §18 are to ensure that:

- Prior to de-commission any SIS from active service, a proper review is conducted and required authorization is obtained.
- The required SIF(s) remain operational during de-commissioning activities.

The paper by Daniels [8], provides an excellent overview of the Lifecycle Activities required and their management for the design and implementation of a Safety Instrumented Systems (SIS) in an Alumina Refinery.

In the case described by Greg Daniels the following number of SIS was implemented:

- 7 Boiler Units
- 2 Low and High Temperature Digestion
- 2 Calcination Units
- 2 Liquor Purification Units.

Also, and especially important, the paper provides an example of how to organize a Master SIF Document Register, so that the documentation is compliant with §19 Information and Documentation Requirements of IEC 61511.

However, as no code or standard can be prepared to cover every industry specific application, sound engineering judgement must be used to ensure "fit for purpose" solutions are implemented. A selected listing of some is provided below. Note the latest version of these must always be used.

13.4.3.4 Equipment Isolation, Testing and Inspection

- API 2217A—Guidelines for Safe Work in Inert Confined Spaces in the Petroleum and Petrochemical Industries
- API 650 Welded Tanks for Oil Storage
- NACE SP0403 Avoiding Caustic Stress Corrosion Cracking of Refinery Equipment and Piping
- NACE SP0198 Control of Corrosion Under Thermal Insulation and Fireproofing Materials—A Systems Approach.

13.4.4 Overpressure Protection Standards

- API 520, Part I—Sizing, Selection, and Installation of Pressure-Relieving Devices
- API 521 Pressure-Relieving and De-pressuring Systems
- API 2000 Venting Atmospheric and Low-Pressure Storage Tanks
- ASME VIII Division I—Rules for Construction of Pressure Vessels
- ASME B31.1—Power Piping, ASME Code for Pressure Piping
- ASME B31.3—Process Piping, ASME Code for Pressure Piping
- NFPA 69—Standard on Explosion Prevention Systems.

13.4.5 Overpressure Protection Systems

Overpressure protection systems guard personnel, equipment and the environment from process transients that deviate the plant away from normal operating conditions. These protection systems may take the form of passive protection (using bursting discs, pressure relief valves and mechanical ratings) or the use of Safety Instrumented Systems (also referred to as protection by system design), or some combination of the two.

A reliance on purely administrative tools and systems will not comply with the various codes and standards listed in 13.3.

Within the alumina refinery, these process transients may comprise:

- Vessel/equipment/piping blockages
- Power Failures (vessels exposed to changing energy states)
- Hydraulic expansion (liquid filled chambers exposed to sensible heat)
- Pressure transients (water hammer → hydraulic shock waves, steam and condensation induced water hammer → collapse of trapped steam pockets).

For the above process transients, the response of the system and resultant relief capacity requirements generally change with time. Salient considerations for each transient are further discussed below.

13.4.5.1 Vessel and Equipment Blockage

Blockages of vessels in the Digestion area flash tank train may result from internal wall scale dislodgment (particularly during equipment startups) or mechanical failure of interposing components.

The provision of passive overpressure protection via simply rating all vessels to the maximum pump shutoff head is generally uneconomic, and more rigorous analyses are required. Independent of the process flowsheet selected, these analyses will be subject to the following variables and assumptions:

1. nominated vessel design pressures
2. selected time frame of transient analysis i.e., 10 min, 15 min, 30 min etc.
3. slurry and liquor supply pump capacities
4. physical interconnecting piping geometry and componentry used.
5. upstream vessel operating pressures during the downstream vessel blockage
6. fluid state temperatures at the inlet to downstream vessel relief valves following a relief event i.e. subcooled, saturated or two-phase mixture
7. dual phase (slurry) or three phase (flashing slurry) fluid movement between interconnecting pressure vessels following the transient condition.

The blockage analysis can multiply in complexity for the high temperature digestion facility as a result of the multiple number of flash tanks employed and sound engineering judgement is required to bound the analysis within practicable limits.

13.4.5.2 Power Failure

A power failure in the digestion facility removes the motive driving force for fluid flows through recuperative heat transfer equipment. As a result, the 'heat sink' for the vessel train is lost in this transient condition.

The motive driving force for slurry flow throughout the vessel train remains in the form of the thermal energy of the slurry.

The hydraulic impedance to slurry flow (and therefore energy flow) is the cumulative result of the components and fittings present in the interconnected piping between vessels. For this transient condition, satisfying the dynamic energy balance results in vessel energy states that continuously change (and generally increase) with time.

Analytical methods using the thermo-hydraulic response characteristics of the equipment and piping are required to assess relieving rates. Once relieving rates are determined, overpressure protection may be provided to ensure that vessel pressures during the transient, do not exceed code allowable pressures.

13.4.5.3 Hydraulic Expansion

Hydraulic expansion relates to the increase in volume of an isolated fluid when exposed to an increased energy state. Most often, this condition is associated with heat transfer equipment in the digestion facility. Hydraulic expansion of a heater tube side fluid will occur when the heater train is isolated at the inlet and outlet ends with the fluid at an initial temperature of the inlet to the first stage heater. Alternately, fluid in the heaters may have been isolated as a result of a short-term operational disturbance and allowed to cool. Subsequent opening of an indirect steam supply and/or recuperative flash vessel vapor will impart sensible heat to thermally expand the isolated tube side fluid. Rapid increases in system pressures may eventuate from relatively small rises in fluid temperature.

The determination of required passive relief may be simplified by neglecting volumetric expansion of the system piping. Liquid relieving rates will then be a function of:

1. The heat input rate, Q (J/h)
2. The fluid heat capacity, Cp (J/kg °C), and
3. The fluid coefficient of volumetric expansion at constant pressure, β ($°C^{-1}$).

13.4.5.4 Pressure Transients

Pressure transients may take the form of water hammer (hydraulic shock waves), steam hammer or condensation induced hammer (collapsing of trapped steam pockets). All types can manifest themselves with an elevation of piping system internal pressures by multiples of the normal operating system pressure. Where closure of valves is concerned, the general method employed to limit the water hammer pressure wave is to restrict the closure time of control valves to not less than

30 s or more. This control, however, is not intrinsically failsafe, nor self-limiting, where manual valves may be inadvertently closed against online process streams.

A complementary approach to overpressure protection employing the layers outlined below (i.e., neither purely passive or safety instrumented systems), may offer a satisfactory level of risk mitigation to this transient condition. Note that the response time of pressure relief valves and bursting discs are generally longer than that required to mitigate the rate of rise of the pressure wave.

- administrative controls incorporating locks on all manual isolators, along with strictly supervised operating procedures.
- surge arrestors. Where positive displacement pumps are employed, the dampener must be sized for both the normal acoustic attenuation and the pressure transient.
- control interlocks on manual isolators to initiate pump shutdown during impending incorrect valve operation.
- modification of manual valve actuators to incorporate an intrinsic characteristic with a limiting low speed gear ratio to limit the valve closure period.

13.5 Emission to the Environmental from Alumina Refineries

The major source of environmental emissions from an alumina refinery is disposal of the Red mud from digestion of the bauxite.

Reference is made to Chap. 6: Bauxite Residue Management, for further readings about modern bauxite residue management technology and best practices.

In 2010 the Department of Environment and Resource Management, Environmental Protection Agency, Queensland Government issued the report [9] concerned with emissions to the atmosphere from alumina refineries titled:

"**Benchmarking Queensland Alumina Limited and Rio Tinto Yarwun (RTAY) Alumina Refineries' Emissions to Australian, Canadian, Irish and USA Refineries**".

The report contributes to the study of **Clean and Healthy Air for Gladstone** examining the possible health impacts related to atmospheric emissions of contaminants from activities in the Gladstone/Calliope Shires.

The report provides atmospheric contaminant emission estimates (as both kilograms per year and grams per ton of alumina) for alumina refineries, drawing on publicly accessible data from Australia, Canada, Ireland and USA.

The pollutants are grouped into (1) Metals and metalloids; (2) Organics and Organohalogens; (3) Inorganics; and (4) Radionuclides. Readers interested in the comprehensive contaminant list for the Gladstone/Calliope Air—shed is referred to study the report.

Table 13.2 is an extract from Table 5 of the report [9] showing the mean specific emission of selected pollutants expressed in gram/ton alumina.

Table 13.2 Mean specific atmospheric emissions from Alumina Refineries in gram/ton alumina [9]

Pollutants source	Observations	VOC	CO	Nitrogen oxides	SO_2	H_2SO_4	PM_{10}
QAL	6	35	470	2.400	1.100	9.7	320
RTAY	3	92	34	360	350	0	53
Australia	30	47	380	700	2400	0.0011	200
Canada	6	22	6400	–	710	–	150

The mean emissions may arise from the storage and burning of fuels to provide energy to the Bayer process operations, from the digestion operations, the clarification, the precipitation process, and emissions from the final calcining process.

The report notes that RTAY utilize the tube digestion process whereby organic non-condensable gases are delivered to the boilers for destruction and that oxides of nitrogen are low due to the use of low-NOx technology.

For the distribution of VOC or odor pollutants emission among the different process units in the Bayer process reference is made to Sect. 13.5.1.4 for further details.

13.5.1 Gas Compounds Emitted into the Air from Bayer Process Units, Calcination and Liquor Burning

A major source of atmospheric emission from an alumina refinery comes from the boilers, calcination and liquor burning units. These emissions comprise mainly particulate matter and gaseous compounds such as NO_x, SO_x, CO, Volatile Organic Compounds (VOC/Odors) and CO_2 to mention the major quantities of flue gas compounds.

The following discussion of flue gas compounds mentioned above will focus on calcination and to some extend on liquor burning.

13.5.1.1 Nitrogen Oxides—NO_x

The term NO_x refers to the content of NO and NO_2 in the gas phase emitted from the stack of a boiler, calciner, liquor burner, or another industrial furnace. Combustion products from industrial furnaces comprises mainly NO.

Formation of NO_x is attributed to three distinct mechanisms:

- Thermal NO_x formation.
- Fuel NO_x formation; and
- Prompt or rapidly formed NO_x.

Thermal NO_x formation in gas-, oil, or coal fired furnaces results from reaction of oxygen and nitrogen in the atmospheric air used in the combustion process.

Early kinetic analysis of NO_x formation in gaseous combustion in pre-mixed flames (that, is the reactants are thoroughly mixed before combustion) by Zeldovich can be approximated by an Arrhenius-type equation where A and b are constants [10] showing the relative importance of time t, absolute temperature T and volume concentration of oxygen O_2 and nitrogen N_2.

$$[NO_x] = A[N_2] \int_0^t e^{-b/T}[O_2]d\theta$$

The Zeldovich relationship cannot be used to determine the NO_x formation, but it does illustrate the impact from the major factors and shows that the thermal NO_x formation increases exponentially with combustion temperature.

Experimentally measured NO_x formation rates near the flame zone are higher than predicted by the Zeldovich relationship. The formation of thermal NO_x may be reduced by reducing the peak temperature and/or the oxygen concentration.

Fuel NO_x is formed by the reaction of nitrogen compounds in the fuel with oxygen. The most common form of fuel nitrogen is organically bound nitrogen present in liquid or solid fuels where individual nitrogen atoms are bound to carbon or other atoms. These bonds break more easily than the diatomic N_2 bonds so that fuel NO_x formation rates can be much higher than those of thermal NO_x.

Fuel NO_x is much more sensitive to stoichiometry (i.e., oxidizing or reducing conditions) than to thermal conditions as seen from the formula, where κ is a constant:

$$[NO_x] = \kappa \int_0^t [C_xH_yN_z][O_2]d\theta$$

Since oxidation of the fuel bound nitrogen is a fast reaction, and the fuel concentration is limited by the required stoichiometry, fuel bound NO_x formation may be reduced only by replacing the fuel i.e. from heavy fuel oil to natural gas and/or reducing the oxygen concentration.

Prompt NO_x refers to the rapidly formed NO from N_2 in the air. Prompt NO in NO_x emission is negligible in comparison to thermal and fuel NO_x. This is so because nitrogen radicals are exceptionally difficult to form from molecular nitrogen.

A more detailed discussions of the various NO_x formation mechanism is beyond the scope of this chapter and can be found elsewhere [11, 12].

Control of NO_x emissions can be made to a certain extent by the design of the calcination furnace as well as the fuel nozzles themselves. It is therefore important to understand the different designs and operation of the calcination furnaces for alumina production as well as the fuel nozzle design to understand its impact on NO_x formation.

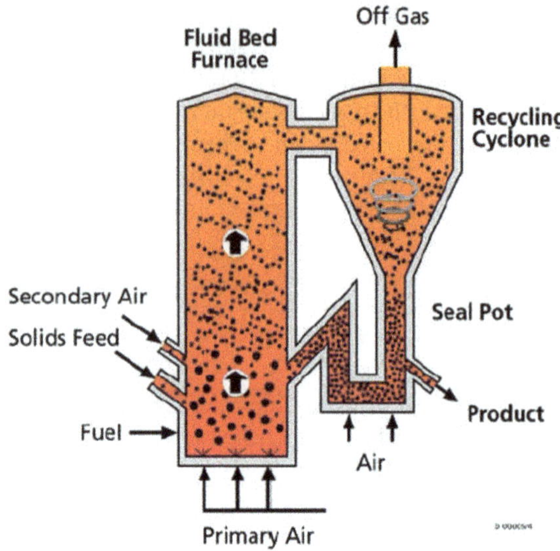

Fig. 13.8 Circulating fluid bed (CFB) furnace design and operating principle

The Expanded or **Circulating Fluid-Bed (CFB)** calciner (Fig. 13.8) was developed by **Lurgi**, now **Metso Outotec, Germany**, in the early 1970. The CFB utilizes expanded fluid-bed technology with two-stage addition of air for combustion of fuel in the calcination furnace.

Dust free primary air is required for fluidization while secondary air for complete combustion is added directly to the circumference of and through the casing of the Fluid Bed Furnace. This means that the initial part of the combustion of the fuel is sub-stoichiometric and thus low in NO_x generation.

The operating temperature is approximately 950 °C of the hot mixture of partly calcined alumina and Smelter Grade Alumina re-circulated many times over the expanded Fluid Bed Furnace/Recycling Cyclone.

The **Gas Suspension Calciner (GSC)** technology was introduced into the cement industry by FLSmidth in 1976. In 1991 two (2) 1850 TPD GSC natural gas fired units for Alumina were commissioned at the Sherwin Alumina Plant, in Texas, USA.

The gas flow into the GSC furnace becomes a confined or enclosed jet with internal re-circulation of gas and solids, and an entrained bed with respect to the net solids flow out (Left of Fig. 13.9). The gas–solid mixture from the GSC furnace passes the Furnace cyclone in one—pass only. This is like in the Alcoa designed Fluid-Flash Calciner, however without a Holding Vessel (HV).

The solids retention time in the GSC furnace is about 3–4 times the gas retention time. This has been verified experimentally in a 3200 TPD GSC unit for cement operating in India [13].

The calcination temperature is 1070 °C in the GSC without a fluidized Holding Vessel (HV) in the Furnace Cyclone to obtain the required degree of calcinations of the SGA. In the GSC unit designed with a fluidized Holding Vessel (HV) into the bottom of the Furnace Cyclone (Right of Fig. 13.9), the nominal solids retention time

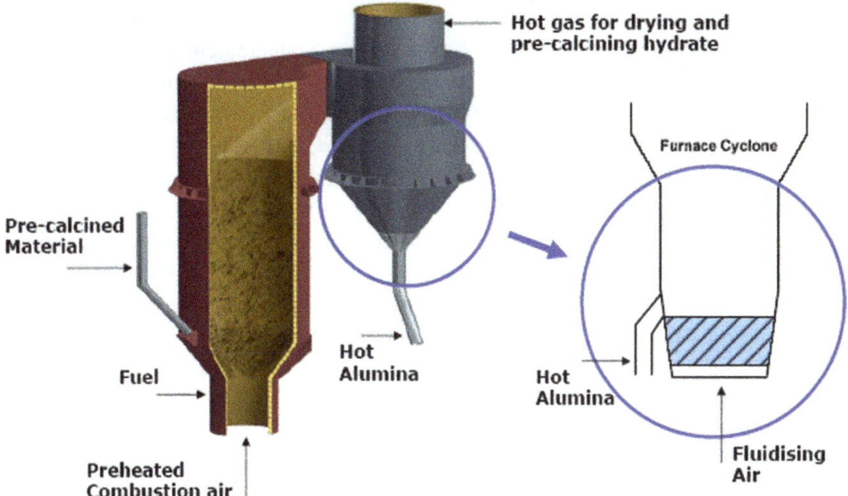

Fig. 13.9 Gas suspension calciner (GSC) design and operating principles without/with holding vessel

in the HV is like in the Alcoa units resulting in a calcination temperature of about 960 °C.

The described calcinations furnaces (Figs. 13.8 and 13.9) are designed to operate with a gas velocity in the range 5–10 m/s to ensure safe entrainment of the solids up through the calcination furnace into the re-cycling/furnace cyclone at full and partial capacity.

There is a high to medium degree of turbulence or "back-mixing" in all the calciner furnaces, all be it for different design reasons. In the CFB the back-mixing is dictated by the circulating mass of solids, which tends to block or reduce the internal re-circulation of gaseous combustion products but as mentioned the initial part of the combustion of the fuel is sub-stoichiometric and thus low in NO_x generation.

The fluid dynamics in the GSC furnace design is a confined jet [13] which results in the setting up of a re-circulating flow of solids and gaseous combustion products down the walls of the cylindrical furnace towards the combustion air inlet where the burner nozzles are located. The GSC furnace volume act as a "large" back-mix reactor with "flue gas" and solids [13] recirculation back into the incoming combustion air.

The "flue gas" acts as a diluent, reducing the local combustion temperature and suppressing the partial pressure of oxygen, and thus reducing NO_x formation. Similar fluid dynamics may be present in the Alcoa calciner furnace.

In general, **NO_x reduction** may be achieved in a number of ways [10] such as: (1) Burners out of services, (2) Derating, (3) Oxygen and combustibles trim, (4) Tramp air reduction, (5) Steam or water injection, (6) Staged combustion, (7) Alternate fuels, (8) Flue gas recirculation and (9) Low NO_x burners.

(**Re 1**) Burners can be taken out of services when multiple burners are used as in both the CFB and GSC furnaces. (**Re 2**) The actual burners used can be derated with the calcination capacity. (**Re 3**) Air/Oxygen trim is used to minimize excess oxygen above the minimum required by the Burner Management System (BMS) for safe operation. (**Re 4**) Tramp or false air may enter the furnace volume through leaks of various kinds. This is not an issue for the CFB furnace as this is operating above the ambient pressure. This may happen to the GSC furnace, but the quantities of false air are relatively small. On the other hand, leaks of hot combustion products are prevented from escaping to the atmosphere which can represent a safety issue to operator or maintenance people passing the furnace area. (**Re 5**) Steam atomization is frequently used for calciners using Heavy Fuel Oil (HFO) rather than air. (**Re 6**) Staged combustion is used in the CFB furnace (Fig. 13.8). (**Re 7**) Change from HFO to Natural Gas eliminates fuel NO_x emissions and SO_x emissions from sulfur in the HFO with an improved alumina quality in addition hereto. (**Re 8**) Internal recirculation of combustion products or flue gas, takes place in the GSC furnace as explained above. (**Re 9**) Low NO_x burners are used in the GSC furnace and is achieved by design of the burner nozzles to ensure low injection velocity of the fuel. NO_x emission from GSC units can be less than 20 ppm, with burner nozzles designed to minimizes the local mixing intensity of fuel and incoming air for combustion.

Reporting of NOx Emission Measurements

The proper weighing of the many factors as discussed above impacting NOx formation and emission is unfeasible, and it is thus the resulting measured NOx emission that lump all the factors into one single value. One must however be incredibly careful when comparing measured NOx values owing to the following issues:

- NO_x is not a regular molecular compound but a mixture of NO and NO_2, with the main component being NO.
- Measured NOx can be expressed in various physical units, but by convention using the molecular weight of NO_2 for the conversion of NO to NO_2 reported in the various measures of NO_x.
- On an absolute basis NO_x can be expressed as flowrate, i.e., g/s.

It is however impractical to compare emissions from different calciners on an absolute basis, as this value represents different calcining capacities with different calciner designs and operating conditions.

Instead of comparing absolute values it is more relevant to compare relative "concentration" values of NO_x either expressed in ppm or mg/Nm^3. If measured in ppm the total content in the wet or dry gas phase of NO_x is simply the sum of NO and NO_2, independent of temperature and pressure but subject to a reference volumetric concentration of oxygen for reference purposes.

To do a proper un-biased comparison of NO_x emission values it must also be defined if the value is reported on a wet or dry gas basis? And what shall be the percentage of oxygen used as reference?

NO_x Sample Calculation

To illustrate the importance of a clearly specified basis for comparison, the below sample calculations are made.

Example 1: Wet versus dry flue gas

Measured Moisture in flue/stack gas = 47 vol%

Measured NO_x = 200 ppm, dry.

$$200ppm(dry) = 200 \times \frac{(100 - 47)}{100} = 106ppm(wet)$$

Example 2: Influence due to Reference Oxygen

Measured NO_x = 200 ppm, dry.

Measured Oxygen = 5%

Oxygen reference 3%:

$$200ppm(5\%O_2) = 200 \times \frac{21 - 3}{21 - 5} = 225ppm(3\%O_2)$$

Oxygen reference 10%:

$$200ppm(5\%O_2) = 200 \times \frac{21 - 10}{21 - 5} = 138ppm(10\%O_2)$$

Example 3: Calculating from ppm to mg/Nm³

Measured NO_x = 100 ppm (dry) as NO_2

$$\mathbf{Y}\,\text{mg}\,NO_x/\text{Nm}^3(\text{Dry Gas}) = (46/22.4)\,\text{Z ppm}\,NO_x$$
$$\text{Y} = 2.054 \times 100 = 205\,\text{mg}\,NO_x/\text{Nm}^3(\text{Dry Gas})$$

Even though NO_x is mainly NO, it is however mostly reported as NO_2 and, it is therefore important to specify that the mg refers to NO_2 and not NO.

13.5.1.2 Carbon Monoxide—CO

There may be two sources of CO emission into the stack from calciners. The major and permanent source of CO is from incomplete combustion of the fuel (HFO, Natural gas or Coal Gas) in the calciner furnace and/or re-cycling/furnace cyclone.

However, in plants where the hydrate to calcination contains organic compounds, i.e., occluded oxalate, expressed as Total Organic Carbon (TOC), CO is generated together with Volatile Organic Compounds (VOC) by pyrolysis of TOC in the pre-calcination stage. Since the solid and gas temperature in the pre-calcination stage is around 400 °C and the gas temperature de-creasing towards the stack, there is not enough retention time for CO and VOC to be oxidized to CO_2 and H_2O before it reaches the stack.

According to Williams [14], CO is formed in flames by the rapid oxidation of the fuel (hydrocarbons) and hydrogen in the reaction zone. Since the CO is formed rapidly but only slowly consumed, the concentration of CO present in the reaction zone is above the water–gas equilibrium values:

$$CO + H_2O = CO_2 + H_2$$

The high temperature oxidation of CO to CO_2 in a turbulent flow reactor [15] resulted in the global kinetic equation below:

$$-\,d[CO]/dt = 10^{14.4 \pm 0.25} \exp[(-48.400 \pm 1200)/RT][CO]^{1.0}[H_2O]^{0.5}[O_2]^{0.25}$$
$$\left(\text{mole}\,\text{cm}^{-3}\,\text{s}^{-1}\right)$$

The equation is applicable at atmospheric pressure, 757–957 °C, water concentration 0.1–3.0 vol% and an equivalence ratio, $\Phi = (Air/Fuel)_{stoic}/(Air/Fuel)$, ranging from 0.01 to 0.5, which indicates a reach mixture with respect to the fuel, CO.

Together with similar global equations for the over-all rate of disappearance of CH_4 and the over-all rate of CO_2 appearance, the global kinetic equation for CO can be used as an approximation to model the combustion of CH_4 in the calciner furnace and/or re-cycling/furnace cyclone in calciners fired with natural gas.

While extraordinarily little CO results from a well-designed calciner furnace operating above 850 °C, the strong temperature dependence of the final oxidation of CO to CO_2 suggest that it is impossible to oxidize the CO formed at low temperatures such as the CO formed in the pre-calciner from organics occluded in the hydrate.

13.5.1.3 Sulfur Oxides—SO_x

Sulfur oxides is formed by combustion of fuels containing sulfur as H_2S, i.e., in Natural or Coal gas, or in fuel oils, notably Heavy Fuel Oil (HFO), containing 2.5–4.5 wt% sulfur in complex organo-chemical compounds.

The product of combustion, SO_x is a gaseous mixture of SO_2 and SO_3 formed in the calciner-furnace. SO_2 is formed simultaneously with the combustion of, say HFO, and subsequently partly oxidized to SO_3 by excess oxygen:

$$SO_2 + {}^{1}\!/_{2}O_2 = SO_3$$

The SO_2 oxidation kinetics are fast at high calciner-furnace temperatures and the SO_3 concentration in the combustion products can conservatively be assumed to be close to the equilibrium concentration. The oxidation rate of SO_2 decreases quickly when the temperature is reduced to below about 1000 °C, which is the case in stationary calciners for alumina.

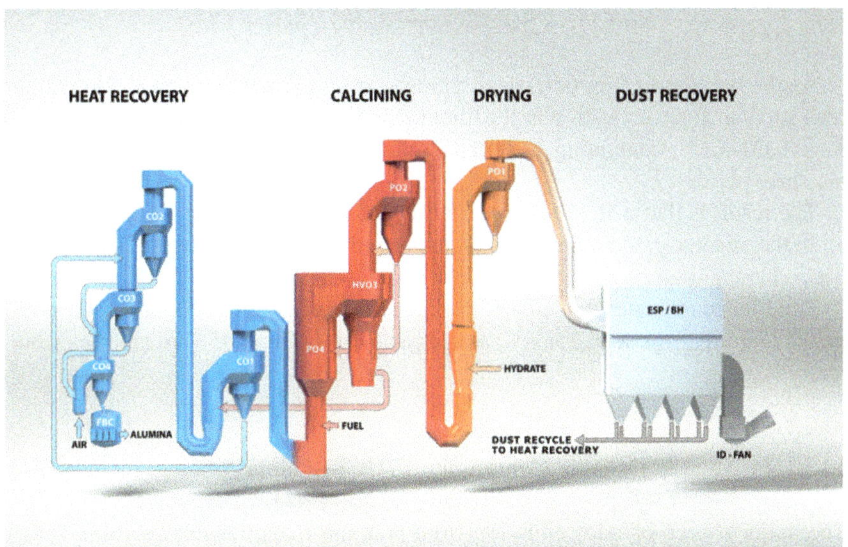

Fig. 13.10 Flowsheet of gas suspension calciner with holding vessel, HVO3

The hot combustion products formed in the Calciner- Furnace, PO4, Fig. 13.10, leaves the Furnace—Cyclone, HVO3, and flows into the riser duct to cyclone PO2 where it is mixed with dry hydrate from cyclone PO1. The dry hydrate is pre-calcined by the heat exchange with the hot combustion products which is cooled to below 400 °C, at which temperature the rate of oxidation of SO_2 is slow.

The formation of SO_3 has two implications for the calcination process with respect to equipment design and alumina purity.

First, the products of combustion and calcination out of cyclone PO2 contains a high concentration of water vapor, which have a strong influence on the sulfuric acid dewpoint. The combined water vapor content from combustion, calcination and drying of hydrate feed reaches about 50 vol% out of cyclone PO1.

Because of the high concentration of water vapor, the SO_3 content of the gas forms sulfuric acid as the produced gas is cooled down below the sulfuric acid dewpoint:

$$H_2O(g) + SO_3(g) = H_2SO_4(l)$$

To avoid corrosion of the equipment such as an electrostatic precipitator or baghouse by the sulfuric acid, the operating temperature of the equipment shall be kept safely above the acid dewpoint. Otherwise, a corrosion inhibitor shall be used or an alternate fuel source supplied.

Secondly, part of the SO_x will diffuse into the pre-calcined and porous alumina particles and react with occluded soda, Na_2O and/or Na_2CO_3:

$$Na_2O + SO_2/SO_3 = Na_2SO_3/SO_4$$

$$Na_2CO_3 + SO_2/SO_3 = Na_2SO_3/SO_4 + CO_2$$

As the hot alumina particles is discharged from the holding vessel, HVO3, into preheated air for combustion in the alumina heat recovery/cooling section, cyclone stage CO1–CO4, containing 21.0 vol% O_2, final oxidation of Na_2SO_3 to Na_2SO_4 may take place.

The result is that a significant part of the sulfur contained in the, say HFO, end up in the smelter grade alumina as an undesired impurity [16] by the smelters, see Chap. 11.

Experience indicates that about 20–25% of the sulfur in HFO is absorbed as SO_2 on the alumina from a stationary calciner, while it is about 45% in alumina from a rotary kiln.

13.5.1.4 Volatile Organic Compounds (VOC)—Odors

The major source of odor emission from alumina refineries comes from TOC in the bauxite feed to the refinery. Part of the TOC is dissolved in the liquor during digestion, see Sect. 13.2.4 above, and accumulates in the circulating liquid in the Bayer process until it reaches an equilibrium level over time.

Queensland Alumina Limited (QAL) and Alcoa World Alumina, Australia, has reported their Refinery Odor Survey [17–19] using Dynamic Olfactometry to determine the odor concentration expressed in odor units per cubic meter (OU/m^3).

Flow rate of gas was either measured or estimated for each potential odor emission point and multiplied with the measured OU/m^3 at the point to estimate the odor emission rate from each point in OU per unit time.

Organic odor emissions emanate from non-condensable gases containing hydrocarbons that break down from the bauxite under contact with caustic soda in digestion. These odors are generated ostensibly from the refinery vent stacks and venting of non-condensable from digestion heater vents. If not vented, these gases would otherwise interfere with the heat transfer process. A high temperature digestion facility may constitute ~80% [17] of the refinery odor emissions.

At QAL the Digestion plant section was the major source of odor emission accounting for 82.0%, while the Calcination plant section only accounts for 3.0%, Fig. 13.11:

Several approaches to minimizing odor emissions are employed throughout industry including thermal oxidation of the vent streams, biological destruction of the organics and natural oxidization of high organic containing condensate streams from digestion [17, 18].

Alcoa World Alumina, Australia, has reported their studies of odor emissions from their Wagerup Refinery in WA made in 1999, 2000 and 2001 [18, 19] (Fig. 13.12).

The calcination including vacuum pump vents and liquor burning process units are seen to be the major contributors of odor emissions from the Wagerup refinery.

The above surveys show that odor sources from different plants are not identical, due to the different bauxite feed sources and many other process factors.

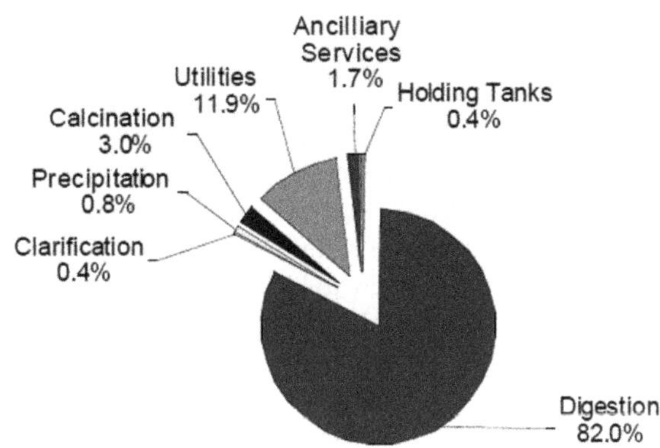

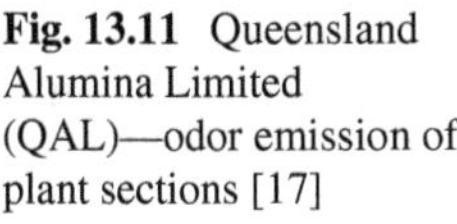
Fig. 13.11 Queensland Alumina Limited (QAL)—odor emission of plant sections [17]

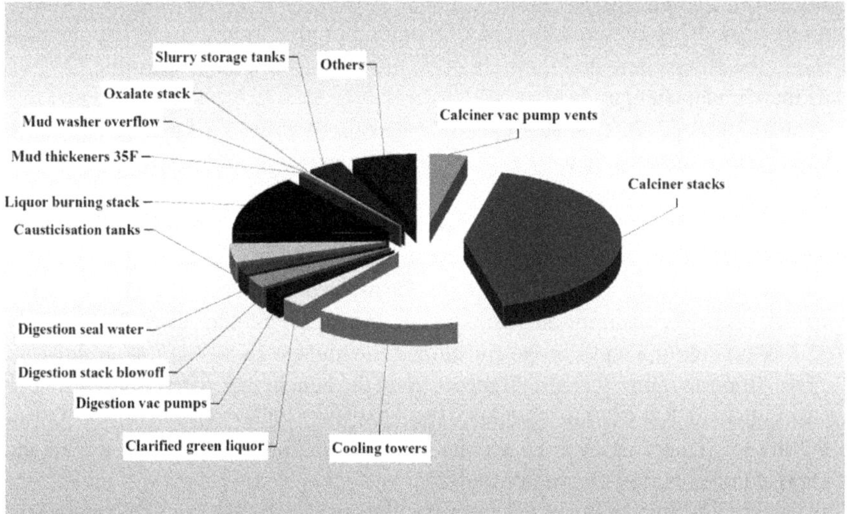

Fig. 13.12 Alcoa Wagerup survey 2000—odor emission of major sources [19]

At Wagerup, the odor from the calciner vacuum pump vents comes from using digestion condensate rather than evaporation condensate lower in odor for hydrate washing. The odor from the calcination stack comes from organics in the hydrate. The concentration of odor in calciner exhaust gas is low, but multiplied with the high exhaust gas flow rate, the total odor emissions form calcination stacks correspond to almost 50% of the total refinery odor emissions estimated in the Wagerup 2000 Survey [18].

The liquor burning process applied at the Wagerup refinery is a modified Showa Aluminum Industries technology using a direct fired rotary kiln [20]. The odor concentration measured in the exhaust gas from the liquor burning stack comes mainly from the organics in the concentrated spent liquor fed to the liquor burning unit together with hydrate, ESP dust or bauxite, or a mixture thereof. Recent studies [21] suggest that the VOC is emitted in the drying zone of the kiln when the temperature exceeds 250 °C in line with authors experience [22].

Reference is made to Chap. 8: Bayer Process Impurities and Their Management for a description of the Liquor Burning and Solid Liquid Calcination (SLC) process flowsheet respectively and their emission of odor.

If the equilibrium TOC in the liquor is 10 gpl or below, a Liquor Burning/SLC control system is unlikely to be justified. But if the concentration is 15 gpl or above, it is highly likely that a Liquor Burning/SLC control system is justified [22].

13.5.1.5 Carbon Dioxide—CO_2

Emission of carbon dioxide has become an extremely sensitive subject nowadays with the increasing accumulation of CO_2 and other greenhouse gases like methane, CH_4, in the atmosphere and their effect on the increasing global warming [23], and thus the climate change.

The alumina production accounts for emission of about 1.8 kg-CO_2 eqv./kg Al, which is slightly more than 10% of the total emission of 17.6 kg-CO_2 eqv./kg Al, assuming coal fired power plants [24].

The specific energy consumption per ton alumina produced is the driver of CO_2 emission from alumina refineries together with the type of fuel used.

In 2009 the International Aluminium Institute published the average regional specific energy consumption with South America as the lowest consuming 9.32 MJ/ton alumina and Europe consuming the highest 18.84 MJ/ton alumina.

The Alunorte refinery in Brazil seems to set the benchmark with a specific energy consumption of 8.0 MJ/ton alumina. The coal fired boilers accounts for 4.8 and HFO fired calciners on average accounts for 3.0 MJ/ton alumina with the balance comprised of imported electric power [25].

Typical CO_2 emission per GJ from burning coal is 0.093-ton CO_2, from HFO 0.079-ton CO_2 and from natural gas 0.058-ton CO_2. From these numbers, a significant reduction in CO_2 emissions can be realized in the short to medium time frame by a shift to natural gas where possible.

Long term fossil fuels can be replaced with "Green-Hydrogen" from electrolysis of water, with the electrolysis plant supplied by power from wind-turbines, sun-panels, or hydro-electric power where available.

In China almost all Gas Suspension Calciners installed since 1987 are fired with coal–gas containing 23–30 vol% CO, 10–15 vol% H_2 and 0.5–5.0 vol% CH_4 so combustion of hydrogen in the fuel mix is already proven.

Conversion to 100% use of "Green-Hydrogen" will require research with respect to the combustion rate of hydrogen, control of peak flame temperatures to limit NO_x and the logistics with respect to transport and storage of "Green-Hydrogen".

However, assuming the technology will be available to scale in the not-so-distant future alumina refining will become a fossil fuel free endeavor.

Fig. 13.13 2 × 2000 tpd GSC units equipped with ESP's—Sherwin Alumina, Texas, USA

Adding hereto that the Alcoa-RTA Carbon-Free Smelting Cell technology now being developed becomes successful and powered by stabilized renewable energy, the future looks "Green" also with respect to producing the "Green-Metal".

13.5.2 Particulate Matter

13.5.2.1 Electrostatic Precipitators (ESP)

Electrostatic Precipitators (ESP), have been used exclusively for dedusting of the exhaust gases from rotary kilns and stationary calciners producing alumina (Al_2O_3) by calcination of aluminiumhydroxide ($Al(OH)_3$).

Figure 13.13 shows two 2000 tpd Gas Suspension Calciners (GSC) equipped with Electrostatic Precipitators.

The main components of an Electrostatic Precipitator are shown in Fig. 13.14.

13.5.2.2 Electrostatic Precipitator (ESP)—Dedusting Principle

The de-dusting of the exhaust gas in an ESP is the result of passing the gas with the dust particles through a high voltage electric field with a discharge electrode

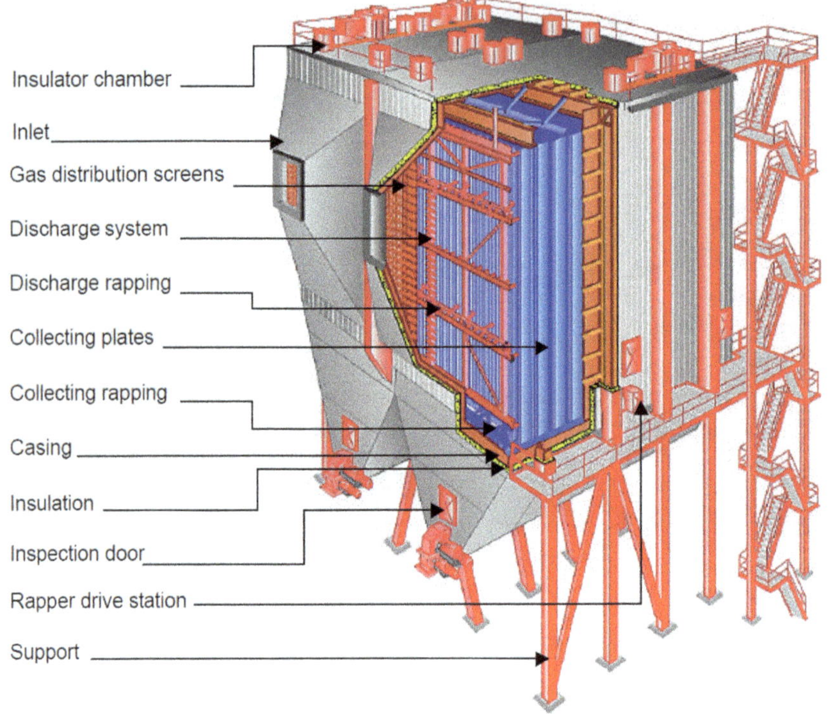

Fig. 13.14 Main components of an electrostatic precipitator (Courtesy of FLSmidth)

charging the dust particles with negative ions, which makes the charged particles move towards the positive collection plate, Fig. 13.15.

The collected particles on the collecting plate are discharged by periodically knocking the collection plates so that the dust particles are released from the collection plate and drops into the discharge hopper by the action of gravity.

The particles are removed from the hopper by a screw-conveyors or a drag-chain.

Electrostatic Precipitators can be designed to particulate emission levels not exceeding 10–20 mg/Nm^3 according to required specification. But in the event of a power failure the collection efficiency of an electrostatic precipitator is close to zero and a plume of particulate matter will be emitted to the atmosphere.

The particulate emission from an ESP dedusting the exhaust gases from a stationary calciner is typically designed not to exceed 50 mg/Nm^3 dry gas.

A more comprehensive description of the de-dusting principles and theory of electrostatic precipitators can be found in the monograph by White [26].

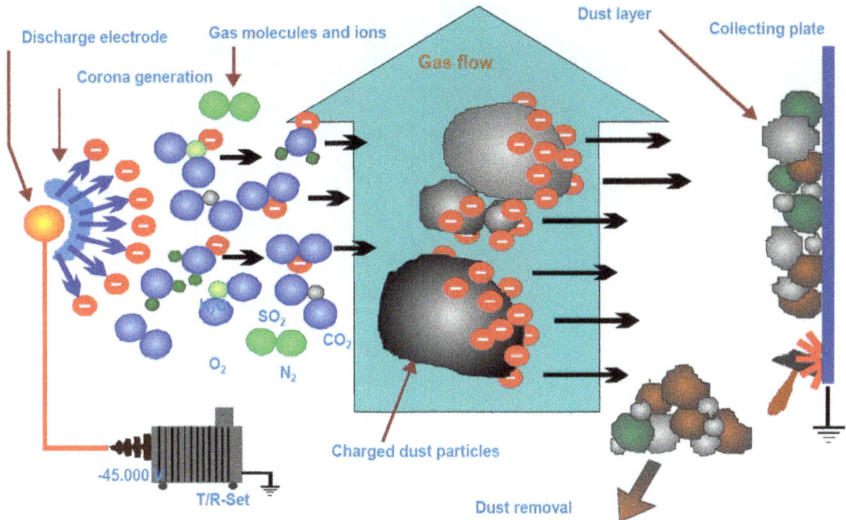

Fig. 13.15 Principle of dedusting the exhaust gas in an electrostatic precipitator (Courtesy of FLSmidth)

13.5.2.3 Bag House/Fabric Filter

In 2001 Queensland Alumina, Australia, placed order for 3 × 4500 TPD Gas Suspension Calciners (GSC) equipped with bag houses or Fabric Filters, Fig. 13.16.

The decision to install bag houses rather than ESP's was because a bag house offers an absolute barrier against dust emission in the event of a power failure, whereas an ESP will allow a plume of non-dedusted exhaust gas to be emitted.

The dust emission from the bag house or fabric filter installed at Queensland Alumina was designed not to exceed 50 mg/Nm^3 (dry) with an experienced operating range of 29–41 mg/Nm^3 (dry) [27]. A three-year bag life has also been experienced. The main components of a bag house or fabric Filter is shown in Fig. 13.17.

13.5.2.4 Bag House/Fabric Filter—Dedusting Principle

The de-dusting of the exhaust gas in a bag house or fabric filter is the result of passing the exhaust gas with the dust particles through a porous filter bag made of fabric and supported by a cylindrical metal cage, Fig. 13.19. As the filter bag is covered by a grooving layer of dust, the pressure-drop across the filter bag increases to a pre-set threshold level (Fig. 13.18).

When the threshold differential pressure is reached, or certain filtration time is exceeded, whatever comes first, the filtration process is "reversed" by blowing compressed air through the filter bag in opposite direction of the exhaust gas flow

Fig. 13.16 3 × 4500 tpd GSC units equipped with bag houses at Queensland Alumina, Australia

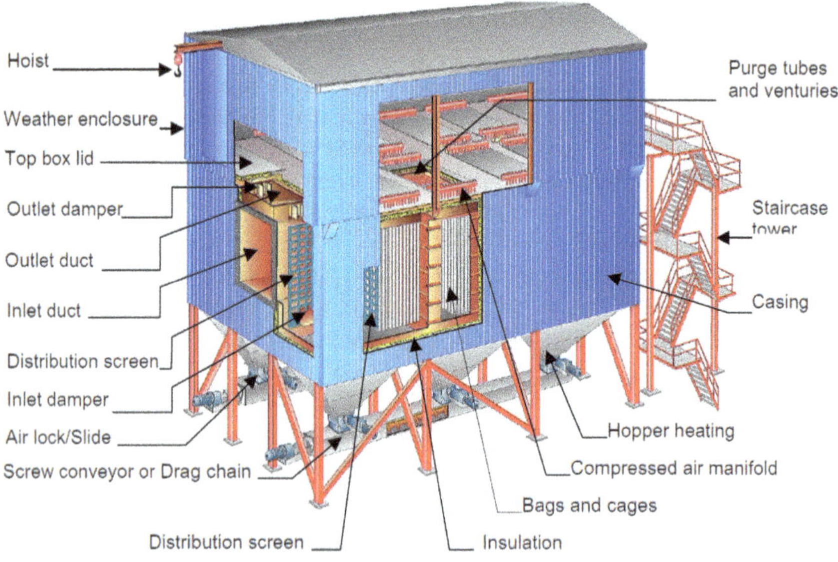

Fig. 13.17 Main components of a bag house or fabric filter (Courtesy of FLSmidth)

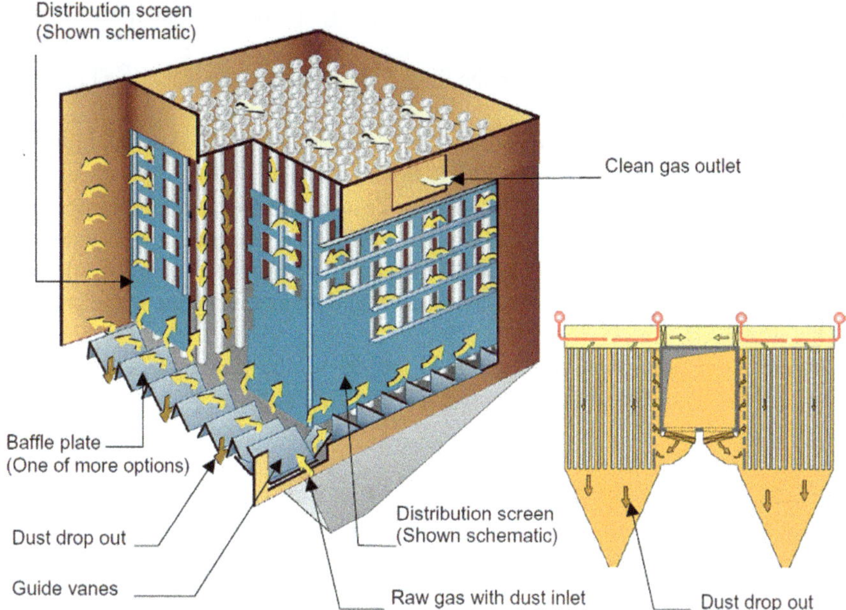

Fig. 13.18 Principle of exhaust gas flow in a bag house or fabric filter (Courtesy of FLSmidth)

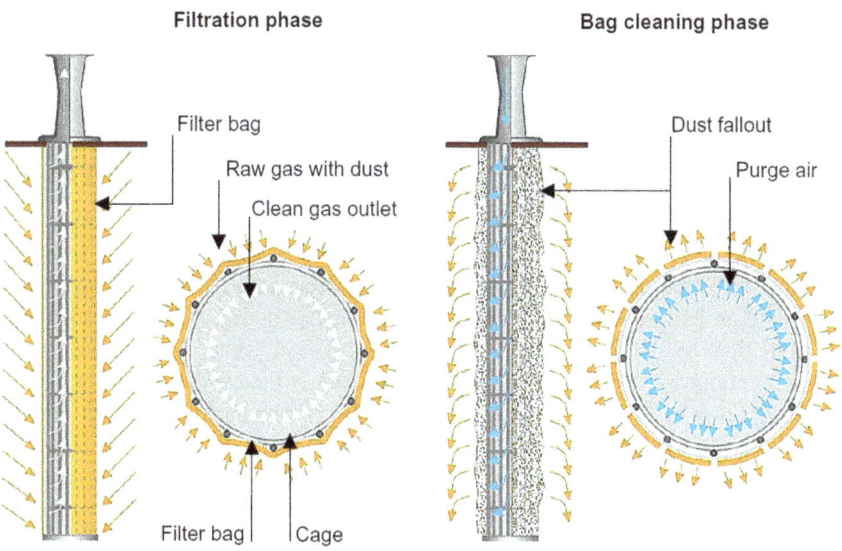

Fig. 13.19 Principle of dedusting the exhaust gas in a bag house or fabric filter (Courtesy of FLSmidth)

through the filter bags thereby releasing the dust particles from the filter bag, Fig. 13.19.

The dust particles released from the filter bags drops into the discharge hopper by the action of gravity.

The particles are then removed from the hopper by a screw-conveyors or a drag-chain.

A bag house or Fabric Filter is typically designed to particulate emission level not exceeding 20 mg/Nm^3 but can also be designed to particulate emission level <5 mg/Nm^3 according to specification. In the event of a power failure the collection efficiency of a bag house or fabric filter is essentially maintained due to the presence of the filter bags.

A more comprehensive description of the de-dusting principles and theory of fabric filters can be found in the monograph by Croom [28].

13.5.2.5 Electrostatic Precipitator Versus Bag House/Fabric Filter

It is not all calciner projects for alumina where the local authorities, like for the Queensland Alumina project in Australia, do not allow temporary emission of dust exceeding the emission of particulate matter designed for, i.e., as result of a power failure.

It is therefore of interest in planning and execution of a calciner project to establish the most economic selection of de-dusting solution between a bag house and an electrostatic precipitator.

Preconditions—technical requirements:

1. Guaranteed particulate emission: 30 mg/Nm^3 (dry).
2. Temporary emissions exceeding the guaranteed particulate emission is allowed.
3. Gas flow 10.900 Am^3/min.
4. Temperature 155–240 °C.

Table 13.3 provides a comparison of general process parameters between a Fabric Filter and an Electrostatic Precipitator which impacts both CAPEX and OPEX of the units.

As mentioned above the Fabric Filter is fail safe in the event of a power failure. A Fabric Filter is designed with an equal number of compartments. Each of them can be isolated by closing the inlet and outlet dampers to the compartment. This way Fabric Filters can be maintained on-line when it comes to replacement of bags and other equipment placed in each compartment. Another remarkable difference when compared to an ESP.

The main characteristics of the two technologies considered is summarized in Table 13.3 for the same set of process and performance conditions, and subject to the same local conditions i.e., wind and earthquake load. The equipment information in Table 13.4 is the main cost drivers for capital, operational and maintenance.

A comparison of CAPEX and OPEX for the same stationary calciner de-dusting duty is shown in Table 13.5.

Table 13.3 Comparison of fabric filter and electrostatic precipitator parameters [29]

Parameter	Fabric filter	ESP
Pressure drop [mmWG] (typical)	150	25
Emission levels [mg/Nm^3, wet]	Typical: 5–20 Possible: <5	Sized to suit Possible: 5
Emission sensitivity to process variations	Low	Medium to high
Temperature resistance	Dependant on bags: Typical: 240–260 °C	Typical: 400 °C
Explosion risk	Low	High
Fire damage risk	High	Low
Absolute filter	Yes	No
Maintainability	High, on line possible	Low, only off line
Service interval	Bags: varies 2–5 years Cages: 2 set of bags	4 years typically

The Fabric Filter is seen to have the lowest CAPEX, but the highest OPEX.

A Net Present Value analysis has shown a breakeven point after nine years of operation [29], whereafter the ESP is the most economic choice.

Since most NPV analysis consider a time horizon of 10 years or more a solution with ESP is preferred, when the pre-condition do accept a temporary release of a plume of dust, i.e., caused by a power failure.

The above analysis implies that there might be a third option: A filter solution combining an ESP with a Fabric Filter for minimum CAPEX and OPEX, that will also provide an absolute filter barrier. This will be briefly outlined in the next sections.

13.5.2.6 New Developments

In the above analysis 6.0 m bags has been considered for the Fabric Filter. They need to be replaced, say each 3 year on-line. This requires significant manpower. In a Fabric Filter with about 3700 bags, this corresponds to about 1725 man-hours.

Now a days Fabric Filters with 10 m bags and cages, operating with low Purging Pressure (<2 barg), have been developed and proven to be much more cost-effective than Fabric Filters with 6 or 8 m bags [30] operating with higher Purging Pressure.

With a weight reduction from 25 to about 16 kg/m^2 filtration area, the new generation Fabric Filters offers a significantly lower CAPEX and OPEX from lower compressed air requirements for purging [30].

Table 13.4 Main equipment data for fabric filter and electrostatic precipitator [29]

	Fabric filter
Steel weight	
Casing	
Support	
Access	
Weather enclosure	
Bags + cages	
Total [ton]	336.4
Insulation surface	
Total [m^2]	1160
AC [m/min]	1.1
Consumption of compressed air [Nm^3/h]	281
Electrical heaters	
– Installed [kW]	48
– Normal operation [kW]	30
	ESP
Steel weight	
Casing and frames	
Support	
Access	
Misc.	
Collecting plates	
Total [ton]	640.8
Insulation surface	
Total [m^2]	2830
Gas velocity [m/s]	0.94
T/R set	
– Installed [kVA]	687
– Normal operation [kW]	184
Electrical heaters	
– Installed [kW]	115
– Normal operation [kW]	58

In relation to the above example, only about 2.220 bags and correspondingly 1.035 man-hours will be needed for replacement of the 10 m bags. This is a significant reduction in maintenance cost also considering a longer bag life (~6 years) experienced.

With proven 10 m bag technology available, a better match to the height of an ESP is obtained when compared to 6 or 8 m bags. This means that a ESP/Fabric

Table 13.5 2002 cost between a fabric filter versus and ESP for the same stationary calciner de-dusting duty [29]

	Fabric filter	ESP
CAPEX (capital expenditure)		
Imported parts	789.000	687.000
Local parts	545.000	633.000
Installation cost	656.000	1407.000
Total CAPEX, US$ flange/flange	1950.000	2727.000
OPEX (operational expenditure per annum)		
Baglife is assumed to be 3 years and cage life is assumed to be 2 set of bags		
Power consumption, fan, kW	302	50
Power consumption, Hopper heaters, kW	24	58
Power consumption, T/R set, kW	NA	184
Power consumption, compressor, kW	25	NA
Total OPEX, US$, based on 0.06 US$/kWh	230.000	138.000
Summary, US$		
CAPEX	1970.000	2727.000
OPEX (per annum)	251.455	138.337

Filter-Hybrid can be designed or a Fabric filter can be retrofitted to an existing ESP [30] to create an absolute filter barrier when required.

A relative comparison is shown in Fig. 13.20 of CAPEX and OPEX for an ESP, ESP/Fabric Filter-Hybrid, and a Fabric Filter (FF) applied to a 3500 TPD Gas Suspension Calciner Unit for Alumina.

The comparison in Fig. 13.20 indicates that the ESP/FF-Hybrid is more competitive than an ESP and that the lower OPEX than the FF will be paid back in just 5.3 years.

Hybrid filter elements of fabric sandwiched with porous ceramic elements impregnated with a catalyst has the potential to oxidize VOC like formaldehyde and remove NH_3 and NO_x [31, 32] from stack gases in one dedicated unit.

The applied catalyst is effective in the 200–400 °C range and may work for a Liquor Burning plant with a relatively high stack temperature [31]. In the future, when a new catalyst is developed that is effective at around 150 °C (the stack temperature of most stationary calciners and the SLC process) application of the hybrid filter element technology will be possible on stationary calciners as well.

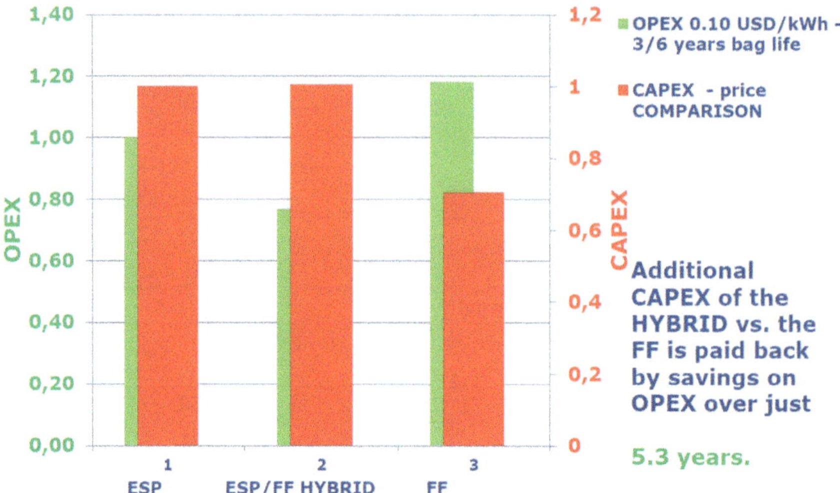

Fig. 13.20 CAPEX versus OPEX comparison between ESP, ESP/FF-hybrid and FF (Courtesy of FLSmidth)

References

1. D. Donaldson, Bayer process description. Presentation at TMS Course in "Alumina Refinery Fundamentals and Practice", in *Light Metals* (2009)
2. J.J. Kotte, Bayer digestion and pre-desilication reactor design, in *Light Metals* (1981), pp. 45–81
3. C.E. Suarez, Process safety management application in alumina refineries. Presentation at TMS Course in "Alumina Refinery Fundamentals and Practice", in *Light Metals* (2009)
4. A. Connor, R. Greenhalgh, P. Mullinger, Prevention of gas related explosions in ESPs. World Cem. 36–40 (1993)
5. G. Bánvölgyi, The red mud pond dam failure at Ajka (Hungary) and subsequent developments. Presentation at ICSOBA Red Mud Seminar, Goa, India (2011)
6. H. Li, P. Johnstone, Full life cycle integrated and risk-based approach for bauxite residue management, in *8th International Alumina Quality Workshop* (2008), pp. 174–178
7. M. Schmidt, Tolerable risk. Chem. Eng. (2007)
8. G. Daniels, Lifecycle activities for safety instrumented systems in an alumina refinery, in *8th International Alumina Quality Workshop* (2008), pp. 280–285
9. D. Campin, *Benchmarking Queensland Alumina Limited and Rio Tinto Yarwun (RTAY) Alumina Refineries' Emissions to Australian, Canadian, Irish and USA Refineries* (Department of Environment and Resource Management, Environmental Protection Agency, Queensland Government, 2010)
10. J. Colannino, Low-cost technology for NOx reduction. Chem. Eng. 30–36 (2020)
11. S.R. Turns, *An Introduction to Combustion: Concepts and Applications* (McGraw Hill Education (India) Edition, Noida, 2012)
12. I. Glassman, *Combustion*, 3rd edn. (Academic Press, 1996)
13. S. Hundebøl, S. Kumar, *Retention Time of Particles in Calciners of the Cement Industry*, no. 8 (Zement-Kalk-Gips, 1987), pp. 422–425
14. A. Williams, *Combustion of Sprays of Liquid Fuels* (Elek Science, London, 1976)
15. F.L. Dryer, I. Glassman, High temperature oxidation of CO and CH_4, in *14th Symposium (International) on Combustion* (The Combustion Institute, Pittsburg, 1973), pp. 7–9

16. A.H.S.P. Morais et al., Studying SO_2 emissions in the calcination area at alumar, in *Light Metals* (2005), pp. 241–243
17. G. Graham, R. Capil, R. Davies, Odour destruction for digestion vent gases, in *6th International Alumina Quality Workshop* (2002), pp. 316–319
18. S.J. Cox, Odour emissions reduction case study at Alcoa's Wagerup Alumina Refinery, in *6th International Alumina Quality Workshop* (2002), pp. 309–315
19. P.S. Coffrey, Assessing odour impact of an alumina refinery by source measurement, dispersion modelling and field odour surveying, in *6th International Alumina Quality Workshop* (2002), pp. 320–326
20. C. Sato et al., New process for removal of organics from Bayer liquor, in *Light Metals* (1982), pp. 119–128
21. I. Galbally et al., Alumina refining and air quality: VOCs and odor, in *Proceedings of 9th International Alumina Quality Workshop* (2012)
22. J.G. Pulpeiro et al., Sizing an organic control system for the Bayer process, in *Light Metals* (1998), pp. 89–95
23. E. Boeker, R. van Grondelle, *Environmental Physics*, 3rd edn. (Wiley, Hoboken, 2011)
24. H. Kvande, Trends in the global aluminium industry, in *Proceedings of XXX I International Conference "ICSOBA"* (2013), pp. 40–43
25. R. Wischnewski et al., Alunorte global energy efficiency, in *Light Metals* (2011)
26. H.J. White, *Industrial Electrostatic Precipitation* (Addison-Wesley Publishing Company, Inc., Boston, 1963)
27. J. Fenger, B.E. Raahauge, C.B. Wind, Experience with 3 x 4500 tpd gas suspension calciners (GSC) for alumina, in *Light Metals* (2005)
28. M.L. Croom, *Filter Dust Collectors* (McGraw-Hill, Inc., New York, 1994)
29. H.V. Pedersen et al., Environmental compliance for stationary calciners for alumina: now and in the future, in *Proceedings of 6th International Alumina Quality Workshop* (2002)
30. C.V. Rasmussen, H.V. Pederersen, Fabric filter operating results with 10m long bags and low purging pressure, in *Proceedings of 12th International Alumina Quality Workshop* (2012)
31. G. Elliot, Developments of Cerafil® filter elements in environmental emission technology for alumina refineries, in *Proceedings of 7th International Alumina Quality Workshop* (2005)
32. C. Poulsen, *Catalytic Filtration Solutions* (FLSmidth AirTech Presentation)
33. G. Roach, E. Jamieson, Effect of bauxite and digestion conditions on iron in SGA, in *6th International Alumina Quality Workshop* (2002), p. 340
34. P. Basu, Reactions of iron minerals in sodium aluminate solutions, in *Light Metals* (1983), pp. 83–97
35. X. Pan, H. Yu, B. Wang, S. Zhang, G. Tu, S. Bi, Effect of lime addition on the predesilication and digestion properties of a Gibbsitic bauxite, in *Light Metals* (2012), pp. 39–42
36. B. Whittington, The chemistry of CaO and $Ca(OH)_2$ relating to the Bayer process. Hydrometallurgy **43**, 13–35 (1996)
37. G. Roach, The equilibrium approach to causticisation for optimising liquor causticity, in *Light Metals* (2000), pp. 228–234
38. A. Lalla, R. Arpe, 30 years of experience with tube digestion, in *COM* (2006), pp. 105–113
39. A. Lalla, R. Arpe, 12 years' experience with wet oxidation, in *Light Metals* (2002), pp. 177–180
40. D. Mulligan, *Environmental Management in the Australian Minerals and Energy Industries, Principles and Practices* (1996)
41. F. Habashi, A hundred years of the bayer process for alumina production, in *Light Metals* (1988), pp. 85–93
42. O. Levenspiel, *Chemical Reaction Engineering*, 3rd edn. (1999)
43. N. Raghavan, G. Fulford, Mathematical modeling of the kinetics of gibbsite extraction and kaolinite dissolution/desilication in the Bayer process, in *Light Metals* (1998), pp. 255–262
44. P. Landry, H. Edwards, Pressure decantation at gramercy alumina, in *Light Metals* (2007), pp. 470–475
45. J. Doucet, C. Hendriks, E. Patz, G. Forté, S. Kumar, R. Paradis, D. Puxley, Pressure decantation technology: the Kaiser gramercy experience, in *6th International Alumina Quality Workshop* (2002), pp. 94–99

46. J.M. Lamerant, Y. Perret, Boehmitic reversion in a double digestion process on a bauxite containing trihydrate and monohydrate, in *Light Metals* (2002), pp. 377–380
47. M. Authier-Martin, R. Girard, G. Risto, D. Verville, Boehmite reversion: predictive test and critical parameters for bauxites from different geographical origins, in *6th International Alumina Quality Workshop* (2002), pp. 109–114
48. Q. Chen, X. Yuming, L. Hepler, Calorimetric study of the digestion of gibbsite, A1(OH)3(cr), and thermodynamics of aqueous aluminate ion, A1$(OH)_4{}^-{}_{(aq)}$. Can. J. Chem. **69**, 1685–1690 (1991)
49. B. Hemingway, R. Robie, J. Apps, Revised values for the thermodynamic properties of boehmite, AIO(OH) and related species and phases in the system AI-H-O. Am. Mineraologist **76**, 445–457 (1991)
50. C. Dobbs, C. Armanios, L. McGuiness, G. Bauer, P. Ticehurst, J. Lochore, R. Irons, K. Ryan, G. Adamek, Mercury emissions in the Bayer Refinery—an overview, in *7th International Alumina Quality Workshop* (2005), pp. 199–204
51. P. Prinsloo, C. Beattie, The application of Cerafil® filter elements on particulate removal on liquor burning application for Worsley Alumina Pty. Ltd., in *Proceedings of 6th International Alumina Quality Workshop* (2002)
52. J.M. Bee'r, N.A. Chigier, *Combustion Areodynamics* (Applied Sciences Publishers Ltd, 1972)

Brady Haneman -Principal Metallurgist (Digestion Technology), High Pressure Metals, Hatch, Canada. Email: brady.haneman@hatch.com

Brady has a Hons Degree in Chemical Engineering from Sydney University Australia and has over 30 years experience in industry. He has extensive experience in conducting feasibility studies, plant and equipment design, engineering management, development of new process technologies and project execution having spent 20 years on feasibility and subsequent detailed engineering phases of major projects.

Brady has designed high temperature and low temperature digestion units totaling 9.0 million tpa of alumina refining capacity and these refinery projects have incorporated Tube Digestion, Double Digestion and Hybrid Tube Digestion flow sheets. He has also acted as a Consultant to numerous refineries globally, in particular in Australia, Ireland, Italy, Ukraine, the United States and Saudi Arabia on design improvements for digestion facilities.

Currently, Brady is the Principal Metallurgist for Hatch's High Pressure Metallurgy practice specialising in Digestion technologies. Previously as the Global Design Lead—Alumina, Brady led major efforts to improve process plant safety by applying a fundamental understanding of refinery dynamic behaviour during process upsets to the design of overpressure protection systems. He has also developed mitigation strategies for complex pressure transients in piping systems and led the detailed design of Digestion Units and their safety relief systems for over 15 yrs.

Benny E. Raahauge Deceased—Owner—Director Raahauge-SGA ApS, Denmark.

Benny received his M.Sc. Chemical Engineering from the Danish Technical University in November 1972. Start working with programming minicomputers controlling the raw material mixing for production of Cement Clinker, December 1972 in Process Technical Department, FLSmidth, Cement Division, Copenhagen, Denmark.

From 1975–76, Benny worked as process and plant engineer in a Danish Sugar Factory before re-joining FLSmidth as R&D engineer and later R&D Manager in the Mining Division in Copenhagen. Benny was assigned the task to develop the stationary Gas Suspension Technology for Smelter Grade Alumina (SGA) to replace the rotary kilns.

After successful commissioning of the first 1000 tpd Gas Suspension Calciner (GSC) unit for alumina at Hindalco Industries, India, in 1986, Benny worked in the role as General Manager-Pyro and Alumina Technology based in Copenhagen. This assignment included the responsibility for design, marketing, sales and commissioning of GSC units for Alumina, culminating with commissioning of 3 × 4500 tpd GSC units at Queensland Alumina 2004–05, the world's largest stationary calciners.

During +44 years employment with FLSmidth & Co. contributed with many papers on calcination to international ICSOBA, Alumina Quality Workshop and TMS meetings until retiring in 2018.

Together with Don Donaldson, Benny was Lead-Editor to "Essential Readings in Light Metals—Volume 1—Alumina & Bauxite" with Fred Williams as Co-Editor, Published by Wieley & Sons. Inc. Copyright © 2013 by The Minerals, Metals & Materials Society (TMS).

In 2018 Benny was editor and contributor to Chapter 12.1 Alumina, in the "SME Mineral Processing & Extractive Metallurgy Handbook", Published by Society for Mining, Metallurgy & Exploration (SME), Copyright © 2019.

Chapter 14
Economics of Smelter Grade Alumina Projects

Peter-Hans ter Weer

Abstract The aim of this chapter is to show the intricate relationship between bauxite resource, refinery technology, sustainability, and economics of (Bauxite and) Smelter Grade Alumina projects, and by doing so generate a better understanding of the underlying factors controlling the technical and economic success of these projects. The chapter opens with an overview of key aspects of project economics, including evaluation criteria, sensitivities, main scope items, project development stages and some of its main economic drivers. It also defines general and economic, and project specific assumptions of a reference project which is used throughout the chapter. The following two sections provide more in-depth analyses of capital and operating costs of a (Bauxite and) Smelter Grade Alumina project, including cost build-up aspects, and a detail breakdown of capital and operating costs of the reference project (based on economic conditions as at end 2016). The chapter concludes with a section covering plant and process capital and operating cost saving opportunities, e.g., through plant productivity and energy savings, Bayer Loop simplification and alternative plant design.

14.1 Overview

14.1.1 Key Aspects of Project Economics and Analysis Method

14.1.1.1 Project Economics Evaluation Criteria

Project proposals in most industries are economically evaluated applying several criteria to ascertain if they meet (company) threshold levels, and for ranking purposes in case a choice between projects needs to be made. These criteria include [1, 2]:

P.-H. ter Weer (✉)
TWS Services & Advice – Bauxite & Alumina Consultancy, Owner-Director, Huizen, The Netherlands
e-mail: twsservices@tiscali.nl

B. E. Raahauge and F. S. Williams (eds.), *Smelter Grade Alumina from Bauxite*, Springer Series in Materials Science 320,
https://doi.org/10.1007/978-3-030-88586-1_14

- **Net Present Value (NPV)**, the sum of a project's annual cashflows at a chosen annual interest/discount percentage (net meaning that both the costs and benefits of an investment are included). The NPV is generally used as the primary economic evaluation criterion of a project, which is calculated as follows:

$$\mathrm{NPV(i)} = \sum \mathrm{a_j} * (1 + \mathrm{i}/100)^{-\mathrm{j}} \tag{14.1}$$

with

i	discount rate, %;
$\sum$	summation from j = 0 to n (project duration, no. of years);
a_j	annual cashflow (after tax) at time (year) j.

Key inputs of the annual cashflows:

- On the revenue side: the amount of product generated and its sales price.
- On the cost side: capital cost, operating cost, sustaining capital, tax payable and depreciation.

The selected discount rate, i in Formula (14.1) above, is generally the appropriate weighted average cost of capital (WACC), reflecting two elements: a risk-free rate and a risk premium which depends on the industry sector and the specifics of a project. Risks in this context include technical risks (e.g., geological, operational), economic risks (e.g., deteriorating fiscal climate), and political risks (e.g. political stability of a country).

In the bauxite/alumina industry the discount rate may range from typically 5–15%. In this chapter on economics of smelter grade alumina projects, a discount rate of 8% per annum has been used, indicated as NPV (8%). NPV measures how much value is added or lost by a project and is often used as prime economic criterion to assess the attractiveness of an investment. It is a central tool in the Discounted Cash Flow (DCF) analysis and is a standard method for using the time value of money to appraise long-term projects. The NPV of a project may be expressed in any currency (USD, EURO, GBP, CNY, etc.) or unit (USD, 000USD, MUSD). The DCF analysis method is used throughout this chapter.

- **Internal Rate of Return (IRR)**, the discount percentage at which the NPV equals zero (illustrated in Fig. 14.4). IRR may be considered a measure of the "quality" of an investment.
- **Value over Investment (VIR)** or **Capital Efficiency Ratio (CER)**, the ratio of NPV and the investment, expressed on a selected basis. This basis may vary e.g. "Money-of-the-day" (MOD, nominal money, no inflation correction applied), "Real terms" (= adjusted for inflation, or zero inflation assumed) or Discounted Money. The terms "constant" and "current" prices are often used instead of "real" and "nominal". The VIR/CER ratio represents another investment quality indicator.
- **Payback Period**. The period for the cumulative cashflow to become zero from a defined point in time. As in the previous bullet point, a project's cashflow may

be expressed in different ways. A correlation exists between IRR and Payback Period (refer Sect. 14.1.1.7, "Correlation between IRR and Payback Period").

In most cases several other criteria are also taken into consideration when evaluating a project, such as availability of resources, strategic reasons, country view, marketing and financing aspects, risks, etc. However, these fall outside the scope of this chapter. Economic robustness of a project with respect to operating and capital costs is discussed in Sect. 14.1.1.4.

The economic criteria mentioned above are also used in the bauxite and alumina industries and may be applied to projects of all scopes.

In this chapter on "Economics of smelter grade alumina projects", the assumptions shown in Table 14.1 have been used as reference project case. Table 14.2 shows the results of the DCF economic evaluation of the reference project. An Excel spreadsheet has been used for the calculations (refer Appendix 4).

14.1.1.2 Impact of Alumina Price, Opex, Capex, and Tax Holiday on NPV and IRR

This section reviews the impact of changes in alumina price, project operating cost ("Opex"), capital cost ("Capex") and tax holiday on NPV and IRR [1, 2].

Alumina Price

Despite being outside the control of a project, alumina price plays a key role in bauxite/alumina project economics. Figure 14.2 illustrates the correlation between NPV (8%) and alumina price for the expansion project detailed in Table 14.1.

A linear correlation exists. The angle of this line is the same for different sets of values for Opex and Capex, however the intercept with the y-axis changes. This correlation is a result of the way the NPV is calculated (refer Formula (14.1) above).

Figure 14.3 provides an illustration of the correlation between IRR and alumina price for the same project. It shows a good fit with a second order polynomial correlation. At different sets of values for Opex and Capex, second order polynomials remain showing a good fit, however their coefficients differ. This is caused by the shape of the curve describing NPV as function of discount percentage (refer Fig. 14.4).

Figures 14.2 and 14.3 illustrate the impact of alumina price variations in the range of about 250–450 USD/tA. During an economic cycle, prices may range from approximately 150–400 USD/tA. As can be seen from these figures, the timing of a capacity expansion project in an economic cycle (a cycle of typically 7–10 year covering a period of high and low prices—refer for instance Fig. 14.14) may have a big impact on its economics.

Opex and Capex

Both Opex and Capex of a project influence project economics. To illustrate their impact, the alumina price has been kept constant at 350 USD/tA in the following examples.

Table 14.1 Reference project assumptions

Variable	Quantification
General and economic	
Evaluation period	Construction time + 30 yr
Construction time	3 years
Construction start	Next year
Production build-up (yr1/2/3+)	70/95/100%
Capex spread (yr1/2/3)	15/70/15%
Tax depreciation period on capex[a]	20 years
Corporate tax rate	30%
Tax holiday	No
Sustaining capex build-up (yr1-7+)	0/20/35/60/75/90/100%
Numbers in 2017 USD	No inflation assumed
Financing method and costs	Not included[b]
Costs of studies, etc	Not considered ("sunk costs")
Project specifics[c]	
Green/Brownfield/De-bottlenecking	Greenfield
Alumina capacity expansion	1.6 Million tonnes/year
Alumina price	350 USD/tA
Total cash operating cost[d,e]	163 USD/tA
Refinery sustaining capex (Susex)	8 USD/tA
Total capital cost[f]	2241 Million USD 1401 USD/Annual tA capacity

[a] Capital Depreciation is the gradual reduction of an asset's value. It is an expense but because it is non-cash, it is often a tax write-off; that is, a company usually may reduce its taxable income by the amount of the depreciation on the asset, thus providing a source of free cashflow
[b] May differ significantly per project; assumed here is that the project is ungeared (the project owner takes the full share of the capital cost; or putting it differently, project financing is not taken into consideration)
[c] Some of these project specifics are used as variable parameter in sections below
[d] Including consumables and alumina handling and transport, and port costs, excl. financing costs (interest, amortization), head office costs, etc. (refer Sect. 14.3.1 and Table 14.35 for more details)
[e] The bauxite quality assumed for the Reference Project is shown in section Appendix 5
[f] Includes mine, refinery, and infrastructure (assuming no Capex required for jetty/wharf); excludes escalation, financing costs, and working capital

Figure 14.5 provides an example of the correlation between NPV (8%) and Opex for the reference project of Table 14.1. The correlation is again linear, however with a negative angle. As with the NPV-Price correlation above, the angle of this line remains the same for different sets of values for price and Capex, while the intercept with the y-axis changes.

Table 14.2 Reference project DCF economic evaluation results

Criterion	Value
NPV (8%)	90.2 Million 2017 USD (MUSD)
IRR	8.4%
VIR/CER	
NPV (8%)/discounted capex	0.047 (or 4.7%)
NPV (8%)/capex (2017 USD)	0.040 (or 4%)
Payback period (refer Fig. 14.1)	
In 2017 (nominal) USD[a]	~9.5 years after start of ops
In discounted money	~25.5 years after start of ops

[a] No inflation assumed

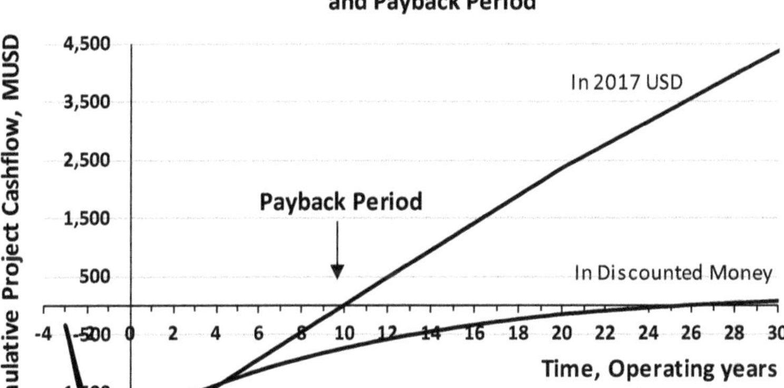

Fig. 14.1 Cumulative cashflows and payback period

It should be kept in mind that the Opex for expansion projects (brownfield, greenfield, de-bottlenecking) may vary significantly, indicatively between 120 and 250 USD/tA (includes sustaining Capex).

The correlation between NPV (8%) and Capex is also linear, with similar characteristics as those between NPV and Opex above.

Figures 14.6 and 14.7 illustrate the correlation between IRR and Opex respectively IRR and Capex (assuming 350 USD/tA price; 1401 USD/Annual tA Capex; respectively 170 USD/tA Opex; and no tax holiday).

Second order polynomials show a good fit, similar as above between IRR and alumina price, however with different coefficients. Note that the Capex for expansion projects may range widely depending on expansion type (de-bottlenecking,

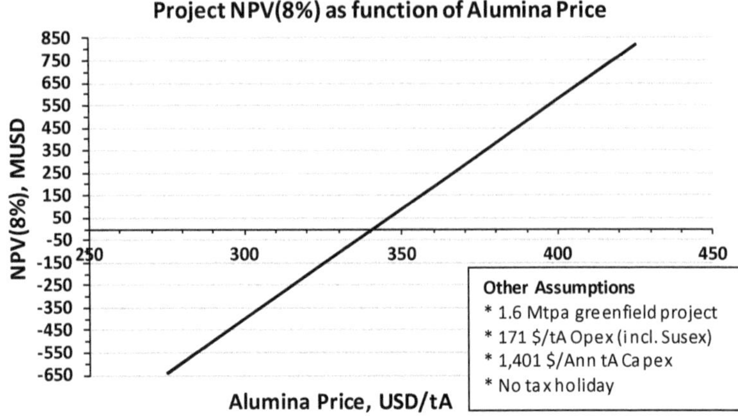

Fig. 14.2 NPV (8%) as function of Aa price

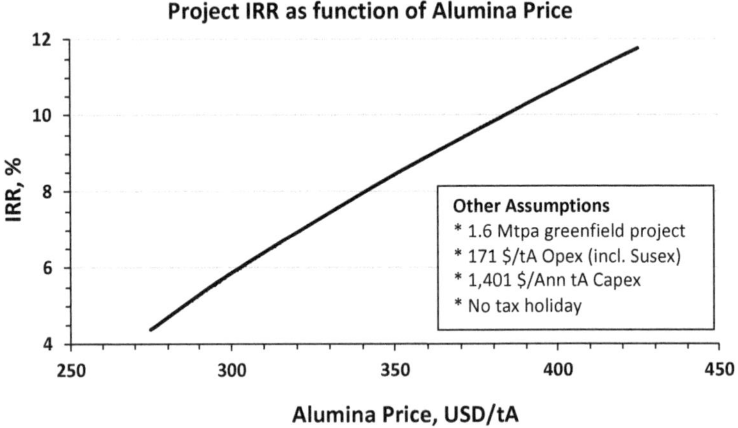

Fig. 14.3 IRR as function of Aa price

brownfield, greenfield), location (close to the coast, shipping lanes), extension of required infrastructure (port, rail, town, roads), etc. Brownfield projects may indicatively range between ~600 and 1200 USD/Annual tA capacity, and greenfield projects between ~900 and 1900 USD/Annual tA (see e.g. [3], slide 17).

Tax Holiday

The impact of a tax holiday on project economics can be significant. and may be granted by governments to provide an incentive for the implementation of a project in their country. It is therefore frequently (but not only) granted to greenfield projects because these often represent the first or a significant investment of a company in that country.

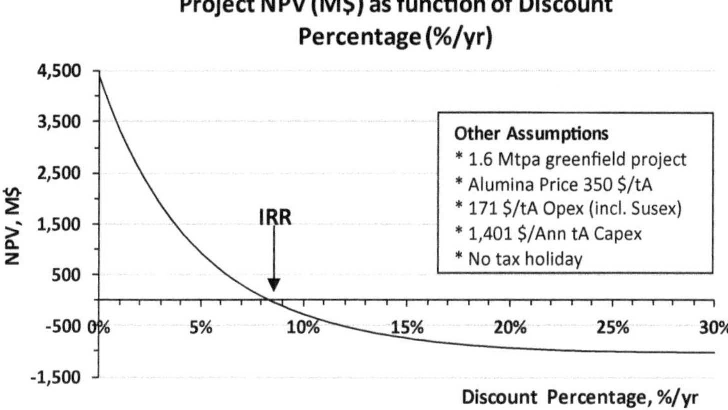

Fig. 14.4 Reference project NPV as function of discount percentage

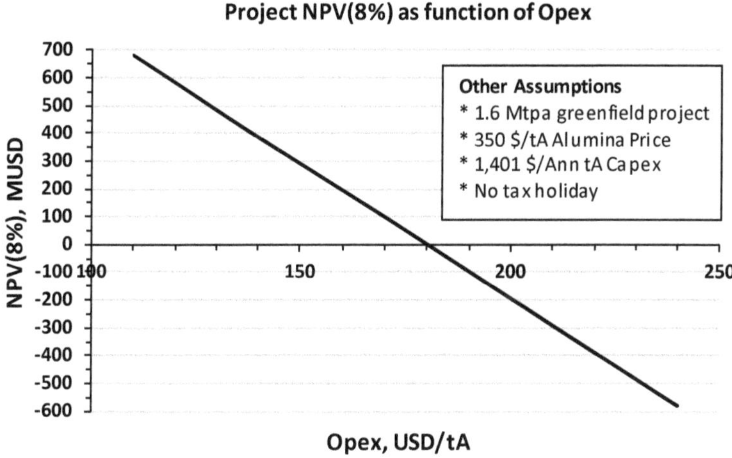

Fig. 14.5 NPV (8%) as function of opex

The duration of a tax holiday may depend on project scale, country norms and agreements between a government and a company, but a 10-year tax holiday on a greenfield project is not uncommon.

Tables 14.3 and 14.4, graphically represented in Figs. 14.8 and 14.9, illustrate the impact of a 10 year tax holiday on NPV (8%) and IRR of the reference project (other assumptions as per Table 14.1).

Tables 14.3, 14.4, Figs. 14.8 and 14.9 illustrate that a tax holiday of 10 years could mean the difference between go and no-go for a project.

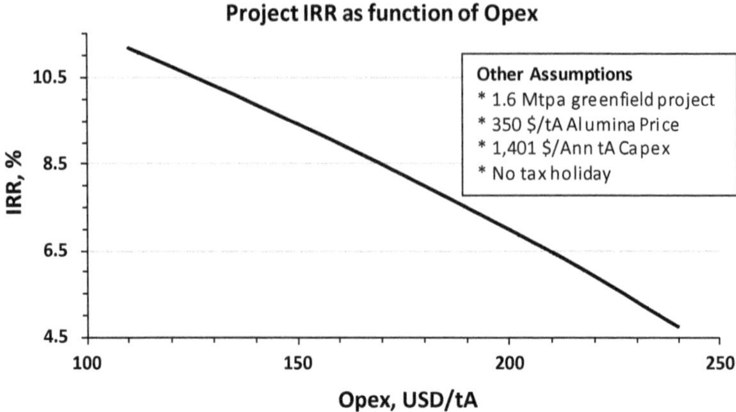

Fig. 14.6 IRR as function of opex

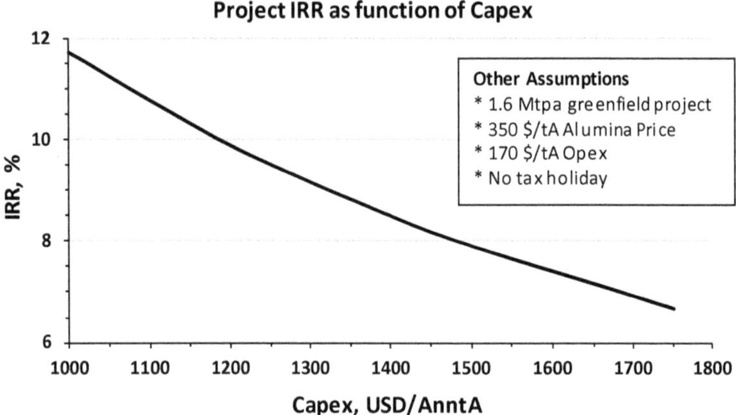

Fig. 14.7 IRR as function of capex

Table 14.3 NPV (8%) versus opex—impact of tax holiday[a]

Opex[b], USD/tA	NPV (8%), MUSD	
	No tax holiday	10-yr tax holiday
110	681	1102
140	389	737
170	97	372
210	−292	−114
240	−583	−479

[a] Other assumptions as per Table 14.1
[b] Includes Sustaining Capex

Table 14.4 IRR versus opex—impact of tax holiday[a]

Opex[b], USD/tA	IRR, %	
	No tax holiday	10-yr tax holiday
110	11.2	13.3
140	9.9	11.7
170	8.5	9.9
210	6.5	7.4
240	4.7	5.3

[a] Other assumptions as per Table 14.1
[b] Includes Sustaining Capex

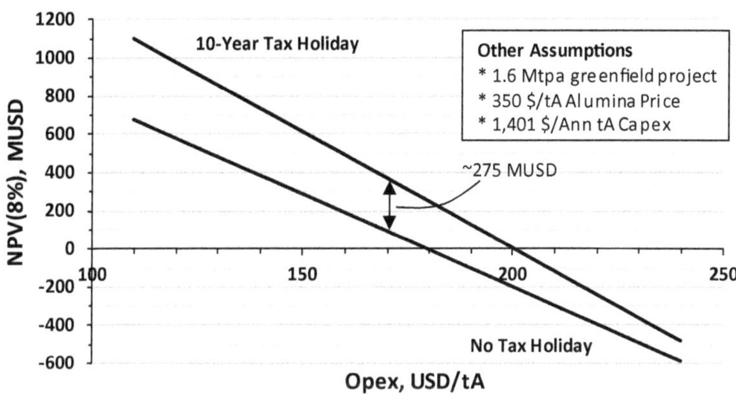

Fig. 14.8 NPV (8%) versus opex—impact of tax holiday

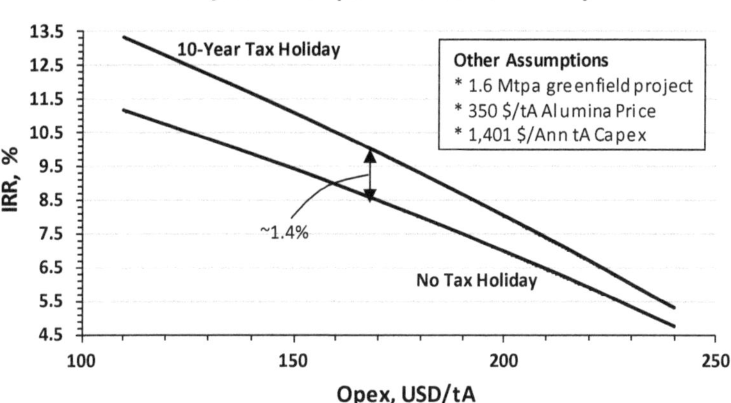

Fig. 14.9 IRR versus opex—impact of tax holiday

14.1.1.3 Exchangeability of Opex and Capex

Both project Opex and Capex represent negative cashflows. Thus a form of exchangeability exists between the two with respect to their impact for instance on NPV and thus on project economics [2], refer Formula (14.1).

Assuming that the project's NPV (8%) should remain unchanged (in this case 90.2 MUSD) the following correlation can be found for the reference project (refer Table 14.1, no tax holiday):

$$\text{Capex} = -8.16 * \text{Opex} + 2794 \quad (14.2)$$

with

Capex in USD/Annual tA expansion capacity, and
Opex in USD/tA.

Correlation (14.2) means that an increase in Opex of 10 USD/tA for the reference project would require a drop in Capex of ~82 USD/Ann tA if NPV (8%) was to remain unchanged. This is illustrated in Fig. 14.10.

The slope of the line in Fig. 14.10 (−8.16) applies to a fixed NPV (8%) target. At other discount percentages this slope would be different (steeper at lower discount percentages, flatter at higher discount percentages).

Similarly for the reference project with a 10-year tax holiday (not uncommon for a greenfield project), the following correlation is found for a fixed project NPV (8%) of 363.3 MUSD:

$$\text{Capex} = -9.25 * \text{Opex} + 2981 \quad (14.3)$$

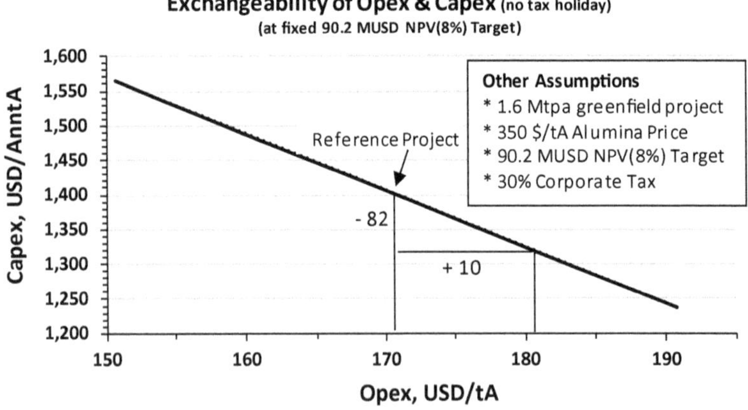

Fig. 14.10 Exchangeability of opex and capex (no tax holiday)

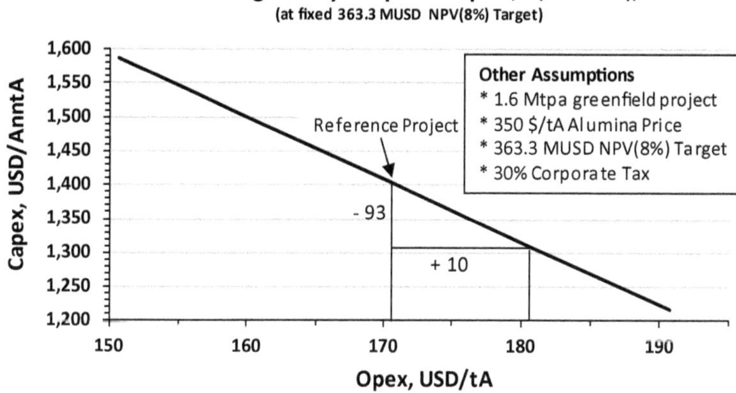

Fig. 14.11 Exchangeability of opex and capex (10-yr tax holiday)

In other words an increase in Opex of 10 USD/tA for the reference project with a 10-year tax holiday would require a drop in Capex of ~93 USD/Ann tA if NPV (8%) was to remain unchanged, as illustrated in Fig. 14.11.

The following conclusion can be drawn from Formulae (14.2) and (14.3): in order to keep NPV (8%) at the same level, an increase in Opex for a project with a 10-year tax holiday would require a significantly larger decrease of Capex than a project without tax holiday.

The difference between Formulae (14.2) and (14.3) is reflected in the different slopes, the consequence of the 10-year tax holiday.

Another way of putting the conclusion above is that under the assumptions mentioned, a change in Opex has a more profound impact on the NPV of a project with a tax holiday than on the NPV of a project without tax holiday (in absolute terms, in relative terms it is the other way around).

This may be illustrated as follows: a decrease of 10 USD/tA in the Opex of the reference project would result in an NPV (8%) of:

- 187.4 MUSD for the reference project without tax holiday, i.e., an increase of 97.2 MUSD (= 187.4 − 90.2 MUSD) or 108%.
- 484.9 MUSD for the reference project with a 10-year tax holiday, i.e., an increase of 121.6 MUSD (= 484.9 − 363.3 MUSD) or 33.5%.

14.1.1.4 Economic Robustness of Structural Low Opex/Low Capex Projects

Projects with structural low operating cost under otherwise the same conditions e.g., the same Capex per annual tonne of alumina production capacity, are inherently more robust with respect to adverse economic conditions such as a drop in the alumina price [2].

Table 14.5 Low opex robustness

Low opex robustness		Low opex project	High opex project
Reference case	Aa price, USD/tA	375	375
	Opex, USD/tA	171	191
	Capex, USD/Ann tA	1401	1401
	NPV (8%), MUSD	333.3	138.8
	IRR, %	9.6	8.7
Price sensitivity	*Aa Price-30 USD/tA*	345	345
	NPV (8%), MUSD	41.6	−152.9
	IRR, %	8.2	7.2

This is illustrated in Table 14.5 for a project based on Table 14.1 (no tax holiday).

Table 14.5 shows that the economics of the low Opex project prove to be more robust with respect to a realistic fluctuation of the alumina price (refer e.g., Fig. 14.14) by remaining attractive whereas the economics of the high Opex project become unattractive.

A related important characteristic is that a low Opex project once constructed remains more robust in periods of low alumina prices which may occur several times during the lifetime of a project (refer Sect. 14.1.1.6 "Alumina Price, Opex, and Margin").

Due to the exchangeability of Opex and Capex (refer previous section), the same applies to projects with structural low capital cost under otherwise the same conditions e.g., the same Opex per tA. This is illustrated in Table 14.6 for a project based on Table 14.1 (no tax holiday).

Table 14.6 Low capex robustness

Low capex robustness		Low capex project	High capex project
Reference case	Aa price, USD/tA	375	375
	Opex, USD/tA	171	171
	Capex, USD/Ann tA	1401	1551
	NPV (8%), MUSD	333.3	154.5
	IRR, %	9.6	8.7
Price sensitivity	*Aa price-30 USD/tA*	345	345
	NPV (8%), MUSD	41.6	−137.2
	IRR, %	8.2	7.4

14.1.1.5 Sensitivities

Generally, a sensitivity ("what-if") analysis is conducted as part of an economic evaluation to assess which parameters have the greatest impact on project economics (i.e. on NPV, IRR, and VIR/CER). A sensitivity analysis indicates how much project base case economics would change if any one of the major parameters were to be higher or lower than originally expected. Parameters that may be considered for a sensitivity analysis include:

- Product alumina sales price.
- Total Opex or elements thereof e.g.
  - Consumables Prices (fuel oil, caustic soda, coal, power)
  - Consumption Rates (power, coal, fuel oil, caustic soda).
- Total Capex or elements thereof e.g.
  - Infrastructure (port, rail, town, roads)
  - Alumina refinery.
- Production capacity in/decreases.
- Currency exchange rate.
- Discount rate.
- Interest rate.
- Tax holiday.
- Other variables.

In this section the first three parameters mentioned above (alumina price, total Opex, and total Capex) are addressed in steps of 5% points over a range of ±15% for the reference project of Table 14.1. The results are presented in Table 14.7 and are graphically shown in Fig. 14.12 [sensitivity of NPV (8%)] and Fig. 14.13 (sensitivity of IRR).

Figure 14.12 illustrates that the reference project is:

Table 14.7 Reference project economics—sensitivities

Factor	Change in factor						
	−15%	−10%	−5%	0%	+5%	+10%	+15%
Sensitivity of NPV (8%), no's in MUSD							
Alumina price	−420	−250	−80	90.2	260	431	601
Total capex	341	257	174	90.2	6.7	−76.8	−160
Total opex	339	256	173	90.2	7.2	−75.8	−159
Sensitivity of IRR, no's in %							
Alumina price (%)	5.7	6.7	7.6	8.4	9.3	10.1	10.8
Total capex (%)	9.9	9.4	8.9	8.4	8.0	7.6	7.3
Total opex (%)	9.6	9.3	8.9	8.4	8.0	7.6	7.2

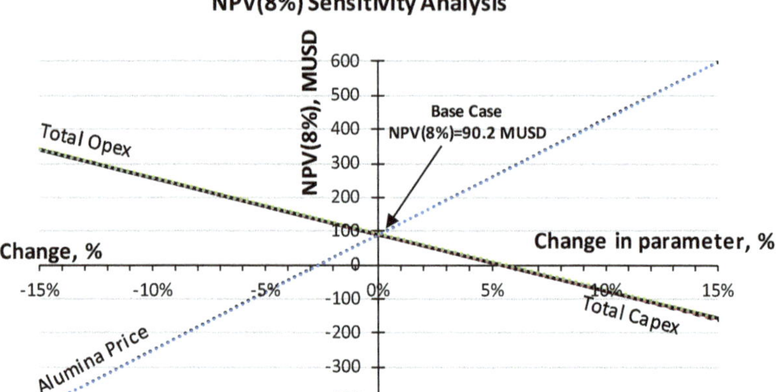

Fig. 14.12 NPV (8%) sensitivity analysis

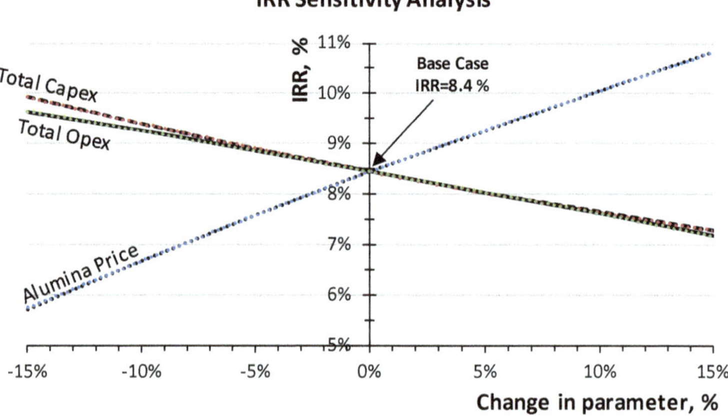

Fig. 14.13 IRR sensitivity analysis

- Very sensitive to alumina price fluctuations: a drop of the alumina price by 3% (~10.5 USD/tA) or more results in a negative NPV (8%).
- Quite sensitive, roughly to the same (relative) extent, to Opex and Capex fluctuations: an increase of total Opex or total Capex by just over 5% results in a negative NPV (8%).

Figure 14.13 shows similar relationships between IRR and the three above mentioned parameters.

The above sensitivity analysis results reflect the findings of Sect. 14.1.1.2 on the impact of alumina price, Opex, and Capex on NPV and IRR.

14.1.1.6 Alumina Price, Opex, and Margin

The development of the alumina price in the period 1980–2016 is shown in Fig. 14.14 [4] (solid line—left axis). To better interpret the fluctuations in the alumina price, Fig. 14.14 also includes the World GDP/capita growth rate (dotted line—right axis), as criterion for the growth of the global economy. Note: the number of people worldwide is growing continuously.

Reason for including the GDP/Capita growth rate is that the usage of aluminium permeates the global economy, and with it the demand for alumina: worldwide aluminium's main end uses include transportation (~25–35%), building and construction (~20–25%), packaging (~12–15%) and engineering (~15–20%, including electrical and machinery and equipment).

Comparison of the two lines shows a reasonable to good correlation: the alumina price has followed World GDP/capita growth rate, with a time lag of one year or less for at least the last fifteen years.

Noticeable in Fig. 14.14 is the significant and consistent increase in the alumina price in the period from about 2003 till 2008 (when the banking/economic crisis occurred), which does only partly seem to be supported by an equivalent trend in the World GDP/Capita growth rate.

As shown in Fig. 14.15, this alumina price increase appears to correspond with China significantly increasing alumina imports over part of that period (dotted line—right axis).

Figure 14.15 also shows the average alumina price between 1980 and 2004 (~182 USD/tA), when the Chinese alumina imports reached a level of close to 6 million tonnes and kept increasing; and between 2006 and 2016 (~318 USD/tA). Although not shown here, the alumina (LME-linked contract) price closely followed changes in the aluminium 3-month LME-price, with a time lag of typically one year.

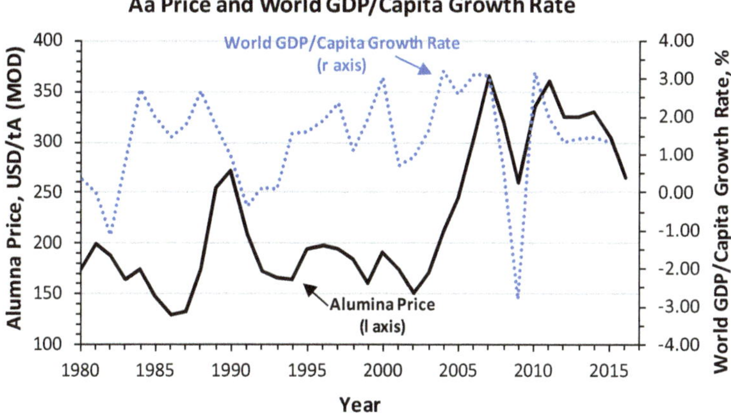

Fig. 14.14 Alumina price and world GDP/capita growth rate. *Sources* [4], USDA and Platts

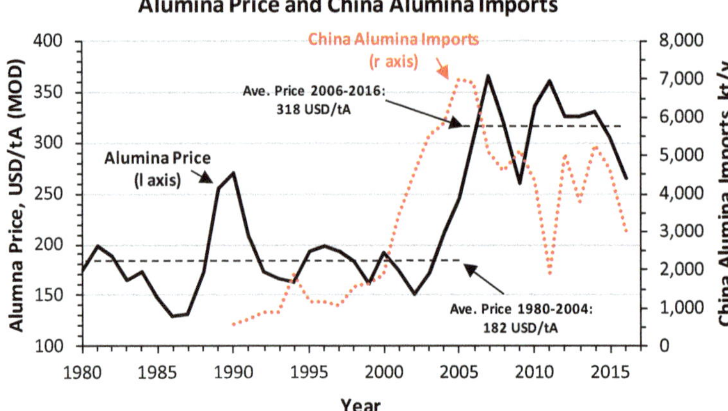

Fig. 14.15 Alumina price and china alumina imports. *Sources* [4], Hydro, Alumina Ltd, CBA

Note: the trend over the past ~7 years has been that the alumina price is becoming more index based than linked to the LME Aluminium price.

Although the total cash Opex (excluding sustaining Capex) differs for an individual greenfield project, in almost all cases total cash Opex of a greenfield project ends up in the first quartile (Q1) (≤25%) of the industry cash operating cost curve of the year in which it starts operations. This is not surprising as greenfield projects normally incorporate new (and normally most efficient) technologies, equipment, and operating procedures. In fact it is essential for their longer term competitiveness because over time Opex tends creeping up the industry cost curve.

For the purpose of the current analysis the average of the first quartile of the annual alumina industry's cash cost curve has been used as greenfield cash Opex.

Figure 14.16 compares the alumina price with the cost curve Q1 cash Opex (excl. sustaining Capex). This figure indicates a reasonable to good correlation between cash Opex (dotted line—right axis) and alumina price (solid line—left axis), with the alumina price following Opex changes with a time lag of around a year.

In summary the following correlations have emerged for the period 1980–2016:

- Changes in the 3-month LME aluminium price followed changes in the global economy as expressed by the World GDP/Capita growth rate with a time lag of a year or less.
- Changes in the alumina price closely followed changes in the aluminium price, with a time lag of typically a year.
- The significant increase in China's alumina imports from 2001 onward was followed by an increase in the alumina price resulting in an average alumina price in the period 2006–2016 of ~135 USD/tA above the average alumina price in the period 1980–2004.
- Changes in the alumina price followed Q1 cost curve Opex (excl. sustaining Capex) changes with a time lag of roughly a year.

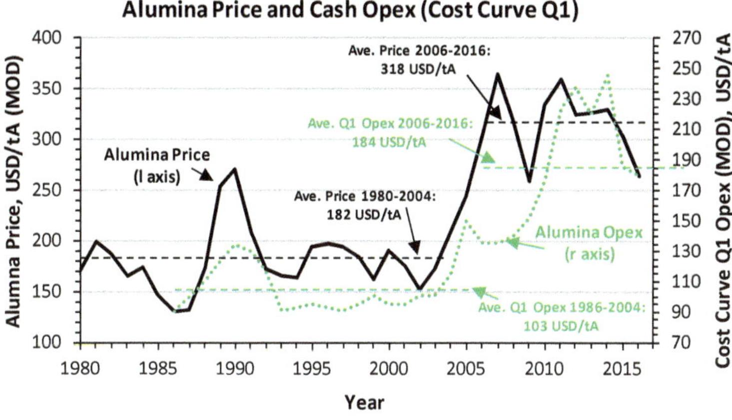

Fig. 14.16 Alumina price and cash (cost curve Q1) opex. *Sources* [4], Hydro, Rio Tinto

As both alumina price and operating cost have been assessed, the margin between the two may be considered, a parameter reflecting the combined effect of price and Opex on economic criteria such as NPV, IRR and payback period. The result is shown in Fig. 14.17, which also includes the average margins in the periods 1986–2004 (~81 USD/tA) and 2006–2016 (~134 USD/tA), the latter showing a rapid drop in the last few years.

In summary the average cash margin fluctuated for many years around 81 USD/tA, increased rapidly from 2004 onward to an average of ~170 USD/tA between 2006 and 2010, and appears to be trending downward since 2011 to a level of ~100 USD/tA.

The margin may also be expressed as percentage of the alumina price. Figure 14.18 shows the result and illustrates that over a long period of time the margin expressed

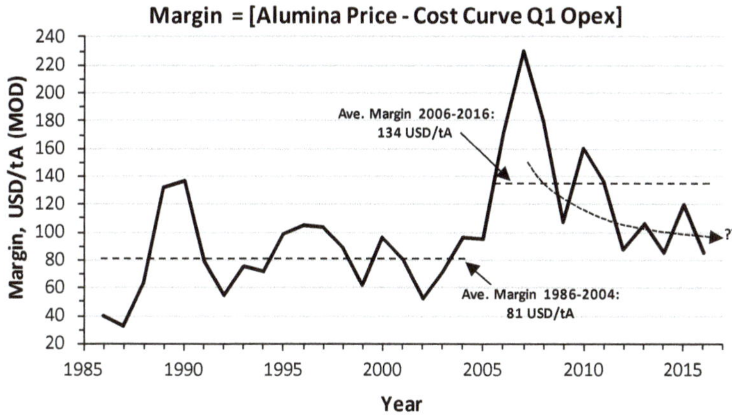

Fig. 14.17 Margin (alumina price—Q1 cash opex)

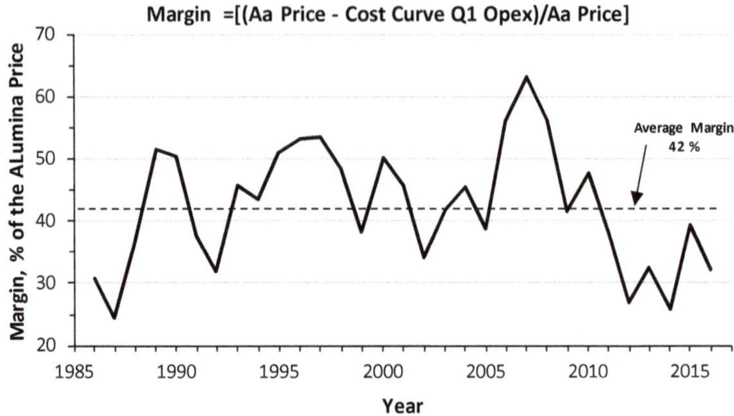

Fig. 14.18 Margin (as percentage of the alumina price)

as percentage of the alumina price has ranged between approximately 25–65%, with a long-term average of ~42%.

At a November 2011 investor seminar Rio Tinto presented the following graph (Fig. 14.19) on historical margins for Cost Curve Q1/Q2 Alumina Assets for the period 2000–2011 which is consistent with Fig. 14.18 w.r.t. the Cost Curve Q1 Alumina Assets.

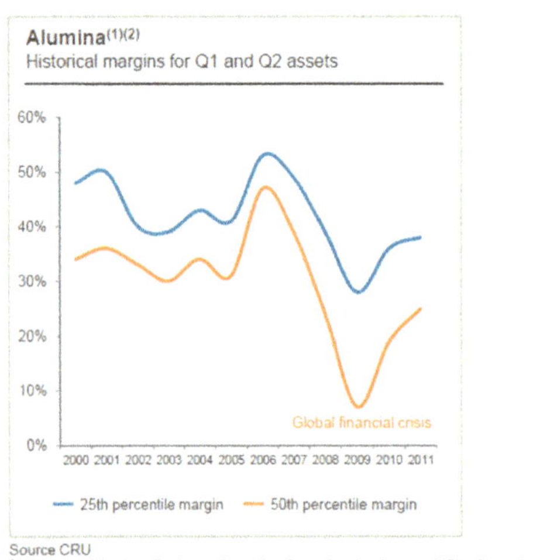

Fig. 14.19 Rio Tinto Alumina margin info (November 2011 investor seminar)

14.1.1.7 Tools for Economic Screening Purposes

Capital Charge—Correlation with NPV and Target IRR

A comprehensive economic analysis of a (Bauxite and) Alumina project requires developing annual discounted cashflows for the lifetime of the project, accounting for Opex and Capex, construction period, production capacity build-up, alumina price, plant depreciation period, tax rate, discount rate for the calculation of the Net Present Value (NPV), etc. However, when an assessment of the economics of an early-phase project or a preliminary comparison of several early-phase project options is required, the necessary time to execute such an exercise is sometimes not available and/or its accuracy not required [5]. This section describes a useful tool to screen a (list of) potential project(s) by testing against a target IRR hurdle rate, when (indicative) Capex and Opex values are known, without the need to generate a full DCF analysis.

As illustration Table 14.8 shows the NPV of the sum of discounted cashflows of 100 Million USD/year for a project with a production capacity of 1 Million t/year and a margin of 100 USD/tA (= [Alumina Price] − [Total Opex, incl. Susex]), at a series of discount rates. Other assumptions:

- Evaluation period: construction time + 30 operating years.
- Construction time: 3 years, starting next year.
- Full production from operating year 1 onward.
- Capital depreciation period: 20 years.
- Corporate income tax rate: 30%, no tax holiday.

Table 14.8 illustrates for instance that the above defined project with net annual cashflows of 100 MUSD/year (net = difference between revenues and total operating costs) starting after 3 years of construction during an operating period of 30 years

Table 14.8 NPV of 100 MUSD/y net cashflows over a 30-year period

Discount rate[a] %	NPV of 100 MUSD/y net cashflows over 30-year period[b] MUSD
6	809
7	709
8	626
9	555
10	496
11	445
12	401
13	364
14	331

[a] Representing project hurdle rate (e.g. target IRR)
[b] After 3 years of construction, and at a 30% tax rate

represents the same value as 496 MUSD today when a discount rate of 10% is applied. Or stated differently: at a discount rate or target IRR of 10%, an expense of 496 MUSD today must be compensated by net annual cashflows of 100 MUSD/year during an operating period of 30 years starting after 3 years of construction.

The same does not apply however to a capital expenditure of 496 MUSD today because a capital depreciation period is included (applicable to most projects). During this period, the capital investment is generally equally spread and tax deductible. In this analysis a capital depreciation period of 20 years has been assumed (typical for many Bauxite and Alumina projects), meaning that 5% of the capital investment per annum is tax deductible for this period of time.

The consequence is that a capital expenditure of ~678 MUSD may be made (instead of 496 MUSD), and still achieve an IRR of 10%. Along the same line's capital expenditures of ~844 MUSD (instead of 626 MUSD) and ~558 MUSD (instead of 401 MUSD) may be made, and still achieve IRR's of 8% respectively 12% (compare numbers in Table 14.8). Note: the same Capex spread is assumed as for the reference project in Table 14.1 i.e., 15/70/15% (year1/2/3).

Or put differently and expressing everything per tonne of product: if [Total Opex USD/tA] + [100 USD/tA "Capital Charge"] = "Full Cost USD/tA" of a project were exactly equal to the (assumed) unit price of the product sold, for instance 678 USD/tA, the IRR of the project would be 10% based on the example above. And if the Full Cost were less than the price for the project's lifetime, the IRR of the project would be more than 10%, equal to saying that the NPV (10%) would be positive.

In other words, by adding a "Capital Charge" in this example of ~15% of the Capex (=100/678 * 100%), expressed in USD per tonne alumina product, to its average annual total Opex per tonne alumina (which includes sustaining Capex), a Full Cost is obtained providing a practical method to check if the project meets (in this example) a hurdle rate of 10% IRR based on the above mentioned assumptions.

The Capital Charge (expressed as percentage of project Capex to be added to the Total Opex—both expressed as USD/tA) required to test a project for screening purposes against a target IRR (project hurdle rate), is shown in Table 14.9 which also includes an approximate capital charge range reflecting the following parameter ranges:

- Tax rate: 25–35%.
- Project evaluation period: 25–30 operating years.
- Capital depreciation period: 15–20 years.
- Investment capital phasing: 1st/2nd/3rd year—15/70/15% and 33/33/33%.

Figure 14.20 graphically illustrates the Capital Charge as function of Target IRR.

The Full Cost (in USD/tA) arrived at by adding the Total Opex (in USD/tA) of a project to the relevant Capital Charge (expressed as USD/tA) may also be used for screening purposes to compare with Full Costs of other project options obtained on the same basis.

Table 14.9 Capital charge and range as function of target IRR

Target IRR %	Capital charge[a] to be added to total opex % of project capex	Capital charge range[b] % of project capex
6	9.2	8.9–9.9
7	10.5	10.1–11.2
8	11.8	11.3–12.5
9	13.3	12.7–13.9
10	14.8	14.1–15.3
11	16.3	15.5–16.8
12	17.9	17.0–18.4
13	19.6	18.6–20.0
14	21.3	20.2–21.7

[a] "Base Case" reflecting assumptions used for Table 14.8
[b] Reflecting ranges of tax rate, evaluation period, capital depreciation period, and capital phasing used for this table

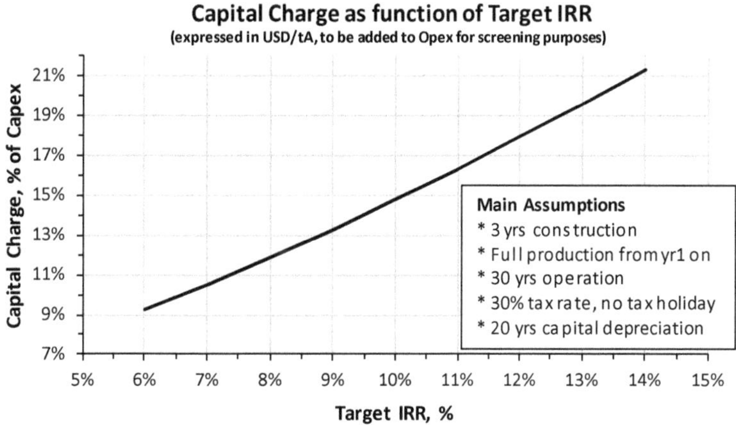

Fig. 14.20 Capital charge as function of target IRR

Use of Capital Charge—Example

Applying a screening hurdle rate of 8% IRR and an alumina price of 380 USD/tA, with other assumptions as above, will the Bauxite and Alumina projects of Table 14.10 meet this hurdle rate and which one is most attractive?

Using the above outlined approach, a "Full Cost" can be calculated as shown in Table 14.11.

Table 14.11 shows that only projects A and C meet the hurdle rate of 8% IRR minimum (Full Cost of both is smaller than the assumed alumina price of 380 USD/tA). And project A ranks higher than C because its Full Cost is less than project C's Full Cost.

Table 14.10 Project examples

Project	Ave. annual total opex USD/tA	Capex USD/Ann tA
A	160	1770
B	180	1730
C	200	1490

Table 14.11 Full cost comparison

Project	Ave. annual total opex USD/tA	Capital charge[a] @ 11.8% of capex USD/tA	Full cost USD/tA
A	160	209	369
B	180	204	384
C	200	176	376

[a] To meet the 8% IRR hurdle rate—refer Table 14.9

Correlation Between IRR and Payback Period

Once a comprehensive economic analysis has been made of a potential (Bauxite and) Alumina project, the correlation between IRR and Payback Period can be established.

Table 14.12 represents general IRR, Capex, and "simple" (see below) and actual payback period values, while Fig. 14.21 provides a graphical representation of the correlation between IRR and actual payback period of the reference project of Table 14.1 (solid line). The latter shows a good fit with a second order polynomial correlation.

In fact, independent of the selected refinery production capacity and combinations of selected values of alumina price, Capex, and total Opex, Fig. 14.21 (solid line) applies to all (Bauxite and) Alumina projects with the following characteristics:

- Primary characteristics
  - Construction period: 3 years
  - Capex depreciation: 20 years
  - No tax holidays
  - Corporate tax rate: 30%
  - Production build-up: 70%/95%/100% (year 1/2/3$^+$)
  - Operational period: 30 years (note: affects IRR, not payback period).
- Secondary characteristics
  - Capex spread: 15%/70%/15% (year 1/2/3)
  - Sustaining Capex build-up: 0%/20%/35%/60%/75%/90%/100% (year 1/2/3/4/5/6/7$^+$).

Table 14.12 Correlation IRR—payback period

IRR %	Capex USD/Ann tA	Margin, USD/tA		Payback period, y		
		a	b	Simple[c]	Actual	Delta
5.9	1401	129.3	111.5	12.6	12.8	0.2
6.5	1401	141.3	119.9	11.7	11.9	0.2
7.2	1401	154.3	129.0	10.9	11.1	0.2
7.8	1401	166.3	137.4	10.2	10.4	0.2
8.0	1401	170.0	140.0	10.0	10.2	0.2
8.4	1401	179.3	146.5	9.6	9.8	0.2
9.2	1401	194.3	157.0	8.9	9.2	0.2
10.1	1401	214.3	171.0	8.2	8.4	0.3
10.7	1401	229.3	181.5	7.7	8.0	0.3
11.3	1035	179.3	141.0	7.3	7.6	0.3
11.9	1401	256.3	200.4	7.0	7.3	0.3
12.4	1401	269.3	209.5	6.7	7.0	0.3
12.7	1401	276.3	214.4	6.5	6.8	0.3
13.0	1401	284.3	220.0	6.4	6.6	0.3
13.5	1401	298.3	229.8	6.1	6.4	0.3
14.0	1401	310.3	238.2	5.9	6.2	0.3

[a] Margin = [Aa Price] − [Total Opex (incl. Susex)]
[b] Margin corrected for 30% tax rate, and 20 yr Capital Depreciation
[c] Simple PP = Capex/Margin corrected for tax rate and Capital Depreciation

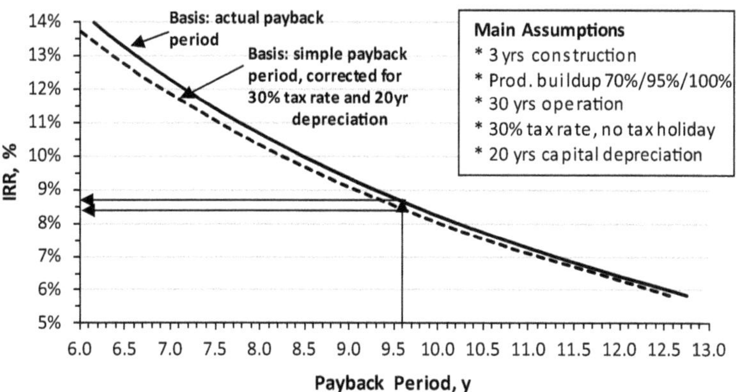

Fig. 14.21 Correlation IRR—payback period

For a preliminary assessment of the IRR and payback period of an early-phase project, a simplified approach may be followed illustrated by the following example:

1. Assume basic primary and secondary project characteristics as mentioned above.
2. Assess/assume values for
   - Capex e.g., 1500 USD/Ann tA
   - Alumina Price e.g., 370 USD/tA
   - Total Opex (incl. Susex) e.g., 180 USD/tA.
3. Calculate
   - Margin = [Aa Price] – [Total Opex] = 370 – 180 = 190 USD/tA
   - Annual capital depreciation = 1500/20 = 75 USD/tA
   - Corrected margin for tax rate and depreciation = 190 – ((190 – 75) * 0.3) = 155.5 USD/tA
   - "Simple payback period" = Capex/Corrected Margin = 1500/155.5 = 9.6 year.
4. Estimate IRR using Fig. 14.21 (solid line): ~ 8.6%.
5. Note that a selected payback period and related IRR can be obtained through several combinations of alumina price, Capex, and total Opex.

To assess the accuracy of this approach, Table 14.12 includes the following items for the Reference Project: margin (a), margin corrected for 30% tax rate and 20 year capital depreciation (b), and the "simple payback period" applying the approach of the example above, as well as the delta between the actual payback and the simple payback period. Figure 14.21 includes the graphical representation of the correlation between IRR and simple payback period of the reference project (dashed line).

Table 14.12 shows that for the investigated IRR-range of 6–14% the "Simple payback period" is optimistic by only 0.2–0.3 years (or ~2–5%).

14.1.2 Project Scope

The scope of a Bauxite and Alumina project may include some or all of the following elements [6].

14.1.2.1 Mine

Generally main scope items for a mine of greenfield Bauxite and Alumina projects are:

- Mining equipment.
- Mine infrastructure.

- Crushing and/or beneficiation/washing plant.
- Bauxite storage and reclaim facilities.
- Bauxite transport system (e.g. conveying systems, trucks).

Brownfield expansions (adding a production "train/line" to an existing plant) and de-bottlenecking capacity expansion projects (indic. 20–100 kt/y per individual D/B project) may not include mine-related scope items if the bauxite feed is purchased from a third party, however greenfield projects will ordinarily include a bauxite mine.

14.1.2.2 Refinery

The main scope items of an alumina refinery cover:

- Materials handling facilities: bauxite stacking and reclaiming, alumina storage and handling, caustic storage and handling, coal storage and handling, heavy fuel oil storage and handling, lime storage and handling, etc.
- Process facilities: bauxite grinding, (pre-)desilication, digestion, clarification, precipitation, calcination, causticisation, etc. This scope item may be broken down differently (see Table 14.13).
- Utilities: steam and power generation, air compressors, etc.
- General facilities: cooling pond, mobile equipment, residue disposal area, communications, etc.

Table 14.13 provides an indicative breakdown of refinery Capex in major plant elements.

Brownfield (Bauxite and) Alumina projects (the addition of an additional production unit/train/circuit) are generally (much) more attractive in economic terms than greenfield projects, which involve, as the word implies, a project starting on a greenfield site i.e., without existing infrastructure etc. (refer Sect. 14.1.5). An important reason is that conventional greenfield plant layouts often include (significant) provisions for future expansions that are not (yet) generating a return. Several of the

Table 14.13 Alumina refinery overall capex breakdown (indicative)

Element	% of refinery capex (%)
Bayer loop[a]	38–49
Utilities	20–26
General facilities	14–19
Calcination	5–10
Materials handling	5–8
Disposal of residues	3–5

Source [6]. Copyright © 2007 by The Minerals, Metals & Materials Society. Used with permission
[a] Including bauxite grinding, desilication, digestion, clarification, security filtration, HID, precipitation, hydrate classification, oxalate control, liquor evaporation

projects constructed in the early 1980s for example were thus able to accommodate additional digestion and/or precipitation production units on the original plant layout (e.g. Wagerup, Worsley, Alumar, Alunorte).

Another aspect to this is that alumina refineries have typically been over-designed due to conservatism by project owner, technology supplier, engineering contractor, and equipment manufacturers. And finally, plant operators once they have familiarized themselves with their specific bauxite feed characteristics find ways to increase plant production capacity ("creep"). To avoid unnecessary negative impact on project economics greenfield plants should therefore aim right from day one to achieve maximum production level. These design aspects have resulted in the past in projects with significant (low cost) de-bottlenecking/brownfield potential as illustrated by their steadily increasing production capacities over the past 30 or more years. In other words the economic potential of these projects had been under-utilized for a significant period of time.

Table 14.13 illustrates that the Bayer loop represents refinery Capex largest cost element. Approximately 20% of total refinery Capex is precipitation-yield related [6].

14.1.2.3 Infrastructure

A second important reason for the difference in economics between greenfield and brownfield (Bauxite and) Alumina projects is the requirement for greenfield projects to construct infrastructure, both "external" (e.g. port facilities, alumina and raw materials transport, housing facilities, roads) and "internal" (e.g. piperacks, power distribution, water supply, buildings). Examples of infrastructural requirements and related Capex ranges are shown in Table 14.14.

In other words the Capex for a greenfield project is penalized even for greenfield projects with limited infrastructural requirements as illustrated by Table 14.15 which shows indicative Capex numbers for a 1.6 Mt/y Bx and Aa greenfield project with limited infrastructure requirements (used for the reference project—Table 14.1).

Table 14.14 Examples of infrastructure elements—indicative capex range

Item	Capex range MUSD[a]
Port facilities[b]	50–150$^+$
Roads and bridges, Aa and raw materials transport (trucks, barges)[c]	40–100$^+$
Water supply (dam, bores, pipelines, pump stations)	10–30$^+$
Town, personnel housing, medical facilities	30–150$^+$

Sources [6] and author

[a] Actual numbers may deviate significantly, also depending on refinery location (mine site, port site)

[b] Excluding wharf/jetty

[c] Excluding railroad transport

Table 14.15 Indicative capex 1.6 Mt/y Aa greenfield project with limited infrastructure requirements

Item	Capex MUSD[a]	Capex USD/Ann tA[a]
Mine	124	77
Refinery	1896	1185
Infrastructure	221	138
Total	2241	1401

[a] Indicative only, actual numbers may deviate significantly

The extent of external infrastructure requirements plays an important role in project economics and the necessity for plant/project production capacity increases (economy of scale effect), refer Sect. 14.1.4.2, "Economy of Scale and Capital Cost", item 2 (External Infrastructure Aspects and Overall Economics).

14.1.3 Project Development Stages

Development of a greenfield (Bauxite and) Alumina investment opportunity from initiation through to actual operation takes the project through several development stages [6]. Table 14.16 provides an overview of typical project development stages, their objectives, some of their main elements, and duration, cost and estimate accuracy indications.

These project development stages require appropriate capital and operating cost estimates as shown in the table, the key distinction between these estimates being the reliability of the information used to prepare the estimate to the targeted accuracy. Cost estimates are therefore classified, each with its accuracy, and are related to the stage of development of a project. Indicative percentages of total engineering completed at each stage and contingency ranges, and type of project risk evaluation are also included in Table 14.16.

Behre Dolbear (mineral industry advisory firm) commented in 2015 that "... it takes 10 years to discover, define, and determine the feasibility of a project and an additional 6 years or more until investors can expect returns from a greenfield mining construction project...". Table 14.16 illustrates the first point while the 6-year return suggests an indicative IRR of ~13–14% (refer Table 14.12). Table 14.16 shows that the cost of preparing a capital cost estimate is related to the stage of development of a project and its targeted accuracy and may be expressed as percentage of the total project cost as illustrated in Fig. 14.22. Note that this graph is indicative only, and that actual estimate costs may vary considerably based on the specifics of a project.

Table 14.16 Greenfield project development stages

Item	Project development stage (typical)			
	Conceptual/order-of-magnitude/scoping	Pre-feasibility	Feasibility	Execution
Objectives	Order-of-magnitude evaluation of investment opportunity	Review alternatives, select preferred option, economic evaluation	Project optimization, economic evaluation for approval to proceed	Project execution, including commissioning and start-up
Duration[a], months	$6–10^{+}$	$8–18^{+}$	$12–18^{+}$	$36–48^{+}$
Study cost[a], MUSD	$0.5–2^{+}$	$3–6^{+}$	$10–30^{+}$	$40–70^{+}$
Location, topography and climate	Basic map	Detailed map	Detailed map showing all claims and boundaries	
Geological description	Preliminary review	Prelim. site-specific analysis	Detailed site-specific analysis	
Resources and reserves[b] (international reporting codes)	Indicated and inferred; but usually inferred	Indicated or above (Inferred as upside)	Measured and indicated	
Mining method	Based on knowledge: assumed either open pit or underground	Specific method identified based on preliminary assessment	Method and mine plan finalized	
Processing and process description	First estimate of production rate and general description	Prelim. mining and processing rates; detail engineering. 1–2% complete	Fixed mining and processing rates; detail engineering. 5–15% complete	

(continued)

Table 14.16 (continued)

Item	Project development stage (typical)			
	Conceptual/order-of-magnitude/scoping	Pre-feasibility	Feasibility	Execution
Tailings	Design costed from similar projects	Detailed site selection study; cost est. ± 30% accuracy	Test work carried out on representative samples; cost est. ± 15% accuracy	
Infrastructure	General overview with types of support facilities described	All required support facilities identified, sizes and quantities estimated	All necessary support facilities identified, sized and costed	
Hydrogeology	Water sources est. using regional data	Preliminary hydrology study	Specific water source(s) identified	
Environmental	Preliminary evaluation of project setting for potentially significant environmental or permitting constraints; EIS/EA not required	Preliminary evaluation of impact on environment; schedule of environmental and/or other permitting requirements; draft EIS/EA	Characteriz. of all potential impacts on environment; finalize schedule environmental and/or other permitting requirements; draft EIS/EA submitted	
Completed engineering[c], %	0–2$^+$	5–20$^+$	20–40$^+$	65–100
Capital cost (capex) est.	Order of magnitude	Preliminary	Budget/detail	Definitive
Capex est. accuracy (typ.) (before contingency), %	±35–50	±20–30	±10–15	±5–10[d]
Capex contingency range[e], %	15–25	15–20	10–15	5–10

(continued)

Table 14.16 (continued)

Item	Project development stage (typical)			
	Conceptual/order-of-magnitude/scoping	Pre-feasibility	Feasibility	Execution
Operating cost estimate	Order of magnitude	Preliminary	Detail	Definitive
Opex est. accuracy (typ.), %	±30–40	±20–30	±10–15	±5–10
Economic evaluation	Preliminary assessment	Assessment principal econ. parameters	Full assessment	
Risk evaluation	General overview	Fatal flaw analysis	Formal Monte Carlo and fatal flaw analysis	

Sources [6] and author

[a] Indicative for a greenfield 1.6 Mt/y Bauxite/Alumina project. Duration time does not include time required for Joint Venture partnership discussions and negotiations, government-related agreement discussions, etc.

[b] International standards include: Australasian Code for Reporting of Mineral Resources and Ore Reserves—prepared by the Joint Ore Reserve Committed (JORC); U.S. Securities and Exchange Commission Industry Guide 7; Canadian National Instrument 43-101 and 43-101 CP; and SME Guide for Reporting Exploration Information, Mineral Resources and Mineral Reserves

[c] Level of project definition, stated as percentage of total engineering

[d] At completion of the definitive estimate

[e] As percentage of (Direct + Indirect) Capital Cost

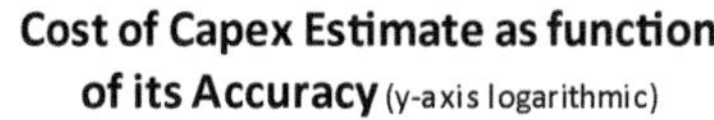

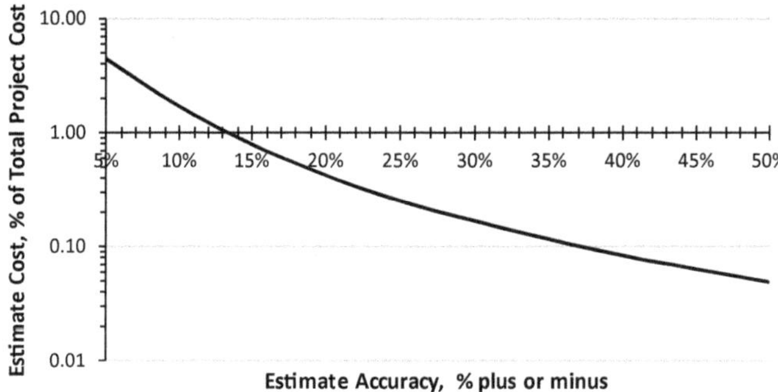

Fig. 14.22 Cost of capex estimate as function of its accuracy (y-axis logarithmic)

14.1.4 Project Economics—Drivers

14.1.4.1 Sustainability ("Profit", "Planet", "People")

Perhaps surprisingly, sustainability is more intricately linked with economic drivers of Bauxite and Alumina projects than expected. In the following sections the relationship is discussed between economics as element of the overall sustainability of a project, and Bauxite Resource Quality respectively Alumina Refinery Design.

Bauxite Resource Quality

In the following the impact of the quality of a bauxite resource is reviewed, including its connections with sustainability, key resource quality criteria, and their impact on project economics [7].

Impact of Resource Quality

Generally basic data on bauxite resources include items such as bauxite horizon thickness, overburden/bauxite ratio, mineralogy, %Available Alumina, %Reactive Silica, %TOC and impurities, etc. Obviously, these are important to perform a resource evaluation. By the same token they do not by themselves provide the information required for an overall understanding of the strengths and weaknesses of a bauxite resource.

An all-encompassing understanding requires a review of the quality of a bauxite resource in its widest sense, a key aspect of a bauxite/alumina project which also has a significant bearing on some of its fundamental sustainability aspects, i.e. "People", "Planet", and "Profit" (refer Appendix 1): its environmental strengths and weaknesses (w.r.t. energy, water, and materials consumption, biodiversity, and waste—overburden, tailings, etc.), its capital and operating costs (Capex and Opex), and its social strengths and weaknesses (impact on communities, resettlements, etc.).

Major resource quality aspects are (refer also Sect. 14.3.1.1):

- **Location** (country, export port, relocation requirements): includes country aspects such as the presence and state of repair of a port (via which raw materials are imported and alumina/bauxite is exported), a rail road (if the deposit is a significant distance from the port), a town site (incl. hospital), roads, and water availability, and the requirement to relocate people living on the deposit. These aspects may be combined in "infrastructure" Capex and Opex. Other country aspects include royalties (on bauxite or alumina), levies, duty on raw material imports, the impact of legislation (e.g., with respect to environmental requirements), taxes, and tax holidays.
- **Logistics**: accessibility of and logistics to the port via which raw materials are imported (e.g., caustic soda, coal, fuel oil, lime) and alumina/bauxite is exported. This element is influenced by port location, ship size (water depth—need to dredge) and the proximity to frequently used sea lanes, as reflected in the cost of raw materials CIF the importing port and/or in freight charges for the export of alumina/bauxite.
- **Accessibility** (distance, height, mountain, river crossings) of the deposit relative to the port, reflected in Capex and Opex related to transportation, either of the raw materials and alumina to the refinery site or of the bauxite to the (refinery at the) port.
- **Deposit characteristics** (uniform vs. pockets, overburden and bauxite horizon thickness, beneficiation requirements), affecting costs of mining, crushing, storage, beneficiation, tailings, rehabilitation, etc. This aspect is reflected in the overall USD/tBx cost CIF refinery (bauxite transportation may either be included—e.g., if the mine is close to the refinery, or be covered by a separate transportation cost) and in the related mine Capex.
- **Refinery feed bauxite characteristics** (hardness, % Available alumina—level and % Gibbsite vs. Boehmite, % Reactive silica, impurities): bauxite mineralogy and chemical composition dictate several process conditions of the Bayer refining process and associated raw material consumption and capital requirements, i.e. the quality of the bauxite feeding an alumina refinery has a significant impact on process conditions and raw materials consumption of an alumina refinery. These aspects are mainly reflected in refinery Capex and Opex related to bauxite, caustic soda, energy, lime and flocculant consumption, and residue disposal.
- **Resource size**: a bauxite resource should be able to support a refinery project for its lifetime (typically 50^{+} years; however, for evaluation purposes a 30-year lifetime is often used).

Bauxite resource quality in its widest sense may affect approximately a third of plant Capex or half of total project Capex if mine and infrastructure Capex are included [6], and has a more profound impact on Opex than technology/design (refer Sect. 14.3.1.6 and [2]). In other words, it is important to take account of the above elements when reviewing a bauxite resource's quality criteria.

Bauxite Resource Quality Criteria

In general terms deposit quality criteria should focus on major issues, provide target values, and should not be applied rigidly. In other words, a resource not meeting one (or perhaps more) of the target values should not necessarily be excluded, but the overall result of a resource review should be considered. In this section strategic criteria which could result in a different outcome of a review have not been taken into consideration (e.g., importance of a presence in a particular country for other reasons than participating in a Bauxite (and Alumina) project).

Table 14.17 presents a set of quality criteria and target values used for bauxite resource review purposes addressing the resource quality aspects discussed above.

Table 14.17 Bauxite resource quality criteria, target values, and sustainability facets

Bauxite deposit quality criterion	Target	Related GRI performance indicator[i]		
		Economic	Environ-mental	Social
1. Country infrastructure capex[a]	250 USD/Ann tA max	EC1, EC4, EC8, EC9		LA1, SO1, SO1-MM9
2. Distance resource—port	Rail (+ other): 150 km max Bx slurry pumping: 100 km min	EC1	EN4, EN29	
3. Disturbed acreage per tA produced[b]	0.35 m^2/tA max[b,h]	EC1	EN12-MM1, EN14	SO1, (SO1-MM9)
4. Material handled per tA, consisting of				
4A material mined[c]	3.4 t/tA max (dry basis)[c,h]	EC1	EN1, EN21, EN22-MM3	SO1, (SO1-MM9)
4B residue to disposal[d]	1.2 t/tA max (dry basis)[d,h]	EC1	EN4, EN12-MM1, EN21, EN22-MM3	
5. Alumina in boehmite	2% max	EC1	EN3, EN16, EN20	
6. Total caustic consumption per tA[e]	75 kg/tA max[e,h] (100% NaOH basis)	EC1	EN1, EN4	
7. Ratio extractable org. carbon/available Al_2O_3	0.002 max[f,h]	EC1	EN3, EN21, EN22, EN22-MM3	

(continued)

Table 14.17 (continued)

Bauxite deposit quality criterion	Target	Related GRI performance indicator[i]		
		Economic	Environ-mental	Social
8. Resource contained alumina	30 years min[g]	EC1		SO1, (SO1-MM9), SO1-MM10

Source [7]. Copyright © 2014 by The Minerals, Metals & Materials Society. Used with permission

[a] Railway, housing, roads, resettlement of villages, and non-refinery related port items (e.g. wharf/jetty, power supply)

[b] At in-situ bauxite SG = 1.85 and mining recovery = 90%

[c] Includes overburden and bauxite beneficiation tailings (if applicable), at mining recovery = 90%

[d] Bauxite residue to disposal, incl. sand and lime products

[e] Incl. chemical soda loss (reactive SiO_2), physical soda losses (bauxite residue), and other losses (e.g. oxalate, product, etc.)

[f] Assuming an Organic C extraction efficiency of about 50%, this means effectively a ratio of %TOC/% Avail. Aa of ~0.004

[g] For a 1.6 Mt/y alumina refinery project: ~50 Mt contained alumina, or in situ bauxite ~ 50 × 3 = 150 Mt

[h] First quartile/Median of global bauxite mines excl. China

[i] Refer Table 14.40 in Appendix 2. In this chapter GRI G3 (version 3) has been used as basis

The table includes references to relevant sustainability-related performance indicators of the Global Reporting Initiative (GRI) (refer Appendix 2 and [8]), illustrating the relationship between bauxite resource quality criteria and their sustainability facets.

Rationale Bauxite Resource Quality Criteria

The rationale behind the bauxite resource quality criteria of Table 14.17 is as follows [7]:

1. **Country infrastructure Capex maximum 250 USD/Annual tA production capacity**: this criterion refers to several elements: "hardware" related (housing, railway, access roads, bridges, port facilities); a country's legal framework; and the requirement to resettle people on the deposit. It affects local employment and local communities. If the Capex for country infrastructure increases well above this number (e.g., when a new port needs to be developed, and a large number of people require resettling), the negative impact on project economics may become unacceptable. Its impact on Opex may range from low (e.g., limited royalty requirements and no physical infrastructure to run) to significant (bauxite levies and extensive infrastructure Opex). Some aspects may be negotiable and therefore more difficult to quantify at an early stage of a project. In addition, a country's government and/or other third parties may be interested to assume responsibility for (some of) the physical infrastructure requirements.
2. **Distance bauxite resource to alumina export port maximum 150 km**: if the transportation distance increases significantly above this number (worldwide range: ~10–360 km for Bx and Aa projects), raw materials and alumina or

bauxite transportation costs increase prohibitively. For bauxite slurry pumping, in which case the bauxite is pumped in slurry form from the mine site to the refinery located at the port, the opposite applies: this technology may become a viable alternative for distances above ~100 km. The impact of bauxite slurry pumping on the refining process such as bauxite de-watering, additional evaporation requirements etc. must be taken into account in the design of the alumina plant. If appropriate design measures are taken, slurry pumping should have less environmental impact than installing a railroad.

3. **Disturbed acreage maximum 0.35 m^2/tA**: the strength of this criterion is that it is a function of a number of characteristics of the resource (bauxite horizon thickness, in-situ SG, % Available alumina and moisture); mining (mining recovery); and processing (beneficiation/washing recovery, refinery alumina recovery). Its global range is ~0.1–1.1 m^2/tA. It is proportional to the inverse of the effective bauxite horizon thickness, and it has an impact on the three sustainability aspects "Profit" (mining and rehabilitation Opex, and beneficiation Capex and Opex if applicable), "Planet" (disturbed and rehabilitated land, and biodiversity), and "People" (impact on communities living close to or on the deposit).
4. **4A. Material Mined (=overburden + [non-beneficiated] bauxite) maximum 3.4 t/tA (dry basis)**: also, a function of a number of characteristics (overburden/bauxite ratio, mining recovery, beneficiation recovery if applicable, % Available alumina and refinery alumina recovery). Global range is ~2–25 t/tA. This criterion takes into account the overburden removed, the bauxite to the refinery, mining losses, and beneficiation tailings if applicable. Bauxite beneficiation/washing has several drawbacks: additional installations are required involving Capex and Opex, and environmental issues (water usage, tailings disposal). However, applying bauxite beneficiation may be appropriate and needs to be considered on a case by case basis.

   4B. Residue to Disposal (= bauxite residue [incl. sand] + lime products) maximum 1.2 t/tA (dry basis): the bauxite residue ("red mud") leaving the refinery after removal of the available alumina in the process (range of available alumina content of globally processed bauxites is ~30–50%).This criterion is affected by %Available Alumina and Reactive Silica in the bauxite feeding the refinery, the alumina recovery in the plant, and lime and other products (e.g. filter aid) ending up in the bauxite residue. It impacts residue handling and disposal facilities, i.e., it has both economic (Capex and Opex) and environmental angles. Bauxite residue represents one of the main environmental issues related to alumina refining because of the large volume involved, its characteristics (very fine), and its alkaline properties (due to the presence of NaOH in the adhering liquor and Na_2O in the de-silication product). Globally bauxite residue ranges from indicatively 0.6–2.3 t/tA.
5. **Alumina in boehmite ($Al_2O_3.H_2O$—alumina monohydrate) maximum 2%**: below this number the large majority of available alumina is present as Gibbsite ($Al_2O_3.3H_2O$—alumina trihydrate) requiring a low temperature (LT) for the digestion operation (~140–150 °C), which has significant advantages: 1.

Required temperature and pressure for steam to the digestion area in the refinery are at a low level, providing the opportunity to first use high pressure boiler steam for co-generation of steam and power positively impacting overall energy cost and thus Opex, lower Capex due to the requirement of lower-pressure vessels, simpler feed pumps, and less flash stages, and lower Opex as a result of a lower power consumption and maintenance costs; 2. The total steam and power energy for a LT digestion plant is typically ~1–2 GJ/tA lower than for a high-temperature (HT) digestion refinery (~240–250 °C), i.e. lower Opex; and 3. Capex and maintenance costs of digestion and power and steam generation equipment are lower than would be required for HT digestion. At an alumina in boehmite of ~6% economic considerations normally favor HT digestion.

Another aspect is that below a certain boehmite in bauxite feed level, processing technology and operating conditions can be chosen such that boehmite does not dissolve in digestion and should not give rise to boehmite reversion (which would otherwise negatively affect efficiencies and operational conditions in the plant). Note that boehmite particle size, morphology, and impurities also affect reversion. In summary: high boehmite levels require HT digestion which negatively affects economics (increasing Opex—energy and maintenance costs, and Capex—high pressure refinery and powerhouse equipment), and the environment (higher energy consumption and emissions of greenhouse gases, NO_x and SO_x).

6. **Total caustic soda consumption maximum 75 kg (100% NaOH basis)/tA**: this criterion represents an important refinery Opex component, mainly influenced by: 1. The ratio of %Available Alumina to %Reactive Silica in bauxite refinery feed which globally ranges from ~6 to 40 (a high ratio meaning a low chemical caustic soda loss)—Reactive silica in bauxite reacts with caustic soda and alumina in solution forming undesired De-Silication Product (DSP—a sodium aluminium silicate) with a significant impact on caustic soda consumption and thus operating cost ("chemical soda loss"); 2. Physical soda losses with bauxite residue (affected by %Available Alumina and alumina recovery in the refining plant); 3. The technology chosen in the bauxite residue wash circuit; and 4. Other soda losses (e.g. with oxalate, organics, and in product alumina). In environmental terms the energy required for the production of NaOH should also be taken into account. Global range: typically, ~45–150$^+$ kg NaOH/tA.

   Note that the advantage of very low reactive silica content in a bauxite w.r.t. caustic soda consumption may present a potential secondary processing disadvantage in LT digestion alumina refineries in that de-silicating the liquor may be more difficult, potentially affecting product quality and equipment scaling conditions. Bauxite test work needs to confirm the processing conditions required to control this aspect which could entail additional holding time in the desilication facility and the necessity of additional equipment descaling, i.e., additional Capex and Opex.

7. **Ratio of %Extractable Organic Carbon (EOC) to %Available Alumina maximum 0.002**: below this number organic impurity removal may either occur sufficiently by natural removal processes (e.g., with bauxite residue) or

by relatively simple removal methods (e.g., concentrating plant spent liquor in a salting out evaporator). At higher ratios extensive facilities may be required (oxalate/organics removal) at significant Opex and Capex. Global range of this ratio is ~0.0003–0.007. In addition, precipitation yield is affected, negatively increasing overall energy consumption and Capex. An environmental aspect is that the removed impurities (e.g., as calcium compound or as mixed "salt" cake) need discarding, e.g. by combining them with the bauxite residue. Note: at an extraction efficiency of typically 50%, the ratio %Total Organic Carbon (TOC)/%Available Alumina is ~0.004.

8. **Resource contained alumina minimum 30^{+} years**: a key criterion for the selection of the production capacity of a greenfield alumina project is the size of the deposit. Although an alumina refinery may actually operate effectively for 40^{+} years (refer e.g., Corpus Christi, Point Comfort, Woodside, Paranam, Stade, QAL, Gove, Kwinana, Pinjarra, etc.), typically a project life of 30 years is applied when evaluating the significant investment of a greenfield Bauxite and Alumina project. In other words, the resource on which the project is based should be able to sustain refining operations for such a period, i.e., for a 1.6 million tonne per year alumina refinery project using 3 tonne bauxite per tonne alumina produced, the bauxite resource required would be of the order of 150 million tonnes. As the optimum refinery capacity may differ from location to location, the same applies to resource size. Due to its long lifetime, a Bauxite and Alumina project often plays an important role in the lives of communities around the deposit and/or the refinery.

Laboratory and field test work is required to establish several of the above discussed quality criteria such as disturbed acreage, material mined, residue to disposal, alumina in boehmite, caustic soda consumption, ratio of extractable to available alumina, etc.

Table 14.17 shows that economic and environmental aspects of bauxite resource quality criteria are intertwined most of the time (in several cases social aspects as well). Putting it differently, economically more attractive deposits are often also more attractive in environmental terms (and sometimes in social terms).

In other words, Table 14.17 and its rationale illustrate that the three pillars of sustainability (economic, environmental, and social) play an important role in the evaluation of the quality of a bauxite resource. Or viewed from another perspective, economics and sustainability in the context of bauxite resource quality are not abstract and isolated from "the real world" but can be qualified and quantified.

Third Party Bauxite Quality Criteria

The development of a potential bauxite project aiming at selling its bauxite to a third party would require applying the same resource quality criteria as those outlined in Table 14.17. However, in view of the fact that the bauxite will be sold, an important success factor will be refinery feed bauxite quality and—related to that—its impact on the alumina refinery operations of potential customers.

Three key aspects in this context are:

1. **Specific (= per tA produced) chemical caustic soda consumption**. Reactive silica extracted from the bauxite at the applied plant digestion conditions resulting in the formation of undesired DSP. At LT digestion conditions it is mainly kaolinitic silica forming DSP, while at High Temperature (HT) conditions quartz will (partly or fully) also react to DSP.

   The soda component of DSP represents a major element of an alumina refinery's total caustic soda consumption, the so-called "chemical soda loss". The extent of caustic soda lost with reactive silica depends on the DSP type formed. This may be translated into a chemical soda loss factor.

   The Reactive Silica content of a bauxite by itself is however an incomplete indicator of its specific chemical caustic soda loss. A more appropriate measure is the ratio %Available Alumina/%Reactive Silica, with a high ratio meaning a low chemical caustic soda loss. The specific (= per tA) "chemical" caustic soda consumption of a bauxite expressed in kg NaOH (100%) per tonne Alumina product is calculated as follows:

   Specific Chemical Caustic Soda Consumption in tNaOH(100%)/tA

   $$= \text{Bx Factor(refer item 2 below)} \times \text{\%Reactive Silica} \times \text{Soda Loss Factor}$$

   Note: This third-party bauxite quality aspect is one of the components of resource quality criterion 6 of Table 14.17.

   Figure 14.23 and Table 14.18 show the specific total caustic soda consumption expressed as tNaOH(100%)/tA of several operating and potential West African, South American, and Asian projects, most of them after bauxite beneficiation/washing (see next section), some of them washed as well as unwashed. The specific total caustic soda consumption values shown are calculated on the

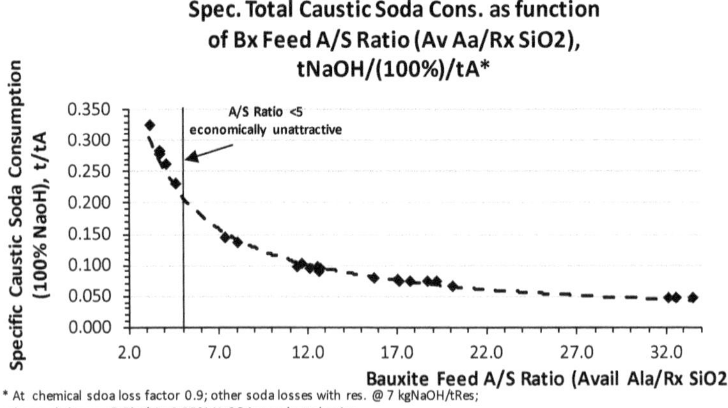

Fig. 14.23 Specific total caustic soda consumption (100% NaOH) as function of A/S ratio

Table 14.18 Specific total caustic soda consumption (100% NaOH) as function of A/S ratio[a]

A/S ratio (Avail. Aa/Rx SiO_2)	Specific total caustic soda consumption, tNaOH (100%)/tA[b]
3.16	0.325
3.67	0.282
3.73	0.276
4.04	0.261
4.58	0.229
7.34	0.143
8.04	0.136
11.35	0.097
11.65	0.101
12.11	0.095
12.53	0.095
12.62	0.089
12.64	0.090
12.68	0.094
15.66	0.078
17.05	0.076
17.13	0.074
17.68	0.072
18.67	0.072
19.22	0.072
20.08	0.065
32.09	0.047
32.50	0.047
33.50	0.045

[a] Ratio %Available Alumina/%Reactive Silica

[b] At 95% Alumina Recovery; chem. soda loss factor 0.9; other soda losses with residue @ 7 kgNaOH/tRes; other soda losses (powerhouse, water treatment) @ 5 kg/tA; 0.35% Na_2O in product alumina

same basis using bauxite %Available Alumina and %Reactive Silica of these projects and include:

- Chemical soda loss as discussed above at a chemical soda loss factor of 0.9.
- Other ("physical") caustic soda losses with residue at 7 kg NaOH/tResidue.
- Other caustic soda losses (e.g., powerhouse, water treatment) at 5 kg NaOH/tA.
- Caustic soda loss with product alumina assuming a content of 0.35% Na_2O in product alumina.

Also indicated in Figure 14.23 is the threshold A/S Ratio of 5, below which the economics of a project (without beneficiation) are generally considered questionable, as also indicated by Australia's Alumina Ltd. Company in their Full Year 2015 presentation. The worldwide actual A/S-range for operating projects is ~6–40. Figure 14.23 and Table 14.18 suggest that at an A/S ratio of 5, the Specific Total Caustic Soda Consumption is about 0.210 t NaOH (100%)/tA. Worldwide the actual specific total caustic soda consumption for operating alumina refineries ranges from ~0.045 to 0.150^{+} t NaOH(100%)/tA.

2. **Specific bauxite consumption**. An alumina refinery's specific bauxite consumption is a function of the %Available Alumina content of the bauxite dissolved at the plant's digestion conditions, and the recovery of alumina in the product. The latter is generally referred to as the "Alumina Recovery", and globally ranges between ~88 and 96%. For this review an Alumina Recovery of 95% is assumed (note: this includes a correction for impurities reporting to the final product).

   The specific bauxite consumption (often referred to as the "Bauxite Factor"), tonnes of Bauxite required (on bone dry—0% moisture—basis) to produce 1 tonne of alumina product, is calculated as follows:

$$\begin{aligned}&\textbf{Bx Factor}\text{ in tBx(dry} - 0\%\text{ moisture basis)/tA}\\&\quad = 10000/(\%\text{Available Alumina} \times \%\text{Alumina Recovery})\end{aligned}$$

   Notes:

   - "Available Alumina" refers to the alumina extracted from bauxite at the refinery's digestion conditions. Generally, a split is made between "Low Temperature" (LT; typ. 140–150 °C) and "High Temperature" (HT; typ. 230–270 °C) digestion operations.
   - This third-party bauxite quality aspect is implicitly covered by resource quality criteria 4, 5, 6, 7, and 8 of Table 14.17.
   - Worldwide actual Bauxite Factor of operating alumina refineries ranges from ~2 to 3.6 tBx (dry basis)/tA.

3. **Specific residue production**. The residue leaving an alumina refinery consists of several components such as bauxite residue, lime compounds, other (organic and inorganic) effluents, and sometimes coal fly ash and bottom ash. The bauxite (digestion discharge) residue, including DSP formed in the digestion stage, generally represents the largest constituent.

   Put differently the corollary of the specific bauxite consumption covered in item 1 above is the specific residue production, tonnes of Digestion Discharge Residue produced (on bone dry—0% moisture—basis) per tonne of alumina product.

   Notes:

   - This third-party bauxite quality aspect is covered by resource quality criterion 4B of Table 14.17.

- Worldwide the actual specific residue production of operating alumina refineries ranges from ~0.65 to 2.3 tRes (dry basis)/tA.

Bauxite Beneficiation/Washing

For the successful development of some bauxite deposits, bauxite beneficiation (or "washing" as it is normally referred to) may be required to improve bauxite product quality characteristics.

The main objective of beneficiation is to reduce the reactive kaolinitic silica (clay) content which in turn reduces the caustic soda consumption in the alumina refining process (a major operating cost item in alumina production; refer above item 2 on specific caustic soda consumption), i.e., the bauxite is upgraded to a quality suitable for use as an alumina plant feedstock. Essentially, beneficiation to lower the level of reactive silica will generally improve its quality and suitability, and its value for sale on the open third-party bauxite market. Resources containing bauxite requiring beneficiation are therefore often expressed on beneficiated/washed bauxite basis, which is the saleable product from the beneficiation plant.

Beneficiation refers to the crushing, wet screening, washing and classification ('sizing') operations separating the finest grained fraction of the raw Run of Mine (ROM) bauxite. The fraction containing much of the clays is higher in kaolinite/reactive silica and lower in alumina than ROM bauxite. Removal of the fines thus results in a reduction of the reactive silica and an increase in the alumina content of the washed product. Ideally, most of the reactive silica (clays) is removed without removing a large proportion of the mined mass, thus obtaining a better product without sacrificing large tonnages.

In summary, the requirement for bauxite beneficiation has the following major consequences:

- **Loss of resource**: a mass yield recovery (on dry—0% moisture content—bauxite basis) for example of 75% means that 25% of the ROM bauxite is lost and considered waste. Worldwide this mass yield recovery may vary between ~35 and 70%.
- **The fines material (tailings) lost through the beneficiation operation require storage**. The settled solids percentage of these tailings is often low (25–40% w/w), consequently large tailings areas are required unless the tailings are subjected to filtration or another treatment stage.
- **Increased capital and operating costs** (personnel, power, water, flocculants, maintenance materials, etc.) associated with mining, the beneficiation plant proper, and handling and storage of additional waste.
- **Overall impact on project economics** due to the combined effect of the above.

Alumina Refinery Design

This section builds on the section above (**Bauxite Resource Quality**) and discusses the relationship between the three sustainability "pillars" ("Profit", "Planet", "People") and ten key design criteria for alumina refineries in the context of applicable

Global Reporting Initiative (GRI) sustainability performance indicators (refer also [12] and [13]).

Bauxite Feed Quality and Alumina Refinery Design

As discussed in the previous section, bauxite quality has a profound impact on the design of an alumina refinery [7], i.e., the choice of a specific bauxite feed affects several important refinery design criteria. On the other hand several plant design criteria are chosen independent of the bauxite feed quality or are based on considerations other than bauxite quality alone, in the following areas [12, 13]:

- **Process conditions**: important process conditions such as plant liquor productivity/yield.
- **Equipment technologies and layout** e.g., for digestion, bauxite residue settling, precipitation and overall plant.
- **Plant location specifics** affecting plant design such as annual rainfall, rainfall fluctuations and net precipitation/evaporation impacting on the method of residue disposal, and the overall plant water balance which plays an important in a (Bauxite and) Alumina project's design; country legal requirements e.g. with respect to emission standards; and local geography e.g. with respect to gravity flow between precipitator tanks.
- **Operating and maintenance philosophies** of the project owners e.g., with respect to outsourcing activities, integration of maintenance and operational activities (multi-skilling), maintenance shop integration, etc. Here too the local conditions may play a significant role.

Alumina Refinery Design Criteria, Benchmarks, and Sustainability Facets.

As it is not possible to cover all alumina refinery design criteria, ten of the most important ones, their benchmarks and sustainability performance indicators are included in Table 14.19 with references to relevant GRI performance indicators—refer [8], illustrating the relationship between these criteria and sustainability. Refer Appendix 2 for a description of the indicators used.

Rationale Alumina Refinery Design Criteria

The rationale behind the refinery design criteria in Table 14.19 is as follows [12, 13]:

1. **Plant Liquor Productivity/Yield** (refer [14] for details). Figure 14.24 provides a schematic of the alumina refining process: caustic liquor is used in Digestion to dissolve alumina from bauxite at a temperature of typically 145–150 °C for so-called Low Temperature (LT) digestion plants processing Gibbsitic bauxites, and 240–270 °C for High Temperature (HT) plants using Boehmitic or Diasporic bauxites—while the dissolved alumina is crystallized from the solution in Precipitation through cooling and seeding. The "spent" solution is recycled to the front end (Digestion). In other words, the higher the productivity/yield of the liquor that is pumped around, the more cost effective the use of installed equipment. A key design objective of an alumina

Table 14.19 Alumina refinery design criteria and sustainability performance indicators

Refinery design criterion	Target/benchmark	Related GRI performance indicator		
		Economic	Environmental	Social
1. Plant liquor productivity/yield	90⁺ kg/m³	EC1	EN1, EN3/EN5, EN4/EN7, EN6, EN16, EN20, EN22-MM3	MM11
2. Digestion temperature	Refer Table 14.17, bauxite deposit criterion 5	EC1	EN3/EN5, EN16, EN20	
3. Digestion technology	Slurry heating ("Single Streaming")	------------- As 1 (Liquor yield) --------------		
4. Bauxite residue settling and washing technology	High rate thickeners (settlers)/washers	EC1	EN1, EN4/EN7, EN8, EN12-MM1, EN21, EN22-MM3	SO1, SO1-MM9, SO1-MM10
5. Heat interchange technology	Direct heat transfer	-------------As 1 (Liquor yield) --------------		
6. Precipitation technology	High solids tanks; seed filtration; inter-stage cooling; green liquor split; split seeding; classification by hydrocyclones	------------As 1 (Liquor yield)-------------		
7. Power and steam generation	– Refer[7], Bx deposit criterion item 5	EC1	EN3/EN5, EN16, EN20	
	– Gas as energy carrier (if available)	EC1	EN1, EN3/EN5, EN16	SO1
	– Off-gas de-sulphurisation		EN16, EN20	SO1
8. Calcination technology	Stationary calciner(s)	EC1	EN1, EN3/EN5, EN16, EN20	MM11
9. Bauxite residue disposal technology	Dry disposal in areas lined with clay or HDPE/PP seal, with underdrains; rehabilitation/re-vegetation afterward; sea water neutralization if applicable	EC1	EN4/EN7, EN12-MM1, EN21, EN22-MM3	SO1 (SO1-MM9), SO1-MM10
10. Overall plant				
10A Occupational health and safety	Maximize personnel safety	EC1		LA8

(continued)

Table 14.19 (continued)

Refinery design criterion	Target/benchmark	Related GRI performance indicator		
		Economic	Environmental	Social
10B Design and layout	Conventional: design accommodates future digestion/process units New approach: dedicated design and layout for a specified production capacity[a] Design for disassembly	EC1	EN1,EN3/EN5, EN4/EN7, EN12-MM1	LA1, SO1 (SO1-MM9), SO1-MM10
10C Production capacity	General: depends on deposit size, plant considerations, economies of scale, infrastructure requirements, and market economics; New approach: compact capacity ~300–600 kt/y alumina[a]	EC1, EC4	EN1,EN3/EN5, EN4/EN7, EN12-MM1, EN16, EN22, EN22-MM3	LA1, SO1 (SO1-MM9), SO1-MM10
10D Equipment and additives	– Mechanical Seal Pumps – Low-NO_x burners – Mechanical vapor compression – Anti-scaling chemicals	EC1	EN1, EN3/EN5, EN20, EN22	
10E Control equipment	Variable Speed Pump Drives	EC1	EN1, EN3/EN5, EN4/EN7, EN16	

[a] Refer Sect. 14.4.4.
Source: [13] Copyright © 2015 by The Minerals, Metals & Materials Society. Used with permission.

refinery is therefore to maximize alumina dissolved in the liquor in Digestion and maximize alumina crystallized from the liquor in Precipitation, i.e. maximize the alumina produced per cubic meter of circulated liquor (liquor productivity/yield).

Increasing plant liquor yield has significant advantages:

- Increased plant capacity, i.e., Capex/annual ton production capacity drops for several process areas (e.g., for digestion, decantation, precipitation, steam and power station).
- Lower consumption per tA of energy (e.g., digestion steam, overall pumping power), labor, maintenance materials, overheads and other fixed costs.
- Alumina product quality control. Although not straight-forward, some aspects of product quality control improve when the conditions for yield increase improve (example: increasing the precipitation fill A/C ratio of

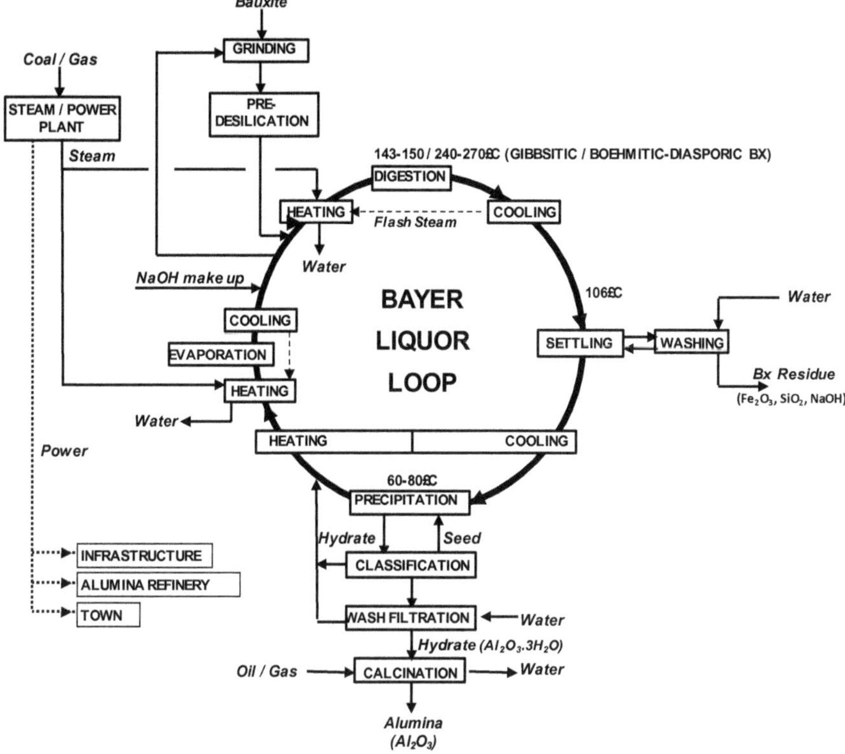

Fig. 14.24 Alumina refinery process schematic

the mixture of green liquor and spent liquor recycled with the seed charge feeding precipitation).

- Lower specific energy consumption also means a drop in greenhouse gas emissions per tA and (if applicable) coal ash residues (bottom ash/fly ash) per tA, i.e., an improvement of direct environmental performance. And the potential to improve alumina quality (previous bullet point) enables lowering alumina fines losses during transport and handling, thus improving environmental performance indirectly.

The focus for maximizing liquor productivity in a refinery is mostly but not solely on precipitation yield because the reaction kinetics of precipitation (alumina tri-hydrate crystallization) are more difficult to control and enhance than those of the digestion reaction (dissolution of alumina from bauxite).

Figure 14.24 also illustrates that heat recovery occurs extensively throughout an alumina refinery by recovering heat from streams requiring cooling down to streams that require heating up. Examples: heat exchanger (HX)/flash vessel combinations in digestion and evaporation, heat exchanged between liquor to precipitation and spent liquor returning to digestion in the

Table 14.20 Precipitation yield improvement options versus capacity increases and energy savings

Plant adaptation[a]	Yield increase[a] kg/m^3	Capac. increase[b] %	Energy savings[c] GJ/tA
Plants using Gibbsitic bauxite			
Seed filtration in precipitation	5–10	7–15	0.4–0.9
Interstage cooling in precipitation—5 units	2–5	3–8	0.2–0.5
Direct cooling in heat interchange	1–3	1.3–5	0.1–0.3
High rate thickeners in decantation	1–2	1.3–3	0.1–0.2
Total (proven technology)	**~9–20**	**~13–31**	**~0.8–1.9**
Plants using Boehmitic bauxite			
Sweetening in digestion	6–10	9–15	0.7–1.1
Seed filtration in precipitation	5–10	8–15	0.6–1.1
Interstage cooling in precipitation—5 units	2–5	3–8	0.3–0.6
Direct cooling in heat interchange	1–3	1.5–5	0.1–0.4
High rate thickeners in decantation	1–2	1.5–3	0.1–0.3
Total (proven technology)	**~15–30**	**~23–46**	**~1.8–3.5**
Proposed new technologies[d]	**Yield increase[d] kg/m^3**	**Capac. increase[b] %**	**Energy savings[c] GJ/tA**
Plants using Gibbsitic bauxite			
Fines destruction in precipitation	4–8	5–12	0.3–0.7
Bauxite residue re-digestion at high solids density	4–5	5–8	0.3–0.5
Plants using Boehmitic bauxite			
Fines destruction in precipitation	4–8	6–12	0.5–0.9
Bauxite residue re-digestion at high solids density	Not applicable	n. a	n. a

Source [14].

[a] Based on proven technology; refer [14] for further details

[b] Using a base case precipitation yield range of 65–75 kg/m^3 for LT plants (digestion typ. 143–150 °C) using Gibbsitic bauxite; and a precipitation yield of 65 kg/m^3 for HT plants (digestion typ. 240–250 °C) using Boehmitic bauxite

[c] Using a steam and power energy requirement of 7.6 GJ/tA for plants using Gibbsitic bauxite (all required power generated in a co-generation facility); and 9.7 GJ/tA for plants using Boehmitic bauxite (part of the power required generated in a co-gen facility, and 0.5 GJ/tA of imported power required to meet remaining power needs). In other words, in both cases alumina calcination energy is not included

[d] Refer [14] for further details

heat interchange area, and the water used in the bauxite residue wash circuit being used first for cooling purposes in a barometric condenser for instance in the evaporation area, thus recovering heat and optimizing energy consumption.

Precipitation yield increases are feasible (refer Table 14.20) of ~9–20 kg/m^3 (or g/l) for LT resp. 15–30 kg/m^3 for HT digestion plants raising typical current precipitation yield levels of ~65–75 kg/m^3 to a benchmark level of ~90 kg/m^3 and beyond, by implementing plant design adaptations—refer [14]. These yield increases are equivalent to ~13–31% production capacity increases for LT resp. 23–46% for HT digestion plants; and equivalent to steam and power energy savings of ~0.8–1.9 GJ/tA for LT resp. 1.8–3.5 GJ/tA for HT digestion plants. These energy savings are significant compared with typical total steam and power energies of ~6–9 GJ/tA for LT resp. 8–11 GJ/tA for HT plants.

Increasing precipitation yield through extra liquor dilution (instead of or in addition to the plant adaptations indicated in Table 14.20) requires additional evaporation i.e. additional energy consumption. The overall economic and environmental gains are not clear-cut especially for LT digestion plants. A similar argument applies to increasing precipitation yield by just increasing precipitation holding time. These options therefore have not been included.

Some proposed new technologies are also shown in Table 14.20. Maximum achievable yield may be constrained by liquor impurities originating from the bauxite feed such as oxalate, carbonate and sulphate. Various liquor removal options may be considered to offset their impact.

2. **Digestion Temperature**. Mainly the consequence of bauxite quality—refer Table 14.17, criterion 5. Other process considerations may play a role as well e.g., related to the water balance.
3. **Digestion Technology**: benchmark digestion technology is Slurry Heating (also termed "single streaming") comprising heating of a combined bauxite/liquor slurry rather than heating the bauxite slurry separately and in parallel with the liquor which are mixed only in the digester vessel (liquor heating or "double streaming").

   Slurry heating has significant advantages over Liquor Heating: **A**. Improved recovery of heat exchanged in the train of flash vessels/HX's between the slurry ex the digester vessel being cooled down and the slurry to the digester vessel being heated up (put differently: a better heat balance between the flows being heated up and cooled down), i.e., lower energy consumption and lower Capex per tA produced. **B**. The slurry flow through the heater tubes keeps them clean from scaling for a longer period time due to the erosive effect of the slurry (scaling inhibitors may have a similar effect)—especially the HX's operating at higher temperatures, i.e. plant on-line time improves (improving plant efficiencies) and less heater cleaning is required (i.e. lower steam consumption due to an overall improvement in heat transfer rates, and less waste, maintenance materials); and **C**. An improved digestion yield potential as there is no Free Caustic (FC) constraint because the Gibbsite in the combined bauxite/liquor slurry rapidly dissolves during the passage through the heater tubes (typically more than 75% of the extractable alumina) thus reducing the extraction liquor's FC (at elevated liquor FC concentrations the risk of stress corrosion cracking—caustic embrittlement—of steel in the heating equipment becomes prohibitive while the allowable liquor FC concentration decreases with increasing liquor temperature—refer e.g. [15]). This means that the caustic concentration of the liquor to digestion and therefore the digestion liquor productivity can be significantly increased impacting positively on Opex, Capex and environment as discussed under point 1 above (Plant Liquor Productivity).

Despite their slightly smaller heat transfer coefficient, vertical HX's are overall more attractive for slurry heating than horizontal heaters because they prevent solids blocking up the tubes and they are easy to maintain using an overhead crane. An alternative, notably in the case of HT digestion, is the use of tube digestion employing jacketed pipes for heat transfer [16].

Although not mentioned as separate item the technology selection for bauxite crushing and grinding should be driven primarily by the requirements to handle the range of expected bauxite feed ore characteristics (e.g. top size ROM ore, ore "stickiness", strength, targeted grind size, etc.). With single streaming digestion, closed circuit grinding would be preferred whereas open circuit grinding is more suited to double streaming digestion. Pre-desilication of the slurry ex the grinding mills at high solids density (50–55% m/m) and ~100 °C is benchmark.

4. **Bauxite Residue Settling and Washing Technology** (refer also Table 14.17, criterion 4B): Bauxite residue slurry discharging from Digestion is subsequently separated in a Solid/Liquid separation/decantation step and the residue ("red mud") is washed counter-currently with water to recover dissolved alumina and caustic soda values in the solution adhering to the residue solids.

   High rate thickening/washing technology (incl. the use of appropriate flocculants, and the recycle of thickener overflow to dilute the tank feed) is benchmark and has the following advantages over conventional large-diameter thickeners: **A**. Greatly increased mud throughput per m^2 thickener area, i.e. lower Capex and Opex (e.g. maintenance costs of thickeners and pumps); **B**. Increased underflow densities, i.e. more effective washing and thus reduced soda and alumina losses with liquor adhering to the residue at the same wash water requirement (or less wash water required at the same soda and alumina losses), and a smaller acreage requirement for the residue (lower Capex, Opex and environmental burden); **C**. Reduced contact time between liquor and mud and hence less potential of alumina reversion/auto-precipitation occurring in the tanks (premature gibbsite crystallization from the super-saturated alumina liquor on non-extracted gibbsite and on goethite in the bauxite residue), i.e. further reducing alumina losses; and **D**. These tanks can handle "sands" (coarser fraction of the bauxite residue), saving Capex and Opex for separate sands separation and washing systems.

   To maximize recovery of caustic soda and alumina values and to further increase solids in the residue to disposal, thus minimizing residue disposal acreage, the inclusion either of bauxite residue filtration downstream of the wash train or of a "super-thickener"/"deep thickener" at the storage disposal site may be considered. Although not mentioned separately here, a proposed plant design modification encompasses by-passing or excluding the Security

Filtration process area (refer Sect. 14.4.3.1 and [17]) which has economic (Capex, Opex) and environmental (scale, cleaning liquor, etc.) advantages.

5. **Heat Interchange Technology**: direct heat transfer (e.g., plate HX's) rather than indirect heating (e.g., by vacuum flash steam) in the Heat Interchange area. Advantages: **A**. Avoids the liquor concentrating effect of flashing, i.e. Digestion can operate at a higher caustic concentration without changing precipitation caustic concentration, enabling an increased digester discharge A/C ratio and thus also Liquor to Precipitation (LTP) A/C ratio (i.e. potential for higher precipitation yield and thus lower Capex, Opex, energy consumption, and greenhouse gas emissions per tA produced); and **B**. Plate HX's are inexpensive and simple to operate and maintain (low Capex and Opex).
6. **Precipitation Technology**: in addition to the process adaptations indicated in Table 14.20, the following features are included in the design of a benchmark precipitation area:

   - *For yield increase reasons* (high surface area): **A**. Precipitation tanks designed for 600–800^{+} kg/m^3 solids and more (current typically ~400–600 kg/m^3), including appropriate agitators; and **B**. Use of additives (e.g., crystal growth modifiers).
   - *For improved product quality control*: **A**. Pregnant/green liquor feed split (control of soda inclusion) and split fine/coarse seeding (control of particle size and strength); and **B**. Agglomeration (mainly quality control) and growth (mainly yield increase) sections.
   - Product/seed classification by hydrocyclones rather than gravity classifiers *for reasons of both yield and product quality control improvement.*

   The advantages of benchmark technology in the precipitation area are covered under point 1 above (Liquor Productivity/Yield).
7. **Power and Steam Generation** (refer also Table 14.17, criterion 5): using gas as energy carrier enables the use of gas turbines in combination with co-generation of steam and power. A gas-fired co-generation facility comprises a gas turbine linked to an electrical generator (producing power required by the alumina refinery), and heat recovery steam generator recovering waste heat from the gas turbine exhaust which is used to produce steam for the refinery, with surplus electricity exported to the local power grid. This electricity has about one-third the emissions intensity of coal-fired electricity. This type of co-generation technology or 'combined heat and power' technology provides greater conversion efficiencies than conventional generation methods, harnessing heat that would otherwise be wasted and reducing greenhouse gas emissions. End result: a reduction of about 26% in the operation's greenhouse gas intensity compared with electricity and steam generated from coal (Yarwun—refer [18]). When applicable: off-gas de-sulphurisation to be included to reduce SO_2 emissions.
8. **Calcination Technology**: alumina hydrate crystals from precipitation are washed and calcined to smelter grade ("sandy") alumina for benchmark performance of aluminium smelters. Calcination in stationary calciners is

benchmark, incorporating effective heat recovery from combustion gases (i.e., improved Opex and less greenhouse gas emissions per tA produced), and improved product quality control over rotary calciners. Benchmark energy consumption is ~2.7 GJ/tA [19]. The selection of energy carrier (natural gas, coal gas, heavy fuel oil) should be based on: **A**. An economic evaluation of availability, price and capital cost (coal gas fired calciners require coal gasification and modifications to the calciner); **B**. The related environmental footprint (CO_2 emission, coal fly ash); and **C**. Alumina product quality (e.g., with respect to coal impurities ending up in the alumina).

9. **Bauxite Residue Disposal Technology**: although both so-called "wet" and "dry" residue disposal technologies are applied worldwide, thickened tailings/"dry" disposal (deposition of bauxite residue layer by layer) is benchmark. Depending on residue characteristics and local conditions (rainfall), dry stacking (residue slope formed "naturally") or slope deposition (discharge onto a slope with an angle equal to the residue's natural angle of repose) may be more appropriate. Dry stacking requires the formed layer of residue to consolidate by solar drying (assisted by so-called amphirolls to plough the residue to promote the drying process) prior to the deposition of the next layer on top. Rotating between residue discharge points with intervals achieves this.

   Perimeter dikes of a Residue Disposal Area (RDA) prevent contamination of the surrounding environment; when an area is filled up a next lift is created on top. An RDA is clay- or HDPE/PP lined to prevent seepage of alkaline solution into the ground water. The lining should be covered with a layer of sand housing a network of porous pipes to collect the alkaline drainage from the residue which is returned to the refinery. Once an RDA has been totally utilized, it may be capped with clay and covered with topsoil for re-vegetation. The run-off from a de-commissioned RDA is returned to the process until its composition is acceptable for return to the environment. Thickened Tailings Disposal of bauxite residue is more attractive than wet mud disposal because **A**. It requires less surface area and results in less rainwater runoff requiring treatment, **B**. It provides the potential for more controlled residue disposal management, and **C**. It is aesthetically more acceptable. Several variations are in use in the world e.g., in Brazil (Alunorte), Australia (Worsley, Gove), Jamaica (Ewarton) and Ireland (Aughinish).

   Sea water neutralization followed by dry stacking may be applicable in case of a refinery location close to the sea (requires test work to confirm applicability). Magnesium and calcium in the sea water neutralize the alkaline components in the bauxite residue. The neutralized slurry is subsequently thickened to a high solid's concentration; the saline overflow is returned to the sea. The thickener underflow is discharged to a conventional dry stacking area. Control of the rainwater runoff from the storage area (solids, pH, heavy metals) would be required because this cannot be recycled to the refinery and needs discharging into the sea. Advantages of sea water neutralization: **A**. Long term liability and management issues related to a storage area and its rehabilitation are significantly reduced; and **B**. The storage area may not require a lining, depending on local ground conditions.

10. **Overall Plant**

10A. Occupational Health and Safety represents an important facet of the design of an alumina refinery, particularly in view of the presence of many areas with potentially dangerous conditions: high temperature and high pressure (digestion, evaporation, and power and steam generation areas), very high temperatures (calcination area), caustic soda liquors (throughout the refinery), acids (digestion and evaporation areas), moving and rotating parts (conveyor belts, pumps, filters, etc.), flammable liquids (diesel, fuel oil, gas, coal) and dust formation (lime, bauxite, fly ash, alumina). Several safety standards may apply in this area such as for example IEC 61511-1 (Functional safety—safety instrumented systems for the process industry sector) in the electrical and electronic fields. The basic process control system (BPCS) as defined in IEC 61511-1 may be considered as a first protection layer around the process, enveloped by following layers such as "Prevention", "Mitigation", "Plant Emergency Response", and "Community Emergency Response". Introducing these safety standards early in the design process enables improving refinery design while maximizing personnel safety.

10B. Overall Plant Design and Layout: conventional designs aim at accommodating additional future digestion and other process units, i.e., plant design incorporates provisions for future expansions resulting in significantly increased Capex of the design/initial production capacity negatively affecting economics (refer [1]). A new approach is based on a dedicated refinery design and layout for a specified production capacity, i.e. tailored to the equipment and infrastructure requirements of the selected production capacity (earth works, power, water supply, pipe racks, roads, cable trays, etc.) (refer [20] and [21]). This enables optimizing plant layout e.g., with respect to positioning similar equipment close to each other; and the use of common spare equipment. The approach results in a focus on a "lean" design positively impacting on commodity volumes: for the same production capacity commodity volumes for steel, concrete, piping, etc. for a greenfield plant designed this way are similar to that of a brownfield expansion of an existing refinery (refer e.g., Sect. 14.4.4.3 and further). In other words, per annual tA production capacity significantly lower amounts of commodities are required for greenfield projects compared with conventional designs. The dedicated design excludes provisions for future expansions, which require their own economic justification. And because commodities represent a significant element of refinery Capex, the consequence is a drop in Capex (indicatively by ~9–10%) and Opex, more effective use of available space for the plant, and avoidance of unnecessary energy consumption (e.g., pumping energy) and maintenance costs.

This approach is independent of preferred/selected refinery technologies and includes several specific design and layout elements (refer Sect. 14.4.4.10 and [21]).

The operational period of an alumina refinery is generally substantial (50^+ years). Its disassembly costs are significant and in some countries project owners are legally obliged to build up financial reserves for a refinery's final disassembly. If an alumina plant is designed for disassembly right from the outset, disassembly costs could be significantly lowered with conceptually little impact on the project's initial Capex. In addition, this approach is environmentally more attractive as it will be easier to return the area to its original situation.

10C. Plant Production Capacity: key selection criteria to decide on the capacity of an alumina refinery typically include: **A**. Bauxite deposit size; **B**. Plant considerations e.g. one or more "trains", and project complexity; **C**. Economies of scale; **D**. Infrastructure requirements (both "external" e.g. port (extension), consumables and alumina transport, personnel housing, roads; and "internal" e.g. pipe racks, power distribution, water supply, buildings), representing a significant part of project Capex; and **E**. Market economics.

A new Dedicated Compact Sustainable (DCS) approach applies a dedicated design to a compact refinery of ~300–600 kt/y alumina, resulting in a project with a simple and limited scope. As a result, plant Capex decreases significantly (indicatively by another ~8–10% on top of the Capex decrease resulting from the dedicated-design approach—refer Sect. 14.4.4.5). To ensure acceptable economics for the overall project, infrastructure capital should be limited. At the same time such a (compact) project has few infrastructural requirements, especially if it is located close to an existing port. Advantages:

- The *smaller project Capex* (lower risk) enables development of bauxite and alumina projects by smaller companies without a need to form (complex) joint ventures, thus increasing the number of companies potentially interested in developing bauxite deposits. Competition increases, resulting in a more efficient use of (capital and bauxite) resources.
- *Small and simple projects carry less risk; require less time to develop, and to construct and start up*, positively impacting economics.
- Alumina refining projects (generally with a long lifetime) based on this approach require only a small deposit (~40 Mt would support a 400 kt/year project for 30 years), i.e. worldwide the *number of deposits lending themselves to development increases, improving the use of resources and employment opportunities.*
- Note that this approach would also apply when *developing part(s) of a large deposit.*
- In some cases this approach would enable *value creation through alumina refining* rather than being limited to bauxite export sales (attractive to the host country and to companies developing bauxite and alumina projects).
- An *adapted version* of the approach may in some cases *enable bauxite deposit development even in locations with little existing infrastructure, albeit at a larger than compact scale* (e.g., ~1.6 Mtpa alumina).
- *The approach is independent of selected refinery technologies.*

An overall process plant layout for a compact 400 kt/year alumina refinery based on the dedicated design approach illustrates that the approach leads to a compact, simple and efficient layout with a small Bayer loop (refer Fig. 14.41), illustrating that the goal to tailor plant design to the equipment and infrastructure requirements of the specified production capacity is achievable: most of the infrastructure is integrated in the process areas and only limited infrastructure is required outside those [21].

10D. Equipment and Additives: equipment related design criteria include: **A**. Mechanical seal pumps instead of pumps with water-purged glands (less process dilution, less infrastructure, lower water consumption); **B**. Low-NO_x burners in power and steam generation (improved NO_x emissions); and **C**. Mechanical vapor compression in case of low-cost power from cogeneration (if no excess steam to be condensed in condensing turbines). Additives related design criteria include using appropriate chemicals e.g., anti-scalants in Liquor Evaporation.

10E. Control Equipment: variable speed pump drives instead of control valves for level control, flow control, etc. This reduces pumping energy, improves erosion and cavitation (i.e., less wear). In other words lower Opex, Capex and greenhouse gas emissions.

Table 14.19 and its rationale illustrate that several of the key criteria of alumina refinery design comprise one or more of the three pillars of sustainable development: economic (“profit”), environmental (“planet”) and social (“people”) aspects. In other words, sustainability in the context of refinery design can be qualified and quantified once bauxite characterization test work has been completed and project size decided.

As Table 14.17 did for Bauxite Resources, Table 14.19 illustrates that economic and environmental aspects are in fact two sides of the same coin, while social aspects are often also integral to refinery design. Putting it differently, optimum refinery design in economic terms is (long term) often also the most attractive environmentally (and to some extent socially).

In view of the above findings, it seems reasonable to assume that the role of sustainability will become more prominent in future decisions on the design of alumina refinery projects. Elements of that trend include a continuing push to increase precipitation liquor yield (item 1 of Table 14.19); finding new approaches and technologies to improve plant and process efficiencies (e.g. item 10 of Table 14.19), including a growing thrust to improve on bauxite residue disposal (item 9 of Table 14.19); and developing a more sustainable energy supply (e.g. Integrated Solar Combined Cycle).

14.1.4.2 Economy of Scale

Economy of Scale—Overview

An alumina refinery consists of a number of unit operations such as grinding, digestion, evaporation, etc. A unit operation generally comprises a string of equipment

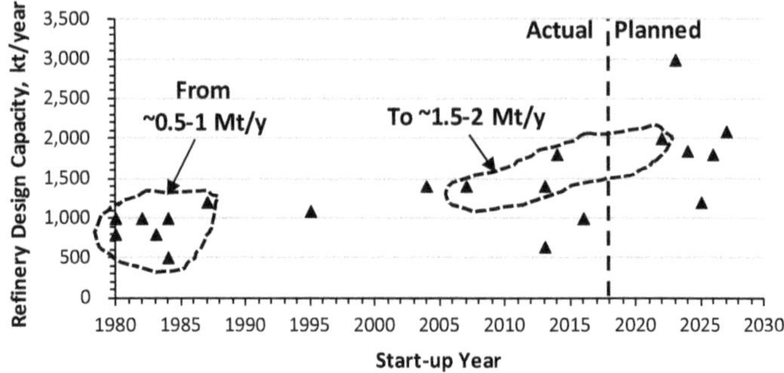

Fig. 14.25 Refinery design capacity versus start-up year (excl. China)

which together performs the desired process step, e.g. digestion with feed tank, heat exchangers, pumps, digester vessel(s), flash vessels, etc. Such a string of equipment is often referred to as a "train", "unit" or "circuit" (e.g., digestion unit, precipitation train, mill circuit). Alumina refinery design generally takes the digestion area as plant bottleneck due to its high unit capital cost and its requirement of constant (optimum) flow for optimum performance.

The design/initial refinery production capacity of greenfield projects outside China has evolved over time from about 0.5–1.0 Mt/y alumina 25–30 years ago (e.g., Worsley, Alumar, Aughinish) to 1.5–2 Mt/y alumina and more for more recently constructed and future planned projects (e.g. Lanjigarh, Yarwun, Ras az Zawr, GAC). Figure 14.25 illustrates this trend for a selected number of refinery projects. Note that actual refinery production capacities of several of the projects indicated in Fig. 14.25 have significantly increased over time as a result of de-bottlenecking, and improved process efficiencies and operations performance.

The rationale of the trend in Fig. 14.25 has been economy of scale: an increased design production capacity should improve the economics of a greenfield (Bauxite and) Alumina project or perhaps more accurate or appropriate: is required to achieve improved economics (refer also [20]).

In this section economy of scale aspects of Operating Cost and Capital Cost are reviewed.

Economy of Scale and Operating Cost

To better assess the effects of the economy of scale on Operating Cost, its major components are considered [2, 20]:

- **Variable costs** (refer Sect. 14.3.2.1 for further details): In USD/year these costs vary with plant production, at least within certain plant production rates (typically $\pm$ 10–15%), examples: bauxite, caustic soda, coal, fuel oil, lime. The overall plant on-line time of an alumina refinery with more than one train/unit/circuit (e.g., a

digestion train), is higher than a plant with one train only, as a result of more flexibility in equipment operation and maintenance. The impact on plant on-line time is generally limited (indic. 0.2–0.5% abs.), however may vary widely and in a specific case could be significant (>1% abs.). As a result the plant operates with less interruptions while operating efficiencies (e.g. bauxite, caustic soda, energy consumption) improve, albeit generally to a limited extent (indic. 0.5–3%).
- **Fixed costs** (refer Sect. 14.3.2.2): In USD/year these costs do not vary with plant production, at least within certain plant production rates (typically $\pm$ 100,000 t/y), examples: labor, maintenance materials, administration, other fixed costs. This is the area on which the economy of scale potentially has the largest impact, i.e., a drop in cost per tonne of alumina produced, due to the "dilution" of "fixed" annual expenses by a larger production volume. This applies particularly to labor and other fixed costs. If the increase in production capacity includes an increase in the number of trains, this positive effect is dampened because not just the size of the equipment involved increases, but also its number. In addition, the requirements of complex and large alumina refineries may result in disproportional increases of overhead costs.

Table 14.21 illustrates the above for a greenfield project evaluated at two different production capacities (1.6 and 3.2 Mt/y). The larger refinery capacity is based mainly on an increased number of production units in several areas such as digestion, resulting in only a limited improvement of the fixed costs per tA. The decrease in variable costs is the result of improvements in operating efficiencies mainly due to operating two production units instead of one.

In another case, a greenfield project increased its design production capacity from initially 2.8 to 3.2 Mt/y without changing the number of production units. This modification represented an increase of equipment size at very similar process conditions. This has two major consequences: virtually unchanged variable operating costs, while fixed costs are diluted. This is illustrated in Table 14.22, which shows a drop in fixed costs (delta fixed costs = 2.8 USD/tA at a production increase of

Table 14.21 Impact of economy of scale on opex—increased no. of operating units

Refinery production capacity, Mt/y[a]	1.6	3.2
Variable costs, USD/tA		
Bauxite	11	11
Energy	61	59
Consumables (caustic soda, lime, etc.)	31	30
Other variable[b] and transport costs	26	25
Total variable	128	125
Fixed costs, USD/tA	35	31
Total operating cost	**163**	**155**

[a] Mt/y = million tonne alumina per year
[b] Flocculants, acids, grinding media, filter cloths, etc

Table 14.22 Impact of economy of scale on opex—increased equipment size

Refinery production capacity, Mt/y	2.8	3.2
Variable costs, USD/tA	126	125
Fixed costs, USD/tA	33.4	30.6
Total operating cost, USD/tA	**159**	**155**

0.4 Mt/y), which, in relative terms, is much larger than the drop in fixed costs of Table 14.21 (4 USD/tA at a production increase of 1.6 Mt/y).

In summary, a significant impact of economy of scale on Opex is on fixed operating costs (expressed per tA), particularly if a production capacity increase is the result of an increase in equipment size rather than equipment number. The impact on variable operating costs is generally small, unless the increased production capacity results in an increase in production trains (as a result of increased operating flexibility and consequently plant on-line time), especially going from one to two trains.

Economy of Scale and Capital Cost

This section discusses the main effects and implications of economies of scale on capital cost ("Capex"), divided in two categories [6, 20]:

1. *Refinery—General Characteristics*

In general, larger equipment, particularly tanks and vessels, are more cost effective per tonne alumina (tA) produced because larger tanks have a smaller surface area over volume ratio than smaller tanks, hence a lower material cost per m^3 stored volume. This effect is sometimes known as the "0.6 factor rule", and potentially represents a significant drop in capital cost per tA (note: this scaling factor may be different for different equipment types and unit operations). Although technological improvements have resulted over time in a general increase in equipment size available for most processing equipment (vessels, tanks, pumps, mills, filters, etc.), there are physical, technical and/or economic limitations to equipment size. In addition, design and/or other considerations may favor in specific cases a larger number of small equipment over a smaller number of large equipment.

When a larger production capacity results in the construction of more production units, plant—"internal"—infrastructure costs (both shared and non-shared) are diluted (e.g., pipe racks, water supply, power distribution), and spare equipment may act as common spare parts. Both aspects result in lower Capex per tA produced. As an illustration: for a refinery with two digestion units/trains, shared facilities (e.g., raw materials handling, general facilities, shared spares, etc.) represent indicatively ~20–25% of its capital cost. Of course, there are limitations: firstly with respect to sharing of spare equipment and secondly capacity increases in plant infrastructure will be required at some stage. The overall effect is a drop in capital cost per tA produced at higher production capacities. A straightforward power factor relationship between these would look like Fig. 14.26.

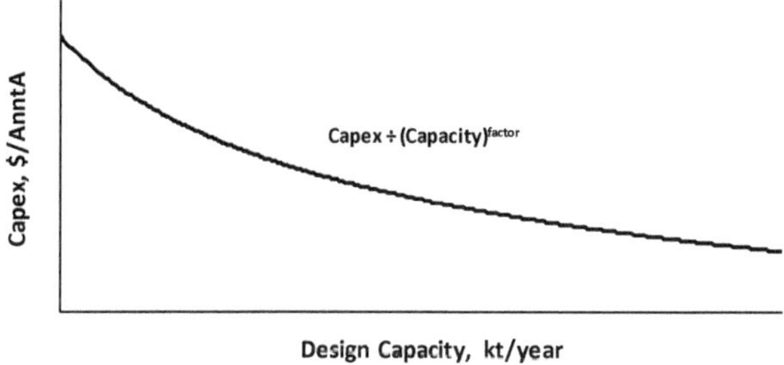

Fig. 14.26 Alumina refinery capex versus design capacity—power factor. *Source* [20]. Copyright © 2011 by The Minerals, Metals & Materials Society. Used with permission

In most cases, however, increases in plant (and thus project) capacity are a combination of increases in equipment size and in equipment number (e.g., as a result of an increase in the number of production trains). In addition, an increased project scope also adds (sometimes disproportionally) to complexity.

As a result, actual capital cost per tA produced may deviate from a smooth curve as shown in Fig. 14.26. Canbäck et al. [22] refer to Bain who found in a study of 20 industries that at the plant level, beyond a minimum optimum scale few additional economies of scale can be exploited.

With respect to the relationship of alumina refinery Capex and plant design capacity, available data suggest that the impact of economy of scale on refinery Capex appears most pronounced for design production capacities up to ~1.5–1.6 Mt/y, with a relatively smaller potential to improve capital cost per tA beyond that point, i.e. a differentiation can be made in two design capacity ranges as illustrated in Fig. 14.27 [20]:

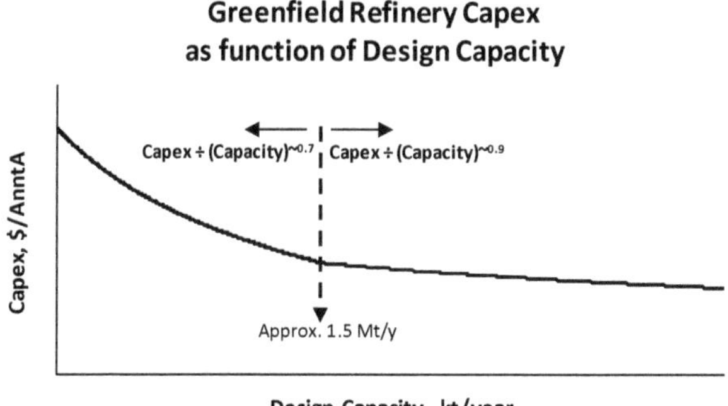

Fig. 14.27 Alumina refinery capex versus design capacity—available data. *Source* [20]. Copyright © 2011 by The Minerals, Metals & Materials Society. Used with permission

- Up to approximately 1.5–1.6 Mt/y: a power factor of ~0.7 (±~0.05).
- Above ~1.5–1.6 Mt/y: a power factor of ~0.9 (±~0.1).

The production capacity of roughly 1.5–1.6 Mt/y is in the range of the current maximum production capacity of ~1.4–1.65 Mt/y for an alumina refinery with one digestion unit/train. This seems reasonable as the digestion area is generally taken as plant bottleneck as mentioned in the "Overview" subdivision of this section. This production capacity might represent the "minimum optimum scale" referred to by Canbäck et al. [22].

In summary, economy of scale effects on alumina refinery capital cost appears most pronounced for design production capacities up to ~1.5–1.6 Mt/y, with a lower potential to improve capital cost per tA at larger capacities. Note that 1.5 Mt/y is not a fixed number, but indicative only (range ~1.4–1.65 Mt/y) and may increase over time as equipment sizes increase.

2. *External Infrastructure Aspects and Overall Economics*

In seeming contrast to the above, several new and future planned projects have design production capacities (well) above the apparent optimum of about 1.5–1.6 Mt/y. An important reason is that greenfield projects have external infrastructural requirements which may include access roads, bridges, railway line, port facilities, personnel housing, etc. (refer Sect. 14.1.2.3 and Table 14.14). In case of extensive infrastructural requirements, the related capital cost is significant and may disproportionally affect the economics of a greenfield project with a relatively small production capacity.

An example illustrates this point for two greenfield project options at the same location: one at a design production capacity of 1.6 Mt/year, the other at 3.2 Mt/year. Assumed infrastructural requirements for this location:

- 150 km roadway between plant site and alumina export port.
- Alumina and raw materials transportation equipment (trucks).
- Port general facilities (jetty/wharf, utilities, general facilities).
- Ship loading/unloading facilities (alumina, raw materials) at the alumina export port.
- Port alumina and raw materials storage and handling facilities.
- Employee housing and living facilities.

Table 14.23 provides indicative numbers for capital, operating and sustaining capital costs for the two options considered in this example and their economics. The table shows that, despite the refinery Capex per annual tA for the two options following the trend illustrated in Fig. 14.27 (within the indicated range), the overall project economics flip from a significant positive NPV (8%), an IRR of 9.7%, a payback period of 8.75 years and a VIR of more than 18% for the 3.2 Mt/year case to a negative NPV (8%), an IRR of 7.9%, a payback period of more than 10 years and a VIR of −1.4% for the 1.6 Mt/year case.

A major contributor to this change of economics between the 3.2 Mt/y and the 1.6 Mt/y project is the disproportional increase in USD/tA of Infrastructure Capex (143 vs. 232 USD/Ann tA). To underpin that: if the Capex scaling factor for the

Table 14.23 Impact of economy of scale on capex and overall economics

Refinery production capacity, Mt/y	1.6	3.2
Capital cost, MUSD[a]		
Mine	124	215
Refinery	1896	3372
Infrastructure	371	456
Total capital cost, MUSD[a]	**2391**	**4043**
USD/Ann tA	1494	1263
Operating cost, USD/tA[b]	**163**	**155**
Sustaining capital, USD/tA[c]	**9**	**9**
Economics[d]		
NPV (8%), MUSD	−29	637
IRR, %	7.9	9.7
Payback period[e]	~10.25	~8.75
VIR/CER[f]	−0.014 (or −1.4%)	0.184 (or 18.4%)

[a] At plant location, December 2016 USD (location factor 1.15—refer Table 14.28)
[b] Refer Table 14.21 (includes infrastructure Opex)
[c] Includes infrastructure Susex
[d] Alumina price 350 USD/tA
[e] After start of operations, in nominal USD
[f] Capital Efficiency Ratio, expressed as NPV (8%)/Discounted Capex, refer Sect. 14.1.1.1

refinery Capex (0.83) would have been applicable to the project overall Capex, the economics of the 1.6 Mt/year project (in that case at a total Capex of 1421 USD/Ann tA) would have looked as follows: NPV (8%) = 58.5 MUSD; IRR = 8.3%; Payback period = ~9.5 years, VIR = 3%.

The reasoning could be turned around: for greenfield projects in locations requiring (extensive) external infrastructure to achieve acceptable economics, a significant (disproportionate) increase in project scale is required. In a similar context, A. Kjar discusses in a paper presented at the TMS 2010 Annual Meeting in general terms the uncompetitive capital cost of recent Western-developed greenfield alumina projects as a result of (among other reasons) large project size and increased project complexity [23].

3. *Implications*

The significant increase in the design/initial capacity of greenfield (bauxite mine and) alumina refinery projects over the past decades as discussed in this section had major consequences:

- Project capital cost has grown in many cases to several billion USD, with project owners reducing risk through project financing and the formation of multi-party joint ventures. Although this is perfectly reasonable it complicates project implementation (e.g., with respect to decision making processes).

- The complexity of these mega projects (projects typically over 1 billion USD) increased significantly, especially w.r.t. project planning and management. Significant infrastructural works are often required, involving extensive government involvement, adding to project complexity.
- Due to the financial commitments involved, globally only a limited number of (very) large companies have the financial and human resources to develop greenfield bauxite and alumina projects.
- For the same reasons (project scope, complexity), only a limited number of engineering and construction companies have the required engineering, construction and project management skills and experience to successfully implement these projects.
- Typically, a project life of 30^+ years is applied to justify the significant investment of a greenfield (Bauxite and) Alumina project. Reason: an alumina refinery can operate effectively for decades (refer above, Table 14.17, rationale of criterion 8). For greenfield Bauxite and Alumina projects with a captive refinery this means that the bauxite deposit on which a project is based should be able to sustain refining operations for such a period of time. Therefore only (very) large bauxite deposits are developed, indicatively 150–300 Mt and above.

Economy of Scale and Capital Cost—Summary

In summary, worldwide only a small number of companies develop mostly very large greenfield Bauxite and Alumina projects, which often take a decade and more to develop.

The question could be asked if the underlying trend, viz. seemingly ever-increasing alumina project design capacities, is inevitable, or if viable alternatives exist. The basic reason for the trend being economics, the question could be reformulated as follows: is it possible to develop smaller greenfield Bauxite and Alumina projects at acceptable economics?

A. Kjar addresses this question and some of the issues discussed above, albeit from a different perspective, in his earlier mentioned paper. He indicates that as a means to overcome some of these issues, attempts were made by others: 1. To gain improved control over the project execution process; and 2. To increase the level of pre-assembly to reduce total costs of on-site construction labor and low productivity—refer also a paper by Valenti and Ho [24]. Mr. Kjar proposes the use of replication of a modern plant design, and small increments of capacity (without quantifying a capacity), in order to quickly and more cost-effectively build a large plant/project.

Although Kjar's paper has a different angle, viz. building a large plant at lower capital cost, there are overlaps with the current subject (investigating the possibility to lower the threshold for the development of—smaller—Bauxite and Alumina projects). This is further explored in Sect. 14.4.4.

14.1.5 Economics of Greenfield and Brownfield Projects

A number of greenfield projects were built in the early 80s, e.g., Aughinish, Alumar, Worsley, Wagerup, Nikolaev, San Cyprian, Ciudad Guayana. However, the growth of the aluminium market, and in its stride the alumina market, did not meet expectations. In addition, these greenfield projects had a huge de-bottlenecking and brownfield expansion potential built into their design. The result was that only a limited number of greenfield projects were required and built in the subsequent two decades (examples: Damanjodi, Pingguo, Alunorte—actual design also early 1980s).

Several of the early 80s projects ended up falling short of the projected economic returns on which basis they were approved. However, it was (and in most cases still is) economically (much) more attractive to develop brownfield than greenfield projects, as illustrated in the following.

Using Table 14.23, and assuming that the capacity increase from 1.6 to 3.2 Mt/y is the result of a brownfield expansion project, the related Capex, Opex, Susex, and economics of this brownfield expansion can be established. The result is shown in Table 14.24.

Table 14.24 Illustration of brownfield economics

Refinery production capacity, Mt/y	Start production capacity 1.6	End production capacity 3.2	Brownfield expansion capacity (delta) 1.6
Capital cost, MUSD			
Mine	124	215	91
Refinery	1896	3372	1475
Infrastructure	371	456	85
Total capital cost, MUSD	**2391**	**4043**	**1651**
USD/Ann tA	1494	1263	1032
Operating cost, USD/tA	**163**	**155**	**148**
Sustaining capital, USD/tA	**9**	**9**	**9**
Economics[a]			
NPV (8%), MUSD	−29	637	666
IRR, %	7.9	9.7	12.1
Payback period[b]	~10.25	~8.75	~7.25
VIR/CER[c]	−0.014 (or −1.4%)	0.184 (or 18.4%)	0.470 (or 47%)

[a] Alumina price 350 USD/tA

[b] After start of operations

[c] Capital Efficiency Ratio, expressed as NPV (8%)/Discounted Capex, refer Sect. 14.1.1.1

Table 14.24 shows that Capex and Opex of the brownfield project are much more attractive than those of either the greenfield 1.6 Mt/y or the greenfield 3.2 Mt/y project, resulting in significantly improved economics as illustrated by its NPV (8%), IRR, and VIR/CEP.

In addition, the greenfield projects of the 80s were conceived as large-capacity projects of which in all cases only one production train/line was initially developed, with a following train not being built at all (e.g., Aughinish, Nikolaev) or many years or even a decade or more later (e.g. Alumar, Worsley, Wagerup). The consequence being that the initial projects had plant layouts with significant provisions for future expansions which were not or only much later exploited.

This meant that the economics of the initial project suffered heavily due to the construction of "internal" infrastructure (e.g., buildings, roads, pipe racks, power distribution, water supply, and alumina and raw materials storage facilities) and external infrastructure (e.g., roads, port facilities, alumina and raw materials transportation facilities) remaining under-used or unused for a long period of time i.e. was not generating a proper return. Refer also to Sect. 14.1.2.3. In the context of operating cost something similar applies: an organizational infrastructure needed to be put in place for a greenfield project (e.g., operations, maintenance, purchasing) while a brownfield project can build upon the existing infrastructure.

Potential options to address the gap in economic attractiveness between greenfield and brownfield projects are discussed in Sect. 14.4.

14.2 Total Investment and Capital Cost

The capital cost ("Capex") of a Bauxite and Alumina project is a key parameter in meeting corporate threshold economic criteria and in many cases may "make or break" a project. Although capital costs of projects tend to be quoted and compared, a consistent basis is often lacking, and "apples" are often compared with "pears". This section addresses elements of capital cost estimates of (Bauxite and) Alumina projects [6].

Like operating cost (refer Sect. 14.3), the Capex of a Bauxite and Alumina project is not a stand-alone item but forms an integral characteristic of a project. It is influenced by the quality of the bauxite resource, country infrastructure, logistics of raw materials import and alumina export, the refinery's alumina production capacity, and the choice of technology and design (refer Sect. 14.3.1.6 for further details).

Capital cost plays a key role in the economics of a Bauxite and Alumina project due to its capital intensity, typically ~1000–1900 USD per annual ton alumina production capacity (USD/Ann tA) for a greenfield project (range quoted by Alumina Limited at its June 2010 presentation—note that the CEPCI plant construction cost index between then and end 2016 are similar). Actual numbers depend on project scope, location, size of expansion, complexity, inclusion of items like owners' costs, etc. Understanding the capital cost of a project and its build-up is therefore essential, especially if comparisons between projects are made, and to assess opportunities to lower the capital cost of a potential project.

14.2.1 Investment Cost Build-up

14.2.1.1 Based on Main Project Scope Items

The capital investment cost build-up of a Bauxite and Alumina project based on its main scope items (mine, refinery, infrastructure) is covered in Sect. 14.1.2.

14.2.1.2 Based on Major Cost Items

The total investment cost of a project (i.e. including mine, refinery and infrastructure) may be broken down as shown in Table 14.25 (numbers consistent with the reference project used in this chapter, refer Table 14.1).

Total installed cost is often referred to, as it is in this chapter, as "Capex". Tax conditions, which are location specific, should be taken into consideration before deciding on the final (optimum) split of the capital costs between the indicated categories.

Table 14.25 Investment cost build-up

Investment cost item	USD/Ann tA
Direct costs	699
Indirect costs	527
Project contingency	175
Total installed cost[a]	**1401**
Escalation	**133**
Financing costs	**7**
Working capital	**60**
Total investment cost	**1601**

[a] Often referred to as "Capex"

Direct Costs

The Direct Costs (sometimes referred to as Direct Field Costs) of the mine, refinery and infrastructure may be split up as shown in Table 14.26. Contractor's mobilization and de-mobilization costs are assumed to be included in the Direct Costs.

Equipment costs cover tanks, process vessels, pumps, agitators, heat exchangers, filters, thickeners, mills, calciners, etc. **Bulk materials** include items such as piping, concrete, structural steel, electrical cables, instrumentation, insulation, valves, etc.

Table 14.26 Typical direct costs build-up

Investment cost item
Direct costs
Equipment
Bulk materials
Concrete
Structural steel
Mechanical
Piping
Electrical
Instruments
Insulation
Paving
Roads
Fencing
Buildings
Construction labor costs
DSC/control systems
Information systems
Direct costs contingency

Source [6].

Indirect Costs

Contractors' cost items such as support and non-productive labor, mobile equipment, craneage (incl. major craneage for heavy lifts), spare parts, logistics, scaffolding, utilities during construction, etc. are considered part of indirect costs or owners' costs. Others may include these items with the direct costs, either explicitly, or in the rates. Owners' costs are also considered part of indirect costs while others may leave them out of installed costs altogether (and thus also out of Capex). The Indirect Costs may be split up as shown in Table 14.27.

Freight costs include handling charges. **Engineering, Procurement and Construction Management** (or alternative contracting) costs may amount to somewhere between ~5 and 20% of the total investment cost, depending on the nature of the project and the engineering/contracting strategy. **Insurances, bonds and guarantees** cover the engineering, construction and warranty maintenance period. **Temporary**

Table 14.27 Typical indirect costs build-up

Investment cost item
Indirect costs
Freight
Engineering, procurement and construction management (EPCM)[a]
Engineering contractor's fee
Commissioning/start-up assistance
Insurances, bonds and guarantees
Temporary construction facilities
Construction equipment/scaffolding
Vendor representatives
Spare parts
Mobile equipment
Utilities during construction
Start-up/commissioning modifications
Outside consultancy fees
NDT and geotechnical testing
Indirect costs contingency
Owners costs

Source [6]. Copyright © 2007 by The Minerals, Metals & Materials Society. Used with permission
[a] Or cost of alternative contracting approach

Construction Facilities include removal and rehabilitation. **Construction equipment** covers items like heavy cranes and trucks and scaffolding. **Mobile equipment** includes items like forklift trucks, pick-up trucks, cranes, trailers, fire engine, front end loaders, etc. **Utilities during construction** cover gas, water, electricity and air.

Indirect Costs may be expressed as percentage of Direct Costs ranging indicatively from ~25 to 100^{+}%, depending on project scope, location, and other project specifics

14.2.2 Contingency and Escalation

Allowances and Contingency

A cost estimate is made with a degree of certainty based on what is known (design criteria, process description, PFD's, P&ID's, equipment lists, general arrangement drawings, etc.). However, there is always a gap between what is known and the expected cost based on previously completed work. Therefore, often allowances are included on quantities, and growth allowances for quotes and expected contractual awards. A quantity allowance on bulk materials may vary between ~3 and 10%, while a growth allowance on bulk materials may vary between ~3 and 5%, both depending on the level of engineering.

Contingencies are included to cover unforeseen (unknown) items of work which will have to be performed, or elements of cost which will be incurred within the defined scope of work of the estimate but that cannot be explicitly foreseen or described at the time the estimate is being prepared because of lack of information. For example the impact of technology risk, risk of contractor insolvency, divergent weather patterns, industrial relations effects, etc. The contingency allowance is thus an integral part of the estimate.

In other words, the contingency is developed by considering all of the factors that could affect the expected final cost of the project, but which have not been able to be quantified as part of the basic estimate. Contingency evaluation is often carried out in combination with project risk analysis and consideration of the high and low range of the estimate. This analysis is generally carried out at two levels.

Firstly a pass is made through the direct and indirect content of the estimate. It is necessary to consider the nature of the scope of each package and the degree of technology risk and design development of each scope item assessed against a range of contingency factors that have been collected from experience in a broad sample of projects. This will generate a package contingency for the project.

The other factors to be considered in the second pass through the contingency analysis are project wide risks that may influence the expected cost outcome of the project. This project wide component of contingency is sometimes termed project contingency to distinguish it from package contingency.

A contingency is not to be considered as a compensating factor for estimating inaccuracy. Nor is it intended to cover items like potential changes in project scope, acts of god, labor disruptions beyond the control of the project manager, currency fluctuations or cost escalation beyond the estimated rates. Depending on the stage of development of the capital cost estimate contingencies may indicatively range from ~5 to 10% (at project execution) to ~15–25% (at concept phase), refer Sect. 14.1.3 and Table 14.16.

Escalation

Escalation is the combined effect of general inflation and market conditions specific to the project (e.g., other planned major projects in the region, the demand for contractors' labor, equipment, materials, etc.). The total impact of inflation on a project depends on the phasing of expenditure during the project. Or putting it differently, the Capex estimate is based on a certain date, while actual expenditures will be made in the future. To get an indication of the actual final expenditure, escalation is added to the Capex estimate.

Overall escalation numbers may vary depending on projected timing and inflation projections. Escalation may typically vary between ~10 and 15% of total Capex.

14.2.3 Contract Types

Different contract types are being used for the implementation of (Bauxite and) Alumina projects [6]. The cost of each of these contracts may differ significantly. However, the cost of the engineering contractor should not be considered in isolation, but in combination with owners' costs and financing costs.

Following are some of the more common ones.

EPCM (Engineering Procurement and Construction Management)

Under EPCM the contractor manages the project on behalf of the project owner. The project risk will therefore largely be borne by the owner, resulting in more efficient project pricing than for EPC.

However, as the risk is largely borne by the owner, he will have to secure financial reserves to meet project risk and contingencies.

EPC (Engineering Procurement and Construction)

Under an EPC contract the contractor is to deliver a complete facility, for a guaranteed price, by a guaranteed date and one that performs to a specific level. If the contractor does not perform, he will face monetary liabilities. In other words, if a problem occurs, the project company will focus on the contractor to fix the problem and provide compensation. The EPC contract provides a single point of responsibility. In other words, in an EPC contract the project company is guaranteed among others a fixed completion date, a fixed completion price, no or limited technology risk, guaranteed output, security from the contractor, which protects the project company if the contractor does not comply with its obligations and restrictions on the ability of the contractor to claim extensions of time and additional costs.

The above factors make it attractive to the project company as it provides a level of certainty for the project owner and its bankers. Plus, the construction risk is borne substantially by the contractor. A drawback is that it allocates most of the construction risk to the contractor, who is then forced to build contingencies into the contract price for unforeseeable events. This leads to higher contract prices than the ones applied for some of the other contractual structures. Another drawback is that the project company has limited ability to intervene when problems occur during construction.

LSTK (Lump Sum Turnkey)

This contract shifts more risk to the contractor: the contractor is paid a predetermined sum for completing a particular stage or the whole contract works. It is a fixed contract price. In case of a change, the sum is not adjusted. The contractor therefore carries the risk of correctly estimating, at the time of contracting, the extent of work required to be carried out. The contractor is selected by competitive bidding. Basic and detail design of the project is the contractor's responsibility.

On the other hand, because the design criteria are given by the owner, problems most frequently arise when the owners' design criteria are open-ended or subject to varying interpretations. Then the question is, who is responsible for the cost and time impact associated with the change.

Cost Reimbursable

Two ways of arranging a contract are by competitive bidding or negotiation. In competitive bidding, contractors bid for the right to execute the project. Instead of competitive bidding the owner may choose to award construction contracts to one specific contractor (e.g. based on its reputation, expertise, etc.).

In a competitive bidding context, lump sum and unit price compensation are mostly used, while cost reimbursement is usually followed in the context of a negotiated contract. In the latter case the contractor is paid his costs including overheads and preliminaries together with a fee which may be either a percentage fee or fixed. This payment mechanism is appropriate when an early start is required but the project lacks sufficient definition to allow the other two payment mechanisms to be adopted. Usually, the owner perceives that he has the necessary construction expertise to make the major decisions relating to method of construction and only requires the contractor's resourcing skills.

By choosing a cost reimbursable contract, the owner is facing two major issues:

- Difficulty in determining how much the project will cost, and
- A lack of incentives for a builder to keep costs down.

A solution is to create a legal relationship which requires the contractor to notify the owner of a possible cost overrun. In the US, the law expects each party to exercise good faith and fair dealing.

Another approach is to share the risk of a cost overrun between the owner and the contactor. This is achieved by using target cost contracts or contracts with a maximum guaranteed price.

The target cost contract specifies a penalty or a reward to a contractor depending on whether the actual cost is greater or lower than the contractors' estimated direct cost.

A maximum guaranteed price arrangement imposes a penalty on a contractor for cost overruns and failure to complete the project on time. Amounts below the maximum are shared between the owner and the contractor. The contractor is responsible for costs above the maximum.

Unit Price or Re-measurement Contracts

This type of contract is used when the design responsibility lies with the owner and the design is to be completed during construction. The advantages for the owner are that by adopting overlapping phases of design and construction, construction can start early.

The overlap of design and construction means that variations to the original outline scheme are likely, but the unit price mechanism provides a ready means of pricing the variations.

A drawback is that the contractor is paid a price or rate for each unit element of work carried out and identified at the time of contracting. Usually, it is done on a monthly basis, based on a monthly valuation. So the unit price payment mechanism thus places the risk with the owner, for estimating the extent of the works at the time of tender.

14.2.4 Financing Structure and Costs

Historically industrial projects such as (Bauxite and) Alumina projects have been promoted by large industrial corporations and financed through the traditional corporate finance scheme: equity allocation along with corporate debt, issued with securities over existing and cash-generating assets.

Several factors result in modifications to the existing scheme and point to a different financial approach: Project Financing. Some confusion exists on its definition as it is sometimes thought to be a kind of corporate financing applied to large industrial undertakings. However, project finance, also referred to as non-recourse or limited-recourse finance, can be defined as the raising of debt aimed at financing an asset with security on the asset and its cash-generating capacity rather than on the sponsor of the project. This means that the lenders will have no or very little recourse against the project owner and thus have to ensure through a due diligence that the project provides satisfactory and low-risk future revenues.

The four main characteristics of project finance over corporate finance are:

1. Risk reduction for the owner (risk being borne by the lenders will be significantly higher, which is reflected in an increased interest rate);
2. Increased leverage (measure of the amount of debt raised to finance a firm's asset, traditionally to a debt-to-total capitalization of 50%; project finance enables greater leverage, on average around 70%, greatly increasing the expected return on capital investment, but also increasing the equity risk, and therefore the interest rate).
3. Legal separation (project finance involves the creation of a legally independent entity—the Project Company—that will finance, develop, construct, own and operate the project.; this is mainly aimed at facilitating the securitization of the project assets), and
4. Contracting structure.

Figure 14.28 provides the standard contracting framework for a project-financed deal.

Although project financing has been applied for more than a century, this approach has recently been gaining broad popularity. Total project-financed investment has increased from 10 billion USD per year in the late 1980s to more than 220 billion USD in 2001. Project finance has a wide range of applications, e.g. power plants, industrial plants, refineries, pipelines, infrastructure, etc.

The reason for this success in many industrial areas is the possibility to develop massive projects with limited capital resources and minimum risk for the sponsor. Besides, project companies, like joint ventures, enable the participation of many

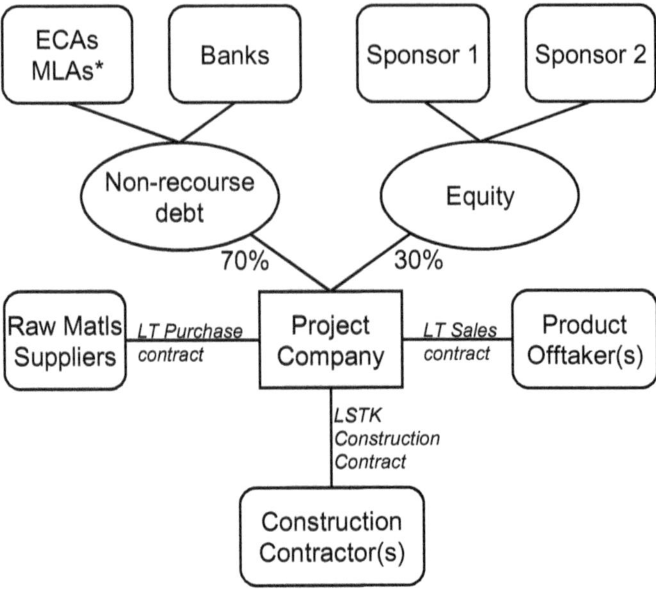

Fig. 14.28 Project finance deal structure (example). *Source* [6]. ECA = Export Credit Agencies; MLA = Multi-lateral and Bi-lateral development finance institutions; LT = Long Term; LSTK = Lump Sum Turnkey. Copyright © 2007 by The Minerals, Metals & Materials Society. Used with permission

industrial and financing partners. Finally, the possibility for complex financing schemes involving several institutions allows the funding of large sums. Commonly found financiers in the project finance sector are commercial banks, export credit agencies that are providers of government-backed loans, guarantees and insurance aimed at supporting their home country's exports and agencies in which more than two nations participate (multi-lateral agencies such as the World Bank and regional Banks for Reconstruction and Development).

Relatively small (Bauxite and) Alumina projects with a unique sponsor and developed in low-risk countries will most likely be financed under a standard corporate finance structure. However, when the political risk, the size of the project or the number of sponsors increase, the project finance approach becomes more relevant.

Financing costs such as interest during construction, costs incurred in contracting loans, taxes, etc. are not included in the scope of this section. As outlined above, they will depend on the project financing construction, the country involved, etc.

14.2.5 Owners' Costs

Owners' costs comprise those activities which are the responsibility of the project owner when implementing a project from the beginning of the study phase to the production of the first product. From then on project costs become the responsibility of operations. They may be categorized in pre and post investment decision as follows.

- **Pre-investment decision**: costs made before the project proposal has been approved. In other words these costs are "sunk costs" by the time the investment is approved (e.g. costs of studies and sampling and testing costs). Although they represent part of the costs to a company to develop a project they are normally not included in the economic evaluation.
- **Post-investment decision**: these costs are part of the investment to be approved and should be included in the economic evaluation of the project.

Owners' costs cover a wide scope. The following provides an outline of components of owners' costs.

- **Staff**: project staff (management, technical support—incl. engineering consultants—and administration and support services), and staff involved with pre-commissioning, commissioning and start-up (operations and maintenance).
- **Pre-Operational Expenses**: recruitment costs, operations training, salaries before commissioning are started, start-up team, first precipitation charge with purchased seed hydrate and excess raw material consumption during initial operation.
- **Indirect staff costs**: travel, accommodation, expenses, training, etc.
- **Office costs**: project office, site office, vehicles, equipment.
- **Insurances**: public liability, material and construction, professional indemnity, vehicles.
- **Fees**: technology, council, state, federal.
- **Bonds and licenses**: e.g. for construction, start-up, reclamation, temporary facilities.
- **Legal**: costs made in the context of putting together construction agreements, agreements for land access and land acquisition, operating agreements, etc.
- **Overheads**: items like advertising, community relations, auditing, soil and geotechnical tests, temporary power usage, marine and land surveys, exchange rate variation allowance, taxes and duties, corporate charges, etc.
- **Land and rights of way**: Cost of land acquisition is sometimes included in owners' costs, while in other cases it is included as part of capital cost.
- **Geology, Mining, Process** covers activities like drilling, exploration, (bulk) sampling, laboratory test work, trials and licenses.

In this chapter on "Economics of smelter grade alumina projects", owners' costs are included in Indirect Costs, and are thus included in project Capex. However, if owners' costs are not capitalized, they should not be included in project Capex.

14.2.6 Working Capital

Working capital includes alumina and hydrate, both in process and in product, raw materials (caustic soda in solution and in stock, heavy fuel oil in stock, bauxite on stockpiles, lime, etc.), warehouse inventories (incl. capital/insurance spares, process and operating supplies), accounts payable/receivable and cash.

14.2.7 General Capex Aspects

The following general Capex aspects are covered elsewhere in this chapter:

- **Capex, NPV, and IRR**: The impact of changes in Capex on NPV and IRR in Sect. 14.1.1.2.
- **Exchangeability of Opex and Capex**: Sect. 14.1.1.3.
- **Economic robustness of Structural Low Capex Projects**: Sect. 14.1.1.4.

14.2.8 Capex Improvement Opportunities

Potential opportunities to lower the Capex of a Bauxite and Alumina project are discussed in Sect. 14.4.

14.2.9 Reference Project Capex Breakdown

Breakdown of the Capex of the reference project used throughout this chapter on "Economics of smelter grade alumina projects" (refer Sect. 14.1.1.1 and Table 14.1) is shown in Table 14.28. Note that the Capex of (Bauxite and) Alumina projects may vary widely (refer Sect. 14.1.1.2, paragraph below Fig. 14.7).

14.3 Operating Cost

Note: cash operating cost ("Opex") as used throughout this chapter covers all on site production costs ("on site" meaning mining, refinery, residue disposal, and all related infrastructure sites), and includes consumables and alumina handling and transport, and port costs, but excludes financing costs (interest, amortization), head office costs, etc. Opex is often regarded as a burden needed to be dealt with or at best as an unavoidable sometimes even bureaucratic management requirement. From a more positive perspective however the Opex and its underlying rationale provides tools to assess the "health" of a project and they may offer opportunities for improvement. This section provides an insight in facets and issues related to the Opex of a (Bauxite and) Alumina project [2]. As discussed in Sect. 14.1.1.1, the operating cost represents a key input on the cost side of the annual cashflows used in the calculation of economic evaluation criteria such as NPV, IRR, etc. In other words, Opex is an important element in the economics of a project, including (Bauxite and) Alumina projects.

Table 14.28 Capex breakdown of 1.6 Mt/y Aa greenfield reference project with limited infrastructure requirements (refer Table 14.1)

Bauxite and alumina project capital cost (End 2016, accuracy approximately −20% to +50% on grand total capex)	Capex	
	MUSD	USD/tA
Mine	**124**	**77**
Refinery		
Materials handling and storage (bauxite, caustic soda, fuel oil, lime, alumina, etc.)	62.1	
Process facilities (bauxite grinding, digestion, etc.)	374.4	
Steam and power generation (incl. sub stations)	212.4	
General facilities (buildings, pipe racks, residue disposal area, etc.)	173.6	
Total direct capital cost	**823**	
Indirect capital cost (EPCM, freight, etc.)	**621**	
Contingency (@25% of direct capex)	**206**	
Total installed cost, basis NW Europe/US Gulf (location factor 1.0)[a]	**1649**	
Total installed cost, basis project site (location factor assumed at 1.15)[a]	**1896**	**1185**
Infrastructure[b]		
Roads and bridges, Aa and raw materials transport	90	
Port facilities (excluding jetty/wharf)	102	
Personnel housing, medical facilities, school, shops, etc	29	
Total installed cost, basis project site	**221**	**138**
Grand total capex	**2241**	**1401**

[a] A location factor accounts for the difference in costs between the case of bringing the project online in NW Europe/US Gulf with the additional costs associated to the actual project location: additional costs associated with availability of qualified personnel, materials (e.g. concrete and steel), and logistics

[b] Numbers at project site; including Direct and Indirect Capex, and Contingency @ 20% of Direct Capex

14.3.1 *Opex Basics*

Conventionally project operating cost may be split in different ways, e.g., "controllable" and "uncontrollable", fixed and variable. This serves a purpose for an existing bauxite/alumina project. However, when considering a greenfield project it makes good economic sense, as outlined in Sect. 14.1.1.4, to focus on identifying projects with structurally low Opex. We therefore firstly review the basics of operating cost.

The Opex of a bauxite/alumina refinery project is not a stand-alone item but is an integral characteristic of the context of the project as it is influenced by:

- Resource quality
- Country infrastructure
- Logistics of raw materials import and alumina export
- Alumina production capacity
- Technology.

(Note: commodity prices such as caustic soda, coal, fuel oil, lime, etc. are not included in the list above because FOB commodity prices are assumed to be predominantly the same for all projects).

Simply put, Opex reflects the efforts required to deal with the "imperfections" of the bauxite resource in the widest sense when converting it to smelter grade alumina with the selected technology, at the desired quality, in a responsible way with respect to safety and environment.

14.3.1.1 Bauxite Resource Quality

Resource quality includes (refer also Sect. 14.1.4.1 for further details):

- **Location** and accessibility of the resource with respect to the port from which the alumina is exported and most of the raw materials are (often) imported (distance, height, mountain and river crossings, etc.). This is reflected in a transportation cost (either of the raw materials to and alumina from the refinery site at the resource or of the bauxite to the refinery at the exporting port). Depending on resource specifics this transportation cost may indicatively range from ~5 to 10^{+} USD/tA.
- **Deposit characteristics** (uniform versus pockets, overburden thickness, bauxite horizon thickness, beneficiation requirement), affecting costs of mining, crushing, storage, beneficiation, rehabilitation, etc. This aspect is reflected in the overall USD/tBx cost CIF refinery and may indicatively range from ~10 to 40^{+} USD/tA (bauxite transportation may either be included—e.g., if the mine is close to the refinery, or be covered by a separate transportation cost referred to above).
- **Bauxite** refinery feed **characteristics** (hardness, available alumina—level and % gibbsite/boehmite, reactive silica, impurities, etc.).

  These aspects are mainly reflected in caustic soda cost (partly) and energy cost (partly) and may indicatively range from ~15 to 65^{+} USD/tA.

Note that if an alumina refinery project does not participate on an equity basis in a (captive) bauxite mine, the first two aspects mentioned above do not apply. In that case a market price will have to be paid for the bauxite, which will be more than the sum of the costs mentioned above for these two elements. For reasons of economics it is unlikely that new greenfield alumina refinery projects outside China would be based on export bauxite. The reason being that China, besides considering economics, takes "strategic" considerations into account.

14.3.1.2 Country Infrastructure

In the context of this paper, country infrastructure includes:

- **A "hardware" aspect**. This covers items such as the presence and state of repair of a port (via which raw materials are imported and alumina exported), railroad (if the deposit is a significant distance from the port), a town site (incl. hospital), roads, and water availability. This aspect is reflected in infrastructure operating and maintenance costs.
- **A "software" aspect**. This refers to elements such as royalty (on bauxite or alumina), levies, duty on imported raw materials, and the impact of legislation (e.g., with respect to environmental requirements).

  Taxes (and tax holidays) are very important, but more so from an overall economics perspective than from the point of view of the Opex per se.

Total infrastructure related costs may indicatively range from ~2 to 12^+ USD/tA.

14.3.1.3 Logistics of Raw Materials Import and Alumina Export

Accessibility of and logistics to the port via which raw materials are imported (e.g., caustic soda, coal, fuel oil, lime) and alumina (or bauxite) is exported. This item is affected by location of the port (e.g., industrial area), ship size (water depth) and proximity to frequently used sea lanes, and it is reflected in the cost of raw materials CIF the importing port and/or in freight charges for export alumina (or bauxite). In other words, it is affected by refinery location in case the refinery is built in the same country as the bauxite resource. Total logistics costs cover a wide range, indicatively ~5–45^+ USD/tA.

14.3.1.4 Alumina Production Capacity

Refinery production capacity has an impact on Opex because it "dilutes" fixed operating and maintenance costs, may result in efficiency improvements, and it affects capital cost (refer Sect. 14.1.4.2 "Economy of Scale"). A larger production capacity generally results in more attractive overall project economics.

Deposit size is an important aspect in this context. In order to capture the Opex (and Capex) advantage of a large production facility, a bauxite resource should be able to support an alumina refinery for its lifetime (typically 50^+ years).

However, a project's initial and final production capacities are also the result of a company's view on alumina demand and supply developments, access to capital, and to a lesser extent selected process technologies.

14.3.1.5 Technology and Design

In terms of operating cost, "technology and design" relates to elements such as plant design and layout, process and equipment technologies and operating and maintenance philosophies. These aspects are reflected in:

- **Energy and raw materials costs** (caustic soda, lime, coal, fuel oil, etc.)—albeit partly in some cases. These are affected by technology choices such as co-generation of steam and power, choice of energy carrier (coal/fuel oil/gas), digestion, residue circuit thickener/washer, precipitation, residue disposal, heat recovery, return of disposal run-off water etc.
- **Other variable costs** (crystal growth modifier, filter cloth, acids, anti-foam, grinding media, etc.).
- **Maintenance materials** and contract services costs, affected by the complexity of the refining plant, choice of layout, outsourcing strategies, etc.
- **Labor and other fixed costs**: similar to maintenance materials, affected by technology and layout choices, outsourcing strategies, etc.

The sum of the costs related to capacity (Sect. 14.3.1.4), and technology and design (this section) may indicatively range from ~60 to 145^+ USD/tA.

14.3.1.6 Summary

Table 14.29 summarizes Sects. 14.3.1.1–14.3.1.5 with indicative ranges of these five elements (numbers include sustaining Capex).

Table 14.29 Ranges of opex elements (indic.)

Element	Range[a] (indic.) USD/tA
Resource quality	$35–115^+$
Country infrastructure	$2–12^+$
Logistics	$5–45^+$
Capacity and technology and design	$60–145^+$
Total opex incl. sustaining capex	**$110–310^+$**

[a] Including sustaining Capex

Table 14.30 Opex ranges of resource quality and capacity and technology (indic.)

Element	Range[a] (indic.) USD/tA
Resource quality	45–170$^+$
Capacity and technology and design	60–145$^+$

[a] Including sustaining Capex

If country infrastructure aspects of the bauxite resource and the location of the exporting port are also considered part of resource quality in its widest sense, the indicative ranges of "Resource Quality" and "Capacity and Technology and Design" costs are as shown in Table 14.30 (incl. sustaining Capex). Table 14.30 shows that the cost range of "Resource Quality" is even wider than that of "Capacity and Technology and Design". Taking the capacity effect out (which may range from ~2 to 15 USD/tA) would narrow the latter range perhaps to ~65–135$^+$ USD/tA.

In other words, "Bauxite Resource Quality" in its widest sense has an as profound or more profound impact on operating cost than technology. Or putting it differently, for a greenfield project it makes good economic sense to primarily focus on identifying the "right" bauxite resource.

This may be done by evaluating bauxite resources on the basis of a set of quality criteria taking the above elements into account. As outlined in Sect. 14.1.4.1 such a set of bauxite resource selection criteria should focus on main criteria, should provide target threshold values, while it is not meant to be applied rigidly. In other words a bauxite resource not meeting one (or possible more) of the threshold values should not necessarily be discarded. The overall result of the evaluation should be considered.

Criteria relating to other than economic considerations (e.g., "strategic" aspects) which could result in a different outcome of a resource ranking exercise, have not been included because those fall outside the scope of this chapter.

Table 14.17 in Sect. 14.1.4.1 presents a set of bauxite resource selection criteria addressing the resource quality elements (Sects. 14.3.1.1–14.3.1.3) and the alumina production capacity element (Sect. 14.3.1.4) discussed above. These criteria address both Opex as well as Capex related aspects as the two are linked. This set of selection criteria may be used as ranking tool for bauxite resource evaluation purposes. The rationale behind these quality criteria are also discussed in Sect. 14.1.4.1.

In summary, for greenfield bauxite/alumina projects the focus should primarily be on identifying the "right" bauxite resource using a limited set of appropriate resource quality criteria.

14.3.2 Opex Build-up

Project Opex may be broken down in the following main components: variable costs, fixed costs and sustaining capital.

14.3.2.1 Variable Costs

In USD/year these costs vary with plant production, at least within certain plant production rates (indicatively ±10–15%). Variable costs comprise items such as bauxite, energy, caustic soda, other raw materials (e.g., lime, flocculants) and process and operating supplies (e.g., crystal growth modifier, grinding media, filter cloth), and they may range from indicatively ~75–215 USD/tA.

Transportation and handling costs of raw materials and product alumina are sometimes accounted for in the raw materials costs (i.e., raw materials costs CIF plant site), but sometimes presented by a separate variable cost item "materials transportation and handling cost".

A significant part of the variable costs is directly or indirectly the result of "Bauxite Resource Quality" in its widest sense (see previous section). They are therefore difficult to improve once the bauxite quality is selected and technology choices have been made for the various processing steps of the alumina refinery.

In other words, structurally improving the variable cost of an existing alumina refinery processing a given bauxite type requires an adaptation or complete change of technology of one or more of the processing steps (e.g. deep thickening technology instead of conventional thickeners, seed recycle instead of solids retention in precipitation).

This also means that the scope of significantly improving the variable costs of an existing refinery without spending significant Capex is generally limited. A good opportunity is sometimes presented when a brownfield capacity expansion project of an existing refinery is considered: it may be feasible to implement new technologies of the brownfield project also in existing units, thus more than proportionally improving project economics.

14.3.2.2 Fixed Costs

In USD/year these costs do not vary with plant production, at least within certain plant production rates (indicatively ±100,000 t/y). They may range indicatively from ~25 to 85 USD/tA. Fixed costs comprise labor, plant maintenance materials, contract services, infrastructure operating and maintenance costs, overheads and other fixed costs.

Plant design and layout affect fixed costs to some extent through simplicity of design, distance to get to facilities, central control room, etc. At the same time plant management controls fixed costs by their choice of operations management systems, number of shifts, maintenance procedures, outsourcing etc.

As mentioned earlier (Sect. 14.1.4.2, "Economy of Scale and Operating Cost") another way of improving fixed costs expressed as USD/tA is "dilution" through increased plant production. In fact, the "phenomenon" of a long term decrease of plant Opex is often to a large extent the result of plant production "creep" (the small but steady increase of plant production on a long term basis).

14.3.2.3 Sustaining Capital

Sustaining capital cost/expenditure, abbreviated as sustaining Capex or Susex, is a project cost item which is easily overlooked and sometimes ignored. To many it is unknown, and it may be (partly) hidden in other operating cost items. The relationship—and the difference—with repair and maintenance costs is not always clear and may inadvertently lead to incomplete cost estimates.

This section explores several aspects of sustaining capital and aims at clarifying some of these Susex issues for alumina refinery projects [25].

Sustaining Capital—General

Susex in general refers to expenditure related to capital asset additions, replacements or improvements required to maintain/sustain existing assets. Required investments covered under Susex contribute to ensuring that the operation's overall reliability is maintained at the existing production level and that all (existing and new) reliability, legislative, regulatory, environmental and safety requirements are met. Considered from an economic perspective: Susex does not generate additional revenues.

Drivers of Susex include:

- Reliability and risk mitigation which includes but is not limited to life extensions, end of life replacements, and system reinforcements (designed and implemented to meet existing and new requirements)
- Regulatory compliance
- Health, Safety and Environment compliance and improvements
- Social requirements
- Product market requirements.

Alumina Refinery Projects

Focusing on alumina refinery projects, Susex is the ongoing capital expenditure required to avoid the eventual discontinuation of the operation, to maintain its production level, its product at target quality and to maintain Health Safety and Environment (HSE) on target/in compliance. It covers items such as:

- Replacing worn out equipment, and major equipment repair
- Developing bauxite residue ("red mud") disposal areas
- Repair of facility structures
- Investments to meet (existing and new) HSE requirements
- Investments to meet (existing and new) alumina product quality standards
- Investments to meet social obligations.

Often an alumina refinery project includes other project elements such as a Bauxite mine and (country) Infrastructure (port, railway, roads, personnel housing, etc.). In that case most of the above applies also to the additional project elements, and supplementary Susex items may include (but are not necessarily limited to):

- Mine mobile equipment replacement

- Mine fleet expansion to maintain plant throughput capacity (if applicable)
- Resettlement costs of local communities (if applicable)
- Port dredging (if applicable)
- Railway rolling stock and locomotive replacement (if applicable).

Operation Closure Costs

The development of a greenfield alumina refinery project may require inclusion of an allowance in the financial model covering the costs of closure of the operation, and the costs of rehabilitation of the affected mine lease and refinery areas at the end of project life. Mine area rehabilitation is normally a progressive process, but final closure of the operational residue area and the refinery site itself will require expenditure at the end of project life.

After operation closure, port and housing facilities and any other functioning infrastructure are sometimes handed over to the country's government without further expense. Operation closure costs are not included in this section's Susex cost.

Difference with Repair and Maintenance Costs

The principle difference between Susex and Repair and Maintenance (R&M) costs (including maintenance materials and contract services) which form part of the alumina refinery cash operating cost (Opex), is that Susex is depreciated (spread) over a period of time while R&M costs are expensed i.e. form part of plant Opex (refer Sects. 14.3.1.5 and 14.3.2.2 above). However, when comparing Susex between different projects, elements of Repair and Maintenance costs should be considered as well because operating plants do not always use the same basis for these two cost items (see below).

Sustaining Capital Cost Values

The actual sustaining capital cost of a project depends on aspects such as its location, age, technologies used, operational and maintenance methods, infrastructure, etc. It should also be remembered that due to its nature actual Susex values may vary (significantly) from year to year, oscillating around a long-term average.

The Susex values in Table 14.31 are based on available global historical data and include typical Bauxite Mine and Alumina Refinery Susex values and ranges

Table 14.31 Susex ranges and values

Susex element	Indicative range USD/tA	Typical value USD/tA
Bauxite mine	~2.5–5.5	~3.5
Infrastructure	~1.5–3.5	~2.5
Alumina refinery	~6–12	~8
Total	~10–21	~14

Source [25]. Copyright © 2016 by The Minerals, Metals & Materials Society. Used with permission (both for low and high temperature digestion plants) of existing bauxite and alumina projects, and Infrastructure values based on greenfield project data.

Figure 14.29 provides a graphical illustration of Table 14.31.

Calculated/Derived Alumina Refinery Susex Values

The approach used in this section to reconcile an alumina refinery's Susex is based on the underlying Capex build-up and the lifetime of its components as shown in Table 14.32.

Table 14.32 illustrates that based on typical values for refinery Capex and the lifetime of equipment and structures, a typical annual long-term alumina refinery Susex valuc may be derived of ~8.7 USD/tA which is consistent with the actual typical value shown in Table 14.31 (~8 USD/tA).

The indicative Susex range derived this way of 7.5–10.5 USD/tA, although narrower than the actual range of 6–12 USD/tA (refer Table 14.31), is consistent as it falls entirely within the range of Table 14.31.

Sustaining Capital—Industry Majors

To provide a context for the above numbers, Susex values in relation to Properties, Plant and Equipment (PP&E) of Alcoa, Hydro, and Rio Tinto have been compared for the years 2013 and 2014 based on their respective annual reports [26, 27] and [28]. The results are shown in Table 14.33. Figure 14.30 provides a graphical illustration of Table 14.33.

Land and land rights (Alcoa, Hydro) and Mining properties and leases (Rio Tinto) have been excluded from PP&E in this comparison under the assumption that no sustaining capital has been spent on these items. Construction Work-in-Progress, respectively Capital Works in Progress (WIP, Alcoa, respectively Rio Tinto) and Plant under Construction (PUC, Hydro) are separately indicated. These are excluded

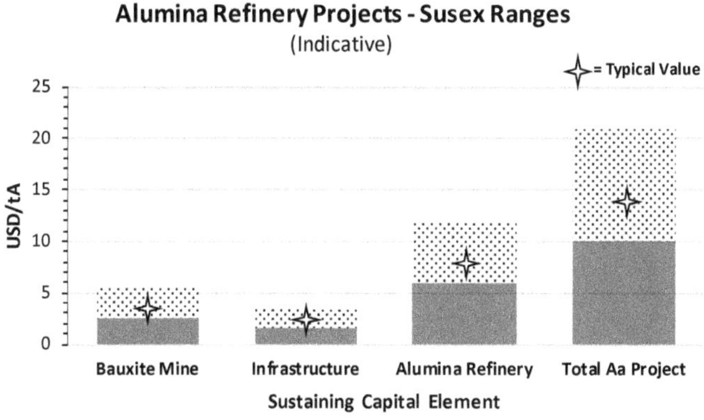

Fig. 14.29 Alumina refinery projects—susex range of project elements. *Source* [25]. Copyright © 2016 by The Minerals, Metals & Materials Society. Used with permission

Table 14.32 Alumina refinery Susex build-up

Alumina refinery	Typical greenfield capex values, USD/Annual tA		Lifetime, Years		Calculated annual long term Susex				
					Greenfield project basis				Mature portfolio basis[d]
Capex element	Indicative range	Typ. value	Indic. range[e]	Typ	Indicative range, USD/tA	Typ., USD/tA	% of direct/total capex (Range)	% of direct/total capex (Typ.)	% of direct, total capex (Range)
Equipment		150	40–60	50	2.5–4	3			
Commodities[a]/structures		370	60–70	65	5–6.5	5.7			
Direct capex		520			7.5–10.5	8.7	1.4–2	1.7	2.8–4
Indirect capex [b]		480							
Total capex	850–1400[c]	1000					0.8–1.1	0.9	1.6–2.2

Source [25]. Copyright © 2016 by The Minerals, Metals & Materials Society. Used with permission

[a] Structural Steel, concrete, piping, wire and cable, raceway (cable trays), instrumentation, electrical equipment, earthworks, etc

[b] Freight, EPCM, temporary construction support, commissioning, insurance, owners' engineering, etc

[c] Dependent on production capacity, location, bauxite quality, technologies, etc

[d] Project portfolio with (Bauxite and) Alumina projects which on average are halfway their expected lifetime

[e] Refer for instance [26]

Table 14.33 Years 2013 and 2014 sustaining capex of Alcoa, Hydro and Rio Tinto in relation to their properties, plant and equipment

Numbers based on 2013 and 2014 annual reports	Alcoa				Hydro				Rio Tinto			
	2014		2013		2014		2013		2014		2013	
	MUSD	As % of cost	MUSD	As % of cost	MNOK	As % of cost	MNOK	As % of cost	MUSD	As % of cost	MUSD	As % of cost
Properties, plant, and equipment[a], at cost[b] (I axis)	33503	100%	34660	100%	102856	100%	96199	100%	80680	100%	79757	100%
Accumulated depreciation, depletion, and amortization	−19091	−57%	−19227	−55%	−50757	−49%	−47274	−49%	−33785	−42%	−33881	−42%
Sub total	14412	43%	15433	45%	52099	51%	48925	51%	46895	58%	45876	58%
Constr. work in progress (WIP)/plant under construction (PUC)	1466	4%	1567	5%	2687	3%	3020	3%	9885	12%	14071	18%
Properties, plant, and equipment, net (I axis)	15878	47%	17000	49%	54786	53%	51945	54%	56780	70%	59947	75%

(continued)

Table 14.33 (continued)

Numbers based on 2013 and 2014 annual reports	Alcoa				Hydro				Rio Tinto			
	2014		2013		2014		2013		2014		2013	
	MUSD	As % of cost	MUSD	As % of cost	MNOK	As % of cost	MNOK	As % of cost	MUSD	As % of cost	MUSD	As % of cost
Sustaining capital (r axis)	735		770		3300		2700		2700		3000	
Sustaining capex as % of "PP&E, at cost" value, excl. WIP/PUC	2.2%		2.2%		3.2%		2.8%		3.3%		3.8%	
Sustaining capex as % of "PP&E, Net" value, excl. WIP/PUC	5.1%		5.0%		6.3%		5.5%		5.8%		6.5%	

Source [25]. Copyright © 2016 by The Minerals, Metals & Materials Society. Used with permission

[a] PP&E

[b] Excl. Land and Mining Properties and Leases

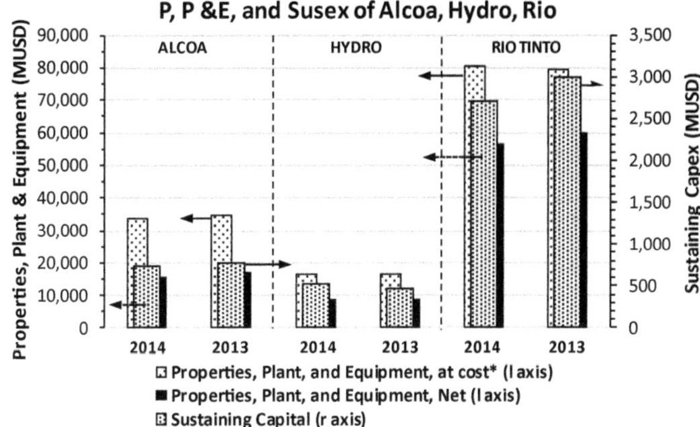

Fig. 14.30 Properties, plant and equipment, and Susex of Alcoa, Hydro and Rio Tinto (2013 and 2014). *Source* [25]. Copyright © 2016 by The Minerals, Metals & Materials Society. Used with permission
* Excluding Land and Mining Properties & Leases)

in the calculation of Susex as percentage of PP&E under the assumption that they have not (yet) required sustaining capital expenditure.

When interpreting the numbers of Table 14.33 and Fig. 14.30, it should be realized that Alcoa, Hydro and Rio Tinto differ significantly in many respects such as scale (2014 Revenue: Alcoa ~ 24 BUSD; Hydro ~ 12.5 BUSD; Rio Tinto ~ 48 BUSD) and business focus (Alcoa 2014, an integrated aluminium company with significant downstream activities: Alumina + Primary Metal ~ 43% of Revenue; Hydro 2014, an integrated aluminium company with limited downstream and other activities: Bauxite and Alumina + Primary Metal + Metal Markets ~ 69% of Revenue; Rio Tinto 2014, a major mining company (iron ore, copper, diamonds) with integrated aluminium activities: Rio Tinto Alcan ~ 25% of Revenue). Refer to Appendix 3 for further details.

The differences between these industry majors are also reflected in the range of Susex as percentage of PP&E on "at cost" basis: 2.2% (Alcoa 2013) to 3.8% (Rio Tinto 2013) and to a lesser extent on "net of depreciation" basis: 5.0% (Alcoa 2013) to 6.5% (Rio Tinto 2013). Table 14.34 provides a comparison of all of the results, and Fig. 14.31 provides a graphical illustration of some elements of Table 14.34.

Noticeable from Table 14.34 and Fig. 14.31 is that Susex as percentage of PP&E (i.e. Total Capex), both on "at cost" ("greenfield") as well as on "net of depreciation" ("mature portfolio") basis is a factor 2–4 times larger as reported in the annual reports of Alcoa, Hydro, and Rio Tinto than it is as derived above and shown in Table 14.32. Unfortunately, due to the availability of only limited detail information on Susex provided in annual reports and company presentations in general, and on alumina refinery operations in particular, it is difficult to reconcile this difference.

Table 14.34 Comparison of Susex values

Alumina refinery sustaining capex					
Source	Indic. range USD/tA	Typ. value USD/tA	% of total capex (typical)	As % of total capex on greenfield/"at cost" Basis (range)	As % of total capex on mature[b]/net of deprec Basis (range)
Actual (Table 14.31)	~6–12	~8			
Calculated (Table 14.32)	~7.5–10.5	~8.7	0.9	~0.8–1.1	~1.6–2.2
Industry majors[a] 2013–2014 (Table 14.33)				2.2–3.8	5.0–6.5

Source [25]. Copyright © 2016 by The Minerals, Metals & Materials Society. Used with permission
[a] Note that sustaining Capex percentages of industry majors relate to their overall portfolio of operations, not just alumina refinery projects
[b] Project portfolio with (Bauxite and) Alumina projects which on average are halfway their expected lifetime

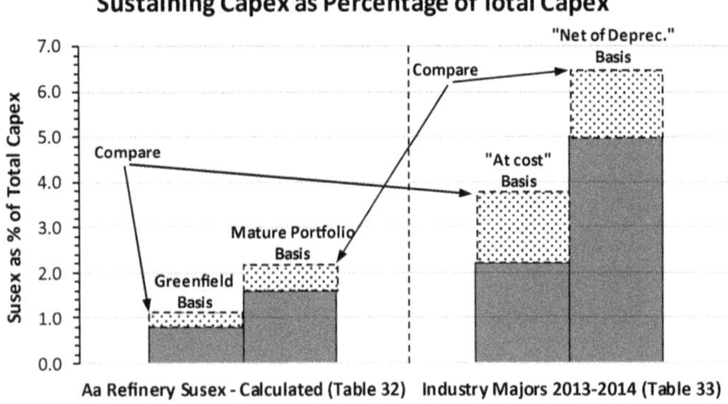

Fig. 14.31 Sustaining capex as percentage of total capex. *Source* [25]. Copyright © 2016 by The Minerals, Metals & Materials Society. Used with permission

Relationship between Susex and Maintenance Materials and Contract Services

The following consideration may explain at least part of the delta found above between Susex as percentage of PP&E (= Total Capex) as reported by the industry majors, and actual and calculated Susex values (Table 14.34).

To properly compare Susex values between projects, costs of "Maintenance Supplies/Materials" and "Contract Services" of an operation's operating cost should also be considered because some operations account for some Susex cost elements (as described in the "*General*" and "*Alumina Refinery Projects*" sections above) under these headings rather than under Susex. The same could well apply to the collection of operations reported on in a company's annual report.

Available information of existing operating alumina refinery projects (both low and high temperature digestion) indicates a range of ~18–30 USD/tA for the total of [Susex + Maintenance Materials + Contract Services] [25].

Conclusions

The actual typical Susex value of an alumina refinery operation (~8 USD/tA) and its range (~6–12 USD/tA) can be reconciled reasonably well by using an approach based on the underlying Capex build-up and lifetimes of major Capex elements.

The discrepancy between the Susex range found by this approach and Susex as percentage of Property, Plant and Equipment in the 2013 and 2014 annual reports of some of the industry majors cannot easily be resolved. This applies both to a comparison on "at cost" basis (0.8–1.1% vs. 2.2–3.8% Industry Majors) as well as on "net of depreciation" basis (1.6–2.2% vs. 5.0–6.5% Industry Majors).

When comparing Susex values between operations, costs of "Maintenance Supplies/Materials" and "Contract Services" should be taken into account: for alumina refinery projects the sum of Susex plus these two items ranges between ~18 and 30 USD/tA.

14.3.3 General Opex Aspects

The following general Capex aspects are covered elsewhere in this chapter:

- **Opex, NPV, and IRR**: The impact of changes in Capex on NPV and IRR in Sect. 14.1.1.2.
- **Exchangeability of Opex and Capex**: The exchangeability of Opex and Capex in Sect. 14.1.1.3.
- **Robustness of Structural Low Opex Projects**: The robustness of projects with structural low Opex in Sect. 14.1.1.4.

14.3.4 Opex Improvement Opportunities

Potential opportunities to lower the Opex of a Bauxite and Alumina project are discussed in Sect. 14.4.

14.3.5 Reference Project Opex Breakdown

The breakdown of the Opex of the reference project used throughout this chapter on "Economics of smelter grade alumina projects" (refer Sect. 14.1.1.1 and Table 14.1) is shown in Table 14.35. The bauxite quality assumed for the reference project is shown in Appendix 5.

14.4 Cost Saving Opportunities

14.4.1 Overview—General Plant and Process Optimization Opportunities

As discussed in Sect. 14.3.1.6, bauxite resource quality in its widest sense has a more profound impact on Opex and Capex than technology. In other words, for an alumina refinery project the emphasis should be on identifying the "right" bauxite resource. In order to enhance the economic success of a project, technology and innovation should be focused on supporting that objective [1, 6].

As outlined in Sects. 14.1.2.2 (Project Scope, Refinery) and 14.1.5 (Economics of Greenfield and Brownfield Projects), greenfield projects used to be designed with a huge de-bottlenecking and brownfield expansion potential in their layout. In other words they were able to accommodate additional trains/lines/units, such as a digestion train or precipitation circuit, on the original plant layout. In addition, plants have typically been over-designed due to conservatism by the project owner or technology supplier, by the engineering contractor, and equipment manufacturers. This often resulted in projects with significant (low cost) de-bottlenecking/brownfield potential.

The above leads to the following general plant and process opportunities to improve alumina refinery project economics:

- Design a project requiring minimum infrastructure, both "external" (port, rail, town, etc.) and "internal" (piperacks, buildings, structural steel, water supply, power distribution, etc.). Another approach might be not to "take the plant to the bauxite", but taking bauxite to a site with existing infrastructure, refer for instance Alunorte's Paragominas bauxite pumped by slurry pipeline to the existing plant. A so-called "split plant" (spent liquor grinding taking place at a different location from the majority of refinery operations) could in certain cases perhaps provide economic benefits.
- In the context of Opex: design a plant requiring "no organization", e.g., by applying automation and "maintenance-less" equipment (as much as possible) and using a simple straight forward plant design (see further down). Outsourcing of activities may be considered in general terms but doesn't by itself address the issue of the necessity of organizational infrastructure.
- Don't build costly provisions for future expansions into a project/plant design. This shouldn't necessarily be taken literally. However, with that mindset, plant design may look different, as the focus is on maximizing the advantage of building a dedicated plant. A next production unit would have to be justified based on its

Table 14.35 Opex breakdown of 1.6 Mt/y Aa greenfield reference project (refer Table 14.1)

Bauxite and alumina project operating cost End 2016 USD, accuracy approximately −20% to + 50% on total opex	Consumption per tA	Price/rate USD/unit	Opex USD/tA
Variable costs, USD/tA			
Bauxite[a], wet basis (11% moisture), t	3.03	3.65[b]	**11.0**
Energy			
Coal (steam and power generation—LHV @ 25 GJ/t), t	0.288	90[c]	26.0
Heavy fuel oil (calcination—LHV @41 GJ/t), t	0.073	410[c]	30.0
Diesel (start-up of powerhouse, calcination, etc.), t	0.0063	790[c]	4.9
Total energy			**60.9**
Consumables			
Caustic soda (100% NaOH basis)[d], t	0.061	420[c]	25.8
Lime (100% CaO basis), t	0.036	130[c]	4.7
Flocculants, kg	0.8	3.5[c]	2.8
Total consumables			**33.3**
Other variable costs[e]			**5.0**
Raw materials and Aa handling and transportation costs, t	1.64[f]	9	**14.7**
Port operating costs, t	1.64[f]	2	**3.3**
Total variable costs			**128**
Fixed costs			
Maintenance materials and contract services		28 MUSD/y	**17.5**
Labor, no. personnel/average annual manyear cost	810	11,200 USD/y	**5.7**
Other fixed costs[g]		18.2 MUSD/y	**11.3**
Total fixed costs			**35**
Total opex			**163**
Sustaining capex			
Refinery			**6**
Infrastructure			**2**
Total Susex			**8**

[a] Delivered to alumina refinery bauxite storage and handling area
[b] Includes all mining related operating costs, royalties and sustaining Capex
[c] CIF port of raw materials import
[d] Note that caustic soda is actually shipped as 50% NaOH solution
[e] Limestone, acids, crystal growth modifier, grinding media, filter cloths, inhibitor, defoamers, BFW additives, etc.
[f] Includes an additional 0.11 t/tA of other raw materials (limestone, acids, etc.)
[g] Insurance, IT, computing, consultants, training, taxes, medical provision, security, social projects, etc.

own (economic) merits. This also relates to the optimum plant capacity (refer Sect. 14.1.4.2—Economy of Scale).

Section 14.4.4 outlines an alternative plant design based on the above three bullet points.

- Related to the previous point: do not over design a project/plant. One angle could be to critically review factors ("fat") built into the design (e.g., with respect to plant operating factor, design allowances, and sparing philosophy). Another could be to make a dedicated effort to "achieve maximum production capacity" right from start-up.

  Section 14.4.2 reviews production capacity increases and energy savings as function of liquor yield increase options.
- Investigate simplifications of the Bayer process, e.g., by deleting or by-passing complete operating steps. Section 14.4.3 explores this option applied to the green liquor (security) filtration facility by using current and if necessary, further developing flocculation technologies.
- Don't re-invent the wheel but look over the fence (e.g., to other industries) to assess if technologies and/or practices that are "proven technology" elsewhere could be applicable and economically attractive to Bauxite/Alumina projects. This may apply both to process related and to project related aspects. An example could perhaps be the application of sustainable energy technologies.
- Investigate other methods of plant construction as applied in other industries (e.g., using skids/platforms/modules [24]).

14.4.2 Plant Productivity/Capacity Maximization and Energy Savings—Liquor Yield

In Sect. 14.1.4.1 (Sustainability, Alumina Refinery Design), the first refinery design criterion mentioned is Plant Liquor Productivity/Yield. The focus for maximizing liquor productivity in a refinery is mostly on precipitation yield because the reaction kinetics of precipitation are more difficult to control and enhance than those of the digestion reaction [14].

Table 14.20 summarizes precipitation yield increase options of ~9–20 kg/m^3 (or g/l) for LT resp. ~15–30 kg/m^3 for HT digestion plants, and includes the following options using proven technology (refer [14] for further details):

- **Seed filtration in precipitation**: installing seed filters drastically reduces spent liquor recycle, increasing the so-called precipitation A/C fill ratio, and thus the driving force for precipitation and precipitation yield. The liquor A/C ratio signifies to what extent a caustic liquor is saturated with alumina, the ratio being that of the alumina concentration in solution (in g/l, expressed as Al_2O_3) and the caustic soda concentration (in g/l, expressed as Na_2CO_3). In addition, an increased fill ratio increases the capacity to agglomerate fines, enabling a finer hydrate seed, increasing specific surface area of the in-tank solids, also resulting in an increased yield.

- **Interstage cooling in precipitation**: the optimum temperature in precipitation associated with the highest crystallization rate depends on the A/C ratio and caustic concentration of the liquor in a precipitation tank. By increasing the number of cooling steps between the tanks it is possible to operate the precipitation area more closely to the optimum temperature profile.
- **Direct Cooling in the Heat Interchange area**: accomplishing heat transfer in the heat interchange area after digestion by direct means such as plate heat exchangers rather than by indirect means (flash steam), thus avoiding the concentrating effect of liquor flashing. As a result the digestion area can operate at a higher liquor caustic concentration without changing the caustic concentration of the liquor to precipitation, enabling an increased precipitation fill liquor A/C ratio.
- **High-rate thickeners in decantation**: high rate decanters/settlers and washers reduce the liquor-to-mud contact time and hence the potential of premature Gibbsite crystallization (so-called reversion) in the liquor coming from digestion (which is super-saturated with alumina) onto un-extracted Gibbsite (or Boehmite) in the bauxite residue.
- **Sweetening in HT digestion**: the HT Boehmitic bauxite digest is used to increase the digestion liquor A/C ratio with a first increment. As the HT digester discharge slurry progresses through flash cooling, Gibbsitic bauxite is added when the slurry reaches a temperature of ~180 °C, thus raising the A/C ratio of the digestion discharge liquor to a level which is more typical for Gibbsitic bauxite digestion.

Also considered in Table 14.20 are the following yield increase options for proposed new technologies:

- **Fines destruction in precipitation** (refer patent by Hiralal [29]): dissolving excess fines controls the amount of fine seed to match the capacity of classification and agglomeration systems in precipitation, and the high A/C ratio liquor obtained after dissolution of fines is utilized to increase the precipitation fill ratio.
- **Bauxite residue re-digestion at high solids density** (refer patent by Den Hond [30]): inclusion of a second low temperature digestion step of the decanter underflow with additional spent liquor, with the objective to maximize alumina extraction from the bauxite residue of the main digestion step.

The above-mentioned precipitation yield improvement options and their related plant capacity increases and energy savings as reported in Table 14.20 are based on the following main assumptions:

- Refinery base production capacity of 2 Million t/year.
- Base case precipitation yields of 65–75 kg/m^3 for Low Temperature (LT) plants, and 65 kg/m^3 for High Temperature (HT) plants.
- Base case steam and power energy requirement of 7.6 GJ/tA for LT plants (all required power generated in a co-generation facility), and 9.7 GJ/tA for HT plants (part of the power required generated in a co-gen facility, and 0.5 GJ/tA of imported

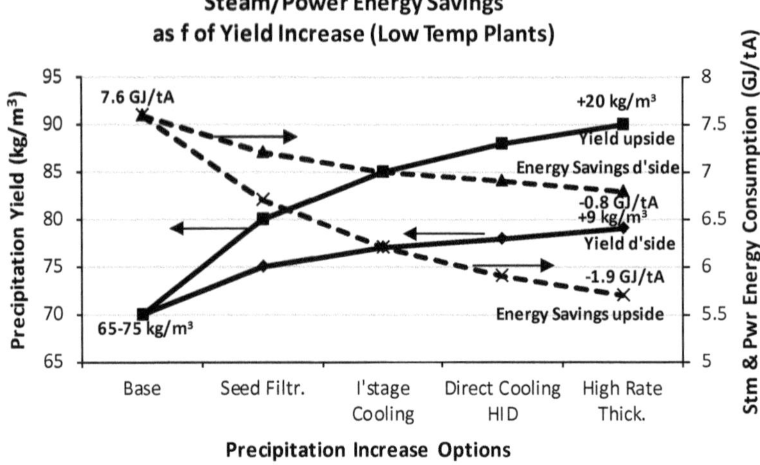

Fig. 14.32 Energy savings by proven technology as function of yield increase (LT plants). *Source* [14]. Copyright © 2014 by The Minerals, Metals & Materials Society. Used with permission

power required to meet remaining power needs). In other words, in both cases alumina calcination energy is not included.

- Heat of Gibbsite dissolution of 0.7 GJ/tA for LT digestion, and 1.0 GJ/tA for HT digestion.
- Yield increases for the various plant adaptations as indicated in [31], at an energy efficiency improvement of 85% of theoretical (a result of an increased number of on-line equipment, and other inefficiencies).

Figure 14.32 provides an illustration for LT plants of the steam and power energy savings by proven technology as function of yield increases (both cumulative) resulting from the above indicated plant adaptations (refer to Table 14.20 for actual values).

Conclusions from the above are that increasing precipitation yield in an alumina refinery by implementing proven technology is a powerful tool:

- To increase production capacity: by (indic.) ~13–31% for LT respectively 23–46% for HT digestion plants.
- To lower steam and power energy consumption per tA: by (indic.) ~0.8–1.9 GJ/tA for LT resp. 1.8–3.5 GJ/tA for HT digestion plants.

The above results in a drop of Capex and Opex i.e., overall in improved project economics.

Interestingly the improved specific energy consumption also means a drop in greenhouse gas emissions per tA and (if applicable) coal ash residues (bottom ash/fly ash) per tA, i.e., an improvement of direct environmental performance. And finally, the potential to improve alumina quality as a result of improved quality control in the precipitation area enables lowering alumina fines losses during transport and handling, thus improving environmental performance indirectly. In other words, projects to increase precipitation yield address two of the three "pillars" of sustainable development (economic, environmental, and social—refer Sect. 14.1.4.1 and Appendix 1).

14.4.3 Bayer Loop Simplification—By-Passing Security Filtration

Several approaches may be considered to address the issue of lowering project Capex and Opex. One of these approaches is to investigate opportunities to simplify the Bayer loop e.g., by deleting or by-passing a process area. This section describes one such opportunity in the security filtration area (sometimes called liquor filtration), if necessary, by further developing flocculation technologies [17].

14.4.3.1 Security Filtration

The purpose of the security filtration area in the Bayer alumina refining process is to lower by filtration the solids concentration in the decanter/settler (or sometimes first washer) overflow to the desired level in Liquor to Precipitation (LTP) to meet alumina product quality requirements.

The criteria for solids in LTP can be derived from the targeted product alumina quality specifications. For instance, the AP 18/30 specification for Fe_2O_3 in product alumina is <0.0165%. Assuming for example 62% Fe_2O_3 in the solids of LTP and a precipitation yield of 75 g/l, the solids in LTP should not exceed 20 mg/l. The solids levels in the filtrate from security filtration typically range from ~5 to 15 mg/l [32–34] (Fig. 14.33).

14.4.3.2 Capex, Opex, and Economics

The installed Capex of the security filtration area for brownfield and greenfield alumina capacity expansion projects depends on a number of factors such as location, technology employed, refinery capacity, etc. and ranges indicatively from ~15 to 25 USD/Annual tA capacity. In other words, for an alumina capacity expansion project of 1.6 Mt/year, the Capex of the security filtration area may be of the order of ~24–40 MUSD.

The Opex of the security filtration area includes fixed costs (e.g., labor, maintenance) and variable costs (e.g., filter aid, filter cloth, power, lime). The total Opex for security filtration differs from plant to plant and ranges indicatively from ~1.2 to 1.8 USD/tA (incl. sustaining capital). Lime for TCA (tri-calcium aluminate—filter aid) often comprises a large element of the Opex of security filtration. To

Fig. 14.33 Horizontal security filter—"Kelly Press". *Source* [17]. Copyright © 2010 by The Minerals, Metals & Materials Society. Used with permission

achieve targeted settler overflow (SOF) solids in a line-up excluding security filtration, alternative (perhaps more expensive) or additional flocculants may be required (see below). This means that net Opex savings would be less than the range indicated above. Assuming that the costs of alternative or additional flocculants are of the order of 0.4 USD/tA, the net Opex savings would amount to 0.8–1.4 USD/tA (incl. sustaining capital).

Indicative NPV's are as follows:

- **Existing refinery** at 1.6 Mt/y capacity fully by-passing Security Filtration, and 1.1 USD/tA net Opex savings as a result of by-passing Security Filtration: NPV (8%) = 10.7 MUSD.
- **Brownfield/greenfield expansion project** of 1.6 Mt/y capacity, 1.1 USD/tA net Opex savings and 32 MUSD net Capex savings as a result of excluding Security Filtration from the design: NPV (8%) = 34.5 MUSD. Although modified clarification equipment may be required (e.g. modified feedwells), and/or some alternative equipment may be considered for inclusion in the scope of a brownfield project (e.g. an extra-redundant-settler, an additional SOF tank or precipitation tank), related additional Capex is expected to be significantly less than Capex saved.

The alumina losses related to the use of filter aid will disappear, ranging from typically 0.4–0.7% Al_2O_3 (absolute). Thus, the alumina recovery will improve by the same absolute percentage, improving bauxite consumption and lowering generation of residue. The effect of this improvement of alumina recovery on Opex is not included in the numbers above.

14.4.3.3 Digestion Blow-Off (DBO) Ratio Limitation

In some alumina refineries the security filtration area puts a limitation on the DBO ratio due to excessive formation of hydrate scale on the filter cloth from too high supersaturation levels for the variability of the operating temperature. Partial or full elimination of security filtration in such a situation would make it feasible to increase the DBO ratio because the liquor remains at a higher and less variable temperature.

The gain in DBO ratio depends on the specific situation of an existing alumina refinery and could be of the order of ~0.02–0.03 A/C ratio points. Depending on the liquor conditions of the plant and the operation of the precipitation area, this could improve precipitation yield by ~2–4 g/l. This improvement could either be used to increase plant production capacity and/or be used to improve energy efficiency, etc.

14.4.3.4 The Opportunity

Decanter/settler overflow solids in current alumina refinery operations may typically range from 100 to 250 mg/l [35]. Significant strides have been made in the design of high-rate thickeners and their feed wells, and similarly in the preparation of modern synthetic flocculants. This is illustrated in a paper presented by Laros [36], which mentions that sizing parameters for settlers assume an overflow clarity of 50 ppm. Another illustration may be AMIRA's project P266F 'Improving Thickener Technology' (see www.p266project.com), which includes a novel feed well design as outlined on the website: www.p266project.com/Pages/About/ProjectAchievements.asp.

Some alumina refineries have reportedly executed tests aimed at SOF solids of <10 mg/l, including stepwise application of a combination of flocculants achieving roughly 5 mg/l solids, allowing part of the clarified liquor to by-pass Security Filtration. In another case, a second clarification step was performed. However, no refinery seems to have implemented by-passing Security Filtration as a routine practice. Perhaps this is understandable from an operator's perspective, but less so from an operating cost perspective (as outlined above).

The current state of technology should enable achieving solids (clarity) levels in the SOF low enough to meet the criteria required for the solids in LTP. This applies to existing refineries as well as to brownfield and greenfield projects. Key issues are the particular nature of a bauxite residue and the scaling issues of its related liquor. Another important aspect is the protection of the precipitation area from upsets in the settler operation which could, for instance, result in too high levels of solids in the LTP. This is addressed below.

Summarizing, developments in thickener design and flocculants as outlined above provide an opportunity to lower alumina refinery operating and capital costs by

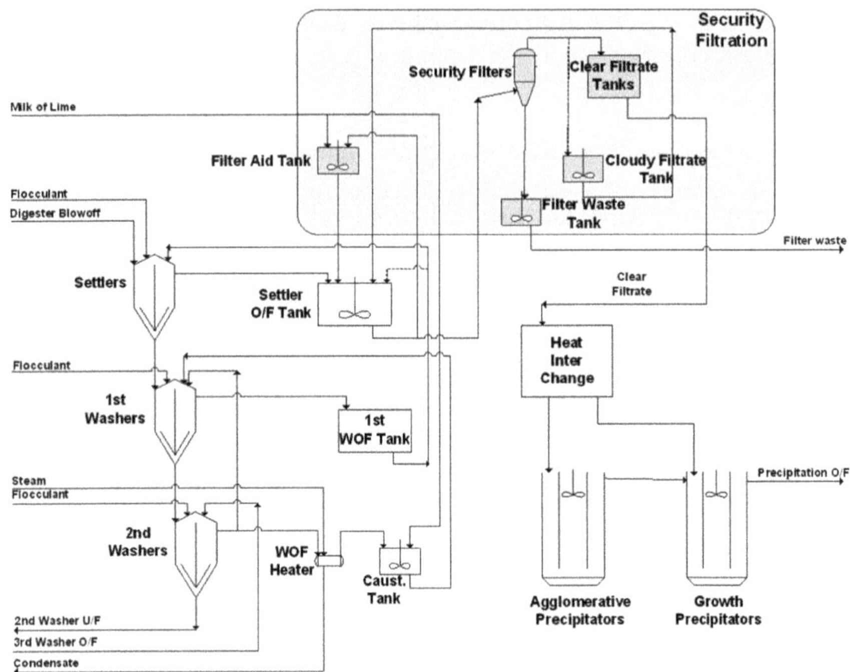

Fig. 14.34 Security filtration, settling and precipitation. *Source* [17]. Copyright © 2010 by The Minerals, Metals & Materials Society. Used with permission

(partly) by-passing (in an existing refinery) or excluding (in a brownfield or greenfield project) the security filtration area. For a brownfield or greenfield project this would mean that the following significant facilities and operations would no longer be required (refer Fig. 14.34):

- Infrastructure requirements of a security filtration area (building, structural steel, pipe racks, air supply, piping, etc.).
- Security filters including controls, other peripherals (valves, motors, piping) and related operator activities such as re-clothing and caustic cleaning.
- Clear filtrate tanks and peripherals.
- Cloudy filtrate tanks and peripherals.
- Filter waste (sometimes called press mud dump) tanks and peripherals.
- Filter aid tank and peripherals. The alumina losses related to the use of filter aid will disappear as mentioned above.
- In general terms, the exclusion of Security Filtration as unit operation from a brownfield or greenfield project represents a simplification of the layout, operation, and maintenance of an alumina refinery's Bayer loop.

This opportunity may be considered from three perspectives: existing refineries, brownfield expansion projects and greenfield expansion projects.

14.4.3.5 Existing Refineries

Existing alumina refineries could consider by-passing (part of) the SOF around the Security Filtration area. This requires sufficiently low solids concentrations in the SOF and in the filtrate from security filtration, and could be implemented in phases as follows:

Phase 1—No Additional Requirements. Example: a plant's current SOF solids are at 100 mg/l, the solids in the filtrate from Security Filtration are at 10 mg/l, and LTP should not have more than 20 mg/l solids for alumina quality reasons. Assuming no other relevant constraints, about 10% of the SOF could be by-passed around Security Filtration while still meeting the criterion for solids in LTP, i.e. without additional requirements. Operating cost savings depend on the specific plant situation and could indicatively be of the order of ~0.1–0.2 USD/tA.

Phase 2—Improve SOF Solids. Once SOF solids have dropped e.g., to 50 mg/l through a combination of using a different mix of flocculants and/or modifications to the settler feedwell, about 25% of the SOF could be by-passed around Security Filtration while still meeting the criterion for solids in LTP. Depending on the mix of flocculants required, Opex savings could be of the order of ~0.2–0.4 USD/tA.

This step would require some laboratory test work encompassing residue settling tests to investigate which (mix of) flocculants results in the required SOF clarity improvement while still meeting other process conditions. The scope of test work depends on a plant's existing line-up, potential line-up modifications, etc. Modifications to the settler feedwell may or may not be required depending on test work results.

Phase 3—Complete By-pass of Security Filtration. The results of Phase 2 could provide the basis to further improve SOF solids to the point that no SOF passes through the security filtration area. Depending on the requirement of the mix of flocculants and the current usage, Opex savings could be of the order of ~0.8 to 1.4 USD/tA as mentioned above.

Once all of the SOF by-passes Security Filtration, digestion blow-off (DBO) A/C ratio may be increased at those refineries where Security Filtration puts a limitation on the DBO ratio due to the excessive formation of hydrate scale on filter cloth. As mentioned earlier precipitation yield improvement may be of the order of ~2–4 g/l. Test work mentioned above could include tests to quantify the potential of a DBO A/C increase. Alternatively, historical plant operating data may provide information required to assess this potential.

Strategy for High Solids in SOF. Obviously, it will be necessary to have a strategy in place how to handle an upset in the operation of the settler(s) which would result in too high LTP solids concentrations and thus deterioration of product quality. Causes: unplanned and relatively fast changes in bauxite composition, control and instrumentation problems, significant flow fluctuations of the DBO, etc. Essentially the operating measures required to remedy this situation in the case of an existing refinery are similar to the operating steps currently in place if such an event was to happen today.

Fig. 14.35 Security filtration facility. *Source* [17]. Copyright © 2010 by The Minerals, Metals & Materials Society. Used with permission

In addition to the steps currently applied in such a situation, or perhaps as an alternative to those steps, the SOF at an undesirably high solids level may be directed to an offline SOF tank until the required number of security filters have been put online and other necessary steps have been taken. Due to a specific refinery design and/or its operating requirements, details of the necessary steps may require tailoring.

Or in other words for existing refineries the fallback position is to re-direct SOF to the existing security filtration area, i.e. reverting to the situation before SOF by-passed security filtration. When the filter building is being by passed it will of course take at least half an hour to an hour to make up a fresh batch of TCA before the filters can be brought back into service. An alternative approach may be to use filter aid from a flocculant supplier which can be turned on immediately when the problem occurs.

Implementation—Existing Refineries. Further details on the implementation of by-passing the security filtration area in an existing refinery can be found in [17] an example of a security facility is shown in Fig. 14.35.

14.4.3.6 Brownfield Expansion Projects

The medium-term perspective is to consider the opportunity as element of a brownfield expansion project of an existing refinery. Capacity expansion projects of existing refineries of ~600–1500 kt/year alumina and above are generally considered brownfield expansions as they often require additional trains/units to be added to some of the major plant areas such as digestion, precipitation etc.

In this case the opportunity is to exclude new (additional) security filtration capacity from the scope of the brownfield project (see above for the relevant facilities and operations involved). This technological development offers an opportunity attractive to alumina producers that are considering a brownfield capacity expansion as well as to technology suppliers offering technology for brownfield expansion projects to the alumina industry.

As mentioned earlier, the design of modern mud settling facilities offered by several technology suppliers enables achieving SOF solids acceptable for direct feed to the precipitation building. Designs by these suppliers may include proprietary feed well designs and/or feed well modifications developed in co-operation with others. As mentioned earlier a key issue is the specific nature of the bauxite residue and therefore laboratory test work is required to investigate which (combination of) flocculants will result in the required SOF clarity.

Strategy for High Solids in SOF. As for existing refineries, a strategy needs to be developed how to handle an upset in the operation of the bauxite residue settler(s) in the case of a brownfield expansion project which does not include a Security Filtration area. Several strategies are possible, and it depends on the combination of the existing facilities and the brownfield project which one would be most appropriate to a specific refinery. One of these strategies may be described as "upstream" because it encompasses design aspects related to facilities upstream of security filtration:

- Include state-of-the-art digestion and settler controls in the design (such as DBO controls, mud level indication and controls, clarity measurement—on-line turbidity meters).
- Include additional SOF tank capacity in the design. Conventionally SOF tanks act as buffer between the settlers and the Security Filtration area, while the Clear Filtrate tanks do the same between security filtration and the precipitation area. However, in this case the SOF tanks act as buffer between the settlers and precipitation. In practical terms this would normally mean that larger SOF tanks are required. Optionally an additional SOF tank is included in the design. The rationale being that sufficient SOF tank capacity is included to enable solving potential operational problems which would otherwise result in too high solids in SOF, ensuring in the meantime that the main plant flow is not (significantly) affected.
- In other words, while the operational problem is being solved, the SOF at too high solids is collected in a separate SOF tank.
- This liquor needs to be stabilized for instance by the addition of a small amount of lime because as it cools auto-precipitation could occur.
- When the problem is solved, bleed the liquor of the "contaminated" SOF tank (too high solids concentration) over a period of time into the settler feed flow. Additional flocculant may be required (requires testing).

An alternative "upstream" strategy could encompass the inclusion of an extra settler in the brownfield design. This extra settler would be brought online in case of an operational problem which has resulted in too high solids in SOF. The extra settler would act as "second stage settler", receiving the normal settler's overflow as feed, with its overflow becoming the "new" SOF to precipitation and the underflow being bled over a period of time into the normal settler's feed.

In addition to the above mentioned "upstream" strategies the brownfield project design could include "downstream" elements (encompassing design aspects related to facilities downstream of security filtration), such as:

- The contaminated SOF tank (refer first upstream strategy covered above) may be bled over a period of time into the LTP flow. The first agglomerative precipitator could then be operated in a "scavenging" (or "scalping") mode. Depending on the specific plant situation, the off-spec product from this tank may either be added to the regular material from the agglomeration step, recycled for instance to the digestion feed flow, or directed to a liquor burner if present.
- Alternatively, an extra agglomerative precipitation tank may be added to the design. As above this extra tank can be lined up as "scavenging" precipitator.

Other strategies may be more appropriate to the brownfield expansion of a particular plant, while for some brownfield expansion projects the existing security filtration area may provide (some) excess capacity that may be put to use in case of upsets of settler operation.

In the quantification of the overall annual plant operating factor for a brownfield expansion project, this aspect (a strategy for high solids in SOF) should be taken into account. It is expected that for a state-of-the-art design brownfield alumina project which excludes a security filtration area the annual plant operating factor should not be significantly different from a similar design which includes one.

Key is that the potential of high solids in SOF is taken into account in the design of the brownfield project. As mentioned earlier, high solids in SOF may be caused by unplanned and relatively fast changes in bauxite composition, control and instrumentation problems, significant flow fluctuations of the DBO, etc. These aspects need therefore extra attention during process design (and mine design—specifically with respect to bauxite quality control).

Implementation—Brownfield Projects. Further details on the implementation of excluding the Security Filtration area in brownfield projects can be found in [17]. An example of a security facility is shown in Fig. 14.35.

14.4.3.7 Greenfield Projects

In this case the opportunity is similar to that of the brownfield expansion case, i.e., excluding Security Filtration from the scope of the greenfield project, and is attractive both to companies considering a greenfield project as well and to technology suppliers offering technology to the alumina industry.

A difference with brownfield projects is that the owner of a greenfield project may not have an operating plant in which design aspects for the settling area can be tested in the field. However, this is not a major issue because properly executed laboratory residue settling test work results these days does not require confirmation by large scale tests.

The implementation of the opportunity in a greenfield project is expected to require 1–3 months in addition to the time normally required when the greenfield project design includes Security Filtration. This does not necessarily mean an extension of the normal greenfield project implementation time: if required information is collected as part of an R&D effort before or in parallel to the brownfield project, project development duration should not be affected at all (refer to [17] for further details).

14.4.4 Alternative Plant Design

Section 14.1.4.2 ("Economy of Scale") discussed the significant increase in design/initial capacity of greenfield (Bauxite Mine and) Alumina refinery projects over the past decades and its implications. It ended asking the question if the underlying trend of seemingly ever-increasing alumina refinery design capacities is inevitable or if viable alternatives exist. And it noted an overlap with A. Kjar's paper [23], although that has a different angle, viz. building a large plant at lower capital cost.

The basic reason for the trend being economics (refer Sect. 14.1.4.2), the question could be reformulated as follows: is it possible to develop smaller greenfield Bauxite and Alumina projects at acceptable economics? To further explore the subject, a closer look at the make-up of a greenfield project's capital cost is required [20, 21].

14.4.4.1 Overview

The capital cost of a greenfield alumina refinery may be broken down as shown in Table 14.36 which shows an indicative breakdown for the reference project's alumina refinery at 1.6 Mt/y production capacity in Direct and Indirect Cost items, and Contingency (compare with breakdown of Table 14.28). Note that numbers for actual projects may deviate significantly as a result of bauxite quality, technology choices, plant location, etc.

14.4.4.2 Commodities and Plant Layout Aspects

Table 14.36 illustrates that Commodities represent a very significant element of refinery Capex. Commodity amounts and their related capital costs reflect plant design and layout.

Table 14.36 Reference project 1.6 Mt/y Aa refinery capital cost breakdown (indic. only)

Refinery cost item, MUSD Basis NW Europe/US Gulf, Dec 2016	Reference project 1.6 Mt/y
Direct costs	
Equipment[a]	231
Commodities[b]	592
Total direct costs	**823**
Indirect costs	
Freight	58
EPCM	252
Temp. construction, start-up, commissioning, etc	184
Owners' engineering and other costs	126
Total indirect costs	**621**
Contingency	**206**
Total refinery capital cost	**1649**

[a] Incl. steam and power generation, sub stations, residue disposal, water supply, communication and info systems
[b] Incl. concrete, steel, mechanical bulks, piping, wire and cable, etc. (refer Sect. 14.2.1.2)

Current alumina refinery layouts are almost always designed to accommodate additional (future) digestion units (and all of the other required process units—e.g., precipitation, evaporation, etc.). The consequence is that plant design is not optimized for its initial production capacity. Plant layout is characterized by an "open architecture", at best compromising between on the one hand the limited layout requirements for the initial/design capacity and on the other the more extensive requirements to accommodate future additional process units. And in the worst case consisting of a layout of a large-capacity plant of which part is built, resulting in an inefficient plant layout for the design/initial capacity. In addition, in some cases plant design includes equipment which at some future stage might be used to its full capacity but operates (well) below design for a considerable part of its lifetime.

14.4.4.3 Alternative Approach Step 1: Dedicated Plant Capacity

A. Kjar's proposal to use replication means developing a design for a dedicated production capacity [23]. Or putting it differently, this alternative design approach aims at designing an alumina refinery for a dedicated production capacity i.e., without provisions for future expansions. This approach enables optimizing plant layout for the targeted production capacity, e.g., with respect to positioning similar equipment close to each other, use of common spare equipment, etc. This more "closed" alternative layout architecture results in a more efficient plant layout, reflected for example in the design of main plant pipe racks.

The above is illustrated in Fig. 14.36 which shows—bottom half—the main Piperack layout of a typical (conventional design) 1.6 Mt/year capacity refinery (i.e. in the expectation that additional production lines in various areas will be added in the future), and the layout—top half—of a dedicated 1.6 Mt/y capacity alumina refinery (same scale).

This alternative approach impacts positively on commodity volumes: for the same production capacity, commodity volumes of a greenfield plant designed along this

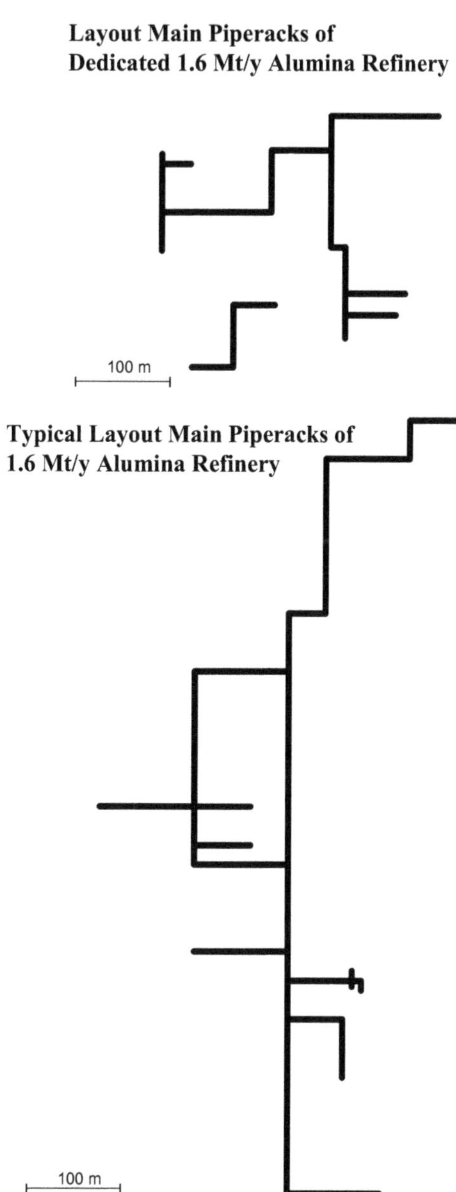

Fig. 14.36 Main Piperack layout comparison. *Source* [20]. Copyright © 2011 by The Minerals, Metals & Materials Society. Used with permission

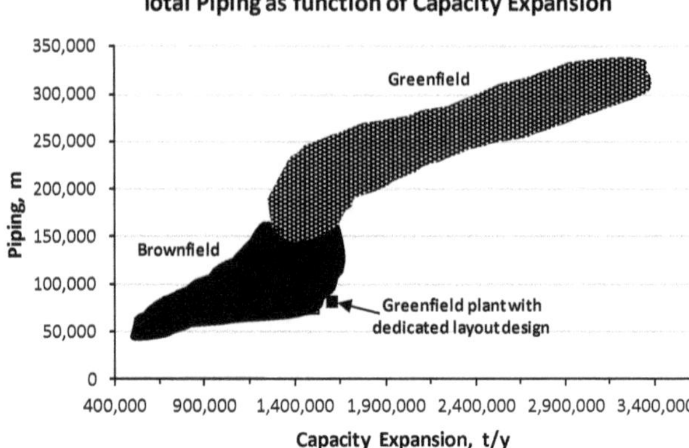

Fig. 14.37 Total piping as function of capacity expansion

alternative approach are similar to that of a brownfield expansion of an existing refinery. This is illustrated in Fig. 14.37 showing the total length of piping of greenfield and brownfield expansion projects as function of plant production capacity, and the requirement of a dedicated plant of 1.6 Mt/y capacity.

14.4.4.4 Impact of Alternative Approach on Commodities Cost

A dedicated greenfield plant design results in lower, in some cases significantly lower, amounts per annual tA produced of commodities such as steel, concrete and piping. This is reflected in lower Commodities costs, resulting in lower Direct Capital Costs, in turn improving Indirect Capital Costs. The overall effect on the capital cost of a greenfield dedicated low-temperature digestion alumina refinery of 1.6 Mt/y is illustrated in Table 14.37 (indicative numbers).

As can be seen in the table, the alternative approach improves total refinery capital cost indicatively by ~9–10%. The capital cost expressed per annual tonne of alumina capacity is approaching that of a current-design refinery at 3.2 Mt/y capacity (925 vs. approximately 916 USD/Ann tA basis NW Europe/US Gulf—refer Table 14.23).

14.4.4.5 Alternative Approach Step 2: Compact Refinery—Simple and Limited Scope

Along the lines of A. Kjar's paper (although he does not quantify "small increments of capacity"), applying the proposed dedicated-capacity approach to a compact alumina refinery capacity for instance of 0.4 Mt/y results in a project with a simple and limited scope. Available data suggest that as a result some Indirect capital cost items decrease more than proportionately, particularly costs related to temporary construction and start-up support, camp and other construction related items, and owners' costs.

Table 14.37 Plant capex comparison reference project with dedicated design approach (indic. only)

Refinery cost item, MUSD Basis NW Europe/US Gulf, Dec 2016	Reference project 1.6 Mt/y	Dedicated design 1.6 Mt/y
Direct costs		
Equipment	231	225[a]
Commodities[a]	592	505
Total direct costs	**823**	**729**
Indirect costs		
Freight	58	52
EPCM	252	223
Temp. construction, start-up, commissioning, etc	184	179
Owners' engineering and other costs	126	112
Total indirect costs	**621**	**566**
Contingency	**206**	**185**
Total refinery capital cost[b]	**1649**	**1480**
USD/Ann tA	1031	925

[a] The more efficient plant layout enables slightly lower equipment cost resulting from a more efficient use of common spare equipment
[b] Basis NW Europe/US Gulf (location factor 1.0—refer Table 14.28 for the description of location factor)

Table 14.38 illustrates the capital cost of a 0.4 Mt/y alumina refinery based on a dedicated design (indicative numbers). The table shows that the capital cost per annual tonne alumina (1343 USD/Ann tA) is higher than that of the much larger 1.6 Mt/y dedicated plant (925 USD/Ann tA—refer Table 14.37), however is at a level which could result in a project with acceptable economics, provided Infrastructure capital cost is limited. The Capex in USD/Ann tA compared with conventional design improves by another indicatively 8–10%.

Table 14.38 also shows that the total capital cost is at a level which would enable many more (relatively small) companies to develop such a project without necessarily requiring the formation of multi-party joint ventures, simplifying overall project management and thus enabling lowering costs (effect not included in Table 14.38).

The 0.4 Mt/y refinery production capacity considered here is not a fixed number but is meant to typify a capacity range of ~0.3–0.6 Mt/y. The higher end of this range is limited by the objective to end up with a total project capital cost (well) below 1 billion USD (to avoid mega projects—refer Sect. 14.1.4.2, Economy of Scale and Capital Cost, item 3—Implications), the lower end is determined by logistical limitations (e.g., with respect to caustic soda and fuel oil shipments) and may vary for different locations.

Table 14.38 0.4 Mt/y dedicated compact design alumina refinery capex (indic. only)

Refinery cost item, MUSD Basis NW Europe/US Gulf, Dec 2016	Dedicated design 0.4 Mt/y
Direct costs	
Equipment	94
Commodities	193
Total direct costs	**287**
Indirect costs	
Freight	28
EPCM	92
Temp. construction, start-up, commissioning, etc	38
Owners' engineering and other costs	34
Total indirect costs	**192**
Contingency	**57**
Total DCS[a] refinery capital cost	**537**
USD/Ann tA	1343

[a] Dedicated Compact Sustainable, refer Sect. 14.1.4.1, Alumina Refinery Design, rationale criterion 10C, Plant Production Capacity

Interestingly the capital cost of a small chemical grade alumina project of 300,000 t/year capacity in West Kalimantan owned by Antam and Showa Denko was reported to have a Capex of ~450 million USD (April 2011). Applying a construction index correction, the Capex of this project would be roughly 426 MUSD in December 2016. Using a capacity factor of 0.7 on the Capex for the DCS plant above (refer Sect. 14.1.4.2, Fig. 14.27) would result in: $(300/400)^{0.70} \times 537$ MUSD = 439 MUSD which seems quite consistent with the 426 MUSD. Or in other words the above described DCS approach appears realistic and feasible.

14.4.4.6 Infrastructure Capex and Possible Project Locations

As mentioned above, in order to realize acceptable economics for a project based on a dedicated compact production capacity, infrastructure capital cost should be limited. Conversely a project based on a compact plant capacity has limited infrastructural requirements and has several advantages over a large plant, particularly if the project is located close to an existing port, e.g., it may be allowed closer to residential areas (i.e. is closer to existing infrastructure); the existing infrastructure may be sufficient for a small plant, but not for a big plant; a suitable location for a small residue disposal area is easier to find than for a large one, etc.

Following are some examples of bauxite resources that may lend themselves to development via the proposed alternative approach (between brackets the potential alumina export port):

- Haden, Queensland, Australia (Brisbane).
- Bindoon, Western Australia (Fremantle).
- El Palmar, Venezuela (Ciudad Guayana).
- Trelawny, Jamaica (Discovery Bay).
- Kibi, Ghana (Tema).

The above list is not exhaustive and meant to be illustrative only.

In addition, some bauxite deposits which in view of their size could support the conventional development approach with large-capacity alumina refining projects, may also lend themselves to stage-wise development applying an adapted version of the proposed alternative approach. This could enable developing bauxite deposits even in locations with little infrastructure, albeit at a larger than compact scale (refer for instance to Table 14.37 for the Capex of a dedicated 1.6 Mt/y capacity project). Example: some of the Eastern Ghats deposits in Orissa and Andhra Pradesh, India, e.g., the Kutrumali deposit (with Visakhapatnam as potential alumina export port).

14.4.4.7 Refinery Technologies

The alternative design approach proposed above is independent of the selected refinery technologies, while at the same time stimulating to focus on design improvements, e.g. positioning similar equipment close to each other, the use of common spare equipment, etc.

14.4.4.8 Replication and Indirect Costs

A. Kjar indicates in his paper that the use of replication of a modern design at small capacity increments has as one of its main advantages far lower indirect capital costs, comprising Project management; Procurement; and Technology and EPCM fees. Although no direct quantification is mentioned in his paper, this appears consistent with the results discussed above for a dedicated plant design at a compact production capacity. Some of the replication-related cost savings mentioned by A. Kjar may come on top of the cost improvements quantified above.

14.4.4.9 Summary

The proposed alternative design approach represents a paradigm shift from "Bigger is Better" to "Smart and Small", and is based on an alumina refinery design with the following key characteristics:

1. First and foremost: a dedicated refinery design and layout for a specified production capacity, "designing a greenfield plant as a brownfield expansion". This means tailoring the design to the equipment and internal infrastructure requirements of the selected production capacity (e.g., earth works, power generation,

water supply, piperacks, roads, cable trays). This approach enables optimizing plant layout for the design capacity, it impacts positively on commodity volumes, and puts the focus on a "lean" design, resulting in ~ 9–10% lower capital cost per tonne of alumina (tA) capacity compared with current plant design. Consequentially the design excludes provisions for future expansions, which should have their own economic justification.
2. Additional potential: applying the dedicated design and layout approach to a compact refinery capacity of ~0.3–0.6 Mt/y (high end limited by the objective to end up with a total project capital cost <1 billion US$—the mega project threshold; the low end by logistical limitations, i.e., dependent on plant location), resulting in a simple and limited scope, further improving capital cost per tA capacity by an additional ~8–10%.
3. To ensure acceptable overall project economics, Infrastructure capital cost should be limited. At the same time such a project has few infrastructural requirements, especially if located close to an existing port.
4. And finally, sustainability is inherent to the alternative approach (optimized plant layout, lean design, effective and efficient use of resources, low capital cost per tA capacity—refer also Sect. 14.1.4.1.

The main advantages of this alternative approach are:

- Due to the significantly smaller project capital expenditure involved (lower risk), this approach enables the development of bauxite and alumina projects by smaller companies without a need to form multi-party joint ventures, i.e., it increases the number of companies potentially interested in developing bauxite deposits. In other words, competition increases, which should result in more efficient use of resources, both in terms of capital resources and in terms of global bauxite deposits.
- Due to the decreased complexity of compact alumina refining projects, the number of engineering companies potentially able to develop these projects increases, again resulting in more competition.
- Small and simple projects carry less risks and require less time to develop, implement and start-up, all of which has a positive impact on economics.
- A long-term alumina refining project based on the alternative approach requires only a relatively small resource (a deposit of ~40 Mt bauxite could support a 0.4 Mt/y alumina project for 30 years). This means that worldwide the number of bauxite deposits increases lending themselves to development.
- The new development model may be applied also to the development of part(s) of a large deposit.
- Furthermore, this approach lowers the threshold to increase value creation through alumina refining rather than being limited to bauxite export sales. This is attractive both to host countries and to companies developing potential bauxite and alumina projects.
- In specific cases, an adapted version of this alternative approach may enable bauxite deposit development even in locations with little existing infrastructure, albeit at a larger than compact scale (e.g. a dedicated 1.6 Mt/year alumina capacity

project, key item again being to tailor the design to the equipment and internal infrastructure requirements of the selected production capacity).

14.4.4.10 Practical Example

The above outlined Dedicated Compact Sustainable (DCS) design approach has been elaborated in a paper presented at TMS 2013 for a 400 kt/y plant [21]. Its overall operating factor and main operating parameters are shown in Table 14.39.

These operating parameters were based on the following design considerations (refer to [21] for further details):

1. **Equipment sparing** aim at finding an optimum between the costs related to plant downtime (loss of production, ongoing fixed operating costs) and the costs of lowering downtime (capital and operating costs). Key items are:

    - The role of equipment cleaning/descaling: strictly following the cleaning schedules and methods for the various pieces of equipment is essential to achieve the overall plant operating factor, and is an integral part of operating and maintaining the plant at design conditions.
    - The use of common spare equipment where possible (e.g., Desilication forwarding/discharge pump; train of heaters for Digestion/Evaporation; CCD washers overflow pumps; Cloudy filtrate/Filter waste tank and pump).
    - If an outage would result in an immediate alumina production loss, a spare is installed (e.g., Digester feed pump; Primary Cyclone Underflow Tank), or equipment bypassing facilities are installed (e.g., for Digester feed tank) or extra capacity in upstream/downstream equipment is included (e.g., Green liquor tank). This also applies to frequently maintained equipment (e.g., many pumps; DSM screens; Desilication feed tank; Green liquor heat exchangers).
    - No sparing is included for the grinding mill; nor for the CCD washers and flash vessels which can be by-passed, accepting transient process efficiency reductions.
    - A spare precipitator is installed, and the third tank can be used both as agglomerating as well as growth precipitator.
    - The refinery operates continuously with planned outages (accounted for in the overall plant operating factor) being used to service equipment. The sparing philosophy assumes no scheduled extended total plant shutdowns.
    - Note that the sparing philosophy may require adjusting to a specific plant location.

2. **Digestion and liquor evaporation areas positioned next to each other**, enabling sharing a common spare bank of heat exchangers. Rationale: using the same hot-end temperatures in Digestion and Evaporation, and an Evaporation cold-end similar to its feed temperature, the Evaporation flow rate is similar to that of Digestion. The design results in the entire spent liquor flow passing through Evaporation. Advantages: equipment standardization, simplified operations and maintenance, less (types of) spare parts (Fig. 14.38).

Table 14.39 Main operating parameters 400 kt/y DCS refinery

Item		
Plant overall operating factor, %		93
Main operating parameters		
Test tank	Liquor flow, m^3/h	508
	Liquor C[a], g/l	336
	Liquor A/C, ratio	0.433
Pre-desilication	Temperature, °C	98
	Retention time, h	24
Digestion	Slurry feed flow, m^3/h	583
	Temperature, °C	150
	Retention time, min	60[b]
	Discharge C, g/l	310
	Discharge A/C, ratio	0.753
CCD	Wash liquor flow, m^3/h	113
	Last washer underflow C, g/l	7
Precipitation	LTP flow, m^3/h	544
	Green liq. to seed reslurry[c], m^3/h	58
	Retention time, h	35
	LTP C, g/l	276
	LTP A/C, ratio	0.738
	LTP C/S, ratio	0.897
	Net precipitation yield, kg/m^3	82
Evaporation	Evaporator feed flow, m^3/h	570
	Hot end temp, °C	150
	Discharge temp, °C	93
Calciner utilization	%	93[d]
Bauxite factor	tBx(bone dry)/tA	2.74
Residue factor	tRes to disposal (bone dry)/tA	1.30

Source [21]. Copyright © 2013 by The Minerals, Metals & Materials Society. Used with permission
[a] Caustic concentration, expressed as Na_2CO_3
[b] It is possible that a retention time of ~30 min would be adequate
[c] Combines with the LTP flow in the agglomeration section
[d] Assuming a calciner overhaul once every 18 months

3. **Bauxite residue settler and washers placed in a horseshoe shape** for easy access to washer overflow standpipes and pumps, etc. (Fig. 14.39). Lime related areas are positioned next to each other (similar operating and maintenance requirements) and close to the washer train (three of the four lime additions are related to the settler/washer train).

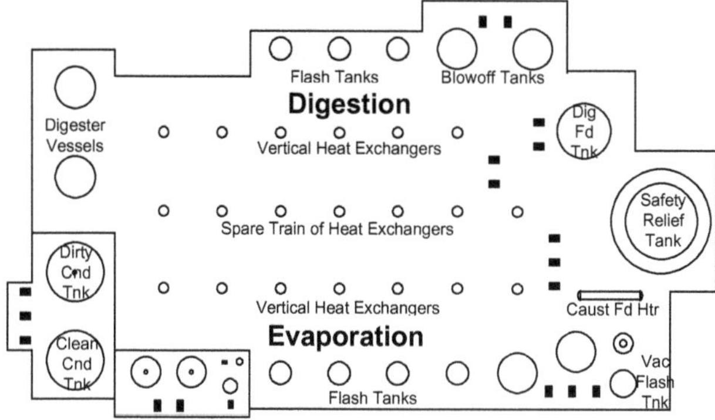

Fig. 14.38 Layout of digestion/evaporation areas. *Source* [21].

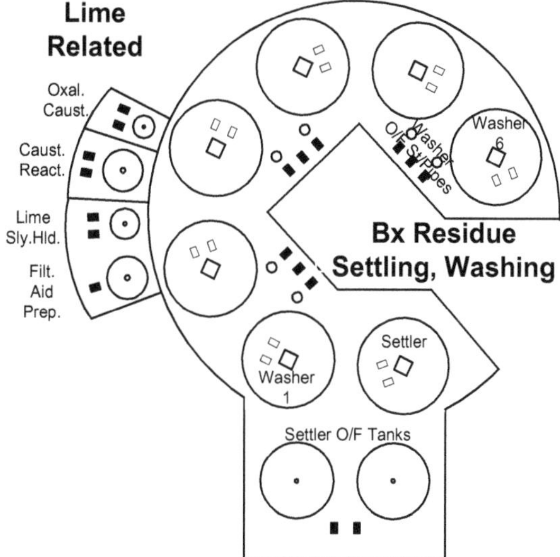

Fig. 14.39 Layout of residue settling, washing and lime areas. *Source* [21].

4. **Bauxite residue discharging from the CCD wash train containing less than ~8 g/l caustic soda** (expressed as Na_2CO_3) in the adhering liquor, enhancing storage/disposal options.
5. **Filters for hydrate to calcination, for fine seed for precipitation and oxalate removal located in one building** (Fig. 14.40). The same comment as above applies with respect to equipment standardization, operating procedures, etc.
6. **Last two on-line precipitators operating with agitators** allowing to vary slurry levels, thus accommodating volume take-up when descaling a (precipitation) tank.
7. **Facility in the center of the plant accommodating plant control room, operations office and plant laboratory**.

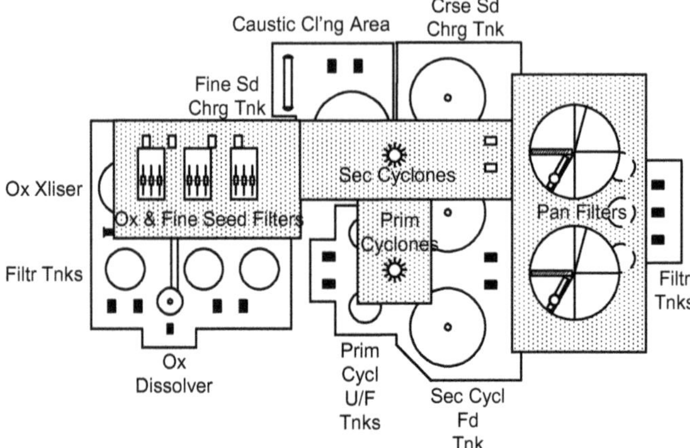

Fig. 14.40 Layout of hydrate classification/filtration and oxalate removal areas. *Source* [21]. Copyright © 2013 by The Minerals, Metals & Materials Society. Used with permission

8. **Key role for equipment cleaning/de-scaling, including mechanical cleaning of the precipitators**. Main advantages: no major plant liquor caustic concentration fluctuations, i.e. better control, allowing a narrower "safety range" for the control of liquor super-saturation (SS), enabling to run it at a higher level. Put differently: tank cleaning and plant liquor concentration control have been separated. Other advantage: steam savings (no caustic cleaning).
9. **A hydrate storage facility between precipitation and calcination**, enabling the Bayer circuit to operate independently and as undisturbed as possible from calcination. Two calciners are installed, both normally in operation. At nominal hydrate production rates, the feed rate will be less than calciner nameplate capacity. When a calciner is off-line for maintenance, the capacity of the remaining unit is maximized, however it will not meet the plant design production rate and hydrate is stored. Stored hydrate is recovered as soon as the second calciner is brought back online and calcination capacity is maximized for both units. Stored hydrate should be kept at a minimum. By uncoupling the Bayer circuit from calcination, product quality control can be optimized.

Refer to [21] for further details. Figure 14.41 presents the resulting overall process plant layout for a 400 kt/y DCS plant, showing that the approach leads to a compact, simple and efficient layout with a small Bayer loop. It also illustrates that the goal to "design a greenfield plant as a brownfield expansion" (tailoring/optimizing plant design to the equipment and infrastructure requirements of the specified production capacity) is achievable: most of the infrastructure is integrated in the process areas and only limited infrastructure is required outside those.

Major advantages of this layout

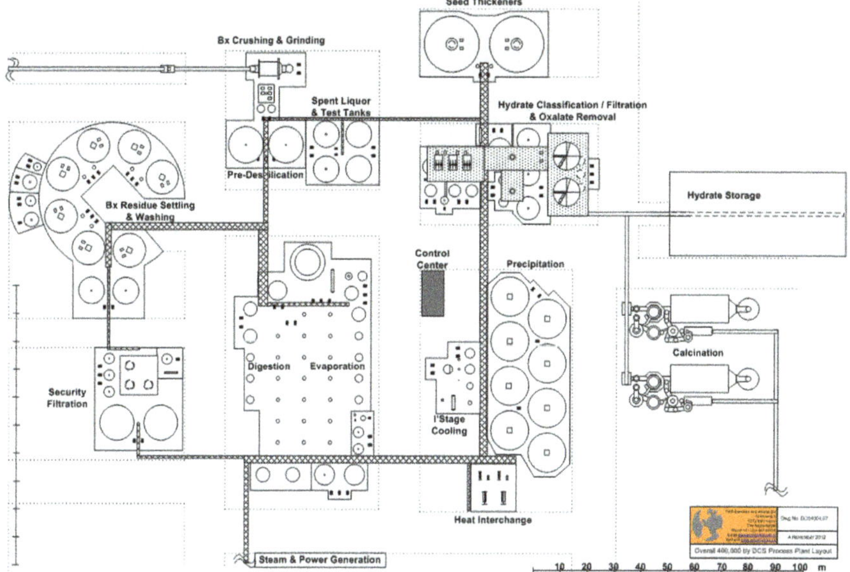

Fig. 14.41 Overall 400 kt/y DCS process plant layout. *Source* [21]. Copyright © 2013 by The Minerals, Metals & Materials Society. Used with permission

- Lower Capital Cost:
  - Lower commodity volumes (concrete, structural steel, piping, etc.): volumes of a greenfield plant designed along this approach are similar to that of a brownfield expansion of an existing refinery, resulting in lower capital cost per tA production capacity.
  - Equipment standardization (e.g., same heat exchangers for Digestion and Evaporation areas, same filters for fine seed and oxalate) results in simpler engineering and design, and less (types of) spare parts.
  - Use of common spare equipment resulting in less spares.
- Lower Operating Cost
  - Lower pumping energy.
  - Increased operating (e.g., filter re-clothing) and maintenance efficiency (e.g. heater cleaning) due to clustering similar types of operations/maintenance.
- Ease of Operations, Maintenance and Control
  - Control center: central control room/office/laboratory.
  - Short distances between process areas.
  - Access to equipment.
  - Limited number of equipment sizes and types due to equipment standardization.

Appendix 1: Sustainable Development: "People, Planet, Profit, Governance" (Refer also [7])

The Global Mining Initiative (GMI) led by companies making up the mining and minerals working group of the World Business Council for Sustainable Development (WBCSD—incl. Alcoa, Rio Tinto, BHP Billiton, Vale, Hydro, and Vedanta Resources) commissioned the independent Mining, Minerals and Sustainable Development (MMSD) project (refer [9]).

This project was conducted by the International Institute for Environment and Development (IIED) between 2000 and 2002. In the Executive Summary of the 2002 MMSD report "Breaking New Ground" it is stated (refer [10]) that

> One of the greatest challenges facing the world today is integrating economic activity with environmental integrity, social concerns, and effective governance systems. The goal of that integration can be seen as 'sustainable development'. In the context of the minerals sector, the goal should be to maximize the contribution to the well-being of the current generation in a way that ensures an equitable distribution of its costs and benefits, without reducing the potential for future generations to meet their own needs.

This builds on the most widely accepted definition of sustainable development by the World Commission on Environment and Development (1987 Brundtland Commission): "Sustainable development is development that meets the needs of the present without compromising the ability of future generations to meet their own needs". The four dimensions of sustainable development thus identified are (refer [10]):

- **Social** sphere sometimes referred to as the "People" aspect.
- **Environmental** sphere ("Planet").
- **Economic** sphere ("Profit").
- **Governance** sphere providing the setting for the other three aspects which are often referred to as the three "pillars" of sustainable development.

In summary sustainable development involves integrating and meeting economic, social, and environmental goals [10]. In their 2012 report MMSD + 10 ("Reflecting on a decade of mining and sustainable development"), the IIED mentions that "*... MMSD helped companies understand that sustainable development is about balancing the needs of society, the environment and economics, in the context of good governance...*" [11]. The Australian Minerals Industry's Framework for Sustainable Development ("Enduring Value") defines Sustainable Development in the mining and metals sector to mean that "*... investments in minerals projects should be financially profitable, technically appropriate, environmentally sound and socially responsible...*".

Prompted by GMI the board of the metals industry's representative organization, the International Council on Metals and the Environment agreed in 2001 to broaden its mandate and transform itself into the International Council on Mining and Metals (ICMM). Currently ICMM members include 22 mining and metals companies (e.g. Hydro, Rio Tinto, BHP Billiton, and Vale) and 34 national and regional mining associations and global commodity associations (e.g. International

Aluminium Institute, and the Minerals Council of Australia). ICMM developed the Sustainable Development Framework consisting of the following three elements which member companies are required to implement (refer website www.icmm.com):

- **Commitments**: 10 principles for sustainable development based on the issues identified in the MMSD project and benchmarked against several leading international standards.
- **Public reporting**: performance reporting against the 10 principles in accordance with the guidelines of the Global Reporting Initiative (GRI) (refer Appendix 2).
- **Independent Assurance**: providing third-party verification against 5 aspects that a company is meeting its commitments to the 10 principles.

Figure 14.42 shows the connections between the above mentioned councils, committees and sustainability related facets.

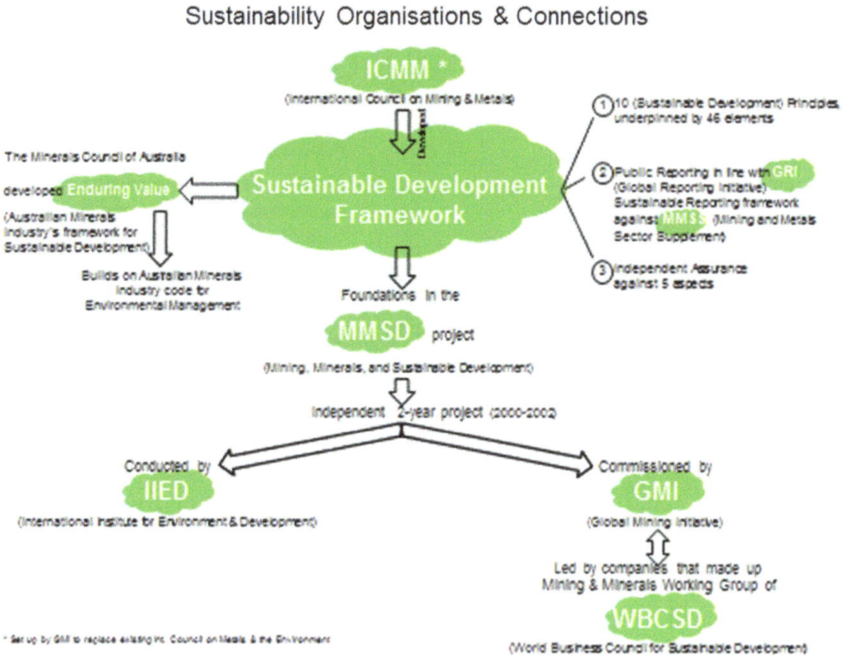

Fig. 14.42 Sustainability organizations and connections. *Source* [7]. Copyright © 2014 by The Minerals, Metals & Materials Society. Used with permission

Appendix 2: Global Reporting Initiative (GRI) Performance Indicators Used

See Table 14.40.

Table 14.40 GRI performance indicators used ("MM" denotes mining and metals sector specific)

Performance indicator	Economic ("Profit")	Aspect: Economic performance
	EC1	Direct economic value generated and distributed, including revenues, operating costs, employee compensation, donations, and other community investments, retained earnings, and payments to capital providers and governments
	EC4	Significant financial assistance received from government
	EC8	Development and impact of infrastructure investments and services provided primarily for public benefit through commercial, in-kind, or pro bono engagement
	EC9	Understanding and describing significant indirect economic impacts, including the extent of impacts
Performance indicator	**Environmental ("Planet")**	**Aspect: Materials**
	EN1	Materials used by weight or volume
		Aspect: Energy
	EN3	Direct energy consumption by primary energy source
	EN4	Indirect energy consumption by primary source
	EN5	Energy saved due to conservation and efficiency improvements
	EN6	Initiatives to provide energy-efficient or renewable energy based products and services, and reductions in energy requirements as a result of these initiatives
	EN7	Initiatives to reduce indirect energy consumption and reductions achieved
		Aspect: Water
	EN8	Total water withdrawal by source

(continued)

Table 14.40 (continued)

		Aspect: Biodiversity
	EN12-MM1	Amount of land (owned or leased, and managed for production activities or extractive use) disturbed or rehabilitated
	EN14	Strategies, current actions, and future plans for managing impacts on biodiversity
		Aspect: Emissions, Effluents, and Waste
	EN16	Total direct and indirect greenhouse gas emissions by weight
	EN20	NO, SO, and other significant air emissions by type and weight
	EN21	Total water discharge by quality and destination
	EN22	Total weight of waste by type and disposal method
	EN22-MM3	Total amounts of overburden, rock, tailings, and sludges and their associated risks
		Aspect: Transport
	EN29	Significant environmental impacts of transporting products and other goods and materials used for the organization's operations, and transporting members of the workforce
Performance indicator	Social ("People")	**Aspect: Labor—Employment**
	LA1	Total workforce by employment type, employment contract, and region
	LA8	Education, training, counseling, prevention, and risk-control programs in place to assist workforce members, their families, or community members regarding serious diseases.
		Aspect: Society—Community
	SO1	Nature, scope, and effectiveness of any programs and practices that assess and manage the impacts of operations on communities, including entering, operating, and exiting
	SO1-MM9	Sites where resettlements took place, the number of households resettled in each, and how their livelihoods were affected in the process

(continued)

Table 14.40 (continued)

	SO1-MM10	Number and percentage of operations with closure plans
		Aspect: Product—Customer Health and Safety
	MM11	Programs and progress relating to materials stewardship

Note In this chapter GRI G3 (version 3) has been used as basis

Appendix 3: Comparison Alcoa, Hydro, and Rio Tinto (Basis: Annual Reports 2014 and 2013)

See Tables 14.41, 14.42 and 14.43.

Table 14.41 Annual revenue comparison Alcoa, Hydro, Rio Tinto 2013–2014

ALCOA			HYDRO			RIO TINTO		
Revenue, MUSD	2014	2013	Revenue, MNOK	2014	2013	Revenue, MUSD	2014	2013
Alumina	3509	3326	Bauxite and alumina	9568	8124	Rio Tinto Alcan	12123	12463
Primary metal	6800	6596	Primary metal	6397	3866			
Global rolled products	7351	7106	Metal markets	37981	29646			
Engineered products and solutions	6006	5733	Rolled products	21345	20286			
Other	240	271	Energy	2492	2830	Iron ore	23281	25994
			Other	124	124	Copper	6282	916
						Diamonds and minerals	4150	4193
						Energy	4308	5454
						Other, intersegment transactions, etc	−2480	-2825
Revenue, MUSD	23906	23032	Revenue, MNOK	77907	64876	Revenue, MUSD	47664	51171
			Equiv. MUSD (@ realized exch. rate)	12406	11052			

Table 14.42 Comparison of upstream business unit revenue as percentage of total revenue Alcoa, Hydro, Rio Tinto 2013–2014

ALCOA			HYDRO			RIO TINTO		
	2014	2013		2014	2013		2014	2013
Upstream[a] revenue, MUS$	10309	9922	Upstream[b] revenue, MNOK	53946	41636	Upstream[c] revenue, MUS$	12123	12463
As percentage of total revenue	43%	43%	As percentage of total revenue	69%	64%	As percentage of total revenue	25%	24%

[a] Alumina, Primary Metal
[b] Bx and Aa, Primary Metal, Metal Markets
[c] Rio Tinto Alcan

Table 14.43 Key operational information comparison Alcoa, Hydro, Rio Tinto 2013–2014

ALCOA			HYDRO[a]			RIO TINTO		
	2014	2013		2014	2013		2014	2013
Bauxite production (kmt)	46300	46900	Bauxite production (kmt)	9481	7567	Bauxite production (kmt)	41871	43204
Alumina production (kmt)	1606	16618	Alumina production (kmt)	5933	5377	Alumina production (kmt)	8134	9307
Primary Al production (kmt)	3125	3550	Primary Al production (kmt)	1958	1944	Primary Al production (kmt)	3361	3555
			Realized Al price LME (USD/mt)	1850	1902			
			Realized Al price LME (NOK/mt)[b]	11624	11160			
			Realized NOK/USD exch. Rate[b]	6.28	5.87			
Engineered products and solutions (MUS$)	6006	5733	Metal products sales, total hydro (kmt)[c]	3305	3164			
Global rolled products (kmt)	2056	1989	Rolled products sales volumes to external market (kmt)	946	941			
Power production (GWh)	13783	12775	Power production (GWh)	10206	10243	Power generating cap. (GWh)[d]	36065	36372

[a] Amounts include Hydro's proportionate share of production in equity accounted investments
[b] Including the effect of strategic hedges (hedge accounting applied)
[c] Sales from casthouses (incl. Neuss), remelters, third party sources and liquid metal
[d] Hydroelectric power for Al production only included here

Appendix 4: Excel Spreadsheet Used for Economics Calculations

Part of the Excel spreadsheet used to calculate NPV, IRR, VIR/CER, and Payback Period throughout this chapter by applying the discounted cashflow (DCF) method.

Real 2017 US$					2017	2018	2019	2020	2021	2022	2023	2024	2025	2026
								Operations –>	Yr1	Yr2	Yr3	Yr4	Yr5	Yr6
Production	ktpa			1,600	0	0	0	0	1,120	1,520	1,600	1,600	1,600	1,600
Production Build-up	%								70%	95%	100%	100%	100%	
Alumina price protocol														
low Aa case					300	300	300	300	300	300	300	300	300	300
Mid Aa case					350	350	350	350	350	350	350	350	350	350
High Aa case					400	400	400	400	400	400	400	400	400	400
Alumina price	$/t	Mid Aa case		2	350	350	350	350	350	350	350	350	350	350
Cash costs														
Raw materials	$/t			11	11	11	11	11	11	11	11	11	11	11
Conversion costs	$/t			152	152	152	152	152	152	152	152	152	152	152
Sustaining capital	$/t			8	0	0	0	0	0	2	3	5	6	7
Sustaining capex Build-up	%				0%	0%	0%	0%	0%	20%	35%	60%	75%	90%
Total cash costs	$/t			171	163	163	163	163	163	164	166	168	169	170
Capital	$/t			1,401	0	210	981	210	0	0	0	0	0	0
Capital Spread	%					15%	70%	15%						
Tax														
Depreciation	$/t		20 yrs		0	0	0	0	70	70	70	70	70	70
Tax	$	30%	0 yrs		0	0	0	0	35	35	34	34	33	33
Total project cashflow														
Operating cashflow	k$		100.0%		0	0	0	0	170,357	229,496	240,230	237,990	236,646	235,302
Capital	k$		100.0%	(2,241,246)	0	(336,187)	(1,568,872)	(336,187)	0	0	0	0	0	0
Total NPV	k$	8.0%	90,205		0	(336,187)	(1,568,872)	(336,187)	170,357	229,496	240,230	237,990	236,646	235,302
Discount Factor					1.00	1.08	1.17	1.26	1.36	1.47	1.59	1.71	1.85	2.00
(Discount Factor)$^{-1}$					1.00	0.93	0.86	0.79	0.74	0.68	0.63	0.58	0.54	0.50
Σ(Discount Factor)$^{-1}$						11.51								
DCF	k$			OK	0	(311,284)	(1,345,055)	(266,876)	125,217	156,191	151,386	138,865	127,853	117,710
Cum DCF	k$	Check: ΣDCF	90,205		0	(311,284)	(1,656,339)	(1,923,215)	(1,797,998)	(1,641,807)	(1,490,421)	(1,351,556)	(1,223,703)	(1,105,993)
Cum MOD (Non-Discounted Cashflows)	k$		4,398,345		0	(336,187)	(1,905,059)	(2,241,246)	(2,070,889)	(1,841,393)	(1,601,163)	(1,363,173)	(1,126,526)	(891,224)
IRR			8.4%											
Payback Period-Nominal Cashflow	y (# operating years)		~9.5											
Payback Period-DCF	y (# operating years)		~25.5											
Masks														
Construction		2018	3		FALSE	TRUE	TRUE	TRUE	FALSE	FALSE	FALSE	FALSE	FALSE	FALSE
Production				**VIR or Cap Eff Ratio**	FALSE	FALSE	FALSE	FALSE	TRUE	TRUE	TRUE	TRUE	TRUE	TRUE
		NPV(x%) or Cum DCF / Discounted Capex		0.047			**Discounted Capex**	(1,923,215)						
		NPV(x%) or Cum DCF / Capex (2017 USD)		0.040			**Disc. Capex/AnntA**	1,202						
		Cum Non-Discounted Cashflows / Capex (2017 USD)		1.962										

Table 14.44 Assumed bauxite quality of the reference project (Table 14.1)

Item	Typical
Chemical (% w/w, dry basis—except moisture)	
Total alumina (Al_2O_3)	43.0
Available alumina[a]	39.2
Boehmitic alumina	1.2
Total silica (SiO_2)	2.85
Reactive silica[a]	1.85
Iron oxide (Fe_2O_3)	28.3
Total organic carbon (as C)	0.1
Total inorganic carbon (as C)	0.2
Impurities (TiO_2, P_2O_5, Cr_2O_3, ZnO, etc.)	2.6
Moisture	11.0
Mineralogical (% w/w, dry basis)	
Gibbsite ($Al_2O_3.3H_2O$)	60.0
Boehmite ($Al_2O_3.1H_2O$)	1.4
Kaolinite ($Al_2O_3.2SiO_2.2H_2O$)	4.0
Quartz (SiO_2)	1.0

[a] At digestion conditions

Appendix 5: Bauxite Quality Used for the Reference Project

In summary, the assumed bauxite quality consists of Gibbsite (major) and Boehmite (minor) as Alumina minerals, enabling low temperature digestion and co-generation of steam and power; and of Iron minerals (Hematite, Goethite, etc.; major), Kaolinite (medium) and Quartz (minor) (Table 14.44).

References

1. P.J.C. ter Weer, Greenfield dilemma—innovation challenges. Paper presented at light metals 2005, San Francisco, California, pp. 17–22
2. P.J.C. ter Weer, Operating cost—issues and opportunities. Paper presented at light metals 2006, San Antonio, Texas, pp. 109–114
3. Alumina Limited Public Announcement 2010—17AWC, 3 June 2010
4. P.J.C. ter Weer, Significance of increased greenfield alumina refinery design capacity. Int. Aluminium J. 20–22 (January/February 2011)
5. P.J.C. ter Weer, Capital charge—tool for economic screening purposes. Int. Aluminium J. 26–27 (July/August 2014)
6. P.J.C. ter Weer, T. McCabe, A. Lavignolle, Capital cost: to be or not to be. Paper presented at light metals 2007, Orlando, Florida, pp. 43–48
7. P.J.C. ter Weer, Sustainability and bauxite deposits. Paper presented at light metals 2014, San Diego, California, pp. 149–154

8. Global Reporting Initiative (website https://www.globalreporting.org/resourcelibrary/G3-English-Mining-and-Metals-Sector-Supplement.pdf), RG & MMSS, Sustainability reporting guidelines and mining and metals sector supplement, RG Version 3.0/MMSS Final Version
9. The Global Mining Initiative, Address to mining 2000, Melbourne September 20, 2000 by George Littlewood, Consultant WMC Resources Limited (Prepared in association with Tony Wells, BHP Minerals)
10. Breaking New Ground, The report of the mining, minerals and sustainable development project. MMSD (2002)
11. MMSD + 10, Reflecting on a decade of mining and sustainable development. International Institute for Environment and Development (IIED), Buxton (2012)
12. P.J.C. ter Weer, Sustainability and alumina refinery design. Int. Aluminium J. 20–25 (July/August 2015)
13. P.J.C. ter Weer, Sustainability and alumina refinery design. Paper presented at light metals 2015, Orlando, pp. 137–142
14. P.J.C. ter Weer, Relationship between liquor yield, plant capacity increases, and energy savings in alumina refining. J. Metals **66**(9), 1939–1943 (2014). https://doi.org/10.1007/s11837-014-1069-x
15. F.A. Champion, Some aspects of the stress corrosion of steel in caustic soda solutions. Chem. Ind. 967–975 (July 13, 1957)
16. R. Kelly, M. Edwards, D. de Boer, P. McIntosh, New technology for digestion of bauxites. Paper presented at light metals 2006, San Antonio, USA, pp. 59–64
17. P.J.C. ter Weer, Redundancy of security filtration. Paper presented at light metals 2010, Seattle, pp. 113–118
18. Website http://bauxite.world-aluminium.org/refining/case-studies/yarwun.html
19. L. Perander et al., Application of optimized energy efficient calcination configuration. Paper presented at alumina quality workshop 2012, Perth, pp. 371–377
20. P.J.C. ter Weer, New development model for bauxite deposits. Paper presented at light metals 2011, San Diego, California, pp. 5–11
21. P.J.C. ter Weer, New development model for bauxite deposits—dedicated compact refinery. Paper presented at light metals 2013, San Antonio, pp. 97–102
22. S. Canbäck, P. Samouel, D. Price, Do diseconomies of scale impact firm size and performance—a theoretical and empirical overview. J. Manag. Econ. **4**(1), 27–70 (2006)
23. A. Kjar, A case for replication of alumina plants. Paper presented at light metals 2010, Seattle, Washington, pp. 183–190
24. R. Valenti, P. Ho, Rio Tinto Alcan Gove G3 experience on pre-assembled modules. Paper presented at the alumina quality workshop 2008, Darwin, pp. 1–5
25. P.J.C. ter Weer, Sustaining capital of alumina refinery projects—important but unloved. Paper presented at light metals 2016, Nashville, Tennessee, pp. 81–83
26. Alcoa Annual Report 2014, "Transforming"
27. Hydro Annual Report 2014, "Better bigger greener"
28. Rio Tinto Annual Report 2014, "Delivering sustainable shareholder returns"
29. I.D.K. Hiralal, Process for the precipitation of aluminium tri-hydroxide from a supersaturated sodium aluminate solution. US Patent Number 5,690,700, November 1997
30. R. Den Hond, Method and apparatus for the extraction of gibbsite from bauxite. PCT, WO 2006/031098 A (2006)
31. R. den Hond, I. Hiralal, A. Rijkeboer, Alumina yield in the Bayer process—past, present and prospects. Paper presented at light metals 2007, Orlando, USA, pp. 37–42
32. M. Normandin et al., Study on the clarification of a red mud slurry during flocculation. Paper presented at light metals 2006, San Antonio, Texas, pp. 23–26
33. G. Bickert et al., Effective Bayer liquor clarification: retrofit or new with DRM Fundabac filter internals. Paper presented at the alumina quality workshop 2008, Darwin, pp. 130–135
34. R. Bott et al., Advanced filtration methods for pregnant liquor purification. Paper presented at light metals 2008, New Orleans, Mississippi, pp. 39–43

35. P.K.N. Raghavan et al., Selection and optimization of synthetic flocculant for high rate decanters (deep cone thickeners). Paper presented at the alumina quality workshop 2008, Darwin, pp. 64–68
36. T. Laros, Selection of sedimentation equipment for the Bayer process—an overview of past and present technology. Paper presented at light metals 2009, San Francisco, USA, pp. 107–110

Peter-Hans ter Weer TWS Services and Advice—Bauxite and Alumina Consultancy, Owner-Director. Master of Science Chemical Technology, University of Technology Delft, The Netherlands, 1978. Email: twsservices@tiscali.nl

Thirty-six years of international exposure to the Bauxite and Alumina industry (22 of which with BHP Billiton), ranging from hands-on experience in plant commissioning, process engineering and production control, to corporate roles in business and technology development and project initiation and evaluation.

Since 2004 independent technical and economic advisor to the bauxite and alumina industry.

Focus: understanding the key factors affecting the technical and economic success of bauxite and alumina projects.

Author of twenty-one articles in industry magazines (International ALUMINIUM Journal, Journal of Metals, Aluminium International Today) and conference proceedings (TMS, ICSOBA, Rusal, Alusolutions); authored a section of the "Alumina" chapter of the 2019 SME-handbook; co-author of a scientific article on Energy efficiency improvement and GHG abatement in the global production of primary aluminium; author and presenter of a TMS short course on Sustainability and Alumina Refinery design.

GPSR Compliance
The European Union's (EU) General Product Safety Regulation (GPSR) is a set of rules that requires consumer products to be safe and our obligations to ensure this.

If you have any concerns about our products, you can contact us on

ProductSafety@springernature.com

In case Publisher is established outside the EU, the EU authorized representative is:

Springer Nature Customer Service Center GmbH
Europaplatz 3
69115 Heidelberg, Germany

www.ingramcontent.com/pod-product-compliance
Ingram Content Group UK Ltd.
Pitfield, Milton Keynes, MK11 3LW, UK
UKHW021007290726
14059UKWH00001BA/10

9783030885854